Integrative Animal Biology

M. Brock Fenton
Western University

Karen A. Campbell
Albright College

Elizabeth R. Dumont
University of Massachusetts, Amherst

Michael D. Owen
Western University

NELSON / EDUCATION

NELSON / E D U C A T I O N

Integrative Animal Biology

by M. Brock Fenton, Karen A. Campbell, Elizabeth R. Dumont, and Michael D. Owen

Vice President, Editorial Higher Education:
Anne Williams

Publisher:
Paul Fam

Marketing Manager:
Leanne Newell

Developmental Editor:
Toni Chahley

Photo Researcher and Permissions Coordinator:
Kristiina Paul

Content Production Manager:
Jennifer Hare

Production Service:
Integra Software Services Pvt.Ltd

Copy Editor:
Rosemary Tanner

Proofreader:
Jay Boggis

Indexer:
Jeanne R. Busemeyer

Production Coordinator:
Ferial Suleman

Design Director:
Ken Phipps

Managing Designer:
Franca Amore

Interior Design:
Dianna Little

Cover Design:
Trinh Truong

Cover Image:
Front cover (rhino), Courtesy of Naas Rautenbach; back cover (sea creature), Courtesy of Geoff Spiby

Compositor:
Integra Software Services Pvt.Ltd

Printer:
RR Donnelley

Library and Archives Canada Cataloguing in Publication Data

Integrative animal biology / M. Brock Fenton ... [et al.].

Includes index.
ISBN 978-0-17-650202-7

1. Zoology—Textbooks.
I. Fenton, M. Brock (Melville Brockett), 1943-

QL48.2.I57 2013   590
C2012-908564-2

ISBN-13: 978-0-17-650202-7
ISBN-10: 0-17-650202-5

For the animals

To James H. Fullard and Donald W. Thomas,
who left the zoology game far too soon
– BF

To Wil and Calum, who understood where
my mind was when it was not with them
– KC

To the study and practice of conservation
– ERD

To Sharon, for her patience throughout this project
– MDO

Brief Contents

Contents

About the Authors

M. BROCK FENTON Brock Fenton received his Ph.D. in 1969 for work in the ecology and behaviour of bats. Since then he has held academic positions at Carleton University (Ottawa, 1969 to 1986), York University (Toronto, 1986 to 2003), and the University of Western Ontario (London, 2003 to present). He has published over 200 papers in refereed journals (most of them about bats), as well as numerous nontechnical contributions. He has written three books about bats intended for a general audience: *Just Bats*, 1983, University of Toronto Press; *Bats*, 1992 (revised edition 2001), Facts On File Inc; and *The Bat: Wings in the Night Sky*, 1998, Key Porter Press. He has supervised the work of 46 M.Sc. Students and 22 Ph.D. students who have completed their degrees. He currently supervises five M.Sc. students and two Ph.D. students.

He continues his research on the ecology and behaviour of bats, with special emphasis on echolocation. He currently is an Emeritus Professor of Biology at the University of Western Ontario.

KAREN CAMPBELL Karen A. Campbell received her Ph.D. in Physiology and Zoology in 1989 from Indiana University. She has spent the last two decades teaching at Albright College, a small liberal arts college in Reading, PA. She has taught over 15 different courses, including the introductory course for Biology majors; advanced Vertebrate Physiology and Comparative Vertebrate Anatomy; seminars in Neuroethology; and numerous courses for non-science majors, including field-based experiences in Pennsylvania and Australia. She was awarded the Lindback Foundation Award for distinguished teaching in 1997. She is a member of the Project Kaleidoscope (PKAL) Faculty for the 21st Century, an informal national alliance taking a lead in the effort to transform undergraduate science teaching. While her primary research interests explore aspects of the behavioral physiology and ecology of bats, many of her undergraduate research students over the past years have been introduced to independent research in very different areas, such as the feeding ecology of owls, the detection of polarized light by painted turtles, the development of a marker library for genome mapping in bats, the application of biochemical techniques in studying enzyme activity in bacteria, and the behavioral physiology of Japanese beetles. She is currently the P. Kenneth Nase Chair of Biology at Albright College.

BETSY DUMONT Elizabeth (Betsy) R. Dumont received a Ph.D. in Anthropology in 1993 from SUNY at Stony Brook. Since that time she has held faculty positions in Anatomy (Northeastern Ohio Universities College of Medicine) and Biology (University of Massachusetts. Amherst). With a strong background in vertebrate morphology, her research is centred on the ecology, evolution, and biomechanics of feeding in mammals. Her work combines empirical and

Left to right: Brock Fenton, Betsy Dumont, Karen Campbell, and Michael Owen

model-based approaches to understanding the links among feeding behaviour, feeding performance, and morphological diversity. She has mentored many undergraduate and graduate research projects and has developed and taught courses in Human Gross Anatomy, Mammalogy, Tropical Field Biology, and the Biology of Social Issues. She is currently Professor of Biology at the University of Massachusetts, Amherst.

MICHAEL OWEN Michael Owen received his B.Sc. in Zoology (specializing in invertebrate marine biology) from Swansea University and a Diploma of the Imperial College (London) in Applied Entomology. He was appointed as a Senior Teaching Fellow at Monash University in Melbourne and studied the stings of some Australian wasps for his Ph.D. He then moved to the United States to begin a postdoctoral fellowship at Harvard for work on the chemistry of wasp and bee stings. In 1970, he joined the Department of Zoology at University of Western Ontario as an Assistant Professor and taught invertebrate biology and marine biology (including field courses) throughout his long career. His research program on bee venom chemistry evolved to work on insect neurochemistry looking at neurotransmitters. Sabbatical projects included researching African bee venom in Kenya, pharmacological techniques in Cambridge, amine brain rhythms in Australian bush flies in Canberra, and neurotransmitter measurements in live brains at the University of Colorado.

In semiretirement, he has worked with the Department of Fisheries and Oceans, looking at meiofaunal invertebrates and taking video transects of benthic fauna in the Bay of Fundy. He has taught courses at the Centre for Enhanced Teaching and Learning at the University of New Brunswick, the University of Western Ontario, and Nipissing University.

Preface

The classic use of separated / independent chapters on individual phyla somehow takes away the "relationship" among those phyla and the questions about why evolution followed so many directions. Evolution cannot be fully appreciated by students and properly taught by their teachers without making explicit the connection between simple and complex forms.

—Pedro Quijon, *University of Prince Edward Island*

As Dr. Quijon suggests, the goal of the integrative approach of this textbook is to make explicit the connections between simple and complex forms. In *Integrative Animal Biology*, we present an integrated view of zoology. As outlined in the Table of Contents, after addressing fundamental topics such as evolutionary processes, larvae and life cycles, body plans, and the fossil record, we cover basic operations from locomotion to neural and endocrine control. Invariably we present information about invertebrate and vertebrate animals. With over 135 years of combined teaching experience, we believe that dividing animals into classification groups overlooks many rich examples of parallel and convergent evolution. If students only experience one course about animals, it should introduce them to as much diversity as possible. Multiple examples of how different animals solve the same problems makes it easier for students to appreciate the diversity of animals.

Organization

Far more than a collection of interesting stories of vertebrates and invertebrates, *Integrative Animal Biology* introduces the foundations of animal biology and diversity in its first five chapters, providing students with the conceptual tools they will need to place the material in the rest of the book in context.

Chapter 1, Tales of Animals, is an introductory chapter that sets the scene for the book and includes examples of animal stories to whet the reader's appetite. Storytelling is an age-old method of education: Aesop knew it; Homer knew it. Humans are storytelling animals! This method of learning lends itself well to the study of animal biology, and so we begin with several stories to illustrate the major themes in this book.

Chapter 2, Body Plans, Skeletons, and Development, introduces the various body plans of animals, an essential framework for what follows. The chapter emphasizes that external features provide scant information on what lies underneath the skin of an animal and explores the ways in which animal bodies are constructed, looking at common elements and how they are combined into a body plan. It also shows how body plans may be conserved within phylogenetic groups and the ways in which they have been modified to meet particular environmental demands.

Chapter 3, The Phylogeny of Animal Groups, and Chapter 4, Evolutionary Processes, explore the phylogeny of animals and the evolutionary processes that have led to our current understanding of animal phylogeny. A phylogeny is an evolutionary history of a species, a group of species, or a higher-level taxonomic group. This chapter reviews current concepts of animal phylogeny and, particularly when used in conjunction with the Purple Pages section of the book, provides access to the evolutionary history of the examples of animals that populate the entire book. The spectacular diversity of animals is a product of evolution. Understanding evolution means understanding how genetic variation is produced and how micro-evolutionary processes such as genetic drift and natural selection act on populations to change their genetic composition over

The purple or grey morph of the pygmy seahorse (*Hippocampus bargibanti*) is found on the gorgonian *Muricella* sp. We chose to open the book with a photo of a pygmy seahorse sitting on a gorgonian because it symbolizes the integration of invertebrate and vertebrate animal biology that informs the philosophy behind this book.

time. Large-scale patterns of evolution involve adaptation, speciation, extinction, and co-evolution.

Closely linked to the evolution of body plans and environmental adaptation is the information in Chapter 5, Larvae and Life Cycles. Development of an egg to a miniature adult form that then grows and becomes sexually mature (the norm in birds and mammals) is relatively uncommon among the animal kingdom. Much more common are either one or more larval forms (with morphologies quite unlike that of the adult) or a nymph that resembles an adult but lacks some adult features and later transitions to the adult form. Larvae and nymphs are often found in different environmental niches from their adult forms, an arrangement that may have evolved for a variety of reasons. A general theme is the adaptation of life cycles to environmental demands.

Conceptual threads introduced in these important foundation chapters are sewn into each of the chapters that follow, with extensive cross-references.

The remaining chapters examine the variety of ways in which animals move (Chapter 7) and feed (Chapter 8), and the similar mechanisms that have evolved in widely different groups of animals. Symbiotic and parasitic associations (Chapter 10) can be viewed as highly specialized modes of feeding. Feeding must be followed by digestion (Chapter 9), a metabolic activity that is linked to gaseous exchange (Chapter 11), circulation (Chapter 12), and excretion (Chapter 13). Like all these activities, reproduction (Chapter 14) must be coordinated. The following chapters (Chapters 15 and 16) address the role of electrical and chemical interactions in coordinating activities within individual animals and between them and their surroundings.

While examples of animal behaviours appear throughout the book, the penultimate chapter (Chapter 17) provides an overview of the diversity of animal behaviours, leading to a final chapter that examines some of the complex interactions between humans and other animals. This final chapter (Chapter 18) demonstrates the importance of animal diversity in our own lives, be it through providing food, spreading diseases, helping to find cures for diseases, or providing recreational activities like birdwatching or fishing. It also describes ongoing threats to animal diversity including pollution, climate change, and habitat loss. The loss of diversity can have devastating effects on the ability of ecosystems to provide the services that we rely on.

Six main themes run throughout the book:

1. the diversity of animal life;
2. the importance of interactions among organisms, whether animals or plants, and how these interactions shape organisms and the ecosystems in which they live;
3. the value of fossils in understanding the history of species and their adaptations to their environment;
4. the evolutionary history of life and interactions among species;
5. the role of biological knowledge in current problems ; and
6. the process of scientific progress.

Elegant Details

In his book *Wonderful Life*, Stephen Jay Gould writes of the scientists and creatures associated with the Burgess Shale, a small quarry in the mountains of British Columbia, which opened a window on our understanding of early multicellular animals. Reflecting on the overwhelming amount of information and detail revealed by the fossils and the evolutionary story that they tell, Gould remarked that, "the beauty of nature lies in detail; the message in generality. Optimal appreciation demands both, and I know no better tactic than the illustration of exciting principles by well-chosen particulars." This book faces the same challenge in selecting from the wealth of amazing examples of animal structures and functions to present a coherent picture of animals and their adaptations.

The diversity of animals—the realm of zoology—is illustrated by the marvellous array of species (and genera, families, orders, classes, and phyla), and the many structures fundamental to the functioning of animals. The examples in this book attempt to identify overarching themes and ideas that put the details of names and structures into functional, evolutionary, and ecological contexts.

We believe that the illustrations are the key to the book. With over 1200 visuals, we have tried to find pictures of animals that may not be familiar to students (or instructors). The pictures, as much as or even more than the words, introduce animals.

Study Tools

Each chapter opens with a "Why It Matters," a story that connects the topic to an important theme in the study of animals. We briefly outline what follows in each chapter and finish by providing an overview of the material. At intervals we provide study break questions, intended to give users a chance to reflect on what they have just read. Self-test questions help to confirm the material that students have learned in the chapter, while questions for discussion should help students to appreciate the many opportunities to learn more about animals.

Within each chapter, separate boxed features explore different topics in more detail. Each chapter highlights a person (a zoologist) and a molecule, both of which play a significant role in the topic of the chapter. These and other boxes present additional material relevant to the text, adding another dimension to the story.

The Purple Pages

We recognize that many students will be intimidated and baffled by the names of animals. So we have created a section in the centre of the book called the **Purple Pages** to allow students to quickly confirm that *Nectocaris* is a cephalopod mollusc or that *Molossus molossus* is a mastiff bat. Every animal in the book is cross-referenced to the Purple Pages, which extend the phylogeny presented in Chapter 2.

Cross-references to the Purple Pages are incorporated everywhere a species is mentioned in the textbook. This allows students to place a specific organism in its phylogenetic context. Using the Purple Pages reference for an animal, students will be able to scan up to find the larger groups (orders, classes) to which that animal belongs. Because the Purple Pages contextualize only the organisms specifically discussed in the book, species are listed alphabetically in groups for easy reference. The in-text reference number for the Purple Pages takes the student to the genus and species. See p. xviii for a guide to using the Purple Pages.

Student Guide to *Integrative Animal Biology*

A. An Integrative Approach: Focusing on the Connections

When we were students, being introduced to animal biology meant one course in vertebrates (animals with backbones) and one in invertebrates (animals without backbones). This approach to teaching zoology implies that there are no connections between these two groups of animals. In fact, the many connections between vertebrates and invertebrates are the main theme of this book. To survive, both flatworms and elephants must solve the problems of acquiring nutrients and disposing of waste. One of the many thrilling aspects of the study of animal biology is that very different animals may have similar approaches to solving the same problem.

Integrative Animal Biology focuses on these connections. After reading this book, we would like you to come away with an appreciation of the diversity of animals and the diverse solutions to the same problems. While it is impossible to cover every animal, we have selected some of the most interesting and exciting examples. Drawing upon examples from fossils to newly discovered organisms, we explore the many evolutionary experiments from past to present, with an eye to understanding why some of these experiments worked and others did not.

Throughout the textbook and in each chapter, we have included examples of both vertebrates and invertebrates. Some of the animal characters will be familiar, while others will be entirely new.

The first five chapters of the book provide some of the foundations of animal biology by exploring key themes. Chapter 1 offers spectacular examples that show how insights from the fossil record, biomechanics, biochemistry, genetics, and ecology help to explain what animals do and how they do it. Chapters 2 and 5 describe the development and organization of animals' bodies and introduce fundamental similarities and differences among groups of animals that will be seen throughout the book. Phylogeny and evolutionary processes (as discussed in Chapters 3 and 4) are inextricably linked: the patterns of relationships among animals are illustrated by phylogenies and are the direct result of evolutionary processes acting over the three billions years since life appeared on Earth.

Some tools are intended to make this book more useful to you. In each chapter, Study Break questions give you an opportunity to identify the important points and encourage you to review the content of the section you have just read. Self-test questions (in multiple choice format) at the end of each chapter give you a way to quickly check your knowledge of important facts and concepts. The Questions for Discussion should help you appreciate that there is a great deal about animals that we have yet to learn.

We think that some of the stories about animals are simply amazing. The diversity of animals continues to astonish us. We hope that you will find the material interesting and engaging.

By the way, whenever you find a mistake, please let us know. We have worked hard to be accurate but recognize that everyone makes mistakes.

B. Names

What's in a name? People are very attached to names—their own names, the names of other people, the names of flowers and food and cars, and so on. It is not surprising that biologists would also be concerned about names. Take, for example, our use of scientific names. The modern system of biological classification grew from the work of Carl Linneaus (1707–1778). He established the framework for **binomial nomenclature**, the practice of formally identifying species with a two-part name. Each species is placed within a classification that reflects its evolutionary history. Scientific names are always italicized and Latinized.

Castor canadensis Kuhl is the scientific name of the Canadian beaver. *Castor* is the name of the genus, while *canadensis* is the species name. Kuhl is the name of the person who described the species. "Beaver" by itself is not enough, because there is a European beaver, *Castor fiber*, and an extinct giant beaver, *Castoroides ohioensis*. Further, common names can vary from place to place (*Myotis lucifugus* is known as the "little brown bat" or the "little brown myotis").

Biologists prefer scientific names because the name tells you about the organism. For example, in *Myotis lucifugus*, *Myotis* means mouse-eared and *lucifugus* means flees the light; hence, the name tells us that this species is a mouse-eared bat that stays away from light. There are strict rules about the derivation, precision, and use of scientific names. Common names are not so restricted and consequently not precise. Common names are lower case, except for birds (see below), geographic names, or patronyms (*geographic* means named after a country, and *patronym* means named after someone; e.g., Pacific dampwood termite or Ord's kangaroo rat, respectively).

Bird common names are an exception to the lower case rule. For example, *Pycnonotus xanthopygos* is the White-spectacled Bulbul. The common names for birds are usually capitalized because they are standardized and accepted in the ornithological world.

In this book, we present the scientific names of organisms when we mention them. We follow standard abbreviations; for example, although the full name of an organism is used the first time it is mentioned (e.g., *Castor canadensis*), subsequent references to that same organism may abbreviate the genus name and provide the full species name (e.g., *C. canadensis*).

In some areas of biology research, the standard representation is of the genus, for example, *Drosophila*. In other cases, names are so commonly used that only the abbreviation is used (e.g., *C. elegans* for *Caenorhabditis elegans*).

Scientific names in the text and figure legends are followed by a number (e.g., *Gadus morhua* (1213)). This number is a cross reference to the position of *Gadus morhua* (Atlantic cod) in the listing in the centre of the book—the Purple Pages—identified by the purple bleed on the page edges (see the Purple Pages Walkthrough on p. xviii). In Chapter 3 (Phylogeny), fish are positioned in the subphylum Vertebrata on a phylogenetic tree, but this gives you no information about where cod belong within the fish. The Purple Page listings tells you that *Gadus morhua* is in the

Family Gadiformes (cod and their relatives)
Order Teleostia (bony fish with moveable maxilla and premaxilla)
Subclass Actinopterigii (ray-finned fish)
Superclass Osteichthyes (the bony fish)
Subphylum Vertebrata

This structured list appears rigid—but this is far from the truth. Often there is little agreement about some classifications reflecting differences of opinion, or the emergence of new information. In any classification some animals cannot be placed with certainty in one group or another. An example is the fossil *Archeopteryx* (1536) that, despite much work (and even more argument), cannot be positioned with certainty as a bird or as a reptile.

C. UNITS

The units of measure used by biologists are standardized metric (or SI) units, used throughout the world in science.

D. DEFINITIONS

The science of biology is replete with specialized terms (sometimes referred to as "jargon") used to communicate specific information. It follows that, as with scientific names, specialized terms increase the precision with which biologists communicate among themselves and with others. Be cautious about the use of terms because jargon can be a veneer of precision. When we encounter a "slippery" term (such as species or gene), we explain why one definition for all situations is not feasible.

E. TIME

In this book, we use CE (Common Era) to refer to the years since year 1 and BCE (Before the Common Era) to refer to years before that.

Geologists think of time over very long periods. A geologic time scale (see the inside back cover of the book) shows that the age of the Earth could be measured in years, but it's challenging to think of billions of years expressed in days (or hours, etc.). With the advent of the use of the decay rates of radioisotopes to measure the age of rocks, geologists adopted 1950 as the baseline—the "Present"—and the past is referred to as b.p. ("Before Present"). A notation of 30 000 years b.p. (14C) indicates 30 000 years before 1950 using the 14C method of dating. In measurements of even longer periods the abbreviation "Ma" (for mega-annum) is used; 1 Ma is a million years.

F. SOURCES

Where does the information presented in a text or in class come from? What is the difference between what you read in a textbook or an encyclopedia and the material you see in a newspaper or tabloid? When the topic relates to science, the information should be based on material that has been published in a scholarly journal. In this context, "scholarly" refers to the process of review. Scholars submit their manuscripts reporting their research findings to the editor (or editorial board) of a journal. The editor, in turn, sends the manuscript out for comment and review by recognized authorities in the field. This process is designed to ensure that what is published is as accurate and appropriate as possible. The review process sets the scholarly journal apart from the newspaper or TV news report.

There are literally thousands of scholarly journals, which, together, publish millions of articles each year. Some journals are more influential than others, for example, *Science* and *Nature*. These two journals are published weekly and invariably contain new information of interest to biologists.

To collect information for this text, we have drawn on published works that have gone through the process of scholarly review. Specific references (bibliographic details) are provided on the website that accompanies the textbook: **www.nelson.com/animalbio**.

A "citation" makes the information accessible. Although there are many different formats for citations, the important elements include (in some order) the name(s) of the author(s), the date of publication, the title, and the publisher. When the source is published in a scholarly journal, the journal name, its volume number, and the pages are also provided. With the citation information, you can visit a library and locate the original source. This is true for both electronic (virtual) and real libraries.

Students of biology will benefit by making it a habit to look at the most recent issues of their favourite scholarly journals and use them to keep abreast of new developments.

THE BIG PICTURE

Each chapter of Integrative Animal Biology is carefully organized and presented in "digestible chunks" so you can stay focused on the most important concepts. Easy-to-use learning tools point out the topics covered in each chapter, show why they are important, and help you learn the material.

Figure 11.1 A spider (*Argyroneta aquatica* (661)) with a web-encased bubble of air (a diving bell).
SOURCE: Claude Nuridsany & Marie Perennou / Photo Researchers, Inc.

11 Gaseous Exchange

STUDY PLAN

Why It Matters
Setting the Scene

11.1 Basic Principles of Gaseous Exchange
11.1a Simple Diffusion
11.1b Pressure
11.1c Respiratory Water Loss

11.2 Sites for Gaseous Exchange
11.2a Surface Layers and Skin
11.2b Gills
11.2c Lungs and Tracheae

11.3 Respiratory Surfaces by Phylum
11.3a Phylum Annelida
11.3b Phylum Mollusca
11.3c Phylum Onychophora
11.3d Phylum Arthropoda
11.3e Phylum Phoronida
11.3f Phylum Echinodermata
11.3g Phylum Chordata

11.4 Some Specific Examples
11.4a Air-Breathing Fish
11.4b Diaphragms and Ventilation
11.4c Atmospheric Gases and Body Size
11.4d Flying High
11.4e Diving Deep

Gaseous Exchange in Perspective

NEL

WHY IT MATTERS

Different species of terrestrial arthropods, such as the diving spider shown here, use bubbles of air when diving. This allows them to stay underwater for relatively long periods of time, like people using diving bells. The diving spider uses specialized silk to contain the air bubble (Fig. 11.2). This approach to diving depends upon a **plastron** (Fig. 11.3), a hydrophobic surface (hydrofuge hairs or cuticular projections) that creates an airspace on the spider's body surface. The plastron allows diving arthropods to take dissolved oxygen from the water, through the thin film of air that creates an incompressible exchange surface, into their gaseous exchange system.

Plastrons provide a large area of water-surface contact with the spiracle opening into book lungs and tracheae of a spider or the tracheae of an insect. Plastrons have evolved repeatedly and independently in terrestrial arthropods, and have also appeared in some arthropods living in habitats subject to flooding, where they are key to survival even for species that do not dive. The amblypygid *Phrynus marginemaculatus* (655) provides an excellent example. This denizen of the Florida Keys that regularly flood has a well-developed plastron (Fig. 11.3).

293

Study Plan The Study Plan provides an overview of all the topics and key concepts in the chapter. Each section breaks the material into a manageable amount of information, building on knowledge and understanding as you acquire it.

Why It Matters Engaging introductory sections capture the excitement of biology and help you understand why the topic is important and how the material you are about to read fits into "the Big Picture."

Study Break Encourages you to pause and think about the key concepts you have just encountered before moving to the next section. The "Study Break" questions are intended to identify some of the important features of the section. The answers are found in the Answers Appendix at the end of book.

STUDY BREAK

1. What three components are involved in the coevolution of the fungal gardens of leaf-cutter ants?
2. What are the differences between high-concentration and low-concentration salt sensors in mice?
3. How does genetic manipulation of a fungus affect its impact on malaria?

BOX 4.4 People Behind Animal Biology
Sean Carroll

What controls morphological diversity, and how do those controls evolve? In the mid-20th century, evolutionary biologists from disciplines as different as paleontology and population genetics came together to answer that question in what is termed the "Modern Synthesis." They concluded that morphological differences among species are the relatively simple products of mutations upon which natural selection acted over very long periods of time.

This idea was rocked in the 1980s when biologists studying fruit flies (*Drosophila* sp. (960)) made inroads into understanding the genetics governing their development. The breakthrough was the discoveries of *HOM-C* genes that regulate *Drosophila* development and the related *Hox* genes of vertebrates (see Box 2.1). The big surprise was that these genes are highly conserved across groups—animals as different as fruit flies and humans showed relatively minor differences in these genes. Understanding how this can be is the focus of the field of evolutionary development, more commonly known as evo-devo.

Dr. Sean B. Carroll is one of the leaders of this new field of evo-devo. After finishing a PhD in immunology from Tufts Medical School, he worked on *Drosophila* genetics and, like many of his contemporaries, soon realized that the genetics underlying morphological diversity were nothing like what earlier researchers imagined they would be. Sean Carroll is a vocal proponent of the idea that all animals share a common, ancestral network of regulatory genes that interact with one another to guide embryonic development. More importantly, small changes in these genes can have large effects on the upregulation (production) and downregulation (turning off) of proteins that determine the timing and

differential growth of various anatomical parts.

It turns out that small changes in regulatory genes do produce remarkable morphological changes. For example, at the same early developmental stage, the bones in the hands of bats and mice are the same length (Fig. 1). Later, the expression of a protein called bone morphogenic protein 2 (bmp2) is upregulated in bats. This protein stimulates the production of chondrocytes that lay down cartilage and cause the digits to lengthen.

In addition to his important contributions to the field of evo-devo, Dr. Carroll has written five popular books about science, including *Endless Forms Most Beautiful: The New Science of Evo-Devo and the Making of the Animal Kingdom* (2005) and *Remarkable Creatures: Epic Adventures in the Search for the Origins of Species* (2008), and has appeared on many television and radio shows. He is currently a Professor of Molecular Biology and Genetics and an Investigator with the Howard Hughes Medical Institute at the University of Wisconsin.

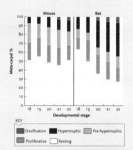

FIGURE 1 The bones of bat wings and mouse hands grow longer by the addition of cartilage by chondrocytes that then ossifies to become bone. This graph illustrates the percentage of a metacarpal (hand bone, y-axis) that is composed of different types of tissues in a mouse (Mus musculus (1891), left) and a short-tailed fruit bat (Carollia perspicillata (1787), right) during developmental stages 18–22. Tissues are divided into resting, proliferative (chondrocytes are increasing in number), pre-hypertrophic (chondrocytes are poised to begin secreting the extracellular matrix in which cartilage forms), hypertrophic (actively secreting extracellular matrix), and undergoing ossification. Beginning at stage 20, the bat metacarpal contains more hypertrophic tissues than does the rodent metacarpal. This reflects the more rapid increase in the length of the bat metacarpal and is controlled by bmp2.

SOURCE: Sears K E et al. 2006. "Development of bat flight: Morphologic and molecular evolution of bat wing digits," Proceedings of the National Academy of Sciences; 103:6581–6586. Copyright 2006 National Academy of Sciences, U.S.A.

People Behind Biology People Behind Animal Biology boxes in each chapter contain boxed stories about how particular people have used their ingenuity and creativity to expand our knowledge of animal biology.

BOX 4.3 Molecule Behind Animal Biology
The Evolution of Opsins

Opsins are molecules that allow animals to perceive light. Some form of opsin is found in animals ranging from cnidarians to annelids, cephalopods, arthropods, and vertebrates. Opsins are the protein components of visual pigment molecules that are found within the membranes of receptor cells in the retina. Different opsin molecules are sensitive to different wavelengths of light, but they all work the same way. Opsin molecules attach to and encase single molecules of **retinal**, a chemical derived from vitamin A. The retinal molecule is forcibly kinked when it fits into the opsin, but the kink straightens out when it is struck by a single photon of light of the correct wavelength. This initiates a series of membrane–protein interactions in the receptor cell that generates a neural signal that is transmitted to the brain, where signals are interpreted as visual images. After releasing the retinal, opsin attaches to another kinked retinal molecule, and the pigment is ready to respond to another photon. This molecular mechanism functions within a narrow portion of the electromagnetic spectrum. No animal can see very far into the infrared or ultraviolet.

Among vertebrates, opsins are located on light-sensitive cells in the retina called **cones**, and each cone is sensitive to a different part of the light spectrum, that is, colour (Fig. 1). Each cone is most sensitive to a specific peak wavelength and less sensitive to wavelengths above or below that peak. The brain perceives colour by comparing the simultaneous firing rates of at least two cells (and usually many more) that favour different colours, and the quality of colour vision depends largely on how many

different classes of cones there are to compare. Vertebrates with two classes of cones are called dichromats, and can see two out of three of red, green, and blue. Vertebrates with three classes of cones are called trichromats; they can see red, green, and blue. Other vertebrates, including fish and reptiles, may be tetrachromats (e.g., red:green:blue:low ultraviolet) and the vision of birds and turtles can be even more sophisticated.

If modern reptiles are any guide, the precursors to mammals were probably active during the daylight hours and had superb colour vision. In the Triassic, mammals left the daylight and became nocturnal, perhaps to avoid predators or to hunt nocturnal prey. In order to navigate the darkness, selection favoured eyes that could gather whatever photons were available, regardless of colour. The result was that colour discrimination degenerated in favour of cells called "light" and "dark," but not separate colours. To this day, most mammals have poor colour vision.

Analyses of the evolution of opsin genes indicate that trichromatic vision evolved independently in primates not once, but twice: first in the group containing Old World monkeys, apes, and humans, and again in a single species of New World primate, the howler monkey (*Alouatta* sp. (1851)). How does colour vision offer such a fitness advantage that trichromacy evolved twice in primates? One possibility is that it has to do with eating fruit. In a predominantly green forest, plants have evolved such that their fruits have bright colours to attract frugivores, which play the vital role of spreading their seeds. As it happens, fruit is an important part of the diet of many Old World monkeys. Another idea is that trichromatic vision helps detect young, succulent leaves, which are often pale green and sometimes even red. Howler monkeys are leaf specialists that prefer young leaves because they are tender and usually less toxic than older leaves. The ability to see young leaves benefits the howler monkeys, although it does not work to the plant's advantage.

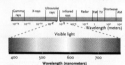

FIGURE 1 The hues we perceive as colours are electromagnetic radiations of different wavelengths that are produced by the sun and pass through our atmosphere. Animals perceive colours that lie roughly between red (wavelength 700 nm), and violet (around 420 nm). For all animals, "light" is a relatively narrow band of electromagnetic wavelengths lying somewhere between ultraviolet and infrared. This is a very narrow part of the whole spectrum of wavelengths. While we can't "see" higher or lower wavelengths, we use them in many different technologies.

SOURCE: Based on Based on Lehninger, A.L. 1965. Bioenergetics. Menlo Park, California.

Molecule Behind Biology Molecule Behind Biology boxes give students a sense of the exciting impact of molecular research. Each chapter the activity of a chemical relevant to topic covered.

SETTING **THE SCENE**

Arguably, the drive to reproduce underlies virtually everything in biology, from structure and function to behaviour and genetic control. Aspects of reproduction are central to the process of speciation (and evolution), as well as to understanding the links between ecology, evolution, and fitness. We will consider sexual and asexual reproduction, and aspects of gender determination, both genetic and environmental. The production of gametes is described as well as the processes around fertilization. The timing of reproduction is considered from the standpoints of environmental cues, synchronization of spawning and mating, promoting the survival of offspring, and resolving timing conflicts around reproduction. We also examine some aspects of sexual selection and how it reflects and influences reproduction in animals.

REPRODUCTION **IN PERSPECTIVE**

This is a good time to consider the links between reproduction and other chapters in this book, particularly evolutionary processes, larvae and life cycles, symbiosis and parasitism, behaviour, and interactions between animals and humans. Consider the impact that the drive to reproduce has on the behaviour of your pet, or on your own behaviour.

Setting the Scene paragraphs provide an overview of the purpose of the content of the chapter, and set it in the context of the textbook as a whole.
In Perspective paragraphs summarize the purpose of the content of the chapter, and set it in the context of the textbook as a whole.

What Are the Purple Pages?

Reference Numbers 100–126

PHYLUM CHOANOFLAGELLATA

- A protistan phylum
- Viewed as the sister group to the Metazoa
- Collar flagellated cells

Reference Number	Genus, Species	Common Name	Page Reference
100	Monosiga brevicollis		41

PHYLUM PORIFERA

- The sponges
- Numerous well-differentiated somatic cell types but no organization into tissues, no nervous system
- Body structured around channels, in which flagellae drive a unidirectional flow of water, and chambers, lined by choanocytes (cells resembling choanoflagellates) that collect food particles

Class Calcarea

- Calcium carbonate (calcite) skeletal spicules
- Asconid, syconid, or leuconid body plan

Reference Number	Genus, Species	Common Name	Page Reference
101	Sycon sp.		50f
102	Sycon coactum		213f

Class Demospongiae

- Leuconid body plan
- Silica spicules
- Often with spongin fibres

Reference Number	Genus, Species	Common Name	Page Reference
111	Agelas sp.		50f
112	Aplysina aerophoba		213
113	Asbestopluma hypogea	a carnivorous sponge	214, 215f
114	Chondrocladia gigantea		17f
115	Halichondria panicea		215f
116	Spongilla lacustris	a freshwater sponge	366
117	Tethya aurantia	orange puffball sponge	272f

Class Hexactinellida

- Leuconid body plan
- 3- or 6-rayed silica spicules that are frequently fused to form a complex skeleton
- Cells are syncytial

Reference Number	Genus, Species	Common Name	Page Reference
125	Acanthascus sp.		50f
126	Euplectella sp.	Venus' flower basket	25f

While your study of animal biology focuses on an understanding of the unique characteristics of the systems of organisms, it is equally important to be able to understand and place each of the organisms you study within their phylogenetic context. The Purple Pages are designed help you easily identify organisms and place them in their phylogenetic context.

Class Pycnogonida

- The sea spiders
- Marine predators on hydroids and bryozoans
- Four pairs of legs with gut diverticulae extending into the legs
- Small body
- Chelifores, at base of proboscis, may not be homologous with chelicerae of rest of the subphylum.

Reference Number	Genus, Species	Common Name	Page Reference
641	Nymphon sp.	a sea spider	134f
642	Pycnogonum	a sea spider	69f

The Pycnogonida (sea spiders) is a group that is probably unfamiliar to you. We will use the positioning of this group to explain the use of the Purple Pages. At different places in the book you will find two pycnogonids, Nymphon sp. (641) and Pycnogonum (642).

PHYLUM ARTHROPODA

- Literally "joint-footed" animals, usually translated as joint legged
- Segmented animals in all environments, with a chitinous exoskeleton that is moulted for growth

†Anomalocarida
- Large predators with a great claw from the Cambrian era

†Marrelomorpha
- Fossil group from the Cambrian to Devonian, soft bodied

SUBPHYLUM CHELICERATA
- Horseshoe crabs, scorpions, harvestmen, spiders, ticks, mites, and their relatives
- No antennae, segment two has pair of chelicera, segment 3 a pair of pedipalps
- Typically a cephalothorax and abdomen (fused in the mites)

Class Pycnogonida
- The sea spiders
- Marine predators on hydroids and bryozoans
- Four pairs of legs with gut diverticulae extending into the legs
- Small body
- Chelifores, at base of proboscis, may not be homologous with chelicerae of rest of the subphylum.

Reptilia or Aves?

Noting the identifying number for each of these species, flip to the Purple Pages in the middle of the book. At the top of each purple page there is a listing of the Purple Page reference numbers that appear on that page.

You will find the number associated with each of the four species listed in the Class Pycnogonida and immediately below that heading are some of the characteristics of the class. A little above the listing for the Class Pycnogonida you will see that this group is a part of the subphylum Chelicerata. Again, under the heading, you will find characteristics of the subphylum. Further up the page you will find that the subphylum Chelicerata is a division of the Phylum Arthropoda with the heading followed by some characteristics of the phylum.

Archaeopteryx (1536) provides an excellent example of the challenges of placing organisms in a phylogenetic context. The first fossil specimen of *Archaeopteryx* was found in 1861, and since then ten additional specimens (and a feather from a possible eleventh) have been discovered. *Archaeopteryx* has a mix of avian (broad wings, flight feathers, wishbone) and reptilian (jaws with teeth, long tail, claws on three toes) characteristics. Early interpretations of this mix of characters placed *Archaeopteryx* as an evolutionary link between the reptiles and the birds. More recent analyses have placed it as either a reptile, or as a primitive bird. The discovery of another fossil *Xiaotingia zhengi* (1535) that resembles *Archaeopteryx* but is slightly more reptilian in its characters adds further complications to the argument. We have treated this problem by placing both Xiaotingia zhengi and *Archaeopteryx* under a heading "Reptilia or Aves?" between the reptiles and the birds.

VISUAL LEARNING

Spectacular photos and illustrations – developed and chosen with great care – help you visualize structures, processes, and relationships.

High quality photos and illustrations play an important role in identifying and understanding the evolutionary relationships among animals. The visuals in this textbook have been carefully chosen and developed to help you place important concepts in context.

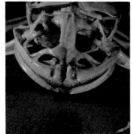

Visualizing Relationships. High quality photographs help you visualize shared attributes and common structures.

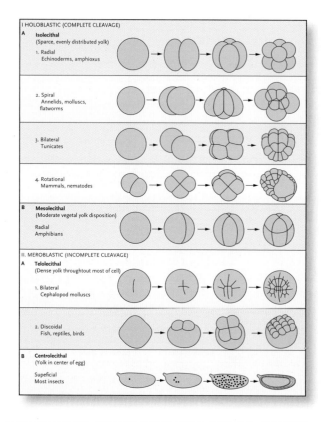

Comparing Systems. Illustrations help you understand and appreciate the variation and technical detail of structures and processes.

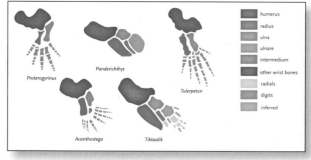

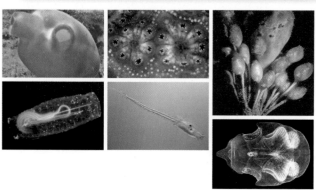

Exploring Diversity. Unique examples expose you to extraordinary animals, and help reinforce the marvelous array of species in the realm of zoology.

Self-Test Questions at the end of each chapter help students test their knowledge of the concepts covered in each chapter.

Questions

Self-Test Questions

1. Which of the following represents the skeletal–muscular interactions typically involved with locomotion?
   a. Animal skeletons transmit the force generated by muscular contraction and translate it into movement.
   b. Muscles always work in opposing pairs with skeletal elements to generate movement.
   c. All animal skeletal–muscular associations are structurally similar.
   d. Only the skeletal elements of limbs are important in locomotion.

5. Which of the following describes features of pulsed jet propulsion systems?
   a. An elastic protein, abductin, plays an important role in the recoil of the squid mantle.
   b. Contraction of rings of muscle around the tubular body wall with the anterior valve closed propels squid forward.
   c. Partial closure of the bell by a shelf of tissue, or velum, creates a jet nozzle to project the squid forward.
   d. Contraction of fast-twitch circular muscle fibres forces water out of the squid mantle cavity.

Nelson Education Ltd. understands that good quality multiple-choice questions can provide the means to measure higher-level thinking skills as well as recall. Recognizing the importance of multiple-choice testing in today's classroom, we have created NETA – the Nelson Education Teaching Advantage program – to ensure the value of our high-quality multiple choice questions.

The NETA program was created in partnership with David DiBattista, a 3M National Teaching Fellow, professor of psychology at Brock University, and researcher in the area of multiple-choice testing.

Answers to the Self-Test Questions are provided in an Appendix at the end of the book. Each answer is accompanied by an explanation to enhance student understanding.

Appendix
Answers to Self-Test Questions

Chapter 1

1. b. Mandibles are not replaced, but rather the roles of the ants in the colony change, so d is incorrect. Choices a and c are not addressed in the story, so b is the correct answer.
2. c. Other animals obtain carotenoids from plants, fungi, and microorganisms, so a and b are incorrect. While the frequency of colours may change in a pea aphid population depending upon selection factors, a single individual cannot change colour, so d is incorrect.
3. a. The other answers illustrate structural or functional characteristics, but only a demonstrates the relationship between structure and function.
4. d. Plantarflexion and the reduced diaphysis angle between metatarsals and plantar surface are important in arboreal primates, rather than in bipedal organisms, so a and b are incorrect. The reduced angle in c would restrict dorsiflexion.
5. b. Structure is described in a and d, and function in c. Only b illustrates the relationship between structure and function.

Chapter 3

1. d. The term *phylogeny* refers to the evolutionary history of an animal, rather than its placement in a taxonomic group, so a is incorrect because it doesn't answer the question. Many of the physical and molecular features of animals are present in organisms from widely diverse taxonomic groups, so b and c are incorrect, confounded by the mollusc DNA from gut contents!
2. b. Polyps, as well as some others, can be considered to have more than one plane of symmetry, so a is incorrect. Not all comb jellies have tentacles, and other features link them somewhat to the Cnidaria. Sea wasps, Scyphozoa, and Hydrozoa are currently considered separate classes in the Medusozoa.
3. a. See text for proposed phylogenies.
4. c. These features define all Ecdysozoa.
5. d. While these terms apply to the number of branches evident in the gut of Platyhelminthes, the specific anatomical feature is not evident. a. These terms describe the location of branchia =

Acknowledgements

Parts of this book owe much to the input we have received from colleagues. In particular we are indebted to:

- Christiane Todt at the University of Bergen, University Museum of Bergen, The Natural History Collections, who participated in the initial planning of the book and provided expert reviews and advice on the chapters on phylogeny and on larvae and live cycles.
- Louise Page at the University of Victoria for her input to a number of chapters and particularly for her contributions to the body plans and reproduction chapters.
- Mick Burt at the University of New Brunswick for his expert review of the chapter on parasitism and symbiosis.
- Erin Fraser at Memorial University for her contribution to the section on the central nervous system.

We are very pleased to thank the many friends and colleagues who allowed us to use images from their work. The list is long but not limited to Greta Aeby, Dennis Anderson, Mark Blaxter, Alexander Bochdansky, Axel Brockman, Mick Burt, Peter Chew, Jenny Clack, Karen Cheney, Beth Clare, Clayton Cook, Jim Cornall, John Dabiri, Nina Fatouros, Philine Feulner, Julian Finn, Diego Fontaneto, John Ford, Peter (Frapps) Frappell, Alex Froese, Clare Hawkins, Ken Haynes, Claire Healey, Ian Hook, Matt Hooge, Robert Jackson, Marshall Johnson, Ryan Kerney, Mike Kinsella, Nikolklai Korniyuk, Liam McGuire, Michael Mares, Steve Marshall, Matt Mason, Donald McMillan, Jeremy N. McNeil, Bill Milsom, Nadja Møjberg, Fernando Montealegre-Z, Cindy Moss, Ryo Nakano, Alan Noon, Mark Norman, Ricardo Ojeda, Rich Palmer, Sheila Patek, Fred Pleijel, Heather Proctor, Paul A. Racey, John M. Ratcliffe, John J. Rasweiler IV, Martin Riddle, Greg Rouse, Mary Rumpho, Aaron Rundus, Robert Schofield, Alexander Semenov, Tom Sherratt, Jake Socha, Dave Stirling, Jeff Stoltz, Jan Storey, Devi Stuart-Fox, Bridget J. Stutchbury, Hannah ter Hofstede, J.G. (Hans) Thewissen, Beverly Van Praagh, George Weiblen, David Williams, Judith Winston, and Jayne Yack. We are grateful to Judith Eger, Kevin Seymour, and Janet Waddington from the Royal Ontario Museum for permitting us to photograph specimens in their care. We thank Marianne Collins for permission to use her reconstructions of *Spriggina* and *Nectocaris*.

Brock Fenton is pleased to thank graduate students for sharing ideas that emerged in the book (Amanda Adams, Matthew Emrich, Erin Fraser, Rachel Hamilton, Colin Hayward, Meredith Jantzen, and Liam McGuire). He also thanks colleagues for discussion and interactions on the topic of animals (Heather Addy, Mark Brigham, Sylvie Bouchard, David and Meg Cumming, Emanuel Mora, Jack Millar, Jeremy McNeil, Bill Milsom, John Ratcliffe, Kevin Seymour, and Hans Thewissen).

Karen Campbell thanks William and Calum Adams for their patience and support, her many colleagues who provided feedback, encouragement, and a sanity check during this project, in addition to countless students who shared thoughts and ideas about animal biology and have shaped her and teaching and learning over the years.

Elizabeth Dumont thanks Tom Eiting and Andy Smith for contributing ideas and content to this text. She also thanks Sean Werle, Beth Jakob, and Penny Jaques for moral support and her many colleagues for sharing their knowledge and love for animals.

Michael Owen's approach to a life in biology was shaped by Wynn Knight-Jones, Alan Osborne, and Teri O'Brien and supported by many colleagues over the years. He thanks Sharon Rich for her patience through this project and the generations of students who have taught him more than he ever managed to teach them.

The authors extend special thanks to Paul Fam who recognized the importance of this project and supported its development, Alwynn Pinard who started us in the right direction, and Toni Chahley who kept us on track. Rosemary Tanner was a master of diplomacy in editing our text; Kristiina Paul did her very best to keep us on budget while locating impossibly concealed sources of images. Jennifer Hare, Carmel Isaac, and Shanthi Guruswamy joined us in the late stages and converted a manuscript to a book.

We were very fortunate to have the assistance of some extraordinary students and instructors of biology across Canada who provided us with feedback that helped shape this textbook into what you see before you. As such, we would like to say a very special thank you to the following people:

Adam Brown, University of Ottawa
Jennifer Gauthier, Dalhousie University
Brenda Hann, University of Manitoba
Janice M. Hughes, Lakehead University
Jon Houseman, University of Ottawa
Drew Hoysak, Brandon University
Brian Leander, University of British Columbia
Deborah McLennan, University of Toronto
Tammy McMullan, Simon Fraser University
Michael O'Donnell, McMaster University
Cynthia Paszkowski, University of Alberta
Pedro Quijon, University of Prince Edward Island
Albrecht Schulte-Hostedde, Laurentian University
Stephen Donald Turnbull, University of New Brunswick
Jane Waterman, University of Manitoba

Student Advisory Boards

University of Western Ontario (pictured above) There are three separate photos for each of these SAB members.

Ellen Denstedt
Lida Zeinali
Tara Pawliwec

Albright College (pictured above)
Jillian Bonitatibus
Travis Dresch
Christopher Hauer

A Special Thank You to the Photographers Who Have Contributed to *Integrative Animal Biology*

We are especially grateful to Naas (I.L.) Rautenbach for his superb wildlife photography and wicked sense of humour, Geoff Spiby for access to his underwater photography, and Sean Werle for his arthropod photos. Bob Bolland shared his collection of Okinawa polyclads; Maria Buzeta and Mike Strong (dive buddies par excellence) were generous with their underwater image collection from the Bay of Fundy; Al Shostak offered the resources of the University of Alberta parasite collection; and Otto Larink and Turston Lacalli supplied amazing pictures of invertebrate larvae.

We also wish to acknowledge in particular the photographers whose work graces the cover of this book:

GEOFF SPIBY Geoff Spiby has been exploring the underwater world since he was a child. He grew up in Cape Town and learned snorkelling and scuba diving in its cold entrancing waters. He started taking underwater photographs when on honeymoon in the Maldives in 1984 and was hooked. Since then, he has generated a huge library of underwater shots from places as diverse as the Red Sea, African east coast, the Maldives, Malaysia and Indonesia, and of course Cape Town. Fortunately, his wife Lyn is as enthusiastic about diving as he is. Geoff works as a small animal veterinarian in Hout Bay, Cape Town, South Africa.

IGNATIUS LOURENS (NAAS) RAUTENBACH PH.D.
Naas is an accredited photographer at Master level and his work has been recognized by election as a Fellow

of the Photographic Society of S.A. He is also an enthusiastic cook, often to the chagrin of his family. His career commenced as field biologist for the Smithsonian's African Mammal Project. The experience later helped him to get a tenured appointment as Curator of Mammals at the Transvaal Museum and later Director. Naas opted for an early retirement from the formal sector to seek his fortunes as a photographer and consultant advising on the environmental impacts new land-use developments may have on the environment. Thus far, he has consulted for 13 years as mammalogist and ecologist. During his collective 46 years as mammalogist, he gained an intimate knowledge of Southern African mammals, which is manifested in numerous scientific publications in peer-reviewed journals, contributions to books, popular articles, public appearances, and hundreds of reports in the Environmental Impact Assessment milieu.

Figure 1.1 Two leaf-cutter ants handling a piece of leaf for transport back to the nest. Note the vein in the leaf, the cut edges, and the way that the ants manipulate the piece of leaf.

1 Tales of Animals

WHY IT MATTERS

The dominant herbivores of New World tropical and subtropical ecosystems, leaf-cutter ants (genera *Atta* (986) and *Acromyrex* (980), Fig. 1.1) cultivate fungi for food. A large colony of these ants can include 8 million workers occupying a nest of 20 m³. Leaf-cutters first appeared in the Eocene and adopted fungal cultivation 50 million years before humans began to practise agriculture. Arguably, fungal cultivation did for leaf-cutter ants what agriculture did for humans, allowing them to achieve very high population densities by manipulating other species to their advantage.

Adult leaf-cutter ants cut and harvest leaves and other plant parts and transport them back to the underground nest where they are further chewed. Although the ants obtain some energy from plant sap, most of the colony's food (fungal derivatives of plant materials) is harvested from symbiotic fungi that, in the nest, grow on the processed plant material. The rate of the fungal harvest determines the rate at which their colony obtains energy.

Each of a leaf-cutter ant's mandibles has several teeth (Fig. 1.2), used asymmetrically during cutting. Opening and closing movements of the leading mandible change the position of the large tooth (the

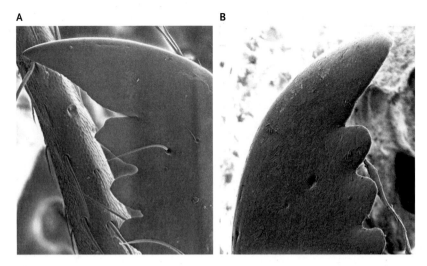

A **B**

Figure 1.2 Scanning electron micrographs of unused (**A**) and worn (**B**) teeth on the mandibles of leafcutter ants (*Atta cephalotes* (987)).

SOURCE: With kind permission from Springer Science+Business Media: *Behavioral Ecology and Sociobiology*, "Leaf-cutter ants with worn mandibles cut half as fast, spend twice the energy, and tend to carry instead of cut," Volume 65, Number 5, 2011, pp. 969–982, Robert M. S. Schofield, Kristen D. Emmett, Jack C. Niedbala and Michael H. Nesson.

V-blade) while the lagging mandible is held in a fixed position. The long distal tooth of the lagging mandible penetrates just beyond the leaf surface away from the ant. Ants sever leaf veins with a cutting action involving both mandibles. About 22% of an ant's cutting time is spent dealing with leaf veins. Cutting time is decreased by a specialized stridulation/vibration produced on the ant's waist (gaster), between thorax and abdomen. The vibratome effect appears to stiffen the leaf material and facilitate cutting.

Tooth wear influences the role that adult ants play in the life of the colony. Leaf cutting dulls the teeth on the mandible (Fig. 1.2), and ants with dull teeth spend twice the energy cutting leaves as do ants with sharp teeth. Thus it is no surprise that individuals with sharper teeth do more cutting, while those with dull teeth do more carrying. Most of us have used scissors to cut fabric and paper in the same way that leaf-cutter ants cut leaves. We know that cutting by holding the scissors "just right" saves time and energy, but additional effort is still required when you hit a seam or a fold.

Leaf-cutter ants make conspicuous pathways (Fig. 1.3), indicating the importance of transportation

costs in the operation of the colony. Having ants with sharp teeth doing the cutting and ants with duller teeth do the carrying may reduce the risk of tooth damage and maximize the amount of food available to the colony relative to the energy expended in harvesting. Individuals with dull teeth are likely also responsible for keeping the transport routes clear. Division of labour among leaf-cutter ants, the influence of jaw (tool) condition on foraging efficiency, and the farming behaviour of these animals present a blend of morphology, physiology, and behaviour that together set the stage for our zoology text.

STUDY BREAK

1. Where do leaf-cutter ants occur? How large are their colonies?
2. How do the tasks of worker leaf-cutter ants change as the teeth on their mandibles become blunt?

SETTING **THE SCENE**

Storytelling is an age-old method of education. Aesop knew it. Homer knew it. Humans are storytelling animals! This method of learning lends itself well to the study of animal biology, so we begin with several stories to illustrate the major themes in this book. Our initial question: "Is life on Earth **monophyletic**?" (in other words, did all species arise from a common ancestor?) Then we will look at the impact of food material on tooth wear in killer whales (*Orcinus orca* (1762)) before moving on to what might be happening in a digestive tract. We also explore how aphids obtain genes from fungi and use them to synthesize carotenoids, and then consider what we can learn about reproduction from fossils. The fossil theme continues with evidence about bipedalism and the importance of feet. We will see how body plans—the organization of animals—are influenced by *Hox* genes. Sperm provide another version of diversity, showing great variation in form reflecting their main task, namely fertilizing eggs under different conditions. Then we will see what underlies the ability to taste salt. We will return to leaf-cutter ants and their fungi as an example of coevolution. We also present another example from genetics, namely the use of transgenic fungi to protect people from malaria. We finish with two stories about animals with solar panels. This sampling mirrors the range of topics and level of detail that you will encounter as you proceed through the book.

Figure 1.3 The cleared transport route of a colony of leaf-cutter ants has two branches merging in the centre before passing under a branch and proceeding out of the picture in the upper right corner.

SOURCE: M. B. Fenton

1.1 Some Stories of Zoology

Common Ancestry

Darwin proposed that all life evolved from a common ancestor, so the idea of **universal common ancestry** is central to the theory of evolution. Most phylogenies are somewhat more focussed, involving specific phyla rather than all life. Evidence of horizontal gene transfer among early species is considered by some to undermine the hypothesis of universal common ancestry (see page 95). However, by using models that did not assume that shared sequences imply genetic relationships, Douglas L. Theobald provided powerful statistical evidence that life on Earth is monophyletic. Although we do not always point out universal common ancestry in the stories that appear in this book, it is fundamental to the relationships among all animals and drives many of the comparisons that we make.

Carotenoids

Carotenoids are widespread molecules that play many roles in the metabolism and ecology of animals. Carotenoids are used for their antioxidant properties and also serve as precursors for visual pigments. In spite of these vital roles, most animals cannot make carotenoids and must obtain these complex molecules from their food: plants, fungi, or microorganisms.

One animal is an exception: aphids can make their own carotenoids (Fig. 1.4). The aphid genome encodes many enzymes for biosynthesis of carotenoids. In several species of aphids, the carotenoid content can vary between colour morphs of a single species (e.g., *Macrosiphum liriodendron* (873) and *Sitobion avenae* (877)). The green forms have α-, β-, and γ-carotene (yellow or yellowish compounds), and the red forms also contain lycopene or torulene (red compounds). In red aphids, a 30-kilobase region of the DNA encodes one carotenoid desaturase, an enzyme missing from green aphids which cannot synthesize the red pigments derived from α-carotene. Pea aphids (*Acyrthosiphon pisum* (865)) can also be red or green in colour, partly influenced by environmental conditions and also by selective predation, with some predators taking more red than green aphids and other predators taking the opposite. Green pea aphids contain mainly γ-carotene, β-carotene, and α-carotene, while red ones have these carotenes as well as torulene and dehydro-γ,ψ-carotene which can be derived from α-carotene (Fig. 1.4). Genetic analysis reveals that the aphid versions of these genes are derived from fungal genes (Fig. 1.5).

Think of animals that are spectacularly coloured, where the colours reflect carotenoid content. For a long time we thought that they always came from the animals' food. The aphid story alerts us to the possibility that other animals may also be able to make their own carotenoids. This story serves to illustrate that we are constantly learning new things about animal form, function, and evolution.

STUDY BREAK

1. What does monophyletic mean?
2. Give an example of horizontal gene transfer. Why could it speak against the concept of universal common ancestry?
3. What roles can carotenoids play in animals?

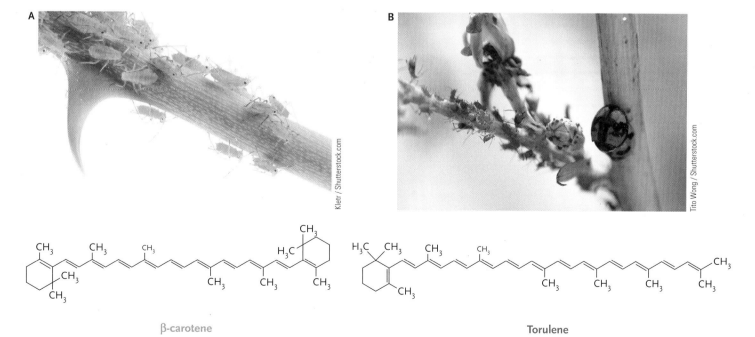

β-carotene

Torulene

Figure 1.4 Variation in carotenoids affect colour in aphids. **A.** Green morphs of pea aphids have beta-carotene, while **B.** red morphs have torulene.

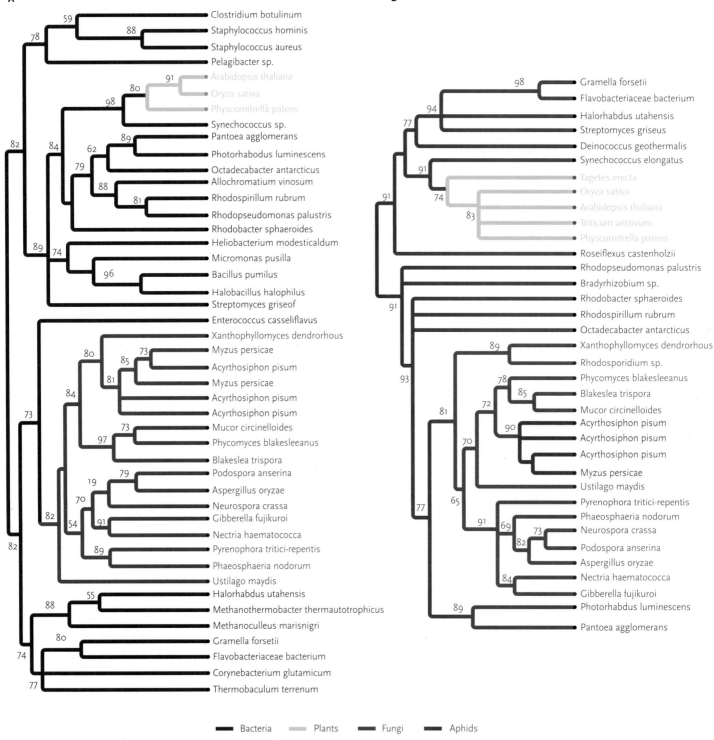

Figure 1.5 Phylogenetic relationships based upon two enzymes involved in carotenoid biosynthesis: **A.** carotenoid desaturases and **B.** carotenoid cyclase-carotenoid synthesases. Genes for the synthesis of these enzymes have been identified in the genome of pea aphids (*Acyrthosiphon pisum* (865). The phylogeny demonstrates that aphids have acquired genetic control over the production of carotenoids by assimilating genes from fungi into their genome. The colors indicate DNA sequences obtained from bacteria (black), plants (green), fungi (brown), and aphids (blue). Statistical analyses (bootstrap values presented as numbers) indicate support >50%.

SOURCE: From Nancy A. Moran and Tyler Jarvik, "Lateral Transfer of Genes from Fungi Underlies Carotenoid Production in Aphids," *Science* 30 Apr 2010; 328: 624-627. Reprinted with permission from AAAS.

Wear and Tear

We started with a story about wear on teeth of the mandibles of a leaf-cutter ant. Many animals use teeth or toothlike structures to obtain and consume food.

Teeth are often a vital first step in food acquisition and digestion. Compared with leaf-cutter ants (individual workers weigh 0.005 g), killer whales (*Orcinus orca* (1762), each weighing about 2 million grams) are at the other end of the size spectrum.

Most mammals have milk teeth (also called baby or deciduous teeth) that are replaced by adult teeth at the time that they have finished growing, making them **diphyodont** (= having two generations of teeth). Like many other odontocete (toothed) whales, killer whales are **monophyodont** and lack milk teeth. Unlike other mammals, killer whales do not use their teeth to chew food, but rather to catch, kill, and tear up their prey. Killer whales are **apex predators** because they are not prey for other animals.

Although there is only one species of killer whale in the world, biologists recognize three different populations: resident, transient, and offshore. Genetically and behaviourally distinct, these populations play different roles in marine ecosystems. The patterns of tooth wear differ substantially among the different populations (Fig. 1.6). Off the coast of British Columbia (Canada), resident killer whales eat mainly fish (salmon) and squid, while transient killer whales prey more on marine mammals such as seals, sea otters, and other whales. The discovery that offshore killer whales often showed heavy tooth wear (Fig. 1.6C) coincided with the recognition that they often eat sharks. At one time, people used shark skin as sand paper because of the abrasive qualities of the dermal denticles (scale-like structures) in the skin. For the same reason, shark skin causes extreme wear on the teeth of killer whales that eat mainly sharks. As we shall see on page 228, other mammals have continuously growing and abrasion-resistant teeth that, combined with different patterns of tooth replacement, protect the owners from starvation arising from worn-out bites.

In the matter of dealing with wear and tear, leaf-cutter ants appear to be ahead of killer whales in that an individual's role and the social structure of the colony change with tooth wear. In studying animals, we will see many examples of different solutions that have evolved in answer to similar challenges.

Figure 1.6 Teeth of killer whales (*Orcinus orca* (1762)) show different patterns of tooth wear associated with differences in diet. **A.** Resident, **B.** transient, and **C.** offshore.

Guts

Guts—digestive tracts—can be filled with food in different stages of digestion, but they also house huge populations of commensal bacteria and other organisms. A typical mammalian gut contains 10^{13} bacteria belonging to two phyla, Bacteroidetes and Firmicutes. The guts of many animals, including mammals such as people, can also be home to a host of parasites from a range of animal phyla (Platyhelminthes, Nematoda, etc.). Commensal bacteria and parasitic worms interact with one another. Successful infection of the large intestine of a mouse by a nematode worm (*Trichuris muris* (592)) depends upon microflora that modulate the host's immune response to the parasite. In mice, a helminth-driven and regulatory helper T-cell response counters infection and repairs damage caused by the parasites. Reducing the number of bacteria in the

mouse's large intestine reduces the number of parasite eggs that hatch. Synergy between parasites and symbionts has not been foremost in the minds of those studying the impact of either group on the host, but it demonstrates the complexity of interactions among many animal groups (see Chapter 10).

STUDY BREAK

1. How are killer whales unlike other mammals?
2. What is surprising about the interactions between a nematode and bacteria in a mouse's large intestine?
3. What is the difference between a symbiont and a parasite?

Fossils and Reproduction

At first glance, you may not think that we can learn much about reproduction from fossilized animals. You, like other scientists, may be surprised! Just as we can learn about the division of labour in leaf-cutter ants by comparing tooth wear and the roles that the ants play in the colony, we can often make inferences about functions from structures. Two specimens of a 16.5-cm-long cartilaginous fish, a holocephalian or ratfish (*Harpagofututor volsellorhinus*) from a Carboniferous deposit in Montana (USA), have multiple fetuses in their uteri. We can be sure that the fossils are of fetuses because they are the same species (the cranial, tooth plate, and pectoral morphology are the same) and there is no sign of digestion that might suggest that the young fish had been consumed by the adult. Each of the two adults has four or five fetuses, one of which is larger than the others in each adult. The young all face anteriorly and their vertebral columns are bent in a U-shape, suggesting that they had been enclosed in an egg. Because there is no evidence of a yolk sac or external gill filaments, the fetuses probably relied on intrauterine absorption of nutrients and gaseous exchange. Therefore, the fossils suggest that the mother provided nourishment for the developing young (**matrotrophy**) as opposed to yolk-based feeding (**lecithotrophy**). It is possible that the fetuses ate eggs (**oophagy**) or mucoid-protein secretions of proteins and lipids from the lining of the uterus (**histotrophy**). Detailed examination of the fossils suggests histotrophy.

In each female, the larger fetus is more developmentally advanced than the smaller ones, suggesting sequential ovulations. Thus, the female is able to carry multiple litters in the uterus at the same time (**superfetation**), a phenomenon reported from some other fish where it may or may not involve matrotrophy.

The data from these fossil ratfish suggest that development of young within the body of the female and associated live birth appeared early in the evolutionary history of vertebrates. Similar situations have been reported in placoderms, another ancient lineage of jawed fishes both earlier (Devonian) and later (Triassic) in the fossil record. This raises the possibility that bearing live young, whether by **vivipary** or **ovovivipary**, could be an ancestral trait among vertebrates (see Chapter 14). As yet we do not have data about reproduction in fossil agnathans, the jawless fishes, to answer that question.

Bipedalism

Lucy (*Australopithecus afarensis* (1852)), one of the most famous hominins, provides a glimpse of one of our ancestors as they appeared about 3.2 million years ago. Originally described from one female specimen, we now have information about male *A. afarensis*, which were considerably larger than the females. Although the pelvis and legs of *A. afarensis* suggested that it was bipedal as we are, other crucial material that supported this hypothesis was not described until 2011.

The discovery of foot bones of *A. afarensis* confirmed that it was bipedal. Specifically, the fourth metatarsal (a long bone in the foot) and arches of the feet are more similar to those of *Homo sapiens* than they are to those of either chimpanzees or gorillas (Fig. 1.7). The relative location of the metatarsal-phalangeal articulation ("diaphysis angle" shown by arrows in Fig. 1.7), as well as the overall angle between the metatarsals and the ground, support the hypothesis that *A. Afarensis* was primarily bipedal. Like in humans, the more dorsally oriented diaphyses and the large angle between the diaphysis and the plantar surface are important in the dorsiflexion of the foot during a bipedal stride. The more plantarly oriented diaphyses of chimps and gorillas, and the small angle between the metatarsals and the plantar surface of the foot, reflect the importance of plantarflexion ("gripping") during arboreal movement. This example illustrates the importance of the morphology of fossils in understanding the functions of the animals they represent.

Other discoveries add to our knowledge about locomotion in our own species. Anyone who has experienced the pain and discomfort of a broken hallux (big toe) will appreciate its importance when it comes to walking, let alone running. People who have lost a big toe can achieve good mobility with a prosthesis. Interestingly, the first records of big toe prostheses came from Thebes, near Luxor in Egypt. Made of a type of papier mâché of linen that had been soaked in animal glue and covered with plaster, the two prostheses discovered to date are shaped like a right big toe (see prosthesis in Fig. 1.8). Both prostheses included a false nail and showed signs of wear.

STUDY BREAK

1. What is the difference between matrotrophy and lecithotrophy? What other patterns of feeding do embryos show?
2. What features of the fourth metatarsal bones occur in bipedal animals? Why are these important?
3. What role does the big toe play in walking and running?

Body Plans and *Hox* Genes

The previous stories have illustrated the importance of recognizing and understanding the relationships between structure and function when studying animal biology. In the past few decades, great strides have been made in understanding genetics, which underlies

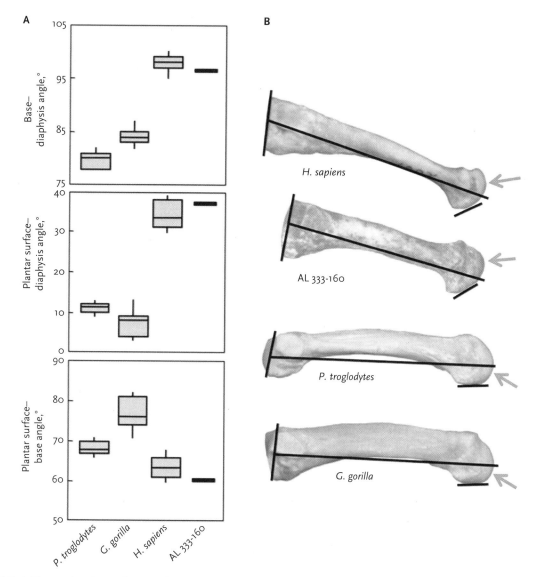

Figure 1.7 **A.** The angular relations between the proximal and distal ends of the fourth metatarsal bone is compared among humans, *Australopithecus afarensis* (1852), chimps (*Pan troglodytes* (1866)) and gorillas (*Gorilla gorilla* (1857)). **B.** Box plots of the data. Together the bones and the angles indicate that the fossil species Lucy was as bipedal as humans.

SOURCE: From Carol V. Ward, William H. Kimbel, and Donald C. Johanson, "Complete fourth metatarsal and arches in the foot of Australopithecus afarensis," *Science* 11 February 2011: Vol. 331 no. 6018 pp. 750–753. Reprinted with permission from AAAS.

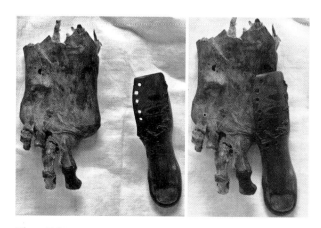

Figure 1.8 At Luxor in Egypt, a mummy missing its right big toe was found with a prosthesis (left). At right is the prosthesis in place. The tomb dates to the 18th dynasty of King Amenhotep II.

SOURCE: MARWAN NAAMANI/Staff/AFP/Getty Images

the morphological and evolutionary concepts. While traditional courses in animal form and function and those in cellular and molecular biology might have separated students into two distinct groups, the boundaries between these types of study are becoming increasingly blurred as we rely more and more upon molecular and genetic data to help explain functional and phylogenetic relationships.

Tetrapods (animals with four legs) have 13 groups of *Hox* genes tightly clustered at four loci, *HoxA* to *HoxD*. The number of repeated body segments along the vertebrate body axis is perhaps best exemplified in tetrapods by the numbers of vertebrae, varying from <10 to several hundreds. Although *Hox* genes show widespread importance in the expression of serial homology (copies of the same structure in a sequence),

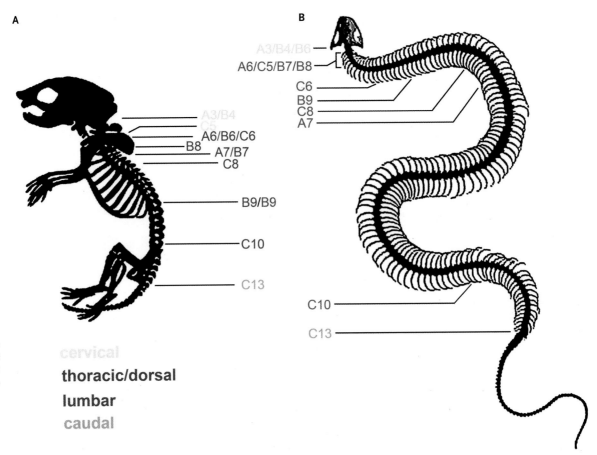

Figure 1.9 A. In the skeleton of a mouse, regional specializations of vertebrae are more obvious than they are in a snake (**B**). Regional specializations include cervical, thoracic/dorsal, lumbar, and caudal. Even in the absence of regional specializations, as shown in the snake, there are similar patterns of expression of *Hox* genes. The approximate anterior boundaries for *Hox* gene expressions are shown.

SOURCE: Reprinted from *Developmental Biology* 332, Woltering, J.M., F.J. Vonk, H. Muller, N. Bardine, I.L. Tuduce, M.A.G. de Bakker, W. Knochel, O. Sirbu, A.J. Durston, M.K. Richardson, "Axial patterning in snakes and caecilians: evidence for an alternative interpretation of the Hox code," Pages 82–89, Copyright 2009, with permission from Elsevier.

there is considerable variation in the number of *Hox* genes in different organisms, as well as in their organization and their patterns of expression (Fig. 1.9).

This is particularly obvious in Squamata, the group that includes snakes and lizards. Variations in the numbers of vertebrae could reflect modification of *Hox* genes. This could involve changes in structure through removal of introns from mRNA that encode the gene product. It also could reflect regulation of gene expression through molecular "clocks" that control the rate at which somites are formed during development. There are major alterations in *Hox13* and *Hox10* expression in the caudal and thoracic regions of corn snakes (*Pantherophis guttatus* (1515)) compared with other reptiles including a turtle (*Trachemys scripta* (1430)), a tuatara (*Sphenodon punctatus* (1458)), a slowworm (*Anguis fragilis* (1462)), a gecko (*Gekko ulikovskii* (1499)), and a green anole (*Anolis carolinensis* (1464)). These alterations in gene expression reflect a higher rate of segmentation in snakes relative to the developmental rate in other reptiles, leading to a greatly increased number of somites.

In their regional specialization of vertebrae, squamates differ strikingly from other reptiles, as well as from mammals, birds, and frogs. On their own, the squamates show considerable variation, demonstrating how the *Hox* gene system has been modified in association with different body plans. For more about the diversity of body plans, see Chapter 2.

Sperm

Sperm (male gametes) are the most diverse type of animal cells, apparently reflecting the variability in the conditions under which sperm operate. Like other features, this diversity can be demonstrated both between and within phylogenetic groups. Sperm of hermaphroditic flatworms (Platyhelminthes) in the genus *Macrostomum* (215) reflect this diversity of cell type. In these flatworms, complex sperm design reflects reciprocal mating (exchanges of sperm both ways between individuals), but the design complexities do not occur in species that use hypodermic insemination (Fig. 1.10). As the name implies, in these species each individual uses a hypodermic "syringe" to inject sperm into the other flatworm's body. The design of these sperm is simpler as is the design of the female genital anatomy.

How are sperm activated? After sperm are rather deposited in the female reproductive tract, the ovarian hormone progesterone initiates gene transcription by binding to a nuclear receptor. For over 20 years we have known that progesterone induces an immediate movement of calcium ions (Ca^{++}) into human sperm. Two papers published in 2011 revealed that in humans, progesterone activates intracellular signalling by acting on CatSper, a sperm-specific Ca^{++} channel. This channel is located only in the plasma membrane of the "principal piece" of the sperm tail (Fig. 1.11). This discovery is interesting because it may offer a means for treating infertility in humans arising from failure

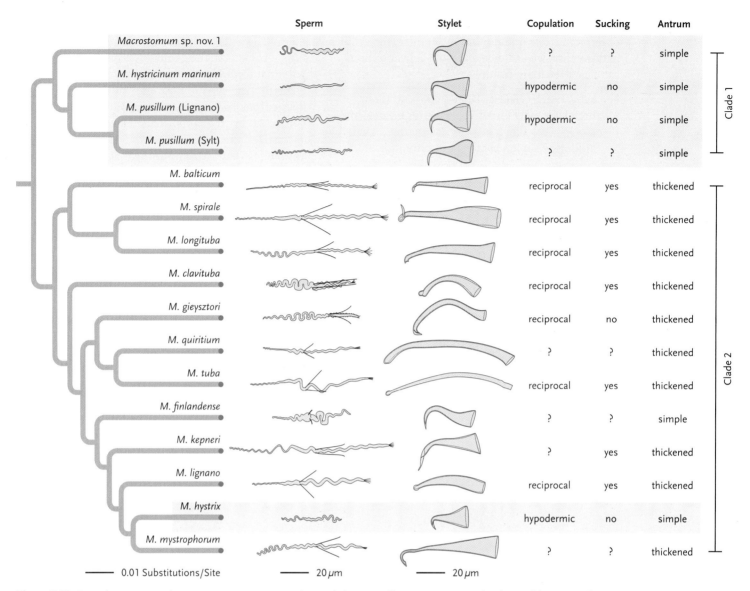

	Sperm	Stylet	Copulation	Sucking	Antrum
Macrostomum sp. nov. 1			?	?	simple
M. hystricinum marinum			hypodermic	no	simple
M. pusillum (Lignano)			hypodermic	no	simple
M. pusillum (Sylt)			?	?	simple
M. balticum			reciprocal	yes	thickened
M. spirale			reciprocal	yes	thickened
M. longituba			reciprocal	yes	thickened
M. clavituba			reciprocal	yes	thickened
M. gieysztori			reciprocal	no	thickened
M. quiritium			?	?	thickened
M. tuba			reciprocal	yes	thickened
M. finlandense			?	?	simple
M. kepneri			?	yes	thickened
M. lignano			reciprocal	yes	thickened
M. hystrix			hypodermic	no	simple
M. mystrophorum			?	?	thickened

— 0.01 Substitutions/Site   — 20 µm   — 20 µm

Figure 1.10 Reproductive traits of *Macrostomum* (215) mapped on a phylogeny to illustrate variation in the shape of the sperm, the stylet (organ for transmitting sperm), and the antrum (female organ for receiving sperm). "Sucking" refers to a worm placing its mouth over its own female genital opening and sucking. This usually occurs right after copulation.

SOURCE: Lukas Schärer, D. Timothy J. Littlewood, Andrea Waeschenbach, Wataru Yoshida, and Dita B. Vizoso, "Mating behavior and the evolution of sperm design," PNAS January 10, 2011.

to activate sperm. Also interesting is the finding that progesterone has a much stronger effect on human sperm than on mouse sperm, suggesting a diversity of sperm activation mechanisms even among mammals.

As we shall see, many other animals show intriguing variety in reproductive behaviour at the cellular level, such as in this story, as well as at the organismal and population levels.

Study Break

1. What is the importance of *Hox* genes in the radiation of squamates?
2. How does the pattern of insemination affect sperm structure in hermaphroditic flatworms?
3. What is a hermaphrodite?

Salt

Often a matter of survival, the sense of taste is widespread in animals and involves a range of sensors. Salt—NaCl—is an interesting substance because it plays a vital role in virtually every fluid in the body. As you may have learned, tasting salt can trigger two quite

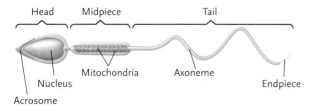

Figure 1.11 The principal piece of the tail of human sperm is the site at which progesterone acts on CatSper, a Ca++ channel through which Ca++ ions enter and activate the sperm tail.

different responses: aversion to a concentrated solution or attraction to a dilute solution. In mice, the cellular pathway involved in detecting low concentrations of salt has three main features:

- activation by NaCl at 10 mM concentration,
- high selectivity for Na versus other cations, and
- blockage of the generation of neural signals by amiloride, an ion channel inhibitor.

The high concentration pathway responds only at 150 mM NaCl or greater, is less selective in that it also responds to other cations, and is not blocked by amiloride. Salt detection in mice is mediated by a specific population of taste-receptor cells with epithelial sodium channels. This discovery demonstrates that each of the five basic tastes (sweet, sour, bitter, umami, and salt) are mediated by independent cellular substrates, and lends insight into the cellular basis of sensory biology. As we shall see, other specialized sensors are involved in communication both within animals and between them (see Chapter 15).

Co-Evolution

The symbiotic association between leaf-cutter ants and fungi goes back more than 50 million years. In this time, leaf-cutter ants have cultivated at least 553 fungi cultivars. Single species of leaf-cutter ants may use more than one cultivar and readily switch among them. The ants' gardens are subject to attack by a fungal parasite in the genus *Escovopsis*, which has the potential to devastate the cultivated fungi. Another element in the ant–fungus relationship is a filamentous bacterium, an actinomycete that is also cultivated by the ants. This bacterium grows in specialized surfaces on the ants' bodies and produces antibiotics that act as antifungal agents to slow the growth of *Escovopsis*.

Thus the ant–fungus system actually involves three mutualists: the ants, the mutualistic bacterium, and the fungal cultivars, all depending on successful cultivation of the fungi. All are potentially vulnerable to *Escovopsis*. A phylogenetic analysis (Fig. 1.12) reveals occasional host-switching by *Escovopsis*, revealing continuous adjustment by the symbionts (ants, bacterium, and fungal cultivar).

The story of leaf-cutter ants contains at least one more surprise to biologists, namely that nitrogen fixation occurs in some ant gardens. Specifically, N_2-fixing bacteria have been isolated from fungus gardens in 80 colonies of leaf-cutters (five species of *Acromyrmex* (980) and three species of *Atta* (986)). Isotopic analysis revealed that the fixed nitrogen enters the ant biomass, representing a previously unrecognized source of nitrogen input into Neotropical communities. This illustrates that animals influence

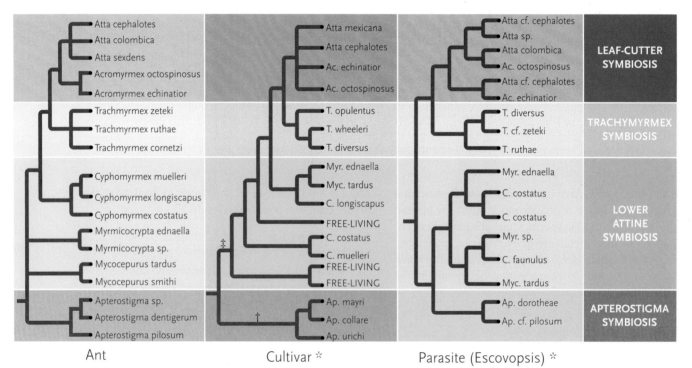

Ant   Cultivar *   Parasite (Escovopsis) *

Figure 1.12 The four panels in this figure portray the tripartite coevolution of leaf-cutter ants, fungal cultivars, a pathogen that affects the fungi, and the various categories (leaf-cutter symbiosis, *Trachymyrmex* symbiosis, lower attine symbiosis, and *Apterostigma* symbiosis). Congruent phylogenetic groups are indicated by colours. Names of the cultivars and parasites correspond to the species name of the ant gardens from which they were obtained. The † identifies derived ants in the genus *Apterostigma* (985) that switched between fungal cultivars. The ‡ indicates cultivars associated with less specialized ants in the genus *Atta* (986), including some free-living fungi.

SOURCE: From Currie, C.R., B. Wong, A.E. Stuart, T.R. Schultz, S.A. Rehner, U.G. Mueller, G-H. Sung, J.W. Spatafora and N.A. Straus, "Ancient tri-partite coevolution in the attine ant-microbe symbiolsis," *Science* 17 January 2003: Vol. 299 no. 5605 pp. 386–388. Reprinted with permission from AAAS.

ecaology as much as ecology imposes selective pressures on animals.

Transgenics and Malaria

As we will see on pages 521–523, malaria is a disease that kills over a million people annually. Almost half of the world's human population is at risk of contracting the disease. The persistence and spread of malaria reflects increased resistance to treatments among both the vectors (mosquitoes) that spread the parasites and the protozoan parasites (*Plasmodium* sp.) that cause the disease.

The fungus *Metarhizium anisopliae* infects mosquitoes by penetrating their cuticle and proliferating in their hemolymph (blood). Using recombinant strains of *M. anisopliae* that attack the sporozoite stage of the parasite (see Figure 18.29 on page 521), researchers reduced counts of sporozoites by 71% to 98% (Fig. 1.13). The key to using fungi such as *M. anisopliae* is that it can be applied by spraying mosquitoes directly on surfaces where they sit, in much the same manner that more traditional pesticides were used against mosquitoes. In this approach, genetic modification of the fungi makes them more effective at reducing the numbers of sporozoites in mosquitoes, thus reducing the chances of a mosquito spreading the sporozoites to humans.

A great potential advantage to this approach is that *M. anisopliae* can infect several species of mosquitoes, including *Anopheles gambiae* (954), *A. funestus* (954), and *A. arabiensis* (954), increasing the impact of the treatment. The animal world contains an undiscovered wealth of systems that humans can benefit from, either directly or indirectly.

STUDY BREAK

1. What three components are involved in the coevolution of the fungal gardens of leaf-cutter ants?
2. What are the differences between high-concentration and low-concentration salt sensors in mice?
3. How does genetic manipulation of a fungus affect its impact on malaria?

Animals with Solar Panels

The solar sea slug (*Elysia chlorotica* (466), Figs. 1.14A and 10.17) extracts chloroplasts from the algae on which it feeds and moves them into their tissues. These solar sea slugs have the genes necessary to control the operation of the chloroplasts, and directly benefit from the photosynthetic activity of the chloroplasts. Other species in the phyla *Porifera* and *Cnidaria* also take up chloroplasts and benefit from at least the maltose and glycerol they produce. Although genetic techniques have allowed recent exploration of the details of the interactions between the animals and the chloroplasts, the association has been known for over 100 years.

Adult spotted salamanders (*Ambystoma maculatum* (1404)) spend most of their lives underground. In spring, they lay their eggs in pools. The eggs develop, hatch into larvae (tadpoles), and metamorphose into adults which leave the pools and move underground.

A symbiosis between algae and the embryos of spotted salamanders was reported in 1888. Individual eggs within large jelly masses appear green because of accumulations of algae around the developing embryo. Only eggs developing in lighted areas acquire algae (Fig. 1.14B), and embryos with algae develop faster, hatch more synchronously, and suffer lower mortality than eggs lacking algae. In 2011, genetic techniques allowed Ryan Kerney and his colleagues at Dalhousie University (Halifax, Canada) to demonstrate the presence of algal DNA in the oviducts of female salamanders as well as an intracellular invasion of salamander tissues by the algae. This situation and the photosynthetic endosymbiosis of protists, sponges, and cnidarians are very similar. For more about symbiotic interactions between animals, see Chapter 10.

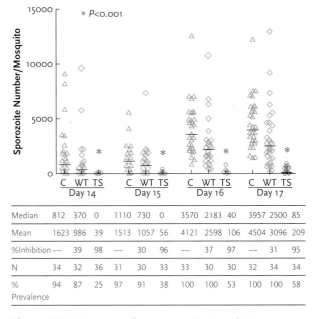

	Day 14			Day 15			Day 16			Day 17		
	C	WT	TS	C	WT	TS	C	WT	TS	C	WT	TS
Median	812	370	0	1110	730	0	3570	2183	40	3957	2500	85
Mean	1623	986	39	1513	1057	56	4121	2598	106	4504	3096	209
%Inhibition	--	39	98	--	30	96	--	37	97	--	31	95
N	34	32	36	31	30	33	33	30	30	32	34	34
% Prevalence	94	87	25	97	91	38	100	100	53	100	100	58

Figure 1.13 The impact of transgenic (TS) *Metarhizium* strains on the prevalence and density of sporozoites in mosquitoes. The TS fungi expressed scorpine and [SM1]$_8$ scorpine. Each mosquito was infected with ~90 fungi 11 days after a blood meal containing *Plasmodium*. The presence of sporozoites in mosquito salivary glands was scored. C indicates control, WT identifies mosquitoes infected with *Plasmodium* and wild-type *Metarhizium*.

SOURCE: From Weiguo Fang, Joel Vega-Rodríguez, Anil K. Ghosh, Marcelo Jacobs-Lorena, Angray Kang, and Raymond J. St. Leger, "Development of Transgenic Fungi That Kill Human Malaria Parasites in Mosquitoes." *Science* 25 February 2011: Vol. 331 no. 6020 pp. 1074–1077. Reprinted with permission from AAAS.

A   B

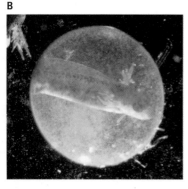

Figure 1.14 A. Solar sea slug (*Elysia chlorotica* (466)). **B.** Developing egg of a spotted salamander (*Ambystoma maculatum* (1404)). The egg has a developing young and is about 3 mm in diameter.

SOURCE: (A) Dr. Mary Tyler & Dr. Mary Rumpho, University of Maine, (2008). "Horizontal gene transfer of the algal nuclear gene psbO to the photosynthetic sea slug *Elysia chlorotica*," *PNAS*, 105 (46), 17868, Copyright 2008 National Academy of Sciences, U.S.A. (B) Gustav Verderber/ Visuals Unlimited, Inc.

STUDY BREAK

1. What do *Elysia chlorotica* (466) and *Ambystoma maculatum* (1404) have in common? How do they differ?

1.2. Teaching and Learning

The wealth of information about animals posed an important challenge in writing this book. On several occasions, we joked about having to stop reading, walking the line between too much material for the space at our disposal, or between too much and not enough detail. We recognize that the level of detail a student needs to know is determined by the instructor, not the text.

The ability to read proficiently is a basic skill for students in particular and citizens in general. Studies of identical (**monozygotic**) and fraternal (**dizygotic**) twins suggest that as much as 82% of an individual's reading skills are genetically determined. Oral reading fluency (ORF) can be tested, and is expressed in terms of gain over first and second grades. When individual performances are compared between twins, the data suggest high intraclass correlation of ORF scores (0.84) in monozygotic twins and dizygotic twins (0.59). These analyses suggest that teachers and family environment can affect reading proficiency.

The implications of ORF levels for students as they progress through the educational system is not clear. Using the internet to access electronic libraries is very liberating, easily allowing students to explore any topic

well beyond the information presented in a class or in a text. But this liberation comes at a cost if a student's performance is assessed based on material presented in class. The question about what is "testable material" is also in the realm of the instructor.

1.3 Generating a Perspective: Links among Chapters

In this introduction we have presented examples that connect to particular chapters in the book. It is important to remember that no chapter should be viewed in isolation because every aspect of an animal's biology is connected. Consider for a moment the cheetah (*Acinonyx jubatus* (1705)) in Figure 1.15. This animal's muscles and limbs are coordinated to allow burst running speeds approaching 120 km/h (Chapter 7) as it pursues prey (Chapter 8), a hunting behaviour (Chapter 17) that is coordinated by an advanced nervous system (Chapter 15). Cheetahs can maintain high speeds only for short distances because oxygen supply to the muscles is limited (Chapter 12) as is the rate of oxygen exchange in the lungs (Chapter 11). A further effect of the extreme metabolic demands of high speed is an increase in body temperature that brings the animal close to a state of heat prostration, a condition that places demands on a complex endocrine system (Chapter 16) to restore a normal body state.

We have made every effort to ensure that this book is rich in facts and new findings, in the hopes that readers will find the material stimulating and, where they want, will provide access to "new and exciting" findings.

Figure 1.15 A running cheetah (*Acinonyx jubatus* (1705)).
SOURCE: Courtesy of Naas Rautenbach

Questions

Self-Test Questions

1. Which best describes the story of the leaf-cutter ants?
   a. It shows the relationship between the metabolic needs of the colony and transportation of food down conspicuous pathways.
   b. It illustrates the interaction between species (ants, leaves and fungi), structure and function (mandibles and cutting efficiency), and animal behaviour within the colony.
   c. It describes the evolutionary history of the ant–fungal interaction.
   d. It explains the need for continual replacement of mandibles in ants as they age, so specialized foragers can provide food for the colony.

2. Why is the ability of aphids to synthesize their own carotenoids significant?
   a. All other animals must then obtain carotenoids from aphids.
   b. It allows plants, fungi and microorganisms that synthesize carotenoids to keep them for their own use, rather than having those carotenoids assimilated by animals.
   c. Identifying aphid genes for carotenoids that are derived from fungal genes illustrates a mechanism by which a group's genome can change over time.
   d. It shows how an individual pea aphid can change its colour depending upon environmental factors or predation.

3. Which of the following best illustrates the relationship between structure and function?
   a. fossil evidence for matrotrophic fetal development in Harpagofututor volsellorhinus
   b. presence of commensal bacteria and parasitic worms in mouse digestive tracts
   c. monophyodont condition of odontocete whales
   d. synthesis of carotenoids by plants, fungi and microorganisms

4. In addition to pelvic and leg structure, what fossil evidence shows that Australopithecus afarensis was bipedal?
   a. great importance of plantarflexion, as shown by a small angle between metatarsals and the plantar surface of the foot
   b. reduced "diaphysis angle" between metatarsals and phalanges to produce arches similar to those of chimpanzees and gorillas
   c. a reduced angle between metatarsals and the plantar surface of the foot, which allows for greater dorsiflexion during a normal stride
   d. foot structure more suited for dorsiflexion during a bipedal stride through a greater diaphysis angle produced by metatarsals that form a foot with more pronounced arches

5. Which of the following best illustrates the importance of studying the relationship between structure and function in animal biology?
   a. Hermaphroditic Platyhelminthes show a wide diversity in sperm structures.
   b. Expression of Hox genes in squamates is reflected in the degree of segmentation and number of somites produced.
   c. Salt detection in mice is mediated by specific taste-receptor cells.
   d. The oviducts of some species of female salamanders express algal DNA, related to the intracellular invasion of salamander tissues by the algae.

Questions for Discussion

1. Why is phylogeny important in putting adaptations of animals in context?

2. How can molecules such as carotenoids affect animals?

3. Why is malaria important? What does it tell us about animals? About people?

4. Use the electronic library to find a recent paper about animals co-opting chloroplasts. How does the story you found differ from those in this chapter?

Geoff Spiby

Geoff Spiby

STUDY PLAN

2   Body Plans, Skeletons, and Development

WHY IT MATTERS

When you look at an animal for the first time, a number of questions may come to mind. Why do animals look the way that they do? For example, what are the two animals shown in the figure above? Do they have the same body plan? Do they belong to the same phylum or to different phyla (plural of phylum)? What is the significance of their colouration? The most important of these questions is: What are they?

The scientific names of the animals in the photos can help. A computer search shows that *Pseudobiceros gloriosus* (222) is a flatworm (phylum Platyhelminthes) and that *Glossodoris symmetricus* (454) is a nudibranch mollusc (phylum Mollusca). Later in this chapter you will learn that the flatworm has a gut with one opening, while the mollusc gut has two openings (a "through" gut). But, if it's a mollusc, where is its shell? So, both look like flatworms, but only one is a flatworm.

What about the animals in Figure 2.2? Do they belong to the same phylum? Which phylum? Again, the scientific names help, identifying the *Chromodoris* (450) species as another nudibranch mollusc. But the other two are both sea anemones (phylum Cnidaria). The basic body design (apparently tubular with tentacles) and living attached to the substratum suggests the same life style (using tentacles

Figure 2.2 These three animals all have generally tubular bodies with what appear to be tentacles. They represent two different phyla, Mollusca and Cnidaria. While **B.** *Metridium senile* (172) and **C.** *Urticina felina* (174) do have tentacles and use them to catch food, the photo of **A.** *Chromodoris* sp. (450) shows the back of the animal, and the apparent tentacles are its gills.

Geoff Spiby

Strong/Buzeta

Strong/Buzeta

to catch passing food particles). Dissecting these specimens would reveal that the mollusc has a through gut, while the cnidarians have only one opening to their guts and that what appear to be tentacles on the mollusc are actually gills.

The five animals in these two figures should convince you that you need more detail than you may get in a single photograph and that you need to know if the photograph is of the entire animal. Some animals may look alike but be quite different, while others in the same group can appear to differ strikingly.

Many animals have worm-like bodies (Fig. 2.3), but closer examination often reveals that "worms" can differ considerably in detail. Figure 2.3 shows

species from two phyla, Annelida and Chordata. The chordates in this case belong to the subphylum Vertebrata (because they have backbones), with the three species representing two classes, Amphibia and Reptilia. Photo A is a burrowing snake (*Leptotyphlops dulcis* (1507)) from Texas, B is a caecilian (*Dermophis mexicanus* (1400), Amphibia), and C is another burrowing snake (*Diadophis punctatus* (1490)). Photo D is a giant earthworm (*Megascolex australis* (359), Annelida). Notice that size is not a good indicator of phylum, as the earthworm is much longer than any of the vertebrate worm-like animals. A closer look at the vertebrate animals shows that the head end may be more obvious than it is in the

LA Dawson

Franco Andreone

Figure 2.3 Are worm-like animals all worms? How many phyla of animals are represented in this figure? **A.** *Leptotyphlops dulcis* (1507), **B.** *Dermophis mexicanus* (1400), **C.** *Diadophis punctatus* (1490), and **D.** *Megascolex australis* (359).

Tony Wear/Shutterstock.com

Copyright INVERT ECO

Figure 2.4 These three animals represent three phyla: Arthropoda, Mollusca, and Annelida. Can you decide which animal is which? What features should you look for to make this decision?

SOURCE: A. Courtesy of Kai Horst George; B. Teguh Tirtaputra / Shutterstock.com; C. Dr. Greg W. Rouse.

earthworm. However, knowing what animal you are looking at still requires more detail than you can get from a photograph.

With the scientific names (*Pteraeolidia ianthina* (459), *Myrianida pachycera* (320) and *Caprella* sp. (789)) and a computer, you can place the animals from Figure 2.4 in their phyla. The name Arthropoda suggests joint-legged animals, perhaps making it obvious which animal is the arthropod (A). What about the annelid? The mollusc? You may be surprised to find that the mollusc is another nudibranch (B)! The only way of placing each of these animals in its phylum is through an understanding of its body plan and life cycle.

Figures 2.5 and 2.6 show pairs of animals that also demonstrate how superficially similar animals from quite different groups may be.

Even more striking than similarities in general appearance are similarities of body parts. Raptorial forelimbs have evolved in four orders of insects, in crustaceans (mantis shrimp), and in arachnids known as tail-less whip scorpions (Fig. 2.7). The

mantid (Fig. 2.7B) demonstrates how the forelimbs function in prey handling.

Similarities among animals may reflect other functions, for example, their value in defense. Sometimes this means camouflage (Fig. 2.8), other times it means looking like something else that is dangerous (Fig. 2.9) (see also page 481).

Particular body plans are closely associated with the ways that biologists group animals into phyla, the major divisions of the animal kingdom. A dictionary definition of a **phylum** is not particularly helpful because the definition calls it a taxonomic unit that ranks below kingdom and above class. A more useful definition is that a phylum is a grouping of animals that share a body plan that was inherited from a common ancestor. A musical analogy is useful: a symphony may be described as being a theme with variations. Similarly, a phylum can be described as a group of animals whose form is derived from variations on a common body plan.

However, this can still leave you bewildered by what the animal is (e.g., Fig. 2.10)!

Figure 2.5 Two animals from the same environment—the sea bed. Despite their similarities, these animals feed in different ways and are almost as far apart on phylogenetic tree as is possible. **A.** One, *Chondrocladia gigantea* (114), is an unusual predatory sponge (Porifera), and **B.** is a filter-feeding chordate, *Molgula pedunculata* (1112) (Subphylum Urochordata) from the Antarctic.

SOURCE: A. Courtesy of the SERPENT Project (www.serpentproject.com); B. Martin Riddle (CEAMARC)

Figure 2.6 Another pair of unrelated animals that feed in the same way (trapping active prey) but come from quite different phyla. **A.** *Actinoscyphia* species (165) is an anemone (Cnidaria, Anthozoa) and **B.** *Megalodicopia hians* (1111), a very unusual urochordate that does not filter feed.

SOURCE: A. Image courtesy of Aquapix and Expedition to the Deep Slope 2007, NOAA-OE; B. Ken Lucas/Visuals Unlimited, Inc.

Figure 2.7 A comparison of the forelimbs that five arthropods use to seize prey (raptorial). These specialized forelimbs have evolved in insects: **A.** a mantispid (Order Neuroptera), **B.** a mantis (Order Mantoidea), and **C.** a fly (Order Diptera). They also occur in **D.** Crustacea (a mantis shrimp, *Hemisquilla californiensis* (785)) and **E.** an arachnid (order Amblypygi).

SETTING **THE SCENE**

Why It Matters has emphasized that observation of external features has limited value in understanding an animal's relationships. As well, external features provide scant information on what lies underneath the skin. This chapter will explore the ways in which animal bodies are constructed, looking at common elements and how they are combined into a body plan. It will also introduce how some body plans are directly related to phylogenetic groups and have been modified to meet particular environmental demands.

2.1 Standard Parts: Animal Tissues

This chapter emphasizes the ways in which animal body plans are built on "standard parts" assembled at hierarchical levels of organization: from cells, to tissues, to organs, to organisms. While some animals have more standard parts and levels of organization than others, the elaboration of the standard parts, and the duplication and subsequent structural and functional divergence of the duplicates, can occur at any level of body organization. Duplication and divergence may occur within an animal's structure or at the level of whole animals within a colony. Underlying any body plan is the way in which an animal develops, so an understanding of general embryological patterns is an essential part of this topic.

The principle that "evolution uses what is already there" is demonstrated by the reality that all members of almost all animal phyla are built from four basic tissue types. A **tissue** is a discrete population of cells that share a set of characteristic anatomical and physiological features. The four tissue types are epithelial, connective, muscle, and nervous. These tissue types, or the individual cells within them, exhibit elaborations and specializations that vary within and between different animals. Despite specializations, the basic tissue categories are always recognizable.

2.1a Epithelial Tissue

An **epithelium** is a sheet of cells that adhere together and have three features: apical-basal polarity, specialized junctions between cells, and a basal lamina. Epithelia form the "floor plan" of all animals. Just as the walls of a house demarcate its outer boundary as well as the number, size, and placement of individual rooms, epithelia form the outer surface of an animal

Figure 2.8 A well-camouflaged scorpion fish.

Figure 2.9 A. The wasp *Polistes* sp. (1006) is a "model": a dangerous species armed with a painful sting. B. It is mimicked by a fly, *Climaciella brunnea* (958), a stingless species.

Figure 2.10 To what (phyla) do these animals belong? A. *Hymenocera picta* (820) and B. *Enteroctopus dolfeini* (495). What visible features of these animals are typical of their phylum?

and delineate its interior compartments. Animal epithelial walls play an active role in exchanges between an animal and its environment and the processes occurring within and between the individual compartments within the animal.

Apical-basal polarity refers to differences in morphological, molecular, and functional properties between the apical and basal (against the basal lamina), surfaces of epithelial cells. Morphological differences include microvilli and cilia that only appear on the apical surfaces. Other differences between apical and basal surfaces include different populations of intrinsic and peripheral membrane proteins within the two regions, including different types of ion channels, passive and active transporters, protein complexes allowing endo- or exocytosis, and receptors for intra- and extracellular signalling molecules.

Intercellular junctions (Fig. 2.11) are formed by clusters of membrane proteins interacting with similar proteins in neighbouring cells. Intercellular junctions may be **adherens** junctions that simply fasten cells together, **gap** junctions that allow small molecules and ions to move from one cell to another or **tight** (septate) junctions that prevent molecules from passing between cells within a layer, restricting movement across the layer to passage through the cells (Fig. 2.11B). Intercellular junctions also prevent membrane-localized proteins from moving between apical and basolateral surfaces.

The **basal lamina** is a meshwork of fibrillar proteins that helps to hold the epithelium together. It is secreted from the basal side of epithelial cells. The basal lamina allows epithelia to resist tensile forces and helps bind an epithelial layer to underlying tissues.

Epithelia are typically classified by morphological criteria (Fig. 2.12). **Squamous** (flat), **cuboidal**, and **columnar** epithelia are identified by the shapes of their cells. The number of layers is also important: **simple epithelium** consists of a single cell layer anchored to the basal lamina, whereas **stratified epithelium** consists of multiple layers of cells in which superficial layers have lost their connection to the basal lamina. Epithelia can also be classified functionally, for example, as **secretory**, **ciliated**, **transporting**, or **sensory**.

Epidermal epithelia actively mediate between an organism and its environment. The epidermis may release mucus onto the exterior (apical) epithelial surface to promote gaseous exchange, or enhance protein production in the apical layers to minimize desiccation or to facilitate ciliary movement. Epidermal epithelia may release defensive chemicals or excess salt ions to the exterior, such as gills of marine fishes.

When epithelia line internal compartments, they can isolate and control physiological processes within those compartments. Epithelial cells in one region of an animal's gut may release digestive enzymes into the gut lumen, while epithelia in another region of the gut may absorb the products of digestion using **transporter proteins** in the apical membranes.

Individual cells or clusters of cells in an epithelium can be highly specialized, both structurally and functionally. For example, secretory cells (secreting

CHAPTER 2 BODY PLANS, SKELETONS, AND DEVELOPMENT

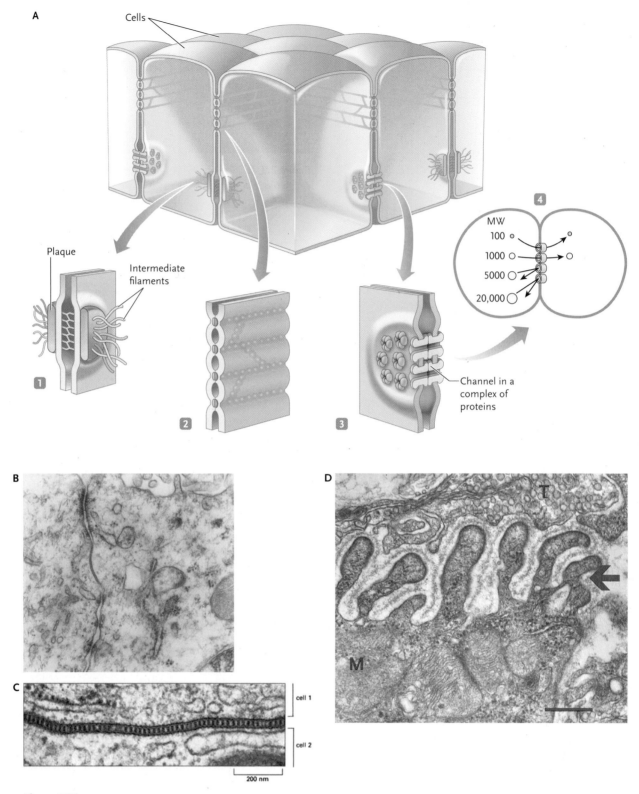

Figure 2.11

A. Adherens junctions surround cells and anchor them to their neighbours. An adherens junction has actin and protein anchors in both cells and bridges of cadherin between the cells (not visible in Figure 2.11A). Gap junctions allow amino acids, sugars, and other small molecules, but not large peptides and proteins, to pass between cells, through channels, each of which is made up of six connexin molecules. **B.** Junctions between cells in a mouse brain. Red arrow: an adherens junction; blue arrow: gap junction. Scale bar: 500 nm.
C. Septate junctions between two molluscan cells. The cell membranes are linked by parallel rows of junctional proteins.
D. Neuromuscular junction from a frog abdominal muscle. Legend: T = axon terminal (containing synaptic vesicles); M = muscle with deep sarcoplasmic infoldings; Arrow = basal lamina. Scale bar = 300 nm.

SOURCE: A. Copyright 2002 from Molecular Biology of the Cell by Alberts et al. Reproduced by permission of Garland Science/Taylor & Francis Books, LLC; B. Courtesy of Dr. Josef Spacek, Atlas of Ultrastructural Neurocytology; C. From N.B. Gilula, in Cell Communication [R.P. Cox, ed.], pp. 1–29. New York: Wiley, 1974. Reprinted by permission of John Wiley & Sons, Inc.; D. Courtesy of Dr. Josef Spacek, Atlas of Ultrastructural Neurocytology

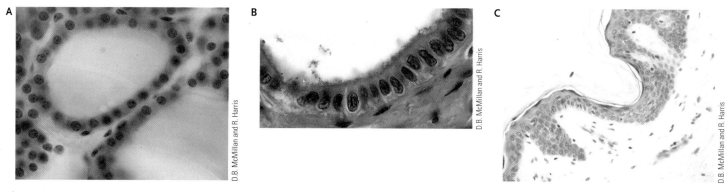

Figure 2.12
A. Cuboidal epithelium (from the thyroid gland of a dog). **B.** Columnar epithelial cells lining the gall bladder of a toad. **C.** Stratified squamous epithelium from the human skin forms a thick, protective layer. Surface cells are lost by abrasion but these cells are readily replaced by cells produced in the underlying basal layers which are mitotically active, and push older cells to the surface as they divide.

mucus or other materials) may be dispersed in an epidermal sheet, or clusters of secretory cells may be invaginated to form a subepidermal gland connecting to the surface via an epithelial duct and pore. Figure 2.13 shows a venom-secreting gland from a honey bee. The highly branched series of epicuticular canals are continuous with the epicuticle that covers the body. Thus the secretory cells are protected from their own toxic secretions by being transported in these impermeable canals.

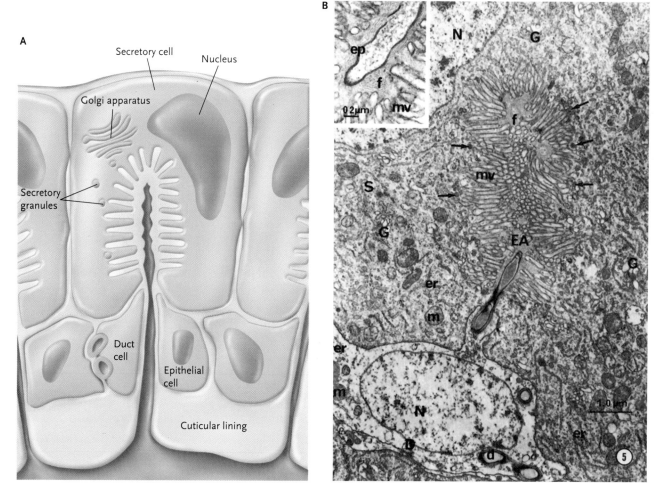

Figure 2.13 A. The diagram presents an idealized view of the secretory cells. **B.** A venom-secreting cell from a honey bee (*Apis mellifera* (984)). The epicuticle (ep) that lines the gland is extended into the secretory cell (S) and connects to the microvilli (mv) of the end apparatus (EA). Venom components made in the Golgi apparatus (G) of the secretory cell are carried as secretory granules (arrows), moved through the fibrous layer (f), and transferred from the cell to the discharge duct (d). Legend: N: nucleus; m: mitochondrion; er: endoplasmic reticulum; D = cell that secretes the duct.

SOURCE: Based on Bridges, A.R. and Owen, M.D. 1984. "The morphology of the honey bee (Apis mellifera l.) venom gland and reservoir," J. Morphology 181:69–86.

During development, epithelia may form from any of the three embryonic cell layers (ectoderm, endoderm, and mesoderm (see pages 29–30)). Epithelial tissue was probably the original animal tissue type, with nervous, muscle, and connective tissues evolving from epithelial tissue.

Sponges, some of the most primitive animals, show only vestiges of epithelial organization. However, the outside of a sponge body is a layer of cells known as **pinacoderm** separated by intercellular junctions and collagen IV, a key element in a basal lamina. Further, **choanocytes** (feeding cells, see page 213) are arranged as a layer (**choanoderm**) with clear apical and basal polarity.

2.1b Nervous Tissue

Nervous tissue consists of interconnected nerve cells (**neurons**) that transmit information between cells. This information is encoded as changes in **membrane potential**, which is a voltage difference across the plasma membrane. Membrane potentials are created by a separation of charged ions on either side of the membrane (see page 415). Animal cells use sodium/potassium pumps to maintain a voltage difference across their cell membranes. Ion channels in nerve cell membranes allow localized reversal of the membrane potential; this reversal can be propagated over the entire membrane of the nerve cell. Excitability depends on a suite of components at the molecular level associated with the morphological features of nerves and nervous tissue.

A **monopolar** neuron consists of a cell body (**soma**) that includes the **nucleus** and **perinuclear cytoplasm** and at least one long **neurite**, a thin cytoplasmic process. Most neurons are **bipolar**, having two neurites—a **dendrite** and an **axon**—extending from the soma. Signals are received by the dendrite, which transduces the signal into a change in membrane potential that is propagated down the dendrite to the soma, and then along the axon. At a **synapse**, a specialized terminal, the axon communicates the depolarization (information) to **follower neurons** or **target cells**. At synapses, the cytoplasm of the pre-synaptic neuron contains a cluster of vesicles containing chemical neurotransmitters, and the electrical signal of the nerve cell is converted into a chemical signal (see Chapter 15).

Nervous tissue is subdivided into three functional categories: **sensory neurons** receive environmental stimuli and transduce them into electrical signals, **interneurons** transport and process the incoming information, and **motor neurons** carry the "decisions" of the interneurons to effectors.

Models of nerve cell evolution (Fig. 2.14) visualize epidermal cells that communicate with their neighbours through gap junctions and have contractile tails acting as muscles. Sensory cells may have evolved from such a multifunctional cell that was electrically isolated from its

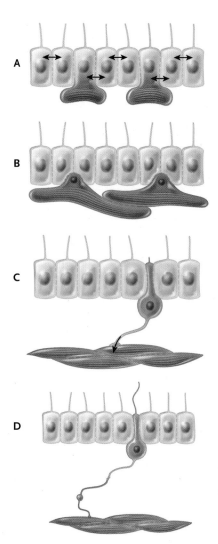

Figure 2.14
A model showing a possible way that neural and muscle cells could have evolved in the primitive metazoa. **A.** Epidermal cells (with muscle tails) that receive inputs from the environment and communicate with its neighbours. **B.** The muscle tails have separated to become specialized contractile cells. **C.** One cell takes over the role of communicating between the several epidermal cells and the contractile cells (epidermal cells are still in continuous communication with each other). **D.** A cell in the epidermal layer is isolated from its neighbours and becomes a sensory receptor cell, communicating with muscle cells through a neuron.

neighbours and received information that it passed through a second conducting cell (which might become a neuron) to a contractile cell (the beginnings of a muscle).

2.1c Muscle Tissue

Muscle tissue is composed of elongate cells specialized for linear shortening (contracting) along the long axis of the cell. Muscles are tension-generating devices whose contractions can generate movement. Often, muscles operate in pairs, for example, when contraction works in combination with skeletal elements to effect movement of a limb. Contraction of one muscle of the pair extends the limb, and this action is opposed by the other muscle that pulls the limb towards the body. Tension generated by muscle contraction also performs tasks not involving the skeleton, for example, delicate oculomotor muscles control eye movements needed to sample different areas in the visual field.

Muscle fibres are formed during development by the fusion of several cells to form a **functional syncytium** of proteins arranged as **myofilaments** within the fibres (Fig. 2.15A). It is this organization that allows the contraction of the entire muscle, as proteins within the myofilaments slide past one another. At rest, the muscle proteins are arranged so that there is some overlap of the component muscle filaments **actin** and **myosin**. Actin myofilaments are thin filaments (5–8 nm in diameter), arranged in a hexagonal pattern around the thick myosin filaments (13–18 nm in diameter). These filaments are arranged such that they form a pattern of alternating dark and light bands that give skeletal muscle its characteristic **striated** appearance. Cardiac muscle also appears somewhat striated, although

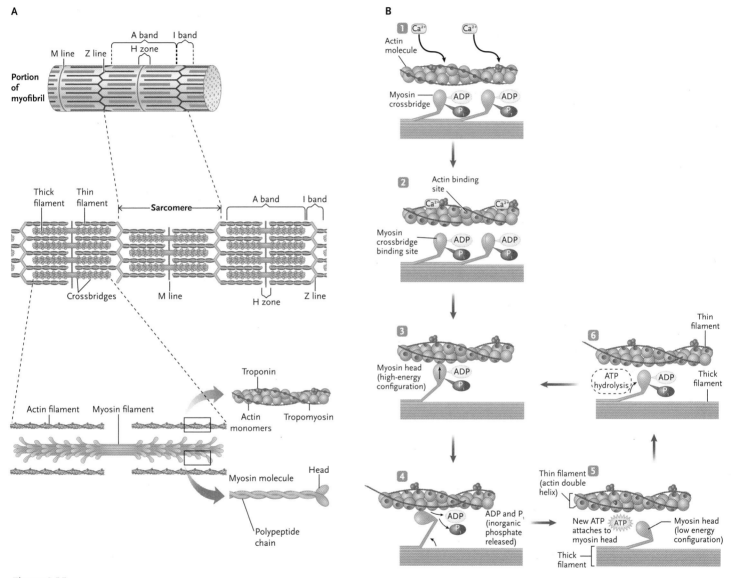

Figure 2.15

Skeletal muscle structure. **A.** Muscles are composed of bundles of cells called muscle fibres; within each fibre are longitudinal bundles of myofibrils. The unit of contraction within a myofibril, the sarcomere, consists of overlapping myosin (thick) filaments and actin (thin) filaments. The myosin molecules in the thick filaments each consist of two subunits organized into a head and a twisted, double helical tail. The actin subunits in the thin filaments form twisted double helices, with tropomyosin molecules arranged head to tail in the groove of the helix and troponin bound to the tropomyosin at intervals along the thin filaments. **B.** Proteins that regulate muscle contraction. 1. In the relaxed state, troponin completely covers the sites of actin binding to myosin, so myosin cannot bind to actin. 2. When calcium ions (released from excited muscle cells) bind to troponin, the binding sites of actin are exposed owing to the shift of tropomyosin. 3. When small numbers of myosin heads bind to the binding sites of actin tropomyosin shifts further, resulting in the complete exposure of the binding sites of actin. 4. With the binding sites of actin completely exposed, many myosin heads can bind to actin, which enables the generation of a large muscle contraction power as the myosin head pivots and bends, pulling on the actin filament. 5. New ATP attaches to the myosin head, breaking the connection with the actin filament. 6. ATP is split into ADP and P, and the myosin head returns to a position in which it will bind to the actin filament 3.

SOURCE: From RUSSELL/WOLFE/HERTZ/STARR. Biology, 2E. © 2013 Nelson Education Ltd. Reproduced by permission. www.cengage.com/permissions

somewhat less organized, as the muscle fibres are arranged to form the walls of the chambers of the heart. Smooth muscle cells also contain actin and myosin filaments, but the arrangement is so different that smooth muscle does not appear striated, even though it relies upon the same molecular mechanisms for contraction (Fig. 2.15).

Like nerve cells, muscle cells are excitable, and their excitability is due to the activity of ion channels in the plasma membrane. Depolarization of muscle cells is initiated by release of neurotransmitter at a **neuromuscular junction**, where axons form synapses with muscle cells. Depolarization then spreads over the membrane of the muscle cell, initiating a signalling cascade that leads to the sliding between the parallel arrays of actin and myosin filaments. Myosin is a **mechanoenzyme**, a molecular motor that converts chemical energy from ATP hydrolysis into mechanical work. Following depolarization of the muscle fibre, an action potential spreads deep into the interior of the muscle fibre along **transverse tubules**, infoldings of the plasma membrane. Where these structures contact the sarcoplasmic reticulum ion channels open, triggering the release of calcium ions (Ca^{++}). These ions bind to the protein troponin-C, which alters its configuration so that another protein, **tropomyosin**, is moved to expose the myosin binding sites on **actin** (Fig. 2.15B). This allows cross-bridge cycling to occur, in which the myosin head hydrolyzes ATP and is converted to its high-energy configuration. In this form, the myosin head can form a cross-bridge and bind to actin. As the myosin relaxes to its low-energy state, it releases ADP and inorganic phosphate, and the actin slides past the myosin. Binding a new molecule of ATP to myosin releases actin. The hydrolysis of ATP returns myosin to its high-energy configuration, where it can bind to a new binding site on another actin molecule farther along the thin filament. This cross-bridge cycling is asynchronous for myosin heads, ensuring that the thin actin filament continues to be moved by some myosin molecules even as other myosin heads go to their low-energy configuration and are detached from the actin. Relaxation of the muscle fibre requires the uptake of Ca^{++} back into the sarcoplasmic reticulum, so that the filamentous protein tropomyosin blocks myosin from interacting with actin.

2.1d Connective Tissue

Cells of connective tissue are positioned in an **extracellular matrix** rather than being closely interconnected. This matrix, which is similar to the basal lamina of epithelia, consists of protein fibres and **ground substance**, a protein-enriched gel in which the fibres and cells are suspended. Differences in the types of fibres, the chemical nature of the ground substance, and the relative proportion of these components typifies the variety of connective tissues with very different material proper-

ties. Loose connective tissue (e.g., cartilage) binds epidermal epithelium to underlying muscle, bone, and cartilage. Dense connective tissues, such as tendons and ligaments, attach muscles to bones or bones to other bones. Adipose connective tissue stores lipids.

Most connective tissues provide structural integrity and support. When biomineralized to form rigid endoskeletal elements, connective tissues also serve in protection and support, and transmit the forces generated by muscle contraction.

Collagen is the most prevalent fibre type within connective tissues, although reticular and elastic fibres also may be present. Tendons and ligaments consist of densely packed, parallel arrays of collagen fibres with little intervening ground substance. The collagen gives the tendons and ligaments high tensile strength and resilience. Conversely, loose connective tissues have a highly hydrated gel formed by highly negatively charged **proteoglycan** molecules that form a voluminous ground substance that attracts water molecules. **Hyaluronic** acid in the ground substance gives cartilage its characteristic material properties and biomineralization of the matrix greatly increases its stiffness in bone. The ground substance of blood, coelomic fluid, and hemolymph—all connective tissues—is a very dilute solution of salt ions and protein.

STUDY BREAK

1. What is meant by apical-basal polarity with respect to an epithelium? Why is apical-basal polarity important for the various types of tasks performed by epithelia?
2. How do nervous and muscle tissues work together?

2.2 Grades of Body Plan Complexity

2.2a   Size, Shape, and Complexity

The early evolution of animals saw several pivotal changes in body plan that profoundly influenced diversification of lifestyles. Four early changes involved:

- progression from a unicellular to a multicellular body
- increase in size
- evolution of a support (skeletal) system
- increase in mobility.

Further diversification of animals was accompanied by five more changes:

- development of a gut
- bilateral symmetry and anterior concentration of neural material
- development of a through gut (presence of a mouth and an anus)

- emergence of a secondary body cavity
- increasing complexity of the nervous system.

These changes and developments are reviewed in the following sections.

2.2a.1 Development of Multicellularity.
The transition from unicellular protists to multicellular animals occurred more than 600 million years ago. Animal multicellularity probably originated from a colony of heterotrophic protists in which daughter cells remained together as a cluster rather than dispersing as independent, single-celled organisms. Some modern protists live in colonies, and a few species have transient phases in which some cells differentiate into gamete-like cells.

A multicellular organism consists of more than one type of somatic (non-reproductive) cell. Within the individual, connected cells differentiate to perform particular tasks. No single cell carries out all of the tasks essential for survival. Component cells of multicellular organisms are subservient to the interests of the organism as a whole, which survives only though cooperation among its individual cells. Cancerous tumours illustrate the consequences when one cell line ceases to cooperate. Here, a renegade cell breaks from normal controls, proliferates by mitosis, and can threaten the survival of the animal.

2.2a.2 Increase in Size.
Growing to a larger body size may have been an important advantage of multicellular organisms over their protist ancestors. The limits imposed on size are a direct consequence of the non-linear relationship between surface area and volume for any three-dimensional object. When the surface area doubles, the volume increases by the power of three. A sphere with a radius of 1 mm has a surface area of 12.6 mm2 ($S = 4\pi r^2$) and a volume of 4.2 mm^3 ($V = 4/3\pi r^3$). A 10 mm radius sphere has a surface area of 1257 mm^2 and a volume of 4189 mm^3. The surface area:volume ratio for the small sphere is 3:1 and for the sphere with a radius of 10 mm it is 0.30:1.

Diffusion is responsible for many exchanges that occur across the cell membrane, the site of all essential exchanges of materials between a cell and its environment, including influx of oxygen and nutrients and efflux of carbon dioxide and nitrogenous wastes. The cellular machinery that consumes and generates these materials is a function of the volume of cytoplasm, while the surface available for exchange is that of the cell membrane. It is critical for a cell to have enough surface area to exchange materials for the volume of cytoplasm it contains, and this places a limit on cell size. On the other hand, multicellular organisms can increase in size by dividing the cytoplasmic mass among many individual cells. This increase in size provides access to a larger food base compared with smaller, single-celled competitors. Larger individuals are also more successful at competing for space. However, growth is still limited by restrictions imposed by ratio of surface area:volume, which limits the rate of diffusion. This constraint acts both on individual cells and on the bodies of multicellular organisms.

Overcoming the constraints of diffusion and surface area:volume imposed on multicellular organisms required further change. Important steps in the development of body plans include evolution of internal compartments for digesting food, specialized areas of the body surface for gaseous exchange, and internal transport systems to deliver nutrients and oxygen to individual internal cells and to collect their metabolic wastes.

2.2a.3 Evolution of Skeletal Systems and Animal Size.
As organisms increased in size and tissue mass, they could not maintain a functional shape without additional rigidity. The need for skeletal support emerged. Perhaps the simplest skeletal support is the incorporation of mineralized (calcium carbonate or silica) spicules into tissues (Fig. 2.16A) as seen in sponges (Porifera) and soft corals (Cnidaria, Anthozoa, Alcyonacea). Even in these simple animals, the silica spicules in some sponges (Class Hexactinellida) are

A

Dr. John D. Cunningham/Visuals Unlimited, Inc.

B

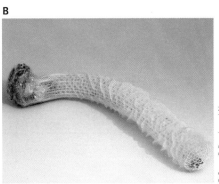

© Susan E. Degginger / Alamy

C

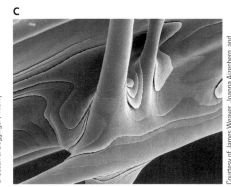

Courtesy of James Weaver, Joanna Aizenberg, and Daniel Morse

Figure 2.16 Skeletal spicules from the Porifera. **A.** Spicules from a member of the Demospongiae. **B.** Venus' flower basket (*Euplectella* sp. (126)) has a meshwork of fused silica spicules (**C.**).

fused into a complex and rigid skeletal support (Fig. 2.16B and C). True corals (Cnidaria, Anthozoa, Scleractinia) are colonial animals that secrete a rigid calcium carbonate skeleton that becomes the foundation of a modern coral reef.

For many invertebrates a **hydrostatic skeleton**—a fluid-filled compartment—provides support. In some animals, this compartment may be connected to the outside world (e.g., sea anemones, Fig. 2.17A); in others it is a single sealed compartment (e.g., Priapulida, Fig. 2.17B), and in still others it is a series of sealed compartments (e.g., oligochaete worms (Fig. 2.17C). In a sea anemone (Fig. 2.17A and 2.18A), the gastrovascular cavity is the hydrostatic skeleton. When contractile tails in the body walls contract, fluid in the cavity is forced out through the mouth as the anemone closes (Fig. 2.18B). Re-expansion of an anemone takes time because ciliary currents pump sea water back into the gastrovascular cavity.

A closed hydrostatic compartment, such as that of a priapulid or those of an earthworm (Fig. 2.17B, C), can change shape but not volume because the fluid content is incompressible. Contraction of the circular muscles around a hydrostatic skeleton lengthens the animal, while contraction of longitudinal muscles shorten it and makes it fatter. See Chapter 7, page 192 for a discussion of burrowing.

Rigid (or almost rigid) skeletons may be in the core of the body (an **endoskeleton**) or may cover the outside of the body (an **exoskeleton**). Endoskeletons are characteristic of Chordata while exoskeletons are found in

Figure 2.18
Urticina felina (174) (Cnidaria, Anthozoa). **A.** When the anemone is expanded, the gastrovascular cavity is fluid filled. **B.** When the anemone contracts, the fluid is expelled through the mouth.

Mollusca and Arthropoda. A hollow tube is more rigid than the same weight of material formed as a rod, a property whose application is obvious in arthropod exoskeletons. This property is also present, albeit concealed, in the long bones of vertebrates. An exoskeleton

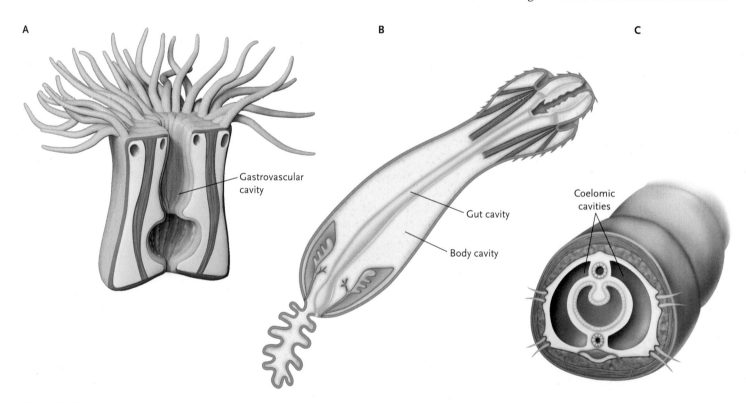

Figure 2.17
Hydrostatic skeletons. **A.** The open hydrostatic skeleton formed by the gastrovascular cavity of an anemone (Cnidaria, Anthozoa). **B.** The single hydrostatic skeleton in the body cavity of a priapulid. **C.** Compartmentalized hydrostatic skeleton in the coelomic cavities of an earthworm (Annelida, Oligochaeta).

has the obvious advantage of surrounding vulnerable soft tissues with a protective cover. But an exoskeleton limits both chemical and information exchange with the external environment, a limitation that can have both advantages and disadvantages. Further, an exoskeleton must be either moulted and rebuilt or continuously modified to allow an animal to grow, while an endoskeleton needs only to be modified to accommodate growth.

The relative mass of skeletal elements is also important when considering support and mobility. Thin-walled exoskeletons are lightweight: they use a relatively small amount of skeletal material but are very strong. If an exoskeleton were built with walls as thick as those of a mammalian femur (thighbone), this exoskeleton would be seven times as resistant to bending as the femur. On the other hand, if a femur were rebuilt as an exoskeleton using the same amount of material, it would be roughly twice as resistant to bending as an endoskeletal element. The advantage of an exoskeleton is particularly apparent in lightweight arthropods with long legs.

However, analysis of bending resistance does not consider impact loading—the reaction forces developed when a moving animal makes a rapid stop such as at the end of a jump or when striking an object. An analysis of impact loading shows that endoskeletal elements can carry much greater loads than an exoskeleton can. Vertebrates are typically larger (more massive) than arthropods and frequently very active, meaning that they develop significant force at the end of each movement. Larger terrestrial invertebrates tend to be slow-moving, and fast-moving species weighing >10 g are rare. This probably reflects the low resistance of exoskeletons to impact loads. Large animals with exoskeletons (e.g., spider crabs and lobsters) live in the aquatic world where their weight is largely offset by the small density difference between their bodies and the surrounding water.

The limitations imposed by an exoskeleton are mainly related to size. An exoskeleton follows the surface of an animal and, as illustrated above, its area increases with the square of linear dimension. Inside the skeleton, the volume of the tissues increases with the cube of linear dimension. Insect tissues are supplied with oxygen through the tracheal system, a series of open airways (see Chapter 11). The volume of the tracheal system scales with the mass of the insect. For example, the space in a beetle leg that is available for tracheae ultimately limits the size to which the beetle can grow. In the late Carboniferous and early Permian eras, partial pressures of oxygen (PO_2) were around 30 kPa, compared to about 20 kPa today (see page 306). Increased oxygen levels in the same volume of tracheal system allowed insects to grow larger than today's insects.

A further size-related issue is discussed in more detail in the consideration of Reynolds numbers in Chapters 7 and 8. This is the effect of the relatively higher viscosity of water on small marine animals when compared to larger ones. To tiny animals and larvae, seawater is a viscous medium (imagine a medium-size vertebrate trying to swim in honey). Small animals use cilia for propulsion and fight against viscosity as they move. At the same time they need to invest little energy to avoid sinking, being supported by the viscosity of the water. Larger body size overcomes viscosity, but demands more active modes of swimming as well as energy investment in buoyancy control, while at the same time allowing for much faster swimming. Copepods (Crustacea) are interesting in that they are small enough animals to be affected by viscosity but at the same time fast swimmers, or rather "jumpers." Sweeps of their enormous antennae produce momentum rapidly, allowing the copepod to overcome the relative viscosity of the surrounding water.

The interplay between size and locomotion in the evolution of animals can be viewed in two ways. In the first view, larger animals have greater access to larger food items than smaller animals and this, combined with systems for collecting and removing wastes, provides greater capacity for locomotion. This, in turn, increases the effective area over which animals can acquire food, while avoiding predators and inhospitable environmental conditions. The other view suggests that locomotor abilities evolved with increased body size because larger animals need more food and must forage further, and because larger animals experience lower predation simply because of their size. There are limits to the number of large animals that can be supported by any ecosystem; in fact, the total number of species in an environment falls as animal size increases. Large animals, produce fewer offspring compared with smaller animals, and their generation time is longer. This combination means that large animals are more dependent on stable environments than are smaller animals.

2.2a.4 The Gut, Mouth, and Anus. Protists and sponges use **phagocytosis** to ingest small food particles (see page 213). Although some sponges can reach large sizes, they do so by developing branching water canals that carry food particles into chambers lined by choanocytes, cells specialized to phagocytose food particles. All other animals have a **gut**: an internal space in which at least the initial stages of food digestion occur. A primary body cavity, the gut is lined by the **gastrodermis**, a specialized epithelium.

While the gastrovascular cavity of a sea anemone (Fig. 2.17) can serve as a skeletal component, its major function is as a compartment into which digestive enzymes are released to break down food. The surrounding gastrodermis takes up the products of digestion. In small animals that are relatively inactive, a simple sac gut can supply nutrients to all parts of the

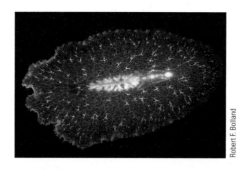

Figure 2.19 In the Polycladida (Platyhelminthes, Turbellaria), the pharynx carries food into a highly branched gut that acts as both a digestive cavity and a transport system.

Robert F. Bolland

body, simply based on its configuration. For example, the branched gastrovascular cavities of ctenophores and the gut of polyclad flatworms (Fig. 2.19) also act as transport systems. However, a simple gut with only one opening for both ingestion and egestion cannot support the metabolic demands of a large, active animal. The development of two gut openings—mouth and anus—allows food to move in only one direction through the gut. This "through gut" allows sequential processing of ingested food with different gut regions specialized for different roles, including mechanical and chemical breakdown of food and active uptake of the breakdown products. The development of muscles in the gut wall uncoupled the movement of food through the gut from the movement of the whole animal.

2.2b Secondary Body Compartments of Bilateria

Most animals have a fluid-filled secondary body compartment located between the body wall and the gut, separating the non-nutritive functions from the gut. A fluid-filled secondary body compartment can accompany increased animal size because the fluid has minimal metabolic activity and circulates close to the epidermal epithelium. This compartment borders both the exterior environment and deep internal tissues, allowing dissolved gases to be rapidly exchanged between the external environment and the internal fluid. Fluid-filled secondary body compartments can also act as hydrostatic skeletons.

There are two types of secondary body compartments: 1. pseudocoels and haemocoels and 2. coeloms (also known as a eucoelom or eucoel). **Pseudocoels** and **haemocoels** are lined by connective tissue and are basically derived from the blastocoel cavity of the early embryo (see below). A **coelom** is completely lined by epithelium (the **peritoneum**) derived from mesoderm, oriented with the apical surface facing the interior of the cavity. This allows the epithelium to control biochemical activities occurring within the secondary body compartment. Coelomic compartments have many functions, forming for example the vertebrate circulatory system and having a critical role in excretion and osmoregulation. Some examples of coeloms include the water vascular system of echinoderms, which also operates their tube feet; the segmental fluid-filled compartments of earthworms, which form their hydrostatic skeleton; and the mammalian pleural cavities, essential for drawing air into the lungs.

The mode of development of the coelomic cavity is a key feature in separating the two major lineages of animals of animals (Protostomia and Deuterostomia, see Sections 2.4d and 2.5).

2.2c Symmetry and Tissue Layers

Sea anemones (Fig. 2.17) and their relatives in the phylum Cnidaria are traditionally described as having radial symmetry, meaning that any vertical section through the axis that runs perpendicular to the mouth is identical (Fig. 2.20A). Radial symmetry means tentacles radiating in all directions, allowing reception of sensory information, detection of threats, and collection of food from any direction.

Anthozoa (sea anemones and corals) are now accepted as the most primitive of the Cnidaria, but do not have a perfect radial symmetry. They show a bilateral radial symmetry because a vertical plane through the oral axis, positioned through the siphonoglyphs, can divide the animal into mirror halves (Fig. 2.20C). Each half has radial symmetry with respect to the gastrovascular cavity and the arrangement of the septa

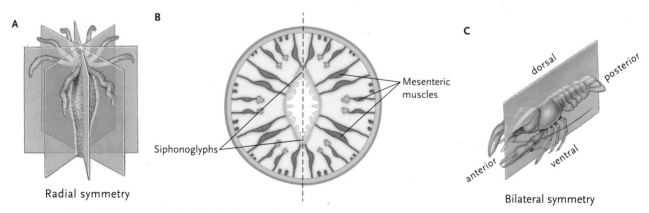

A

Radial symmetry

B

Mesenteric muscles

Siphonoglyphs

C

dorsal

posterior

anterior

ventral

Bilateral symmetry

Figure 2.20 Symmetry. **A.** The hydroid is radially symmetrical, a pattern that may have been secondarily simplified from a bilateral radial pattern. **B.** The section through an anemone shows the way in which the siphonoglyphs and the orientation of mesenteric muscles create a bilateral radial symmetry. **C.** A crayfish shows bilateral symmetry.

dividing the gastrovascular cavity. These data suggest that Cnidaria are primitively bilaterally symmetrical and that their radial symmetry is a secondary development, an interpretation supported by the arrangement of *Hox* genes (Box 2.1). In cnidarians, some members of the *Hox* gene family are arranged along the primary axis through the mouth, while others are arranged asymmetrically along an axis perpendicular to the primary axis.

An apparent secondary radial symmetry is a recurring phenomenon in animals that are attached to, or buried in, the substratum. Examples include many of the tube-living polychaete worms, phoronids, and sea cucumbers (Fig. 2.21) where the body of the animal retains its normal bilateral symmetry but a crown of feeding tentacles allows the animal to sample the environment and capture food from any direction.

The Cnidaria are **diploblastic**, which means their bodies are built from two epithelial layers (epidermis and gastrodermis) that sandwich the **mesoglea**, a layer of connective tissue. Cnidarian muscles are long tails, packed with actin and myosin filaments (Fig. 2.14A). Nerves are arranged as a neural net with no sign of any

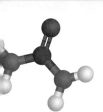

BOX 2.1 Molecule Behind Animal Biology
Hox Genes: Regional Specification along the Anterior–Posterior Axis

The new appreciation and understanding of the role that *HOM/Hox* genes play in animal development may be the most important breakthrough in developmental genetics since the discovery of the genetic basis of inheritance. *Hox* genes now include both *HOM-C* genes of fruit flies (*Drosophila* (960)) and *Hox* genes in vertebrates. They regulate development at the executive level where their expression controls major aspects of body patterning. *Hox* genes regulate regional specification of structures along the anterior–posterior axis. For example, expression of *Hox* genes in *Drosophila* development ensures that antennae develop on the head, legs and wings develop on the thoracic segments, and no legs develop on the abdomen. When one of the genes is mutated, structures develop in inappropriate places along the anterior–posterior axis. For example, legs may develop on the head where antennae ought to be.

Hox genes encode transcription factors, proteins that bind to enhancer regions of DNA to control the expression of downstream genes. In insects, the eight *Hox* genes may have arisen by duplication and divergence of two original genes. Remarkably, these genes are encoded on chromosome 3 in the order of their expression along the anterior–posterior axis of the body. That is, the *Hox* gene that lies closest to the 3' end of the chromosome is expressed in anterior structures, whereas the *Hox* gene closest to the 5' end of the chromosome is expressed in posterior areas of the developing body. This correspondence between gene order on the chromosome and expression along the anterior–posterior axis is called **co-linearity**. In essence, every segment of the metameric body of *Drosophila* has a *Hox* address, and expression of an appropriate overlapping set of Hox genes in that segment directs development of structures appropriate for that segment. In 1995 this discovery won a Nobel Prize in Physiology for E.B. Lewis, C. Nüsslein-Volhard, and E. Wieschuas.

Flies are very different from mammals, so genes controlling development in a fly would not be expected to also control development in a mouse. However, homologues of fly *Hox* genes are present in mammals and perform the same function in specifying regional features along the anterior–posterior axis (Fig. 1). While the actual structures produced in flies and mice are very different, the ultimate genetic address system that specifies where structures will form is shared by these animals, suggesting that they evolved in a common ancestor over 600 million years ago. Sponges (Phylum Porifera) do not have *Hox* genes that evolved after the divergence of the Porifera–Eumetazoa lines, although a gene cluster known as NK in sponges appears to have been the ancestor of the *ProtoHox* gene which evolved from this cluster after the sponge-metazoan branch point.

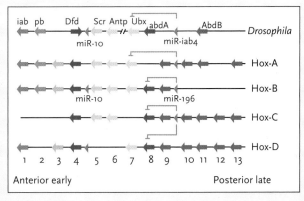

FIGURE 1
Hox complexes of Drosophila (top line) and mammals (lower four lines). The Hox complex has been duplicated twice in mammalian genomes and comprises 39 genes. Note that microRNA genes, which inhibit translation of more anterior Hox mRNAs, have been conserved between Drosophila and humans.

SOURCE: Reprinted from *Cell*, Volume 132, Issue 2, E.M. De Robertis, "Evo-Devo: Variations on Ancestral Themes," Pages 185–195, Copyright 2008, with permission from Elsevier.

Figure 2.21 Animals that appear to be secondarily radially symmetrical. **A.** *Myxicola infundibulum* (336) (Annelida, Polychaeta); **B.** *Phoronis vancouverensis* (577) (Phoronida); and **C.** *Neothyonidium magnum* (1097) (Echinodermata, Holothuroidea).

concentration of nerve tissue to form a "brain." This nerve net is the first appearance of a nervous system (see Figure 2.14 for a conceptual model of the possible development of a nervous system).

Animals that are **triploblastic** (having three layers of cells, Fig. 2.22C) and bilaterally symmetrical are referred to as **Bilateria**. Triploblastic animals have a third population of cells lying between the epidermal and gastrodermal epithelia. These cells, known as **mesoderm**, are formed during embryogenesis. Mesoderm gives rise to muscle tissue, most of the vertebrate skeleton, and cells that differentiate as reproductive organs and accessory digestive organs. In the phyla Platyhelminthes and Gastrotricha (Fig. 2.22A, B), mesoderm fills the space between the epidermal epithelium on the outside and the gastrodermal epithelium lining the gut on the inside. These animals lack a secondary body cavity and are known as **acoelomates**. Acoelomates have one internal compartment, the gut lumen. The mesoderm-derived cells of acoelomates differentiate into muscle tissue, gametes, the epithelial walls of various chambers of the complex reproductive system, and parenchyma cells.

Bilateral symmetry (Fig. 2.20B) allows active animals to move forward with an anterior end always leading. The sensory receptors, and the processing of sensory inputs, of these animals can be concentrated at the anterior end, an evolutionary change called **cephalization**. This allows the animals to monitor the environment into which they are moving. **Cephalization** is the first step in the differentiation of an anterior head. Mobile benthic animals typically also have ventral surfaces specialized for locomotion and dorsal surfaces specialized for protection or defense, further differentiating a dorso-ventral body axis.

As triploblastic animals increased in size, they still had to deliver nutrients and oxygen to internal tissues efficiently. This led to the development of the coelom, an epithelium-lined, fluid-filled, secondary body cavity between the gut and body wall (see Section 2.4d).

STUDY BREAK

1. What is the difference between true radial symmetry and bilateral radial symmetry?
2. How can an animal have a mouth and pharynx but no anus?

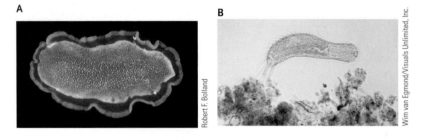

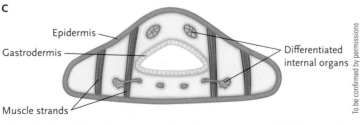

Figure 2.22 A. A free living platyhelminth (*Pseudoceros* sp. (223)), **B.** an unidentified gastrotrich, and **C.** cross section through an acoelomate animal showing the epidermis, gastrodermis, and mesoderm.

Epidermis

Gastrodermis

Differentiated internal organs

Muscle strands

2.3 Duplication and Divergence

Evolutionary successes at any level are selected for in many different lineages (i.e., evolve convergently), whether the focus is genes, biochemical pathways, cells, tissues, whole body parts, and even whole organisms. This may happen within a single body (serial repetition, metamerism, and segmentation) or at the level of the whole body (colony formation).

2.3a Metamerism and Segmentation

The terms "segmentation" and "metamerism" are frequently used interchangeably, but the terms are really applied to different levels of organization. In **serial repetition,** a single organ or structure is repeated along an axis of an animal's body (Fig. 2.23A). **Metamerism** describes a body plan in which two or more sets of organs or structures are repeated along the body axis (Fig. 2.23B). **Segmentation** involves metameric repetition of most of the organs of the body, with each repeat clearly separated from its neighbours (Fig. 2.23C).

Only Annelida and Arthropoda appear to include fully segmented animals. In less specialized members of both phyla (e.g., *Nereis* (321) and a remipedian), every segment behind the head appears to be the same (Fig. 2.23C and 2.24A). But when you compare *Nereis* and a remipedian with *Chaetopterus* (331) (Annelida, Polychaeta, Fig. 2.24B) and *Homarus* (819) (Arthropoda, Crustacea, Decapoda, Fig. 2.24C), the segmentation remains obvious but not all segments are the same. Different segments and different body regions are specialized for different functions, a process known as **tagmatization,** where a **tagma** is a single region specialized for one function. Tagmatization may be accompanied (e.g., in *Homarus*) by fusion of segments (Fig. 2.25).

Because of their similar segmentation, Annelida and Arthropoda were once considered to be closely related phyla. Currently accepted phylogenies, however, place the Arthropoda in the Ecdysozoa and Annelida in the Lophotrochozoa (Fig. 2.26) so they are quite separated in evolutionary history (see Chapter 3). Two opposing possibilities emerge from an examination of the phylogenetic distribution of animals with segmented, metameric, and serial repetitive patterns. First is the potential for development of these patterns very early in metazoan evolution, a potential exploited in some groups but not in others (and perhaps secondarily lost in some groups). Second, metamerism and segmentation could have evolved independently several times in different evolutionary lines.

An increasing body of evidence supports the first model. *Hox* gene clusters mediate anterior–posterior body organization, and proto-*Hox* clusters can be traced back into the Cnidaria where they are involved in differentiation into three or four regions along an anterior–posterior axis. *Hox* genes are fundamental to the organization of metameric and segmented body plans. Another gene, *engrailed,* is associated with seriality and metamerism in animals such as chordates, arthropods, onychophorans, and annelids. *Engrailed* is not involved in serial repetition in molluscs but is expressed around shells and spicules; this means that in chitons it is expressed serially around the shell plates. Although *engrailed* plays different roles in different groups, it also is expressed in non-segmented groups (e.g., ophiuroid echinoderms), where it is associated with skeletal development.

At this time, the available evidence supports the idea that ancestral gene regulatory networks have been co-opted to control development of metamerism and segmentation in different ways in different animal groups.

2.3b Colonial (Modular) Animals

A colony is formed when a multicellular animal reproduces asexually by budding or fission and generates a whole new individual that does not completely separate from its parent. The colony enlarges as asexual reproduction is repeated. The definition of a **colony** requires an organic connection between the asexually produced individuals along with some level of coordination in their activities. This definition excludes colonies of social insects where the individuals are closely related

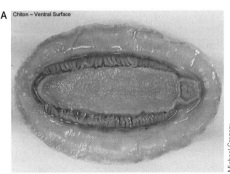
A Chiton – Ventral Surface

Michael Gregory

B

Michael Owen

C

Alexander Semerov

Figure 2.23 A. Ventral view of a chiton (Mollusca, Polyplacophora) showing serial repetition: multiple pairs of gills (on either side of the foot) but no repetition of other organs or body divisions. **B.** Representatives of the phylum Kinorhyncha shows metamerism (division of the cuticle, musculature, and nervous system into 13 zonites) but not segmentation. **C.** Like many annelids, *Nereis* (321) (Annelida, Polychaeta) has a segmented body plan. *Nereis* is an active predator, burrowing through soft substrata in search of prey. Each segment of *Nereis* bears a pair of parapodia (lateral extensions used in burrowing, walking, and swimming). Internally, each segment is separated from its neighbours by a partial septum (in many annelids the septum completely separates segments) and has a separate coelomic cavity, a pair of metanephridia, a segmental nerve ganglion, and segmental nerves that coordinate the activity of segmental muscle units.

Figure 2.24 **A.** *Godzilliognomus frondosus* (710) (Crustacea, Remipedia), an evenly segmented, cave-dwelling crustacean. Each segment bears a pair of paddle-like swimming legs. Internally the segmented pattern is expressed in the nerve ganglia, segmental nerves, and the arrangement of the muscles associated with the swimming legs. **B.** *Chaetopterus* sp. (331) (Annelida, Polychaeta, the parchment worm). This annelid lives in a U-shaped tube of parchment-like material secreted by the worm. Water is pumped through the tube by the action of three fans (labelled 1), formed by the fusion of the dorsal lobes of three pairs of parapodia. Food is trapped from the water current in a mucus bag (2) supported by a pair of aliform (wing-like) parapodia (3). When the bag is clogged with food it is collapsed and passed forward to the mouth and a new bag secreted. **C.** Ventral view of a male *Homarus americanus* (819) (Crustacea, Decapoda, the American lobster). The head appendages are specialized for sensory functions (antennules and antennae) and food handling. Appendages on the thorax are the chelae, which handle and crush prey, and four pairs of walking legs. The pleopods on the abdomen (also known as swimmerets) are used in slow swimming and the uropods (tail fan) are used in the fast backward swimming escape response.

and live cooperatively but without direct physical connections.

Pyrosomes, pelagic urochordates, show coordinated behaviour (Fig. 2.27). Pyrosome zooids are all morphologically and functionally similar, using cilia to transport water through the body to filter the micro-algae that they eat. The cylinder-shaped colonies may be up to 20 m long and include thousands of zooids. Water from each zooid enters the hollow lumen of the cylindrical colony and out of its open end. The resulting current powers the animal and is an example of non-pulsed jet propulsion (see Chapter 7). Disturbances trigger each zooid to produce a flash of bioluminescence and stop its cilia, causing the whole colony to sink. Disturbance to any individual zooid generates both ciliary arrest and a light flash. This flash is received by a photoreceptor in neighbouring individuals, which

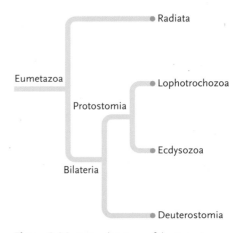

Figure 2.25 Tagmatization. The bodies of arthropod ancestors were composed of a head followed by a series of individual segments (top). In insects these segments came together to form three distinct regions: the head and associated appendages (sensory appendages and mouthparts, red), the thorax (blue), and the abdomen (yellow). The process was accompanied by fusion of segments, for example, an insect head may involve the fusion of seven segments.

Figure 2.26 Major divisions of the Eumetazoa.

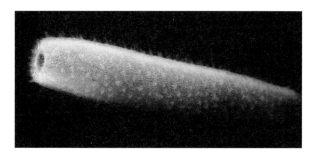

Figure 2.27 *Pyrosoma* sp. (1115) (Urochordata, Thaliacea). The barrel-like colony is closed at one end.

SOURCE: Carver Mostardi/age fotostock/Getty Images

initiate ciliary arrest and a light flash in these zooids and so on throughout the colony.

In many attached animals, colonies of the same species occasionally encounter one another. Closely related colonies usually fuse, but more distantly related colonies generate a natural histoincompatibility and the two colonies aggressively compete for space (Fig. 2.28). Here, at the base of the metazoan phylogenetic tree, we see a simple immune defence system. Allelorecognition genes in the hydroid *Hydractinia* (136) involve a hypervariable protein from the immunoglobulin superfamily. In some *Hydractinia* rejection responses, contact with an unrelated colony triggers nematocyst discharge.

Colonial organisms are frequently **polymorphic**, exploiting the capacity of a single genome to generate more than one body morphology. Polymorphic colonies are found in the Cnidaria, Bryozoa, and Chordata (in the subphylum Urochordata). The best-known

siphonophore (Cnidaria, Hydrozoa), the Portuguese man-o'-war (*Physalia* sp. (139)), is an example of a polymorphic colony, although an atypical siphonophore because it uses a gas-filled bladder (the gas includes a large amount of carbon monoxide) to float on the ocean surface (Fig. 2.29.A). Most siphonophores live submerged, such as *Muggiaea atlantica* (137) (Fig. 2.29.B). A generalized siphonophore colony (Fig. 2.29C) has a number of specialized zooids,

A

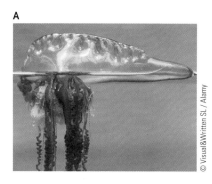

B

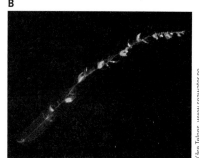

C

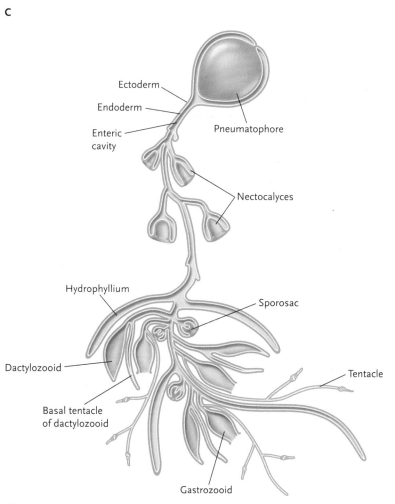

Ectoderm
Endoderm
Enteric cavity
Pneumatophore
Nectocalyces
Hydrophyllium
Sporosac
Dactylozooid
Tentacle
Basal tentacle of dactylozooid
Gastrozooid

Figure 2.29 Polymorphic colonies. **A.** The Portuguese man-of-war (*Physalia* sp. (139)) has a gas-filled surface float and clusters of tentacles (cormidia) that include feeding, reproductive, swimming, and stinging polyps. The tentacles may be 10 m in length. **B.** *Muggiaea* (137) (Cnidaria, Hydrozoa) is a typical siphonophore, living in the mid water. A single stolon carries different polyp types. **C.** Part of a generalized siphonophore colony showing the different types of specialized medusae and polyps (see text).

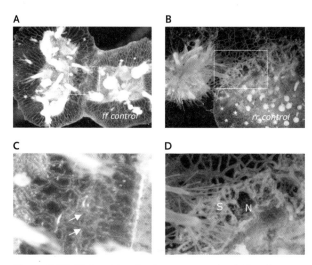

Figure 2.28 *Hydractinia* rejection response. **A., C.** Closely related *Hydractinia* colonies (ff strain in this experiment) fuse on contact with the fusion of the gastrovascular cavities of the two colonies (arrows in C). **B., D.** An ff strain colony touches and is rejected by an rr strain colony. Notice in D. that stolons from the ff colony (S) cause necrosis (N) and loss of tissue.

SOURCE: Poudyal, M., S. Rosa, et al. (2007). "Embryonic chimerism does not induce tolerance in an invertebrate model organism." *Proceedings of the National Academy of Sciences of the United States of America* 104(11): 4559–4564. Copyright 2007 National Academy of Sciences, U.S.A.

some derived from a medusa. Included are floats (pneumatophores), swimming bells (nectocalyces), reproductive zooids (sporosacs), nematocyst-bearing defensive structures (hydrophyllia), feeding (gastrozoids) and defensive (dactylozooids) components, and tentacles (tentilla).

Bryozoa (moss animals) (Fig. 3.53A & B) may also form polymorphic colonies that include zooid forms specialized for different functions. Feeding zooids (**autozooids**) have a ring of ciliated tentacles that trap food particles and transport them to the mouth. In many bryozoan colonies, a number of zooids cooperate to set up feeding and discharge currents on a larger scale than an individual zooid might create (Fig. 2.30). Colonies may also include specialized reproductive **ooecia** (Fig. 2.31.A), defensive **avicularia** (Fig. 2.31.B), as well as **vibracula** that clean the colony surface (Fig. 2.32). Some bryozoan colonies spread over a substratum through the growth of stolons, **kenozooids**, or with another type of zooid (Fig. 2.33).

STUDY BREAK

1. Why do we need carbon monoxide detectors in our houses while this gas is not toxic to *Physalia*?
2. What is meant by polymorphism?
3. What is a segment?

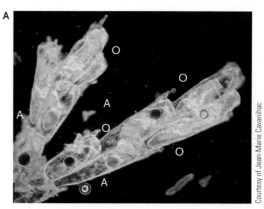

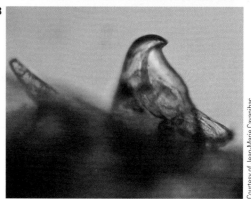

Courtesy of Jean-Marie Cavanihac

Figure 2.31 **A.** Part of a colony of *Bugula* sp. (280) showing reproductive ooecia (O) and avicularia (A). **B.** An avicularium, a modified zooid with a jaw-like morphology. Avicularia protect the colony against invaders (nematodes and polychaetes) and against the settling of larvae of fouling organisms. The tuft of hairs between the jaws may be the sensory triggers that cause the jaw to snap shut.

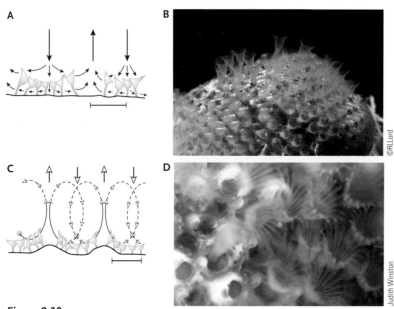

Figure 2.30

Cooperative feeding in bryozoans. **A.** A group of zooids of a flat-surface bryozoan colony sets up cooperative feeding currents. Scale bar: 2 mm. **B.** *Watersipora* species (287), a flat-surface colony. **C.** Feeding current circulation in a mound-shaped colony. Scale bar: 4 mm. **D.** *Celleporaria sherryae* (282), a mound-shaped colony.

2.4 Development and Evolution of Animal Body Plans

2.4a Cleavage and Body Orientation

Every animal body plan can be traced back to a single cell: a fertilized egg or zygote. The phylogeny of body plans is deeply rooted in the patterns and changes that can be seen when one follows the development from egg to adult. **Embryos** are multicellular organisms under construction. With few exceptions, body cells are genetic clones of the fertilized egg. Selective gene transcription and post-translational modifications in the lineages of proliferating embryonic cells generate the regionally differentiated body form and the diversity of cells, tissues, and organs of a multicellular animal.

Fertilization is the fusion of male and female gametes (sperm and ovum). A mature sperm is a motile nucleus typically with a beating flagellum along with mitochondria to provide energy. Unlike sperm, mature ova typically have extensive cytoplasm with mitochondria, ribosomes, yolk proteins, and stored mRNA transcripts. While sperm have completed

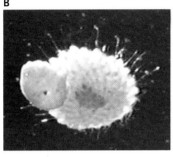

Figure 2.32 A. Vibracula (seen in a scanning electron micrograph of *Nematoflustra* sp. (285)) are reduced zooids that form tentacle-like structures that are used in cleaning the surface of the colony. **B.** Frames from a video sequence showing the removal of a shell fragment by the vibracula of a colony of *Cupuladria* sp. (283)

SOURCE: Judith Winston, "Life in the Colonies: Learning the Alien Ways of Colonial Organisms," *Integr. Comp. Biol.* (2010) 50 (6): 919–933, by permission of v Oxford University Press.

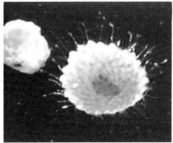

meiosis and are haploid, mature ova do not normally complete meiosis until after fertilization. The time of fertilization relative to the stage of meiosis of the ovum is species-dependent and not correlated with phylogeny. In many species, the first event after sperm fusion with the ovum is release of one or two **polar bodies** (and sometimes a third) as the ovum completes meiosis. This is followed by consolidation of the haploid nuclei of the ovum and sperm to produce a diploid zygotic nucleus.

The cytoplasm of a fertilized egg includes encoded information that determines a central axis that will become the anterior–posterior axis of the zygote. This axis lies between two poles, the **animal pole** and the **vegetal pole**. The animal pole typically becomes the anterior end of the developing animal, the vegetal pole the posterior. Egg polarity may be established early in oogenesis, long before fertilization, when cytoplasmic constituents within the egg acquire a gradient of

distribution relative to the future embryonic axis. In *Clytia* (132) (Cnidaria, Hydrozoa) the axis is determined as the oocyte matures and depends on a group of microtubule-forming genes; this axis can be disrupted experimentally by low-speed centrifugation. In *Synaptula* (1098) (Echinodermata, Holothuroidea) axis orientation is determined even earlier and is set by the position of the contact between the developing oocyte and the basal lamina of the germinal layer, with the apical-basal polarity of oocyte becoming the animal-vegetal axis of the ovum. However, in many other species, egg polarity is determined at the time of the ovum's meiotic divisions, and the location where the polar bodies are extruded from the ovum becomes the animal pole. An example of this mode of axis determination is the limpet *Lottia* (405) (Mollusca, Gastropoda), where physical deformation of the egg can shift the meiotic apparatus so that the animal–vegetal axis is similarly rearranged.

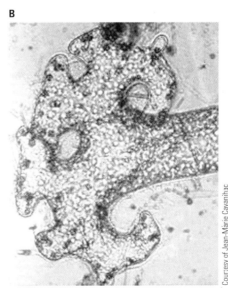

Courtesy of Jean-Marie Cavanihac

Courtesy of Jean-Marie Cavanihac

Figure 2.33
A. The stolon from which the zooids of *Zoobotryon* species (288) arise is another modified zooid type, a kenozooid (notice the protist, *Vorticella*, that has colonized the kenozooid). **B.** The tip of a kenozooid forming an attachment to the substratum.

2.4b Cleavage Patterns: Radial, Spiral, and Others

Fertilization is followed by **cleavage**, the mitotic division of the zygote. The pattern of cleavage planes is an early and important step in the translation of encoded information and resources into an embryo with its parts in the right places and its nutrient resources distributed according to the needs of its different parts. In turn, the cleavage planes depend on the orientation of the mitotic spindle within the fertilized egg and in all the subsequent cell progeny. The cell progeny produced by embryonic cleavage divisions are called **blastomeres**.

Patterns of embryonic cleavage are correlated with animal phylogeny. Deuterostomes (chordates and echinoderms) have radial cleavage while lophotrochozoans (Platyhelminthes, Annelida, Mollusca, etc.) typically have spiral cleavage. Ecdysozoans (Arthropoda and Cycloneuralia) have a variety of different cleavage patterns that are probably highly derived.

Radial cleavage (Fig. 2.34) occurs when the mitotic spindle is oriented either parallel or perpendicular to the primary embryonic axis. Alternation between these two orientations produces tiers of blastomeres, and the blastomeres of each tier are in line with blastomeres above and below. Radial cleavage may be the ancestral cleavage pattern for all bilaterians.

Spiral cleavage (Fig. 2.35) occurs when the first two cleavages divide the egg longitudinally into four blastomeres, and at the third cleavage, the mitotic spindle is oriented at an oblique angle relative to the primary embryonic axis. A spiral pattern develops after the third cleavage as the mitotic spindle within each blastomere is positioned to divide the blastomere unequally, generating a tier of four small blastomeres at the animal pole and a tier of four larger blastomeres at the vegetal pole. Since the spindle is tilted relative to the embryonic axis, the small animal blastomeres (now called **micromeres**) are offset relative to the larger vegetal blastomeres (now called **macromeres**). (Note: Sometimes the size difference between micromeres and macromeres is negligible.) Repeated cleavage of macromeres eventually generates four tiers of micromeres, most giving rise to ectoderm, while one micromere gives rise to much of the mesoderm (see Section 2.4f). Macromeres give rise to endoderm.

Different groups of animals exhibit other modes of cleavage. Only part of the ovum cleaves in some, for example discoidal cleavage in fish, reptiles, and birds, and centrolecithal cleavage in insects (Fig. 2.36).

2.4c Cell Fate Specification

The genome of a fertilized egg contains all of the information required for the development of an animal. Some of this information is differentially distributed

Figure 2.34 Radial cleavage.

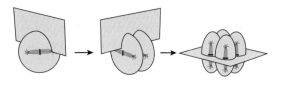

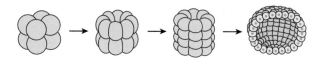

Figure 2.35 Spiral cleavage. The letters and numbers on the blastomeres are the codes used to follow the fates of individual blastomeres.

SOURCE : Hejnol, A., "A twist in time—the evolution of spiral cleavage in the light of animal phylogeny," Integr Comp Biol. 2010 Nov;50(5):695-706, by permission of Oxford University Press.

A

VIEW FROM ANIMAL POLE

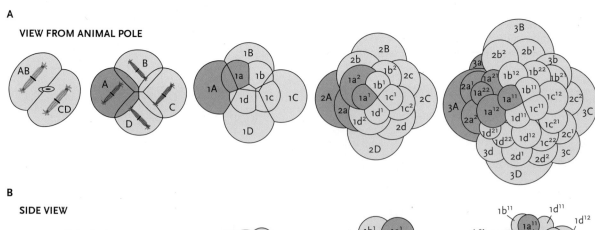

B

SIDE VIEW

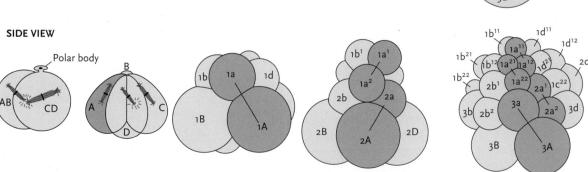

in the egg when polarity (the specification of animal and vegetal poles) is established. As cleavage progresses, this information may be divided so that all blastomeres retain a complete set of information, or may be differentially divided so that different blastomeres include only parts of the genome. Investigations of two marine snail species show that different cytoplasmic mRNAs and their transcripts are sorted among different blastomeres during cleavage, providing a molecular basis for differentiation.

When the fate of early blastomeres is largely determined by the differential segregation of "cytoplasmic determinants," this mechanism is called **autonomous specification** and the pattern of embryonic development is **determinate (or mosaic) development**. Mosaic development is characteristic of spirally cleaving eggs of lophotrochozoans including Annelida, Mollusca, and Nemertea. In these groups experimental separation of the blastomeres produces cells that will continue dividing but are incapable of producing a complete embryo (Fig. 2.37A).

If early cleavage does not differentially segregate essential cytoplasmic determinants, then all early blastomeres retain the potential to form a whole embryo. Differential development of early blastomeres depends on signalling between the blastomeres, not on differential sorting of special cytoplasmic determinant factors. This mode of development is described as **dependent (or conditional) specification**, and the pattern of embryonic development is called **indeterminate (or regulative) development**. Regulative development is characteristic of the radially cleaving embryos of deuterostomes. In these groups experimental separation of the blastomeres produces cells that are each capable of developing to a fully formed embryo (Fig. 2.37B).

Spiral and determinate cleavage in Lophotrochozoa and radial indeterminate cleavage in deuterostomes are traditionally regarded as defining fixed boundaries. However, many embryos incorporate a mix of both strategies during embryonic development. For example, the eggs of frogs and salamanders have typical deuterostome radial and indeterminate cleavage, but a small crescent of cytoplasm on the future dorsal side of the uncleaved embryo is irreversibly committed to form the invaginating dorsal lip of the blastopore and initiates subsequent embryonic inductions during and after gastrulation (see Section 2.4d). Further, studies of mesoderm specification among gastropods show that the mechanism for specifying blastomere fate may change during evolution.

2.4d The Blastula and Gastrulation

Very early in development, animal embryos pass through two landmark stages: blastula and gastrula. A simple **blastula** is a hollow ball of cells enclosing a cavity, the **blastocoel**. The blastula stage then forms the **gastrula** by **gastrulation**. Gastrulation occurs in

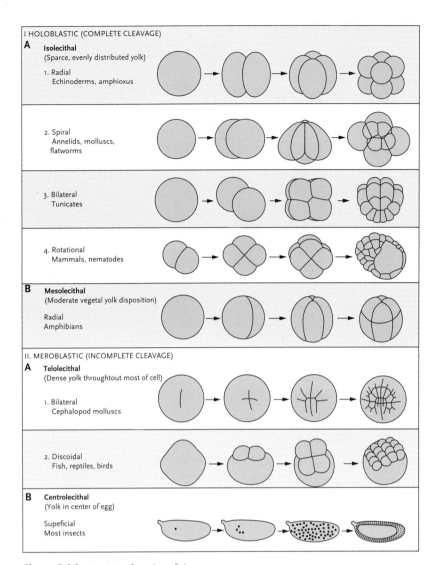

Figure 2.36 A variety of modes of cleavage.

SOURCE: Gilbert SF. 2000. *Developmental Biology*, 6th edition. Sunderland (MA): Sinauer Associates.

different ways in different animal groups but is the process by which a single-layered embryo becomes two layered. Figure 2.38 shows gastrulation in a sea urchin. Notice how the wall of the blastula at the vegetal pole invaginates as a closed cylinder of cells moving into the blastocoel. As a result, the body of the embryo becomes a tube within a tube. The **blastopore** is the site where the inner tube invaginated. The outer layer of embryonic cells, the **ectoderm**, eventually forms the epidermal epithelium of the body wall. The inner layer, the **endoderm**, forms the gastrodermal epithelium or inner lining of the gut. In a gastrula it is called the **archenteron**, literally the "primitive" or "original" gut.

Embryos are multicellular organisms under construction. With few exceptions, body cells are genetic clones of the fertilized egg. Selective gene transcription and post-translational modifications in the lineages of proliferating embryonic cells generate the regionally differentiated body form and the diversity of cells, tissues, and organs of a multicellular animal. Ectoderm and endoderm are embryonic cell layers. Cnidarians

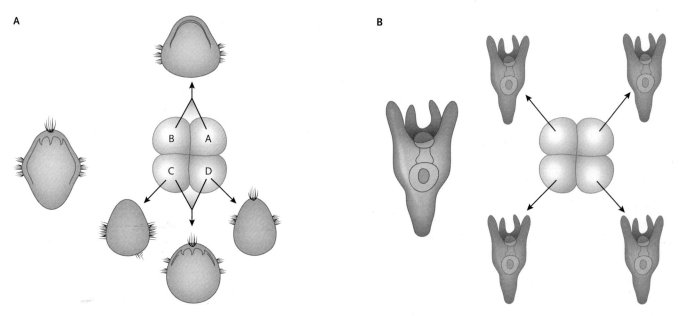

Figure 2.37 An experimental demonstration of the consequences of determinate and indeterminate cleavage. **A.** The four blastomeres of a spirally cleaving snail embryo are separated; none is able to continue development to form a complete trochophore larva (sketched at left). **B.** The same experiment is performed on a four-cell embryo of a sea urchin; each separated blastomere can form a complete (albeit small) pluteus larva.

(including sea anemones, jellyfish, and corals) are diploblastic animals with bodies derived entirely from ectoderm and endoderm. Most other animals differentiate **mesoderm** and are triploblastic. Mesoderm is the layer of embryonic cells situated between ectoderm and endoderm (Fig. 2.38).

2.4e Origin of Mesoderm and the Development of Body Cavities

The secondary body cavities of bilaterians are spaces between the epidermal epithelium and the gastrodermal epithelium lining the gut (see page 19). There are two types, the persistent blastocoels (pseudocoels and haemocoels) and the coeloms. These body cavities form during embryogenesis.

The blastocoel, the hollow interior of the blastula stage embryo, remains after gastrulation forms the archenteron, and is bordered by the basal lamina under the ectoderm (future epidermal epithelium) and endoderm (future gastrodermal epithelium). The blastocoel may also remain after mesoderm has been generated in it. In nematodes and some other Bilateria, mesoderm differentiates as muscle cells underlying the epidermal epithelium. The blastocoel is now enclosed by muscle tissue on the body wall side and gastrodermal epithelium on its inner surface. This persistent blastocoel is called a pseudocoel and plays an important role as a hydrostatic skeleton and as a medium for circulation of dissolved nutrients and gases.

The blastocoel may contain connective tissue that forms interconnected channels through the body. Fluid circulates through these channels to deliver gases and nutrients to bodily tissues. This persistent blastocoel is called a **haemocoel** and the individual channels are **haemocoelomic** sinuses. The haemocoel is never lined by an epithelium oriented with its apical side facing the interior of the haemocoel (an important distinction separating the haemocoel from a true coelom). In many invertebrates, the body plan includes both persistent blastocoels and coelomic compartments.

Mesoderm, the third embryonic cell layer in triploblastic animals, is present in all bilaterians. Mesodermal cells form the morphological elements between the epidermis and the gastrodermis. These elements are basic to the evolution of complex circulatory, reproductive, digestive, and excretory systems, and the attainment of large body size and rapid locomotion that have evolved in different groups of Bilateria. Mesoderm originates in different ways in deuterostomes and protostomes, perhaps because mesoderm evolved more than once, or perhaps because mesoderm differentiation has undergone evolutionary change in at least one of these groups.

In most deuterostomes, mesoderm originates from the endoderm soon after gastrulation, arising as outfoldings of the archenteron that detach as coelomic vesicles. The lumen of these vesicles enlarges as the left and right coelomic compartments, which may subsequently duplicate to form multiple pairs. The epithelial lining of the coelomic vesicles becomes the peritoneum. Cells ingress from this epithelium to differentiate into muscles of the gut and body wall and cells that will contribute to the formation of internal organs. Therefore, in deuterostomes, the embryological

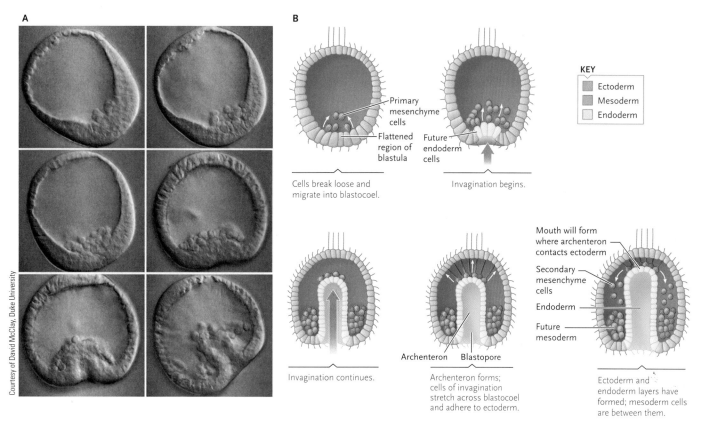

Figure 2.38 Gastrulation in a sea urchin (Echinodermata, Echinoidea). **A.** Micrographs taken at the time intervals showing gastrulation in *Lytechinus* (1089) **B.** The development of structures in the developing gastrula.

origin of mesoderm and the embryological formation of the coelomic compartments constitute the same process, called **enterocoely**, the out-pocketing and pinching-off of left and right coelomic vesicles from the archenteron (Fig. 2.39).

In lophotrochozoans, the mesoderm that will form the epithelial lining of the coelomic compartments originates from the 4d micromere of the embryo. This "mesoderm mother cell" undergoes mitoses to eventually produce two cellular masses on either side of the archenteron. Each of these clusters of mesodermal cells becomes organized as an epithelium surrounding a central space that forms the coelomic compartments. This pattern of coelom formation is known as **schizocoely** (Fig. 2.40). The mesodermal mother cell is labelled as the 4d cell in the notation system of numbers and letters identifying each micromere generated during the development of a spirally cleaving embryo (see Fig. 2.35). Interesting experiments on *Tubifex* (360) (Annelida, Oligochaeta) show that the 2d and 4d micromeres (Fig. 2.41) are vital to axis determination. In their absence, no anterior–posterior axis develops. If the 2d and 4d cells from a second embryo are grafted onto an intact embryo, a second head-tail axis develops at right angles to the natural axis.

A true coelom (eucoelom) is a secondary body compartment completely lined by epithelium derived from mesoderm. This epithelium is oriented with the apical surface facing the interior of the cavity. Deuterostomes and lophotrochozoans have coelomic secondary body compartments, sometimes co-existing with hemocoelomic compartments, but the way in which the coelom forms differs in the two major animal lineages.

Whether a coelom forms by enterocoely or schizocoely, the result is a bilateral pair of coelomic compartments to the left and right of the future gut. These two epithelial compartments meet along the mid-sagittal plane of the developing animal, and the boundary between the two forms a double layer of mesothelium at the dorsal and ventral side of the gut (Fig. 2.39 and 2.40). These double layers of mesothelium are the dorsal and ventral **mesenteries**.

The mesothelium lining the coelomic compartments is often ciliated to circulate fluid, but many animals require a more efficient transport system to deliver oxygen and nutrients to their tissues. In vertebrates the circulatory system is a specialized coelomic compartment, lined by endothelial cells. Most invertebrates depend on a persistent blastocoel for fluid circulation, for example, the dorsal and ventral blood vessels of an annelid are essentially tunnels in the dorsal and ventral mesenteries and are delineated by the basal lamina sides of mesothelium.

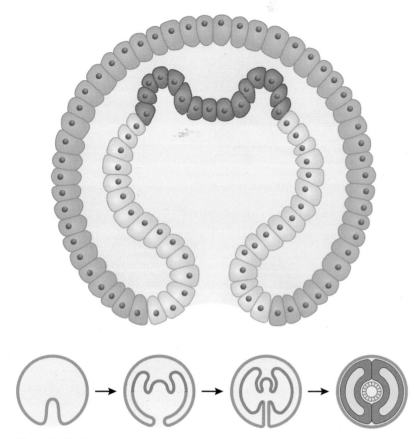

Figure 2.39 Enterocoely.

2.4f Fate of the Blastopore: Two Theories

Gastrulation is a key event in development because it establishes the primary body cavity and the future gut, and sets the stage for differentiation of mesoderm. Morphogenesis depends on successful completion of gastrulation, and the site and fate of the blastopore have traditionally been viewed as basic to reconstructions of evolutionary history and phylogenetic relationships. Traditionally, zoologists proposed that the mouth and the anus evolved simultaneously during the early transition from radial to bilateral symmetry. In this view, the lophotrochozoan blastopore became the mouth, and the anus formed secondarily. In the traditional view of the deuterostomes, the blastopore became the anus, and the mouth was formed secondarily when the tip of the archenteron fused with the ectodermal wall of the gastrula.

More recent information shows that the fate of the blastopore and the origin of mouth and anus with respect to the blastopore are more variable than previously expected. As well, recent analysis of genes associated with patterning of the foregut and hindgut of *Convolutriloba longifissura* (201) (Acoela) offer a different interpretation, suggesting that a through gut (with mouth and anus) evolved independently in different lineages of animals.

Study Break

1. How would you determine the polarity of an egg?
2. Name three phyla of animals that have a schizocoel and three that have an enterocoel?
3. What is meant when cleavage is described as "determinate"?

2.5 Are Protostome and Deuterostome Body Plans Inverted Copies of the Same Thing?

In 1822 Étienne Geoffroy Saint-Hilaire proposed that the body plans of arthropods, with their ventral nerve cord and dorsal blood vessel, were simply an inverted version of the body plan of a craniate vertebrate with its dorsal nerve cord and ventral major blood vessel. (Fig. 2.42.A). Saint-Hilaire illustrated his argument with a famous picture of a lobster in an inverted position (Fig. 2.42B). Notice that the only obvious feature that does not fit into the inversion model is the position of the mouth (it is on the upper side in Saint-Hilaire's lobster and has been moved in the diagram in Fig. 2.42A). Saint-Hilaire's concept was strongly opposed by Georges Cuvier, whose standing in the French scientific community condemned Saint-Hilaire's idea to obscurity.

While anterior–posterior axis features are regulated by *Hox* genes (see Box 2.1), other genes control

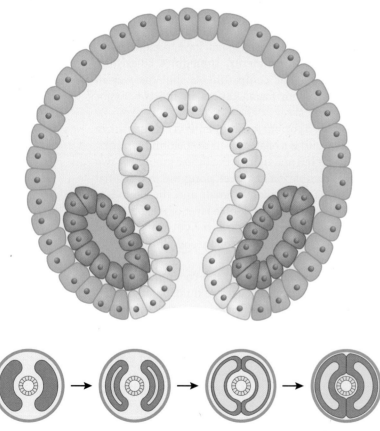

Figure 2.40 Schizocoely.

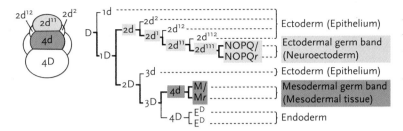

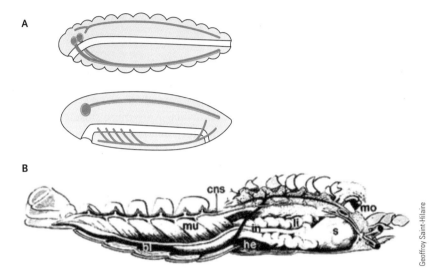

Figure 2.41

The 2d and 4d micromeres in *Tubifex* have a role in axis determination. The position of the "d" micromeres and "D" macromere and the code that identifies the descendants of the "D" macromere.

SOURCE: Reproduced with permission from Ayaki Nakamoto, Lisa M. Nagy and Takashi Shimizu. 2011. "Secondary embryonic axis formation by transplantation of D quadrant micromeres in an oligochaete annelid," *Development* 138, 283–290. The Company of Biologists Ltd.

the development of the dorsal-ventral axis. In *Drosophila* (960) a gene known as *dpp* (decapentaplegic) regulates protein synthesis in embryonic cells that form dorsal structures in the embryo. Another gene, *sog* (short gastrulation), antagonizes *dpp* and regulates the development of the ventral nerve cord and associated structures, meaning that these two genes oversee the division of the embryonic ectoderm into neural and non-neural domains. The dorso-ventral axis in *Xenopus* (1392) (Amphibia) is regulated in a similar way, in this case by an interaction in which *chordin* controls the neural domain and is opposed by *bmp4* (bone morphogenetic protein) controlling the ventral region (Fig. 2.43). Isolation of the proteins coded by *sog* and *chordin* led to the recognition that the two were homologous. From this finding the obvious experiment was to test the effect of injecting the mRNA for *sog* in *Xenopus* and that for *chordin* in *Drosophila*. The result showed that the two proteins were interchangeable; *sog* could determine a dorsal nerve cord in an amphibian and *chordin* a ventral nerve cord in the fly. If these molecular determinants of the dorso-ventral axis have been conserved from a common ancestor, the body plans of protostomes and deuterostomes may indeed be inverted versions of each other.

2.6 Embryological Patterns Mapped onto Animal Phylogenetic Relationships

In this chapter we have repeatedly mentioned the ways that different features of body plans and the embryological development of body plans differ among phyla or groups of phyla. Rearranging the data allows body plan features and developmental processes to be mapped against animal phylogeny.

At the base of the tree, there is strong evidence that Eumetazoa (all multicellular animals) are a sister group to the **Choanoflagellata**, a group of protists. The cells of choanoflagellates are similar in form to choanocyte cells found in sponges (see page 213). In these cells, the flagellum creates water currents, and the surrounding microtubular collar traps food particles for the cell to take up by phagocytosis. The genome of the choanoflagellate *Monosiga brevicollis* (100) (Fig. 2.44) has been sequenced and includes genes for **cadherins** (cell adhesion molecules) and **tyrosine kinases** (cell

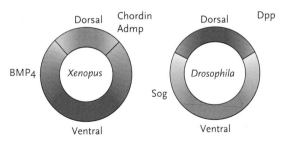

Geoffroy Saint-Hilaire

Figure 2.42 A. A simplified view of protostome (above) and deuterostome (below) body plans. The nerve cord is blue, the gut yellow, and the major blood vessel red. **B.** Geoffroy Saint-Hilaire's view of an inverted lobster forming a deuterostome body plan except for the mouth (mo) which would be dorsal.

signalling molecules) that are homologs of animal genes and quite different from those in plants, fungi, and microorganisms.

Studies using molecular data and cladistic analysis to determine the phylogenetic branching order for early clades of animals yield inconsistent results. Morphological and some molecular phylogenies suggest that Porifera are the most basal lineage of animals with living descendants. Cnidaria and Ctenophora were

Figure 2.43 Diagrams of these controls with the neural region in red and the non-neural in blue.

SOURCE: Reprinted from *cell*, Volume 132, Issue 2, E.M. De Robertis, "Evo-Devo: Variations on Ancestral Themes," Pages 185–195, Copyright 2008, with permission from Elsevier.

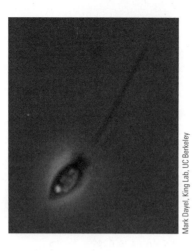

Figure 2.44
A microscope image of the choanoflagellate *Monosiga brevicollis.* A single flagellum is surrounded by a collar of microtubules. Each cell is about 10 µm long.

Mark Dayel, King Lab, UC Berkeley

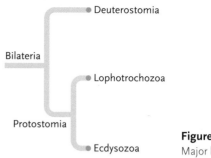

Figure 2.46
Major bilaterian clades.

early branches from the animal family tree. Some evidence suggests that Cnidaria are more closely related to the main line of metazoan evolution, with Ctenophora as a separate branch. Other evidence suggests that Cnidaria and Ctenophora are more closely related to each other and share a branch from the metazoan line (Fig. 2.45).

Hypotheses for phylogenetic relationships among the remainder of the animal phyla (the huge group known as the Bilateria (Fig. 2.46)) are explored in Chapter 3. Two major clades of bilaterians, the **Deuterostomia** and **Protostomia**, are recognized with the Protostomia further subdivided into the **Lophotrochozoa** and **Ecdysozoa**.

While this tree is based on molecular evidence, the separation of Deuterostomia and Protostomia was established long ago on the basis of embryological characteristics (Table 2.1). A major difficulty in this traditional arrangement is that only some of protostomes, notably Annelida and Mollusca, fit these "classic" protostome embryological characteristics. Arthropods display none of these embryological characteristics and were classified in Protostomia because their metamerism is so similar to annelid metamerism that the two groups were assumed to be closely related. A number of other bilaterians, including Nematoda

and a number of phyla with a pseudocoelom, have embryological features that do not fit with either protostomes or deuterostomes.

These problems were resolved by reconstructing phylogenetic relationships among representative bilaterians by comparing 18S ribosomal DNA nucleotide sequences coding for the RNA that forms the small subunit of the eukaryotic ribosome coding for 18S ribosomal DNA. This gene is essential and has tolerated only rare changes during evolutionary time, so it can be used to reconstruct divergent relationships among animals. The reconstruction identified a clade within the bilaterians whose members share two main characteristics: 1. they periodically moult an exoskeleton to allow body growth, and 2. they lack motile cilia. This group included arthropods, nematodes, and many of the small pseudocoelomate phyla. This research simultaneously resolved uncertainties about the phylogenetic placement of both arthropods and the pseudocoelomate phyla and showed that the Protostomia includes two clades. Lophotrochozoa exhibit the embryological characteristics formerly attributed to all protostomes, and Ecdysozoa exhibit various forms of highly derived embryological development.

The picture remains far from clear in some areas. For example, Lophotrochozoa are presumed to exhibit spiral cleavage, but in a number of the phyla in this group, spiral cleavage has been either lost or modified (Fig. 2.47). Improved techniques and study of more examples are gradually improving our understanding of phylogenetic relationships.

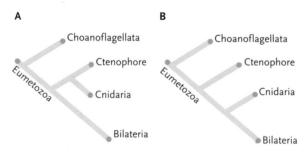

Figure 2.45
Two possible arrangements for the relationships between phyla at the bottom of the phylogenetic tree. A number of additional possibilities would emerge if the Placozoa were included since they might have evolved before or after the Ctenophora–Cnidaria branch in **A**, before the Ctenophora, or either before or after the Cnidaria in **B**.

Table 2.1	A Summary of a Classical View of the Embryological Characteristics of the Protostomia and Deuterostomia	
	Protostomia	Deuterostomia
Cleavage	Spiral determinate	Radial indeterminate
Mesoderm Formation	From 4d micromere	From endoderm
Coelom Formation	Schizocoel	Enteroceol
Mouth	From blastopore	Not from blastopore

BOX 2.2 People Behind Animal Biology

Ernst Haeckel: Famous for Being Wrong!

Ernst Heinrich Philipp August Haeckel (or von Haeckel) was a towering figure in late 19th and early 20th century biology. Initially trained as a physician Haeckel found an aversion to patients. After flirting with the possibility of life as an artist, he spent three years at the University of Jena to attain a doctorate in zoology. He remained there for most of his career. As a zoologist Haeckel was a comparative anatomist and specialized in a number of invertebrate groups, particularly the radiolarian protists, the Cnidaria (Coelenterata in Haeckel's time) and the segmented worms. He described and elegantly illustrated (e.g., Figure 1) many species of radiolarians, including samples from the H.M.S. *Challenger* expedition in the late 1800s—the first major and systematic investigation of the fauna of the deep ocean.

Had Haeckel restricted himself to systematics and comparative embryology, he might have appeared in biological histories as a significant, although not very exciting, figure. However, the death of Haeckel's wife, after only eighteen months of marriage, turned him

against religion. After reading Darwin's *Origin of Species* he argued the case for evolution in a way that put him in direct conflict with the church. Indeed it has been suggested that Haeckel was responsible for the polarization of views of religion and evolution that persists to this day.

While Haeckel was a fierce proponent of evolution, he did not accept the Darwinian concept of natural selection; rather he argued the Lamarckian concept of progressive evolution through the inheritance of acquired characteristics. This view of evolution conformed to Haeckel's concept of animal development, capsuled in the phrase, "ontogeny recapitulates phylogeny," meaning that stages in the development of a species repeat stages in its evolutionary history. Haeckel argued this theory on the basis of his observations of the development of different vertebrates, in one publication challenging readers to see the difference in embryos of a dog, a chicken, and a turtle. This challenge was not likely to succeed since the printer, whether by accident or design,

used the same woodcut for the pictures of all three embryos! While this error was corrected in reprinted editions, there was still controversy over the accuracy of Haeckel's drawings, with some claims that he biased his drawings to support his theories.

Haeckel produced a number of versions of animal evolutionary trees (e.g. Figure 2) with differing levels of detail for different groups. In a number of these trees Haeckel presented his view that the races of man evolved separately; based on studies of language groups he suggested that there are ten races of man, with Caucasian as the most highly developed. Unfortunately Haeckel's views were later adopted by the Nazi party (despite Haeckel having placed Jews in his top tier of races) and also influenced the arguments of the American anthropologist Carleton Coon in his views of racial evolution.

Haeckel defended his theories with vigour and his reputation has suffered in consequence. He might share the descriptor applied to some politicians, "frequently wrong but never in doubt!"

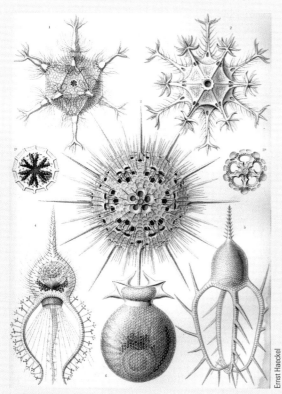

FIGURE 1
Illustrations of radiolarian species drawn by Ernst Haeckel.

Ernst Haeckel

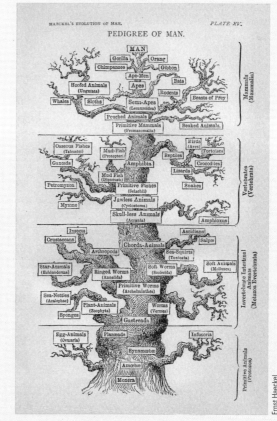

Ernst Haeckel

FIGURE 2
One of the evolutionary trees drawn by Ernst Haekel.

Figure 2.47

Phylogenetic relationships of the Spiralia (a major grouping within the Protostomia). Taxa showing spiral cleavage are marked with a white square. Dotted squares are taxa that are not fully studied. L = the loss of spiral cleavage. The position of the Polyzoa remains uncertain.

SOURCE: Hejnol, A. (2010). A Twist in Time--The Evolution of Spiral Cleavage in the Light of Animal Phylogeny. *Integrative and Comparative Biology*, 50(5), 695-706.

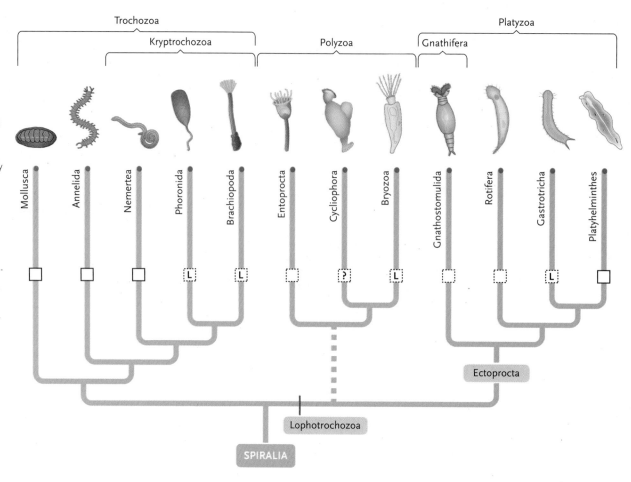

BODY **PLANS IN PERSPECTIVE**

This chapter has introduced the variety of body plans found in different animal groups, considered the ways in which those body plans develop, and introduced some of the linkages between body plans and phylogeny. Not explicitly stated in the chapter, but important, is the realization that "advanced" and "primitive" are not useful words in a discussion of body plans. A body plan may be recent or ancient and might be simple or complex, but an ancient and simple body plan should not be viewed as primitive. A body plan that has survived unchanged through the generations and evolutionary pressures that separate today from the Precambrian is clearly highly successful. The next chapter (Phylogeny) will build on these links as it looks at the relationships between animal groups. Chapter 4 (Evolutionary Processes) will examine some of the ways differences in body plan evolve.

Questions

Self-Test Questions

1. Which of the following tissue characteristics is correctly matched with its primary function?
   a. Fusion of several cells forms a functional syncytium of proteins with the ability to transmit electrical potentials.
   b. Apical–basal polarity of epithelial cells isolates and controls physiological processes in body compartments.
   c. Contractile myofilaments are suspended in a protein-enriched gel, or ground substance, in which fibres and cells are suspended.
   d. Clusters of invaginated secretory cells release transmitters at neuromuscular junctions.

2. Which of the following evolutionary changes in body plan is most directly influenced by the relationship between surface area and volume?
   a. Increased complexity of the nervous system and development of an anterior concentration of neural material
   b. Increased mobility and skeletal support
   c. Increase in size and multicellularity
   d. Development of a through-gut and bilateral symmetry

3. Which of the following body compartments border both the exterior environment and gut tissues, allowing gas exchange between the external environment and the internal fluid, as well as serving osmoregulatory and circulatory functions?
   a. Hydrostatic skeletons
   b. Pseudocoels and coeloms
   c. Gastrovascular cavities
   d. Interstitial spaces

4. Which of the following is involved in establishing an anterior–posterior body axis during embryonic development?
   a. Cytoplasmic determinants present in the egg before fertilization
   b. Location at which the polar bodies are extruded from the egg
   c. Synthesis of microtubules early in development
   d. All of the above

5. According to the text, what currently forms the strongest evidence that links the development of body form to phylogenetic relationships between animal groups?
   a. Classical embryological patterns of cleavage and mesoderm formation
   b. Comparison of 18S ribosomal nucleotide sequences
   c. Molecular data and cladistic analysis of Porifera, Cnidaria, and Ctenophora
   d. Embryological mechanism of coelom formation and origin of the mouth

Questions for Discussion

1. What limits the size of an animal that has an exoskeleton? How can crustaceans reach much larger sizes than insects?

2. What are the advantages of a segmented body plan? If segmentation is advantageous, then why are segments quite different in the bodies of many segmented animals?

3. What problems are raised by accepting the concept that deuterostome body plans are an inverted version of the protostome body plan? What answers have been offered as explanations of these problems?

4. Use the electronic library to find a recent paper that explores some aspect of animal body plans. Does the new information revolutionize any of the information in this chapter?

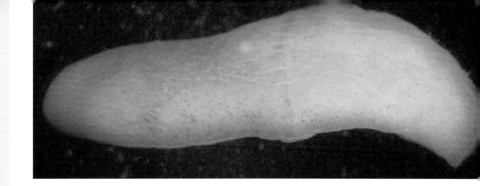

Figure 3.1 *Xenoturbella* (210) is 2–3 cm long. The band around the middle is a sensory structure.

SOURCE: Graham Budd/Wellcome Library, London

3   The Phylogeny of Animal Groups

WHY IT MATTERS

Knowing the identity of the animal under study is central to all fields of Biology. Many published works have later embarrassed the authors when their identification of an animal proved to be wrong, or when their sample included more than the species that they named. An understanding of **phylogeny** (the evolutionary history of groups of animals) provides a framework for **taxonomy** (the naming of animals and their placement in related groups).

A complication for biologists who do not deal with animal phylogenies every day is that phylogeny is a live science. Phylogenies are not fixed—they change as new information emerges. This means that, just as taxonomists may move a species from one genus to another, some animals may be moved from the group in which they were described to a completely different position on the phylogenetic map.

Xenoturbella (210) (Fig. 3.1) is an example of an animal that, over time, has been placed in several different positions in different animal phylogenies. *Xenoturbella* is a ciliated marine flatworm, first described in 1949. It was then considered to be a primitive member of the Turbellaria (phylum Platyhelminthes), since only the gonads were noticeably different from those of the primitive turbellarians. Later studies of the

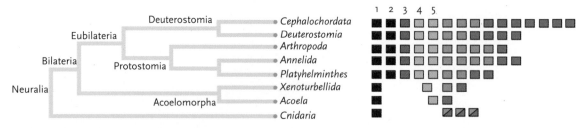

Figure 3.2 *Hox* genes of major eumetazoan groups. The anterior (red), group 3 (orange), and the anterior central (*Hox* 4 and 5) (yellow) genes are arranged according to presumed orthologies (orthologies, or orthologous genes, are genes in different species that originated by vertical descent from a single gene of the last common ancestor). Only the numbers of the posterior central (green) and posterior genes (blue) are indicated. (The *Nematostella* genes are blue-green to emphasize the uncertainty about their orthology.)

SOURCE: Nielsen, C. (2010). "After all: Xenoturbella is an acoelomorph!" *Evolution & development* 12(3): 241-243. © 2010 Wiley Periodicals, Inc.

ciliated and the mucous secreting cells of the *Xenoturbella* epidermis supported this conclusion. More detailed electron microscopy of the epidermal features, sperm morphology, and the structure of the statocyst led to a firm conclusion that *Xenoturbella* did not belong in the Platyhelminthes. Taxonomists then suggested that it should be classified in the hemichordates or echinoderms.

In 1997 this positioning was challenged by molecular evidence that suggested that *Xenoturbella* is actually a bivalve mollusc, albeit one that has lost its diagnostic features. However, a study in 2003 showed that *Xenoturbella* eats the eggs and larvae of bivalve molluscs and that the molluscan DNA came from its gut contents! More recent molecular work supports the earlier suggestion that *Xenoturbella* is a deuterostome, related to hemichordates and echinoderms. Interesting support for the early studies suggesting a relationship between *Xenoturbella* and the Acoela comes from the suggestion that the Acoela are actually primitive deuterostomes (see Fig. 3.72D). Analysis of *Hox* genes (Fig. 3.2) supports moving the Acoela and *Xenoturbella* back to a basal position. Clearly the phylogeny of *Xenoturbella* remains far from settled.

Another animal that has been moved from one phylum to another is *Nectocaris* (see page 143).

SETTING **THE SCENE**

A **phylogeny** is an evolutionary history of a species, a group of species, or a higher-level taxonomic group. Organismal biology texts are commonly based on a phylogenetic sequence. In such texts, the material and ideas found in single chapters in this book are divided and scattered throughout the various chapters. This chapter reviews current concepts of animal phylogeny and, particularly when used in conjunction with the Purple Page section of the book, provides access to the evolutionary history of the animal examples populating the entire book. Pictures and their captions introduce animal examples from the groups included in the phylogenies.

3.1 Introduction

Humans are intrinsically fascinated with the history of life on earth. Identifying groups of characteristics that indicate close relationships among species is fundamental to an understanding of evolutionary history. This understanding is encapsulated in phylogenies, which provide a framework for asking questions about large-scale patterns of evolutionary change. Phylogenies are the tools we use in this historical reconstruction.

Phylogenies (or phylogenetic trees) are built using numerical techniques that compile data from any aspect of an organism's genotype or phenotype: anatomy, DNA sequences, behaviour, and/or physiology. Some phylogenetic trees show relationships among genera or species, others illustrate relationships among higher taxonomic groups or even populations based on their genetic similarities (Fig. 3.3). A common ancestor and all of its descendants is a **monophyletic group**, also referred to as a **clade** (Fig. 3.4). A single clade may contain as few as two species but its size is limited only by the number of descendants of the common ancestor. Conventionally, single branches are referred to by the taxonomic unit they represent (a family, genus, species, population, etc.). The clade that defines the kingdom "Animalia" contains millions of species.

Beginning in the 1990s, molecular tools and techniques were applied to questions of animal phylogeny, a subject that had previously been the exclusive domain of palaeontologists, morphologists, and embryologists. This change opened a Pandora's box of new concepts and ideas. Molecular techniques improved rapidly, the data sets they examined grew in size, and their results, combined with existing data, often produced phylogenies that differed substantially from those previously accepted.

Elements of current thinking on animal phylogeny were introduced in Chapter 2. In this chapter we will examine the history and the current ideas of the relationships among animal phyla. Given the current pace of the appearance of new findings, it is almost certain that, by the time this book is in print, some of the material in this section will be viewed as out of date (see Box 3.1).

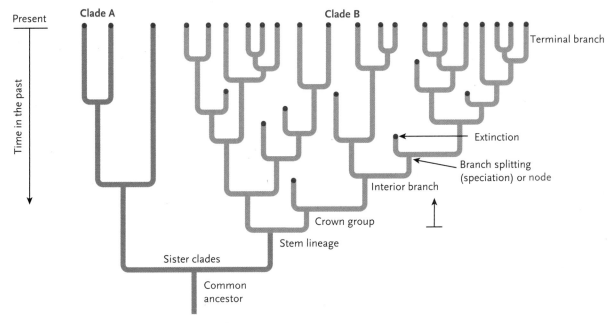

Figure 3.3 A phylogenetic tree. Clades A and B diverged from a common ancestor as stem lineages, each of which gave rise to more lineages (or branches). Clades are defined by having two or more independent lineages (branches) at any one time. Branch splitting (speciation) occurs at nodes that represent the last common ancestor of the descendent lineages. Within this tree in the present, Clade A includes three species (terminal branches or lineages) and Clade B contains 18 species. Eight species in Clade B are extinct.

Note: The phylogenies drawn in green in this chapter were created by the authors and are usually simplified versions of phylogenies in the literature. The lines leading to groups are not scaled to give any indication of the time at which groups separated.

3.2 Phylum Porifera

The problems of interpreting the phylogenetic relationships of the Porifera, Cnidaria, and Ctenophora were introduced in Chapter 2 and alternative phylogenies offered (see Fig. 2.45). Despite many morphological studies as well as the application of different molecular approaches, there remains no consensus on the phylogeny of these groups. Results from molecular analyses are challenged by the original separation of these groups having occurred 600+ million years ago and consequent long-branch attraction (see pages 78–79).

There is general agreement that the single-celled Choanoflagellata (see Fig. 2.44) are the sister group to the rest of Metazoa (Fig. 3.5). The presence of food-collecting choanocytes (see page 213), which closely resemble choanoflagellates, suggests a close relationship between Porifera and Choanoflagellata, although such a relationship is not supported by available evidence. Choanocytes are found in adult sponges but

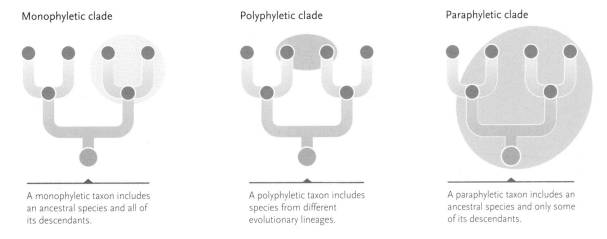

A monophyletic taxon includes an ancestral species and all of its descendants.

A polyphyletic taxon includes species from different evolutionary lineages.

A paraphyletic taxon includes an ancestral species and only some of its descendants.

Figure 3.4 The goal of building phylogenetic trees is to identify monophyletic groups, based on shared characteristics that are derived from a common ancestor. Ideally, the taxonomy of an organism reflects its membership in a series of nested monophyletic clades (groups). In practice, many taxonomic groups are polyphyletic or paraphyletic.

The starting point for many studies in biology is a phylogeny showing the evolutionary relationships among species or groups of organisms. For many years phylogenies have been derived from objective, statistical analyses of data about characteristics based on a variety of data from morphology to genetic, but including behavioural and physiological data.

Willi Hennig (1913–1976), an entomologist, developed a coherent theory for investigating and presenting relationships among species. This became the foundation for phylogenetic systematics first presented in a German publication, "Grundzüge einer Theorie der phylogenetischen Systematik" (1950) that became more widely known when published in English as "Phylogenetic Systematics" in 1966. He is widely considered to be the founder of cladistics, a common name for phylogenetic systematics.

There are four guiding principles to cladistics:

1. Sister lineages are used as clade relations.
2. Shared, derived characters (synapomorphies) are major features of organisms used to determine sister lineages.
3. Objective evidence (the most synapomorphies) is used to determine the most logically consistent portrayal of historical relationships (phylogeny) among organisms.
4. The taxonomy (classification) of organisms must be consistent with the evolutionary relationships presented by the phylogeny.

The Willi Hennig Society, founded in 1980, promotes the field of phylogenetic systematics. Members support the view that groups of organisms (taxa or species) should only be recognized and formally named when the evidence indicates that they are monophyletic because they have evolved from one taxon (i.e., they have a single common ancestor). The society publishes *Cladistics*, a journal that presents theoretical and empirical peer-reviewed contributions to the area of phylogenetic systematics.

The August 2012 issue includes papers on the phylogeny of Poaceae (grasses), Cyanobacteria, aconitate sea anemones, and floral evolution in plants in the order Fabales. You can appreciate the importance of Hennig's contribution if you recognize that it underlies all of the phylogenies considered in this book. The importance of basing phylogenies on quantitative data also should be clear.

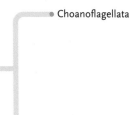

Figure 3.5 Proposed phylogenetic relationship between Choanoflagellata and Metazoa.

never in sponge larvae. Ultrastructural studies reveal significant differences between choanocytes and choanoflagellates.

Three classes, Calcarea, Demospongiae, and Hexactinellida, are generally recognized (Fig. 3.6) as making up the Porifera, although Demospongiae may not be a **monophyletic** group (a group of organisms which forms a single clade).

There is general agreement that the Demospongiae and Hexactinellida share a monophyletic origin, but the

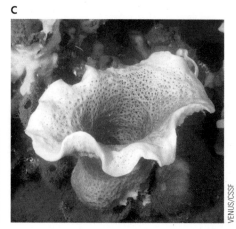

Figure 3.6 Phylum Porifera includes three classes: A. Calcarea (*Sycon* sp. (101)), B. Demospongiae (*Agelas* sp. (111)), and C. Hexactinellida (*Acanthascus* sp. (125)). The Calcarea have calcium carbonate spicules and may have an asconoid or syconoid body plan (see Fig. 8.16). The Demospongiae and Hexactinellida have a leuconoid body plan and silica spicules (see Fig. 2.16). The Hexactinellida have a syncitial choanocyte layer. The examples of the three classes in this figure look superficially similar—sponge identification requires specimen destruction in order to investigate the shape and chemistry of the spicules and the choanocyte arrangement.

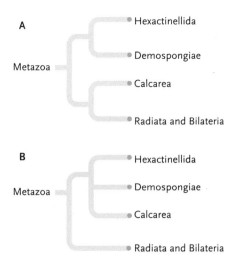

Figure 3.7 Possible phylogenetic arrangements of sponges relative to other animals. Current evidence supports arrangement **A** over **B**.

position of the Calcarea is less certain. One recent study, based on 128 different protein-coding genes, separates the Calcarea (as well as the Homoscleromorpha from within the Demospongiae) as having a separate origin (Fig. 3.7A). Another study using data from 18S, 28S, and mitochondrial 16S ribosomal DNA sequences concludes that the Porifera are monophyletic (Fig. 3.7B).

3.3 Radiata

Radiata traditionally includes the phyla Cnidaria and Ctenophora, both traditionally described as having a primary radial symmetry around an oral-aboral axis. However, recognition of the polyp body plan of Anthozoa (Cnidaria) as the primitive cnidarian body plan replaces the traditional view. The primary symmetry has either one or two planes of mirror image symmetry (see page 28), a form of bilateral symmetry. Radial symmetry, seen in the Medusozoa (Cnidaria), is secondary and is best viewed as adaptive to efficient locomotion and environmental interaction in all directions in swimming forms.

Figure 3.9 Class Cubozoa. Cubozoans such as *Chironex fleckeri* (155) are medusae with tentacles arising from four corners of the bell. A shelf-like velarium extends inwards from the lower margin of the bell and aids swimming by jet propulsion. Cubozoans (known as sea wasps) have potent nematocysts that can have serious effects on humans. Cubozoan reproduction produces a short-lived polyp stage, which then buds off a single medusa. Cubozoans have both simple and compound eyes and navigate using visual cues.

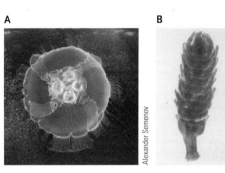
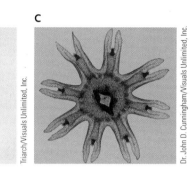

Figure 3.10 Class Scyphozoa. **A.** Scyphozoans such as *Aurelia aurita* (150) have a life cycle with a dominant medusoid stage. In this adult, the four white horseshoe shapes are the gonads. The four frilled tentacles—extensions of the mouth—can be seen through the bell. Typically a planula larva settles and develops as a polyp. **B.** The polyp buds off ephyra larvae. **C.** which grow to the adult form. Notice the sensory structures (rhopalia) on the arms of the ephyra and edge of the adult bell. These are compound sensory organs with photo-, chemo-, and gravity-sensing functions.

3.3a Phylum Cnidaria

Old phylogenies described the Cnidaria as being radially symmetrical and diploblastic, and composed of three classes, the Hydrozoa, Scyphozoa, and Anthozoa. The Hydrozoa were considered the most primitive class because of their apparent simplicity compared with the more complex Anthozoa. The Cubozoa (box jellies) were included in the Scyphozoa.

More recent phylogenies recognize the Cubozoa as a separate class and have revised the phylogeny, placing the Anthozoa (Fig. 3.8) as the first group to have evolved. The Cubozoa, Scyphozoa, and Hydrozoa (grouped as the Medusozoa, Figs. 3.9, 3.10, and 3.11)

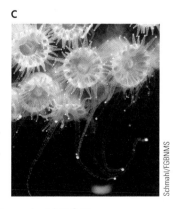

Figure 3.8 All anthozoans have a polyp form which may be single (e.g., **A.**, the anemone *Urticina* (174)), or colonial (B & C).
B. Mutiple polyps on branches of a sea pen (*Ptilosarcus gurneyi* (161)) and **C**. individual polyps of a star coral (*Monastrea* sp. (184)).

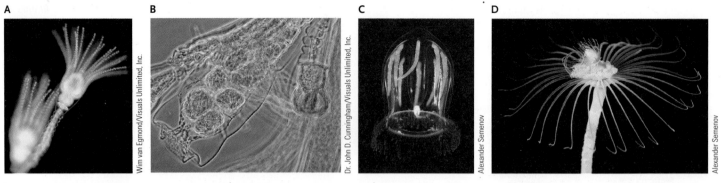

A Wim van Egmond/Visuals Unlimited, Inc.
B Dr. John D. Cunningham/Visuals Unlimited, Inc.
C Alexander Semenov
D Alexander Semenov

Figure 3.11 Class Hydrozoa. Hydrozoan forms and life cycles show great variability. Many hydrozoans are colonial (e.g., *Obelia* (138)) and may have **A.** differentiated feeding polyps and **B.** reproductive polyps which bud off to form medusoid stages. **C.** Hydrozoan medusae have a shelf-like velum projecting inward from the edge of the bell (as seen in *Aglantha* (130)). Life-cycle modifications range from being a polyp (e.g., *Hydra* (135)) with no medusa, to being only a medusa with no polyp stage (e.g., *Aglantha*). **D.** In *Tubularia* (140), a feeding polyp is surrounded by a number of reproductive buds. In each bud a medusa is formed, but remains attached. Eggs produced by the medusa are fertilized and develop to a planula larva. Further development of the planula produces an actinula larva that resembles a small swimming polyp. This larva settles to found a new colony.

are recognized as appearing later (Fig. 3.12). This means that the primitive form of the Cnidaria was a polyp, with medusoid stages evolving later. In the Hydrozoa, considerable variation in life cycles has

evolved with the polyp totally eliminated in some (the trachyline hydrozoans) and the medusa secondarily lost in others (e.g., *Hydra* (135)).

3.3b Phylum Ctenophora

Ctenophora (comb jellies, Fig. 3.13) are traditionally classified with the Cnidaria (Fig. 3.14A) because they appear to be diploblastic and radially symmetrical. Current views question this interpretation. Ctenophores are biradially symmetrical about both a sagittal plane (determined by the position of the anal pores relative to the oral-aboral axis) and, on a separate plane, through arrangement of the tentacles. Unlike the contractile tails of epithelial cells in Cnidaria, muscle cells in the ctenophores are separate from both endodermal and ectodermal layers—an arrangement leading to the suggestion that Ctenophora are triploblastic.

A recent study, based on 150 genes from 77 taxa, concludes that the ctenophores are the sister group to the rest of the Metazoa (Fig. 3.14B). This arrangement requires that nerve and muscle cells, as well as the development of a gut, evolved twice. Some evidence

Figure 3.12 A current view of the phylogeny of the Cnidaria and the life cycle stages (polyp and medusa) present in each cnidarian group.

SOURCE: Redrawn from *Molecular Phylogenetics and Evolution*, Volume 24, Issue 3, Mark Q. Martindale, John R. Finnerty, Jonathan Q. Henry, "The Radiata and the evolutionary origins of the bilaterian body plan," Pages 358–365, Copyright 2002, with permission from Elsevier.

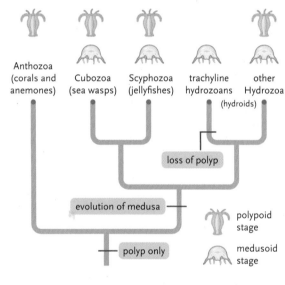

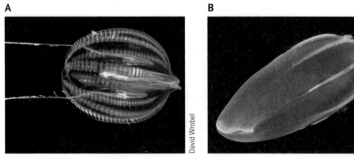

A David Wrobel
B Alexander Semenov

Figure 3.13 Phylum Ctenophora. Ctenophores have gelatinous bodies and swim by the action of eight rows of ctene plates (each plate is composed of fused cilia). **A.** Tentacles are present in some (e.g., *Pleurobrachia* (195)) but are in sheaths and not around the mouth like the tentacles of cnidarians. The mouth (at the bottom) opens into a branching gut cavity that has anal pores at the aboral end, alongside a complex sensory structure that coordinates their swimming. Ctenophores are predatory and capture prey by discharging sticky threads (colloblasts). **B.** The red colour of *Beröe* (196) comes from a consumed prey item.

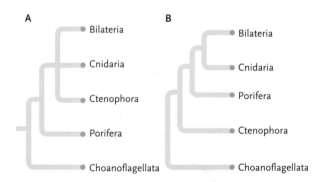

Figure 3.14 Possible phylogenetic relationships among Bilateria, Cnidaria, Porifera, Ctenophora, and Choanoflagellata.

suggests that Cnidaria are more closely related to the rest of the Bilateria than to Ctenophora and a more likely phylogeny allows for the possibility that Cnidaria and Ctenophora are primitive bilaterians.

STUDY BREAK

1. What features are important in determining the group to which a sponge (Phylum Porifera) specimen belongs?
2. Why are there problems in determining the relationship between the Porifera, the Cnidaria and the Ctenophora?

3.4 Platyhelminthes and Acoela

The heading of this section, in itself, reflects changes to ideas of platyhelminth phylogeny in that, until recently, the Acoela were viewed as being within the Turbellaria in the Platyhelminthes (Fig. 3.15A). This arrangement recognized a clade uniting the Nemertea and Platyhelminthes, based on the absence of a body cavity (hence acoelomates). The mesoderm surrounding the proboscis cavity and blood vessels of Nemertea led to suggestions that the nemerteans are actually coelomates and that the acoelomate form is secondary. Molecular techniques have confirmed the position of the Nemertea in the Lophotrochozoa (Section 3.6c), and they will be considered there (Fig. 3.15B).

Within the Platyhelminthes (Fig. 3.16), notice the division splitting the Turbellaria (mostly free-living; the Polycladida, Tricladida and Urastomida) from the parasitic groups in the Neodermata, a name ("new skin") referring to the syncytial epidermis of the Aspidogastrea, Digenea, Monogenea, and Cestoda. The parasitic Digenea, Monogenea, and Cestoda have

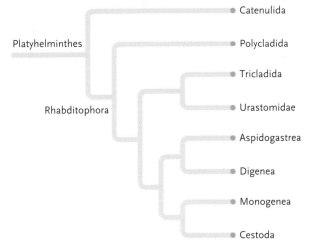

Figure 3.16
A possible phylogeny of the Platyhelminthes. The Acoelomorpha have been removed from this phylogeny and the Catenulida separated from the rest of the turbellarian orders.

traditionally been treated as classes within the phylum Platyhelminthes (alongside the Turbellaria). They are now included, with other turbellarian orders, in the Rhabditophora.

Turbellaria has been divided into a number of orders (Fig. 3.16), some of which will be omitted from this discussion, while the parasitic groups (Monogenea, Cestoda, and Digenea) are discussed in Chapter 10 (Symbiosis and Parasitism).

Acoelomorpha, one historically recognized group, includes Acoela and Nemertodermatidae. Once thought to be the most primitive of the platyhelminths, Acoelomorpha (e.g., *Praeconvoluta castinea* (203), Fig. 3.17) is a possible basal group sister to the Bilateria. It is not now included in the Platyhelminthes. More recent suggestions propose a major rearrangement to this plan and place Acoelomorpha with *Xenoturbella* (210) in the Deuterostomia (see pages 47 and 72).

Species in the order Catenulida (Fig. 3.18) are considered to be an early branch in Platyhelminthes. Older taxonomies showed the catenulids branching from the

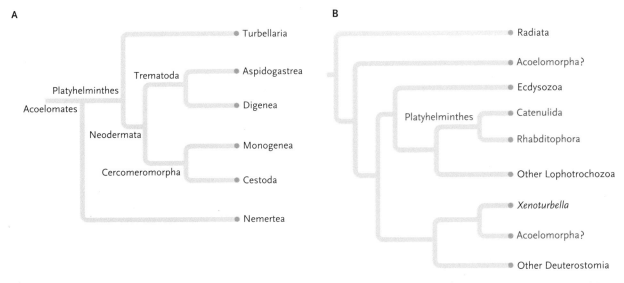

Figure 3.15 A. A traditional view of the phylogenetic relationships among the Platyhelminthes. **B.** A current view of the position of the Platyhelminthes, showing the separation of the Catenulida from the other platyhelminths and the current uncertainty over the position of the Acoelomorpha.

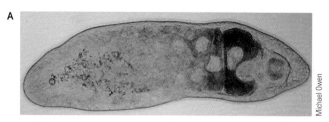

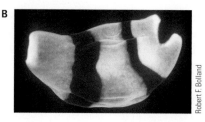

Figure 3.17 Acoelomorpha, such as **A.** *Praeconvoluta castinea* (203) and **B.** *Amphiscolops* sp. (200), take their name from their lack of gut cavity. A simple pharynx conveys food (algal cells, microorganisms, and detritus) into a central digestive region of loosely packed cells that take up food by phagocytosis. Some acoels (see *Convoluta* on page 271) have symbiotic algal cells in their epidermis and benefit from the algal photosynthesis. Acoels have no differentiated gonads; gametes form in the mesenchymal tissue of the animal. *Praeconvoluta* is about 400μm long.

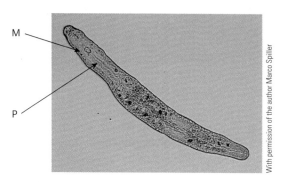

Figure 3.18 The catenulids (e.g., *Stenostomum unicolor* (208)) are small (~1 mm long) platyhelminths found in fresh water. They are omnivores, feeding on detritus, algal cells, and small animals encountered as they swim. The mouth (m) opens into a long pharynx (p) that joins a gut (g) running the length of the body. Asexual reproduction by transverse fission can produce the appearance of a chain of individuals.

platyhelminth line before the separation of the Acoelomorpha. The removal of the Acoelomorpha from the Platyhelminthes places the position of the catenulids in question. Evidence from the sequencing of the 18s and 28s rDNA genes places the Catenulida as the sister group to the Rhabditophora (which includes the rest of the Platyhelminthes) and creates a phylogeny with the Platyhelminthes as the sister group to the rest of the Lophotrochozoa. (Notice that the phylogeny in Figure 3.16B also shows the two possible positions of the Acoelomorpha).

Molecular data have done much to resolve the complexities of group relationships within the Rhabditophora. The phylogeny in Figure 3.16—although much simplified—shows the relative positions of the Catenulida and the Rhabditophora, and the positions within the Rhabditophora of the Tricladida (Fig. 3.19), Polycladida (Fig. 3.20), and Urastomidae (Fig. 3.21) relative to the parasitic groups (Digenea, Monogenea, and Cestoda).

An important point emerging from the new phylogeny is that the parasitic groups (Digenea, Monogenea, and Cestoda) are monophyletic and are now embedded within the turbellarians. Similarly, other groups with extreme morphological adaptations (common in parasitic forms) that had been proposed as separate phyla have now been classified within other phyla. Other examples include Acanthocephala (see pages 56–57), Pentastomida (once considered a separate phylum and now placed among branchiuran crustaceans, Section 3.8b.2), and Pogonophora (at one point actually suggested to be two phyla and now classified within polychaete annelids).

3.5 Platyzoa

There is increasing support for a grouping, the Platyzoa, that places the Platyhelminthes with the Phylum Gastrotricha and the phyla united in

Figure 3.19 A. The Tricladida (Platyhelminthes, Rhabditophora) are named for the three branches to the gut, one running forward from the pharynx, the other two posterior (shown in this stained specimen). **B.** The group includes *Dugesia* (230), commonly known as *Planaria* (232), a carnivore and scavenger in freshwater. **C.** A few members of the group are terrestrial, e.g., *Microplana terrestris* (231) seen on a termite nest where this animal preys on termites.

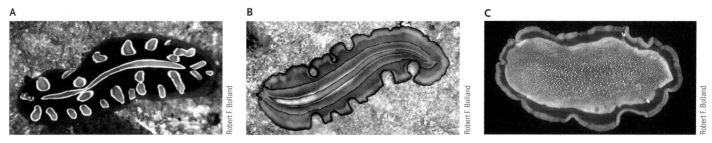

Figure 3.20 Polycladida (Platyhelminthes, Rhabditophora) are named for their many-branched gut (see Fig. 2.19) that serves both to take up food and to transport it throughout the body. These animals feed by inserting the pharynx into a prey item, secreting digestive enzymes, and then taking in their food in liquid form. The group includes many brightly coloured species, for example **A.** *Pseudoceros* sp. (223), **B.** *Prostheceraeus giesbrechtii* (221) and **C.** an unidentified polyclad.

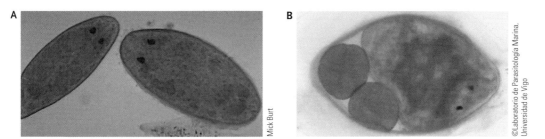

Figure 3.21 Urastomatida (Platyhelminthes, Rhabditophora). **A.** *Urastoma cyprinae* (235) is found as a commensal, living on the gills of mussels where it feeds on the mucus that coats the gill filaments. The species is also reported from oysters (*Crassostrea virginica* (535)) and giant clams (*Tridacna* (550)). **B.** In the life cycle of *Urastoma*, the adults leave the mussel host and produce a cocoon around themselves. They lay eggs that develop to new adults before emerging to invade a new host. The adult appears to die after the cocoon opens.

the Gnathifera (see below) (Figure 3.22). There is strong support for the Platyhelminthes as the sister group to the Trochozoa but the heterogeneous body plans (acoelomates and pseudocoelomates) included in the Platyzoa leaves the recognition of the Platyzoa open to question.

3.5a Phylum Gastrotricha

The gastrotrichs (Figure 3.23) are abundant marine and freshwater animals, living either among the meiofauna (the interstitial fauna living between sand grains) or on the surface of detritus or aquatic plants. Most of the roughly 500 species of gastrotrichs are less than 1 mm in length. The gastrotrichs have traditionally been placed in the cycloneuralian branch of the Ecdysozoa (see Fig. 3.57) but such a placement is not supported by 18S DNA analysis.

3.5b Gnathifera

Gnathifera (Fig. 3.24) is a recently created grouping based on a combination of molecular evidence and the morphological characteristics of the epidermis and internal jaws. Included are the Syndermata, a group that combines the Rotifera (Monogononta, Bdelloidea,

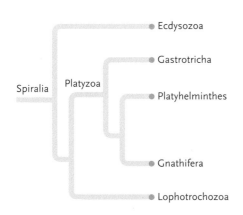

Figure 3.22 A possible phylogeny of the Platyzoa and their position relative to the Ecdysozoa and the Lophotrochozoa.

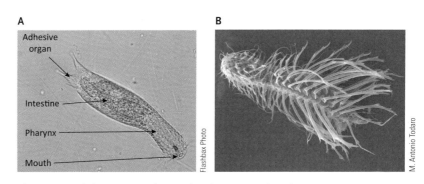

Figure 3.23 Phylum Gastrotricha. **A.** Like other gastrotrichs, *Chaetonotus* (270) moves on a carpet of ventral cilia. Adhesive organs in the forked posterior allow the animal to attach to its substratum. The mouth opens into a muscular pharynx, leading to a posterior intestine. **B.** A scanning electron micrograph shows the spines that cover the body surface of *Chaetonotus*. Other gastrotrich species may have spines, scales, or plates covering the body.

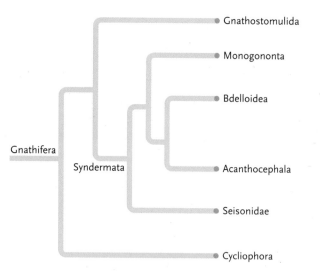

Figure 3.24 Phylogeny of the Gnathifera.

and Seisonidae) and Acanthocephala (previously treated as unrelated phyla at a pseudocoelomate level of organization), Gnathostomulida (previously either included in the Platyhelminthes or viewed as a separate acoelomate phylum), and Micrognathozoa, a newly discovered group. The Cycliophora (Section 3.5.b.5) may be allied with the Gnathifera although some recent phylogenies suggest a grouping that places the Cycliophora with the Bryozoa (Section 3.6d) and Entoprocta (Section 3.6e) in a group known as the Polyzoa.

3.5b.1 Phylum Gnathostomulida.
Gnathostomulida (Fig. 3.25) are small (most are < 2 mm long) members of the meiofauna, living in spaces between grains of sand. Their single body opening, serving as mouth and anus, and the simple gut cavity suggests affinities with the Turbellaria. It was not until their jaws (Fig. 3.25B) were studied in detail that their obvious similarities with rotifer jaws were discovered. Further examination showed the jaws to have the same internal cuticular support rods as rotifers. Consequently, they were moved to a position within the Gnathifera. Molecular studies have provided additional support for this phylogenetic position.

3.5b.2 Phylum Rotifera.
Rotifers get their name from the anterior crown of cilia (the "corona") that functions in feeding and locomotion. Males are not found in the Digononta (Bdelloidea) (Fig. 3.26), and reproduction is asexual; males are rare in the Monogononta (Fig. 3.27). Males have no gut, while females have a gut with a complex set of pharyngeal jaws (e.g., Fig. 8.37B) in an apparatus known as the mastax. Rotifers are eutelic, meaning that they have a fixed number of cells in each species (typically about 1000 cells). Eutely is characteristic of many animals living in ephemeral environments. Rotifers have a syncytial epidermis and this, along with the morphology of their sperm, has led to their being grouped with the Acanthocephala as the Syndermata (Fig. 3.24).

The Seisonidae (Fig. 3.28) are a small group of ectoparasitic rotifers found on the gills of the crustacean *Nebalia* (783) (and perhaps other hosts). Their jaw apparatus (Fig. 3.28B) has clear homologies with that of other rotifers, although their body form has lost most of the other characteristics of the phylum.

3.5b.3 Acanthocephala.
The Acanthocephala (Fig. 3.29) are parasitic worms found in the gut of vertebrate hosts. They attach themselves by the spiny proboscis from which the group takes its name (Acanthocephala = thorny headed). Because acanthocephalan life cycles (see Chapter 10)) involve an arthropod intermediate host, we presume that in their evolution they were

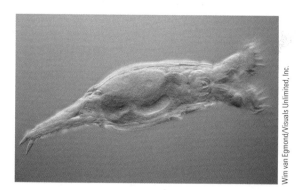

Figure 3.26 Phylum Rotifera. *Philodina* sp. (305) (Digononta (Bdelloidea)). The Digononta are only known as females.

Figure 3.25 Phylum Gnathostomulida. **A.** *Rastrognathia* sp. (294) is about 2 mm long. The jaws are about a quarter of the way back from the anterior end and are associated with a mid-ventral mouth that serves as both mouth and anus. **B.** The jaw apparatus (from *Gnathostomulida* sp. (293)) is complex but has clear homologies with the jaws of other gnathiferan groups (see Figs. 8.36B and 8.37B).

A

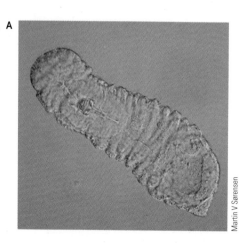

B

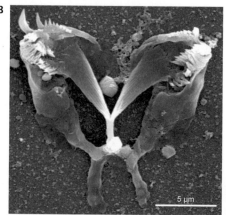

5 µm

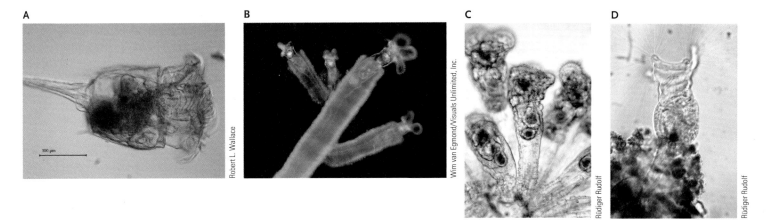

Figure 3.27 Phylum Rotifera. Populations of the largest rotifer group, Monogononta, are dominated by females and parthenogenetic reproduction, but dwarf males do occur and fertilize resting eggs that survive difficult environmental conditions. Monogonontans are found mostly in fresh water and are mostly benthic with some free-swimming forms such as **A.** *Epiphanes* (300). **B.** Some benthic species build tubes (e.g., *Floscularia* sp. (301), within a tube composed of detritus); **C.** a few are colonial (e.g., *Sinatherina socialis* (302)); and **D.** a small number have become predatory (e.g., *Collotheca ornata* (299), in which the corona is modified to a four-jawed trap).

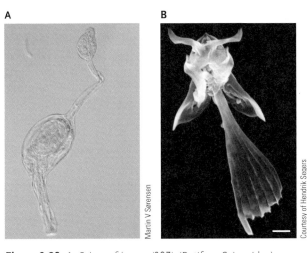

Figure 3.28 **A.** *Seison africanus* (307) (Rotifera, Seisonidae), an ectoparasite. **B.** Despite the specialized form of the adult *Seison*, the jaw apparatus is clearly homologous with that of other rotifer groups.

initially parasites of arthropods and then extended their life cycle to vertebrates that feed on arthropods.

The syncytial epidermis and sperm morphology of Acanthocephala suggest a close relationship to rotifers, a suggestion that has since received strong support from molecular evidence and led to the recognition of the clade Syndermata. The older placement of the Acanthocephala as a phylum is another example of a group with highly specialized adaptations to a parasitic way of life being accorded a higher taxonomic ranking than their placement within the Syndermata (Fig. 3.30).

Notice that acanthocephalans are placed closer to the monogonont and bdelloid rotifers than to the Seisonidae. The evidence suggests that seisonids were an early offshoot of the Syndermata.

3.5b.4 Micrognathozoa. The Micrognathozoa includes a single species, *Limnognathia maerski* (290) (Fig. 3.31), found in Greenland in 1994 and since then at a site in the sub-Antarctic Crozet Islands. The jaw apparatus of

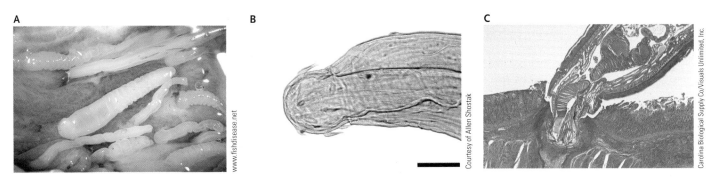

Figure 3.29 Phylum Acanthocephala. **A.** Unidentified acanthocephalans in the intestine of a fish. **B.** The extended proboscis of an acanthocephalan, with hooks for attachment to the gut wall. **C.** A section through an acanthocephalan and the gut of its host showing the proboscis embedded in the gut wall.

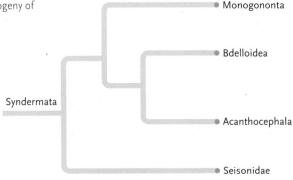

Figure 3.30 The phylogeny of Syndermata.

Monogononta
Bdelloidea
Syndermata
Acanthocephala
Seisonidae

Limnognathia is complex (Fig. 3.31B) but with elements that appear homologous to parts of the jaw apparatus in rotifers. Furthermore, a section through micrognathozoan jaw elements reveals the same cuticular support rods present in rotifers and gnathostomulids. This morphological evidence, combined with molecular data from nuclear loci, supports an association of Micrognathozoa with Gnathifera, perhaps as a sister group to the Syndermata. Data from the mitochondrial cytochrome c gene supports a linkage between Micrognathozoa and Entoprocta.

Figure 3.31 Micrognathozoa.
A. *Limnognathia maerski* (290) from a mossy pool in Greenland. The animal is about 150 μm long.
B. Its complex jaw apparatus (a reconstruction of the isolated jaws) has elements that are homologous with rotifer jaws. Only female *Limnognathia* have been found, and these produce two egg types, one thin-shelled that hatch rapidly, the other with a thick protective shell and designed to be frozen through the long winter.

SOURCE: Gonzalo Giribet, Martin V. Sørensen, Peter Funch, Reinhardt Møbjerg Kristensen and Wolfgang Sterrer, "Investigations into the phylogenetic position of Micrognathozoa using four molecular loci," Cladistics (2004) 20:1–13. © 2004, John Wiley and Sons.

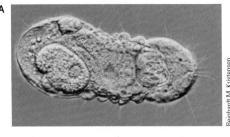

A

Reinhardt M. Kristensen

B

Figure 3.32 Phylum Cycliophora. *Symbion pandora* (275) occurs as a commensal on the mouthparts of the Norway lobster (*Nephrops* (824)). An anterior feeding funnel is suggested to have some homologies with the feeding apparatus of members of the Syndermata. *Symbion* has a complex life cycle that includes periodic shedding and replacement of the feeding funnel.

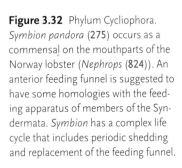

© Peter Funch, Aarhus University

3.5b.5 Phylum Cycliophora. The Cycliophora (Fig. 3.32) are a recently discovered group of ectoparasites on the mouthparts of *Nephrops* (824) (the Norway lobster) and may be associated with Gnathifera. Their inclusion in a possible gnathiferan phylogeny is shown in Figure 3.33.

STUDY BREAK

1. Adaptations to a parasitic lifestyle are often dramatic. Which taxonomic groups have been rearranged as a result of work using modern phylogenetic techniques?
2. What advantage do rotifers gain from either eliminating or reducing the role played by males in their life cycles?

3.6 Molluscs, Annelids, and Nemerteans

The Lophotrochozoa include the Platyhelminthes and Gnathifera (Fig. 3.33) and a number of other phyla. Among them are two major phyla, Annelida and Mollusca, as well as a number of smaller phyla including the Bryozoa, Nemertea, Brachiopoda, and Phoronida.

3.6a Phylum Mollusca

A 2011 phylogenomic study strongly supports placing seven of the eight classes of Mollusca on a tree (Fig. 3.34) that identifies two major lines of molluscan evolution: Aculifera that have calcareous spines or spicules, and Conchifera that have shells. Aculifera

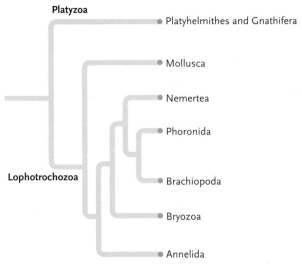

Platyzoa

Platyhelminthes and Gnathifera
Mollusca
Nemertea
Phoronida
Lophotrochozoa
Brachiopoda
Bryozoa
Annelida

Figure 3.33 A phylogeny of the Lophotrochozoa showing the Platyhelminthes and Gnathifera (Platyzoa) and the lophotrochozoan phyla (Annelida, Mollusca, Bryozoa, Nemertea, Brachiopoda and Phoronida).

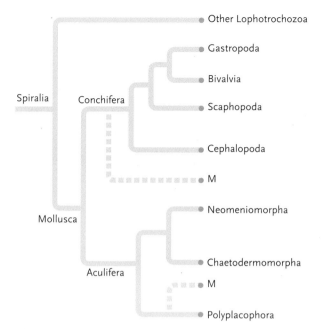

Figure 3.34 Phylogeny of Mollusca. Dotted lines leading to "M" indicate possible positions of the Monoplacophora.

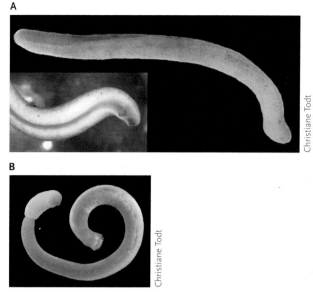

Figure 3.36 Neomeniomorpha (Mollusca, Aplacophora). **A.** Solenogasters, such as *Alexandromenia crassa* (390), can be up to 3 cm long and bears short needle-like sclerites all over their mantle. The foot sole is pink and quite broad (insert), but can be completely retracted into the pedal furrow. The dorsal view has the head to the left. The insert is a ventral view with the head to the right. The tip of the head bears the roughly triangular vestibular sense organ and the mouth cavity, which are clearly separated from the prepedal-ciliary pit at the frontal end of the foot sole. *A. crassa* lives in water depths of 90–150 m, where it crawls over rocks and gravel to hunt for its prey, cnidarian polyps. **B.** Chaetodermomorpha (Mollusca, Aplacophora). *Chaetoderma* (392) may be up to 4 cm long and is covered in scale-like sclerites that are longest on the tail. In the photo, the anterior end of the animal is to the left and the tail end is curled anti-clockwise. On the head end, the mouth shield is partly everted. The neck region and trunk are clearly separated by a deep constriction of the mantle. These animals live in soft sediments, where they burrow using the mouth shield and peristaltic movements of their whole body. They feed on sediment particles rich in organic matter and microorganisms.

includes Polyplacophora (chitons, Fig. 3.35) along with the Aplacophora, which are divided into the Neomeniomorpha and the Chaetodermomorpha (Fig. 3.36). The position of the eighth class, the Monoplacophora (Fig. 3.41), is uncertain—two possibilities are illustrated in Figure 3.34.

Conchifera include Gastropoda (snails and slugs) (Fig. 3.37), Bivalvia (clams and their relatives) (Fig. 3.38), Scaphopoda (tooth shells) (Fig. 3.39), and Cephalopoda (octopus, squid, and cuttlefish) (Fig. 3.40). Fossil cephalopods (e.g., ammonites and belemnites) had heavy shells. The interpretation of *Nectocaris* as a Cambrian squid (see page 143, Chapter 6) raises an interesting question because the evolution of the modern cephalopods has been viewed as involving a progressive reduction of the shell.

Figure 3.35 Polyplacophora (Mollusca, Aculifera). Polyplacophorans, known as the chitons, have a dorsal shell made up of eight plates. In *Acanthopleura granulata* (395), a spicule-covered mantle extends beyond the edge of the shell. In some chitons the mantle extends outwards and folds up over the shell. A ventral view of a chiton is seen in Figure 2.23A.

Possible positions of the Monoplacophora, limpet-like molluscs, are indicated in Figure 3.34. Monoplacophorans have a serial repetition of gills, retractor muscles, nephridia, gonads, and parts of the heart (Fig. 3.41). Monoplacophora may be basal to Conchifera (the more popular view), or perhaps are a sister group to Polyplacophora.

3.6b Phylum Annelida

The Annelida might be described as the not-so-little phylum that grew! Originally described with three classes—Polychaeta, Oligochaeta, and Hirudinea—the phylum was simplified to two classes when oligochaetes (earthworms) and hirudineans (leeches) were brought together as Clitellata. The Siboglinidae (Pogonophora), Echiura, and Sipunculida, previously viewed as separate phyla, have now been incorporated into the Polychaeta. Older taxonomies listed Polychaeta as a

Figure 3.37 Gastropoda (Mollusca, Conchifera). By far the largest molluscan group, this class includes three subclasses: **A.** the Proso-branchia (e.g., *Strombus lentiginosus* (**429**)), **B.** the Opisthobranchia (a paraphyletic grouping) in which many species (e.g., *Nembrotha kubaryana* (**456**)) have lost or reduced shells, and **C.** the Pulmonata (e.g., the giant African land snail *Achatina glutinosa* (**469**)) in which the vascularized mantle cavity acts as a lung, unlike the gills of the other subclasses.

Figure 3.38 Bivalvia (Mollusca, Conchifera). Adults have two shell valves, but juveniles have a single shell that has a figure-eight appearance. As they approach adulthood, the two lobes harden and become lateral, with the cross-over point of the "8" becoming the dorsal hinge. Adductor muscles close the valves, and a spring element in the hinge opens them. In the scallop, *Placopecten magellanicus* (**538**), the edges of the mantle have numerous simple eyes that detect the light changes that could signal an approaching threat. The tentacles between the eyes are chemosensory. Bivalves have greatly increased their gill area to form a food-collecting filter (see pages 214–215).

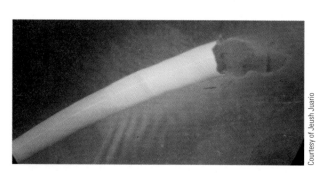

Figure 3.39 Scaphopoda (Mollusca, Conchifera). Tooth or tusk shells have a tapered cylindrical shell open at both ends. The animal (e.g., *Dentalium* sp. (**525**)) shown here extends a conical foot from the larger end of the shell. The foot is used to burrow into soft substrata. A cluster of tentacles at the base of the foot extends to capture protists and other food items from the sediment.

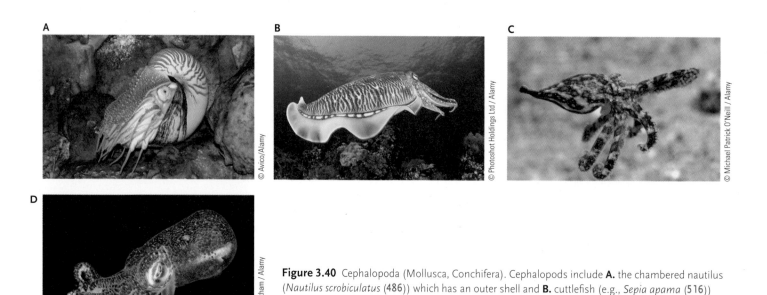

Figure 3.40 Cephalopoda (Mollusca, Conchifera). Cephalopods include **A.** the chambered nautilus (*Nautilus scrobiculatus* (**486**)) which has an outer shell and **B.** cuttlefish (e.g., *Sepia apama* (**516**)) which has a reduced internal shell. **C.** Octopus (e.g., *Hapalochlaena maculosa* (**500**), the blue-ringed octopus) have no shell, and **D.** squid (e.g., *Euprymna scolopes* (**496**), the Hawaiian bobtailed squid) have reduced the shell to an internal flexible cartilaginous pen.

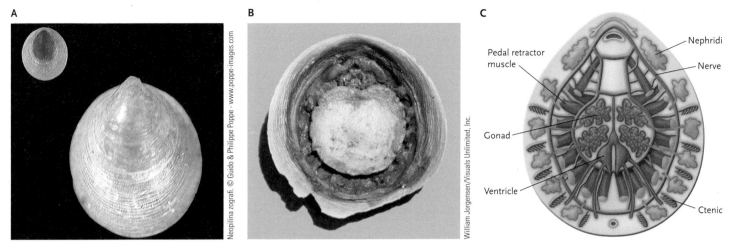

Figure 3.41 Class Monoplacophora (Mollusca). **A.** In dorsal view the monoplacophoran *Neopilina* (400) looks like a limpet. **B.** The ventral view, however, shows features not seen in any gastropod. The head is clearly separated from the foot and a series of gills (something like those of a chiton) lies in the groove between the edge of the foot and the mantle. Monoplacophorans were well known from the fossil record, however, the discovery of *Neopilina* (which looks very similar to *Pilina* from the Silurian) in a sample from the Costa Rican trench in 1952 was a surprise. A number of other species, all from deep water, have been found since. **C.** The internal anatomy of *Neopilina* shows serial repetition of organs and structures (see page 31). The heart has two pairs of ventricles, there are two pairs of gonads, eight pairs of pedal retractor muscles, five pairs of gills (ctenidia) and five pairs of nephridia (the numbers of gills and nephridia vary between species). The nervous system has paired lateral connectives.

class with about 80 families and no intermediate divisions although a behavioural separation of these families divided them into the Errantia (active) and Sedentaria (tube building) assemblages.

A 2011 phylogenomic analysis provides a new picture of the phylogeny of Annelida. This clade now shows three groups branching basally (Figure 3.42) from the remainder of the annelids. These are the parchment worms (*Chaetopterus* (331), see Fig. 2.24B), the Myzostomidae (Fig. 3.43), and the Sipunculida (Fig. 3.44).

The remainder of the annelids are divided into two major clades (Fig. 3.45). One includes the families previously included in Errantia (e.g., *Nereis virens* (323), Fig. 3.46A), *Hermodice carunculata* (318) (Fig. 3.46B) and *Glycera* (317) (Fig. 8.34A). The other

clade, the Sedentaria, combines the sedentary annelids (Terebellidae, Fig. 3.47, and Sabellidae, Fig. 3.48) with the Clitellata (Fig. 3.49) and the Echiura (Fig. 3.50). The Sedentaria also includes the Sibcglinidae (Fig. 3.51).

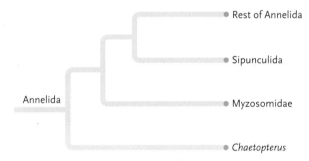

Figure 3.42 A possible phylogeny of basal annelids.

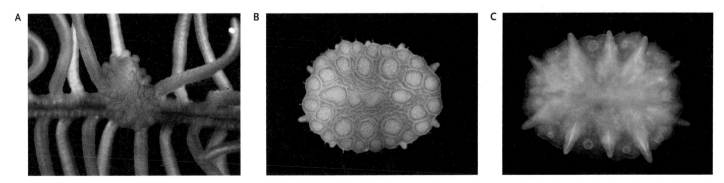

Figure 3.43 A. Myzostomidae live as commensals on crinoids (Echinodermata). The body of *Myzostoma* sp. (337) is flattened and disc-like. **B.** The dorsal surface is plain. **C.** Five pairs of parapodia, equipped with attachment hooks, are visible on the underside. Some species are mobile on the host's surface, others may be encapsulated in the host tissue.

SOURCE: Greg Rouse

Figure 3.44 **A.** Sipunculida (e.g., *Sipunculus* sp. (375)) are commonly known as the peanut worms, and include about 150 species. These benthic marine worms can withdraw the anterior region of the body (the introvert) back inside their trunk. An adaptation to a burrowing life is the gut doubling back forward, with the anus at the base of the introvert. **B.** When the introvert is extended, a cluster of anterior tentacles (shown in *Themiste* sp. (376)) extend to either suspension feed or to collect benthic detritus.

A

B

Ken Lucas/Visuals Unlimited, Inc.

Stan Elems/Visuals Unlimited, Inc.

Figure 3.45 Clades within the Annelida.

- Hirudinea
- Oligochaeta — Clitellata
- Echiura
- Terebellidae
- Siboglinidae
- Sabellidae
- Errantia

Sedentaria

A

David Cowles at http://rosario.wallawalla.edu/inverts

B

© john t. fowler / Alamy

Figure 3.46 Errantia (Annelida). The body segments of Errantia polychaete annelids are similar, typically with well-developed parapodia. The well-differentiated head bears sensory palps and lobes. Many species have eyes that may be simple light detectors but in some appear to be image forming. **A.** *Nereis virens* (323), known as the clam worm, has well-developed pharyngeal jaws. **B.** The bristle worm *Hermodice carunculata* (318) has tufts of hollow, venom-filled chaetae on each segment. These chaetae break off and penetrate skin if the worm is handled and, from the pain they produce, must be an effective deterrent against predatory fish.

A

Alexander Semenov

B

© Mark Conlin / Alamy

Figure 3.47 Terebellidae (Annelida) are polychaetes that live in tubes and feed on deposits. **A.** *Thelepus crispus* (345) lives in a tube of fine sediment that may either be attached to a hard surface or buried in the mud. The long white tentacles extend across the substratum surface to collect detritus particles. These are carried by mucociliary currents along each tentacle to the mouth, and passed through the gut where digestible material is extracted. The anterior red tentacles are the gills. **B.** *Pectinaria koreni* (338) builds a tapered tube of sand grains. The worm lives head down in soft substrata and scrapes a tunnel using the gold-coloured bristle setae. Long tentacles extend into the substratum to collect detritus particles.

Figure 3.48 Sabellidae (Annelida) are also suspension-feeding polychaetes. When feeding, they extend a crown of ciliated tentacles into the water column. **A.** *Filogranella elatensis* (333) secretes a meshwork of delicate calcareous tubes. **B.** *Myxicola infundibulum* (336) forms a mucus tube in cracks between rocks or in the substratum. A giant axon coordinates a rapid withdrawal of the worm into its tube. The central individual in this photo is fully expanded, the worm to its left is contracting, and above and slightly right of the expanded individual is the opening of a tube into which a worm has fully withdrawn.

A

Robert Fenner

B

Strong/Buzeta

A
B

Figure 3.49 Clitellata (Annelida). This groups includes earthworms (e.g., **A.** *Lumbricus rubellus* (357)) and their relatives (e.g., **B.** the medicinal leech, *Hirudo medicinalis* (367)). The clitellum, easily visible as the red collar in *Lumbricus*, is a glandular region that lubricates gamete transfer when the worms copulate. It then forms a cocoon around the eggs that hardens and closes as it is shed from the worm. All leeches have 33 segments but the segmentation is obscured by the fusion of segments to form the anterior and posterior suckers and by the annulations (variable in number) on the outer surface. In leeches, the clitellum is only visible at the time of reproduction. Historically the medical leech was used indiscriminately for many forms of illness. Medical leeches have recently come back into use to reduce swelling, caused by venous blood accumulation, after microsurgery.

A
B

Figure 3.50 Echiura (Annelida). Echiurans are treated as a clade within Annelida rather than a separate phylum. Echiurans live in burrows in soft marine sediments and extend their proboscis over the surface to collect food particles. **A.** Echiurans are also known as spoon worms because of the shape of the retracted proboscis as it appears within an *Echiurus* sp. (372). **B.** *Bonellia viridis* (371), a 10-cm-long worm, may extend the green-forked proboscis up to 1 m; this can leave a series of traces on the sediment surface, forming a rosette as the worm searches for food in all directions.

3.6c Phylum Nemertea

The Nemertea (ribbon worms, Fig. 3.52) were traditionally considered alongside the Platyhelminthes as an acoelomate phylum. The pilidium larva of nemerteans is somewhat similar to turbellarian larvae. At the same time, the cavity around the nemertean proboscis (see page 233) meets the definition of a true coelom, as do the cavities of the thin-walled blood vessels. Phylogenomic analyses place Nemertea among the Lophotrochozoa and suggest that

A
B

Figure 3.51 Siboglinidae (Annelida), once classified as Pogonophora. One example of these polychaetes, *Riftia* (383), is discussed on page 263. **A.** *Osedax* sp. (381) is a worm found consuming whale bone. **B.** *Lamellibrachia* sp. (380) depends on symbionts in its life associated with cold methane seeps in the deep ocean.

A

B

Figure 3.52 Phylum Nemertea. **A.** Ribbon worms, are unsegmented, soft-bodied, predatory marine worms such as *Nipponnemertes pulcher* (565). **B.** Nemerteans have an extensible proboscis housed in a fluid-filled cavity surrounded by circular muscles that contract to extend the proboscis (e.g., *Cerebratulus lactuca* (561). In primitive nemerteans, the proboscis and its cavity (the rhynchocoel) are some distance from the mouth. In more complex groups the proboscis and rhynchocoel have moved to a position that allows the proboscis to be extended through the mouth. Nemertean species such as *C. lactuca* may reach lengths of 1 m. Nemerteans have a through gut and a simple blood circulatory system.

Figure 3.53 Phylum Bryozoa. "Moss" animals include some 4000 species of living and 15,000 fossil species. Most bryozoans are marine, with one class living in fresh water. The great majority of bryozoan species are colonial. **A.** The phylum includes both independent colonies such as *Plumatella* sp. (286) and **B.** encrusting forms such as *Membranipora membranacea* (284). Some bryozoan colonies are polymorphic, with zooids specialized for different roles in the life of the colony. (See also Figs. 2.30–2.33).

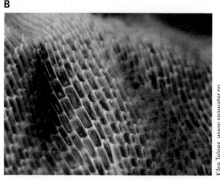

Nemertea are a sister group to the brachiopod–phoronid clade (Fig. 3.33).

STUDY BREAK

1. What features are thought to have characterized an ancestral mollusc? How have they changed in the adaptation of the Bivalvia to a burrowing lifestyle?
2. Why are the nemerteans no longer regarded as being acoelomate?
3. Describe the body form of typical errant and sedentary polychaete worms. How do the differences relate to their way of life?

3.6d Phylum Bryozoa

Bryozoa (or Ectoprocta) (Figs. 3.53 and 2.30–2.33) have traditionally been included in the lophophorates (see above). While some argue to maintain this arrangement, an increasing body of molecular evidence supports a positioning as a sister group to the grouping that includes the Nemertea, Phoronida, and Brachiopoda. A further complication in bryozoan phylogeny is their relationship to the Entoprocta (Section 3.6e).

3.6e Phylum Entoprocta

The Entoprocta (or Kamptozoa, Fig. 3.54) have been grouped with Bryozoa in some phylogenies and placed as a sister group to the Cycliophora in others. While the morphology of an entoproct appears superficially similar to that of a bryozoan, there are significant differences. Entoprocts have the anus positioned within the ring of tentacles (hence their name) while in the bryozoan the anus is outside the tentacular ring. Entoproct tentacles retract by curling down while bryozoan tentacles are pulled directly down into the body.

3.7 Lophophorata

Older taxonomies positioned lophophorates (including Phoronida, Brachiopoda, and Bryozoa) close to the divergence between the protostome and deuterostome

lines of evolution. This arrangement reflected disagreement about whether these phyla were protostomes or deuterostomes. More recently, molecular evidence provides clear support for including the Bryozoa, Brachiopoda, and Phoronida in Lophotrochozoa. Recent phylogenies have separated the Bryozoa from the other two phyla (as in Figure 3.33).

3.7a Phylum Brachiopoda and Phylum Phoronida

There are some 300 species of living brachiopods and around 12,000 species are known from the fossil record. Brachiopods (Fig. 3.55A), phoronids (Fig. 3.55B), and bryozoans use a lophophore, a double-horseshoe-shaped structure that supports a series of ciliated tentacles, to take up oxygen from the water and to collect food (Fig. 3.56).

3.8 Ecdysozoa

Ecdysozoa (Figure 3.57) was first recognized as a major animal grouping in the early 1990s. The proposal generated considerable controversy because it contradicted what had been interpreted as a clear evolutionary line linking the metamerically segmented annelids with the segmented arthropods. The characteristics of the Edysozoa include growth that depends on moulting of

Figure 3.54 Phylum Entoprocta ("nodding animals") are solitary or colonial zooids, and most of the 150 species are marine. The example here is unidentified but is probably a species in the genus *Pedicellina* (278).

A

B

Figure 3.55 Brachiopoda and Phoronida. **A.** A brachiopod (*Laqueus californianus* (570)) attached to the substratum by its pedicel. In this natural position, the dorsal valve is down and the ventral valve above (compare with bivalve molluscs in which the valves are lateral). **B.** Phoronids (*Phoronis australis* (576)) with lophophores extended for feeding. The worm has a slightly bulbous body that is buried in the substratum. Phoronids live in very low oxygen levels and have a complex circulatory system that includes cells containing hemoglobin as an oxygen transport system—the only example of this adaptation in the invertebrates.

SOURCE: A. Lovell and Libby Langstroth © California Academy of Sciences; B. Gary Robinson/Visuals Unlimited, Inc.

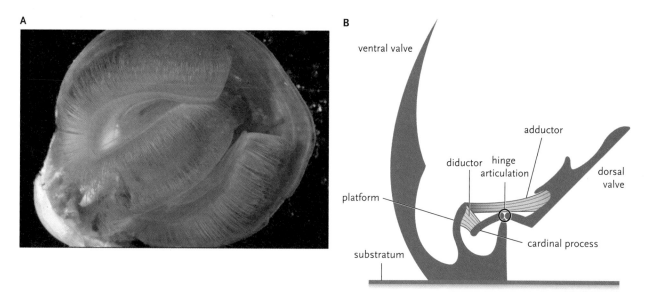

Figure 3.56 A. A brachiopod (*Laqueus californianus* (570)) with one valve removed to show the lophophore, the two-armed structure bearing a series of ciliated tentacles. **B.** Around the hinge, a complex platform structure provides attachment for muscles that both open (diductor muscle attached to the cardinal process) and close (adductor muscle) the shell valves (again compare this with the adductor muscle and hinge spring system in the bivalves discussed on page 201).

SOURCE: A. Lovell and Libby Langstroth © California Academy of Sciences; B. Based on Rudwick, *Living and Fossil Brachiopods*, Hutchinson University Library, 1970.

the secreted cuticle and not having motile cilia in their larvae and adults.

The phylum Chaetognatha (Fig. 8.61), at one time viewed as being in the Deuterostomia, is now included in the Ecdysozoa although its placement in the group is uncertain. Because of this uncertainty, the chaetognaths have been omitted from the phylogenies in this section.

3.8a Cycloneuralia

Recognition of Ecdysozoa as a group places all animals that moult their exoskeleton together. This means that the Panarthropoda (which includes Arthropoda, Tardigrada, and Onychophora) are grouped with Nematoda (Fig. 3.58), Nematomorpha (Fig. 3.59), Kinorhyncha (Fig. 3.60), Priapulida (Fig. 3.61), and Loricifera (Fig. 3.62). The latter five phyla are grouped as the Cycloneuralia on the basis of the structure of their central nervous systems (Fig. 3.57).

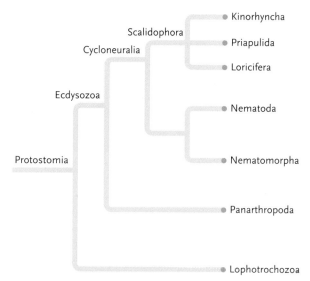

Figure 3.57 A phylogeny of the Ecdysozoa.

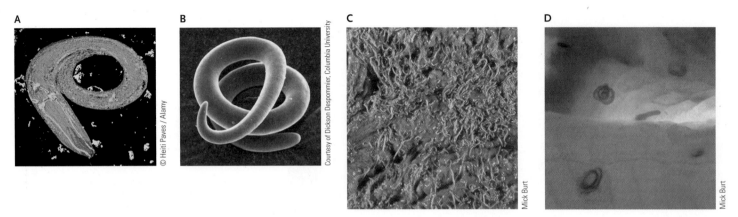

Figure 3.58 Phylum Nematoda (round worms) is the second-largest phylum of animals. It includes a huge number of as yet undescribed species. Nematoda is unusual because it includes large numbers of both free-living and parasitic species. **A.** *Caenorhabditis elegans* (585), one of the best known free-living nematodes, is the metazoan with the most completely described genome. **B.** Parasitic forms include *Trichinella* sp. (591), the parasite responsible for trichinosis. **C.** The opened stomach of a seal showing the density that *Phocanema decipiens* (590) (the cod worm) may reach in this host. **D.** A fillet of cod infested with larvae of *P. decipiens*.

Figure 3.59 Phylum Nematomorpha (horsehair worms) are parasitic worms. The adults are very long and thin, reaching lengths of up to 1 m while being only 2–3 mm in diameter. Adults are short lived and have a non-functional gut. **A.** Adult male and female *Nectonema* (596). **B.** The larvae are parasitic in arthropods, mostly insects, as illustrated by the newly emerging adult *Spinochordodes* sp. (597) leaving its grasshopper host. **C.** A few species are found in marine hosts (e.g., *Nectonema* with the shrimp host from which the adult has just emerged).

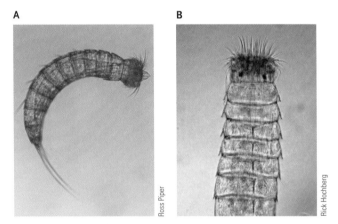

Figure 3.60 A. Phylum Kinorhyncha, such as *Echinoderes* sp. (600) are sometimes known as mud dragons. They are <1 mm long and occur in and on marine benthic sediments. An adult kinorhynch body is divided into 13 zonites, making the animal metameric in body plan (see page 31). Larval kinorhynchs have 11 zonites. Kinorhynch movement depends on the scalids, the flattened spines around the head, which are pushed forwards into the substratum and then swung back to the position seen in A. to pull the animal forward. **B.** Scalids can be pulled back into the anterior region. Notice the red eye spots and also the muscular pharynx, seen in the interior of the body.

Figure 3.61 Phylum Priapulida are burrowers, using the bulbous anterior introvert as a terminal anchor. The thin-walled, branched caudal appendage is thought to be part chemosensory, part an accessory excretory structure, and mainly a surface for oxygen exchange (its lumen is continuous with the internal body fluid). The priapulids are predators, feeding on small worms captured as they burrow. *Priapulus caudatus* (605), seen here, is the largest living member of the group and is about 10 cm long.

A

B

Figure 3.62 Phylum Loricifera are part of the meiofauna, living between sand grains. They cling firmly onto particles and are difficult to extract. **A.** *Nanaloricus* sp. (610) is said to have been accidentally discovered when sand grains were rinsed with fresh water rather than seawater! The animal can withdraw its body within the lorica (the word means a case or sheath). **B.** *Spinoloricus* sp. (611) is one of three new species found in the benthos of the Mediterranean. These animals have attracted considerable attention as they are the first metazoans to be found in a totally anoxic environment.

SOURCE: A. Reinhardt M. Kristensen; B. Danovaro R., Dell'Anno A., Pusceddu A., Gambi C., Heiner I. & Kristensen R. M. (2010). "The first metazoa living in permanently anoxic conditions". BMC Biology 8: 30.

Moulting in Panarthropoda is controlled by ecdysteroids (see Box 5.2). Moult control in the other ecdysozoans has still to be resolved although finding an ecdysteroid receptor in a nematode suggests the possibility that ecdysteroids are also involved in nematode moult control.

Within the Cycloneuralia, nematodes and Nematomorpha are viewed as sister groups. Priapulida, Kinorhyncha, and Loricifera are grouped together as Scalidophora; the name coming from the flat stylets found on the anterior region of animals in each of these phyla.

3.8b Panarthropoda

3.8b.1 Phylum Tardigrada and Phylum Onychophora.
The three phyla grouped in the Panarthropoda (Tardigrada, Fig. 3.63; Onychophora, Fig. 3.64; and Arthropoda) are segmented animals that have paired segmental appendages.

Molecular evidence supporting the arrangement of the panarthropod phyla is mixed. Onychophora and Arthropoda are clearly sister groups and in some older phylogenies, Onychophora were included within the Arthropoda. But the placement of the Tardigrada is open to question. A site-heterogeneous–mixture model favours the arrangement shown in Figure 3.65A, while an amino-acid–replacement model suggests that the Tardigrada are a sister group to the Nematoda and Nematomorpha (Fig. 3.65B). One recent study, based on two independent genome data sets involving expressed sequence tags and microRNAs, strongly supports the arrangement in Figure 3.65A. The data suggest that arrangement Fig. 3.65B reflects the result of long-branch attraction (see Fig. 4.5).

A
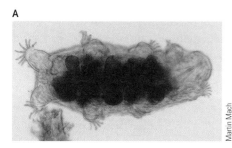

B
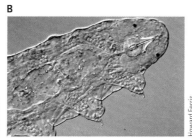

C

Figure 3.63 Phylum Tardigrada. **A.** Tardigrades such as *Echiniscoides* (617) are less than 1 mm long. Each of four segments has a pair of lobe-like legs ending in claws (as in **B.** *Panagrolaimus* (621)), hooks (as in A. and **C.** *Cornechiniscus* (616)). In A. and B., the stylets used to puncture plant cells for food can be seen inside the mouth. B. shows the muscular, sucking pharynx just behind the base of the stylets. Tardigrades have remarkable survival abilities (see pages 138–140).

A

B

Figure 3.64 Phylum Onychophora. Velvet worms have a pair of lobe-like legs on each segment. Each leg is tipped by a claw (*onych* means claw or nail). **A.** The head bears a pair of sensory antennae, each with a pair of simple eyes at its base. **B.** Oral papillae on the head are nozzles that discharge a sticky secretion to trap prey. The onychophorans are the champion spitters of the animal kingdom, being able to squirt their secretion up to ten times their body length.

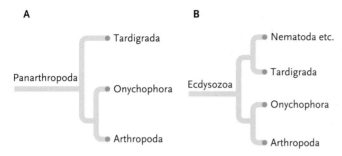

Figure 3.65 Two possible positions of the Tardigrada within the Ecdysozoa.

Some older texts discussed the Pentastomida (Fig. 3.66) as another phylum within the Panarthropoda. Molecular analysis now places the pentastomids with the Branchiura as parasitic crustaceans.

3.8b.2 Phylum Arthropoda. The major groups included in the Arthropoda are Chelicerata, Crustacea, Hexapoda (Insecta), and Myriapoda. The phylogeny of Arthropoda as a phylum, as well as the major groupings within it, has a history of changing views. Traditionally, Arthropoda was considered monophyletic with two evolutionary lines, one leading to the Chelicerata and Pycnogonida and the other to the Myriapoda,

Insecta, and Crustacea (Fig. 3.67A). A variation on this arrangement treated Crustacea as a sister group to chelicerates and pycnogonids.

Some embryological and morphological studies provided a basis for viewing the arthropods as polyphyletic. These studies concentrated on the form of the appendages (triramous in the trilobites, biramous in the crustaceans, and uniramous in other groups) and the evolutionary origin of the jaws. These studies proposed that the arthropoda were really four independently evolved groups—the chelicerates + pycnogonids, the crustaceans, the fossil trilobites, and a grouping called the uniramia (based on having unbranched legs)—from the stem arthropods (Fig. 3.67B). The Uniramia includes the Onychophora, Myriapoda, and Insecta, and some supporters of this arrangement viewed the Insecta as being polyphyletic (shown by their repeated appearance in the diagram).

The arguments for polyphyletic arthropod origins reflect the differences among animals in this group that are considered too great to be derived from a common ancestor. Support for the polyphyletic theory weakened as first the mandibles of onychophorans and other uniramian groups were shown to be not homologous (they develop on different segments), and then the jaws of crustaceans and uniramians were found to be homologous, having the same musculature and both developing from leg bases. The final blow to the polyphyletic theory came with the demonstration that in some fossil insects, the legs had outer branches (exites), making them biramous and destroying support for the Uniramia as a group.

A current possibility is that Onychophora may be grouped with *Anomalocaris* (see pages 153–154) to form a phylum Lobopodia.

Application of molecular techniques has produced some clarification in the relationships of arthropod groups and has introduced some surprises. A current view (Fig. 3.67C) supports the separation of Chelicerata (Fig. 3.68) from the rest of the arthropod groups and maintains the separation of Pycnogonida from other chelicerates (grouped in the Euchelicerata). Chelicerata is a sister group to the Mandibulata, a grouping that includes Myriapoda (Fig. 3.69), Crustacea (Fig. 3.70), and Insecta (Fig. 3.71).

The challenge to traditional phylogenies comes in the latter two groups. Insecta and Crustacea form a sister group (Pancrustacea) to Myriapoda, with Insecta being monophyletic and evolving from a crustacean stem. Crustacea are paraphyletic with Malacostraca and most other groups on one branch. The other branch includes Insecta as a sister group to other crustacean groups with Remipedia (and/or Branchiura) appearing as probable sister groups to Insecta. Current evidence provides strong support for this arrangement but, given its dramatic variation from traditional views, a number of researchers are working to provide further clarification.

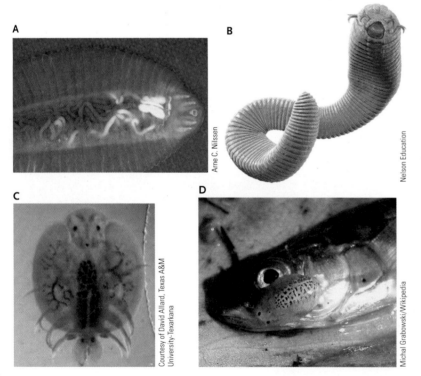

Figure 3.66 Pentastomids, the tongue worms, were named for the way that two pairs of reduced appendages near the mouth give the appearance of five mouths. **A.** A female *Linguatula arctica* (745) (the coiled central structure is the ovary) came from the sinuses of a reindeer (*Rangifer tarandus* (1767)). The mouth and appendages are at the right. **B.** In *Sebekia oxycephala* (746), a parasite from an Amazonian fish, *Phalloceros harpagos* (1206), the appendages are reduced to hooks. **C.** *Argulus japonica* (722), an ectoparasite shown on fish **D.**

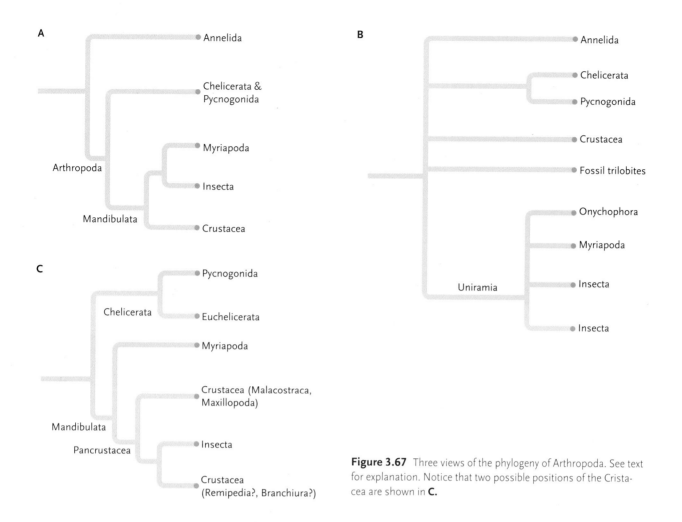

Figure 3.67 Three views of the phylogeny of Arthropoda. See text for explanation. Notice that two possible positions of the Crustacea are shown in **C.**

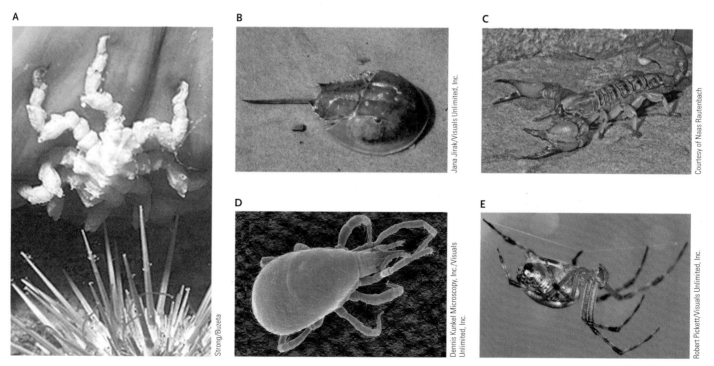

Figure 3.68 Chelicerata are characterized by a prosoma, formed by the fusion of head and thorax. Chelicerates lack antennae, have chelicerae and pedipalps on the 2nd and 3rd segments respectively, and then four pairs of walking legs. The rear part of the body is the opisthosoma. **A.** Pycnogonida (sea spiders, e.g., *Pycnogonum* (642)). **B.** Xiphosura (horseshoe crabs, e.g., *Limulus* (645)). Arachnida includes **C.** scorpions (e.g., *Opistophthalmus glabrifrons* (692)), mites, **D.** ticks (e.g., *Ixodes* (650)) and **E.** spiders (e.g., *Araneus diadematus* (659)).

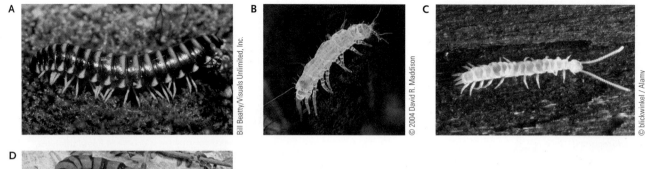

Figure 3.69 Myriapoda. **A.** Diplopoda (millipedes, e.g., *Apheloria virginiensis* (705)) are nocturnal terrestrial detritivores with paired segments, giving the appearance of having two pairs of legs per segment. There may be 80 000 species of millipedes. **B.** Pauropoda and **C.** Symphyla occur in terrestrial leaf litter and are also detritivores. **D.** Chilopoda (centipedes, e.g., *Scolopendra* (700)) are predators, and all species have a pair of poison claws. There are about 3000 species of centipedes and large species may be up to 30 cm long.

STUDY BREAK

1. Explain why the Arthropoda are now no longer seen as being closely related to the Annelida.
2. What are the features that separate the Chelicerata from the Mandibulata and the Insecta from the higher Crustacea?
3. In Figure 3.71, identify an example from each of the following orders: Coleoptera, Dictyoptera, Diptera, Hymenoptera, Lepidoptera, Odonata, Orthoptera, and Siphonaptera.

3.9 Deuterostomia

Given the attention paid to the origins of the vertebrate classes, it might be expected that the phylogeny of deuterostomes is firmly established. However, before the recognition of Ecdysozoa and the later inclusion of Chaetognatha as ecdysozoans, Chaetognatha was viewed as an early branch from deuterostomes. Traditionally, relationships between the deuterostome groups were those seen in Figure 3.73A. A combination of morphological and molecular evidence led to a small but significant shift in this arrangement, with recognition that hemichordates were more closely related to echinoderms (forming the group Ambulacra) than to the line leading to the chordates (Fig. 3.72B).

This arrangement changed a little with the recognition that *Xenoturbella* (210) (Fig. 3.1) was not a mollusc but perhaps a deuterostome (Fig. 3.72C). More recently, data from molecular studies were used to challenge the arrangement in Figure 3.72C. While some studies supported placing Cephalochordata as the sister group to vertebrates, other results suggested

Figure 3.70 Class Malacostraca (Arthropoda, Crustacea) includes Decapoda (crabs, shrimp, and lobsters, e.g., **A.** *Crangon crangon* (814) and **B.** *Pagurus acadianus* (826)), **C.** the Amphipoda (e.g., *Gammarus* (792)), and **D.** Isopoda (e.g., *Ligia* (805)). Class Maxillopoda (Arthropoda, Crustacea) includes **E.** barnacles (Cirripedia, e.g., *Semibalanus* (751)) and **F.** Copepoda (e.g., *Calanus* (727)). **G.** Class Remipedia (Arthropoda, Crustacea; e.g., *Lasionectes entricoma* (711)) may be a sister group to the insects.

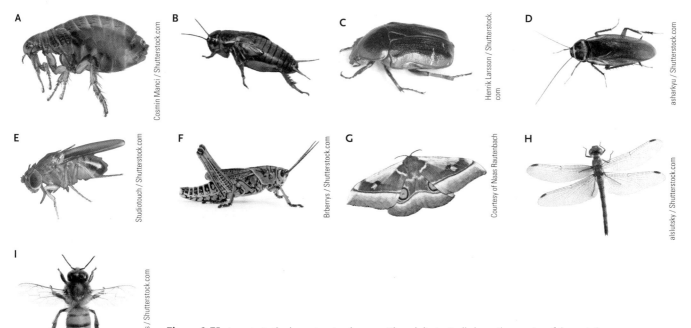

Figure 3.71 Insecta is the largest animal group. The adults typically have three pairs of thoracic legs, a single pair of antennae, and, in most, two pairs of wings. The thumbnail photographs show examples from some of the more than 30 orders in the Insecta.

that Urochordata are the sister group to Vertebrata (Fig. 3.72D).

An additional complication to current understanding of deuterostome phylogeny comes from microRNA data evidence that Acoela and Nemertodermatida are not, as previously thought, primitive and a sister group to the Bilateria (see Section 3.4). Rather, they share microRNA with *Xenoturbella* (Fig. 3.1), which also shares microRNA with Ambulacra (echinoderms and hemichordates). Investigations of *Hox*

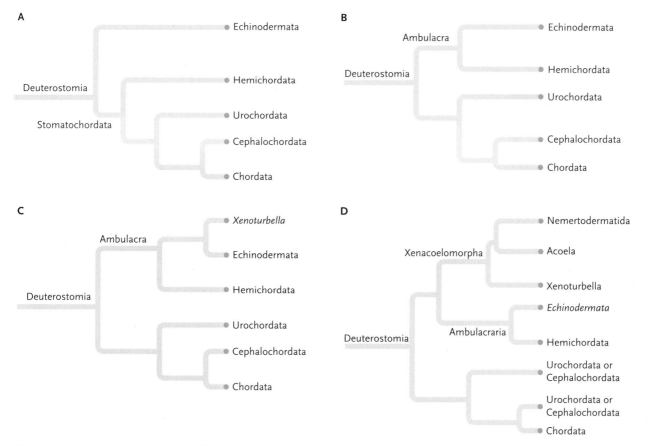

Figure 3.72 Concepts of the phylogeny of deuterostomes.

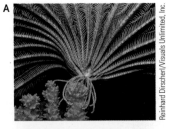

Figure 3.73 Echinoderms (literally "spiny skin") are characterized by bilaterally symmetrical larvae that transform to penta-radial adult forms. This approximation to radial symmetry gives sedentary or slow-moving animals the ability to sample and react to their environment in all directions. The water vascular system is unique to echinoderms. **A.** The phylum is divided into the attached Crinoidea (sea lilies, e.g., *Antedon* sp. (1065)) (although many modern forms have lost the attachment) and Eleutherozoa—four mobile classes, namely **B.** Asteroidea (sea stars, e.g., *Henricia* (1074)), **C.** Echinoidea (sea urchins, e.g., *Strongylocentrotus* (1090)), **D.** Holothuroidea (sea cucumbers, e.g., *Cucumaria* (1095)), and **E.** Ophiuroidea (brittlestars, e.g., *Ophiothrix* (1081)).

genes (Fig. 3.2) now suggest that perhaps the Xenacoelomorpha (the grouping of acoels and *Xenoturbella*) should be removed to a basal position in the bilateria.

Deuterostomia includes the Echinodermata (Fig. 3.73), Hemichordata (Fig. 3.74) and the three classes in the phylum Chordata: Urochordata (Fig. 3.75), Cephalochordata (Fig. 3.76), and Vertebrata (Fig. 3.77).

Traditional views placed Cephalochordata (e.g., *Branchiostoma* (1122), Fig. 3.76) as the sister group to the vertebrates. Cephalochordata have historically been considered in biology courses and textbooks about vertebrates, while Urochordata and other deuterostomes have more often been treated in courses about invertebrates. The Craniata constitute the rest of the Chordata and include the Vertebrata (Fig. 3.77) as well as the lampreys and hagfish.

STUDY BREAK

1. Explain the different functions of the podia (tube feet) in the echinoderms. How do the podia perform each of these functions?
2. What are the characteristics of the Phylum Chordata? How are they expressed in the Urochordata?

Figure 3.74 Phylum Hemichordata (literally "partial notochord"). These animals have a forwardly directed gut pocket, the stomochord, that was originally interpreted as a partial notochord. The phylum is divided into Enteropneusta (e.g., *Saccoglossus* (1101) (**A** & **B**)) and Pterobranchia (e.g., *Rhabdopleura* (1103) (**C** & **D**)). Enteropneusts, known as acorn worms, have an anterior proboscis, a collar that hides the mouth, an abdominal region divided into a branchiogenital region with gill slits (the anteriormost gill slits can be seen in **B**), and an intestinal region (grey coloured in **A**). Pterobranchs are colonial (**C**) with numerous zooids. **D.** Branched tentacles on the zooids are homologous with the enteropneust collar.

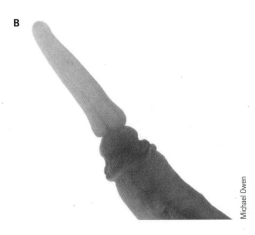

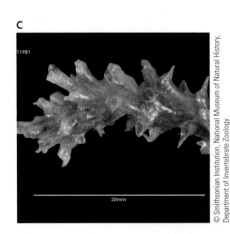

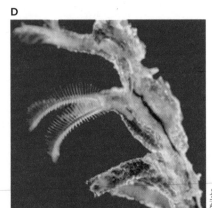

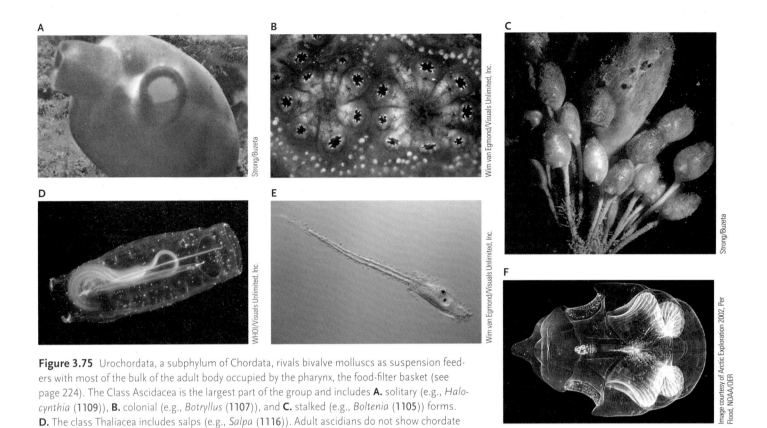

Figure 3.75 Urochordata, a subphylum of Chordata, rivals bivalve molluscs as suspension feeders with most of the bulk of the adult body occupied by the pharynx, the food-filter basket (see page 224). The Class Ascidacea is the largest part of the group and includes **A.** solitary (e.g., *Halocynthia* (**1109**)), **B.** colonial (e.g., *Botryllus* (**1107**)), and **C.** stalked (e.g., *Boltenia* (**1105**)) forms. **D.** The class Thaliacea includes salps (e.g., *Salpa* (**1116**)). Adult ascidians do not show chordate characteristics, but these are present in the larva, such as **E.** an ascidian tadpole, and in **F.** the class Larvacea (e.g., *Oikopleura* (**1120**)).

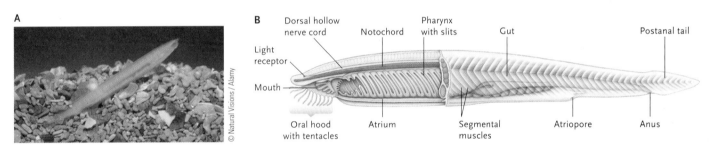

Figure 3.76 Cephalochordates belong to the phylum Chordata, and include animals such as *Branchiostoma* (**1122**) (formerly known as *Amphioxus*). All chordates have a dorsal notochord, a dorsal hollow nerve cord, and pharyngeal gill slits. Many have a postanal tail.

SOURCE: B. From RUSSELL/WOLFE/HERTZ/STARR. *Biology*, 2E. © 2013 Nelson Education Ltd. Reproduced by permission. www.cengage.com/permissions

Figure 3.77 Phylum Chordata, subphylum Vertebrata. Vertebrates include **A.** fish (e.g., *Hippocampus denise* (**1340**)), **B.** amphibians (e.g., *Tomopterna cryptotis* (**1391**)), **C.** reptiles (e.g., *Crocodylus niloticus* (**1437**)), **D.** birds (e.g., *Ardea melanocephala* (**1565**)), and **E.** mammals (e.g., *Panthera leo* (**1726**)). Vertebrates share all of the characteristics of the Chordata (see Figure 3.76) with the addition of a vertebral column.

PHYLOGENY **IN PERSPECTIVE**

A generation ago, biologists and biology texts worked with a phylogenetic scheme that had been in place for some time. While there was argument about the positioning of some phyla, the positions of the major groups were fixed and subject to little argument. Today we are working with a phylogeny that is totally different from that of the 1980s. While major groups appear settled in this phylogeny, the positions of a number of the small phyla are uncertain. When the phylogeny of groups as large as the insects and crustaceans remains open to question, and groups such as the Xenacoelomorpha challenge our understanding, we must recognize that animal phylogeny is still in a state of flux. It is likely that some currently accepted ideas will change within the next few years.

Questions

Self-Test Questions

1. Why has it been so difficult to place *Xenoturbella*[210] in a specific taxonomic group?
   a. Phylogenies are not fixed; they change as new information emerges.
   b. All ciliated marine flatworms are included in the Platyhelminthes based upon their physical characteristics.
   c. *Xenoturbella* has genes in common with flatworms, molluscs, and hemichordates.
   d. Physical characteristics and molecular evidence must be examined together to provide a framework for taxonomy, yet this evidence may yield conflicting results.

2. Which of the following statements describes the Radiata?
   a. All members of this group are diploblastic and radially symmetrical, and so they are readily differentiated from the bilaterally symmetrical organisms.
   b. The medusa body plan may have evolved secondarily, and biradial symmetry of these forms suggests that Cnidaria and Ctenophora are primitive bilaterians.
   c. Ctenophora are included in this group because they have tentacles, like Cnidaria.
   d. Cubozoans are included in the same class as all the other radial medusoid forms.

3. Which class of Molluscs is most difficult to place phylogenetically in relation to the other molluscan classes?
   a. Monoplacophora
   b. Gastropoda
   c. Chaetodermomorpha
   d. Scaphopoda

4. Which of these features place Nematoda in the Ecdysozoa?
   a. Physical adaptations for a parasitic lifestyle
   b. Segmental animals with paired, segmented appendages
   c. Growth depending on moulting of the secreted cuticle, and not having motile cilia
   d. Biramous appendages and jaws homologous with those of arthropods

5. Which of the following terms based on physical characteristics are LEAST explicit about the anatomy of the organism?
   a. Prosobranchia and opisthobranchia
   b. Gnathifera
   c. Echinoderm
   d. Tricladida and polycladida

Questions for Discussion

1. Which groups, previously regarded as separate phyla, have been incorporated in the Annelida? Why has this change taken place? Do you think that there is further change to come in the phylogenetic arrangement of these groups?

2. Sidney Manton devoted her life to anatomical studies of the Arthropoda. Based on her findings, she viewed the arthropods as being polyphyletic and arranged the arthropod groups in a way that is very different from a current phylogeny. What are the differences, and what are they based on? Can you explain how such a difference could have occurred?

3. If a phylum is equated to a symphony and its subgroups to themes and variations, then which phyla have the greatest number of themes and variations?

4. Use the electronic library to find a recent paper that explores some aspect of animal phylogeny. Does the new information revolutionize any of the information in this chapter?

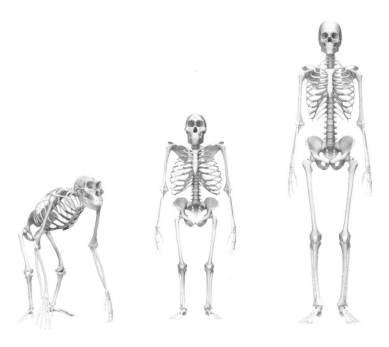

Figure 4.1 These skeletons of a chimpanzee (*Pan troglodytes* (1866)) (left), the *Australopithecus afarensis* (1852) specimen known as Lucy (middle), and a modern human (right) illustrate many changes in body form that occurred during the evolution of humans. The most obvious is the evolution of bipedality (walking on two legs), which is clearly reflected in the changing shape of the pelvis. The *A. afarensis* pelvis is far more similar to the human pelvis than to the chimp pelvis and Lucy was certainly bipedal. However, the width of the pelvis and its bowl-like shape suggest that she walked a little differently from the way we do. Lucy's relatively long arms, along with the hand and shoulder bones of other individuals, suggest that she was a better climber than we are. The presence of some human-like traits in *A. afarensis* and the lack of others illustrates that the process of human evolution proceeded in very small increments over many millions of years.

4 Evolutionary Processes

WHY IT MATTERS

Just after dawn a small family moves quietly through the African forest searching for fruits, young leaves, nuts, grubs, and (if they are lucky) other small animals for their morning meal. Even the oldest member of the group is just 1.2 m tall and weighs perhaps 45 kg. As they walk along the forest floor, their gait is less graceful than ours, but they are excellent climbers. They are remarkably agile in the trees where they search for food and build large, leafy nests where they find safety at night. This animal is *A. afarensis* (1852): your ancestor, or at least your distant cousin (Fig. 4.1).

Like all life on Earth, we are the product of evolution. Over many generations, evolutionary processes—including natural selection and genetic drift—slowly change the genetic composition of a **population** of organisms (a group of individuals of the same species that interbreed). Some of these changes are random, but others are in response to selective pressures imposed by, and opportunities provided by, the environment. For example, the ability of *A. afarensis* to walk upright could have been selected for because individuals who were able to move efficiently on the ground were better at gathering food and, because they were well-nourished, had more offspring who survived to produce offspring of their own.

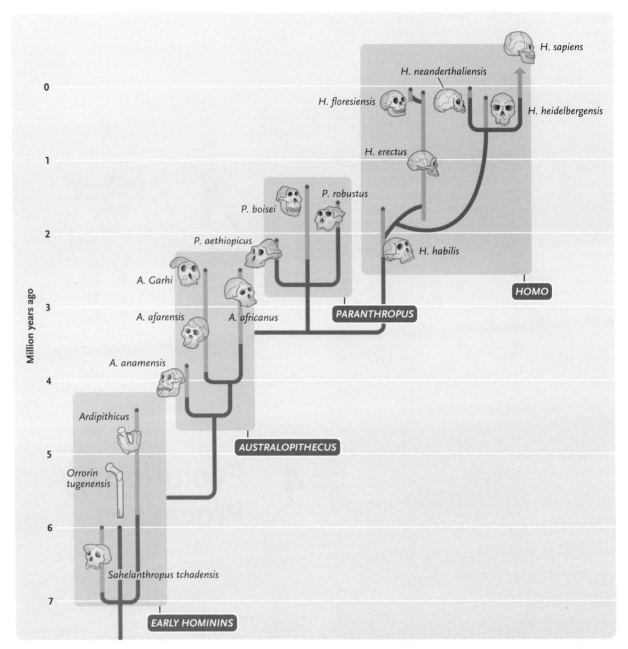

Figure 4.2 Modern humans are the sole survivors of a lineage that may extend as far back as seven million years, and includes ancestors, sister species, and distant cousins. All of the species that we know from pelvis, leg, or foot bones walked upright, though most did not walk in the same way we do. The limbs of our oldest ancestors were far better suited to climbing trees than ours are.
SOURCE: Based on Who's Who in Human Evolution, NOVA, pbs.org.

At some point in our evolutionary history, significant changes in our DNA gave rise to physical features that make us unique primates (Fig. 4.2). For example, over 500 genes that had changed very little between other mammals and chimpanzees (*Pan* spp. (1866)) disappeared from our genome. These regulatory genes control steroid hormone signalling and neural function, and their loss resulted in the loss of, among other things, penile spines and sensory vibrissae (Fig. 4.3). The loss of some of these genes may also have led to the expansion of the neocortex, the part of the brain involved in "higher functions" including language.

Evolution not only shaped our bodies, there is also evidence that it shaped our social systems. All primates live in communities that range in complexity from small family groups to large numbers of individuals living together, but only human societies are based on multi-level, nested alliances (e.g., families within villages, villages within territories, territories within states, etc.). This may have resulted from the evolution of long-term pair bonds that linked social groups.

Anatomically modern humans first appeared in Africa roughly 200 000 years ago. They likely subsisted on small animals including fishes and turtles that they gathered, the meat of larger animals that they likely scavenged, and a wide variety of terrestrial and aquatic plants. By 125 000 years ago, humans were beginning to move out of Africa, as evidenced by stone tools

Figure 4.3 Vibrissae, or whiskers (arrows), are specialized hairs found on the faces of most mammals and are exquisitely sensitive to touch. Mechanosensory neurons at the base of each vibrissa respond to the slightest movements and provide information about what is close to the mouth. **A.** Lesser mole rats (*Spalax leucodon* (1904)) and **B.** giant otters (*Pteronura brasiliensis* (1729)) rely on information from vibrissae to navigate and hunt fast-moving prey in dim lighting. **C.** Our closest relatives, the bonobos (*Pan paniscus* (1865)) have vibrissae but we do not. **D.** Human (*Homo sapiens* (1862)) facial hair lacks specialized mechanoreceptors.

excavated from a rock shelter at Jebel Faya, in what is now the United Arab Emirates. The stone tools are very similar to older tools from Africa, so these people were likely not expanding because of technological innovations but simply moving into a region that then provided fresh water, small animals, large game such as Persian fallow deer (*Dama dama mesopotamica* (1750)), water buffalo (*Bubalis arnee* (1746)), and edible plants.

Not only are our bodies and societies shaped by evolution, our existence is utterly dependent on the products of evolution. Everything from our food supply to the global cycling of oxygen and carbon are the products of millions of years of evolution.

This book is about the diversity of animals that evolution has produced, the fundamentals of how their bodies work, and how they interact with their environments and with one another. Human evolution is often viewed as a separate subject, but the mechanisms driving it are the same for the many different animals you will meet in this book.

SETTING **THE SCENE**

The spectacular diversity of animals (Chapter 3, Purple Pages) is a product of evolution. This chapter reviews concepts behind taxonomy) (naming species) and reconstructing phylogenetic (evolutionary) relationships among species. We then turn to how genetic variation is produced and provide examples of how micro-evolutionary processes such as genetic drift and natural selection act on populations to change their genetic composition over time. With these basics in place, we investigate large-scale patterns of evolution (macro-evolution) including adaptation, speciation, extinction, and co-evolution. Evolution is a central theme in every chapter of this book,

whether the topic is the feeding appendages of acorn barnacles (*Semibalanus balanoides* (751)) (page 209), how the unique gills of decapod crustaceans excrete ammonia (page 373), or the migration of Arctic terns (*Sterna paradisaea* (1560)) (pages 483–484).

4.1 Classification and Phylogeny

Classification is the process of naming species and placing them into hierarchical groups, while **phylogenetics** refers to the study of evolutionary relationships among species. Both are powerful tools for animal biologists. Without a universally recognized name for each animal, it would be impossible for people who speak different languages to talk about them. Similarly, without knowledge of a group's evolutionary history, it would be impossible to identify its adaptations, origins, extinctions, or even its place in the tree of life. The classification and phylogeny of animals is a work in progress and is constantly being revised as we learn more. To complicate matters, sometimes classification reflects an accepted phylogeny and sometimes it does not.

4.1a Classification

Carl Linneaus (1707–1778) is famous for positioning animals and plants within a taxonomic hierarchy based on their observable characteristics. Today we place animals in hierarchical groups based on their relatedness. This is because both similarities and differences can fool you. Closely-related animals are usually quite similar in appearance, but sometimes they look very different. For example, larvae, adult males, and adult females of deep-sea whale fishes (family Cetomimidae) look so different that they were once placed in three different families. In addition to extreme **sexual dimorphism** (differences between males and females), the larvae and adults also have radically different morphologies and ways of life (pages 351–353). In contrast, Juliana's golden mole (*Neamblysomus julianae* (1703)) and the Southern marsupial mole (*Notoryctes typhlops* (1696)) are superficially similar but belong to two different subclasses—one is a placental mammal and the other a marsupial (Fig 7.43 on page 194; see also Chapter 2, pages 15–18)! As we learn more about the evolutionary history of animals, species are often moved from one taxonomic rank to another, as we saw on pages 47–48.

To keep track of all these name changes, taxonomists place the name of the "authority" (the person who first described the species) after the generic and specific name, along with the date of the published description. For example, when Linnaeus described the dog whelk as *Buccinum lapillus*, the name would appear as *Buccinum lapillus* Linnaeus 1758. Since then, the species *B. lapillus* has been moved, first to *Thais lapillus* (Linnaeus 1758) and later to *Nucella lapillus* (424) (Linnaeus 1758). The parentheses around the authority name and date indicate that the species was moved from the original genus and placed in another. The name of the authority must always be spelled out, except for Linnaeus, who can be referenced by his initial (L. 1758), because he is the founder of modern taxonomy.

4.1b Phylogenies: Reconstructing Evolutionary History

Phylogenies (phylogenetic trees) were introduced and discussed on pages 47–48. Taxonomists use three common approaches to building phylogenies, each rooted in different assumptions about how evolution works. The three are maximum parsimony, maximum likelihood, and Bayesian inference.

The oldest approach, **maximum parsimony** (MP), assumes that evolutionary change occurs in the fewest possible steps. Using this method, species are grouped on the basis of shared characteristics derived from a common ancestor (**synapomorphies**). MP analyses attempt to compare all possible combinations of evolutionary steps in the search for the shortest sequence of changes. One criticism of the method is that it is often impossible to search all possible phylogenies, so there is no guarantee that the final phylogeny is the most parsimonious (i.e., requires the fewest steps). Another perhaps more fundamental criticism is that MP analyses do not work well if there are no synapomorphies in the data entered into the analysis, which, of course, cannot be known in advance.

A second approach to reconstructing phylogenies is **maximum likelihood** (ML). ML differs from MP in that instead of searching for the phylogeny (or phylogenies) with the fewest changes, ML predicts the most likely tree given the structure of the input data. ML trees are appealing because they provide statistical estimates of error and are commonly used to analyze DNA data. Like MP, critics of ML note that the method is only as good as the model of evolution inherent in the input data, which can vary in quality.

The third common method of reconstructing phylogenies, **Bayesian inference**, is similar to ML except that analyses are weighted by probabilities based on prior knowledge of relationships drawn from the results of previous research. Bayesian analyses are appealing because the logic parallels the process of testing hypotheses and using the results to formulate new hypotheses. One potential problem is that the analysis is only as good as the prior information entered into it. Despite their differences in philosophy and methods, MP, ML, and Bayesian phylogenetic analyses often yield quite similar results, especially at higher levels of the taxonomic hierarchy.

As we will see, knowing the timing of the branching events in a phylogeny allows us to ask questions about things such as rates of speciation and patterns of evolution in phenotypes that are hypothesized to be adaptations. The most common way of estimating the timing of branching events is with a **molecular clock**. The fundamental assumption behind molecular clocks is that neutral mutations, which are silent and neither harmful nor beneficial, evolve at a constant rate. Once these neutral mutations are identified, their rate of evolution can be calibrated using information about time gathered from fossils that represent specific branching points in a phylogeny (see Fig. 4.5, see also page 144). Some methods for using molecular clocks assume a constant rate of evolution across an entire phylogeny; other methods allow rates to vary in different parts of the tree. In either case, the more fossils that can be used to calibrate a phylogeny, the more confidence we have in the estimated branching times.

Even when the data used to create and calibrate a phylogeny are excellent, none of the methods of building phylogenies are perfect. For example, a common problem in phylogenetic analysis is long-branch

A

B

Courtesy of Craig Jackson

© Auscape International Pty Ltd / Alamy

Figure 4.4 Despite their similarities, **A.** Juliana's golden mole (*Neamblysomus julianae* (1703)) is a placental mammal in the order Afrosoricida, while **B.** the Southern marsupial mole (*Notoryctes typhlops* (1696)) is the sole member of the marsupial order Notoryctemorphia.

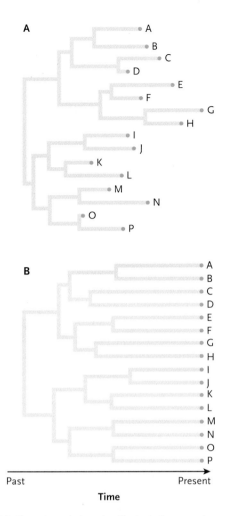

4.1c Strengths and Limits of Phylogenies

Phylogenies provide frameworks for testing hypotheses about patterns of evolution. These hypotheses range from the presence or absence of adaptations to co-evolutionary relationships, in which species are so dependent upon one another for survival that they evolve together. Without a phylogeny it would be impossible to trace the evolution of important characteristics without relying solely on the fossil record, which is often very incomplete. More significantly, many argue that one cannot do statistics that compare species without knowing the phylogeny of the group. This is because one of the fundamental requirements of traditional statistics is that the samples being compared (species) are independent of one another. However, we know this is not true because species are more or less dependent on one another by virtue of their evolutionary history (Fig. 4.6). Traditional statistical comparisons among species fail to meet the requirement of independence, so the results may be biased. This problem has given rise to a whole new branch of statistics called **comparative methods**, which factor phylogenetic relationships into statistical analyses. These analyses are exciting because they can be used to ask questions about the rate and timing of evolution.

Useful as they are, it is important to remember that phylogenies are hypotheses about evolutionary relationships. They are not statements of fact. Indeed we have no way of recognizing a true phylogeny even if we saw it. Phylogenies can and do change as more data accumulate and new methods of reconstructing phylogenies are developed. This means that all of the analyses relying on phylogenies as a scaffold are subject to change when the phylogenies change. In

Figure 4.5 These two phylogenies illustrate the same branching sequences among species A–P. The branch lengths on Phylogeny **A** represent the rate of evolution along each branch, and the tips of the branches are in the present. Short branches indicate rapid evolution; long branches indicate slow evolution. Phylogeny **B** has been calibrated using a molecular clock with information about time derived from the fossil record. Here, branch lengths represent time. Most phylogenies in this book are like phylogeny B.

SOURCE: Based on http://abacus.gene.ucl.ac.uk/software/indelible/manual/files/random_branch_length_trees.JPG

attraction, in which unrelated pairs of taxa are placed together as an artifact of their both having long branches (i.e. they independently acquired many evolutionary changes and/or experienced a long period of independent evolution). This problem led to bizarre results such as removing guinea pigs from rodents (the order containing mice, groundhogs, and porcupines), to which there is no doubt that they belong and to which they have been returned. Several more examples of the problems created by long-branch attraction were mentioned on page 49. Methods of recognizing and avoiding the problem of long-branch attraction are being developed and implemented. Indeed, techniques for generating phylogenies are constantly being invented and refined but, like all techniques, they will never work perfectly all the time.

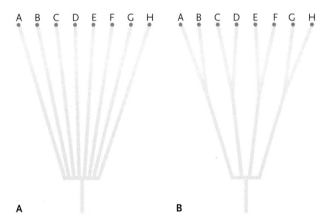

Figure 4.6 These two phylogenies illustrate two different kinds of relationships among species (labelled A–H). Phylogeny **A** assumes that all species are equally related, but in reality that is almost never the case. Usually, any given species is closely related to some other species and distantly related to others. In phylogeny **B**, for example, species A is more closely related to species B than it is to species C. Not recognizing the uneven relationships among species can bias the results of statistical analyses.

SOURCE: Based on Joseph Felsenstein, "Phylogenies and the Comparative Method," The American Naturalist, 1985.

practice, this is not a big problem for scientists who work on groups that have been studied in detail, but it is a significant problem for those working on lesser-known groups. For many clades of animals there is simply not enough information to generate a phylogeny in which we can have confidence.

STUDY BREAK

1. Compare and contrast taxonomy and phylogenetics.
2. What is the primary assumption behind molecular clocks and how are molecular clocks used in the construction of phylogenies?
3. What are comparative methods? How are they different from traditional statistics?

4.2 Micro-Evolutionary Processes

The simplest definition of evolution is a change in the genetic composition of populations over time. The mechanisms that drive evolution are called **micro-evolutionary** processes. These processes can only function in the presence of **genetic variation** among individuals within populations. Each of the brilliant blue-skippers (*Paches loxus* (1042)) in Figure 4.7 has a different **genotype** (genetic composition) due to different combinations of **alleles** (variants of genes) within at least some of its genes. We know this in part because some (but not all) genes affect **phenotypes** (physical, biochemical, or behavioural characteristics), and these butterflies look different from one another. If all of them shared exactly the same genotype (i.e., they were clones), there would be no variation for evolution to act upon, so evolutionary change would not take place. Like these butterflies, most populations of most animals contain many different genetic variants upon which micro-evolutionary

Figure 4.7 The variation in colour between these brilliant blue-skippers (*Paches loxus* (1042)) from the same population is based on underlying genetic variation.

processes can act. Some of these processes are random (genetic drift), some are governed by the movement of genes between populations (gene flow), and some are the result of selection. In the latter case, individuals that inherit phenotypes that allow them to have the largest number of offspring that reach reproductive age are said to have the highest **fitness** (Box 4.1). These individuals will pass more of their genes to the next generation than those with lower reproductive success.

4.2a Sources of Genetic Variation: Recombination and Mutation

What is the source of genetic variation, the fuel for the evolutionary fire? By far the most important source of genetic variation in animal populations is good old-fashioned sex. **Diploid** animals carry two homologous sets of chromosomes, one from their male parent, and the other from their female parent. A diploid organism produces gametes through meiosis, which cuts its diploid number of chromosomes in half (gametes are thus called **haploid**). During meiosis, another important event—**genetic recombination**—or crossover (Fig. 4.8) can introduce variation. Prior to meiosis, homologous chromosomes line up next to each other, and parts of them switch places (cross over). In this process, chunks of genetic material from one chromosome move to the other chromosome, and vice versa (Fig. 4.8). The result is gametes that contain an assortment of the genes from the organism's parents. Recombination is a huge source of genetic variation, because it means that the genes of diploid animals are shuffled like a deck of cards every time reproduction happens. Almost every individual animal carries a unique random combination of genes.

Mutations are the second most important source of genetic variation in animal populations. **Mutations** are changes in genetic information encoded in an organism's DNA. When DNA is replicated during cell division, the DNA molecules "unzip" along their centres each rung of the ladder breaks in the middle, and a new "other half" is built for each side, resulting in two identical copies of the original molecule (Fig. 4.9 on page 84). This process takes place with an amazing degree of fidelity, and mistakes rarely occur. But with the huge amount of information involved, mistakes inevitably happen. The vast majority of mutations occur in somatic cells whose DNA is not passed on to offspring. These have little or no effect because either they occur in a region of the DNA that does not code for anything (so-called "junk DNA"), or they happen in a way that does not affect the product of the gene (a protein). But sometimes mutations do have an effect.

Mutations occurring at the level of the base pairs are called **base pair**, or **point**, mutations because they are changes in a single nucleotide within a DNA

BOX 4.1
Fitness

What do evolutionary biologists mean by "fitness"? In evolution, an individual is more "fit" if it is better at producing descendants than other individuals. Fitness is a concept dependent upon circumstance, and thus you cannot assess the fitness of an individual unless you can watch it over time and see how it does at the game of reproducing relative to other individuals. Genes conferring fitness in one set of circumstances may lead to lack of fitness in another set of circumstances. Only by tracking reproductive success over generations can you determine the fitness of a genetic variant.

There are many colour variants of the European land snail (*Cepaea nemoralis* (475)) and the underlying genetics of the different morphs are well understood (Fig. 1). The distribution of colour morphs within populations has evolved rapidly in response to human disturbance of the snails' habitats. Light-coloured morphs are found in open, recently disturbed habitats while darker morphs occur in shaded habitats that have had time to recover from disturbance. The results suggest that dark colours are more fit in shaded habitats because they are better at absorbing heat. Conversely, coloured snails are less likely to overheat in open, sunny habitats. Other studies of the same species of snails argue that predation is the selective agent for snail colour because light-coloured snails are harder for birds to find in bright, open habitats while snails with darker colours are more hidden in shaded areas. Both proposals can be correct: different selective pressures can act on different populations of the same species or on the same populations at different times.

Robert Cameron

Figure 1 *Colour morphs of the European land snail (*Cepaea nemoralis (475)*).*

Interestingly, the snails achieved "light" and "dark" colours by several different combinations of colors and banding patterns. This is an excellent example of how selection can produce similar results (light or dark) with different combinations of genes, depending on the genetic variation that is present in each population.

---

molecule. Sometimes point mutations change important parts of genes and result in either a small or large change in the phenotype of an organism. Point mutations can simply occur randomly or be induced by exposure to radiation (either cosmic radiation or some terrestrial source) or environmental toxins. Most point mutations have no effect, but some can have a large effect on phenotype. For example, a well-known point mutation in fruit flies leads to the antennae being replaced by legs. Many cases of skin cancer are thought to be caused by point mutations in somatic cells acquired during a person's lifetime through exposure to solar radiation.

Inversions are mutations involving larger pieces of DNA. In this case small bits of the DNA ladder copy properly, but whole regions of the ladder are turned around and put back in the opposite direction. This technically should not affect the genes that remain intact, although now they are arranged differently on the DNA molecule. In reality, inversions can seriously affect the phenotype of the organisms that carry them, so they are an important component of the pantheon of mutations that can lead to evolutionary change (Box 4.2).

A mutation that affects a phenotype can only enter the fitness game if it occurs in a cell that is destined to divide into gametes through meiosis. This is a very small chance. We can take an example from juvenile topshell snails (*Trochus niloticus* (413)) that live near the Great Barrier Reef alongside dwarf turban snails (*Turbo brunneus* (414)) and predatory knobbed rock shell snails (*Thais tuberosa* (430)). Juvenile topshell snails and dwarf turban snails have different anti-predator defenses: the topshells secrete white mucus when a predator snail approaches while the dwarf turban snails move away quickly (for a snail anyway). The predatory snails eat more dwarf turban snails even though the dwarf turban and topshell snails contain the same amount of nutrients. This means that secreting mucus is a more effective anti-predator strategy than fleeing.

Now imagine that a mutation occurs in a cell in a male topshell snail. For this mutation to have any effect on the evolution of topshells, it needs to have occurred in the extremely small fraction of that individual's cells that are destined to become sperm. Then, one of these sperm must succeed in fertilizing a female topshell's egg. After all of this, the great likelihood is that the mutation does not affect its carrier (the juvenile topshell that grows from this egg) in any way. If, however, the mutation happens to allow the juvenile to secrete mucus that is even better at repelling predators, then that carrier may survive while its siblings are eaten. Thus, the carrier is more likely to reproduce. If so, the mutation would be passed on and could have a lasting effect on the

Figure 4.8 The process of meiosis produces four haploid nuclei (gametes) through two meiotic divisions. For simplicity, just one pair of homologous chromosomes is followed through the divisions. Recombination commonly occurs when chromosomes are replicated just prior to the first meiotic division.

SOURCE: From RUSSELL/WOLFE/HERTZ/STARR. *Biology, 2E.* © 2013 Nelson Education Ltd. Reproduced by permission. www.cengage.com/permissions

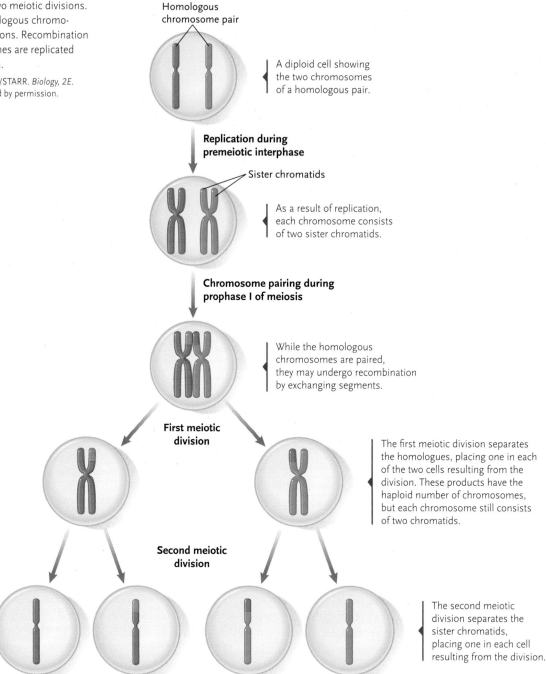

Homologous chromosome pair

A diploid cell showing the two chromosomes of a homologous pair.

Replication during premeiotic interphase

Sister chromatids

As a result of replication, each chromosome consists of two sister chromatids.

Chromosome pairing during prophase I of meiosis

While the homologous chromosomes are paired, they may undergo recombination by exchanging segments.

First meiotic division

The first meiotic division separates the homologues, placing one in each of the two cells resulting from the division. These products have the haploid number of chromosomes, but each chromosome still consists of two chromatids.

Second meiotic division

The second meiotic division separates the sister chromatids, placing one in each cell resulting from the division.

evolution of topshells. As you can imagine, passing on such a mutation is very rare! The more usual source of genetic variation on which evolution can act is recombination.

4.2b Genetic Drift

Genetic drift is one way in which genetic variation can lead to evolutionary change. In **genetic drift**, the genetic compositions of populations fluctuate randomly and, depending on the gene frequencies and population size, can lead to non-adaptive evolution. We can think about genetic drift by considering random effects on the frequency of a single **allele** of a hypothetical gene within a population. Figure 4.10A illustrates the results of computer simulations of a random change in the frequency of an allele in each of 50 generations. Each simulation begins with an allele frequency of 0.5 and the results illustrate the potential for many different evolutionary paths. By the 50th generation the frequency of the allele ranges from 0.0 (loss) to 1 (fixation)—simply by

BOX 4.2
Chromosome Inversions

For many years, inversion mutations were thought to be selectively neutral or detrimental, but recent evidence indicates how inversions can undergo positive selection. One plausible explanation for this, suggested by Theodosius Dobzhansky in 1970, is that inversions cause groups of well-adapted genes to tend to remain together because recombination between chromosomes that are heterozygous for inversions (sometimes called heterokaryotypes) are reduced.

For homologous heterokaryotype chromosomes to line up (synapse) during meiosis, they must form loops such as those shown in Figures 1 and 2. For mechanical reasons, any single crossover that occurs within a heterozygous inversion results in a cell that cannot complete division into daughter cells with complete complements of genes. Therefore they cannot form viable gametes. Double crossovers (or in fact any even number of crossovers) within an inversion can succeed, but odd numbers of crossovers within an inversion are not viable. In this way, inversions can protect groups of genes from recombination. Whatever the reason, it has become clear that inversions are an important source of genetic variation that natural selection acts upon in evolving populations.

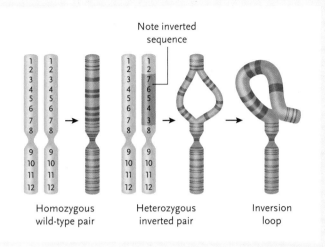

Note inverted sequence

Homozygous wild-type pair — Heterozygous inverted pair — Inversion loop

Figure 1 *The making of a chromosome inversion.*

SOURCE: Based on illustration by Sean F. Werle

Sean F. Werle

Figure 2 *An inverted chromosome.*

chance. Figure 4.10 also shows that the effects of genetic drift vary depending on population size. It has the strongest influence in small populations (4.10A), where it is called the **founder effect**. But even in larger population (4.10B & C), genetic drift can be a powerful driver of evolutionary divergence.

It is important to point out that this example of genetic drift is hypothetical. In real life, many different evolutionary processes are happening at once, and it is difficult to separate the effects of genetic drift from other micro-evolutionary processes. A good example comes from a recent investigation of the evolution of seahorses (*Hippocampus* spp. (1345)) which are found in shallow coastal waters around the world. Seahorses have grasping tails that they use to hold on to stationary objects and they typically live relatively sedentary lives (Fig. 4.11). Given their stay-at-home habits, how did they come to have such a broad geographic range? Researchers suggest that the answer is that the different species evolved from small, isolated founder populations that diverged due to genetic drift via founder effect.

There are two ideas about how small populations became isolated. One possibility is that small populations of seahorses could have been isolated by **vicariance**, the development of physical barriers which, in this case, resulted from the opening and closing of connections among ancient seas. The second, and perhaps more likely, alternative is that small numbers of seahorses dispersed over long distances and founded new populations. Although seahorses live near shore, they are often found far out to sea hanging onto rafts of floating seaweed. A single pregnant male seahorse can carry a brood of several hundred offspring (Fig. 4.11). If he were to end up in a new environment, his offspring could breed and found a new population. Both of these scenarios agree that founder effects and subsequent genetic drift probably played a role in speciation, but they cannot discount the possibility that gene flow and/or selection also influenced the evolution of the isolated populations in their newfound homes. This example serves to illustrate the complexity of evolution and the challenge of uncovering the mechanisms behind it.

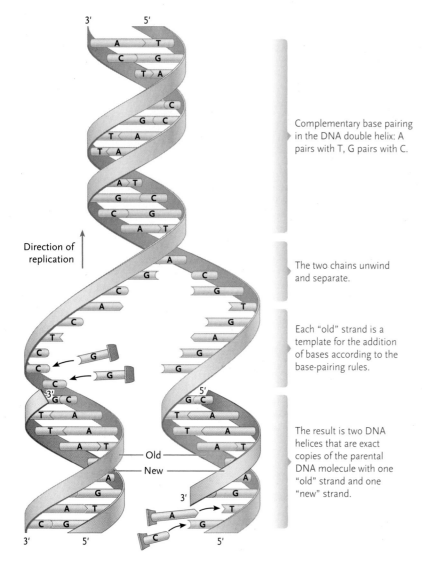

Figure 4.9 DNA replication.

SOURCE: From RUSSELL/WOLFE/HERTZ/STARR. *Biology*, *2E*. © 2013 Nelson Education Ltd. Reproduced by permission. www.cengage.com/permissions

Complementary base pairing in the DNA double helix: A pairs with T, G pairs with C.

The two chains unwind and separate.

Each "old" strand is a template for the addition of bases according to the base-pairing rules.

The result is two DNA helices that are exact copies of the parental DNA molecule with one "old" strand and one "new" strand.

Direction of replication

Old

New

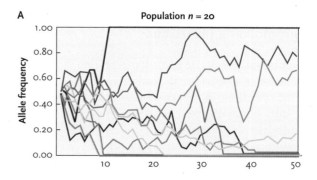
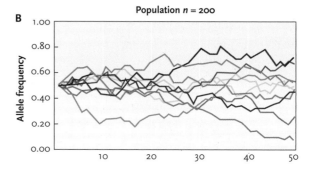
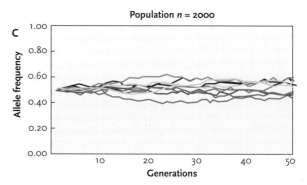

Figure 4.10 Simulations of random genetic drift in a single allele at a population sizes of 20 (**A**), 200 (**B**) and 2000 (**C**). Each simulation is a different colour and each begins with an allele frequency of 0.5. Allele frequency diverges less quickly and less widely as population size increases.

SOURCE: Based on concept found in Figure 3.1 of "Darwinian Detectives", Norman A. Johnson, 2007, Oxford publishers, p. 48. Professor Marginalia

4.2c Gene Flow

Gene flow is the movement of genes between populations and can be an important source of genetic variation. If a population of animals is **monomorphic** for a particular gene, then only one allele for that gene exists in the population. This is also known as **fixation**. If a gene is fixed within a population, there is no variation in that gene upon which evolution can act. If, however, an individual or small group of individuals from another population who carry a different allele join the "fixed" population, then the mixing in of their genes has introduced genetic variation.

Gene flow between populations is regulated by many variables but perhaps most commonly by life history strategies related to reproduction and dispersal. Take, for example, the topshell snail, *Margarella antarctica* (412) and the limpet *Nacella concinna* (406). Both are marine species that feed on the same types of foods and coexist over large regions of the Antarctic coast, but their reproductive strategies are very different. The topshells

Figure 4.11 Seahorses use their grasping tails to hold onto stationary objects. This large-bellied individual is a pregnant male. Male seahorses release sperm into the water at the same time that females deposit their eggs into the males' brood pouches. Males carry the eggs in the pouch until they hatch.

deposit egg capsules that stick to underwater substrates. The larvae can crawl but don't venture far from where they hatched. In contrast, the limpets reproduce by broadcast spawning (releasing gametes into the water column), and the resulting free-swimming larvae are carried by ocean currents for up to two months.

Not surprisingly, tissue samples from five populations of topshells and limpets that span the length of the Antarctic coast show different levels of genetic variation. The five populations of topshell snails were genetically distinct from one another, indicating that there was very little gene flow between populations. Meanwhile, the five populations of limpets were genetically similar—a reflection of extensive gene flow between populations. This example illustrates the complexity of the links among animal ecology, biology and evolution.

Don't let the story of the topshells and the limpets leave you thinking that all marine invertebrates that engage in broadcast spawning exhibit high levels of gene flow across very large areas. The snake-skinned chiton, *Sypharochiton pelliserpentis* (398), inhabits inter-tidal to estuarine habitats on the coasts of New Zealand and eastern Australia. Like the limpet, it is a broadcast spawner, releasing gametes into the ocean during certain times of year. A comparison of the mitochondrial gene cytochrome oxidase from populations at 27 different locations around New Zealand revealed seven different haplogroups (groups based on similarity in suites of adjacent alleles), and most populations contained examples of more than one haplogroup (Fig. 4.12). The primary division between haplogroup frequencies was between populations from North Island plus two populations at the north end of South Island, and the remainder of South Island. Clearly there is some gene flow between the two islands, but why is it so limited? The answer may lie in seasonal upwellings of cold water that coincide with the timing of the spawning. Whether this influx of cold water kills some proportion of the larvae or the associated currents sweep them offshore is not known, but it does appear to form an intermittent barrier to gene flow in the snake-skinned chiton. Interestingly, several other sessile marine invertebrates in the area exhibit a similar pattern of genetic separation between northern and southern populations, but it is not as strong. Those species either spawn at a different time of the year, when there are no upwellings of cold water, or spawn several times a year, again avoiding the seasonal appearance of the cold water upwellings. This is a good example of how gene flow can be limited by a geophysical barrier, in this case seasonal, cold ocean currents.

Physical barriers are not the only forces that can restrict gene flow. When closely related populations have begun to evolve toward becoming separate species, other factors can keep them from interbreeding. These include divergent mating displays, differences in reproductive anatomy, or reduced fitness in hybrids.

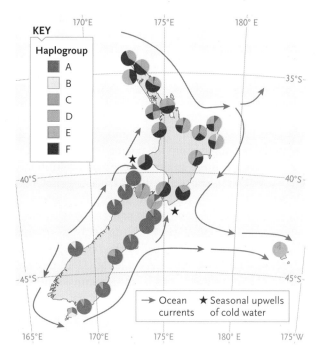

Figure 4.12 A map of New Zealand with pie charts illustrating the distribution of haplogroups (A–F) of the cytochrome oxidase gene in populations of the snake-skinned chiton (*Sypharochiton pelliserpentis*). The grey arrows indicate typical ocean currents, and the asterisks indicate the locations of seasonal upwellings of cold water that may impede gene flow.

SOURCE: Redrawn from A.J. VEALE and S.D. LAVERY. 2011. "Phylogeography of the snakeskin chiton Sypharochiton pelliserpentis (Mollusca: Polyplacoph-ora) around New Zealand: are seasonal near-shore upwelling events a dynamic barrier to gene flow?" *Biological Journal of the Linnean Society*, 104: 552–563. © 2011 The Linnean Society of London.

4.2d Natural Selection

Natural selection was first described in 1858 in two papers presented at a meeting of the Linnaean Society of London, one by Charles Darwin and the other by Alfred Russell Wallace. Both Darwin and Wallace had travelled the world as naturalists and made extensive observations and collections of all kinds of animals and plants.

Darwin in particular drew two important inferences from his observations:

1. that individuals within populations compete for limited resources
2. that inherited characteristics may allow some individuals to survive longer and reproduce more than others.

Taken together, these inferences describe the process of evolution by **natural selection**: that characteristics of individuals within populations change over generations as advantageous, heritable characteristics become more common. Darwin's theory of natural selection was meticulously laid out in his book, *On the Origin of Species by Means of Natural Selection* (1859).

In the more than 150 years since the publication of the *Origin of Species*, technological innovations have given us insights into genetics that Darwin may never have imagined, but his theory of natural selection has remained intact. Restated in modern language, if there is variation in genotypes within a population that produces phenotypes with various levels of fitness, then natural selection will, over many generations, increase the frequency of the genotypes that confer higher fitness. The end result of natural selection is that some genes (or groups of genes) and their phenotypes increase in frequency while others decrease in frequency or vanish from the population completely.

There are three main patterns of evolutionary changes in phenotypes and their associate genotypes: directional, stabilizing, and disruptive selection (Fig. 4.13). Let's illustrate these patterns by imagining an ancestral population of birds with a normal distribution of tail lengths (bell-shaped curve, Fig. 4.13, top row). The most frequent tail length is 10 cm, but tail length ranges between 2 and 18 cm within the population as a whole. Under directional selection (Fig. 4.13A, left curve), selection favours a trait value from one end of the original distribution, in this case longer tails. Over time the peak frequency of tail length increases to 15 cm and few individuals have a tail shorter than 10 cm, but the range of variation is similar to that of the ancestral population. In contrast, stabilizing selection acts to reduce the variation within a trait (Fig. 4.13B, middle curve). In this example, the location of the peak frequency of tail length is the same as in the ancestral population (top row), but the distribution is compressed and the range of tail lengths is restricted to between 5 and 15 cm. Stabilizing selection can occur when fitness drops dramatically as a trait diverges from an optimal value. Disruptive selection occurs when selection favours traits at the extremes of the original distribution and/or there is strong selection against the intermediate phenotype (Fig.4.13C, right curve). Here the result is a frequency distribution of trait values with two peaks; one for individuals with short tails, and the second for individuals with long tails. These are perfect, hypothetical examples of directional, stabilizing, and disruptive selection, but evidence for each can be found in nature.

Body size evolution in the Sociable Weaver Bird *Philetairus socius* (1610) presents a good example of stabilizing selection. These communally nesting birds inhabit savannahs and scrublands of southern Africa.

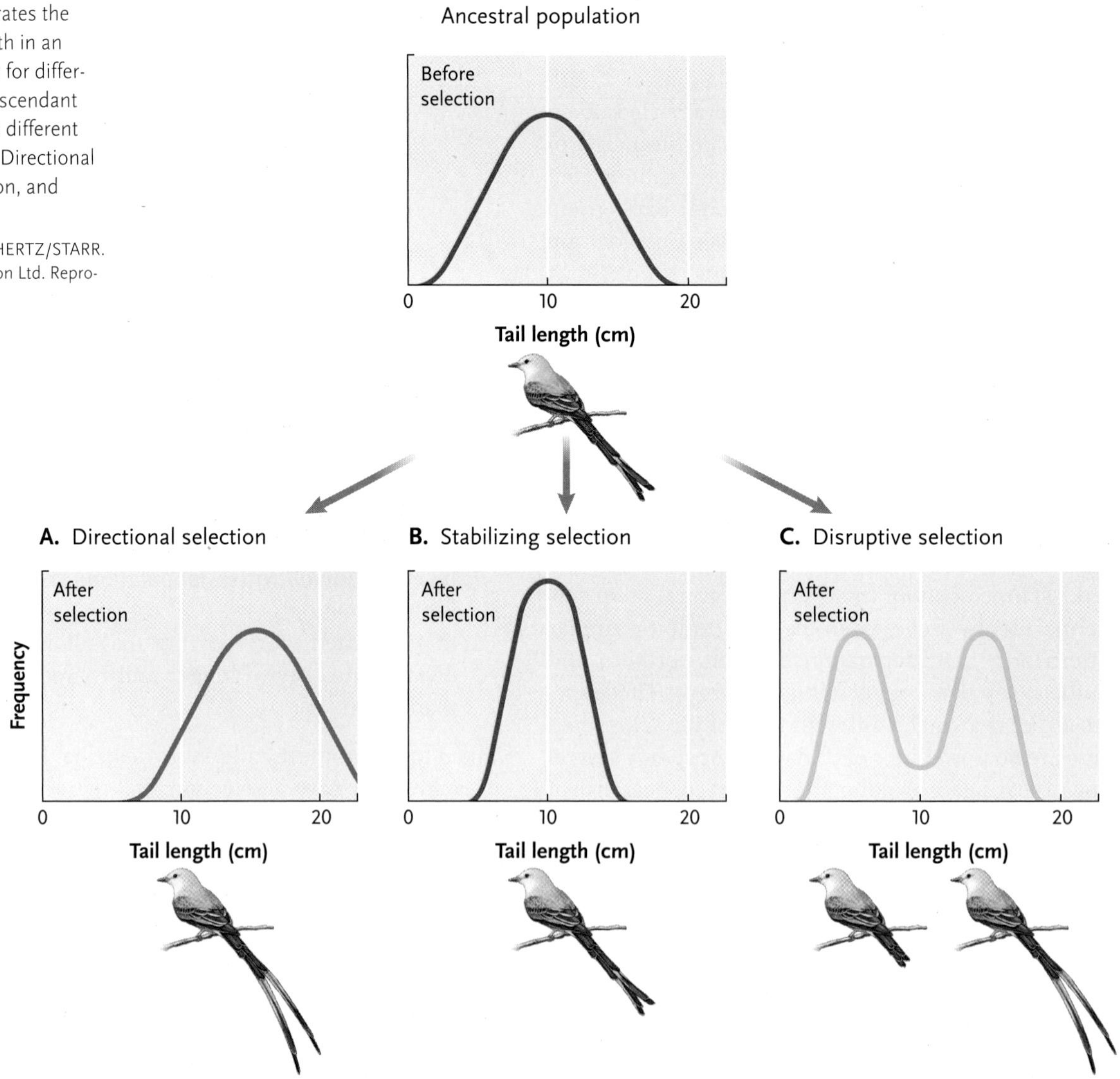

Figure 4.13 The top row illustrates the normal distribution of tail length in an ancestral population. Selection for different lengths (arrows) lead to descendant populations (bottom row) with different distributions of tail lengths. **A.** Directional selection, **B.** stabilizing selection, and **C.** disruptive selection.

SOURCE: From RUSSELL/WOLFE/HERTZ/STARR. *Biology*, 2E. © 2013 Nelson Education Ltd. Reproduced by permission. www.cengage.com/permissions

Ancestral population

A. Directional selection   B. Stabilizing selection   C. Disruptive selection

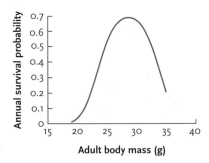

Figure 4.14 Annual survival probability versus body mass in the Sociable Weaver (*Philetairus socius* (1610)). Note that the probability of survival decreases at both the low and high ends of the body mass distribution, indicating strong stabilizing selection around the mean body mass.

SOURCE: Covas R, Brown CR, Anderson MD, Brown MB. 2002. "Stabilizing selection on body mass in the sociable weaver Philetairus socius." *Proc Royal Soc B*, 2002, 269(1503):1905–1909, by permission of the Royal Society.

Food availability is unpredictable in this environment, so it is advantageous for Sociable Weavers to have high fat reserves, which come in handy when food is scarce or when disease or parasites strike. However, being too heavy can have negative consequences. Heavier birds fly more slowly and are less manoeuvrable than lighter birds. This makes them easier targets for predators, such as the Gabar Goshawk *Micronisus gabar* (1581). In fact, data collected over several years demonstrate that Sociable Weavers at both extremes of the body mass distribution exhibit reduced survival ability (Fig. 4.14), illustrating that stabilizing selection is operating on the population. Over time, this selection should lead to reduced variance in body mass.

Pronounced differences in reproductive strategies and sperm size among male Bleeker's squid (*Loligo (Doryteuthis) bleekeri* (504)) is an interesting example of disruptive selection. There are two distinct types of males (Fig. 4.15). Large "consort" males compete with one another and court females using elaborate colour displays. When a female accepts a consort male, he deposits a spermatophore in her oviduct and guards her against other males until she releases strings of eggs. The other type of male is called a "sneaker" because it sneaks up to the female while she is being guarded by a consort and mates with her. Sneakers are smaller than consorts and do not compete with other males or display to attract females. Instead they achieve reproductive success by placing a spermatophore in the female's seminal receptacle (sperm storage organ), which is located on the external surface of her body just under her mouth. When the female produces egg strings, some of the eggs are fertilized in the oviduct with consort sperm, and some are fertilized by sneaker sperm as she pulls the egg string in front of her mouth before depositing it on the ocean floor. The sperm of sneaker males are substantially larger than consort sperm. In some animals with sperm of different sizes, there is clear evidence that the size classes are competing with one another for fertilization (i.e., there is sperm competition). Although this may have happened in the history of Bleeker's squid, it now appears that the size difference between the two types of sperm may be adaptive for different fertilization environments (i.e., the internal oviduct and external seminal vesicle). Whether the result of sperm competition or adaptation to different environments, the two size classes of sperm can be attributed to selection for phenotypes from the extremes of the ancestral distribution.

An elegant example of directional selection comes from a study of hatchling survival in the common snapping turtle, *Chelydra serpentina* (1422), which is common in lakes, ponds, and rivers throughout temperate eastern North America. Adults can be very large (23 kg or more) and aggressive, often preying on fish, frogs, snakes, and small birds and mammals. Like most turtle species, however, snapping turtles rarely survive to adulthood. Instead, many eggs are eaten, and many hatchlings are eaten within the first few

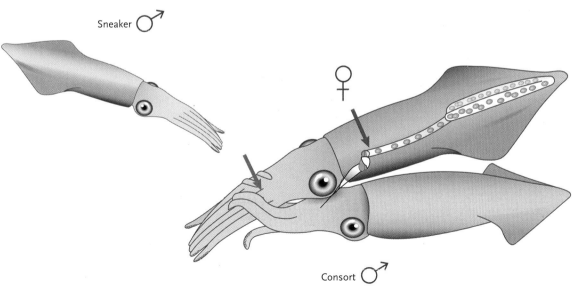

Figure 4.15 Male consorts compete with one another and are larger than sneaker males. Consorts place their spermatophore inside the female's oviduct (red arrow), while sneaker males place their spermatophore in the female's seminal receptacle (blue arrow).

SOURCE: Redrawn from Iwata et al., "Why small males have big sperm: dimorphic squid sperm linked to alternative mating behaviours," *BMC Evolutionary Biology* 2011, 11:236.

CHAPTER 4 EVOLUTIONARY PROCESSES

weeks. Thus, heritable features that promote survival in hatchlings should be under strong selection. Hatchling body size is one such feature; it is moderately heritable in snapping turtles and heavier individuals are often in better condition and less susceptible to predation. In this particular experiment, eggs from several *Chelydra serpentina* nests near the Mississippi River in Illinois were brought into a laboratory where they were incubated and allowed to hatch. After a few weeks, individuals were weighed, marked, taken back to the Mississippi River, and released from an abandoned nest site about 50 m from the water's edge. Over the next couple of days, individuals that succeeded in reaching the river were recaptured and analyzed. The results confirmed the prediction that, on average, larger hatchlings had a greater chance of survival. These results, when combined with the study on the Sociable Weaver *Philetairus*, show that body size can be subject to different types of selection.

These examples illustrate that natural selection is a powerful evolutionary force that has resulted in remarkably effective adaptations of animals to their environments. Just one such example is **mimicry**, which we will see many times in this book. Sometimes mimicry is used as a defence. For example, in New England the wood frog, *Rana sylvatica* (1385), is so well matched to its woodland environment that it is virtually invisible when it sits on a background of dry leaves (Fig. 4.16A). In another example, a harmless robber fly resembles a bumble bee, which defends itself by a painful sting (Fig. 4.16B). Mimicry can also help animals capture their prey. The Costa Rican pitbull katydid, *Lirometopum coronatum* (900), is a sit-and-wait predator. The colouration of its nymphs closely matches the appearance of the small flower heads where they wait for prey (Fig. 4.16C). (See Section 16.4c for a more detailed discussion of mimicry.)

4.2e Sexual Selection

Sexual selection is another important source of animal diversity, and causes some of the most spectacular examples of diversity. **Sexual selection** is a form of within-species natural selection and can be divided into two broad categories, intra- and inter-sexual selection. Intra-sexual selection involves contests between individuals of the same sex, usually males, for access to mates. In inter-sexual selection, one sex, usually females, chooses from among many males based on her assessment of a heritable trait(s). Sexual selection can result in dramatic sexual dimorphism in the appearance and/or behaviour of males and females of the same species.

The most common form of intra-sexual selection is male–male competition, in which males compete to defend resources that are directly or indirectly related to mating opportunities. These resources are things that females need, such as food or places to lay eggs or raise young, but their distribution is patchy (confined to small areas that are widely dispersed across the landscape). When females congregate at resource patches, males compete with one another for access to the females. For example, a study of fiddler crabs (*Uca capricornis* (841)) demonstrated that males defend many empty burrows within their territories as a means of attracting females. Females are allowed to move into the empty burrows and mate with the territory holder, but other males that enter the territory are driven away. Contests between male fiddler crabs involve a series of ritualized displays of their enlarged claws that sometimes escalates to pushing, grappling, and ultimately flipping an opponent (Fig.4.17).

In another example, Japanese rhinoceros beetles (*Trypoxylus dichotomus* (950)) feed on sap and congregate around freshly wounded trees. Males make use of this opportunity to mate with as many females as possible by squaring off against one another and grappling with their horns until one of the males is evicted from or leaves the area. Males with the longest horns usually win the competition and have higher reproductive success than males with shorter horns.

Sometimes males compete directly for or defend females from other males. Male tusked wasps (*Synagris cornuta* (1010)), for example, patrol and guard nests that may contain female larvae with the goal of mating with adult females as soon as they emerge. Here, contests between males are usually won when the male spreads his mandibles to show the intruder his "tusks." If an intruder does land, which is not often, the guarding male grasps the intruder with his tusks and flings him away from the nest. Male Spanish ibex (*Capra pyrenaica* (1749)) also defend females from other males. Males and

Figure 4.16 Some examples of mimicry include cryptic colouration in **A.** a wood frog (*Rana sylvatica* (1385)), **B.** a harmless robber fly (*Laphria* sp.(964)) disguised as a dangerous bumblebee and **C.** a pitbull katydid larva (*Lirometopum coronatum* (900)) disguised as a flower so it can sneak up on prey.

Figure 4.17 A. Male fiddler crabs (*Uca annulipes* (840)) in a contest to determine access to females. The males with the biggest claws usually win. **B.** Female fiddler crabs are smaller than males and do not have enlarged claws.

females typically live in separate herds for most of the year and only come together in the winter to mate. During the mating (rutting) season, males defend small groups of females by pushing one another and clashing with their horns. Males and females are similar in appearance, but males are larger and have much larger horns. The most successful males defend larger groups of females and mate with them almost exclusively. Although serious injuries are uncommon, males do occasionally break their horns in battle.

Whether males compete for mates or resources, the evolution of male weaponry is favoured when the benefits of growing, maintaining, and using those weapons outweigh their costs. As in the examples outlined above, male–male combat rarely results in death or even serious injury. Rather, weapons largely function as signals by which males can evaluate their relative size, strength, or status. This explains why most combats end after just brief encounters and usually with one male retreating. It may also explain why evolution so often favours large, conspicuous weapons—they allow males to assess one another without the need for risky combat. That is not to say that weapons are free. The energy required to grow them can be significant, and the energy required to monitor and defend females is often substantial. It is worth noting that some females do exhibit weaponry. In the case of bovids (cows and their relatives), horns appear to be the result of natural selection for defense against predators rather than sexual selection. Male weaponry has evolved independently hundreds of times in Animalia and can be found in many vertebrate and invertebrate lineages (Fig, 4.18). Not surprisingly, it is a common source of sexual dimorphism (Fig. 4.17).

Among many animals (and plants), gametes (usually sperm) compete with one another. Because sperm are usually motile, they can compete for the goal of fertilizing the larger and usually immobile female gamete, the ovum or egg. This competition occurs in animals with polyandrous mating systems, where a single female mates with two or more different males during the same reproductive bout. The sperm from the different males compete to fertilize the eggs.

Sperm competition can take a variety of forms. A relatively simple strategy is for males to increase the number of sperm in their ejaculate so that they have more "dogs in the race" so to speak. This is a common strategy in insects and is triggered by environmental factors. For example, male false garden mantids (*Pseudomantis albo-fimbriata* (890)) transfer more than twice the number of

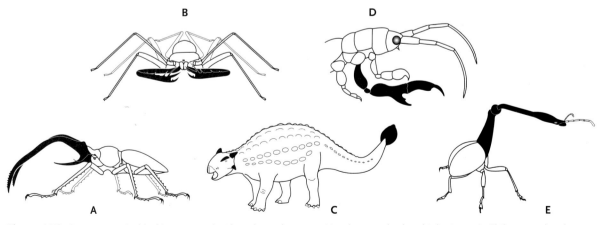

Figure 4.18 Armaments (in black) associated with male–male competition have evolved multiple times. **A.** Chilean stag beetle (*Chiasognathus grantii* (928)), **B.** a whip scorpion (*Damon variegatus* (654)), **C.** the dinosaur *Ankylosaurus magniventris* (1442), **D.** an amphipod (*Ericthonius punctatus* (791)), and **E.** giraffe weevil (*Lasiorhynchus barbicornis* (938).)

SOURCE: Based on Emlen, D.J., "The evolution of animal weapons," *Annual Review of Ecology Evolution and Systematics* 39:387–413, 2008.

sperm to females in the presence of other males that are likely to mate with the same females than when no other males are present. Similar responses have been shown in many different insects and spiders.

Things get more complex when females can also influence sperm competition by choosing among sperm from different mates. In some polyandrous ants and bees, the queen mates with multiple males within a very short period of time (hours to days) and then stores the sperm for use throughout her lifetime. The transferred sperm is suspended in a secretion that the males produce in an accessory gland. This secretion both prolongs the life of their own sperm and negatively affects the sperm of competitors. The best defended sperm are more likely to survive the short time that they are inside the female before she stores them. Since the queens mate only this one time, it is to her advantage to keep all the sperm healthy and available for use. To that end, her sperm storage organ produces a fluid that neutralizes the negative effects of the males' accessory gland secretions on one another's sperm. As is often the case in evolutionary biology; it's a complicated game!

Inter-sexual selection takes many forms but it's usually the females that choose among males. Typically, females choose on the basis of a heritable trait that signals the males' quality that would, in turn, enhance the fitness of her offspring. These traits are often ornaments that come with a cost. For example, in a classic study of sexual selection in Swallows (*Hirundo rustica* (1602)), researchers artificially extended the length of the tail feathers in some males and shortened them in others. The mating success of these males was compared with those of two experimental controls: birds whose feathers were not cut and birds whose feathers were cut and re-glued into place. Males with artificially elongated tails spent fewer days waiting to mate than males with short tails (and controls). They also engaged in more extra-pair copulations and ulti-

mately sired more fledglings. Males with artificially shortened tails were significantly more likely to be rejected by females.

However, during the next moult, when feathers are shed and replaced with new feathers, the birds with artificially elongated feathers often replaced them with defective feathers that broke easily. These defects, called fault lines, form in developing feathers when birds are under stress from lack of adequate nutrient and energy resources, suggesting that elongated feathers come with a cost. More importantly, feather breakage is associated with decreased foraging efficiency and increased predation by hawks, so males with artificially elongated tails experienced higher predation when the next set of feathers grew in. This illustrates another theme in sexual selection via female choice—there is an upper limit to the cost of male ornaments and there is pressure on males to evolve mechanisms that compensate for these high costs.

Stalk-eyed flies are another classic example of sexual selection via female choice (see also page 430). Stalk-eyed flies forage by themselves during the day but roost together at night in small aggregations clinging to root tips under the banks of streams. Typically composed of single males with multiple females, these aggregations show a positive correlation between the number of females and the span of the males' eye stalks. Experiments in the laboratory have shown that there is a genetic basis for the link between female preference and eye-span in males, and that males with wide eye-spans have higher fertility (Fig. 4.19A). These results provide strong evidence for the role of female choice in the evolution of exaggerated eye-span.

In terms of the cost of a wide eye-span, models of flight dynamics indicate that the males' larger eye-spans could make it more difficult for them to manoeuver during flight. However, there is also evidence that the wings of males have evolved to be

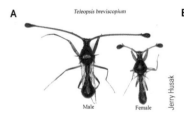

A   *Teleopsis breviscopium*

Male   Female

Jerry Husak

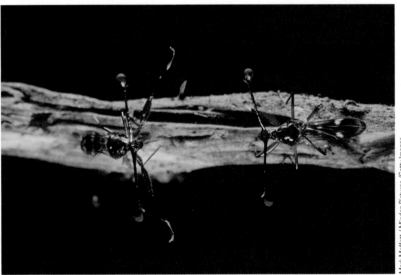

Mark Moffett / Minden Pictures /Getty Images

Figure 4.19 A. Male stalk-eyed flies are large than females. **B.** Two male stalk-eyed flies (*Teleopsis dalmanni* (971)) face off to compete for access to females. The male with the widest eye-span usually wins.

relatively larger than those of females and have a slightly different shape, which appears to compensate for adverse effects of having a wide eye-span. To add complexity to the story of the stalk-eyed fly, there is also strong evidence that wide eye-span is selected for via male–male competition. Males with wide eye-spans are more aggressive and win more interactions (Fig. 4.19B). The prize is the ability to defend and mate with more females. In the end, long eye stalks in males are likely the result of both female choice and male–male competition. This is another excellent example of how many different selective pressures can act at the same time to yield very complex evolutionary processes.

There is debate surrounding the origins of features associated with female choice. What sends the whole process into motion? One possibility is that males vary in a feature that provides an honest signal of genetic quality and that a new mutation in females allows them to perceive it. Another is that males vary in their ability to provide a direct benefit to females, such as providing parental care, and simultaneously vary in a feature that signals that ability. A third possibility is that females are intrinsically predisposed to certain signals and, over time, choose males that satisfy that affinity. This "sensory bias" could either be a by-product of the organization of the central nervous system or linked to other stimuli that females have been selected to recognize because they are associated with food, high-quality breeding sites, predator avoidance, etc. There is certainly evidence for sensory bias in female choice. For example, female stalk-eyed flies prefer models with eye-spans than are wider than any that occur in nature. Female zebra finches (*Taeniopygia guttata* (1618)) prefer males with artificial white crests despite the fact that no finches in the wild have crests. Males are known to exploit sensory biases as well. Water mites (*Neumania papillator* (652)) are sit-and-wait predators that orient toward and grab nearby copepods by sensing the vibrations created as they swim. When a male water mite finds a female, he mimics those vibrations and, in response, the female approaches and grasps him. Whatever their origins, it is clear that female choice, sometimes in concert with male–male competition, has had a significant impact on the evolution of many animals.

STUDY BREAK

1. Why is genetic variation important? What processes produce genetic variation?
2. What is "fitness," and how does it relate to natural selection?
3. Think of some examples of animals that exhibit sexual dimorphism. Generate a hypothesis about how it evolved in that animal.
4. Is everything about an animal's phenotype a direct result of natural selection? Explain.

4.3 Macro-Evolutionary Patterns

The processes of genetic drift, gene flow, and natural and sexual selection act on genetic variation at the level of populations and are often called micro-evolutionary processes. When we step back and look at a broader picture of evolution within clades, larger patterns in evolution begin to emerge; these are referred to as **macro-evolutionary** patterns. These include the evolution of adaptations, the presence of homologies, speciation, and extinction events, co-evolution between large groups of organisms, and explosive speciation events (called adaptive radiations) that are linked to ecological changes.

4.3a Adaptation

Adaptations are traits that result from the process of selection. Not everything about an animal's phenotype is an adaptation, and it is important to avoid the trap of thinking of an animal as simply a conglomeration of adaptations. As we mentioned earlier, processes such as random genetic drift can drive changes in phenotype that have no selective advantage. In addition, **pleiotropic** genes are genes that affect more than one phenotype. Selection for one phenotype controlled by a pleiotropic gene can influence a different phenotype controlled by the same gene, even if selection is not acting on it. Animals also have traits that were selected in the past but no longer serve a functional purpose. The vestigial pelvic bones of whales are a good example.

So how do we recognize adaptations? We often suspect the presence of an adaptation when a trait fits a prediction derived from a model of how it functions. For example, mammals such as echidnas, porcupines, armadillos, and pangolins are all protected by spines or scales (Fig. 4.20). These mammals, from different orders, have lower metabolic rates than mammals of similar size that run away from their pursuers. Comparative analyses confirm that having a tough, prickly covering appears to be an adaptation for conserving energy.

We also suspect adaptation when we see the same traits evolve independently and multiple times within a clade. For example, the forelimbs of scincomorph lizards (night lizards, skinks, whiptails, and relatives) have been lost 11 times, and the hindlimbs have been lost on six different occasions, each time in association with habitats suited for burrowing. This suggests that limb loss in scincomorphs is an adaptation to these niches. In other lizards, loss of limbs is associated with a snake-like existence on the ground. (See also Box 4.3)

Once we hypothesize that a trait is an adaptation, that hypothesis needs to be tested. If a trait is an adaptation, it should increase fitness, and fitness can be tested experimentally. A study of Bahamian mosquitofish (*Gambusia hubbsi* (1203)) demonstrates

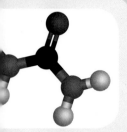

BOX 4.3 Molecule Behind Animal Biology

The Evolution of Opsins

Opsins are molecules that allow animals to perceive light. Some form of opsin is found in animals ranging from cnidarians to annelids, cephalopods, arthropods, and vertebrates. Opsins are the protein components of visual pigment molecules that are found within the membranes of receptor cells in the retina. Different opsin molecules are sensitive to different wavelengths of light, but they all work the same way. Opsin molecules attach to and encase single molecules of **retinal**, a chemical derived from vitamin A. The retinal molecule is forcibly kinked when it fits into the opsin, but the kink straightens out when it is struck by a single photon of light of the correct wavelength. This initiates a series of membrane–protein interactions in the receptor cell that generates a neural signal that is transmitted to the brain, where signals are interpreted as visual images. After releasing the retinal, opsin attaches to another kinked retinal molecule, and the pigment is ready to respond to another photon. This molecular mechanism functions within a narrow portion of the electromagnetic spectrum. No animal can see very far into the infrared or ultraviolet.

Among vertebrates, opsins are located on light-sensitive cells in the retina called **cones**, and each cone is sensitive to a different part of the light spectrum, that is, colour (Fig. 1). Each cone is most sensitive to a specific peak wavelength and less sensitive to wavelengths above or below that peak. The brain perceives colour by *comparing* the simultaneous firing rates of at least two cells (and usually many more) that favour different colours, and the quality of colour vision depends largely on how many different classes of cones there are to compare. Vertebrates with two classes of cones are called dichromats, and can see two out of three of red, green, and blue. Vertebrates with three classes of cones are called trichromats; they can see red, green, and blue. Other vertebrates, including fish and reptiles, may be tetrachromats (e.g., red:green:blue:low ultraviolet) and the vision of birds and turtles can be even more sophisticated.

If modern reptiles are any guide, the precursors to mammals were probably active during the daylight hours and had superb colour vision. In the Triassic, mammals left the daylight and became nocturnal, perhaps to avoid predators or to hunt nocturnal prey. In order to navigate the darkness, selection favoured eyes that could gather whatever photons were available, regardless of colour. The result was that colour discrimination degenerated in favour of cells called **rods**, which are sensitive to degrees of "light" and "dark," but not separate colours. To this day, most mammals have poor colour vision.

Analyses of the evolution of opsin genes indicate that trichromatic vision evolved independently in primates not once, but twice: first in the group containing Old World monkeys, apes, and humans, and again in a single species of New World primate, the howler monkey (*Alouatta* sp. (1851)). How does colour vision offer such a fitness advantage that trichromacy evolved twice in primates? One possibility is that it has to do with eating fruit. In a predominantly green forest, plants have evolved such that their fruits have bright colours to attract frugivores, which play the vital role of spreading their seeds. As it happens, fruit is an important part of the diet of many Old World monkeys. Another idea is that trichromatic vision helps detect young, succulent leaves, which are often pale green and sometimes even red. Howler monkeys are leaf specialists that prefer young leaves because they are tender and usually less toxic than older leaves. The ability to see young leaves benefits the howler monkeys, although it does not work to the plant's advantage.

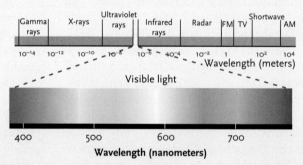

FIGURE 1 *The hues we perceive as colours are electromagnetic radiations of different wavelengths that are produced by the sun and pass through our atmosphere. Animals perceive colours that lie roughly between red (wavelength 700 nm), and violet (around 420 nm). For all animals, "light" is a relatively narrow band of electromagnetic wavelengths lying somewhere between ultraviolet and infrared. This is a very narrow part of the whole spectrum of wavelengths. While we can't "see" higher or lower wavelengths, we use them in many different technologies.*

SOURCE: Based on Based on Lehninger, A.L. 1965. *Bioenergetics*. Menlo Park, California.

A

B

C

D

Figure 4.20 Spines and scales have evolved multiple times in mammals as an anti-predator defense. **A.** The short-beaked echidna (*Tachyglossus aculeatus* (1678)) is a monotreme while **B.** the North American porcupine (Rodentia, *Erethizon dorsatum* (1884)) is a placental mammal. Scales evolved independently in **C.** the nine-banded armadillo (Cingulata, *Dasypus novemcinctus* (1827)) and **D.** the Cape pangolin (Pholidota, *Manis temminckii* (1845))

how this prediction can be tested. Populations of Bahamian mosquitofish have been isolated in "blue holes" (water-filled caves) for the past 15 000 years. Some blue holes have predators, others do not. Mosquitofish that live with predators have larger tails, and individuals with larger tails can accelerate more quickly from a stationary position. These "fast starts" give fish with larger tails a better chance of escaping from predators which, in turn, is linked to higher fitness (Fig. 4.21). A study of the genetics of the different populations supports the hypothesis that they have diverged in genetic composition; in other words, evolution has occurred.

Adaptations are not always morphological features. Consider the case of *Daphnia pulex* (720), a small crustacean that lives in small pools, ponds, and lakes throughout the world (Fig. 4.22A). All species of *Daphnia* feed on tiny bacteria, algae, and detritus. However, the population of *D. pulex* that lives in Star Lake (Vermont, USA) is also exposed to the cyanobacterium *Anabaena affinis*, which is usually toxic to *Daphnia* (Fig. 4.22B). To determine whether the ability to tolerate *A. affinis* is typical of *D. pulex* or unique to the Star Lake population, samples of *D. pulex* from Star Lake and a nearby lake without *A. affinis* were brought into the lab and exposed to varying concentrations of *A. affinis*. The Star Lake *D. pulex* survived and reproduced at constant levels, regardless of the concentration of *A. affinis*. In contrast, the survival and rate of reproduction of the *D. pulex* that had not evolved in the presence of *Anabaena* decreased with higher concentrations of *A. affinis*. These results suggest that

A

B

Figure 4.21 Male mosquito fish (*Gambusia hubbsi*) from a predator-free blue hole (**A**), and a blue hole that contains predators (**B**). The larger tail of **B.** is associated with more rapid acceleration and has evolved in populations of mosquito fish that live with predators.

SOURCE: Courtesy of Brian Langerhans

A

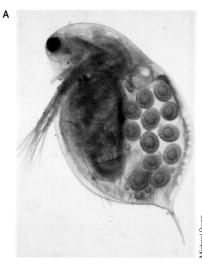

B

Figure 4.22 A. A female water flea, *Daphnia pulex* (maximum length ~3 mm; note orange eggs). **B.** The filamentous cyanobacteria, *Anabaena affinis*.

CHAPTER 4 EVOLUTIONARY PROCESSES |

BOX 4.4 People Behind Animal Biology

Sean Carroll

What controls morphological diversity, and how do those controls evolve? In the mid-20th century, evolutionary biologists from disciplines as different as paleontology and population genetics came together to answer that question in what is termed the "Modern Synthesis." They concluded that morphological differences among species are the relatively simple products of mutations upon which natural selection acted over very long periods of time.

This idea was rocked in the 1980s when biologists studying fruit flies (*Drosophila* sp. (960)) made inroads into understanding the genetics governing their development. The breakthrough was the discoveries of *HOM-C* genes that regulate *Drosophila* development and the related *Hox* genes of vertebrates (see Box 2.1). The big surprise was that these genes are highly conserved across groups—animals as different as fruit flies and humans showed relatively minor differences in these genes. Understanding how this can be is the focus of the field of evolutionary development, more commonly known as evo-devo.

Dr. Sean B. Carroll is one of the leaders of this new field of evo-devo. After finishing a PhD in Immunology from Tufts Medical School, he worked on *Drosophila* genetics and, like many of his contemporaries, soon realized that the genetics underlying morphological diversity were nothing like what earlier researchers imagined they would be. Sean Carroll is a vocal proponent of the idea that all animals share a common, ancestral network of regulatory genes that interact with one another to guide embryonic development. More importantly, small changes in these genes can have large effects on the upregulation (production) and downregulation (turning off) of proteins that determine the timing and differential growth of various anatomical parts.

It turns out that small changes in regulatory genes do produce remarkable morphological changes. For example, at the same early developmental stage, the bones in the hands of bats and mice are the same length (Fig. 1). Later, the expression of a protein called bone morphogenic protein 2 (bmp2) is upregulated in bats. This protein stimulates the production of chondrocytes that lay down cartilage and cause the digits to lengthen.

In addition to his important contributions to the field of evo-devo, Dr. Carroll has written five popular books about science, including *Endless Forms Most Beautiful: The New Science of Evo-Devo and the Making of the Animal Kingdom* (2005) and *Remarkable Creatures: Epic Adventures in the Search for the Origins of Species* (2008), and has appeared on many television and radio shows. He is currently a Professor of Molecular Biology and Genetics and an Investigator with the Howard Hughes Medical Institute at the University of Wisconsin.

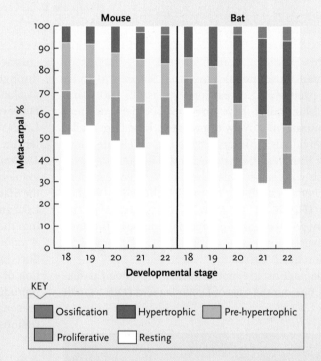

FIGURE 1 *The bones of bat wings and mouse hands grow longer by the addition of cartilage by chondrocytes that then ossifies to become bone. This graph illustrates the percentage of a metacarpal (hand bone, y-axis) that is composed of different types of tissues in a mouse (Mus musculus (1891), left) and a short-tailed fruit bat (Carollia perspicillata (1787), right) during developmental stages 18–22. Tissues are divided into resting, proliferative (chondrocytes are increasing in number), pre-hypertrophic (chondrocytes are poised to begin secreting the extracellular matrix in which cartilage forms), hypertrophic (actively secreting extracellular matrix), and undergoing ossification. Beginning at stage 20, the bat metacarpal contains more hypertrophic tissues than does the rodent metacarpal. This reflects the more rapid increase in the length of the bat metacarpal and is controlled by bmp2.*

SOURCE: Sears K E et al. 2006. "Development of bat flight: Morphologic and molecular evolution of bat wing digits," Proceedings of the National Academy of Sciences; 103:6581–6586. Copyright 2006 National Academy of Sciences, U.S.A.

physiological adaptations have evolved in the *D. pulex* from Star Lake that allow them to tolerate higher concentrations of toxic *A. affinis*.

Adaptations are the result of an evolutionary process so it goes without saying that we cannot identify them with certainty without knowing the phylogeny of the clade (or populations) of interest. Sometimes, certain traits that we think were adaptations for certain functions are, in fact, not. The classic example is feathers. Feathers were originally thought to have evolved for the purpose of flight. However, based on phylogenetic analyses and new fossils, we now know that birds are a clade of theropod dinosaurs and that feathers were likely present in the common ancestor of all coelurosaurian theropods (a clade that also includes tyrannosaurs; see also pages 160–161). We now suspect that feathers originated as an adaptation for insulation and that early feathers may also have been used for catching insects, camouflage, regulating the temperature of eggs, or even mating displays. In this sense, the use of the feathers in flight is an **exaptation**—it evolved for another use (insulation) and was later co-opted for a new function (flight). Certainly, aspects of feathers that improved flight were selected for, but the evolution of flight does not explain the origin of feathers.

4.3b Homology and Homoplasy

Homology, similarity due to common ancestry, is a fundamental concept in evolutionary biology. Traits derived from the same developmental sequence, and thus presumably derived from a common ancestor, are considered homologous. Phylogenies aid the identification of homologies. If we have a well-supported phylogeny, we can map a trait onto it to determine whether it represents a homology.

Homoplasy, sometimes called analogy, occurs when animals share a trait that appears to be similar but is not derived from a common ancestor. The classic example of the difference between homology and homoplasy is the comparison between the wings of birds, bats, and pterosaurs (Fig. 4.23). The presence of forelimbs in birds, pterosaurs, and bats is a homology—the forelimbs of all vertebrates are inherited from a common ancestor that had forelimbs. However, the wings of birds, bats, and pterosaurs had independent origins. These animals do not share a common ancestor that had wings, so the presence of wings is an example of homoplasy. This example illustrates that whether a trait is viewed as a homology or a homoplasy depends on how the trait is defined and on what level of the taxonomic hierarchy it is studied.

4.3c Speciation and Extinction

Life on Earth is not static, and species are constantly arising and going extinct. The primary drivers of speciation (the evolution of new species) are natural selection (including sexual selection) and genetic drift. By harnessing the power of phylogenies, we can identify speciation events and test hypotheses about what triggered them. Extinction, the death of an entire species, can result from processes ranging from

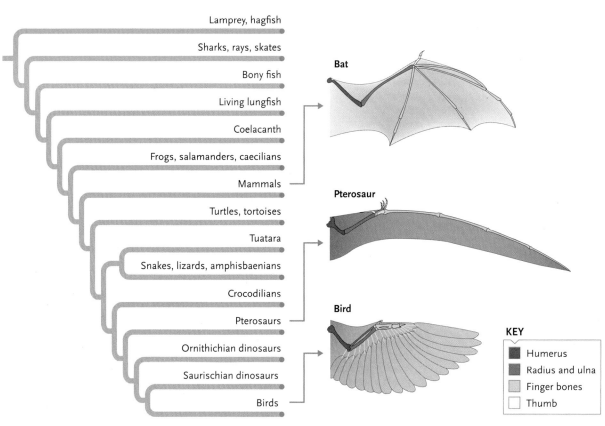

Figure 4.23 The forelimbs of birds, bats and pterosaurs are homologous because they are composed of skeletal elements inherited from a common ancestor. The wings themselves are an example of homoplasy because they have independent evolutionary origins.

SOURCE: Based on http://www.biology.ucr. edu/people/faculty/ Garland/Tedtree.jpg and http://ncse.com/book/ export/html/2192

KEY
- Humerus
- Radius and ulna
- Finger bones
- Thumb

volcanism to predation and climate change. In each case, extinction results from a change in an environmental condition to which animals cannot, or do not have time to, adapt.

Let's first consider speciation. We discussed the role of genetic drift in speciation earlier, so here we focus on ecological speciation and sexual selection.

Ecological opportunity—when natural selection favours individuals that are successful in moving into an available niche—is a potential trigger for speciation. Once populations are separated by occupying different niches, divergent selection drives them apart, and they eventually form new species. An exciting example of ecological speciation in the making is found in each of three distantly-related species of lizards in the desert of White Sands, New Mexico. The region contains two types of habitats that are adjacent to one another: the older habitat is on dark soil, and the other one is on bright white gypsum deposits that formed very recently (no longer than 5000 years ago). In each of species of lizard, the populations that moved onto the white gypsum soil are lighter in colour (Fig. 4.24). Moreover, an analysis of other morphological traits indicates that characteristics other than colour are also under divergent selection. The evolution of two differently coloured subpopulations in each of three independent lineages of lizards is strong evidence of ecological speciation.

We can also use phylogenies to detect the presence and rates of ecological speciation that occurred in the past. North American Wood Warblers (*Setophaga* spp. (1615)) are small songbirds that differ from one another in feeding and foraging behaviour. They appear to have undergone rapid ecological diversification in the Late Miocene or Early Pliocene. Researchers used a dated phylogeny of these birds to show that speciation rates were high when the birds began to diversify, but slowed down as the available ecological niches were filled.

The power of well-supported, dated phylogenies and comparative statistical methods has also allowed researchers to challenge and dispel traditional examples of ecological speciation. One such example was the rise of mammals following the extinction of dinosaurs at the Cretaceous–Tertiary (K/T) boundary, roughly 65 million years ago. Until quite recently, researchers believed that the loss of dinosaurs opened ecological niches that mammals rapidly speciated to fill (much like the niche-filling seen among Wood Warblers). However, modern phylogenetic and comparative analyses demonstrate that the peak rate of diversification in mammals occurred much earlier (Fig. 4.25).

Another traditional example of ecological speciation now disproven is the link between the evolution of angiosperms (flowering plants) and the speciation of beetles, many of which are herbivorous. As it turns out, beetles diversified long before angiosperms did. They are probably so diverse now because their extinction rates are low, and they have since evolved to fill a multitude of ecological niches.

Speciation via sexual selection differs from ecological speciation in that competition within and between the sexes is the agent of selection, rather than the availability of ecological niches. Sexual selection may play a role in speciation by creating **pre-zygotic barriers** that reduce the chances of a mating that would result in a hybrid that has reduced fitness or may be sterile. A recent analysis of crickets (*Laupala* spp. (899)) from the island of Hawaii may be an example. Many species of crickets are **sympatric** (live in the same place) and similar in both morphology and ecology. However, the mating calls of males from different species have different pulse rates, and females respond most strongly to males of their own species. Phylogenetic analyses show that species of *Laupala* evolved very rapidly. That they diverge only in mating calls, and

Soil colour

Holbrookia maculata

Sceloporus undulatus

Aspidoscelis inornata

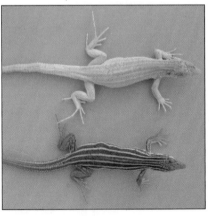

Figure 4.24 In each of these species of lizard, the populations that moved onto light-coloured soils (top) have diverged in both phenotype and genotype from populations that live on darker soils (bottom). This is a clear example of the beginnings of ecological speciation.

SOURCE: Courtesy of Erica Bree Rosenblum

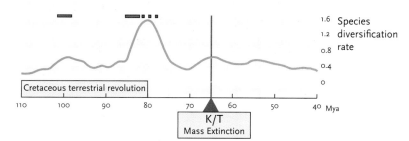

Figure 4.25 Analysis of the patterns of species origination and extinction in the evolution of mammals indicates that a peak in species diversification rate occurred at the end of the Cretaceous Terrestrial Revolution, a period characterized by the rapid evolution of flowering plants. This runs counter to the hypothesis that mammals underwent rapid speciation following the demise of dinosaurs during the K/T mass extinction.

SOURCE: From Robert W. Meredith, et al. 2011. "Impacts of the Cretaceous Terrestrial Revolution and KPg Extinction on Mammal Diversification." *Science* 334, 521–524. Reprinted with permission from AAAS.

not in morphology or ecology, suggests that sexual selection could have been the primary evolutionary driver in their diversification.

Another potential example of sexual selection can be found in the variation in the colouration and the courtship displays of male jumping spiders (*Habronattus pugillis* (670)) in four different areas of Arizona (Fig. 4.26). During courtship, males perform visual displays for females, who then choose a mate. Females are less likely to mate with males from another population, and when they do they often have fewer offspring (a sign of reduced fitness). While the phenotypes of males are largely fixed within each population and evolve quickly, neutral genes inherited from females (i.e., mitochondrial DNA) evolve at a much slower pace. This difference in the rate of evolution in males and females suggests sexual selection. Unlike the Hawaiian crickets which were sympatric species (at least currently), the spiders are **allopatric** (geographically separated) populations. Therefore selection associated with ecological or environmental differences cannot be ruled out. However, the argument for the influence of sexual selection in the divergence of these populations is quite strong.

Currently, scientists are debating whether sexual selection alone can lead to speciation. There is a growing consensus that sexual selection, natural selection, and other genetic processes may all contribute to any one speciation event.

Extinction, like speciation, is an ongoing, natural process. The balance between speciation and extinction rates determines the range and composition of animal diversity at any one time. By using the record of animal diversity through time preserved in geological records, it is possible to reconstruct **background extinction rates**. These are estimated to range around 0.01–1 extinctions per million species per year in marine animals and 0.2–0.5 extinctions per million species per year in mammals. It is not clear why the lowest background extinction rates occur among marine animals. One idea is that living in the oceans buffers some of them from extinction by providing a relatively stable environment. Another possibility is that the extinction rate of marine animals is underestimated because we have underestimated their diversity.

Over Earth's history, there have been five mass extinctions, periods when extinction rates exceeded background rates (Fig. 4.27). The geological record indicates that these events were triggered by abiotic factors including volcanism, the movement of Earth's crustal plates, the growth and recession of polar ice caps, and

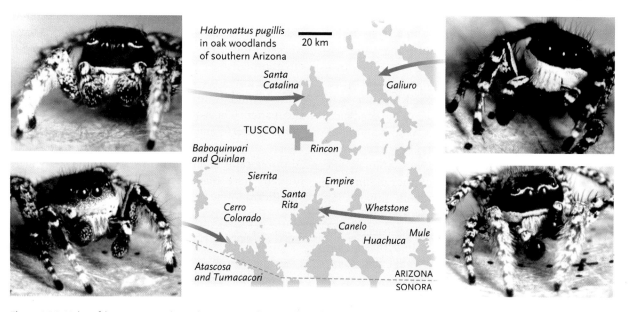

Figure 4.26 Males of the jumping spider *Habronattus pugillis* (670) from four separate locations differ in appearance and in their courtship displays. This is associated with sexual selection in the form of females' preference for the courtship display of males from their own population.

SOURCE: Masta, S. E., Maddison, W. P. 2002. Sexual selection driving diversification in jumping spiders. *PNAS* 2002;99:4442–4447. Copyright 2002 National Academy of Sciences, U.S.A.

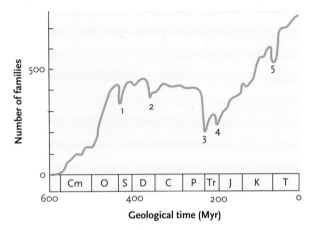

Figure 4.27 Each of the five mass extinctions was characterized by a significant drop in the number of families of animals known from the fossil record (y-axis). The extinctions occurred in the Late Ordovician (1), Late Devonian (2), Late Permian (3), Late Triassic (4), and Late Cretaceous (5). The x-axis is time in millions of years from past (left) to present (right). Time periods: Cm = Cambrian, O = Ordovician, S = Silurian, D = Devonian, C = Carboniferous, P = Permian, Tr = Triassic, J = Jurassic, C = Cretaceous, and T = Tertiary.

SOURCE: From Raup, David M., and J. John Sepkoski, Jr. 1982. "Mass Extinctions in the Marine Fossil Record," *Science* 215:1501–03. Reprinted with permission from AAAS.

perhaps even catastrophic meteor impacts that led to global changes in temperature, rainfall patterns, habitat availability, sea level, and atmospheric conditions. From our limited window into geological time, mass extinctions appear to have happened relatively quickly. In fact they played out over thousands or even hundreds of thousands of years. Extinctions as shown in the fossil record are discussed on pages 148–149.

Extinction is a normal part of the evolutionary process, and each of us will witness the extinction of many animal species in our lifetimes. Most will go unnoticed, but some are sure to draw attention. For example, western gorillas (*Gorilla gorilla* (1857)), wild horses (*Equus ferus* (1841)), northern and southern bluefin tuna (*Thunnus thynnus* (1292), and *T. maccoyii* (1291)), and elkhorn coral (*Acropora palmata* (181)) are currently listed as **critically endangered**, meaning that their numbers have either decreased or are predicted to decrease by 80% within three generations. Without significant investment of time and money in conservation, there is a high probability that these animals will become extinct soon, perhaps within your lifetime.

While extinction is an inevitable product of evolution, there is strong evidence that current extinction rates are 100–1000 times higher than background extinction rates. Many researchers believe that we are currently witnessing another mass extinction event, perhaps the biggest in the history of Earth.

As in the past, modern extinctions are associated with changes in temperature, rainfall patterns, habitat availability, sea level, and atmospheric conditions. The difference is that these changes are happening rapidly and are largely driven by human activity. Global

warming, which today is largely anthropogenic, plays a strong and complex role in these environmental changes. Climate change will rapidly alter the distribution of habitats, thereby placing strong selective pressures on many animal species, some of which will become extinct (Box 4.5). In addition to the threats imposed by climate change, humans are directly responsible for the conversion of animal habitats for agricultural production, housing, and the extraction of natural resources. Finally, there are documented examples of humans hunting animals to extinction. Among these are the iconic Dodo (*Raphus cucullatus* (1572)), Stellar's sea cow (*Hydrodamalis gigas* (1915), an Arctic relative of dugongs), the quagga (*Equus quagga* (1842), a subspecies of zebra), and the Tasmanian tiger (*Thylacinus cynocephalus* (1681), a marsupial carnivore) (Fig. 4.28). Dodos, Stellar's sea cows, and quaggas were hunted for their meat. Tasmanian tigers were thought to be sheep-killers (they weren't) and were slaughtered when a bounty was placed on their pelts. The species was granted protected status on July 10, 1936; the last known individual died in a zoo 59 days later.

Globalization has facilitated the spread of invasive species, which is another significant cause of extinction. For example, the introduction of rats has led to the extinction of many frogs, birds, and lizards in New Zealand. A famous example is the purported extermination of the Stephens Island Wren (*Xenicus lyalli* (1620)), a small, mouse-like, flightless, semi-nocturnal bird by Tibbles, the lighthouse keeper's cat. Although historical documents suggest that more than one cat was responsible for the mayhem, there is no doubt that the bird was extinct within ten years after cats were first seen on the island. In these cases invasive species either consumed or outcompeted the endemic fauna.

Recently, a group of researchers identified biodiversity as one of nine planetary systems that have worked together to maintain a stable global environment over the last 10 000 years. Sadly, they concluded that the rate of biodiversity loss is already well beyond safe limits for continued human occupation. The same is true for climate change and human interference with the nitrogen cycle. Some people advocate naming a new geological period, the Anthropocene, to refer to the period of time over which humans have made, and will continue to make, a significant impact on global ecosystems.

4.3d Adaptive Radiation

Like the word "adaptation," the term "adaptive radiation" has a very specific meaning to evolutionary biologists. A clade can be diverse without having undergone an adaptive radiation, and it is important to distinguish diversity from adaptive radiation. There is debate about exactly how to define "adaptive radiation" and how one might recognize it. A recent

BOX 4.5

Climate Change and Habitat Contraction

Many of the world's most endangered species either have very specific habitat requirements, are found only in small areas, or both. Consider polar bears (*Ursus maritimus* (1731)). While they occupy a vast region of the Arctic, they require summer sea ice in order to travel the distances needed to find food. On the other extreme, the Black and Gold Cotinga (*Tijuca atra* (1619)), a small songbird, lives in a small region of mountaintop cloud forests in eastern Brazil. Polar bears and Black and Gold Cotingas are very different animals with very different needs, but both are threatened by climate change.

Climate change will affect these species because each requires a specific habitat that will inevitably be altered as global weather patterns change. In the case of the Cotingas, its cloud forest habitat will move to higher and higher elevations as global temperatures increase (Fig. 1). As the forests retreat further up the mountains, they will become smaller and more fragmented as the valleys between the mountaintops become warmer. Many species are expected to move to smaller ranges in higher elevations as the climate warms.

Figure 1 *Areas of forest (dark green) and cleared land (buff) are visible in this satellite view of the coast of eastern Brazil. Black and Gold Cotingas (*Tijuca atra* (1619)) currently live in the high-altitude cloud forests shown in blue. As global temperature increases, the cloud forests will shrink and become fragmented, occupying only the areas shown in light green.*

SOURCE: Reprinted from *Current Biology*, Volume 18, Issue 3, Pimm, "Biodiversity: Climate Change or Habitat Loss — Which Will Kill More Species?," Copyright 2008, with permission from Elsevier.

A recent estimate suggests that polar bears may lose up to 68% of their preferred summer habitat and 17% of their preferred winter habitat by 2099 (Fig. 2). During the summer the bears use sea ice as a platform from which to hunt seals, which make up the bulk of their diet. The loss of summer sea ice is likely to force the animals to travel over longer distances to find summer hunting areas. Although it's impossible to tell exactly how that will impact polar bear populations, the increased energy required to travel further to hunt or to respond to increased competition as individuals are forced into smaller areas is more likely to adversely affect mothers and cubs than adult males.

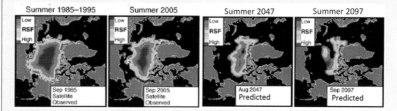

Figure 2 *Complex mathematical models called resource selection functions (RSFs) were good predictors of polar bear distribution in the summer of 2005. Red areas indicate the best polar bear habitats and blue areas are poor habitat, though some bears did occur there. RSF models predict that summer habitats will decline precipitously by 2047 and 2097 as global temperature increases.*

SOURCE: Durner et al. (2009) "Predicting 21st-century polar bear habitat distribution from global climate models." *Ecological Monographs*, 79(1): 25–58.

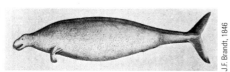

Figure 4.28 Animals that have been hunted to extinction include **A.** the quagga (*Equus quagga* (1842), **B.** the Tasmanian tiger (*Thylacinus cynocephalus* (1681), **C.** the Dodo (*Raphus cucullatus* (1572), and **D.** the Stellar's sea cow (*Hydrodamalis gigas* (1915).

and widely-cited definition states that an **adaptive radiation** must meet four specific requirements.

1. The species comprising an adaptive radiation must share a common ancestor. A diverse fauna that is composed of species from many distantly related groups is not evidence that an adaptive radiation occurred.

2. A strong association should exist between the new, genetically determined phenotypes of the species within the radiation and the ecological niches they inhabit. In other words, there should be evidence that ecological opportunity was the driver of the speciation event.

3. Phenotypic traits must be shown to be linked to ecological niches, confer a fitness advantage and are, in fact, adaptations. This is difficult, but is often achieved by demonstrating that a trait is associated with a species' ability to perform an ecological task that is related to fitness. Two common measures of performance in vertebrates are sprint speed (related to escaping predators or catching prey) and bite force (which limits the foods an animal is able to eat).

4. The speciation events that compose an adaptive radiation must have been rapid. Again, the availability of well-supported and dated phylogenies allows scientists to develop statistics that predict the pace of speciation events.

Given this rather strict definition of adaptive radiation, it should not be a surprise that adaptive radiations are not very common.

West Indian anoles (*Anolis* spp. (1469)) are a classic example of an adaptive radiation in vertebrates. Ancestral *Anolis* species from the mainland independently colonized the islands of Cuba, Hispaniola, Puerto Rico, and Jamaica. On each island the ancestor speciated rapidly into a similar set of ecomorphs, body sizes, and shape phenotypes that occupy similar sets of microhabitats (Fig. 4.29). The convergent evolution of these ecomorphs in itself suggests that they represent adaptations. Even better, the ecomorphs are associated with aspects of performance that, in turn, are linked to fitness. Field studies demonstrate that anoles run at their maximum speeds only when they are fleeing predators. Large-bodied and long-limbed species are most often found on large perches and can sprint faster on them than they can on narrow perches. In contrast, small, short-limbed species that typically live in grass or on small twigs can sprint equally well on perches of all sizes, but are less likely to stumble on small perches than are larger species.

An interesting example of an adaptive radiation among invertebrates is the land snails of the genus *Mandarina* (479) that inhabit the Bonin islands off the coast of Japan. The Bonin islands consist of three separate archipelagos that were uplifted from the ocean floor and have never been connected to one another or with the mainland. The ancestors of *Mandarina* currently on the islands arrived there relatively recently, between 1.8 and 0.9 My, so diversification has occurred rapidly. Three main ecomorphs inhabit each of the three archipelagos: terrestrial palm snails, terrestrial broad-leaved litter snails, and arboreal snails. Each has a distinctive shell shape and colour that appear to be responses to predation pressure and thermoregulation (see also Box 4.1).

The story of the snails is a bit more complex than that of *Anolis* because the same species can take the form of different ecomorphs on different islands. However, this could result from strong selection for a particular ecomorph, the influence of genetic drift on small founder populations, and the fact that, as in other snails, shell colour and striping are likely

Figure 4.29 *Anolis* lizards found on different islands in the Caribbean have evolved to have similar sets of ecomorphs in similar environments. In this example based on Puerto Rican anoles, *A. cooki* (1463) and *A. gundlachi* (1465) are large, "trunk-ground" ecomorphs that spend time on large perches or on the ground, while *A. poncensis* (1467) and *A. krugi* (1466) are smaller "grass" ecomorphs that spend time on thin twigs and vegetation that is low to the ground.
SOURCE: From RUSSELL/WOLFE/HERTZ/STARR. *Biology*, 2E. © 2013 Nelson Education Ltd. Reproduced by permission. www.cengage.com/permissions

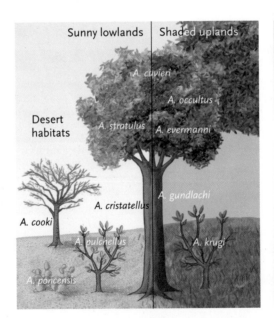

A. krugi

Sunny lowlands Shaded uplands

Desert habitats

A. cuvieri

A. occultus

A. stratulus

A. evermanni

A. gundlachi

A. cristatellus

A. cooki

A. pulchellus

A. krugi

A. poncensis

regulated by a very small set of genes. It is also possible that because the shape of the shell is relatively simple, it too may be regulated by only a few genes.

4.3e Co-Evolution

Co-evolution, an evolutionary process in which two distantly-related clades impose selective pressures on each other, provides a powerful illustration of the relationship between evolution and community-level ecological processes. Co-evolutionary relationships are typically mutualistic or based on predator-prey or host-parasite interactions. Relatively weak forms of each type of relationship are common and often lead to loose associations among groups of species. However, in a few instances, the co-evolutionary relationships are strong enough to produce clear cases of **co-speciation**, in which the phylogeny of one clade is literally a reflection of the other.

Perhaps the best-known example of co-evolution through mutualism is the relationship between figs and the fig wasps that pollinate them. Within this mutualism, the fig provides the wasp a protected place to reproduce, and the wasp provides the fig with pollination services (Fig. 4.30, Box 10.1). The intimate relationship between fig wasps and figs has led to tight co-speciation between them: most fig species are pollinated by a single species of wasp (Fig. 4.31). The fig–fig wasp story is an exception, since most examples of co-evolution through mutualism are characterized by a complex network of mutualistic interactions among multiple species. Even though there is not a one-to-one relationship between the mutualists, significant associations are often found between them when the relationships are examined in a phylogenetic context. This kind of association is called a diffuse co-evolutionary relationship. Importantly, all mutualistic networks have been assembled over long periods of evolutionary time and are important sources of stability in ecosystems.

Co-evolution between predators and prey is common. Think of it as an arms race. When a more effective means of catching prey evolves in a predator, it triggers the evolution of more effective ways for the prey to avoid capture. The presence of anti-predator

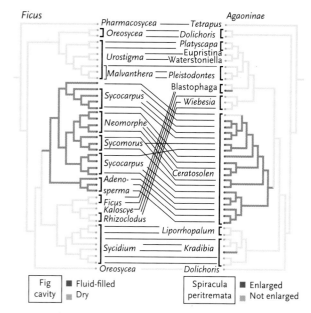

Figure 4.31 The phylogenies of the fig species and their wasp pollinators match almost perfectly, providing strong evidence of co-speciation of the two lineages. These phylogenies also illustrate the correlated evolution of a fluid-filled fig cavity and which wasps of the genus *Ceratosolen* (992) have an enlarged spiracular peritremata (extensions of the respiratory system that allow the wasps to live in the fluid-filled fig).

SOURCE: Weiblen, 2002. "How to be a fig wasp." *Annu. Rev. Entomol.* 47:299–330. ANNUAL REVIEW OF ENTOMOLOGY by ENTOMOLOGICAL SOCIETY OF AMERICA Reproduced with permission of ANNUAL REVIEWS in the format Republish in a book via Copyright Clearance Center.

adaptations and the predator responses produce some remarkable examples of reciprocal evolution, such as the relationship between echolocating bats and moths (Fig. 4.32).

Moths have evolved "ears" independently at least five times as a means of avoiding bats (see pages 478–479), but different bats use different tactics for catching moths. One remarkable example is the barbastelle bat (*Barbastella barbastellus* (1785)), which uses specialized stealth tactics to sneak up on eared moths in flight. Most bats that hunt aerial prey use high-amplitude (loud) echolocation calls to maximize the range over which they can detect prey. Researchers monitored the flight paths of the barbastelles relative to the paths of the

Figure 4.30 Fig wasps (subfamily Agonidae) emerging from the ostiole (opening) of an Indo-Australian fig (*Ficus* sp.) in which they hatched. Wingless males (brown) fertilized the winged females (black) inside the fig and chewed the ostiole to enlarge it to allow females to exit and lay eggs in a new fig.

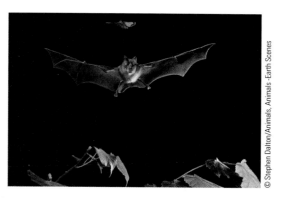

Figure 4.32 Greater horseshoe bat (*Rhinolophus ferrumequinum* (1814)) chasing a moth.

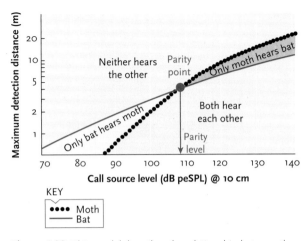

Figure 4.33 This model describes the relationship between the amplitude of the calls emitted by bats (x-axis) and the maximum detection distances of both bats and moths (y-axis). With low-amplitude calls, bats can detect moths over longer distances without being detected by the moths (yellow area). Moths hear higher amplitude calls over longer distances than the bat can detect the moths (grey area).

SOURCE: Reprinted from *Current Biology*, Volume 20, Issue 17, Goerlitz et al., "An Aerial-Hawking Bat Uses Stealth Echolocation to Counter Moth Hearing," Pages 1568–1572, Copyright 2010, with permission from Elsevier.

moths while recording the bats' echolocation calls and the activity of neurons in the moths that respond to sound. It turns out that barbastelles use lower amplitude calls than other moth-eating bats, which allows them to get closer to the moths before they are detected (Fig. 4.33). Other tactics that bats use to avoid being detected by eared moths are to use very low frequencies, very high frequencies, or very short calls.

Because any given community may include several different bat species, moth hearing responds to a wide range of bat calls. Although moths cannot discriminate between different sound frequencies, the range of their sensitivity (20–50 kHz) includes the echolocation frequencies of many bat species. Upon hearing a bat, most moths attempt to avoid capture by diving or assuming an erratic flight path. Moths in the family Arctiidae have evolved a different strategy. They are distasteful and have evolved the ability to make ultrasonic clicks that the bats can hear. Bats quickly learn to associate the clicks with bad taste and avoid these moths. In another twist of evolution, some moths that are not noxious use clicks to mimic noxious moths, thus fooling the bats.

As with predators and prey, the "evolution race" can be between parasites that maximize fitness by taking resources from the host, and the host whose fitness often decreases with increased parasite load. In other cases, the life cycle of a parasite can be so dependent upon their host(s) that it either cannot survive elsewhere or has limited chances to move. Its evolution is inextricably linked to that of its host. A good example of this range of relationships can be seen between New World (North and South American) doves (Family Columbidae) and their lice.

The birds harbour two distantly related genera of lice that eat their feathers: *Physconelloides* sp. (913) live on body feathers, and *Columbicola* sp. (910) live on wing feathers. Each louse species exhibits adaptations that help them escape from being removed when the birds preen. Body lice hide in the birds' soft belly feathers, while wing lice are long and thin and hide between the barbs of the flight feathers. Both louse species complete their entire life cycle on the birds and are transmitted when birds come into contact with one another. Despite these similarities, the strengths of the co-evolutionary relationships between birds and the two types of lice are very different.

The co-evolution of doves and their body lice is strongly supported, but there is little evidence of co-evolution between doves and wing lice. How can the evolutionary histories of lice with similar life histories that live on the same animals be so different? The answer may lie in differences in their hardiness and the way they disperse. Both types of lice move from one dove to another when the doves are in physical contact with one another. Since most doves regularly contact only members of their own species, the lice are most often transmitted between doves of the same species. Wing lice, however, have other ways of moving from bird to bird. Wing lice leave dead hosts more rapidly than body lice, and live longer than body lice when removed from a bird. Wing lice are also known to disperse on detached feathers or in shared dust baths, giving them yet another opportunity to move to other bird species. Finally, wing lice also are known to leave a dying host by attaching to parasitic flies, which visit many different species of doves. Overall, wing lice have many more opportunities to move to a new species of dove than do body lice. This story of doves and their lice begins to illuminate the huge variety of factors that can promote or deter co-evolution.

STUDY BREAK

1. What are the three types of co-evolutionary relationships? Give examples other than those in the text.
2. What is co-speciation? What kinds of interactions promote it?
3. How might the loss of biodiversity affect mutualistic networks and ecosystem stability?

EVOLUTIONARY PROCESSES IN PERSPECTIVE

Evolution is the mechanism that produced the diversity of animals, both living and extinct. Knowing evolutionary relationships among species is interesting in itself and can help us classify animals into meaningful groups. More importantly, this chapter illustrates that phylogenies provide the framework that allows us to test evolutionary hypotheses using statistically robust methods. Without a phylogeny, we would not be able to say with any degree of certainty that, for example, the spines of echidnas and porcupines are adaptations for conserving energy, that the current extinction rate is 100 to 1000 times the background rate, or that there is a co-evolutionary relationship between figs and fig wasps. As you progress through the following chapters, it is important not to lose sight of the fact that everything you will study is the product of a long, and often very slow, evolutionary process.

Questions

Self-Test Questions

1. Which of the following factors is an essential element to the process of evolution?
   a. Point mutations introduced by environmental toxins
   b. Genetic variation
   c. Genetic drift
   d. Gene flow between populations

2. What factors influence gene flow between populations?
   a. Barriers to gene flow between populations are always physical geographic features.
   b. Organisms with offspring that are distributed more widely throughout a geographic region will show greater genetic variation between individuals in different locations than more sedentary organisms whose offspring form separate populations.
   c. Reproductive barriers must involve anatomical differences that prevent mating.
   d. Differences in animal behaviour and life history strategies can form effective barriers to gene flow.

3. Which of the following does natural selection favour?
   a. Genes that are the most numerous within a population
   b. The same genes within different portions of a divided Population, even if they are subjected to different selection pressures
   c. Genetic variants within a population that have the greatest reproductive success
   d. Genes that are most likely to increase reproductive success in future generations

4. Why are phylogenies useful in representing our understanding of evolutionary relationships between organisms?
   a. They represent our best hypothesis of the relationships between organisms, and can be modified as new information becomes available.
   b. They are proven, unchanging illustrations of the relationships between organisms.
   c. They are based upon molecular clocks that provide information about the rate at which genetic changes have occurred in populations.
   d. Species are grouped on the basis of shared characteristics derived from a known common ancestor.

5. Which of the following pairs of terms correctly links the factors on which natural selection acts with a realistic outcome?
   a. monophyletic group; homology
   b. pleiotropic genes; specific adaptations
   c. genetic variation within specific ecological niches; adaptive radiation
   d. species in unrelated ecosystems; coevolution

Questions for Discussion

1. Phylogenies provide the framework for investigating macro-evolutionary patterns. How might you use a phylogeny to test the hypothesis that the evolution of two traits is correlated? (For example, the hypothesis that the evolution of head shape and fin shape is correlated in fishes, or that the evolution of colouration and social behavior is correlated in shrimp.)

2. Many different evolutionary processes are at work within populations at the same time. Can you imagine an example of how natural and sexual selection might act on the same population at the same time?

3. There have been five mass extinctions in the history of the planet. Does the current spike in the extinction rate qualify as another mass extinction? Is the modern spike in extinction any different from those that occurred in the past? Will the effects be the same or different?

4. Use the electronic library to find a recent paper that identifies a micro-evolutionary process in a population or species of animal. How does the paper help to explain the causes of evolution within that group?

5. Use the electronic library to find a recent paper about a macro-evolutionary process in animal evolution. What does it tell you about how evolution has shaped the number or diversity among species in that group?

Figure 5.1 An Atlantic salmon (*Salmo salar* (1316)) travelling upriver to breed. The "salar" species name means jumping.

SOURCE: Kirk Norbury / Shutterstock.com

5   Larvae and Life Cycles

WHY IT MATTERS

Figure 5.1 captures the essence of a wild river—an Atlantic salmon (*Salmo salar* (1316)) fighting its way upstream to breed. As it travels, the fish is the target of a sport fishery that is a major contributor to local economies where guide fees can be several hundred dollars a day plus the costs of travel and accommodation. Governments and private organizations both have major investments in salmon hatcheries. The fish produced there are used to restock rivers that have salmon runs and to reintroduce salmon to rivers from which wild salmon have disappeared. Pages 517–518 include a discussion of salmonid aquaculture and the economic importance it has in today's world. The sport fishery, hatchery operations, and salmonid aquaculture would all be impossible, however, without an understanding of the life cycle of the fish.

Salmon are **anadromous** fish, born in fresh water, swimming to the sea where they spend most of their lives, and returning to fresh water to breed (Fig. 5.2). Male Atlantic salmon in breeding condition change colour and develop a characteristic hooked jaw known as a kyte. Males establish territories and wait for females who select a territory and excavate a shallow dip in the gravel of the river bed where

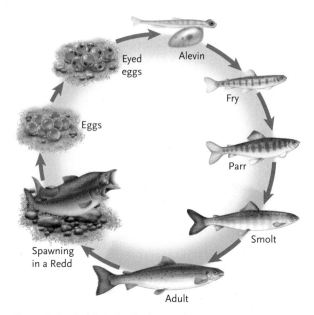

Figure 5.2 The life cycle of Atlantic salmon.

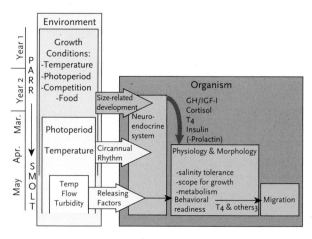

Figure 5.3 Factors involved in the transition from parr to smolt in Atlantic salmon.

SOURCE: Stephen D. McCormick, Lars P. Hansen, Thomas P. Quinn, and Richard L. Saunders, "Movement, migration, and smolting of Atlantic salmon (Salmo salar)," *Canadian Journal of Fisheries and Aquatic Sciences* 55 (Supp 1) 77-92, 1998. © 2008 Canadian Science Publishing or its licensors. Reproduced with permission.

they lay their ~5000 eggs. The eggs are immediately fertilized by sperm (milt) from the male.

The eggs laid in the fall hatch the following spring as **alevin** (Fig. 5.2). Hatching time depends on water temperature, and in cold waters eggs may take about eight months to hatch. The alevin grow using food reserves in the yolk sac. At 3–4 cm long, they are known as **fry** and feed on small plankton and insects. As they grow, the fish take on the camouflage colouring that characterizes **parr**. Parr continue to feed and grow in fresh water, typically for 2–4 years, and reach a length of about 15 cm. Then comes a remarkable transition, timed by the appropriate combination of environmental and physiological factors, when the parr change to silvery coloured **smolt** at the start of their downstream journey to the sea (Fig. 5.2).

A summary of the interacting factors involved in **smoltification** appears in Figure 5.3. Parr must achieve a critical size before smoltification. The time taken to reach this size depends on environmental conditions, including food availability, competition, water temperature, and photoperiod. Once the critical size is reached, spring photoperiod and temperature conditions trigger neuroendocrine changes that control the physiological changes of smoltification. These are followed by behavioural changes that reverse the direction of swimming (parr swim against the current, smolt swim with the current to move downstream) and reduce agonistic and territorial behaviour as the fish start to school.

The change from camouflage to reflective silver during smoltification results from guanine and hypoxanthine crystals deposited in the skin. The reflective appearance is typical of pelagic fish and adaptive for predator avoidance in the ocean (see Chapter 8). There are also changes in visual pigment and the hemoglobin isoform.

The most dramatic changes are associated with salinity tolerance. Smolt move from water with a salinity of ⩽1:1000 to water at 35:1000 within a few days. This change demands changes to the gill physiology (particularly to Na^+K^+ ATPase and to chloride cells, which are mitochondria-rich cells in the gill that regulate the acid–base balance) and to the kidney. These changes are regulated by changes in thyroid and pituitary gland activity, leading to changed levels of thyroxin, growth hormone, prolactin, cortisol, and insulin.

On reaching the sea, salmon populations from eastern North America and northern Europe migrate to feeding grounds off the west coast of Greenland (Fig. 5.4). Here they feed on small fish and shrimp that provide the carotenoids that give salmon flesh its characteristic pink colour.

Salmon are imprinted on the river in which they developed and use olfactory cues to return there. Some salmon (usually males) return to fresh water after a year at sea when they weigh 2–3 kg and are known as **grilse**. More commonly, the fish spend two years at sea and return weighing 4–6 kg. A few fish may spend longer in the ocean and return at higher weights. Unlike Pacific salmon species, where all breeding fish die, a few (up to 10%) of breeding Atlantic salmon survive, of which most (85%) are females that return to the sea and then come back to the river to breed again.

The separation of salmon larvae and adults to live in different environments, with a hormonally regulated physiological transition akin to a metamorphosis, are life cycle features that will re-appear in different examples in this chapter.

SETTING **THE SCENE**

Development of an egg to a miniature adult form that then grows and becomes sexually mature

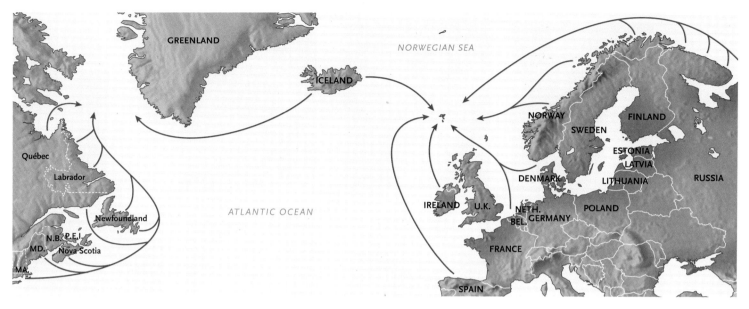

Figure 5.4 Migration of Atlantic salmon to their feeding grounds.

SOURCE: Data from Atlantic Salmon Federation and WWF, The Status of Wild Atlantic Salmon—A River Assessment, 2001.

(the norm in birds and mammals) is relatively uncommon among the animal kingdom. Either one or more larval forms (with morphologies quite unlike that of the adult) or a nymph that resembles an adult but lacks some adult features and later transitions to the adult form are much more common. Larvae and nymphs are often found in different environmental niches from their adult forms, an arrangement that may have evolved for a variety of reasons, some of which are considered in this chapter. A general theme of the examples in the chapter is the adaptation of life cycles to environmental demands.

5.1 What Is a Larva?

At first this appears to be an easy question to answer, but, in reality, there is no simple definition. A **larva** is the developmental stage that follows hatching and precedes metamorphosis to an adult, implying that larvae occur in the life cycles of many marine invertebrates and holometabolous insects, as well as some fish and amphibians (Fig. 5.5). Larvae usually have specific larval organs that serve a particular function during larval life but are reduced or disappear during metamorphosis. On the other hand, **nymphs** are the juvenile stages of animals that hatch as miniature adults (e.g., nematodes and many insects, see Section 5.7c). In most vertebrates, the term **juvenile** is more appropriate (Fig. 5.5).

The characteristics of a larva include:

- the absence of functional reproductive structures
- a morphology that differs significantly from that of the adult

- the presence of either, or both, undifferentiated cells (e.g., imaginal discs in insects) that will develop later and morphological elements (larval characters) that will be eliminated during development.

Figure 5.6 shows some different types of larvae found among invertebrate marine organisms.

One problem with a simple definition for "larva" is that, even in related species, the stages between hatching and metamorphosis may differ substantially (Fig. 5.7). In the Crustacea, the first-stage larva is a **nauplius**, a free-swimming larva in the Copepoda, Ostracoda, Cirripedia, Mysidacea, and the penaeid shrimps. In lobsters and crabs, a nauplius stage forms but does not hatch from the egg (indicated in Fig. 5.7 by the oval outline in the box). A **cypris** larva is found

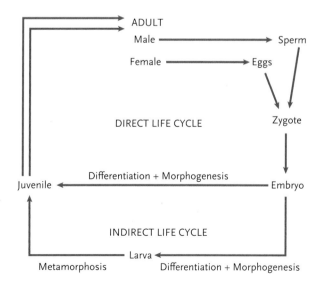

Figure 5.5 Direct and indirect development.

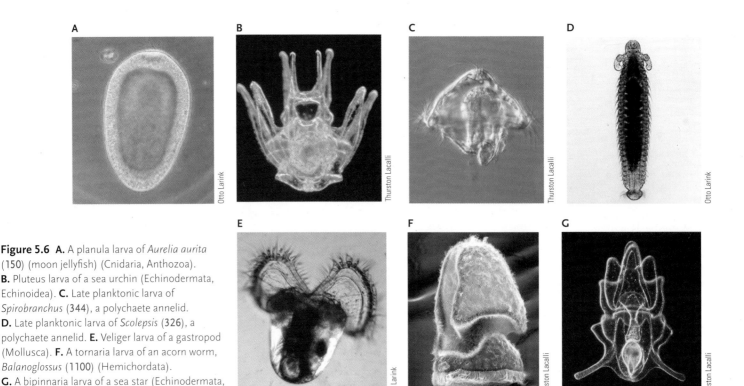

Figure 5.6 **A.** A planula larva of *Aurelia aurita* (150) (moon jellyfish) (Cnidaria, Anthozoa). **B.** Pluteus larva of a sea urchin (Echinodermata, Echinoidea). **C.** Late planktonic larva of *Spirobranchus* (344), a polychaete annelid. **D.** Late planktonic larva of *Scolepsis* (326), a polychaete annelid. **E.** Veliger larva of a gastropod (Mollusca). **F.** A tornaria larva of an acorn worm, *Balanoglossus* (1100) (Hemichordata). **G.** A bipinnaria larva of a sea star (Echinodermata, Asteroidea).

only in ostracods and barnacles. Other groups have a protozoea larva, free swimming in the mysids and shrimps, and retained in the egg in the lobsters and crabs. **Zoea** larvae are free swimming in crabs, shrimps, and mysids but are retained in the egg in the lobsters. An additional larval stage, the **megalopa**, appears in crabs. Clearly both actual hatch point and the timing of the metamorphosis to an adult differ among groups of crustaceans. Later in this chapter we present examples of different timing of hatching, relative to development, in some polychaete annelids and gastropod molluscs.

One thing a larva is almost certainly not, despite a recent and controversial theoretical approach, is the genome of one animal laterally transferred to become a part of the genome of another group, with the transferred genome then being expressed as a larva. This theory has been applied to barnacles and sea urchins (where evidence suggesting hybridization with ascidians was suggested but not confirmed) and to the idea that the genomes of onychophorans and cockroaches were hybridized to produce the modern Lepidoptera. Among the evidence against the latter are the sizes of the genomes involved. A hybridized genome might be expected to have a DNA content roughly equal to the sum of the DNA in its ancestors, but the genomes of species of Lepidoptera are smaller than those of either proposed ancestor.

For decades, much of biology was adultocentric. Based on body plans, phylogenies generally focused on adult forms. Developmental studies focused on the origins of adult features such as bat wings and whale flippers. Modern studies of phylogeny are based on molecular characteristics that have the advantage of not being adultocentric and which avoid the problem, always present in morphologically based phylogenies, of placing different values on different characters. Most genome studies have concentrated on model organisms that have direct development (e.g., *Caenorhabditis* (585), zebra fish, frog, and mouse), with *Drosophila* (960), a holometabolous insect, as an exception. In *Drosophila* the focus has been on development to the adult form rather than on the larva.

All of the examples in Figure 5.6 display morphologies and life styles that are quite different from their adult forms. They are typical of the majority of animal phyla because they are marine and have a biphasic life cycle. A survey of the animal kingdom reveals that most groups are aquatic with a **pelago-benthic** life cycle; in other words, they have a planktonic larva and a benthic adult. This leads to the question of which phase of the pelagobenthic life cycle appeared first. Was the current adult form an addition to the planktonic larval life cycle, or was the planktonic larva interposed between embryology and a benthic adult form?

The **modified trochaea** theory, proposing that planktonic larvae are ancestral, developed from Haeckel's 19th-century concept of recapitulation (ontogeny repeats phylogeny). While Haeckel's theory has been rejected, Jägersten and Nielsen more recently proposed that a ciliated trochaea developed from a gastraea (an adult, planktonic gastrula), which is proposed as

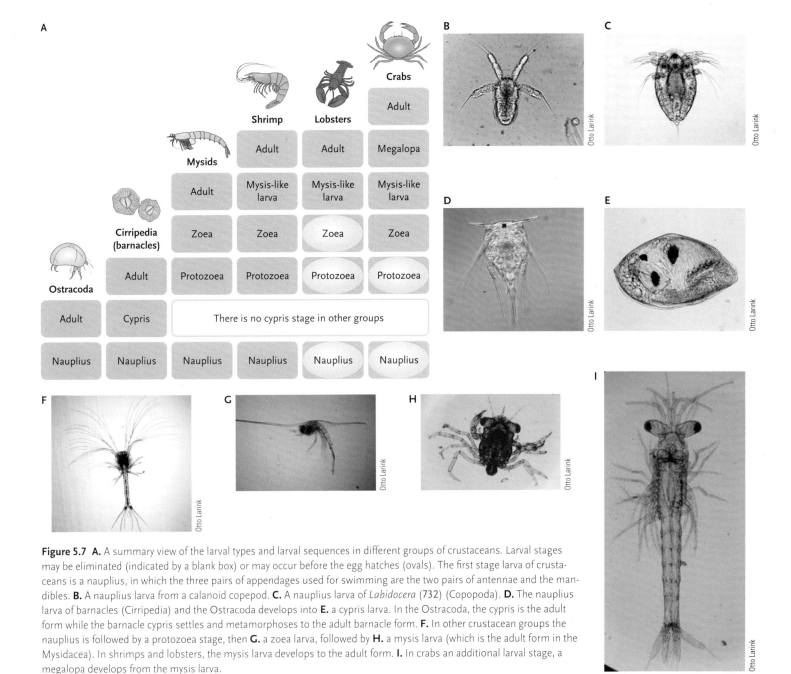

Figure 5.7 **A.** A summary view of the larval types and larval sequences in different groups of crustaceans. Larval stages may be eliminated (indicated by a blank box) or may occur before the egg hatches (ovals). The first stage larva of crustaceans is a nauplius, in which the three pairs of appendages used for swimming are the two pairs of antennae and the mandibles. **B.** A nauplius larva from a calanoid copepod. **C.** A nauplius larva of *Labidocera* (732) (Copopoda). **D.** The nauplius larva of barnacles (Cirripedia) and the Ostracoda develops into **E.** a cypris larva. In the Ostracoda, the cypris is the adult form while the barnacle cypris settles and metamorphoses to the adult barnacle form. **F.** In other crustacean groups the nauplius is followed by a protozoea stage, then **G.** a zoea larva, followed by **H.** a mysis larva (which is the adult form in the Mysidacea). In shrimps and lobsters, the mysis larva develops to the adult form. **I.** In crabs an additional larval stage, a megalopa develops from the mysis larva.

the ancestor of protostomes and deuterostomes. The modified trochaea theory (also known as the maximal indirect development theory) proposes that adults form from cells that are set aside in the larvae and form the adult during metamorphosis. This theory involves repeated convergent evolution of *Hox* genes to control the anterior-posterior axis as well as the evolution of the set-aside cells whose functions evolve later. The requirements of this theory are not well supported by current evidence.

The **intercalation theory** proposes that the metazoan ancestor was an acoel-like benthic animal that had developed the basic organ systems of a bilaterally symmetrical triploblastic animal and the regulatory genes required for bilaterian development. Some part of the genome was then co-opted to regulate develop-

ment of a free-swimming larva. This might have happened in different ways in different animal lineages, explaining, for example, why different genes control the development of an apical ciliary tuft in the larvae of a sea urchin and a mollusc. The intercalation theory (larva second) fits better with recent molecular phylogenies than does the modified trochaea theory. Figure 5.8 diagrams the evolutionary events invoked by each of the two theories.

STUDY BREAK

1. What is the difference between a larva and a nymph?
2. What is a pelagobenthic life cycle?

BOX 5.1 People Behind Animal Biology

Walter Garstang

Walter Garstang (1868–1949) had planned a career in medicine until, in his undergraduate years at Oxford, he was influenced by Professor H.N. Mosley, one of the three naturalists who had laid the foundations of deep-sea biology during the three-year global travels of HMS Challenger. As a new graduate, Garstang's first position was on the staff of the Plymouth Marine Biology Laboratory when it opened in 1888.

Garstang had a distinguished career as a marine biologist, based at Plymouth, Oxford, and Leeds, England. His work included studies of the warning colouration of nudibranch molluscs, the behaviour of burrowing crabs, and the development of siphonophores and tunicates, as well as a period of applied work on the population biology of North Sea fish. Apart from these studies, Garstang was to influence zoology in three quite different ways.

Garstang seems to have been the originator of what is now an element of almost all undergraduate biology programs, a field course. After returning to Oxford from Plymouth, Garstang arranged a marine field course, taught in Plymouth over the Easter vacation, for a group of undergraduates (several of the students in his first course also went on to distinguished careers in zoology).

In the latter part of the 19th century and the early years of the 20th century, Biology was under the influence of what was known as the "fundamental biogenetic law," propounded by Ernst Haeckel in 1866 and popularly restated as "ontogeny recapitulates phylogeny." Haeckel supported Darwin's theory of evolution but added to it the older concept that life progressed through a series of stages, each more advanced than its predecessor, and with all of the previous stages being a part of the embryological development of the new form. Embryological studies had increasingly questioned the validity of Haeckel's law, while supporters of the law passed off these questions as "exceptions." There was even a suggestion that Haeckel himself "fudged" some studies of limb development in echidnas to make his results fit with his law. It fell to Garstang, in a 1921 address "The theory of recapitulation" (delivered to the Linnean Society of London), to move Haeckel's law into history through a review of his own studies on animal development and those of others. Garstang included in his address a new key phrase, "Ontogeny does not recapitulate phylogeny; it creates it."

Garstang's third contribution to zoology lies in the surprising medium of verse. A published offering in this form was a collection of verse, accompanied by musical notation, titled *The Songs of Birds*. Through his later life he compiled a series of verses that, in a whimsical manner, recount features of the development of many animals. There is an account, perhaps apocryphal, that Garstang wrote these verses as he rode on the top level of a double-decker bus, commuting between his house and office. The developmental verses were published after Garstang's death as *Larval Forms and Other Zoological Verses*. The flavour of the verses may be captured in a short extract from "The Trochophores."

The Trochophores are larval tops the
    Polychaetes set spinning
With just a ciliated ring—at least at
    the beginning—
They feed, and feel an urgent need to
    grow more like their mothers,
So sprout some segments on
    behind, first one, and then the
    others...
...Then setose bundles sprout and
    grow, and the sequel can't be hid;
    The larva fails to pull its weight,
    and sinks—an Annelid.

An apt description of a trochophore larva and its development to the form that leaves the plankton and continues its development on the sea bed (Fig. 5.25).

Figure 5.8 Three evolutionary stages of the larva-first and intercalation hypotheses for the origin of indirect life cycles among bilaterians. Under the larval-first hypothesis, Stage A consists of holoplanktonic feeding adults (the trochaea hypothesis posits that the protostome and deuterostome lineages diverged at this stage); Stage B has added a terminal, benthic adult stage, and the original planktonic stage persists as a sexually immature larva that feeds; Stage C is a transition of the planktonic larvae from feeding to non-feeding. Under the intercalation hypothesis, Stage A consists of holobenthic adults; Stage B has added a planktonic larval stage to early development; and Stage C is a transition of the planktonic larva from non-feeding to feeding.

SOURCE: Figure 1 from Page L. R. 2009. *Biol. Bull.* 216: 216–225. Reprinted with permission from the Marine Biological Laboratory, Woods Hole, MA.

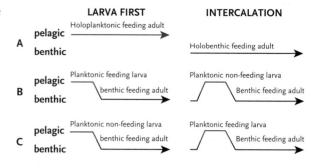

5.2 Loss and Modification of Larvae in Marine Groups

A free-swimming larva is a part of the life cycle of all marine phyla. However, in many groups, the ancestral larval form has been highly modified or lost, and closely related species may develop in quite different ways. For example, in the gastropod genus *Littorina* (the periwinkles), some species have planktotrophic larvae while others, sometimes living on the same shore, complete their development within an egg capsule and hatch as miniature adults (direct development).

The evolutionary modification of larvae is well illustrated in the echinoderm class Asteroidea (the sea stars) where there has been sufficient work to develop a larval phylogeny. This larval phylogeny is complex, and when it is superimposed on a phylogeny of asteroid families developed from adult characteristics, there is only limited correlation between the two.

How can a species evolved from a line that has lost its planktotrophic larva reacquire one? The evolution of the Calyptraeidae, a family of slipper limpets (Fig. 5.9A) (Mollusca, Gastropoda) includes a change from planktotrophic development to direct development. Most members of the calyptraeid genus *Crepipatella* (422) have direct development, while one species, *C. fecunda*, has planktotrophic development. Analysis of its DNA shows that *C. fecunda* is nested within samples from a direct developing congener, *C. dilatata*, suggesting that the latter is paraphyletic (Fig. 5.9B). Following development of *C. dilatata* in the egg capsule, its veliger larvae are retained in the egg capsule but have features required for planktotrophic feeding and swimming. This suggests that *C. fecunda* evolved from *C. dilatata* through a shift in the time (heterochrony) of hatching from the egg capsule.

5.3 Dispersal

Planktonic larvae are presumed to use surface water currents for dispersal because this requires limited energy input from the animal. Larval life in the plankton may be short-term, non-feeding, and purely for dispersal. More commonly, however, larvae in the plankton have access to food sources not available to the adult, and combine feeding with dispersal. Planktonic life is highly competitive, with small zooplankton competing for phytoplankton resources and larger zooplankton competing for smaller zooplankton as food. This generates pressure for rapid growth because the smallest animals are most vulnerable to predation (Table 5.1).

Many zooplankton species have spines (Fig. 5.10) that can serve a dual purpose. They slow the sinking of the animal by increasing its surface area (think of a free-fall parachute jumper flaring to slow the rate of descent) and thus reduce the swimming energy required to stay near the surface. On small animals the viscosity effect (see page 175) converts the spines to what is effectively a continuous parachute. Spines are also a low-energy way to increase the apparent size of the animal, thereby reducing the number of predators able to eat it (Fig. 5.10).

Tidal currents generate local oscillations and may distribute larvae over a period of hours, overriding oceanic currents that allow longer distance dispersal.

If the larval stage facilitates dispersal, then why does it disappear in some species? Species that develop more completely in the egg are called **lecithotrophic** because the egg has more yolk, used by the embryo for energy. Distance dispersal requires food or the ability to feed, so loss of a larval stage is unlikely to occur in species lacking access to an abundant food reservoir. This suggests that newly hatched young compete with neighbouring adults and other hatchlings for food, a situation that places a premium on large size at hatching to provide a competitive advantage. Larval loss is only likely in species that have some degree of mobility so that some dispersal is possible.

A

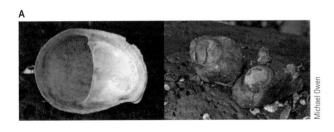

Michael Owen

B

Figure 5.9 A. The slipper limpets (Mollusca, Gastropoda, Calyptraeidae) , e.g. *Crepidula fornicata* (421) are filter feeders, distinguished by the shelf inside their shell. In the genus *Crepipatella* some species develop directly while other species have reacquired a veliger larva. **B.** A phylogeny of *Crepipatella* species based on molecular data. Species shown in blue develop directly; species in red have a planktotrophic larva. Notice that *C. fecunda* is nested between two clades of *C. dilatata*, a clear indication of the paraphyly of *C. dilatata*.

SOURCE: B. Figure 2 from Collin, R., et al. 2007. *Biol. Bull.* 212: 83–92. Reprinted with permission from the Marine Biological Laboratory, WoodsHole, MA.

Table 5.1	Mortality decreases with age and size in a fish, the plaice *Pleuronectes platessa* (1307). Female plaice may produce 600 000 eggs a year, and these metamorphose (to both eyes on the same side of the head, see Fig. 5.35) at 4–6 weeks of age. In this table the number metamorphosing in an experimental population has been set at 100 000 (the point between "Metamorphosis" and 2 months of age).		

Age (months)	Size (cm)	Deaths/month
Hatching	0.7	
Metamorphosis	1.3	1 550 000
2	2.7	80 000
3	3.9	14 000
4	4.6	3 500
5	5.4	1 300
6	6.4	550
7	6.8	220
8	7.3	100
9	7.4	50
10	7.5	20
11	7.6	17
12	7.7	16
Number alive after one year		130

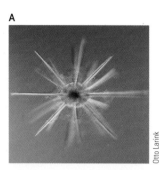

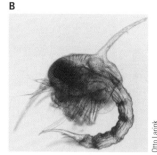

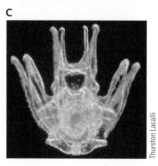

Figure 5.10 Spines are a common feature of larvae in the plankton. **A.** Notice how they make the much smaller body of *Phyllostaurus* (Protista) a much larger target for a potential predator. **B.** A crab zoea and **C.** a pluteus larva of a sea urchin are also both protected from predation by their spines.

The ratio of planktotrophic to lecithotrophic species is very high in tropical waters, but drops in temperate waters where phytoplankton food becomes available only in seasonal blooms. Lecithotrophic species dominate in the deep sea and in polar waters which have greater distances between the breeding adult and planktonic food. A puzzling side note to this is that in the Antarctic, about 95% of species have lecithotrophic larvae, but the most abundant species in the region are in the 5% that have planktotrophic larvae.

STUDY BREAK

1. What conditions might favour direct development (implying loss of a larval form)?
2. What problems must species with planktotrophic larvae overcome?

5.4 How Does a Larva Know Where to Settle?

A dispersed planktonic larva faces the risk of being unable to find a settling place that will allow its survival to the adult stage. Larval settlement is of supreme importance, not just for the survival of the individual and species, but because it underlies the distributional ecology of benthic species. Species that forgo a larval stage benefit by the young remaining in habitats appropriate for their survival.

Physical factors are intrinsic to almost all planktotrophic larval behaviours. Newly hatched larvae typically exhibit some combination of photopositive and geonegative behaviours and swim toward the surface. At the completion of development, these behaviours reverse and the settling larvae find their way to the substratum by becoming photonegative and/or geopositive. On reaching the substratum, larvae respond to the nature of its surface (mud, sand, pebble, rock, shell, algae, etc.) and, in some cases, to its rugosity (a measure of surface roughness).

Interacting with the physical cues are chemical signals, some of which are species specific, others general; some triggering a single event, others conveying more general messages. These chemical signals are typically involved either in settlement in association with conspecifics (gregarious settlement) or in the enhanced or specific settlement of one species with (or on) another (associative settlement). Gregarious settlement has obvious advantages for immobile benthic species in that it places young individuals in a location suitable for their survival and increases the chances of reproductive success for broadcast spawners. Less obvious advantages include those for juvenile sand dollars settling in an established bed of adults (e.g., *Dendraster* (1086) or *Echinarachnius* (1088)), where the reworking of the sediment by adults tends to exclude predatory tanaid shrimp.

Associative settlement underlies many different types of species relationships. Many phoretic barnacles settle on only one or two species of whales, and some parasite larvae locate and settle on their hosts (e.g., cypris larvae of *Sacculina* (750) on crabs). Predators may settle on their prey (e.g., *Onchidoris bilamellata* (457) on barnacles) and other organisms onto their preferred substrata (e.g., *Clava leptostyla* (131) on the stipes of *Ascophyllum nodosum*) (Fig. 5.11).

Settlement of the red abalone (*Haliotis rufescens* (411), Fig. 5.12) involves planktotrophic larvae that settle on coralline red algae. The life cycle of *Haliotis* (Fig. 5.13) starts with simultaneous spawning of sperm and eggs (Fig. 5.13, 1). This synchronicity is coordinated by a chemical signal released by the first spawner. This signal has been identified as a **prostaglandin** hormone. An aid to laboratory investigations of *Haliotis* reproduction is that the last step in prostaglandin synthesis involves an enzyme reaction with arachidonic acid. This reaction can be mimicked by the presence of hydrogen peroxide, which supplies free oxygen for the arachidonic acid conversion.

Fertilized eggs hatch in about 14 hours to release veliger larvae, which then spend seven days developing in the plankton (Fig. 5.13, 2), using food reserves from their yolk sac. At seven days, development is arrested, at which time the larvae are said to be competent, meaning that they are ready to settle to the benthos. In this state of arrested development, they can survive in the plankton for up to a month, searching for a settlement site (Fig. 5.13, 3). However, the larvae will not settle without contact with an exogenous trigger: a red coralline alga (Fig. 5.13, 4). *Haliotis* metamorphoses, continues development, and feeds on this alga (Fig. 5.13, 5). Under experimental conditions, competent larvae of *H. rufescens* remained in the plankton until an extract of red coralline alga was added to the culture. At that point, the larvae settled.

A small peptide was isolated from the coralline algal extract (molecular weight about 1000, or about ten amino acids). The peptide includes at least one unusual, unidentified amino acid. One of the identified amino acids in the chain is γ–aminobutyric acid (GABA), which mimics the settlement-inducing action of the natural peptide.

Using radioactive labelled GABA and competitive inhibitors of GABA binding, researchers identified GABA receptors on the velum of *Haliotis* veliger larvae. Binding of the inducer peptide activates adenyl cyclase within the cell (Fig. 5.14), leading to the synthesis of cyclic AMP (a universal second messenger) from ATP. In the presence of Ca^{++}, cyclic AMP activates protein kinase A (PKA) which adds a phosphate group to a cytoplasmic protein. This initiates the opening of a chloride channel in the cell membrane, which

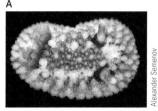

Figure 5.11 **A.** Veliger larvae of *Onchidoris bilamellata* (457) (Mollusca, Gastropoda, Nudibranchiata) are attracted to settle on **B.** barnacles, their prey. **C.** Planula larvae of *Clava leptostyla* (131) (Cnidaria, Hydrozoa) are attracted to *Ascophyllum nodosum* where they settle and then move to an attachment point on the stipe of the alga.

Figure 5.12 *Haliotis rufescens* (411).
SOURCE: Photographer: Dr. Dwayne Meadows, NOAA/NMFS/OPR.

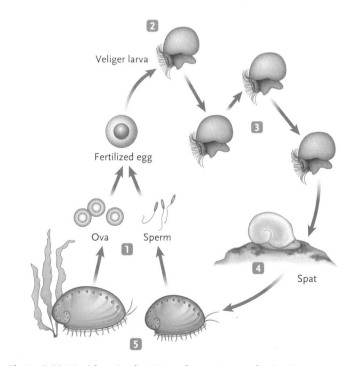

Veliger larva

Fertilized egg

Ova   Sperm

Spat

Figure 5.13 The life cycle of *Haliotis rufescens*. See text for details.

CHAPTER 5 LARVAE AND LIFE CYCLES

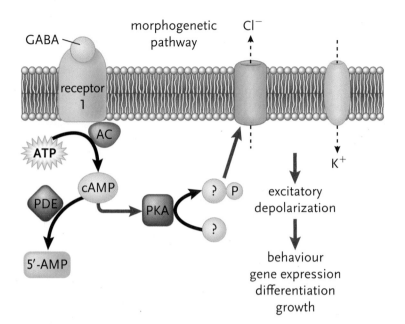

Figure 5.14 The cyclic AMP second-messenger pathway that is activated when GABA binds with the receptor.

generates an electrochemical signal that is mediated through the larval nervous system and that activates the further development of the larva.

Lysine and other diamino acids are also involved in the regulation of the morphogenetic pathway leading to settling. In their presence, the GABA morphogenetic pathway is more sensitive. Higher levels of diamino acids in the dissolved organic material (DOM) in seawater indicate higher levels of available nutrients, thus signalling an appropriate area for larval settlement. Lysine (or DOM) by itself does not stimulate settling, but in the presence of DOM the larvae are much more sensitive to the coralline algal peptide. The diamino acid/DOM regulatory pathway has been shown to operate through a different receptor system from that for GABA (Fig. 5.15). Binding lysine to its receptor activates a G protein (a signal-transducing

molecule) that activates a phospholipase. This enzyme releases diacylglycerol from the inner face of the cell membrane, which then activates protein kinase C (PKC) to phosphorylate another protein. It appears that this phosphorylated protein regulates the GABA morphogenetic pathway.

Controls on other larval settling behaviours may be either simpler or more complex than those described for *Haliotis*. For example, the tubeworm *Hydroides elegans* (335) (Annelida, Polychaeta, Figure 5.16) settles on surfaces covered by a natural biofilm (Fig. 5.17) that includes multiple bacteria species, as well as fungal and diatom species. When individual species from the biofilm are isolated, they remain somewhat attractive to *Hydroides* larvae, but less so than in the natural complex. It appears that a fatty acid and a simple hydrocarbon are the compounds involved. The larvae of a

Figure 5.15 The G protein regulatory pathway triggered by a diamino acid.

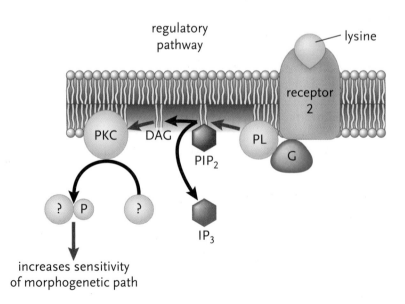

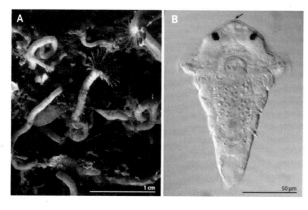

Figure 5.16 *Hydroides elegans* (335): **A.** adults, **B.** larvae that are competent to settle (arrow points to sensory apical cilia)

SOURCE: Republished with permission of Annual Reviews, Inc., from Hadfield, "Biofilms and Marine Invertebrate Larvae: What Bacteria Produce That Larvae Use to Choose Settlement Sites," *Annual Review of Marine Science*, Vol. 3: 453–470, 2011; permission conveyed through Copyright Clearance Center, Inc.

related species, *Hydroides dianthus* (334), exhibit a divided strategy. Some of each clutch of larvae respond to chemical signals from adults and settle in the same area, while some of each clutch settle immediately after reaching competence and then search for a benthic biofilm, thus allowing the colonization of new habitat.

Animals and plants that settle on nets, wharves, boats, and other structures and impede water flows are referred to as fouling organisms. Barnacles are major fouling organisms, and their settlement has long been known to be a response to a compound called arthropodin. It appears that this compound is a glycoprotein released by adult barnacles, although the same compound is present in cyprid larvae where it may be involved in larva–larva settlement interactions.

The planula larvae of reef-building corals settle on crustose red algae and on bacterial films. Metamorphosis of *Acropora millepora* (180) planulae to a polyp has been demonstrated to be triggered by a compound (a tetrabromopyrrole) isolated from species of the bacterial genus *Pseudoalteromonas*, cultured from the biofilm. Despite metamorphosing (induced by levels of the tetrabromopyrrole at concentrations as low as 0.1–1.0 % of the bacterial levels naturally present), the new polyps do not settle, showing the need for separate developmental and settlement cues in this species.

The surfaces of many benthic animals are free of any epibiotic fouling organisms. This is frequently attributed to the secretion of negative settling compounds, although it is not clear what compounds are involved. For example, tests of the larvae of polychaetes, barnacles, and bryozoans showed that they would not settle on colonies of the colonial ascidian *Botryllus schlosseri* (1107) (Fig. 5.18). Natural antifoulants have obvious commercial application in that they might replace toxic antifouling paints based on tin and copper compounds. One example of research in this area has shown that compounds extracted from marine *Pseudomonas* bacteria species from the surface of a nudibranch can be incorporated into paints that greatly reduce the settling of both barnacles and *Ulva* sp. (a green marine alga). Sponges are another source of compounds that have natural antifouling activity.

Sometimes fully developed and competent larvae that have received appropriate chemical cues are still unable to settle. Parts of the coast of California experience variable strengths of upwelling currents. When currents are sustained and high, the barnacle larvae are unable to settle in the three-week span of

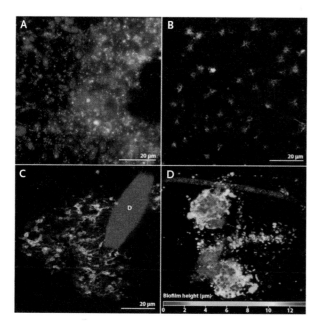

Figure 5.17 Marine biofilms are complex multicellular communities composed of both microorganisms and extracellular polymeric substances (EPS). **A.** Biofilm from the sea tables at Kewalo Marine Laboratory (Hawaii) labelled with 4,6-diamidino-2-phenylindole and photographed with UV fluorescence. Bacteria and diatoms dominate. DNA in bacterial cells fluoresces blue, and chlorophyll in diatoms autofluoresces red. **B., C.** Scanning confocal micrographs of bacteria and EPS in a biofilm. Bacterial DNA is labelled with propidium iodide and fluoresces red, whereas EPS containing α-L-fucosyl residues fluoresces green due to binding of lectins from *Tetragonolobus purpureas*. In **C**, a large autofluorescent diatom (D) is also present. **D.** Marine biofilms form complex topologies. A depth-projection of stacked scanning confocal images of a bacterial biofilm showing 12-µm-tall towers of both bacterial cells and EPS.

SOURCE: Republished with permission of Annual Reviews, Inc., from Hadfield, "Biofilms and Marine Invertebrate Larvae: What Bacteria Produce That Larvae Use to Choose Settlement Sites," *Annual Review of Marine Science*, Vol. 3: 453–470, 2011; permission conveyed through Copyright Clearance Center, Inc.

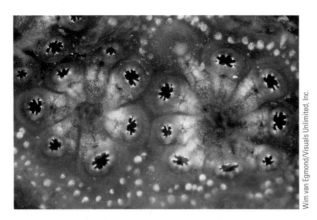

Figure 5.18 The colonial ascidian *Botryllus schlosseri* (1107) has been shown to repel settlement by larvae of several different phyla.

their planktonic life, resulting in very low settlement levels compared to periods of reduced upwelling currents.

STUDY BREAK

1. List examples of three conditions under which the larvae of different species might fail to settle.
2. What would be the ideal characteristics of a molecule that is to be a settlement cue for the larvae of a particular marine species?

5.5 Larvae in Fresh Water

Larval ecology in fresh water is very different from the marine environment. While small, ciliated planktonic larvae are abundant in the sea, they are uncommon in fresh water. This may be explained, in part, by the absence or scarcity in fresh water of the groups whose larvae abound in the marine plankton (e.g., echinoderms, prosobranch gastropods, polychaete annelids). Small larvae that are present in fresh water are mostly those of crustaceans whose appendage-powered swimming gives them significantly greater mobility than ciliated larvae have. These larvae, along with adult copepods, ostracods, other small crustaceans, and rotifers, are the primary consumer level in a freshwater trophic pyramid.

Downstream transport of larvae by the current is an obvious disadvantage in rivers and streams. This may underlie the loss of typical larval types in some groups. For example, Cnidaria in general and most species of Hydrozoa have a ciliated planula larva. Freshwater hydroids, *Hydra* spp. (135), have no planula larva and no free-swimming medusa stage in their life cycle. Instead, adults retain the fertilized eggs which develop directly into attached juvenile polyps.

Limited nutrient levels in fresh water may also influence the loss of larval stages. Many freshwater species whose marine relatives have planktonic larvae produce large, yolk-rich eggs that support development to a miniature adult form.

While some freshwater clams and mussels have a veliger larva (as in marine bivalves), in many freshwater bivalves the veliger stage has been lost and replaced by a parasitic larva called a **glochidium** (Fig. 5.19A). Many have evolved remarkable lures to attract fish that are then parasitized as the glochidia attach to either their fins or gills. Figure 5.19B shows a fish-like lure produced by the mantle edge of a mussel. Predatory fish attacking the lure become hosts to the glochidia. Fish respond to an attached glochidium by tissue growth around it to form a cyst. Phagocytic cells in the mantle of the glochidium feed on fish tissue as the larva develops to a miniature adult. This breaks out of the cyst and settles to the lake or stream bed. Individual large freshwater clams may produce as many as 17 000 000 glochidia in a year.

Short-lived parasitic larvae are relatively common in fresh water. For example, the miracidia of trematodes (see page 285), whose intermediate host lives in fresh water, must hatch and find their host in fresh water. Similarly the coracidia of tapeworms (Cestoda) that spend all, or an intermediate part, of their life cycle in fresh water, normally hatch from eggs shed into fresh water.

Despite the negative tone of the preceding paragraphs, freshwater environments are rich in larvae. These include the larvae of crustaceans and the more numerous nymphs of insects. Some insect species are fully aquatic, with both nymphs and adults living in fresh water (Fig. 5.20). In some species the adult phase of the life cycle is minimal, with almost the entire lifespan being spent as an aquatic nymph, e.g., mayflies (Order Ephemeroptera, Fig. 5.21). Most mayfly nymphs are herbivorous, some are detritivores, and a

A

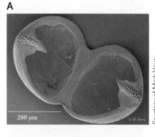

B

Figure 5.19
A. A glochidium larva of *Lasmigona compressa* (541) (Mollusca, Bivalvia).
B. The mussel *Lampsilis reeveiana* (540). What at first sight might be a small fish with its head resting on the shell is the lure, an ovisac filled with glochidia (their outlines are visible through the wall of the ovisac).

A

B

Figure 5.20 A. The Belostomatidae (giant water bugs) complete their life cycle in fresh water. B. Females attach their eggs to the back of the male and they remain there until hatching as nymphs that resemble adults but have no wings.

Barbara Strnadova / Photo Researchers, Inc.

© Daniel L. Geiger/SNAP / Alamy

few are carnivores. Nymphs live for 1–2 years, while the terrestrial adults are short lived, from a few hours to about a month.

Terrestrial adults with aquatic larvae can be compared to benthic marine animals with planktotrophic larvae. In each case the species is taking advantage of the food supply and conditions of an alternative environment.

STUDY BREAK

1. Explain why larvae (excluding nymphs) are relatively uncommon in freshwater species in comparison to related marine species.
2. What advantage may a parasite of a terrestrial host gain from having a larval form in fresh water?

5.6 Terrestrial Larvae

If the freshwater environment poses problems for larval forms, then the terrestrial world is the ultimate challenge. Holometabolous insects are the only group of animals with larvae that have truly con-

A

B

Figure 5.21 Mayflies (Ephemeroptera) from the genus *Siphlonurus* (847). A. A larva. B. Adult.

© blickwinkel / Alamy

Steven Russell Smith Photos / Shutterstock.com

quered the terrestrial environment. (Insect life cycles are discussed in Section 5.7c.) Not all insects have larvae; hemimetabolous insects have nymphs (as defined at the start of the chapter). In holometabolous insects, a larva hatches from the egg and goes through a series of moults before entering a transitional pupa and a complete (sometimes called catastrophic) metamorphosis.

Terrestrial larvae of insects can be split into several categories. Some are symbiotic or parasitic, a number in the Hymenoptera and Isoptera are raised in colonies of social insects and are totally dependent on the colony structure for their food, and many are free living. This last group includes detritivores, herbivores, and predators. Among herbivorous larvae are many serious agricultural pests (e.g., the light brown apple moth (*Epiphyas postvittana* (1034)), discussed on page 261). Larvae of the familiar cabbage white butterfly (*Pieris brassicae* (1047)) cause multi-million-dollar crop losses to both food (cabbage, broccoli, etc.) and decorative plants.

Some predatory insect larvae are used commercially to control insect pests. For example, the larvae of many coccinellids (the ladybird beetles, Fig. 5.22) are voracious feeders on aphids. They are bred and sold by horticultural suppliers to gardeners trying to protect their roses from aphid attacks. Although most larval Lepidoptera (caterpillars) are herbivorous, a predatory caterpillar from Hawaii, *Hyposmocoma molluscivora* (1039), uses its silk glands to spin threads that anchor its prey, a snail, to a leaf to prevent its escape (Fig. 5.23).

5.7 Metamorphosis

5.7a Introduction

At a meeting in 2006, 14 scientists working on animal development attempted to define metamorphosis. The group produced a list of nine characteristics that might be part of a definition of metamorphosis (Table 5.2). Only two of the participants were in agreement on these characteristics of metamorphosis, each selecting the same seven of the list of nine characteristics.

Some of the differences of opinion revealed by Table 5.2 undoubtedly result from the animal group (or

A

B

photofun / Shutterstock.com

SvenButstraen / Shutterstock.com

Figure 5.22 Ladybird beetle (*Harmonia axyridis* (935); Coleoptera, Coccinellida). A. Larva and B. adult.

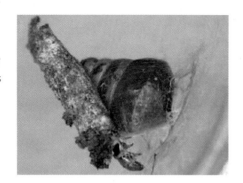

Figure 5.23 The larva of *Hyposmocoma molluscivora* (1039) (Lepidoptera) lives in a case (the anterior region of this larva is visible outside its case) and preys on snails.

SOURCE: From Rubinoff, Haines, "Web-spinning caterpillar stalks snails," *Science* 309 no. 5734 p. 575. Reprinted with permission from AAAS.

groups) on which each participant based their definition. Despite the lack of consensus, four features of metamorphosis are generally accepted:

1. a movement from one habitat to another
2. a major change in the morphology of the animal
3. a change in the environment to which the animal is adapted
4. a change to a life cycle in which a post-embryonic stage* precedes metamorphosis

Metamorphosis may involve endogenous controls (as in amphibian tadpole development and the transitions from a pupa to an adult butterfly) or may depend on exogenous factors (e.g., the development of coral planulae to polyps, without settlement, as in *Acropora millepora* (180) discussed above). In general, where endogenous controls are involved, the process of metamorphosis is slow, while exogenous controls are typically linked to a rapid metamorphosis.

Table 5.2 | **Features that might characterize metamorphosis.**

Characteristic	Selected by n/14 participants
Habitat shift	10
Major morphological change	12
Change in adaptive landscape	13
Rapid	1
Change in feeding mode	8
Pre-metamorphic stage is post-embryonic	13
Usually a pre-reproductive to reproductive stage transition	8
Transition is generally hormone regulated	5
Is plant flowering metamorphic?	Yes: 5; No: 9

* Described elsewhere in this chapter as a larva.

5.7.b Evolution of Metamorphosis

Planktonic marine larvae are, in general, small. This may be linked to their use of cilia for locomotion and flotation, a way of life that limits their size to less than about 1 mm (see discussion of Reynolds number on pages 27 and 207–209). While a larva of this size is potential food for many larger members of the zooplankton (copepods in particular), the hazards awaiting it on the seabed are even greater. Apart from the risk of settling on an inappropriate substratum or settling on a surface only to be eaten by a grazing animal, the seabed is home to not only a new range of predators but is also likely to be carpeted with an array of filter-feeding animals, all able to catch, hold, and eat a small larva. This generates a strong selection pressure for the ability to settle, metamorphose, and grow (which demands a start to feeding) as rapidly as possible.

Examination of the development of many of these planktonic larvae shows that this metamorphosis is not quite as rapid as first sight suggests. The trochophores of many annelid species, for example, develop a number of segments as well as segmental setae before leaving the plankton (Fig. 5.24). Molluscan veliger larvae form a shell and develop a foot before settling (Fig. 5.25). In the phylum Nemertea (and in many echinoderms), an adult rudiment forms within the planktonic larva, and metamorphosis really only involves the "hatching" of a juvenile adult form from the larva (Fig. 5.26).

This premetamorphic development supports the concept that metamorphosis may have evolved through heterochrony (changes in the timing of developmental events). This concept is shown in Figure 5.27 and explained in the figure legend.

5.7c Insect Metamorphosis

Insect metamorphosis is considered separately from that of other groups for several reasons:

- it occurs in the largest animal orders in the world
- insect larvae and metamorphosis are secondary adaptations to the original insect life cycle
- it has been more studied than any other metamorphic process.

Three different patterns of development are recognized in different insect groups (Fig. 5.28):

1. ametabolous development in which the egg hatches as a miniature adult (pronymph)
2. hemimetabolous development in which the first five nymphal stages resemble the adult but lack wings, with wings being formed at the fifth moult to produce the adult form
3. holometabolous development in which the insect hatches as a larva (caterpillar or grub) that grows in size through a series of moults, then pupates, to emerge from the pupa as an adult (Fig. 5.29).

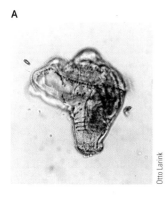

A

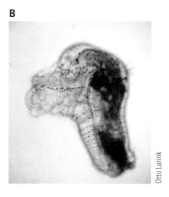

B

C

Figure 5.24
Developing trochophore larvae of polychaete annelids. **A., B.** Larvae of the tubeworm *Pectinaria* sp. (338). Notice the formation of an additional segment in **B**, a 14-day-old larva. **C.** A late trochophore of *Scolelepis* at a stage when it is ready to settle.

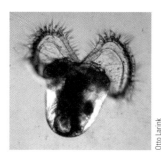

Figure 5.25 A veliger larva of a gastropod showing the shell already formed before settlement and metamorphosis.

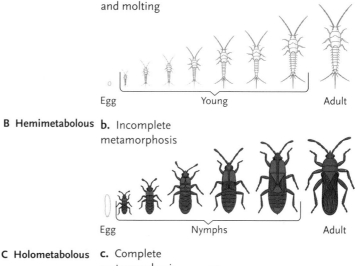

Non-metamorphic ontogeny

Metamorphic ontogeny

Larval development — Larval period — Metamorphosis — Adult period

Figure 5.27 A hypothesis of the way in which metamorphosis may have evolved through heterochrony. Symbols represent particular ontogenetic events. In the metamorphic ontogeny, some events (+ symbol, green hexagon, cyan triangle, and blue square) have been accelerated and occur together as metamorphosis. New developmental features (black circle and square with arrows) have appeared to characterize a larval form; these features are destroyed, as the larva enters the metamorphic period (grey oval).

SOURCE: Jason Hodin, "Expanding networks: Signaling components in and a hypothesis for the evolution of metamorphosis," *Integrative and Comparative Biology*, 2006, volume 46, number 6, pp. 719–742, by permission of Oxford University Press.

Given the origin of the insects within a crustacean ancestry (see Chapter 3), it is clear that the larval stage and metamorphosis evolved after the first radiation of the Insecta. Some aspects of insect development parallel those of echinoderms and nemerteans in that precursors of adult structures can be identified in larval stages. In insects these appear as clusters of undifferentiated cells, known as imaginal discs.

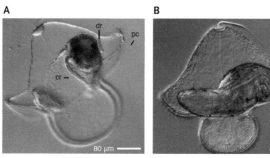

A

dr

pc

cr

B

80 µm

110 µm

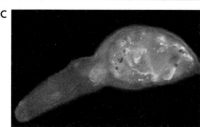

C

Figure 5.26 Development of the pilidium larva of *Micrura alaskensis* (564) (Nemertea). **A.** In this 17-day-old larva, the rudiments of the adult body have already formed (cd = cephalic rudiment; dr = dorsal rudiment; cr = ciliated ridge of esophagus; pc = posterior cirrus). **B.** A complete juvenile within the larva. **C.** The juvenile after metamorphosis. The pigment granules are the remains of the pilidium, which is eaten by the juvenile.

SOURCE: Svetlana A Maslakova, "Development to metamorphosis of the nemertean pilidium larva," *Frontiers in Zoology* 2010, 7:30.

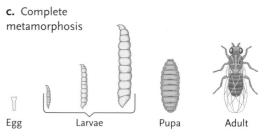

A Ametabolous **a.** Growing and molting

Egg — Young — Adult

B Hemimetabolous **b.** Incomplete metamorphosis

Egg — Nymphs — Adult

C Holometabolous **c.** Complete metamorphosis

Egg — Larvae — Pupa — Adult

Figure 5.28 Modes of insect development. **A.** Ametabolous development in a silverfish, **B.** hemimetabolous development in a hemipteran, and **C.** holometabolous development in a dipteran.

SOURCE: From RUSSELL/WOLFE /HERTZ/STARR. *Biology*, 2E. © 2013 Nelson Education Ltd. Reproduced by permission. www.cengage.com/permissions

Figure 5.29 **A.** Larval (caterpillar), **B.** pupal (chrysalis) and **C.** adult stages in the holometabolous development of a swallowtail butterfly (*Papilio* sp. (1043).)

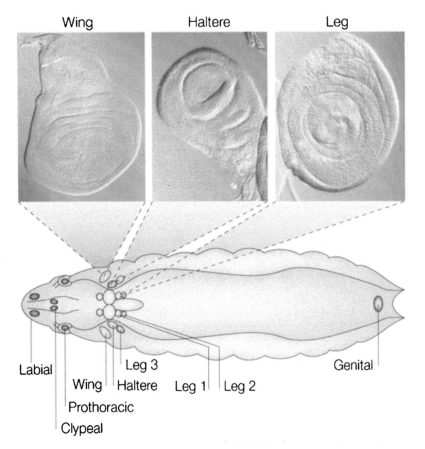

Figure 5.30 The positions of imaginal discs in a third-instar larva of *Drosophila* showing the structures they will form in the adult. The upper panels are micrographs of three of the imaginal discs.

SOURCE: Reprinted by permission from Macmillan Publishers Ltd: *Nature Reviews Molecular Cell Biology* 2, Ginés Morata, "How *drosophila* appendages develop," pp. 89–97, copyright 2001.

In *Drosophila* (960), for example, imaginal discs that will form a number of adult structures can be identified in a third-instar (third-stage) larva (Fig. 5.30).

Hemimetabolous and holometabolous development are not as different as first sight suggests. This supports the concept that metamorphosis evolved through heterochrony, as suggested in the model in Figure 5.27. A key regulatory gene in pupal development of holometabolous insects is a transcription factor known as "broad" (*br*). Figure 5.31 shows that *br* is expressed in hemimetabolous insects throughout the nymphal instars, the time at which wing buds show progressive development. In holometabolous insects, *br* is expressed at only the final stage of development, the time at which the imaginal discs for wings are transformed into wing buds and other adult structures are differentiated.

Development and moulting in insects is under a complex set of endocrine interactions, mediated and controlled through the central nervous system. These interactions have been best studied in lepidopterans and are summarized in Figure 5.32. The two circulating hormones that target cellular processes are ecdysone and juvenile hormone. Ecdysone (20-hydroxy-ecdysone, considered in more detail in Box 5.2) is the hormone that stimulates the processes involved in each moult. Juvenile hormone (JH) regulates the type of moult by inhibiting the development of adult characteristics. High levels of JH result in larva–larva moults, reduced JH is associated with the larva–pupa moult, and absence of JH characterizes the moult leading from pupa to adult (imago). The production of ecdysone (by the prothoracic glands) is regulated by the secretion of prothoracicotropic hormone (PTTH) from

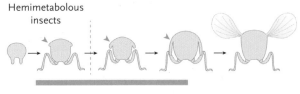

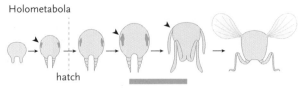

Figure 5.31 Diagrammatic cross-sections through a hemimetabolous and a holometabolous insect through development. Gray arrowheads point to developing wing buds in the hemimetabolous nymph. Black arrowheads show the imaginal discs for wing buds in the holometabolous insect. Gray bars show the period during which the transcription regulator gene *br* is expressed.

SOURCE: Deniz F. Erezyilmaz, Lynn M. Riddiford, and James W. Truman, "The pupal specifier broad directs progressive morphogenesis in a direct-developing insect," *PNAS* May 2, 2006 vol. 103 no. 18 6925–6930. Copyright 2006 National Academy of Sciences, U.S.A.

the lateral neurosecretory cells in the cerebral ganglia. JH is synthesized in, and released from, the paired *corpora allata*, close to the cerebral ganglia. The *corpora allata* are closely associated with the paired neurohemal *corpora cardiaca*. JH production and release are regulated by two neuropeptides from the medial neurosecretory cells: allatotropin that stimulates JH production and allatostatin (or allatoinhibin) that inhibits JH production. The cerebral ganglia integrate environmental and physiological information and regulate the hormonal pathways on the basis of these inputs.

Ecdysone and JH can act at the level of imaginal discs. Figure 5.33 shows the hormone levels associated with the growth of the imaginal discs that develop to form the horns of a dung beetle.

5.7d Metamorphosis in Vertebrates

Figure 5.34A shows a larva of *Reinhardtius hippoglossoides* (1309), the Greenland halibut or turbot. With its yolk sac, it resembles the larva of most teleost fish. A remarkable metamorphosis occurs around 800 degree days (since fish development rates are a function of both time and temperature, they are calculated as accumulated degrees, i.e., a larva developing at a steady 10° temperature would reach 800 degree days in 80 days). This larva is transformed to the adult form shown in Figure 5.34B. The stages in the metamorphosis of another flatfish (*Paralichthys dentatus* (1305), the summer flounder) are explained in Figure 5.35. As in amphibian tadpole metamorphosis (described below), fish metamorphosis is hormonally controlled.

The eggs of amphibians (frogs, toads, salamanders, newts, and caecilians) are embedded in a jelly-like substance and lack the protective membranes that allow amniote eggs to be laid on dry land. Most amphibians lay their eggs in fresh water but some have evolved other tactics that provide adequate moisture while avoiding predators. Many salamanders and caecilians lay eggs in damp soil or leaf litter, some tree

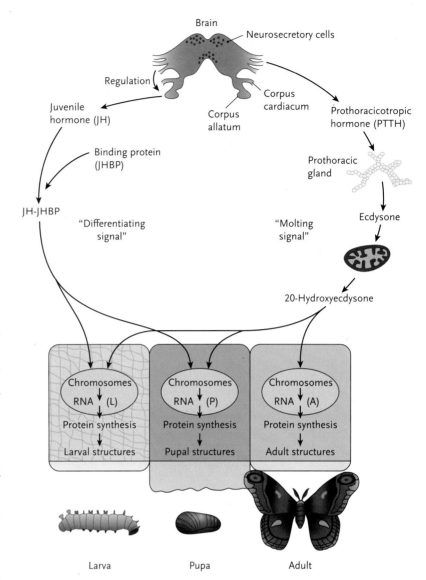

Figure 5.32 A schematic representation of the hormones and pathways involved in development regulation in the wax moth, *Galleria mellonella* (**1035**). See text for explanation.

SOURCE: Gilbert SF. *Developmental Biology*, 6th edition. Sunderland (MA): Sinauer Associates; 2000. Reprinted by permission.

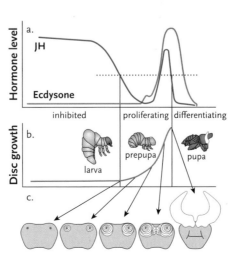

Figure 5.33 The growth of imaginal discs that will form dung beetle horns is inhibited in the larval stage when JH levels are high and ecdysone levels low. When JH levels fall below a genetically mediated threshold (red line), the imaginal discs grow as ecdysone and JH levels rise, until the JH levels fall below the threshold for the second time. This inhibits disc growth and leads to differentiation of the imaginal discs to form the beetle horns (seen in the inset picture of *Onthophagus taurus* (**943**)).

SOURCE: Douglas J. Emlen, Cerisse E. Allen, "Genotype to Phenotype: Physiological Control of Trait Size and Scaling in Insects," *Integrative and Comparative Biology*, (2003) 43 (5): 617–634, by permission of Oxford University Press.

BOX 5.2 Molecule Behind Animal Biology

Ecdysone

OH

ecdysone

A

HO

HO

H

H

OH

O

FIGURE 1
A. *Ecdysone prohormone*, **B.** *20-hydroecdysone*.

OH

HO

OH

B

HO

HO

H

H

OH

O

20-hydroxyecdysone

Ecdysone (Fig. 1A) is the steroidal **prohormone** of the moulting hormones produced by arthropods, best studied in crustaceans and insects. It was first isolated in crystalline form from the common silkworm, *Bombyx mori* (1028), in 1954, and its chemical structure was elucidated in 1965. Like other steroid hormones, **ecdysteroids** act by influencing the transcription of DNA, the first step in protein synthesis in cells. While this is now an obvious action of developmental hormones, the first evidence that steroid hormones act at the transcriptional level came from studies of the effects of ecdysone on dipteran polytene chromosomes in the 1960s.

Ecdysteroid moulting hormones are produced by the prothoracic glands of immature insects and by crustacean Y-organs located at the base of the antennae or near the mouthparts. Ecdysone is converted to the active form, 20-hydroxyecdysone (Fig. 1B), in the fat body or epidermis by a cytochrome P-450 enzyme in the mitochondria.

Insects cannot synthesize steroids. Consequently sterols, usually cholesterol or a closely related structure, are essential dietary constituents. This is important because ecdysteroids also appear in many plants, mostly as protective agents (toxins or feeding deterrents) against herbivorous insects. Phytoecdysteroids have been identified in a wide range of plants, and some have been shown to have medicinal value as components of herbal remedies.

Ecdysone might have been chosen as the molecule associated with Chapter 3 (Phylogeny) since moults, controlled by ecdysteroids, characterize one of the major branches (the Panarthropoda, and perhaps the Ecdysozoa as a whole) of the phylogenetic tree. Its linkage to this chapter on larvae and life cycles is the importance of ecdysteroids in regulating the life cycles and developmental stages in the biology of arthropods. There is a wide variety of active ecdysteroids found in arthropods, with a greater variety of ecdysteroids found in the hemolymph of crustaceans than in insects. This may reflect the differences in insect and crustacean life histories. The insect life strategy, for both hemimetabolous and holometabolous forms, is compartmentalized into larval and adult stages. Larval insects do not reproduce, and most adult insects do not moult. Many crustaceans, on the other hand, both reproduce and continue to molt well into adult stages. Because insects partition growth and reproduction into different stages, fewer hormones may be able to serve more functions while the more complex adult stage of crustaceans may require a greater variety of ecdysteroids.

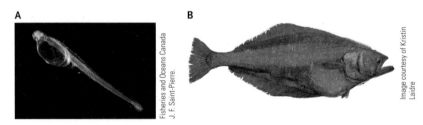

Figure 5.34 A. Larva (a few mm long) and **B.** adult (up to 200 kg) of *Reinhardtius hippoglossoides* (1309), the Greenland halibut or turbot.

frogs deposit eggs in a frothy mixture of surfactants and proteins that bind carbohydrates and have antimicrobial qualities, and in many caecilians and salamanders, the eggs hatch inside the mother and are born either as larvae or fully formed (Fig. 5.36). No matter where the eggs are deposited, almost all amphibians pass through a larval stage before metamorphosing into their adult forms.

Metamorphosis in frogs is particularly spectacular because larval frogs (tadpoles) and adult frogs differ dramatically in anatomy, physiology, and ecology. The differences are accomplished in three distinct phases: premetamorphosis, prometamorphosis, and climax metamorphosis (Fig. 5.37). You may have seen tadpoles in a pond or puddle on a warm spring day. They are fully aquatic, usually (but not always) herbivorous, exchange oxygen through gills, and have long, laterally compressed tails used to generate propulsion for swimming. Frogs, on the other hand, are largely terrestrial, eat other animals, use lungs for oxygen exchange (and skin when it is moist), and have elongated hind limbs allowing them to jump and/or swim effectively. The transition from tadpole to adult

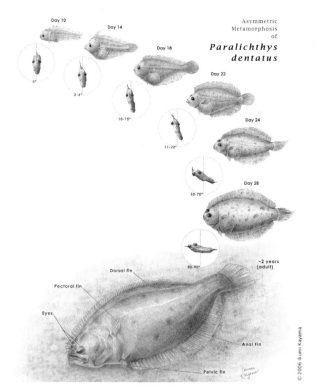

Asymmetric
Metamorphosis
of
*Paralichthys
dentatus*

Day 10
Day 14
Day 18
Day 22
Day 24
Day 28

0°
3-6°
10-15°
17-22°
50-70°
80-90°
~2 years (adult)

Dorsal fin
Pectoral fin
Eyes
Anal Fin
Pelvic fin

© 2006 Ikumi Kayama

Figure 5.35 Asymmetric metamorphosis of *Paralichthys dentatus* (1305), the summer flounder. All flatfish undergo dramatic changes mediated by thyroid hormones (T3). Summer flounders are born with one eye on each side of the head. However, typically between day 10 and day 28 post hatching, the right eye migrates over the top of the head to the left side. Before metamorphosis, the frontal and parietal bones of the cranium the bones are symmetrical. As the fish develop, these bones undergo remodelling as shown in circled figures. The right frontal and parietal bones become progressively thinner to accommodate the migrating right eye. At the same time, the dorsal fin migrates anteriorly and ventrally. It continues to extend over the dorsal canal and eventually passes the eyes. The pectoral fins undergo changes from paddle-shaped cartilaginous fins to narrow ossified fins. The pelvic fins develop in a similar manner. Shortly before metamorphosis, the summer flounder starts to change its swimming orientation from free swimming to a tilted angle, until it swims and lies parallel to the ocean floor. The left side (with two eyes) develops normal skin colouration, and the bottom (right) side becomes white. Flounders normally burrow into sand for camouflage, but occasionally they emerge out of the sand to relocate for hunting or to escape from predators.

SOURCE: (c)2007 Ikumi Kayama

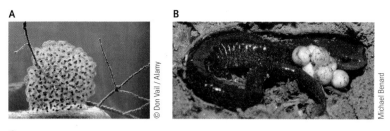

A
© Don Vail / Alamy

B
Michael Benard

C
"Hatch on attack" -- Red-eyed treefrog. *Agalychnis callidryas*, embryos hatching prematurely during an attack by a cat-eyed snake, *Leptodeira septentrionalis*. Photo taken in Corcovado National Park, Costa Rica. Copyright Karen M. Warkentin.

Figure 5.36 A. An egg mass of wood frogs (*Rana sylvestris* (1380)) in a pool. **B.** A female red-backed salamander (*Plethodon cinereus* (1412)) guarding eggs laid under a rock. **C.** Larvae of a neotropical red-eyed tree frog (*Agalychnis callidryas* (1363)) emerge from an egg mass and drop into the water below. In this picture they are hatching prematurely because they sense the vibrations made by the snake.

upregulates GH expression and action, but in frogs the actions of GH and TH appear to be decoupled. This means that the relative timing of growth and development can change independently, allowing for a great deal of plasticity in larval form as well as the timing of metamorphosis.

STUDY BREAK

1. What do you think are the important characteristics that define metamorphosis?
2. How does the hormonal control of metamorphosis differ between insects and vertebrates? In what ways are the two processes similar?

requires extensive remodelling of the entire body. While some structures grow and change, others are destroyed.

Metamorphosis in frogs consists of growth followed by profound changes in the form and function of the body. Growth is likely regulated by **growth hormone (GH)** and possibly **prolactin**, both produced in the pituitary gland. The dramatic proliferation, death, and remodelling of body tissues are largely under the control of thyroid hormone (TH), also produced in the pituitary, which acts through thyroid hormone receptors located in cells throughout the body. In fish and mammals, TH

Pre-   Pro-   Climax metamorphosis

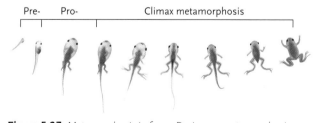

Figure 5.37 Metamorphosis in frogs. During premetamorphosis (left), the tadpole grows, and the cells that will grow into the hindlimbs begin to aggregate. During prometamorphosis (middle), the hindlimbs develop. Climax metamorphosis (right) marks the growth of the forelimbs and resorption of the tail.

SOURCE: Reprinted from *Trends in Ecology and Evolution* 20(3), Rose C.S., "Integrating ecology and developmental biology to explain the timing of frog Metamorphosis," Pages 129–135, Copyright 2005, with permission from Elsevier.

5.8 Larval Polymorphism

In polymorphic colonies, individuals are adapted to different functions (see Chapter 2). Polymorphism (the development of individuals having different morphologies from a single genome) can play other roles in animal life cycles.

5.8a Poecilogony in Nudibranchs and Polychaetes

Poecilogony (meaning "varied offspring") describes the production of two different larval morphs from eggs of the same genome. A population of California sea slugs (Mollusca, Gastropoda, Nudibranchiata), *Alderia willowi* (465) (Fig. 5.38A), may produce either small, planktotrophic larvae or large lecithotrophic larvae (Fig. 5.38B). About half the population produces larvae of each type, thereby combining local survival with maximal dispersal due to the different larval strategies. The dual strategy becomes poecilogony when a single parent produces both larval types. When adult *A. willowi* producing lecthotrophic larvae are starved for a few days, they switch strategies and produce either all planktotrophic larvae or broods of mixed planktotrophic and lecithotrophic larvae (Fig. 5.38C). It appears that the trigger of local food shortage initiates the production of larvae that will disperse farther, thus improving the chances of survival of the genome.

The intertidal tubeworm *Boccardia proboscidea* (330) (Annelida, Polychaeta, Spionidae; Fig. 5.39A) provides a different example of poecilogony. Larvae of this species may disperse in the plankton or remain benthic, and may be planktotrophic or **adelphophagic** because they eat their fellow offspring. Notice that two types of egg capsule are produced, one from which all eggs hatch and become planktotrophic, and the other in which some eggs become food for the larvae developing within the capsule (Fig. 5.39B). These larvae remain longer in the capsule before emerging at a more advanced state of development (Fig. 5.39B), and complete their development either in the plankton or the benthos.

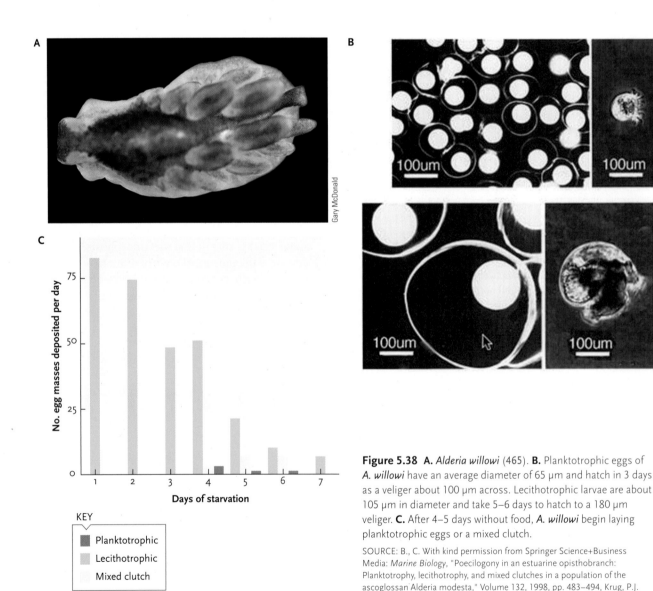

Figure 5.38 **A.** *Alderia willowi* (465). **B.** Planktotrophic eggs of *A. willowi* have an average diameter of 65 μm and hatch in 3 days as a veliger about 100 μm across. Lecithotrophic larvae are about 105 μm in diameter and take 5–6 days to hatch to a 180 μm veliger. **C.** After 4–5 days without food, *A. willowi* begin laying planktotrophic eggs or a mixed clutch.

SOURCE: B., C. With kind permission from Springer Science+Business Media: *Marine Biology*, "Poecilogony in an estuarine opisthobranch: Planktotrophy, lecithotrophy, and mixed clutches in a population of the ascoglossan Alderia modesta," Volume 132, 1998, pp. 483–494, Krug, P.J.

A

B

Ip
(20d)

Hatching (6d)

Metamorphosis
(26d)

Type I
No nurse
eggs

IIIp
(15d)

Type III
Extra-embryonic
provisioning
(nurse eggs)

Hatching
(11d)

IIIa

(0-2d)

Hatching and
Metamorphosis (11d)

→ Benthic
→ Planktonic
→ Varies within
cohort

Days since spawning

Dr G. Read NIWA Wellington, NZ

Figure 5.39 A. *Boccardia proboscidea* (**330**), a poecilogonic worm. **B.** The life cycle of *B. proboscidea*. Types I and III identify egg capsules without (I) and with (III) nurse eggs to provide additional nutrition. Note the different times at which the larvae leave the capsule and the additional development of the adelphophagic larvae (IIIa) that remain associated with the benthos for a short period before they settle.

SOURCE: B. Based on Gibson & Gibson, "Heterochrony and the Evolution of Poecilonogy: Generating Larval Diversity," *Evolution* 58(12):2704–2717, 2004.

5.8b Soldier Rediae in Trematodes

The complex life cycles of parasitic flukes (Platyhelminthes, Trematoda) are reviewed in Chapter 8 (Symbiosis and Parasitism). The normal life cycle goes from egg to a miracidium larva that invades a molluscan secondary host, to redia, metacercaria and/or cercaria larvae (all asexual reproductive multiplier stages), to the final host. In the near-shore marine environment, many species of trematode parasitize the numerous species of sea bird. A relatively limited number of gastropod mollusc species are available as possible intermediate hosts, meaning that competition between parasite species might be expected.

A recent and surprising finding is the development of a caste of "soldiers" in a trematode. (A feature of the life cycles of some ant and termite species is the development of a soldier caste, usually larger than other workers and with well-developed jaws; these soldiers defend the colony.) The trematode genus *Himasthla* (**262**), parasites in the gut of shore birds, is one of several trematode genera that invade the snail, *Cerithidea californica* (**416**), as an intermediate host (Fig. 5.40A). The miracidia larvae that infest the snail develop to form redia, a stage that multiplies asexually. Asexual multiplication produces two morphologically distinct forms of rediae. One is large and

reproductive, the other smaller, more active, and with well-developed jaws (Fig. 5.40B). This small morph is non-reproductive and attacks both conspecific rediae from a different parent and heterospecific larvae (Fig. 5.40C). The recognition of soldier rediae extends the concept of eusociality to the Platyhelminthes and is a remarkable adaptation for life cycle success.

5.8c Cannibal Frogs

The growth and development of frog larvae can also respond to changes in environmental conditions as well as to competition and the presence of predators. Among vertebrates, the ability to produce different morphs from a single genetic background is called **polyphenism**.

Spadefoot toads (*Spea* spp. (**1389**)) are native to dry regions of western North America from Canada to southern Mexico. Adults are burrowers and spend most of their lives underground, but they emerge to lay eggs in water where the tadpoles develop. Most spadefoot toad tadpoles are omnivores, eating algae and organic particulates suspended in the water. But sometimes an omnivore transforms into a carnivore that eats small shrimp and insects and cannibalizes other spadefoot tadpoles (Fig. 5.41). It is not entirely clear what triggers omnivores to morph into carnivores, but it is probably linked to water temperature, food availability, and interactions with other spadefoot

A

Colonies of parthenitae (soldier and reproductive rediae)
in 1st intermediate host snail

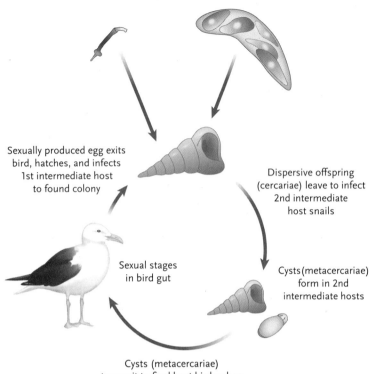

Sexually produced egg exits
bird, hatches, and infects
1st intermediate host
to found colony

Dispersive offspring
(cercariae) leave to infect
2nd intermediate
host snails

Sexual stages
in bird gut

Cysts (metacercariae)
form in 2nd
intermediate hosts

Cysts (metacercariae)
transmit to final host birds when
birds prey upon 2nd intermediate
hosts

SOURCE: Reprinted from *Current Biology*, Volume 20, Issue 22, Philip Newey and Laurent Keller, "Social Evolution: War of the Worms," Pages R985-R987, Copyright 2010, with permission from Elsevier.

B

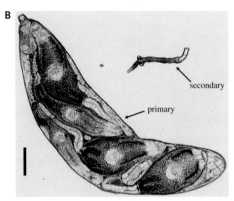

SOURCE: Ryan F. Hechinger, Alan C. Wood and Armand M. Kuris, Social organization in a flatworm: trematode parasites form soldier and reproductive castes," *Proc. R. Soc. B* 7 March 2011 vol. 278 no. 1706 656–665, by permission of the Royal Society.

C

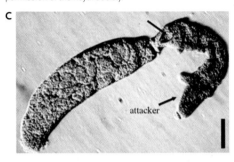

SOURCE: Ryan F. Hechinger, Alan C. Wood and Armand M. Kuris, Social organization in a flatworm: trematode parasites form soldier and reproductive castes," *Proc. R. Soc. B* 7 March 2011 vol. 278 no. 1706 656–665, by permission of the Royal Society.

Figure 5.40 A. The life cycle of *Himasthla* sp. (262) **B.** Redia stage of *Himasthla* showing a large reproductive individual (notice the cercaria within the redia) and a smaller soldier redia. **C.** A solder redia of *Himasthla* attacking a redia of another trematode, *Euhaplorchis californiensis* (260). Scale bar: 0.2 mm.

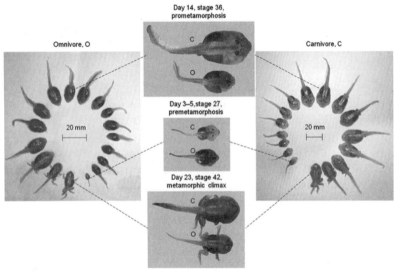

Figure 5.41 Development of omnivorous (O) and carnivorous (C) tadpoles in a spadefoot toad, *Spea multiplicata* (1389). The two morphs are similar at premetamorphosis, very different during prometamorphosis, and similar again during metamorphic climax. The days indicate the number of days since hatching.

SOURCE: With kind permission from Springer Science+Business Media: *Oecologia* Vol. 141, No. 3, 2004 year of publication, pp. 402–410, Brian L. Storz, "Reassessment of the environmental model of developmental polyphenism in spadefoot toad tadpoles."

tadpoles. The omnivore and carnivore morphs diverge early in premetamorphosis, are most different during prometamorphosis, and become similar again at metamorphic climax. They do not differ as adults. Carnivorous tadpoles develop quickly and can escape rapidly drying ponds, while the slower-developing omnivores have higher post-metamorphic survival rates in long-lived ponds. The story of the spadefoot toads highlights the fact that larvae experience different opportunities and challenges from adult frogs. In addition, developmental plasticity can allow individuals to

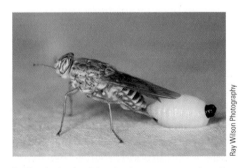

Figure 5.42 The birth of the third instar *G. morsitans* larva.

move into niches where they may experience higher fitness.

STUDY BREAK

1. Give two examples of plasticity in larval development and explain the advantage they confer to their species.
2. When and why might larval cannibalism be advantageous?

5.9 Vivipary

Chapter 1 reviewed the evidence for vivipary in a fossil (Carboniferous) cartilaginous fish that died with embryos in utero. Viviparous animals retain the zygote and developing embryos within their bodies, and some provide food and waste management until the birth of live young. Zoologists recognize three general levels of interaction between the mother and developing young in viviparous animals:

- Ovovivipary: zygotes are retained within the body of the female (or occasionally the male, e.g., sea horses) but there is no trophic link between parent and embryo.
- Histotrophic vivipary: embryos develop within the parent, feeding on either other eggs (oophagy) or sibling young (adelphophagy).
- Hemotrophic vivipary: embryos develop in the uterus and obtain their food from the mother, typically through a placenta.

With the exception of living monotremes (spiny anteaters and duckbilled platypus), modern mammals use hemotrophic vivipary. Among vertebrates, there are examples of ovovivipary, histotrophic vivipary, and hemotrophic vivipary in all other groups of gnathostomes (Chondrichthyes, Osteichthyes, Amphibia, Reptilia). Only birds (Aves) and turtles (Chelonia) are entirely oviparous, laying eggs from which young (miniature adults) hatch directly. In birds, ovipary probably reflects the limitations of loading and flight, while in turtles, living inside a shell may impose problems on giving birth. Vivipary is a recurring theme in some invertebrates.

In truly viviparous animals, the **placenta** is the interface between the developing embryo and (usually) the mother. The term "placenta" is a functional label and is applied to structures that have the same function but quite different physiologies and morphologies in different groups. At least 50 loci regulate the development of the placenta in mammals, including gene families that produce protein hormones and hemoglobin but that are only expressed in the fetus. In some mammals, a molecule that enables circulating leucocytes to bind to blood vessels (L-selectine) is commandeered by the **blastocyst** to initiate interactions with the lining of the uterus, preparing it for implantation.

Among Mollusca, snails in the genus *Viviparus* (480) are ovoviviparous, with females brooding up to 16 young over a developmental period lasting about 8 months. They then give birth to fully developed small snails.

A number of insects are viviparous, for example, the tsetse fly (*Glossina morsitans* (963)) broods a single egg at a time within the uterus (Fig. 5.42). The egg hatches, and the larva passes through its first moult stages within the uterus. As it develops, the larva is nourished by a milky secretion from glands in the uterus.(Fig. 5.43). This larva does not feed after birth but pupates and emerges as an adult. At 25°C a tsetse fly produces a third-stage larva every 7–9 days. Tsetse flies are described as having adenotrophic vivipary, meaning "gland fed, live birth."

The phylum Onychophora includes just over 100 species (*Peripatus* (628) and its relatives). The group includes species that are oviparous, some that are ovoviviparous, and a number that use hemotrophic vivipary. Here, zygotes and developing embryos are nourished via a placenta within the uterus, and live young are born (Fig. 5.44).

Among boney fish, guppies and their relatives provide an interesting perspective on variations in patterns of development, with some species (and genera) being entirely oviparous and others using hemotrophic vivipary (Fig. 5.44). An index of matrotrophy (supply of food from the mother) is based on levels of investment that females make in their young, with higher values translating into hemotrophic vivipary. These guppy data demonstrate just how dynamic biological systems can be.

Fish in the family Sygnathidae—the sea horses, pipefish, and sea dragons (Fig. 5.45)—offer a different approach entirely. In these species, it is the male that "gets pregnant," carrying the developing young in a brood pouch. In some of these animals, the brood pouch provides aeration, osmoregulation, and nutrition, a full-service uterus! In *Syngnathus scovelli* (1347), the male's pregnancy is affected by the size of the mother, the number of eggs she transfers, and the male's sexual responsiveness. Embryo survival in one pregnancy is negatively affected by survivorship in previous pregnancies. In *S. scovelli*, sexual selection and conflict are mediated by choices made by the males as indicated by rates of abortion of offspring from less attractive mothers. In another pipefish, *Syngnathus typhle* (1348), males keep young in the pouch for several weeks, and over this period the number of embryos in the pouch diminishes. Radioactive (^{14}C) labelling of embryos revealed that some of the amino acids ended up in the male: in his liver, the lining of the brood pouch, and his muscle. This is another example of recycling.

While the stomach may seem to be an unusual place to serve as a uterus for raising young inside the

A

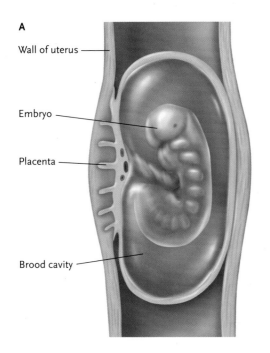

- Wall of uterus
- Embryo
- Placenta
- Brood cavity

B

Rod Morris

Figure 5.43 A. A developing embryo, nourished through a placenta, in the uterus of an onychophoran. **B.** *Peripatoides novaezelandiae* (626) with two new-born young.

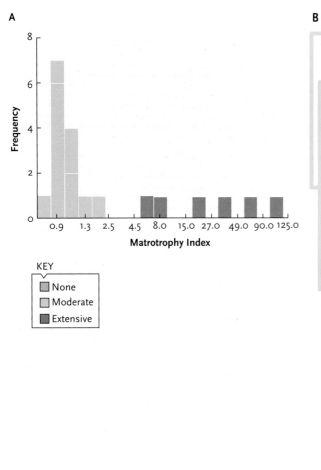

A

B

Gambusla affinis
Poecilia butleri
Heterandria formosa
Phalichthys tico
Neoheterandria umbratilis
Priapkhthys festae

Outgroups

80
56
100

P. paucimaculata (7.8)
P. elongata (68.9)
P. retropinna (117)

Subgenus *Aulophallus*

100
69

P. fasciata (0.81)
P. latidens (0.81)
P. baenschi (1.58)
P. infans (0.86)
P. lucida (1.34)
P. occidentalis BD (1.12)
P. prolifica Moc
P. prolifica Nay   (5.4)
P. prolifica May
P. monacha (0.61)
P. viriosa (0.93)
P. gracilis TB (0.69)
P. hnilickal (0.86)
P. catemaco (0.68)
P. gracilis Coa (0.69)
P. presidionis (21.5)
P. turneri (41.4)
P. scarili (0.87)
P. turrubarensis (0.66)

Subgenus *Poeciliopsis*

96
100
100
98
75
96
96
65
100
93
100
90
100
100
100
100
100

Figure 5.44 A. The distribution of values for matrotrophy index and **B.** a phylogeny illustrating the variation among guppy-like fish. Colour codes in both figures are the same.

SOURCE: From Reznick, D.N., M. Mateos and M.S. Springer. 2002. "Independent origins and rapid evolution of the placenta in the fish genus Poeciliopsis'" *Science*, 298:1018–1020. Reprinted with permission from AAAS.

Figure 5.45 Syngnathid fish, including **A.** a pipe fish (*Syngnathus scovelli* (1347)), **B.** a dragon fish, and **C.** a sea horse.

body, the Australian frog *Rheobatrachus silus* (1386) uses gastric brooding (Fig. 5.46). During development, the young secrete a prostaglandin that inhibits the secretion of gastric acid, perhaps ensuring that the female does not recycle her brood. Unfortunately, it is likely that this frog is now extinct in the wild.

Other amphibians take a more conventional approach to vivipary. The toad *Nectophrynoides occidentalis* (1373) lives on Mount Nimba in West Africa. This 15–26 mm toad gives birth to live young, miniature adults (200–220 μg each), after they develop in a uterus. These tiny amphibians are pregnant for nine months. The long period of gestation coincides with seasonal changes, ensuring that young are born after the start of the rainy season.

Vivipary is also a recurring theme among many species of reptiles, having evolved about 100 times in this group. A Brazilian skink, *Mabuya heathi* (1508), produces the smallest known eggs for a reptile (~ 1 mm diameter), and transport from the placenta is responsible for over 99% of the dry mass of young at birth. This situation demonstrates that vivipary in reptiles is more than just females retaining fertilized eggs within their oviducts or uteri, and that it can involve interactions between mother and young that are as complex as those of placental mammals (or the Mount Nimba toad). In another lizard, *Pseudemoia entrecasteauxii* (1518), placental nourishment appears to provide about half of the organic molecules used in growth of the

Figure 5.46 This female frog (*Rheobatrachus silus* (1386)) raises its young in its stomach. In this picture, a froglet is about to leave its mother's body. The mouth serves as the birth canal.

embryos. This situation is accompanied by a general reduction in the quality of yolk in the eggs.

Maintaining embryos within the body of a parent can have important repercussions. In poikilothermic reptiles (= cold blooded), sex determination by temperature is relatively common. In some lizards (e.g., *Niveoscincus ocellatus* (1514)) the time a pregnant female spends basking in the sun influences the sex of her offspring. This may provide females considerable control over the ratio of sons and daughters in the young she produces. This situation may be more acute in viviparous reptiles (e.g., the gekkonid *Hoplodactylus maculatus* (1500)), where females can delay parturition (birth) of young by up to nine months. As in the Mount Nimba toad, the timing of birth appears to reflect seasonal conditions.

In the original (1893) description of the Cretaceous mososaur *Carsosaurus marchesetti*, A.G. Kornhuber proposed that small skeletal elements in the body cavity were gut contents. More recent studies of the original specimen have revealed that the skeletal remains were those of developing embryos (see also Chapter 1). *C. marchesetti* is an early mosasaurid, suggesting that bearing live young was a trait that appeared early in the diversification of these marine reptiles.

In many but not most cases, vivipary is associated with internal fertilization, which may increase the chances of ensuring paternity. Multiple paternity has been demonstrated in the clutches of a large variety of reptiles, including snakes, lizards, and turtles. Therefore the discovery of monandry (mating with a single partner) in six species of viviparous sea snakes is important. Genetic analysis involving 10 highly variable microsatellite loci on 76 embryos obtained from 12 pregnant sea snakes (six species in the *Hydrophis* (1502) clade) revealed that each brood arose from the female mating with one male.

Mammals are often considered to be the epitome of vivipary, setting aside the oviparous monotremes. We do not know at what point in their diversification that mammals adopted vivipary. It is perhaps likely that the Mesozoic diversification of mammals included the appearance of vivipary, presumably after the monotremes had diverged. Other than the monotremes, the rest of the mammals are classified in two groups: marsupials and placentals. There are two obvious

CHAPTER 5 LARVAE AND LIFE CYCLES

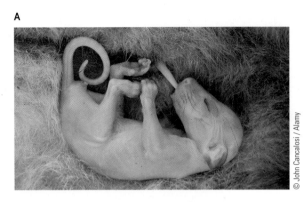

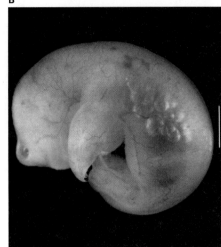

Figure 5.47 Marsupial young. **A.** Brush tail possum (*Trichosurus vulpecula* (1691)) suckling. **B.** A newborn dunnart is the size of a grain of rice (4 mm long, 17 mg), compared to the adult (**C.**). Image B was prepared using computed tomography (CT), using phase contrast X-ray absorption spectroscopy, and generated by taking 1000 images whilst the sample was rotated by 180 degrees.

SOURCE: A., B., and C. Printed with permission of the authors (J.P. Mortola, P.B. Frappell, P.A. Woolley; original publication: *Nature* 397, 161, 1999).

differences between the two groups. First, marsupial young are born at an early stage of development (Fig. 5.47A, B, & D) after a short period of gestation while placentals are born at a more advanced stage after longer *in utero* development. After birth, marsupial young migrate to a pouch where they attach to a nipple (Fig. 5.47C) and complete their development and growth. Second, male marsupials have a bifurcate penis which is distinct from the unbranched penis in male placentals. Both of these differences derive from the pattern of development of ureters, the tubes that conduct urine from the kidney to the bladder to the outside (Fig. 5.48).

During marsupial development, the ureters migrate inside and above the genital ducts, while in placentals the migration is outside and below these ducts. This means that marsupial females have two vaginae while placental females have only one (Fig. 5.48). This arrangement means that during birth, only a very small fetus can pass through the marsupial birth canals. The situation also means that during copulation, the male must deposit sperm in both vaginae, hence the bifurcate penis. The developmental pattern in marsupials probably reflects initial pressure to make the discharge of nitrogenous wastes more effective.

In placentals, blastocysts implant in the wall of the uterus, and the placenta is the interface between mother and young, with mother providing nutrients and removing metabolic wastes from the developing embryo. In marsupials, egg yolk supplies the developing young through the blastocyst stage. After that, the embryo remains encased within three egg coats, and uterine secretions are absorbed across the wall of the blastocyst to fuel its growth.

The ability of a pregnant female to prolong gestation occurred in viviparous amphibians and reptiles, and it is a recurring theme among mammals. Delayed development in mammals can involve the timing of implantation of the blastocyst or the rate of growth and differentiation of the embryo. In marsupials, suckling by a young in the pouch inhibits development beyond the blastocyst stage. Removal of the suckling young reactivates development of the corpus luteum (the hormone-producing follicle from which an egg has been released) and further development of the embryo.

Delayed development is widespread in placental mammals. In short-tailed fruit bats (*Carollia perspicillata* (1787)), development can be halted at the primitive streak stage and suspended for at least six months (Fig. 5.49). The behaviour of the female appears to

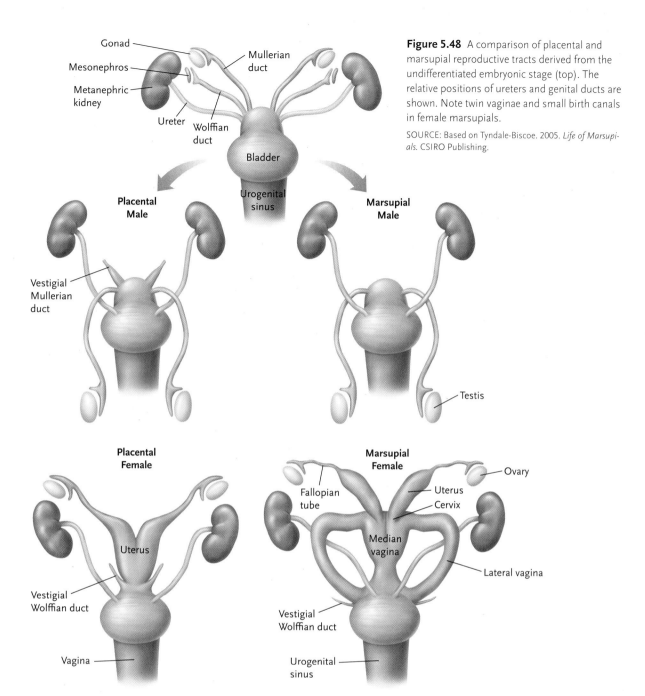

Figure 5.48 A comparison of placental and marsupial reproductive tracts derived from the undifferentiated embryonic stage (top). The relative positions of ureters and genital ducts are shown. Note twin vaginae and small birth canals in female marsupials.

SOURCE: Based on Tyndale-Biscoe. 2005. *Life of Marsupials.* CSIRO Publishing.

influence the delay, and under inclement or stressful conditions there is no further development of the young. These bats are the only mammals known to suspend development at the primitive streak stage.

In species of flying foxes (Pteropodidae) and free-tailed bats (Molossidae), the placenta of females approaching the end of gestation become liver-like in appearance (Fig. 5.50). This specialized part of the placenta is rich in energy and may be used by females to reduce the time they spend flying as they approach the end of the term of pregnancy. This situation could be similar to the disposition of tissues from embryonic pipefish (*S. typhle*, above) into the tissues of their pregnant fathers.

Parturition (birth) is the final phase of vivipary. In mammals, young typically emerge head first from the birth canal; the opposite, tail first, is known as a breech presentation (Fig. 5.51). In bats and cetaceans, breech presentations are the norm. In bats, a breech presentation may minimize the chances of the birth process being halted if the wings of the being-born became tangled. In cetaceans, a breech birth ensures that the mother can continue to provide oxygen to the young until it is virtually clear of the birth canal. At this point the young can be guided to the surface to get its first breath. In most other mammals, head-first presentation ensures that the newborn can breathe during birthing once its nostrils are clear of the birth canal.

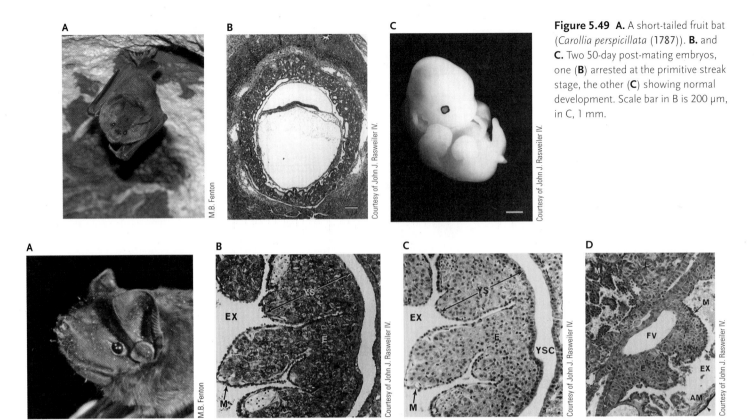

Figure 5.49 **A.** A short-tailed fruit bat (*Carollia perspicillata* (1787)). **B.** and **C.** Two 50-day post-mating embryos, one (**B**) arrested at the primitive streak stage, the other (**C**) showing normal development. Scale bar in B is 200 μm, in C, 1 mm.

Figure 5.50 **A.** Black mastiff bat (*Molossus ater* (1799)) and details of yolk sac showing glandular development. **B.** Dark cells reflect high concentrations of glycogen which has been digested away in **C. D.** Arrowhead at bottom right shows glycogen-rich epithelium. YS = surface of the yolk sac wall; YSC = yolk sac cavity; M = mesothelial cells; EX = exocoelom; E = endoderm; and FV = foetal blood vessel. Magnification is the same in B, C, and D.

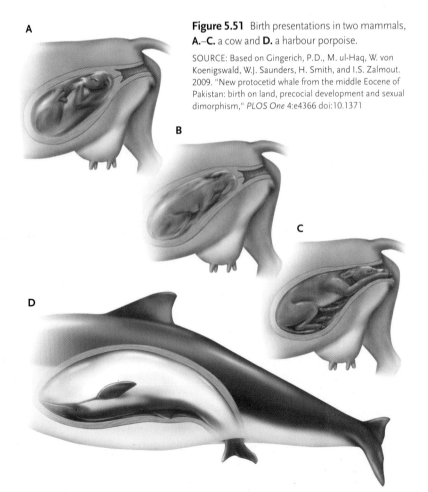

Figure 5.51 Birth presentations in two mammals, **A.–C.** a cow and **D.** a harbour porpoise.

SOURCE: Based on Gingerich, P.D., M. ul-Haq, W. von Koenigswald, W.J. Saunders, H. Smith, and I.S. Zalmout. 2009. "New protocetid whale from the middle Eocene of Pakistan: birth on land, precocial development and sexual dimorphism," *PLOS One* 4:e4366 doi:10.1371

The pattern that emerges in the evolution of vivipary in animals is one of diversity and of the importance of environmental and behavioural factors.

STUDY BREAK

1. What are the important differences between monotreme reproduction and eutherian reproduction?
2. Define vivipary. Name three animals (from different groups) that are viviparous.

5.10 Parental Care

Parental care is fundamental to the success of birds and mammals, but is also a feature of the life cycle of animals in many different groups. Parental care may be viewed as a logical extension to the development of K-selection, the reproductive strategy in which resources are invested the production of a small number of well-provisioned offspring. (Fig. 5.52). If the animal commits major resources to producing just a few offspring, then investment in parental care maximizes the survival of those offspring and the passing of the parents' genes onto the next generation.

Many animals simply release eggs and sperm and parental investment in reproduction is little beyond

Figure 5.52 Hatching eggs.

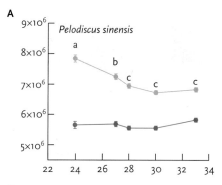

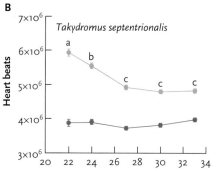

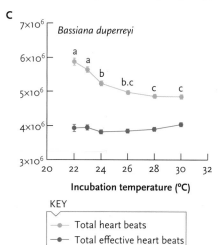

Figure 5.53 A comparison of the numbers of heartbeats to hatching in a turtle (**A.**), and two lizards (**B., C.**). At cooler temperatures, more heartbeats occur before hatching, while at warmer temperatures, cumulative heartbeats is consistent. Closed circles = total heartbeats, open circles = heartbeat rates.

SOURCE: Reproduced with permission from Du, W-G., R.S. Radder, B. Sun and R. Shine. 2009. "Determinants of incubation period: do reptilian embryos hatch after a fixed total number of heart beats?" *Journal of Experimental Biology*, 212:1302–1306. The Company of Biologists Ltd.

the cost of gametes. Eggs, larvae, and young are vulnerable to predators, and sometimes the predators are conspecifics, even their parents. While some vertebrates spend >50% of their lifespan as eggs (e.g., some chameleons), Komodo dragons (*Varanus* sp. (1528)) spend <2%. The time of hatching of poikilothermic reptiles appears to be set by the number of heartbeats, in turn dependent upon prevailing temperature (Fig. 5.53).

A simple parental investment is protecting, and perhaps aiding, egg development. Examples include octopus (Fig. 5.54A) that lay eggs in an enclosed niche and then guard them against predators. The female expels a stream of water from the mantle cavity over the eggs to maintain a flow of oxygenated water around them. Female pycnogonids release eggs into the water where a male fertilizes them. The male then collects the eggs and stacks them into a ball on a modified pair of ovigerous legs (Fig. 5.54B). Pycnogonid males frequently mate with more than one female, and the egg clusters carried by a male may reflect this. Hatched larvae may also be protected, for example in the pycogonids (Fig. 5.54C). Many first instar scorpion larvae do not have a hardened cuticle. These larvae remain with the mother (Fig. 5.54D) and are provided with both protection and food.

By their incubation behaviour, birds can control the general pattern of hatching of the eggs they lay. If a bird lays a clutch of four eggs, one egg a day, and starts incubation on day 1 (first egg), the young will hatch over a four-day period. If it starts incubation on day 4 (last egg), all the young will hatch at about the same time. A spread in the age of young can be advantageous under adverse conditions because older chicks have a competitive advantage over their younger (smaller) siblings. This can ensure that at least one chick fledges.

The bromeliad land crab *Metopaulias depressus* (822) raises its young in water trapped at the leaf bases of bromeliads such as *Aechmea paniculigera* (Fig. 5.55). About 50 larvae are released into an axil, and they take 9–10 days to develop into young crabs. If unprotected, more than half of these larvae would be lost to preda-

tion by nymphs of *Diceratobasis macrogaster* (853), a damselfly (Odonata) larva that also develops in the bromeliad's trapped water. The female crabs destroy the damsel fly larvae, reducing the crab larval loss to predation to less than 25%. Female crabs also manage the physicochemical conditions in the bromeliad axils. In the absence of parental care, the water becomes hypoxic and acidic, and contains low levels of calcium. Females oxygenate the water and maintain both pH and calcium levels by collecting empty snail shells and adding them to the water in the axil.

In even the least complex social insect colonies (e.g., *Polistes humilis* (1006), a paper wasp, Figure 5.56), larvae are totally dependent on adult members of the colony for food and care. Adults forage for a broad variety of foods, including meat (often caterpillars), fruit, and nectar. Droplets of regurgitated food are supplied to the larvae as they grow.

Honey bee (*Apis mellifera* (984)) colonies (Fig. 5.57) are much larger and more complex than the paper wasp

Figure 5.54 A. *Enteroctopus dofleini* (495) (the giant Pacific octopus) guarding strings of eggs.
B. Male *Nymphon* sp. (641) carrying an egg mass and C. the mass of young after the eggs hatch.
D. Scorpion *Parabuthus* sp. (693) carrying young.

colonies. There may be up to 80 000 worker bees in a colony with a single queen, and worker bee activities are regulated by their age (see page 492). Newly emerged worker bees spend about their first three weeks of adulthood without leaving the hive. Older bees guard the entrance to the hive and act as foragers, collecting nectar (concentrated in the hive to become honey) and pollen. These food reserves are stored in hive regions away from the cells in which eggs are laid and larvae develop. When they first emerge, young adult bees spend their first few days cleaning cells in the hive; after that they take on the duties of brood care. During the first two weeks of adulthood, worker bees have functional hypopharyngeal glands that secrete protein- and carbohydrate-rich royal jelly. These glands degenerate in older bees. Some royal jelly is fed to worker and drone (male) larvae, and it is the sole food of larvae destined to become new reproductive queen bees. Other larvae are fed mostly on pollen, supplemented with honey. Food-deprived larvae signal their need for food to workers, although the signal mechanism (behavioural or pheromonal) has yet to be identified.

The consistent provision of high-quality food to developing young can increase their growth and development rates, perhaps minimizing their vulnerability

Figure 5.55 A. A bromeliad crab, *Metopaulias depressus* (822).
B. *Aechmea paniculigera*, a bromeliad host to the bromeliad crab's reproduction.

Figure 5.56 A *Polistes humilis* (1006) colony. The nest is founded by a single queen who lays eggs and feeds the first batch of larvae as they develop. Newly emerged wasps are potentially fertile but act as workers, raising the subsequent batches of brood and enlarging the nest. If the queen is lost, one of the workers becomes fertile and takes over as queen.

Figure 5.57 A. A worker bee (*Apis mellifera* (984)) cleaning a cell in a colony. **B.** A queen bee, surrounded by worker bees. **C.** Honey bee larvae at different stages of development in the cells of a colony.

to predators. It is common for bird species in which adults eat fruit or nectar to feed their young protein-rich food such as insects to promote growth. While most birds bring food to their young, female mammals provide milk, a rich source of food that represents a high cost investment by mothers. Although we associate milk with mammals, many animals provide milk-like food to their young. Discus fish (Fig. 5.58) produce milk-like mucus on their skin, which is eaten by their young. The same is true of at least 30 other species of cichlid fish, and it also occurs in some catfish (Siluriformes, Bagridae).

Caecilians (Amphibia, Apoda) that show direct development (eggs hatch as miniature adults) feed their young by dermatophagy (Fig. 5.59). At birth, these caecilians are relatively large even though their eggs have relatively small supplies of yolk. The young use specialized teeth to feed on the mother's skin, and, in some cases, the young also consume oviductal fluid after birth. It is interesting to note that dermatophagy appears in both African and South American caecilians, even though the two lineages have been isolated for at least 100 million years.

Pigeons, birds in the order Columbiformes (~300 species), are the only birds known to produce "milk," produced by the germinal epithelium in the birds' crops. Crop milk is produced by both males and females. It is rich in proteins and lipids and contains a factor that promotes growth. For the first four days post hatching, young pigeons are fed only crop milk. Another distinctive feature of columbiform birds is that females always lay two eggs. Research involving

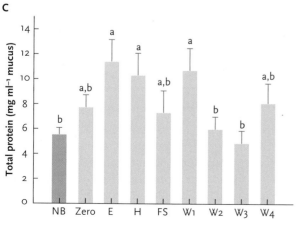

Figure 5.58 A. Discus fish (*Symphysodon* spp. (1289)) produce mucus on their skin surfaces that **B.** is consumed by their young. **C.** Total production of protein in mucus of nonbreeding (diagonal striped bar) and breeding discus fish (open bar) over the breeding cycle. Different letters above the bars denote significant differences. At week 3 (W3), protein in the mucus drops as young become free-swimming. Letters on X axis: NB = no reproduction; E = embryo; H = hatch; FS = free swimming.

SOURCE: C. Reproduced with permission from Buckley et al., "Biparental mucus feeding: a unique example of parental care in an Amazonian cichlid," 2010, *Journal of Experimental Biology*, 213, 3787–3795. The Company of Biologists Ltd.

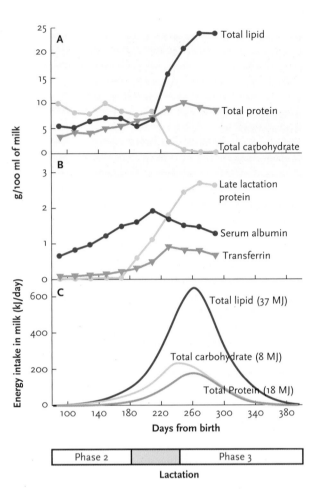

Figure 5.59 After birth, young caecilians are dermatophagous, taking skin and secretions from their mother.

SOURCE: Reprinted by permission from Macmillan Publishers Ltd: *Nature*, Alexander Kupfer, Hendrik Muller, Marta M. Antoniazzi, Carlos Jared, Hartmut Greven et al., "Parental investment by skin feeding in a caecilian amphibian," Vol. 440, pp. 926–929, copyright 2006.

Figure 5.60 Changes in milk composition during lactation in *Macropus eugenii* (**1689**), the tammar wallaby. Shown are changes in (a) carbohydrates, (b) major components of whey proteins, and (c) energy content.

SOURCE: Figure 2.19 in Tyndale-Biscoe, 2005, Life of Marsupials, CSIRO Publishing, http://www.publish.csiro.au/pid/4781.htm, based on published data from KR Nicholas 1988 and SJ Cork and H Dove 1989. Used by permission.

manipulation of clutch size demonstrated that adding a third nestling to Mourning Dove (*Zenaida macroura* (1573)) nests decreased growth rates of all nestlings. This evidence indicates that increased investment (crop milk) means a lower clutch size in birds. Kiwis of New Zealand (Apterygiformes, four species) are the only other birds with a consistent clutch size of two, but there is no evidence of Kiwis producing crop milk.

In mammals, the initial release of milk, **colostrum**, is rich in cytokines, growth factors, hormones, and immunoglobulins in addition to the nutritive content. The milk of monotremes and marsupials differs markedly in composition across the period of lactation (Fig. 5.60). Following the release of colostrum, milk composition in placental mammals is quite consistent across the period of lactation.

Lactation in mammals is usually considered to be a trait of females, and its costs mean that, compared with males, females pay higher costs of reproduction. There are records of male placental mammals lactating, usually domesticated mammals. Dyak fruit bat males (*Dyacopterus spadiceus* (1789)) from Borneo are an exception. While these males lactate, they produce much less milk than females. When discovered, the phenomenon attracted a great deal of attention, but we remain unclear about the behavioural implications of lactating males in this species.

Placental mammals show at least two patterns of nursing. In many species young are nursed more or less continuously, and nursing behaviour is initiated by the young. In species that leave their young so they can feed or hunt, e.g., some hares, females nurse the young for a brief period (3–5 min) once a day. Behaviour of nursing young is associated with rises in the level of **ghrelin**, a hormone associated with initiating feeding behaviour.

STUDY BREAK

1. How do bromeliad crabs demonstrate parental care?
2. How does the milk fed to young vary between mammalian groups?
3. Why do females often invest more parental care in their offspring than males do?

5.11 Dormancy

Dormancy during adverse environmental conditions is a recurring phenomenon in animal life cycles. Dormancy is any form of resting stage, regardless of how it is induced or terminated. It involves a suspension in normal life activities, reduced levels of metabolism, and suspended development. Dormancy under endogenous control is called **diapause**, and when under exogenous controls, **quiescence**.

Many organisms regularly face environmental conditions ranging from inclement to harsh. There can be seasonal changes associated with winter or dryness, or daily changes such as tidal cycles that leave some animals high and dry and perhaps frozen for parts of every day. Some animals avoid harsh conditions by moving away from them, perhaps by migration (see page 482). This allows the migrant to continue to enjoy favourable weather. Other animals do not migrate, but seek shelter from harsh conditions. Animals facing prolonged periods of subfreezing temperatures may simply move below the frost line. By staying dormant in places such as the mud in a lake bottom, many animals reduce levels of activity and metabolism but avoid freezing. Some animals that are exposed to subfreezing conditions produce antifreeze, usually carbohydrates such as glycerol, that prevents their body fluids and tissues from freezing.

For many animals, dormancy means surviving a period of actually being frozen. This often means supercooling, which works only if there are no nucleators to precipitate the formation of ice; water as ice expands in volume and can fracture cell membranes and components of cells. Animals that use supercooling must have mechanisms for eliminating nucleators from their cells.

Other animals become dormant to avoid heat and dessication. Aestivation is a means of surviving long hot and dry periods. For some animals, aestivation means living inside a waterproof cocoon to minimize loss of water to the environment. Large stores of fat may allow aestivating animals to use metabolic water to offset any losses to dehydration.

As we will see in Chapter 17, many animals remain active throughout the year regardless of inclement conditions of temperature and humidity. These are mainly homeothermic (endothermic) or warm-blooded animals such as birds and mammals. Remaining active during winter is expensive and is only viable for animals with a dependable food supply. Other warm-blooded animals hibernate.

5.11a Insect Diapause

Diapause is typically a suspension of activity and development in insects, and can occur in embryos, larvae, pupae, or adults. Diapause may be directly related to environmental conditions (facultative diapause) or may have become a fixed part of the life cycle (obligate diapause).

Facultative diapause is not a direct response to environmental conditions; its onset is triggered by a cue that precedes environmental changes to come. For example, in many temperate insect species, diapause is adaptive to survival through the harsh conditions of winter. In the fall, before cold conditions

begin, the days shorten. This change in day length triggers the onset of diapause. Such a photoperiod trigger is species specific and genetically determined (Fig. 5.61).

Significant activity may continue after the diapause trigger's reception. A dramatic example is the migration of monarch butterflies (*Danaus plexippus* (1031)) after the reduced day length of early fall. A different example is the larva of *Chironomus riparius* (959) (Diptera) that continues activity while the imaginal discs (see above) pause their development at a particular stage.

Diapause can be divided into three phases:

1. **prediapause:** the period between the cue and the onset of actual diapause
2. diapause
3. **postdiapause:** a period in which an appropriate trigger initiates normal activity and development.

If a day-length cue has initiated prediapause and the insect has entered diapause before the onset of cold conditions, then exposure to warm conditions will not end diapause. In postdiapause, such an environmental temperature trigger will initiate the termination of diapause.

While hormones are involved in diapause control, there is considerable variation in different species. In the silk moth (*Bombyx mori* (1028)), a 24-amino acid neuropeptide, released from the suboesophageal ganglion, regulates embryonic diapause. Gypsy moth (*Lymantria dispar* (1040)) diapause comes when the first instar larvae are ready to hatch from the eggs. It is accompanied by high levels of the ecdysteroid moulting hormones. Larval and pupal diapause frequently involves interruption of the brain–prothoracic gland pathway, resulting in the prothoracic gland not

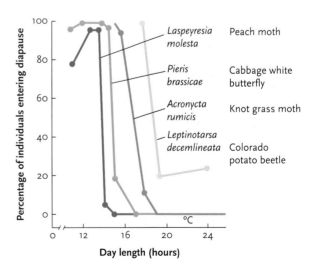

Figure 5.61 Insect photoperiod responses. These four insect species show an abrupt change from active to diapause forms at day lengths between 14 and 17 hours.

SOURCE: Gilbert SF. *Developmental Biology*, 5th edition. Sunderland (MA): Sinauer Associates; 1997. Reprinted by permission.

producing the ecdysteroids needed for development. In other lepidopterans, pupal diapause involves elevated juvenile hormone levels.

One of the roles of diapause is to tailor the life cycle of an insect to match periods of activity and reproductive demand to the time of the greatest abundance of nutrients. At the same time, metabolic activity does not completely stop during diapause (Fig. 5.62). Metabolic rate during diapause is temperature dependent (Fig. 5.62B). The temperature at which a larva is maintained through diapause may also affect the metabolic rate of larvae emerging from diapause.

Most insects do not feed during diapause, which may last periods of months to years. Therefore, nutrients must be stored in the prediapause period to supply the demands of both the diapause period and the immediate postdiapause resumption of development and activity. In prediapause, the fat body increases the synthesis of proteins that are normally made and used immediately; these are now released and stored in the hemolymph. This protein store appears to be the source of amino acids required for the metabolic synthesis in the postdiapause period. Trehalase activity and synthesis of the disaccharide trehalose are features of the prediapause period in many insects. While trehalose has been suggested to be a cryprotectant, it seems more likely that it is converted to glycogen, a source of glycerol and sorbitol in the diapause period. The other major activity in the prediapause period is the synthesis of heat shock proteins; these are interpreted as being chaperones, protecting other proteins and also, in some species, as cryoprotectants.

5.11b Cryptobiosis

Cryptobiosis (literally "hidden life") occurs in many ways in different groups of animals. It has been described in nematodes, rotifers, insects, crustaceans, and tardigrades. Cryptobiosis is distinguished from other forms of quiescence by the extent of reduction of metabolism. While torpor or dormancy may reduce an animal's metabolic rate to between 30% and 5% of active levels, an animal in the cryptobiotic state has a metabolic rate of well under 1% of normal (one experimental measure in tardigrades reports O_2 consumption of 1/600 normal).

A number of different environmental triggers may induce cryptobiosis, which has been subdivided into **anhydrobiosis**, **cryobiosis**, **anoxybiosis**, **osmobiosis**, and **chemobiosis**. Of these, anhydrobiosis and cryobiosis are the best studied. The problems that surface area:volume ratios introduce as animals get larger are considered on page 25. Small terrestrial animals face

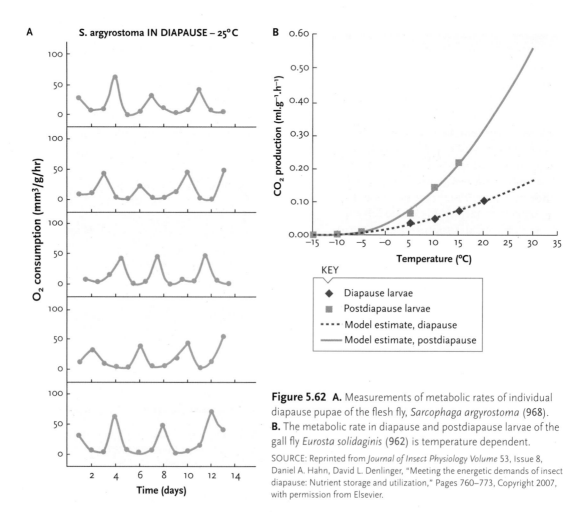

Figure 5.62 A. Measurements of metabolic rates of individual diapause pupae of the flesh fly, *Sarcophaga argyrostoma* (968). **B.** The metabolic rate in diapause and postdiapause larvae of the gall fly *Eurosta solidaginis* (962) is temperature dependent.

SOURCE: Reprinted from *Journal of Insect Physiology Volume* 53, Issue 8, Daniel A. Hahn, David L. Denlinger, "Meeting the energetic demands of insect diapause: Nutrient storage and utilization," Pages 760–773, Copyright 2007, with permission from Elsevier.

a rather different problem; they have a large surface-to-volume ratio and so, under dry conditions, face dehydration. Investigation of anhydrobiosis has focused on *Artemia* (brine shrimp, Crustacea), nematodes, and tardigrades (Fig. 5.63). An animal the size of a small tardigrade, with a cuticle having similar water permeability to a cockroach (unlikely since the tardigrade cuticle would be much thinner), at 90% relative humidity, has been calculated to lose 40% of its water (a lethal level) in five hours. What this means is that in the real world, a small terrestrial animal must either live in a permanently saturated microhabitat or have anhydrobiotic ability.

Many texts include the erroneous observation that when tardigrade specimens in dried moss were rehydrated, they showed full recovery 120 years after the moss was collected. All that was actually reported was movement of a leg. In reality, tardigrade survival in an anhydrobiotic state is species related and does not exceed 10 years, with 1–2 years being more typical (Fig. 5.64). It seems likely that the limit to survival lies in the level of accumulated damage to DNA (single strand breaks) exceeding the ability of the system to effect repair.

The anhydrobiotic state of a tardigrade, known as a tun (Fig. 5.63), is formed after controlled permeability changes in the cuticle. As a tun forms, the cuticle becomes opaque, and the body shortens and thickens as the legs are withdrawn and areas of flexible procuticle (which are significantly more permeable than the harder regions) are folded inwards. Tuns are not formed if tardigrades are dehydrated too quickly, placed under anoxic conditions, or anaesthetized.

In the tun state, metabolic rates are barely measurable (Fig. 5.65). There is argument as to whether this low metabolic rate is genuine or a result of other oxidative reactions. Evidence favouring the latter view is the survival of tuns after cooling to –272.8°C (a temperature at which molecular free energy is negligible) or after vacuum drying (which removes any water that would be needed for metabolic activity).

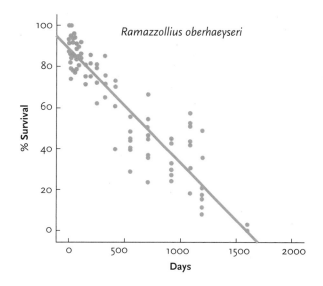

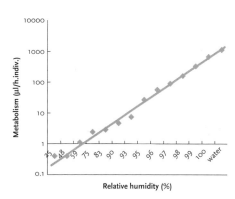

Figure 5.64 Survival of different species of moss-dwelling tardigrades under anhydrobiotic conditions.

SOURCE: L. Rebecchi, T. Altiero and R. Guidetti, "Anhydrobiosis: the extreme limit of desiccation tolerance," *Invertebrate Survival Journal* 4: 65–81, 2007.

The absence of free water and metabolic activity underlies the resistance of anhydrobiotic tardigrades to extreme cold and to agents such as H_2S, CO_2, methyl bromide, and organic solvents. Tuns of *Macrobiotus* (620) and *Echiniscus* (618) are resistant to pressures of 600 MPa (95 and 80% survival respectively), compared with 90% survival of active specimens after exposure to

Figure 5.63 Tardigrades (*Paramacrobiotus richtersi* (622)). **A.** Free living and **B.** tun stage.

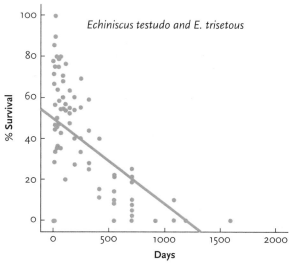

Figure 5.65 The effect of relative humidity on metabolic activity of tuns of *Macrobiotus hufelandi* (620).

SOURCE: Reprinted from *Zool. Anzeiger* 240, Jonathan Wright, "Cryptobiosis 300 Years on from van Leuwenhoek: What Have We Learned about Tardigrades?" Pages 563–582, Copyright 2001, with permission from Elsevier. Zoologischer Anzeiger by Carus, Julius Victor ; Deutsche Zoologische Gesellschaft Reproduced with permission of URBAN UND FISCHER VERLAG in the format reuse in a book/textbook via Copyright Clearance Center.

100 MPa pressure. Adaptation to anhydrobiosis appears also to confer considerable resistance to radiation in tardigrades, nematodes, and the brine shrimp *Artemia* (715). A dose of 5 Gy is normally lethal to humans. (The Gy unit, the gray, is 1 joule of radiation absorbed by 1 kg of tissue. It replaces the older unit, the rad; 1 rad is 0.01 Gy). Tests on *Macrobiotus* show an LD_{50} to X-rays of around 5000 Gy, with anhydrobiotic animals having a slightly higher tolerance than active animals, and a survival to X-ray levels of 1000 to 7000 Gy (although doses above 1000 Gy sterilize the animals).

Given the ability of tardigrades to survive adverse environmental conditions, they are obvious candidates for tests of life forms to survive in either simulated or real conditions of space. These tests have shown an ability to survive high doses of UV radiation and some ability to survive cosmic radiation and microgravity.

Trehalose metabolism is closely linked to tardigrade tun formation (paralleling its involvement in insect diapause). Trehalose levels increase before a tun is formed and may continue for several hours after tun formation. In different species, trehalose levels of 23× base and from 0.2% dry weight to 13% dry weight have been recorded. Trehalose is rapidly metabolized on rehydration. Stress-induced proteins, including heat shock proteins, are another accompaniment to tun formation.

STUDY BREAK

1. What are the characteristics of (a) diapause and (b) cryptobiosis?
2. How are the formation and maintenance of a tun stage, and the recovery from it, regulated?

LARVAE AND LIFE CYCLES **IN** PERSPECTIVE

The wide range of topics and examples in this chapter illustrate both the variability and importance of larval development to the success of species in their environments. One of the features of animal life cycles is the amazing plasticity of larval development; species as closely related as being in the same genus may have radically different development paths (e.g., direct development in one *Littorina* species compared to a planktotrophic larva in a congeneric living in the same place). There are many points at which material in this chapter links with examples in Chapter 14 (Reproduction). You might look for these links when you have studied the material in the two chapters.

Questions

Self-Test Questions

1. Which of these characteristics belong to true "larvae"?
   a. Juvenile forms that hatch as miniature adults, such as those of many ecdysozoans
   b. Morphological elements that may be eliminated during development
   c. Functional reproductive structures that are modified in the adult form
   d. Modifications of the adult form, triggered by environmental or physiological changes

2. What cues are involved in determining where planktonic larvae eventually settle?
   a. Physical features of the substratum, including the presence of conspecifics or other species
   b. Small peptides that may act to trigger further development of the settled larva
   c. Biofilms that include bacterial, fungal and diatom species
   d. All of the above

3. Which of the following larval types is LEAST related to a mechanism of providing nutrients to support its development?
   a. Glochidia
   b. Herbivorous insect larvae
   c. Lecithotrophic larval forms
   d. Nauplius larvae

4. How does metamorphosis in holometabolous insects differ from that which occurs in flounder?
   a. There is a profound change in form and function of the body in insects.
   b. In flounder there is a change in the environment to which the animal is adapted.
   c. Ecdysone and juvenile hormone control insect metamorphosis, whereas thyroid hormone modulates the changes in flounder.
   d. Only exogenous factors influence the process in flounder.

5. Which of these statements applies to viviparity?
   a. Internal fertilization must occur.
   b. In non-mammalian species, viviparity consists simply of the eggs being retained to develop within the mother, without any further interaction.
   c. Either parent may be involved in providing a body chamber in which gestation of the young occurs.
   d. All vertebrate groups have at least some members that demonstrate ovoviviparity, histotrophic viviparity, or hemotrophic viviparity.

Questions for Discussion

1. What factors lead to the development of complex life cycles?
2. How should larval and adult features be combined in the development of ideas of phylogeny?
3. What are the advantages of males being involved in care and feeding of eggs and young? If paternal involvement is advantageous, why is it not more common?
4. Use the electronic library to find a recent paper that explores some aspect of the biology of animal larvae and life cycles. Does the new information revolutionize any of the information in this chapter?

Figure 6.1 *Euhelophus* (1454), a long-necked sauropod dinosaur.

SOURCE: DiBgd

6 History of Animals: Fossils

WHY IT MATTERS

Understanding history is said to be the key to understanding the present and predicting the future. Surely the history of extinctions, and of their associations with climate change, is of obvious importance in today's world. But in the following example we go the other way and apply modern knowledge of physiology and energetics, and of their associations with morphology, to examine a puzzle from the past.

Long-necked sauropod dinosaurs, such as *Euhelophus* (1454) (or the more famous *Brachiosaurus* (1452), previously known as *Brontosaurus*), are thought to have browsed on the upper leaves of trees, giving them access to plant biomass unavailable to shorter herbivores. Evidence from morphology (neck length and tooth and jaw morphology) demonstrates how different species of sauropods could have exploited different feeding zones and avoided interspecific competition. But this "heads up" hypothesis for sauropod feeding is not supported by other interpretations of neck vertebrae and associated flexibility. The estimated costs of maintaining adequately high blood pressure to the brain when the head was elevated (up to 9 m above the heart in the largest forms) appears to support the view that these animals had to keep their heads down when feeding.

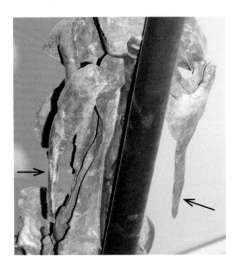

Figure 6.2 The diplodocid sauropods had long necks and long tails. Seen in ventral view, a pair of prominent cervical ribs (arrows) flanked the cervical vertebrae, for example those on *Barosaurus lentus* (1451). The black pipe supports the vertebral column of this specimen on display at the Royal Ontario Museum.

SOURCE: M.B. Fenton

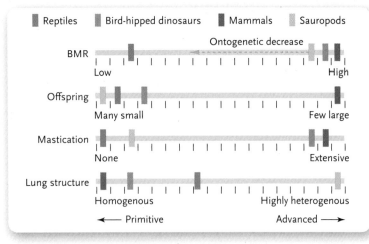

Figure 6.3 The relative size of two of the largest sauropods (*Argentinosaurus* (1450) and *Brachiosaurus* (1452)) compared with an African elephant, a giraffe, an extinct mammal *Indricotherium*, two other dinosaurs (duck-billed *Shantungosaurus* (1456) and horned *Triceratops* (1446)) and a Galapagos tortoise. The inset table (colour coded to silhouette) compares BMR (basal metabolic rate), offspring production, mastication, and lung structure among reptiles, bird-hipped dinosaurs, mammals, and sauropods.

SOURCE: From Sander, P.M. and M. Clauss, "Sauropod gigantism," *Science* 322, 2008:201–202. Reprinted with permission from AAAS.

Biomechanical modelling indicates that *Euhelophus* had a relatively inflexible neck. Consequently, like modern giraffes, long-necked sauropods probably kept their necks straight while raising their heads. Very long cervical ribs (Fig. 6.2) would have allowed efficient transmission of forces over a long distance, allowing concentration of neck muscle mass toward the base of the neck. This arrangement meant that when walking, *Euhelophus* and *Brachiosaurus* kept their heads low, but when feeding could have raised their heads to 40° above the horizontal, contradicting the earlier suggestion. In *Euhelophus* the estimated energetic cost of walking 100 m equalled the cost of maintaining blood pressure in the brain when browsing for 11.6 min with its head up compared to 32.2 min with its neck horizontal. Equivalent numbers for *Brachiosaurus* were 3.8 min and 12.9 min. These animals appeared to have browsed with their heads up!

Unlike herbivorous mammals and duck-billed and horned dinosaurs (Fig. 6.3), very large sauropods lacked an efficient means for grinding up plant food, either the specialized grinding teeth or the gastroliths (see page Chapter 8) present in other herbivorous dinosaurs. Their huge size would have meant a larger gut capacity and longer times to digest food. Size and time for digestion, in turn, would have required the animal to eat large amounts of plant material. Therefore, in situations where food was more readily available farther above the ground, reaching up to eat would have given long-necked sauropods access to the large amounts of food they required. This would have been especially true if other herbivores were consuming foliage closer to the ground. This interpretation supports the heads-up model.

SETTING **THE SCENE**

In this chapter we view the history of animals through the lens of the fossil record. Our coverage includes the ages and types of fossils, as well as the importance of different techniques in revealing the details of animal structure. We consider extinctions and their role in evolution. Then we sample fossil communities from the Ediacaran, Cambrian, Carboniferous, Jurassic, and Eocene to illustrate interactions among animals. We also use the time scenarios to explore events in evolution. We finish with five specific examples from the fossil record, namely animals moving onto land, insects mimicking leaves, Cretaceous mosquitoes, animal size, and the evolution of birds. As you proceed through the chapter, be alert for breakthrough events and structures and interactions involving animals and plants, or among animals. Be aware of the impact that the fossil record has had on the development of phylogenies of animals.

6.1 Introduction

Paleontologists are people who study fossil organisms. Some of their discoveries attract press coverage because they are so spectacular. Many people are entranced by dinosaurs and other charismatic fossils, especially children, an obvious reality to anyone visiting a museum during school break and trying to get close to exhibits of dinosaurs and other fossils. Cartoonists also use fossils to their advantage, and fossils provide biologists with often astonishing glimpses of the histories of the animals and plants they study today.

The word **fossil** comes from the Latin *fossus*, meaning "dug up," and describes any trace or preserved material from the past. Although living organisms have inhabited Earth for over three billion years, most fossils that have been found and described are much, much younger. The diversity of fossils only increases sharply about 600 Ma (million years ago). Fossils can inform biologists about the global distributions of animals and plants, and allow us to track changes in morphology and appearance that have occurred over time. Details of the morphology of fossils often allow biologists to assemble and test ideas about the evolutionary history of specific groups of animals and plants.

6.1a Ages of Fossils

The age of a fossil is a critical piece of information. Many fossils occur in sedimentary rocks. The layer of rock in which it is found can indicate a fossil's relative age (younger than fossils in layers below it, older than those in layers above it). More absolute information (approximate age, usually in millions of years or Ma) about the age of a fossil comes from radiometric analysis of isotopes (Box 6.1). Isotopic analysis also can provide information about the temperatures under which sediments were deposited. Studies of sequences of rock strata collected from around the world have provided a **geological time scale** (inside back cover) that can be associated with both the relative and absolute ages of fossils. Throughout this chapter we will provide information about the age of fossils, placing them in the context of the geological time scale.

6.1b What Paleontologists Do

Paleontologists work with fossils, using their wide knowledge of organisms to describe and understand the all-too-often fragmentary fossil record. This fragmented nature is illustrated by the case of *Nectocaris pteryx* (510) which was originally known from a single specimen. The original specimen (Fig. 6.4A) appears to have a delicate fin supported by rays as well as a possible notochord and a chordate-like gut. These features, combined with apparent muscle segments

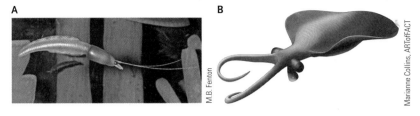

Figure 6.4 An early (**A.**) and a more recent (**B.**) reconstruction of *Nectocaris pteryx* (510).

(myomeres), a large pharyngeal cavity, and tentacles around the mouth, suggested that *N. pteryx* was a chordate. However, the creature had two stalked (pedunculate) eyes which are rare among chordates and some paleontologists suggested that *Nectocaris* was a crustacean.

Later, information from 90 additional specimens of *Nectocaris* revealed that it had an open axial cavity housing paired gills, wide lateral fins, a pair of long tentacles, the eyes on stalks, and a flexible anterior funnel (Fig. 6.4B). There was no notochord. *Nectocaris* is now classified as a cephalopod mollusc, extending the fossil record of cephalopods by 30 million years and challenging the view that the ancestral members of this group were shelled.

The story of *Nectocarus* demonstrates how more specimens and information can change interpretations of the evolutionary record and reminds us that a thorough knowledge of anatomy is essential for knowing what to look for in fossils. The *Nectocaris* example is not unique: the first fossil ichthyosaurs (marine reptiles) were originally thought to be the remains of fish, and the interpretation of *Anomalocaris* is discussed in Section 6.3a. The challenges of accurately identifying fossils, along with the gaps in the fossil record, are sometimes cited as evidence contrary to the phenomenon of evolution. In fact, the discovery of new information often changes and refines our view of natural systems, but does not undermine the reality of evolution.

STUDY BREAK

1. What is the difference between absolute and relative age of fossils? How can this confuse our knowledge of the fossil record?
2. How does *Nectocaris* illustrate the challenges of interpreting fossil material?

6.2 Types of Fossils

Although many fossils occur in sedimentary rocks, fossils have formed in a variety of ways and situations.

Permineralization occurs when ground water containing dissolved minerals penetrates a buried

BOX 6.1
Molecular Clocks and Thermometers

How are dates in geological time established? For example, the Cambrian era began 540 Ma, dinosaurs became extinct 65.5 Ma, and Lucy (*Australopithecus afarensis* (1852)), a fossil hominid, lived 3.2 Ma. The answer lies in one (or sometimes more) techniques of radiometric, or radio-isotope, dating.

Carbon-14 (^{14}C) dating may be the most familiar example of using a molecular clock to measure time. However, ^{14}C dating has more application in putting dates to human changes than to the longer time span of the fossil record.

Carbon-14 is formed in Earth's atmosphere by the collision of radiation-generated neutrons with a nitrogen atom (7 protons, 7 neutrons); the release of a proton generates ^{14}C (6 protons, 8 neutrons). The ^{14}C atoms can combine with oxygen to produce carbon dioxide that may be taken up by plants, through photosynthesis, and passed up through the food chain. The ratio of normal carbon ^{12}C to ^{14}C in air and in life forms is a constant at any moment (with a ^{12}C to ^{14}C ratio of about a trillion to one). Once an organism dies, the ^{14}C in its body gradually decays back to ^{12}C, with a half-life of 5730 years (a half-life is the time taken for half of the molecules in a given sample to decay). Compared with some other isotopes, this is a fairly rapid rate of decay: after several half–lives, very little ^{14}C remains in a sample. Therefore the older the biological material, the greater the error in measuring $^{12}C:^{14}C$ ratios. In practice, ^{14}C dating is limited to dates <50 000 years.

Potassium-argon dating is also applied to the fossil record. Potassium has two stable isotopes (^{41}K and ^{39}K) and a radioactive form (^{40}K) that has a half-life of 1250 million years, decaying to form ^{40}Ar and ^{40}Ca. While 89% of the decay product is ^{40}Ca, this isotope is too common in possible contaminants to be useful. Therefore the

11%—the inert gas, argon—trapped in rocks when they solidify can be usefully measured and provide dates (Ma). The ^{40}Ar accumulates with age rather than being reduced, and the potassium-argon method can be used on rocks from a few 100 000 years to 4 billion years old.

A variant on this technique is the $^{40}Ar:^{39}Ar$ method in which a sample is put in a neutron beam to convert ^{39}K to ^{39}Ar. As ^{39}Ar has a short half-life, it cannot be present in the sample. The $^{40}Ar:^{39}Ar$ ratio allows calculation of the sample's age. This method, with the refinement of the use of a laser beam to melt single rock crystals, was used to establish the age of the Ethiopian rocks from which Lucy was collected.

In 2011, a report extending the use of laser ablation to *in situ* uranium–lead (U–Pb) dating provided evidence that some dinosaurs survived into the Paleocene. Specifically, two dinosaur bones (one fragment, the other from a sauropod *Alamosaurus sanjuanensis*) from New Mexico were dated at 73.6±0.9 Ma and 64.8±0.9 Ma, respectively. The first date agrees closely with a previous $^{40}Ar/^{39}Ar$ date of 73.04±0.25 Ma. The second date, placing the sauropod in the Paleocene, agrees with other data, including palynologic, paleomagnetic, and evidence from fossil mammals.

Oxygen comes in two isotopes, ^{16}O and ^{18}O. The relative amount of these two isotopes, present in water, varies with climate. Molecules of H_2O containing ^{16}O evaporate less readily than those with ^{18}O. Conversely, ^{18}O molecules condense more readily than ^{16}O, the lighter isotope. Measurements of $^{16}O:^{18}O$ ratios are made relative to the ratio of the isotopes in oceanic water collected 200–500 m below the surface and are expressed as percentage change [$\partial^{18}O$] relative to this standard. $\partial^{18}O$ decreases in annual precipitation compared with

the average annual temperature, reaching about −5% $\partial^{18}O$ in Antarctica. Remember that ^{16}O water evaporates more readily, so vapour taken up near the tropics is ^{16}O rich. As this vapour cools and moves towards the poles, ^{18}O is preferentially condensed. By the time the vapour reaches the poles the ^{18}O has been reduced. Therefore, the $^{16}O:^{18}O$ ratio in water trapped in glaciers, ice sheets, and the remains of plants and animals preserves a record of climatic conditions at the time (Fig. 1).

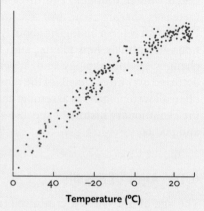

Figure 1
The concentration of ^{18}O in precipitation is lower in cold climates. Cold sample sites (Antarctica and Greenland) have about 5% less ^{18}O than samples from warm sites.

SOURCE: Adapted from Jouzel, et al. From Paleclimatology: the Oxygen Balance, NASA Earth Observatory

The **clumped isotope method** is an extension of this principle used to measure paleoclimate from isotope ratios in carbonate materials. Heavier isotopes, ^{14}C and ^{18}O, have a thermodynamic tendency to form bonds with each other at greater than random frequency. This tendency is increased at low temperatures, meaning that the ratio of heavy isotopes to lighter ones, say in CO_2 released from a carbonate by acid treatment, can provide a measurement of the temperature at which the carbonate (perhaps the shell of a bivalve or brachiopod) was formed.

BOX 6.2 People Behind Animal Biology

Stephen J. Gould and Simon Conway Morris

The fragmented nature of fossils lends itself admirably to the process of science. Whether we pick *Anomalocaris* (630) or *Nectocaris* (510), the first descriptions used the information available at the time. As additional specimens were discovered and described, our views of these animals changed. Changes in the classification of *Anomalocaris* and *Nectocaris* do not reflect poorly on the original workers, rather they demonstrate that it is easy to be misled by partial information, however experienced, expert, and well-intended the researchers. Each of us brings our history, experience, and view of life into our interpretation of fossil (or living) organisms. This can put individuals and their ideas into conflict.

The Burgess Shale has provided astonishing glimpses of life as it was ~500 million years ago. In 1998, Stephen Jay Gould and Simon Conway Morris published a sharp exchange of views on the Burgess Shale in the magazine *Natural History* [107:10 p. 48 (1)]. These differences were evident in their books about the Burgess Shale: Gould's *Wonderful Life: The Burgess Shale and the Nature of History* (1988) and Conway Morris's *The Crucible of Creation: The Burgess Shale and the Rise of Animals* (1998).

Gould, well known for his wide-ranging essays, was a paleontologist by profession and wrote widely on evolutionary theory. In *Wonderful Life* Gould looked at the diversity of body forms and found it difficult to fit forms described from the Burgess Shale into the body plans associated with currently recognized phyla. Gould argued that if the Cambrian "explosion" were repeated (replay the tape of evolution), then some of these unusual body forms may have become dominant and modern groups might not have succeeded, or even survived. Gould offered a series of speculations about what a modern world might look like if the replay resulted in different groups becoming successful. For example, he envisioned reefs made of extinct archaeocyathids rather than modern corals or a world in which fish never evolved a bone structure that could become limbs, leaving the land to the insects and flowering plants.

Conway Morris wrote from a different perspective, including information from the Lagerstätte of Chengjiang (in China) and from Sirius Passet (in Northern Greenland), as well as developments in the interpretation of some fossils not available when Gould wrote *Wonderful Life*. Developments such as the association between the previously enigmatic fossil *Wiwaxia* with annelids allowed Conway Morris's book to remove some of the evidence from which Gould argued but did not affect his alternative view. Conway Morris's position was that if the evolutionary tape were replayed, then environmental constraints and forces would lead to the same, or similar, selective constraints that had operated originally, dictating the selection of the same forms that evolved at the first playing.

Some of the disparity in Gould's and Conway Morris's views arose from different interpretations of some of the bizarre animals of the deposits, whether in the Burgess Shale or comparable deposits elsewhere. The case of *Anomalocaris* (Fig. 6.22) illustrated how, seen in isolation, different parts of this creature were classified in different animal phyla. Between the times of publication of the two books, it had become clear that anomalocarids had jointed appendages and are associated with the arthropod stem.

Biologists should always be open to changing their minds, particularly when new evidence is uncovered. The give and take of the process of peer review and scientific publishing is a fertile ground for exchange of ideas and points of view. Ultimately, decisions about which interpretation is correct must be based on evidence. We are fortunate to have such an array of tools, from descriptive to statistical, to assist us in finding and interpreting evidence.

---

animal's remains (such as shell or bone). When the water evaporates, the minerals remain. A clam shell or a mammal bone seen under a microscope is not solid but penetrated by innumerable pores of various diameters. Silica (commonly found in petrified wood) can be deposited in this way, as can insoluble carbonates and pyrite (iron sulphide). Once a bone has been buried, anaerobic bacteria become involved in decomposition, and the water in bone openings becomes anoxic and alkaline. Bacterial reduction of sulphate produces hydrogen sulphide that reacts with ferrous ions to precipitate iron monosulphide, which further reacts with hydrogen sulphide to produce pyrite.

Sometimes, a chemical constituent of a tissue (hard or soft) of a body disappears and is **replaced** by another (typically insoluble) material. **Recrystalliza-tion** is a similar process through which a mineral (e.g., aragonite in the shell of a mollusc or brachiopod) is replaced by calcite, a similar but harder mineral.

Despite the enormous number of known dinosaur fossils and the amount of attention they have received, there is limited understanding of the changes that took place during conversion of original bone to fossil. In some fossils, the original bone was a matrix of collagen between crystals of carbonate hydroxyapatite. As the collagen decayed it allowed the entry of groundwater. Then some of the crystals along the newly created channels were replaced by francolite (carbonate-fluorapatite) that later crystallized to fill the channels.

A cast fossil is formed when new minerals fill a **mould fossil**, the impression left after the decay of a

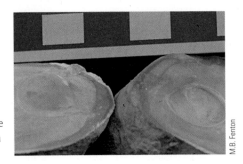

Figure 6.5 A comparison of the cast (left) and mould (right) of a bivalve fossil.

Figure 6.8 A roach and a wasp preserved in Baltic amber. The roach is about 5 mm long. Note the details on the legs of the roach and on the wasp.

dead organism. The mould of a scallop-like mollusc (Fig. 6.5) shows the fine details of the shell. Paleontologists often take casts of fossils and then fill them with plaster of Paris or a similar material to make a replica of the original fossil. This approach was used in making the reproduction of a placoderm fish (Fig. 6.6).

A **trace fossil**, also known as an ichnofossil (from the Greek *ikhnos*, trace or track), is a geological record of biological activity or something left behind by an animal that is not a part of the animal itself. Examples of trace fossils include a coprolite and a footprint (Fig. 6.7).

Many plants (particularly conifers) secrete resins, either as a protection against insects or as a sealant to prevent infection through a wound on the plant's surface. These resins (usually a mixture of terpenes plus dissolved, non-volatile, compounds) lose their volatile components over time and, when hardened, are commonly called **amber** (or resinite). The oldest amber dates back to the Carboniferous (see Section 6.4d). Insects and other organisms (Fig. 6.8) as well as feathers and fur may be trapped in fluid plant resins. Organisms preserved in amber may be protected from decay and, in some cases, recoverable DNA has been extracted from arthropods preserved in amber.

The German word "**lagerstätte**" translates as "storage place" and refers to fossil beds that display exceptional (and unusual) preservation of organismal forms, with soft body tissues preserved as well as shells and skeletal material. Exceptional preservation of soft body tissues usually occurs in fine-grained deposits. In the Burgess Shale lagerstätte (see Section 6.4c.2), some 86% of the fauna lacked mineralized structures in life and consequently would not have been seen in more typical fossil beds. Lagerstätten are found in many geographical locations and from different geological time periods.

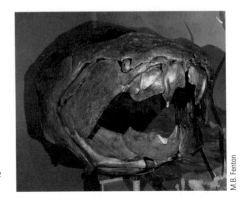

Figure 6.6
A cast of the skull of *Dunkleosteus* (1130), a fossil placoderm fish. The skull was more than 1 m long.

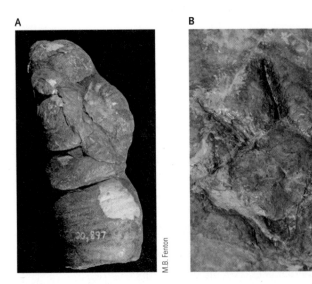

Figure 6.7
Trace or ichnofossils. **A.** A coprolite from an alligator; **B.** a dinosaur footprint.

STUDY BREAK

1. What is amber? How does it contribute to our knowledge of fossils?
2. Name three kinds of trace fossils. How does each of them inform us about animals (and plants) we know as fossils?

6.3 Modern Techniques in Paleontology

The traditional image of a fossil hunter is of the rugged paleontologist heading out into the wild with hammer and pick and later painstakingly excavating her/his finds with dental picks and paintbrushes. Back in the laboratory, either mechanical or chemical (usually an acid bath) techniques are then used to prepare the material for examination. While these images and techniques are still very much valid, they have now been joined and supported by technological advances from other branches of science.

6.3a   Serial Reconstruction (Tomography): *Acaenoplax*

Cutting thin "sections" through a fossil allows serial reconstruction of the organism at a scale reflecting the thickness of the sections. This approach (**tomography**) has been used by paleontologists for many years. The limitation was the thickness of the saw blades used to cut the sections, restricting the technique to larger, thicker specimens. The story of a worm-shaped mollusc, *Acaenoplax* (388) (Fig. 6.9), derives from a combination of sectioning by cutting and modern computer-graphic technology.

About 425 Ma in the Silurian, a series of volcanic events deposited fine layers of volcanic ash over a muddy sea floor community living at depths of 100–200 m in what is now Herefordshire in England. The result was a lagerstätte in which soft-bodied animals were preserved in calcareous nodules. Comparison of the fossil fauna with the rate of decay of modern organisms (e.g., the polychaete worm *Nereis* (321)) suggests that the initial preservation of the fauna must have taken place in less than six days. The fauna assemblage includes a chelicerate, a polychaete, radiolarian protists, other arthropods, and sponges, as well as other organisms of unknown affinity. A common fossil in this assemblage is *Acaenoplax*.

The brute force part of the approach to *Acaenoplax* involved grinding and polishing the fossil in its matrix rather than extracting the fossil. The section through the worm, created by grinding, was photographed (using a digital camera) after each 30 μm of material was ground away. Images were stored and then reassembled to produce a three-dimensional video image of the fossil (Fig. 6.9A & B). The technique shows the valves (V4 and 5) and ridges that are serially repeated along the length of the specimen in a high level of detail.

Acaenoplax was originally placed in the Phylum Mollusca and presumed to be somewhat like Caudofoveata (Aplacophora) and Polyplacophora (the chitons). Later, other workers suggested that *Acaenoplax* was more like a polychaete annelid than a mollusc, but at this time it remains classified as a mollusc, although perhaps not in one of the currently recognized groups.

6.3b   X-Rays

We think of X-rays in association with inspection of live tissues but the use of X-ray imaging of fossils goes back to the late 19th century, immediately after its invention. Fossils that are particularly well revealed by X-rays are

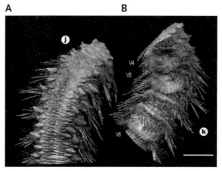
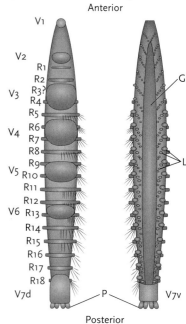

Figure 6.9 Dorsal (**A.**, **C.**) and ventral (**B.**, **D.**) views of a reconstruction of *Acaenoplax* (388). **A.** and **B.** show 7 mm of the anterior region of an *Acaenoplax* taken from 243 "slices" at 30 μm intervals at V4 and V5 in C. **C.** and **D.** are overviews of the animal.

SOURCE: SUTTON, M. D., D. E. G. BRIGGS, et al. (2004). "Computer reconstruction and analysis of the vermiform mollusc Acaenoplax hayae from the Herefordshire Lagerstätte (Silurian, England), and implications for molluscan phylogeny." *Palaeontology* 47(2): 293–318. The Palaeontological Association.

those that have been pyritized (converted to iron sulphide, FeS_2) because iron is an excellent absorber of X-rays.

In beds such as the Hunsrück slate (Lower Devonian), X-rays have revealed both skeletonized animals and soft-bodied animals lacking hard parts. This has worked even for Ctenophora (comb jellies) which are jelly-like and consequently less likely to be preserved in the fossil record.

The value of X-ray examination is clear in Figure 6.10. Figure 6.10A shows the visible surface of a 3–4 mm thick rock slab with an asteroid (*Taeniaster beneckei* (1076)), while the X-ray of the same slab (Fig. 6.10B) shows more detail of the asteroid as well as an excellent image of *Mimetaster hexagonalis* (635), a marrellomorph arthropod, not visible in the photo. Marrellomorpha are discussed in Section 6.4c, the Cambrian.

Combined with computerized tomography, the power of X-ray imaging increases even further. This approach allows resolution of minute morphological

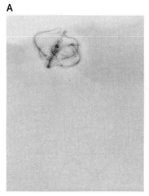

Figure 6.10
A. Photo and **B.** X-ray of a 3–4 mm thick rock slab from the Lower Devonian.

SOURCE: Republished with permission of British Institute of Radiology, from P. Hohenstein, "X-ray imaging for palaeontology," *British Journal of Radiology* (2004) 77, 420–425; permission conveyed through Copyright Clearance Center, Inc.

detail, including internal morphology, and produces a three-dimensional image that can be sectioned at any point or viewed from any angle. An example of the application of the technique is a study of a minute (about 1 mm long) spider preserved in amber about 53 Ma during the Lower Eocene.

6.3c Synchrotron X-Ray Tomographic Microscopy

In the future we can expect even more detailed information from fossil specimens examined with high-energy (hard) X-ray techniques and the very-high-energy hard X-rays produced by a synchrotron. Tuned high-energy X-rays have, for example, now been used to produce three-dimensional images of yeast cells (about 5 μm in diameter) that show detail of the nucleus and intracellular vacuoles at less than 100 nm scales. Higher resolution X-rays require more energy, achieved through synchrotron X-ray tomographic microscopy (SRXTM), that allows non-destructive, submicrometre, three-dimensional analysis of a specimen. SRXTM has been applied to the embryos inside the eggs of *Tianzhushania*, an Ediacaran fossil from China. Reconstruction of an external view of a developing embryo and slices through the developing embryo reveal details of nuclei in the cells (Fig. 6.11).

6.3d Chemical Derivatives

While most paleontology is based on either physical fossils or the physical traces that the organism created (burrows, footprints, etc.), occasionally the past existence of a particular organism may be inferred from the chemical signature it left in the stratum where it lived. In general, chemical signatures can only point to

a fairly large group of organisms. It is rare that a compound is limited to a small group of organisms and can be pin-pointed in the geochronological record. One example is the presence of high levels of 24-isopropyl-cholestane in sedimentary rock from the South Oman Salt Formation. Sedimentary 24-isopropylcholestanes are the hydrocarbon remains of C_{30} sterols, compounds produced in high concentrations only by some groups of Demospongiae (Porifera). The presence of these compounds indicates that demosponges occurred in the South Oman formation in the Neoproterozoic era (1000–542 Ma), well before the end of the Marinoan glaciation (about 635 Ma) and the Ediacaran Period (about 635–542 Ma).

STUDY BREAK

1. Describe different ways in which paleontologists use tomography. What are SRXTM reconstructions?
2. How are X-rays used in the study of fossils?

6.4 Ancient Communities

In this chapter, our sampling of knowledge about the history of animal life through the fossil record begins with the Ediacaran fauna (635–542 Ma), moving to the Cambrian (543–488 Ma), the Carboniferous (359–299 Ma), and finishing with the Eocene (55–34 Ma). These periods demonstrate increasing biodiversity and more and more evidence of interactions among animals and between animals, plants, and other organisms from fungi to bacteria and protists.

As you progress through the chapter, note obvious links to other chapters. Be aware of dramatic changes in faunas. These often occur when animals change their diet, for example switch from eating other animals to eating photosynthetic organisms such as algae and plants. In ecological terms, this involves a switch from secondary to primary consumer and access to a greater biomass. Herbivory has evolved in many terrestrial groups, such as insects, reptiles, birds, and mammals. A vital recurring theme in evolution is that the appearance of more different kinds of organisms (such as herbivores) generates opportunities for predatory species.

6.4a Extinctions

Mass **extinctions** have been recurring events in the history of life on Earth, and life continues in their wake. During mass extinctions, literally hundreds or thousands of species disappeared; some groups of animals and plants (e.g., trilobites and dinosaurs) disappeared completely. The causes of mass extinctions are rarely

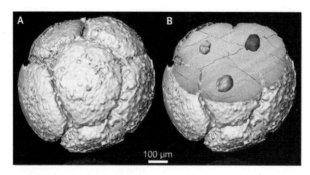

Figure 6.11 SRXTM reconstructions of Ediacaran embryos of *Tianzhushania* from the Doushantuo Formation in China show an 8-cell stage embryo. **A.** The rendering of the surface shows six of the cells. **B.** Three nuclei are shown in a slice through the embryo.

SOURCE: From Huldtgren, T., J.A. Cunningham, C. Yin, M. Stampanoni, F. Marone, P.C.J. Donoghue and S. Bengtson, "Fossilized nuclei and germination structures identify Ediacaran "animal embryos" as encysting protists," *Science* 23 December 2011: Vol. 334 no. 6063 pp. 1696-1699. Reprinted with permission from AAAS.

completely known, but often involve catastrophic events such as massive volcanic eruptions or a large meteor striking Earth. The extinction event that included the disappearance of dinosaurs coincides with unusual amounts of iridium in strata from that time. Iridium is an element found in meteorites, suggesting that a meteor had collided with the Earth.

Over at least 520 million years, global biodiversity, measured as the richness of families and genera of animals and plants, has been directly related to temperature in both terrestrial and marine systems. During warm phases, global biodiversity has been relatively low, but these were times of extinctions of some lineages and originations of new ones. Changes in temperature alone do not predict changes in biodiversity. Periods of glaciation represent extreme environmental changes, but these may be no more damaging to the biota than prolonged periods of heat and drought. Extinction events were usually followed by diversification of some evolutionary lines because the disappearance of some species usually provides opportunities for others. (In a diversification, two (or more) sister groups evolve from the basal form.) Some lineages persisted through every major extinction. For example, one group of mammals, the Multituberculata (Fig. 6.12), radiated at least 20 million years before the Cretaceous-Palaeogene (KT) extinction that affected dinosaurs and many other taxa. Clearly, multituberculates depended upon resources not affected by the KT extinction, and they persisted until the late Eocene when they disappeared.

6.4b Ediacaran Fauna

About 2.5 billion years ago (Ba) the evolution of photosynthesis meant the production of oxygen and its release into Earth's previously anoxic atmosphere. By 2 Ba there is fossil evidence of large (centimetre scale) bacterial colonies, but it took almost 1.4 billion years from then before we have evidence of complex multicellular animals. Their appearance marks the beginning of the **Ediacaran Period**, named after hills in southern Australia where some of these deposits occur. The Ediacaran lasted from 635 to 542 Ma. There are over 30 known deposits of Ediacaran age on five continents. The boundaries of the Ediacaran are set by the end of the Marinoan global glaciation period and the first appearance of the complex fossil *Treptichnus pedum* (607) (discussed below in the Cambrian). Figure 6.13 shows the periods of glaciation, the dating of the Ediacaran period (and the Ediacaran biota), and the changes in the ratio of the isotopes carbon 13 and carbon 14 ($\partial^{13}C$ and $\partial^{14}C$). This carbon isotope ratio indicates climate change with negative $\partial^{13}C$ values correlating with periods of low oxygen levels.

Knowledge of the biota of the Ediacaran is sketchy because of the age of the deposits and the degree to which they have been modified. Some organisms from the period have living representatives (e.g., stromatolites, present-day Cyanobacteria) but most do not. Complex multicellular animals' appearance in the fossil record is a feature of the Ediacaran. The first actual fossil remains of an animal come from China where fossil eggs and embryos (Fig. 6.11) and perhaps

Figure 6.13
A time series showing the dating of the Ediacaran period (in green), periods of glaciation (and their effect on $\partial^{13}C$), and the appearance of the Ediacaran fauna. The Avalon Assemblage (see below) is dated at roughly 575 Ma.

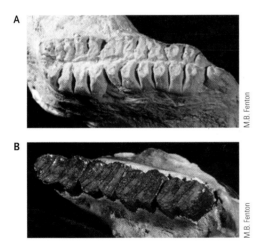

Figure 6.12 A. The cheek teeth of multituberculate mammals (*Taeniolabus* spp.) had a characteristic cusp pattern (many tubercles). **B.** Other mammals also have specialized cheek teeth (e.g., a giant beaver, *Castoroides ohioensis* (1878)). Specialized teeth often reflect diet.

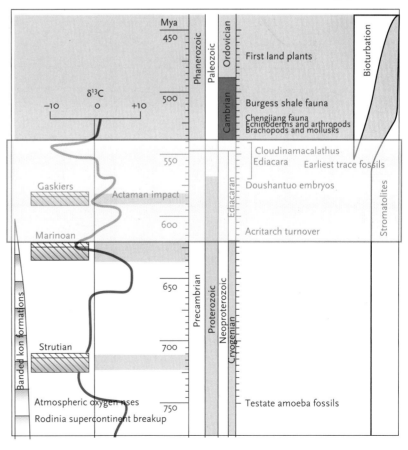

other groups have been identified in rocks dated between 595 and 584 Ma.

Despite their absence from the Ediacaran fossil record, modern groups of animals must have been present through that period and originated before the Marinoan glaciation (Cryogenian period). Molecular clock data, based on seven genes from 118 taxa, place the origins of the Cnidaria and the Demospongiae in the Cryogenian. This suggests that separation of the basal groups in the Deuterostomia and Protostomia had occurred by then (Fig. 6.14).

But many of the Ediacaran fossils from the Doushantou deposits are blob-like and showed progressive increases in cell number without increase in cell size (**palintomy**), reminiscent of embryonic development (see Chapter 2). More recent SRXTM imaging (Fig. 6.11) suggests that the fossils are neither embryos nor animals. A proposed life cycle of *Tianzhushania* (Fig. 6.15) has been used to infer that these fossils are in fact the remains of animals that were outside the evolutionary line that led to the diversification of the Metazoa.

6.4b.1 Avalon Assemblage. The oldest Ediacaran macrofaunal fossils were found in the Avalon Assemblage in Newfoundland and date between 575 and 560 Ma. These are deep-water fossils and, in spite of extensive searches, no shallow-water fossils from the same

Figure 6.14

Divergence estimates for 13 lineages of animals. Yellow and blue show fossil material, while hatching indicates stem lineages (animals in one phylum). Thick black lines are known fossil records from the Cryogenian through the Ordovician.

SOURCE: From Erwin, D.H., M. Laflamme, S.M. Tweedt, E.A. Sperling, D. Pisani and K.J. Peterson. 2011. "The Cambrian conundrum: early divergence and later ecological success in the early history of animals," *Science* 334:1091–1097. Reprinted with permission from AAAS.

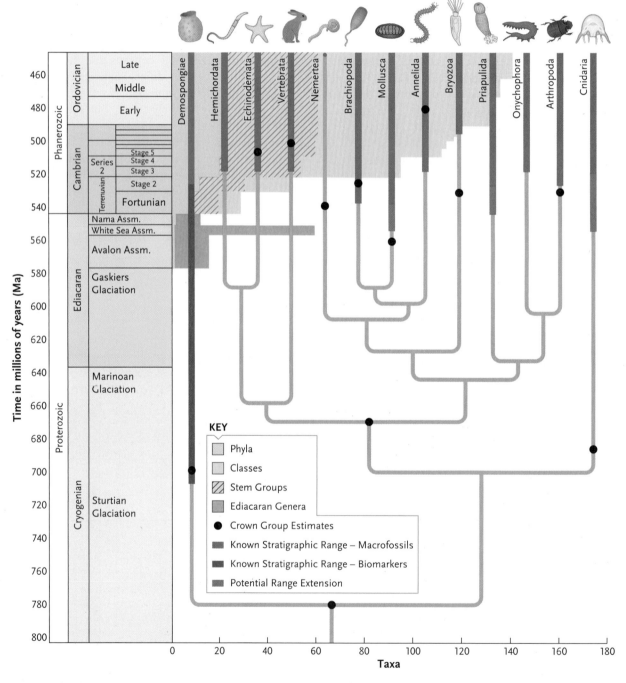

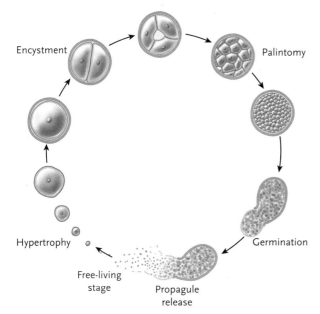

Figure 6.15 The life cycle of the doushantuoan fossil *Tianzhushania* shows hypertrophic grown, encystment in a wall composed of several layers. Palintomic cleavage produced a mass of pre-propagules which germinated after opening of the outer wall of the cyst. Black shows the inner wall of the cyst which is seldom preserved in fossils.

SOURCE: Based on Huldtgren, T., J.A. Cunningham, C. Yin, M. Stampanoni, F. Marone, P.C.J. Donoghue and S. Bengtson, "Fossilized nuclei and germination structures identify Ediacaran 'animal embryos' as encysting protists," *Science* 23 December 2011: Vol. 334 no. 6063 pp. 1696–1699.

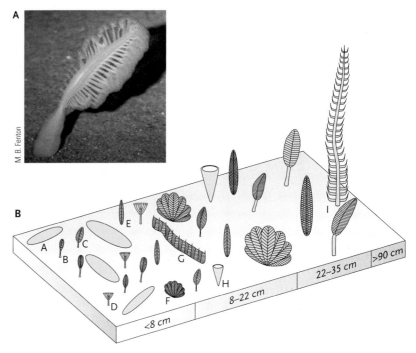

Figure 6.17 Species in the Avalon Assemblage can be arranged into groups by height (scale numbers indicate the height to which species grew). **A.** Spindle, **B.** Ostrich feather, **C.** *Charniodiscus*, **D.** duster, **E.** *Charnia*, **F.** *Bradgatia*, **G.** comb-shaped, **H.** *Thectardis*, **I.** *Frondophyllas grandis*. Some rangeomorphs may have resembled modern sea pens (inset: Cnidaria, order Pennatulacea) like *Ptilosarcus gurneyi* (161).

SOURCE: Republished with permission of Annual Reviews, Inc, from Narbonne, THE EDIACARA BIOTA: Neoproterozoic Origin of Animals and Their Ecosystems *Annual Review of Earth and Planetary Sciences* Vol. 33: 421-442, 2005; permission conveyed through Copyright Clearance Center, Inc.

period have been found. This casts doubt on the idea that the Ediacaran animals might have evolved as photoautotrophs, animals having a symbiotic association with algal cells in shallow, well-lit waters. Most of the Avalon fossils belong to the Rangeomorpha, a group that did not survive into the Cambrian (Fig. 6.16). The size and positioning of the animals of the Avalon Assemblage allow some reconstruction of their arrangement on the seabed (Fig. 6.17). Shorter species were more abundant in shallower waters, taller species

in deeper waters. This arrangement is comparable to that of a population of modern filter-feeding organisms rather than a community in which species competed for light or space.

6.4b.2 White Sea Assemblage. The White Sea Assemblage from Russia dates from 560 Ma and extends to the start of the Cambrian. This assemblage is best known from the White Sea and from Australia. This is a shallow-water fauna, occurring between the subtidal wave depth and the depth affected by storm waves. Some species in the assemblage may have extended into deeper waters. The major change from the Avalon Assemblage is the appearance of bilaterally symmetrical, burrowing animals (Fig. 6.18).

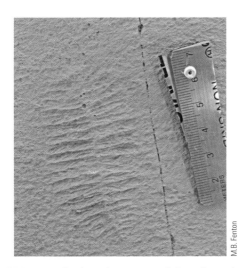

Figure 6.16 A spindle-shaped rangeomorph from the Avalon Assemblage of Newfoundland.

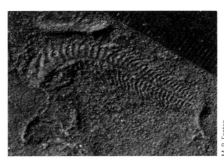

Figure 6.18 Fossil *Spriggina floundersi*, which was probably an errant polychaete or perhaps a primitive arthropod from Australia.

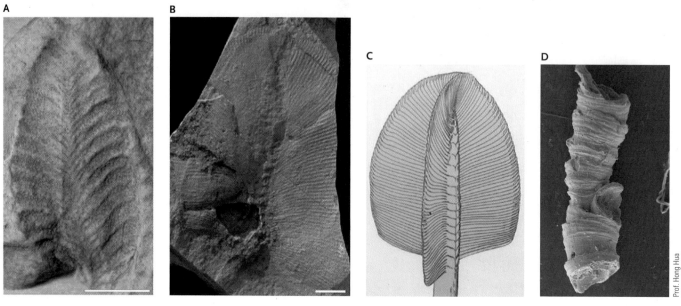

Figure 6.19 Fossils (and a reconstruction) from the Nama Assemblage. **A.** *Rangea*, a rangeomorph; **B.** *Swartpuntia*, a rangeomorph; **C.** reconstruction of *Swartpuntia*. (Scale bar = 2 cm.) **D.** Budding structure in *Cloudina*.

SOURCE: A., B., and C. Republished with permission of Annual Reviews, Inc, from Narbonne, THE EDIACARA BIOTA: Neoproterozoic Origin of Animals and Their Ecosystems *Annual Review of Earth and Planetary Sciences* Vol. 33: 421–442, 2005; permission conveyed through Copyright Clearance Center, Inc.

In contrast to the Nama Assemblage (below), none of the fossils shows evidence of a skeleton although *Spriggina* (Fig. 6.18) may have been a primitive arthropod, although it is usually viewed as an annelid.

6.4b.3 Nama Assemblage. The Nama Assemblage (Fig. 6.19, named for the Nama formation in Namibia) dates between 549 and 542 Ma. Like the White Sea Assemblage, the Nama Assemblage is a shallow-water fauna that includes fronds of rangeomorphs, burrows of bilaterians, and, importantly, the first calcified metazoans. The Nama Assemblage is separated from Cambrian faunas by the presence of rangeomorphs and by evidence of predation from bore holes in the calcareous tubes of the possible annelid, *Cloudina*.

The animal forms that appeared during the Ediacaran represent three important stages in metazoan evolution:

1. diversification of form and epithelial organization, implying sophisticated intercellular communication and development controls
2. appearance of bilateral symmetry and mobility, which would have demanded a reasonably developed neuromuscular system as well as transport and excretory systems
3. biomineralization and predation, the beginning of the evolutionary arms race that continued into the Cambrian.

STUDY BREAK

1. What is the Ediacaran fauna? Where have Ediacaran deposits been found?
2. What are rangeomorphs? What do we know about them?

6.4c Cambrian Seabed

Many of the animals that lived on Cambrian seabeds are preserved in fine-grained lagerstätte sediments that provide many details of both external and internal structures. Some Cambrian fossils are readily identified to Phylum, others are problematic. The rich fossil record of the Burgess Shale in particular, and the Cambrian in general, includes the appearance of animals with skeletons.

The first appearance of a trace fossil, *Treptichnus pedum* (607), marks the beginning of the Cambrian, ~543 Ma. This fossil is a series of burrows that appear to come to the surface at regular intervals. Modern work, in which burrow patterns were created and examined in the laboratory, strongly suggests that these burrows were made by priapulids coming to the surface to prey on small epibenthic organisms.

The change in trace fossils from being two dimensional (and probably associated with a diploblastic fauna) to three dimensional (involving vertical burrows, trace fossils, and remains) suggest bilaterally symmetrical, triploblastic animals (labelled "trace fossils" in Fig. 6.20). This change is part of the rationale for distinguishing Ediacaran and Cambrian periods. Estimates obtained from molecular clocks suggest that triploblastic burrowing animals had originated much earlier in the Ediacaran, but we still lack fossils to indicate when they first appeared. The first true fossils of the Cambrian are small (many measure only a few microns) calcareous plates. These have been interpreted as the remains of early molluscs, brachiopods, worms such as the Machaeridia (see Section 6.4d), and of many unidentified, and probably unknown, taxa. Fossils with shells suggest an early Cambrian arms race with predatory animals evolving and their activities triggering the evolution of protective coverings in their prey, in turn selecting for predators with stronger claws and jaws.

In the first 15–20 million years of the Cambrian, the fossils are traces or shells which increase in number and diversity but provide limited insight into the events that had occurred. Trilobites, the first animals with a full external covering (armour), appeared about 520 Ma. At this time there was a major increase in the numbers of body forms and species (Fig. 6.20), the Cambrian "explosion," or better, the Cambrian "radiation." Shortly after this we have a detailed record of the early Cambrian biota in the Chengjiang lagerstätte, deposits in Hunan province of SW China.

6.4c.1 Chengjiang Biota.
More than 100 species have been described from Chenjiang from 10 modern phyla, four phyla now extinct or with questionable affinities, and several as yet not assigned to any group. Most of the species are arthropods (>50); none of the other groups has more than 10 species.

The exceptional preservation in the fine mudstones of the Chengjiang deposits gives a clear picture of the fauna at the time. Trilobites (Arthropoda) dominated the fauna, and *Hallucigenia* (625) (Lobopodia, probably related to modern Onychophora) also occurred. Among the Chengjiang deposits are species of soft-bodied animals (including a sea anemone, a ctenophore, a nematomorph worm and a priapulid) that can easily be compared to genera living today.

6.4c.2 Burgess Shale Biota.
In the mid-Cambrian, disparity and diversity further increased, which is revealed by a ~520 Ma lagerstätte, the Burgess Shale, in the Canadian Rocky Mountains. This Cambrian seabed lay under the edge of the cliffs of what is now the Cathedral Range and fauna was trapped and preserved in the sediments under occasional falls from the cliffs.

Figure 6.21 shows reconstructions of five fossils from the Burgess deposit. *Marella splendens* (Fig. 6.21A), an arthropod, is the most common species

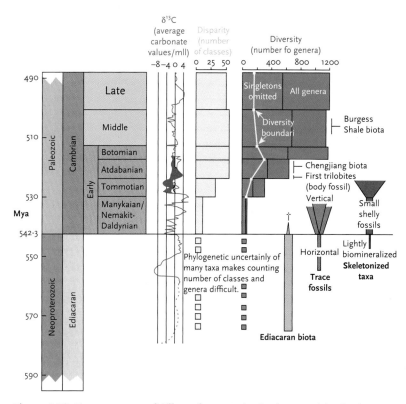

Figure 6.20 The appearance of different faunas in the Cambrian and the development of different body forms and more diversity of species.

SOURCE: Republished with permission of Annual Reviews, Inc, from Marshall, EXPLAINING THE CAMBRIAN "EXPLOSION" OF ANIMALS, *Annual Review of Earth and Planetary Sciences* Vol. 34: 355–384, 2005; permission conveyed through Copyright Clearance Center, Inc.

among the Burgess Shale fossils (>5000 specimens). *Marella* does not fit into any existing arthropod line and appears to represent a stem group arthropod—an early offshoot from the line giving rise to all other arthropod groups. *Marella* was bottom dwelling and appears to have used the pair of appendages behind its antennae to collect particulate food. Fossils of *Marella* are frequently linked to a dark stain interpreted as being gut contents that had been squeezed out of the animal as it was flattened, prior to preservation. *Opabinia* (631), a segmented animal with five eyes and a probe-like proboscis, has been suggested to be related to the anomalocarids (placing it among the stem arthropod groups). *Aysheaia* is a lobopodian, most likely to be placed in the phylum Onychophora. The Cambrian appearance sheds important light on the ancestry of the group because living onycophorans are terrestrial. *Pikaia* is an important member of the Burgess fauna because it is one of the earliest chordates. The reconstruction shows the repeated muscle blocks (myomeres) and fin. *Pikaia* has a notochord and could be related to the modern Cephalochordata (e.g., *Amphioxus*).

6.4c.3 Anomalocaridae.
Anomalocaridae is a variation on the situation of *Nectocaris* (Fig. 6.4). *Anomalocaris*

Figure 6.21 Reconstructions of five Cambrian animals from the Burgess Shale, from an exhibit at the Royal Tyrrell Museum, Drumheller, Alberta. **A.** *Marella*, an arthropod; **B.** *Opabinia* (631), an animal of unknown phylum; **C.** *Aysheaia*, an onycophoran; **D.** *Canadia*, a polychaete annelid; and **E.** *Pikaia*, a cephalochordate.

A

B

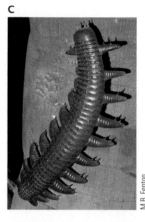
C

D

E

was first described in 1892 and, as its name suggests, was presumed at the time to be an unusual (anomalous) phyllocarid shrimp (Fig. 6.22). The Burgess fossil on which this was based was segmented, resembling the abdomen of a shrimp, apparently with ventral, spike-like, appendages. The anomaly was the lack of segmentation in these appendages. In the early 20th century, Charles Walcott described *Peytoia*, a cnidarian medusa or jellyfish with 32 subumbrella lobes. Another fossil from the site, *Tuzoia*, was identified as the carapace of the crustacean of which *Anomalocaris* (630) was the abdomen. The final element in the history was a fossil, originally described as a sea cucumber (Echinodermata, Holothuria), called *Laggania*. The parts of the puzzle were assembled when a Cambridge-based group collected new Burgess material and re-examined the fossils collected earlier. They determined that *Anomalocaris* was actually a frontal appendage (the "great appendage") of a large arthropod. The former "jellyfish," *Peytoia*, was a series of plates surrounding this animal's mouth. The "holothurian," *Laggania*, was the poorly preserved body of another anomalocarid that had a smaller great appendage. *Tuzoia* is the fossil of an unrelated arthropod with a bivalved carapace resembling that of modern Ostracoda (a group of Crustacea).

Anomalocaris may have reached a metre in length and eaten trilobites (suggested by jaw damage found on the fossil exoskeletons of some trilobites). This interpretation has been challenged because the anatomy of the mouth of *Anomalocaris* is not suited to this type of predation, but the "great appendages" were probably used to hold and break the carapace of shelled prey. Recent evidence indicates that *Anomalocaris* had well-developed compound eyes, providing an early date for the presence of a visually hunting predator.

All currently recognized phyla had appeared by the end of the Cambrian in addition to some others now extinct. The transition from Cambrian to Ordovician is marked by a major extinction event that eliminated ~60% of Cambrian genera. However, the Ordovician was an era of greatly increased animal diversity that included many animals recognized in the Cambrian. A lagerstätte in Morocco has yielded anomalocarids and other animals of Middle- and Late-Cambrian deposits.

STUDY BREAK

1. How similar were Cambrian and Ediacaran faunas?
2. Why have so many fossils known from the Burgess Shale changed classification since their discovery and description? Present one example in detail, other than *Anomalocaris*.

6.4d Carboniferous

The Carboniferous (359–299 Ma) (in the United States and Canada known as the Mississippian followed by the Pennsylvanian) was a time of climate and habitat

M.B. Fenton

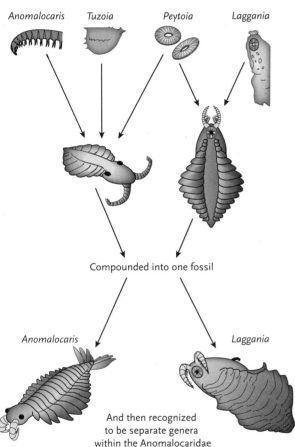

Anomalocaris   Tuzoia   Peytoia   Laggania

Compounded into one fossil

Anomalocaris   Laggania

And then recognized
to be separate genera
within the Anomalocaridae

Figure 6.22

This diagram illustrates how our knowledge of *Anomalocaris* developed and emerged from the discovery of better fossils. In the end, *Anomalocaris* (630) is a composite, including parts originally described as a sea cucumber (*Laggania*), a schyphozoan medusa (*Peytoia*), and a bivalve arthropod (*Tuzoia*).

oxygen in the atmosphere approached 35%, compared with 20% today (see also page 306). Although levels of atmospheric carbon dioxide had been higher earlier in the Paleozoic, coinciding with higher global temperatures, by the Carboniferous they had declined to modern-day levels. Isotopic analyses reveal that in the Carboniferous (Box 6.1), sea surface temperatures were similar to those of today. In some parts of the world, glaciation events began in the middle of the Mississippian and continued through the Pennsylvanian. During these times, high-latitude glaciations were linked to environmental changes in the tropics.

6.4d.1 Forests and Insects. On land, widespread forests meant new habitats for plants and animals. These Carboniferous forests are the source of much of today's supply of coal. The first "trees" had appeared in the Devonian, 8-m-tall tree-fern-like cladoxylopsids (Pseudosporochnales). The prevalent trees of the Carboniferous were lycopsids that were up to 40 m tall (Fig. 6.23A), represented today as club mosses and quillworts. The oldest of the vascular plants, lycopsids had diverged from the other main lineage of vascular plants (leading to ferns, horsetails, and seed plants) in the Silurian or early Devonian. Arborescent lycopsids apparently appeared in the middle Devonian and by the middle of the Carboniferous were notable for thickened stems or trunks indicating cambial growth or wood (secondary tissues). Lycopsids must have been capable of rapid growth, probably achieving maturity in a few years. At maturity, lycopsids developed and released spores and then died. Short generations (which may have made them vulnerable to changes in climate) and the accumulation of large amounts of standing biomass set the stage for coal deposits. In the Permian, tree ferns replaced lycopsids as the trees of swamp forests.

By the Carboniferous, insects were diverse and included winged forms with some dragon flies whose wing spans exceeded 1 m. At least 16 orders of insects are known as fossils from the Carboniferous, 11 that did not survive into the Mesozoic. The wings of mid-Carboniferous dragonflies (Odonata) showed specializations such as automatic depression of the trailing edge of the wing that permit the high performance flight by living odonates. Relationships between plants and insects had developed by the Late Carboniferous when some insect larvae formed galls in the internal tissue of tree-fern fronds, whose fossilized tissues include diagnostic histological and cellular details (Fig. 6.23B) typical of gall-forming plant-insect interactions today. There also is evidence of Carboniferous amber, demonstrating that some trees, probably preconifer gymnosperms, had the biosynthetic mechanisms necessary to produce complex polyterpenoids used by modern plants, in part in defense against insects.

changes associated with higher levels of oxygen as well as the spread of forests and the insects that exploited them. It was also the time of the early evolution of amphibians. Animals and plants had successfully moved onto land by the Carboniferous and levels of

6.4d.2 Radiation of Amphibians. The diversification of insects provided a food base for other terrestrial animals including vertebrates. Exploitation of terrestrial prey required, among other things, mobility on land. *Pederpes* (1351), an early Carboniferous (348–344 Ma) vertebrate, had feet with asymmetrical metacarpals (Fig. 6.24), more suited for locomotion on land than the feet of its more amphibious contemporaries.

Eucritta melanolimnetes (1353), whose name means true (*Eu*) critter (*critta*) from the black (*melano*) lagoon (*limnetes*), was described from the lower Carboniferous in Scotland. *Eucritta* illustrates an early diversification of terrestrial vertebrates, combining characteristics usually found in lineages leading to living Amphibia (frogs, salamanders, caecilians) and Amniota (mammals, turtles, crocodiles, birds, lizards, and snakes). *Eucritta* was a short-bodied animal with hind limbs noticeably larger than the forelimbs. The skulls of the four known specimens range from 30 to 90 mm in length. *E. melanolimnetes* had a short snout, large orbits, a primitive pattern of skull bones, and short legs. While the skulls of living amphibians (Lissamphibia) are more open because of large vacuities in the palates (Fig. 6.25), *Eucritta* had a closed palate and skull like that of larger predatory amphibians such as anthracosaurs, early reptile-like amphibians, and baphetids, an extinct group of amphibians. The ventral ribs (gastralia) are elongated as they are in other baphetids. Baphetids such as *Eucritta* evolved from small to medium-sized terrestrial forms that probably preyed on arthropods. The radiation of these early amphibians convergently gave rise to large, aquatic piscivores that lived in coal swamps. *Eucritta* lacked a middle ear bone (stapes) but had a large tympanic notch that could be the basis of the ears of temnospondyls. The blend of features in *Eucritta* compared with other amphibians suggests that an early Carboniferous rather than a Devonian divergence gave rise to this form.

6.4d.3 Plates of Armour. Chitons (Fig. 2.23) are molluscs (Class Polyplacophora) specialized to adhere to rocks and shells. Their fossils first appear in the Late Cambrian. Both Cambrian and modern polyplacophorans have shells consisting of eight

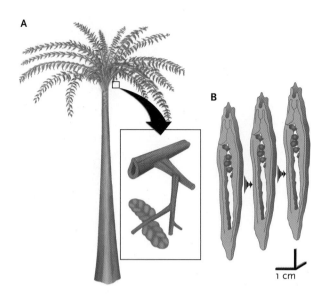

Figure 6.23 A. A 7-m-tall tree fern (genus *Psaronius*) and **B.** a reconstruction of a plant gall in a leaf petiole.

SOURCE: Labandeira, C.C. and T.L. Phillips. 1996. A Carboniferous insect gall: insight into early ecologic history of the Holometabola. *PNAS*, 93:8470–8474. Copyright 1996 National Academy of Sciences, U.S.A.

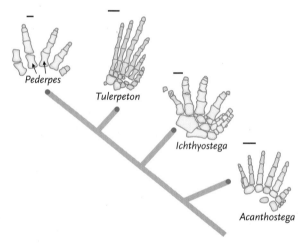

Figure 6.24 The hands of early tetrapods, moving from more aquatic (*Acanthostega* (1352)) to more terrestrial (*Pederpes* (1351)), illustrate progressive asymmetry among digits (black arrows). *Pederpes* is considered to have been a tetrapod. Scale bars = 10 mm.

SOURCE: Based on Figure 6 in Long et al. 2004. "The Greatest Step in Vertebrate History: A Paleobiological Review of the Fish-Tetrapod Transition." *Physiological and Biochemical Zoology* 77:700–719.

Figure 6.25 While the skulls of amphibians from the Mesozoic and before (**A.** *Thoosuchus yokovlevi* (1360)) were relatively solid, those of modern amphibians (e.g., **B.** *Ambystoma mexicana* (1404) or **C.** *Rana catesbiana*) (1381) are more open.

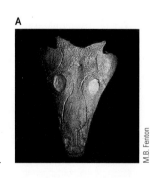
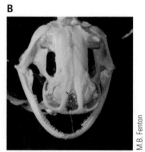

overlapping transverse valves surrounded by scales or spicules. Multiplacophora of the Carboniferous (Devonian to Permian) had more valves than living and Cambrian Polyplacophora. Multiplacophora had head and tail valves as well as five dorsal valves on each side of the body, flanking five central valves. The valves were surrounded by large spines. These details are clear in *Polysacos vickersianum* (397), an exceptionally well-preserved specimen from the Carboniferous of Indiana (U.S.). The Multiplacophora are now considered to be an order within the class Polyplacophora. Chitons were more diverse in the Paleozoic than today.

STUDY BREAK

1. What conditions could have accounted for the "giant" insects of the Carboniferous?
2. What plants were the "trees" of Carboniferous forests? How did forests change the ecological opportunities for other organisms?

6.4e Jurassic

The Jurassic Period (206–144 Ma) started immediately after the mass extinction that marked the end of the Triassic. The Triassic had been marked by an apparent four-fold increase in atmospheric concentrations of carbon dioxide that could have generated a 3–4°C warming. This, in turn, would have raised the temperatures above the lethal limits for leaves, accounting for the >95% turnover of megaflora across the Triassic-Jurassic boundary.

Some animals (e.g., horseshoe crabs, sea spiders, and some bivalves) show little evidence of differentiation in over time; others, including cartilaginous fish (sharks and rays) and plankton-feeding fish were diversifying. On land, amphibians showed both diversification and consistency between Jurassic and living forms. Dinosaurs and mammals provide further examples of diversification, while the appearance of feathers forced biologists to recognize that they are not diagnostic features of birds.

The appearance and rise of angiosperm plants had a huge impact on terrestrial communities. Angiosperms extended the impact of forests that had opened new spatial opportunities, as well as providing food in the forms of vegetation, nectar, pollen, and fruit. While we lack many details of the origin and diversification of angiosperms, they had a positive effect on the diversity of terrestrial animals. Insects provide excellent examples of the relationships between plants and animals.

6.4e.1 Adaptive Radiation: Plants and Insects. Long, tubular mouthparts (proboscices) of some Late Jurassic insects allowed them to feed on ovular secretions and pollen of gymnosperm plants. From the Late Jurassic to the early Cretaceous, scorpion flies epitomized these feeding specializations (Fig. 6.26). These forms disappeared from the fossil record at the same time as contemporary gymnosperms. The extent of specialization of scorpion flies indicates an independent coevolution of associations we now find between angiosperms and their insect pollinators. Brachyceran (short-horned flies, Diptera) were diverse by the Late Jurassic, including some species with short, stout proboscises as well as others with long and slender ones designed for obtaining nectar from tubular flowers. Brachyceran flies with long proboscises may indicate the appearance of the first angiosperms. In any event, long proboscises are a recurring feature of flower-visiting insects (Fig. 6.26), one shared with other flower-visiting animals.

A 165-million-year-old fossil katydid (*Archaboilus musicus*) from the Jurassic shows that these orthopteran insects were "singing" even then (Fig. 6.27). Katydids produce sound by rubbing a toothed area on one wing against a plectrum on the other wing (stridulation). The fossil is preserved well enough that the stridulatory surfaces have been analyzed to derive the sound spectrum they produced. *Archaboilus* song had a frequency of about 6.4 kHz, about half the frequency of modern katydids and about the same as modern bush crickets. Use of low frequency sound is optimal for distance communication at ground level.

6.4e.2 Adaptive Radiation: Neoselachians and Plank-tivorous Fish. The radiation of modern sharks and rays (neoselachians) occurred in the oceans of the early Jurassic (Fig. 6.28), coinciding with the aftermath of the mass extinctions that marked the end of the Triassic. From a basal group there was rapid diversification of the Neoselachia reflecting a combination of life-history features such as small body size, short lifespans, and ovipary. This suite of traits enabled neoselachians to respond more quickly to changes in environmental conditions and biological communities.

Large-bodied, planktivorous animals are the largest creatures known on Earth, including modern giants such as blue whales and whale sharks. In the Triassic, *Shonisaurus* spp. (1440), the largest of the ichthyosaurs, was a planktivore that may not have survived the extinction event at the end of the Triassic. By the Middle Jurassic, pachycormids (bony fish) occupied the niche of the large planktivore and survived at least until the end of the Cretaceous. Modern lineages of large planktivores only appear after the extinction of the pachycormids (Fig. 6.29). The largest of pachycormids were probably not more than 10 m long, less than half the size of 20–25-m-long *Shonisaurus* or blue whales.

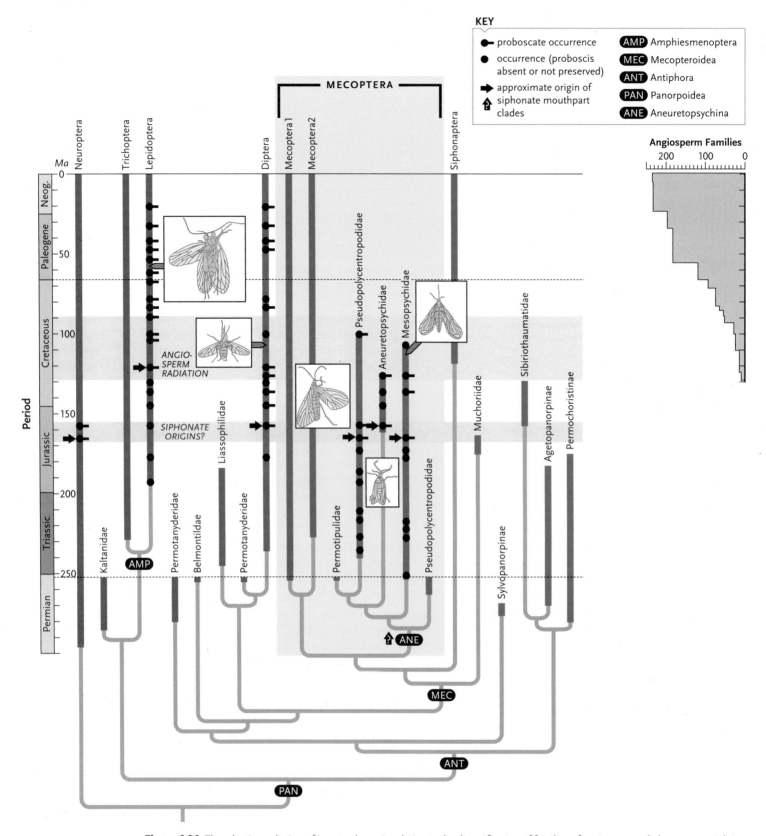

Figure 6.26 The adaptive radiation of insects shown in relation to the diversification of families of angiosperms (light green on right). Mecoptera are scorpion flies; Neuroptera, lacewings; Trichoptera, caddisflies; Lepidoptera, butterflies and moths; and Siphonaptera, fleas. Siphonate mouthparts allowed insects to feed at flowers. Black arrows show repeated origins of these mouthparts. Note several extinct orders of insects.

SOURCE: From Ren, D., C.C. Labandeira, J.A. Santaigo-Blay, A. Rasnitsyn, C. Shih, A. Bashkuev, M.A. V. Logan, C.L. Hotton, and D. Dilcher. 2009. "A probable pollination mode before angiosperms: Eurasian long-proboscid scorpionflies," *Science* 326:840–846. Reprinted with permission from AAAS.

A

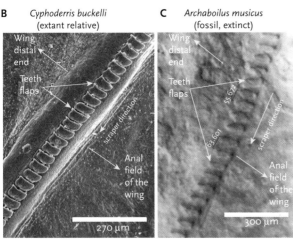

B **Cyphoderris buckelli**
(extant relative)

Wing distal end

Teeth flaps

scraper direction

Anal field of the wing

270 μm

C **Archaboilus musicus**
(fossil, extinct)

Wing distal end

Teeth flaps

scraper direction

Anal field of the wing

300 μm

Kevin Judge

Figure 6.27 A. A living katydid, *Cyphoderris buckelli* (895), and **B.** its toothed vein used in producing sound. **C.** The file of the Jurassic fossil *Archaboilus musicus* (894). The details allowed recreation of the fossil katydid's song.

SOURCE: B. and C. Gu et al., "Wing stridulation in a Jurassic katydid (Insecta, Orthoptera) produced low-pitched musical calls to attract females," *PNAS* February 6, 2012.

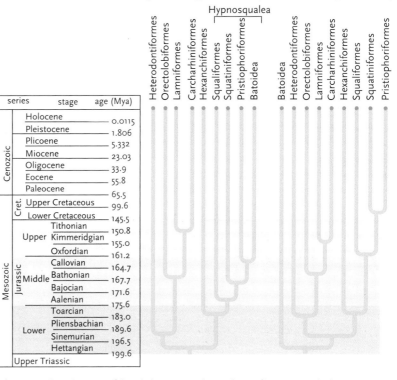

Figure 6.28 Two views of the phylogenetic relationships of living neoselachians are shown with information about relative ages. Thick bars represent confirmed fossil occurrences; thin bars, gaps in the fossil record. The left phylogeny is based on morphological features and proposes that batoids (skates and rays) are highly evolved sharks. The right phylogeny is based on molecular data and indicates that batoids are the basal sister group to sharks.

SOURCE: Jürgen Kriwet, Wolfgang Kiessling, Stefanie Klug, "Diversification trajectories and evolutionary life-history traits in early sharks and batoids," *Proc. R. Soc. B* 7 March 2009 vol. 276 no. 1658 945-951, by permission of the Royal Society.

6.4e.3 Organisms Showing Little Change over Time.

Some organisms in other lineages show little evidence of change between the Jurassic and the present: for example, horseshoe crabs and pycnogonids (sea spiders). An exceptionally well-preserved fossil horseshoe crab (*Mesolimulus* (646)) from the Upper Jurassic in Germany provides details of musculature that differ little from those of living horseshoe crabs (e.g., *Limulus polyphemus*). The details of this fossil even extend to the microbes that formed biofilms. Biofilms are usually postmortem phenomena, and their presence, combined with the detailed preservation of the muscle details, suggest mineralization soon after death.

Sea spiders (Chelicerata, Pycnogonida) apparently have changed little since the Jurassic. Sea spiders resemble arachnids and are recognizable by their long legs, prominent proboscis, a pair of legs used to transport eggs (ovigers), and vestigial abdomens. Sea spiders are small marine forms found from shallow waters to great depths. Pycnogonids were probably established by the Cambrian, but their fossil record is spotty. Jurassic pycnogonids from France were part of deep-water communities and were larger than Silurian species and slightly larger than Devonian ones. Among modern species, some Antarctic pycnogonids have a leg span of up to 30 cm. Fossil pycnogonids generally resemble living species, but some taxa are more similar than others. Living pycnogonids evolved from a common ancestor, and the diversity today reflects an adaptive radiation in the Mesozioc. The long legs of most species allow them to walk on muddy surfaces, while the long proboscis is used in feeding.

6.4e.4 Basal Origin of Amphibian Radiation.

The fossil record of amphibians provides a different perspective on evolution and adaptive radiation because of the similarity between fossil and living forms. The Middle Jurassic amphibian *Chunerpeton tianyiensis* (1409) occurs in volcanic deposits in China. This species belongs to the family Cryptobranchidae (Fig. 6.30), and the extremely well-preserved specimens include one of a larva with obvious external gills. Modern species of cryptobranchids include living species in China and North America. Comparison of Middle Jurassic and modern cryptobranchids suggests little change in salamander morphology over 150 million years. Other Upper Jurassic salamanders from China along with *Chunerpeton* appear to have been basal to the radiation of modern salamanders and indicate an Asian origin for this group.

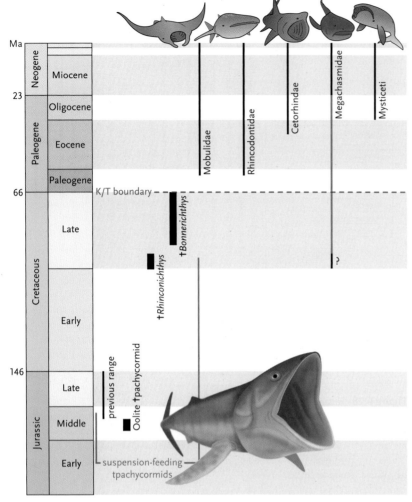

Figure 6.29 Distribution in geologic time of giant pachycormid fishes and modern plankton-eating cartilaginous fish and whales (not to scale). Recent finds (thick vertical lines) have extended the known fossil record of pachycormid fishes from the Middle Jurassic to the Cretaceous/Tertiary (K/T) boundary. The figure shows four lines of cartilaginous fishes (Mobulidae, Rhincodontidae, Cetorhinidae, and Megachasmidae) and the mysticete whales. The † denotes extinct lineages.

SOURCE: From Friedman, M., K. Shimada, L.D. Martin, M.J. Everhart, J. Liston, A. Maltese and M. Triebold. 2010. 100-million-year dynasty of giant planktivorous bony fishes in the Mesozoic seas. *Science*, 327:990–993. Reprinted with permission from AAAS.

6.4e.5 Adaptive Radiation: Dinosaurs. The adaptive radiation of dinosaurs (Fig. 6.31) was an impressive feature of the Mesozoic. However, when the timing of the diversification of dinosaurs is presented (Fig. 6.32), it is clear that they were much less diverse in the Jurassic than in the Cretaceous. On land dinosaurs present another variation on the themes of origin and adaptive radiation, especially when birds are added to the picture.

Figure 6.30 This hellbender, *Cryptobranchus alleganiensis*, belongs to the Crytobranchidae, which has been reported from the Middle Jurassic of China. The fossils and the hellbender belong to the same family of amphibians.

The two main lineages of Dinosauria—saurischians and ornithischians—are readily distinguished by differences such as the structure of their pelvic girdles (Fig. 6.33) but the dinosaur lineage is monophyletic. Five derived characters unite the Dinosauria:

1. an elongated deltopectoral crest on the humerus
2. a brevis shelf on the ventral part of the ilium
3. an extensively perforated acetabulum
4. a transversely expanded subrectangular distal end to the tibia
5. an ascending astragular process on the front of the tibia.

Saurischians such as the theropod *Tyrannosaurus rex* (1445) were carnivorous and bipedal with hollow bones and modified fore and hind feet (Fig. 6.33A). The feet of typical theropods had enlarged second, third, and fourth digits. The digits of the hind limb bore most of the weight, and those on the forelimb usually had sharp claws. Birds are living descendants of theropods. The distribution of theropods and patterns of latitudinal differences among saurischian faunas suggest intercontinental dispersal of these animals.

Some sauropods (ornithischians) (Fig. 6.33C) were enormous, quadripedal dinosaurs that were the dominant vertebrate herbivores of the Jurassic. These animals showed considerable variation in body form, from short- to long-necked, armoured to unarmoured. The very large herbivorous sauropods exceeded the size of the modern megaherbivores such as elephants and rhinos. Food evaluations based on living herbivores suggest that sauropods would have had to have eaten plants such as horsetails (*Equisetum*), monkey puzzle trees (*Araucaria araucana*), ginkgos (*Ginkgo biloba*), and mule's foot fern (*Angiopteris* spp.) because other plants of the Jurassic, such as cycads, tree ferns, and podocarp conifers, were poorer sources of food. Slow fermenting, energy-rich monkey puzzle trees were globally distributed in the Jurassic and may have been the forage of choice for sauropods.

The Theropoda lineage of dinosaurs diverged into two main groups: the Ceratosauria that included Coelophysoidea, and the Tetanurae with many lineages from tyrannosaurs to birds. The Tetanurae included Maniraptora such as *Haplocheirus sollers* (1455), an Early-late Jurassic fossil from China. *H. sollers* provided evidence of an early occurrence of the clade that includes birds, 63 million years earlier than the first specimen of *Archaeopteryx lithographica* (1536). There are obvious similarities in specializations of the forelimb in *H. sollers* and early birds represented by *Archaeopteryx*.

6.4e.6 Feathers. Traditionally, biologists considered feathers to be a diagnostic feature of birds. Our view has changed with repeated discoveries of feathers in other groups, especially theropod dinosaurs ancestral to birds (the Coelurosauria). Troodontidae are the

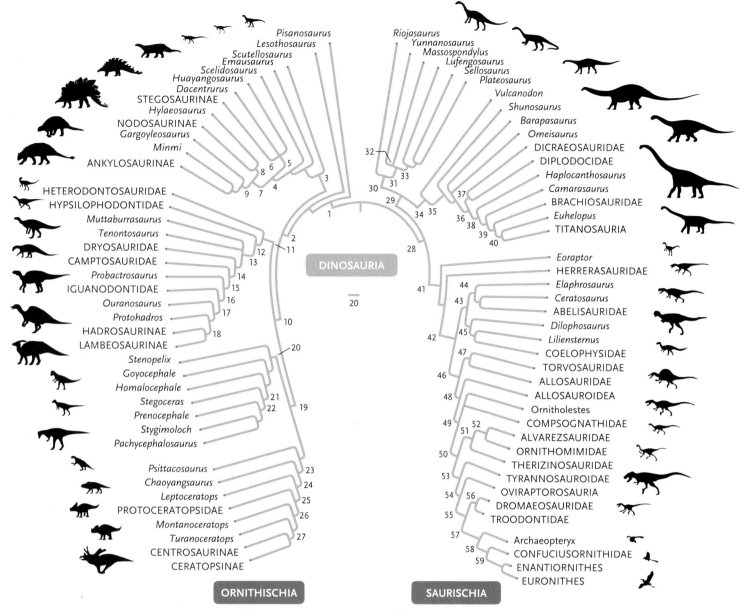

Figure 6.31 The adaptive radiation of Dinosauria is depicted in this 1999 phylogeny, which illustrates the relationships among major groups. The phylogeny is based on analyses of about 1100 characters and shows the numbers of synapomorphies (see Section 6.1b). Scale bar = 20 synapomorphies.

SOURCE: From Sereno, P.C., A.L. Beck, D.B. Duthell, H.C.E. Larsson, G.H. Lyon, B. Moussa, R.W. Sadleir, C.A. Sidor, D.J. Varricchio, G.P. Wilson and J.A. Wilson. 1999. Cretaceous sauropods from the Sahara and the uneven rate of skeletal evolution among dinosaurs. *Science*, 286:1342–1347. Reprinted with permission from AAAS.

theropods most closely related to birds, but although they were not birds, many were feathered, for example, *Anchiornis huxleyi* (Fig. 6.34) from the earliest Late Jurassic in China. This animal had long, pennaceous feathers attached to its legs and feet.

The major radiation of the maniraptoran theropods that produced birds took place in the Middle Jurassic. The distribution of fossil feathers among theropods along this lineage has been used to infer the presence of protofeathers in coelurosaurs, the precursors of maniraptorans. Melanosomes in some fossil feathers would have imparted colour to them. Specifically, quantitative comparisons of melanosomes in feathers of *Anchiornis huxleyi* were used in the reconstruction

of plumage colour (Fig. 6.34). Within-feather pigmentation and patterns such as spots, stripes, and spangles coincides with the appearance of elongate feathers in Maniraptora (*Caudipteryx* in Fig. 6.35).

6.4e.7 Adaptive Radiation: Mammals. Animals classified as mammals are first known from the Late Triassic and Early Jurassic; they lived in the shadow of the dinosaurs. Today we use two physiological features such as endothermy and lactation to distinguish living mammals from reptiles, but dental and skeletal features are more practical for identifying fossils as mammals. Of particular importance is the modification of teeth and bite to enhance chewing and access to

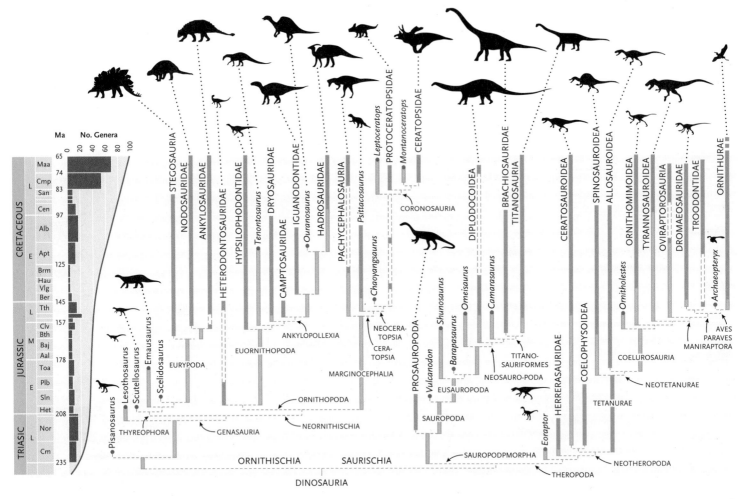

Figure 6.32 The phylogeny of dinosaurs presented in the context of time makes it obvious that the group reached its maximum diversity in the Cretaceous (numbers of genera shown as bar graph on left side of the figure.

SOURCE: From Sereno, P.C., A.L. Beck, D.B. Duthell, H.C.E. Larsson, G.H. Lyon, B. Moussa, R.W. Sadleir, C.A. Sidor, D.J. Varricchio, G.P. Wilson and J.A. Wilson. 1999. Cretaceous sauropods from the Sahara and the uneven rate of skeletal evolution among dinosaurs. *Science*, 286:1342–1347. Reprinted with permission from AAAS.

energy. Changes in the articulation of the lower jaw (mandible) with the skull also influenced mammals' hearing abilities. The auditory ossicles (malleus, incus, and stapes) of mammals are derived from bones that had been involved in the jaw articulation.

Even in terrestrial ecosystems dominated by dinosaurs, mammals diversified and, in the Jurassic,

docodonts showed body forms and life styles comparable to those of modern mammals (Fig. 6.36). Today's mammals include amphibious species with flat tails (e.g., beavers (*Castor canadensis* (1877)) and duck-billed platypus (*Ornithorhynchus anatinus* (1677)), which also appeared in *Castorocauda* (1833), a Jurassic docodont. Today's gliding mammals include several

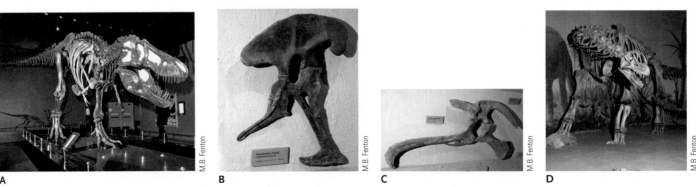

Figure 6.33 **A.** *Tyrannosaurus rex* (1455), a saurischian (theropod) dinosaur and **B.** a "typical" saurischian pelvic girdle. **C.** *Camarasaurus supremus* (1453), an ornithiscian (sauropod) dinosaur, and **D.** a "typical" ornithiscian pelvic girdle. Both exhibited in the Royal Tyrrell Museum in Drumheller, Alberta.

Figure 6.34 Reconstruction of *Anchiornis huxleyi*.

Matt Martyniuk

species of marsupials (sugar gliders, *Petaurus* spp., Marsupialia) and at least three lineages of placental mammals (flying squirrels and flying lemurs, Cynocephalia). The Jurassic *Volaticotherium* (1928) was a much earlier gliding mammal whose phylogenetic relationship to living mammals is unclear. Today the pangolin digs to find termites and ants. The Jurassic *Fruitafossor* (1930) showed similar adaptations of teeth and forelimbs, as do other living mammals such as armadillos and anteaters (Xenarthra).

In the Mesozoic the triconodont mammal *Repenomamus giganticus* (1925) was half again larger than a living Virginia opossum (*Didelphis virginiana*

(1683)), making it much larger than most contemporary mammals which were more like living shrews and rats in size. The remains of a juvenile ceratopsian dinosaur (*Psittacosaurus* sp. (1443)) in the stomach of a well-preserved *R. giganticus* reveal that these mammals preyed on young dinosaurs. *Repenomamus giganticus* demonstrates that some mammals of the Mesozoic were larger than expected, significantly extending their ecological and behavioural potential.

STUDY BREAK

1. Which mammals occurred in the Jurassic? What ecological roles did they play?
2. What evidence suggests that fossil animals are "basal" to the evolution of a group?

6.4f Eocene

The Eocene Epoch lasted from ~54 to ~38 Ma. The transition from Paleocene to Eocene coincided with abrupt climate warming (5–10°C), apparently linked to

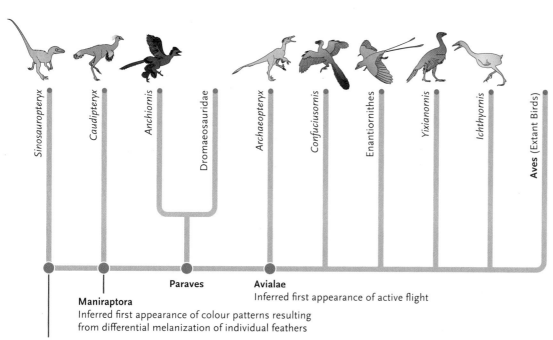

Figure 6.35 Protofeather-like structures appeared by the base of the Coelurosauria, the ancestral line that gave rise to birds. Colour patterns in feathers may have appeared at the same time. Pinnate feathers and within-feather colour patterns are first known from the Maniraptora, in particular the elongated tail feathers of the oviraptorosaur *Caudipteryx* and the troodontid *Anchiornis huxleyi*.

SOURCE: Based on Li, Q., K-Q. Gao, J. Vinther, M.D. Shawkey, J.A. Clarke, L. D'Alba, Q. Meng, D.E.G. Briggs and R.O. Prum. 2010. "Plumage color patterns of an extinct dinosaur," *Science*, 327:1369–1372.

Paraves

Avialae Inferred first appearance of active flight

Maniraptora Inferred first appearance of colour patterns resulting from differential melanization of individual feathers

Coelurosauria Inferred first appearance of colour patterns resulting from variation among proto-feathers

Sinosauropteryx *Caudipteryx* *Anchiornis* Dromaeosauridae *Archaeopteryx* *Confuciusornis* Enantiornithes *Yixianornis* *Ichthyornis* **Aves** (Extant Birds)

 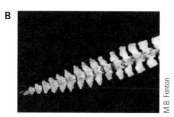  

Figure 6.36 Living mammals similar to Jurassic mammals include amphibious species such as **A.** the beaver which has a flattened tail like **B.** the duck-billed platypus. **C.** Other mammals glide (e.g., a sugar glider), while **D.** others are diggers (e.g., the aardvark).

M.B. Fenton

M.B. Fenton

© Andy Harmer / Alamy

Courtesy of Naas Rautenbach

the release of 1050 to 2100 gigatonnes of carbon. Evidence from carbon isotopes at the boundary indicates more humid conditions in mid-latitude areas. The floral fossil record shows widespread forests (Fig. 6.37). Tropical forests occurred mainly within 15° of the equator, and there is no evidence of xerophytic vegetation types. There is some indication of wet and dry seasons caused by variation in amounts of precipitation. Vegetation types that are adapted to cooler conditions indicate altitudinal gradients in mid-latitudes. Figure 6.37 also shows how the positions of continents differed from today's arrangements.

In this section, we consider plant-insect interactions, the protection provided by shells of bivalves and other animals, and the ways predators circumvent these hard defenses as examples of diversification. The change in feeding behaviour of ray-finned fishes, and their move down the food chain to eat algae, also illustrates diversification. We will also consider how fossils change our view of the evolutionary relationships among birds, and how birds themselves provide feeding opportunities for feather lice. The Eocene appearance of bats and whales is a further demonstration of movement of animals into new ecological situations, as is how *Osedax* worms have exploited feeding opportunities provided by whale bones.

6.4f.1 Plants and Insects. In the Jurassic, plant diversification was paralleled by insect diversification as plant consumers and pollinators. Eocene strata from Patagonia (Argentina) include a wealth of plant fossils from the Early Eocene. A leaf of one of these plants illustrates the damage caused by insects (Fig. 6.38). An example of the insect damage to 6521 leaves, 66% of which represented 186 species of plants, is shown in Figure 6.38. Insect attacks that are evident from frass-filled blotch mines in conifer leaves are strongly reminiscent of the activities of some modern moths in New Zealand or of some beetles in South America (Fig. 6.38A). Other indications are angulate-serpentine leaf mines resulting from the activities of a nepticulid moth in a dicotyledonous plant (Fig. 6.38C). Again the damage resembles that of nepticulid moths on bladderworts (Urtiacaceae) in Micronesia and Polynesia. Some insects make galls in leaves (Fig. 6.38B) and on other parts of plants. Significantly higher rates of damage to fossil leaves from Patagonia compared with leaves from younger Eocene deposits in North America suggest a long history of specialized plant-insect interactions in warm climates. These plant-insect relationships could be the foundation of the current biodiversity of plants and insects in South America.

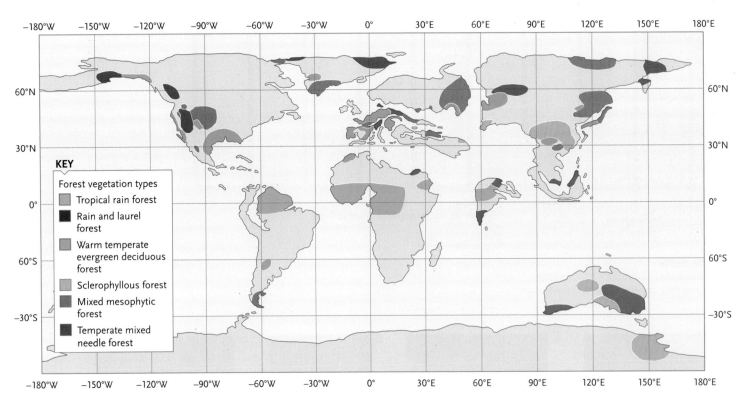

Figure 6.37 Global distribution of forest vegetation types in the Eocene, including tropical rain forest (blue-green), rain and laurel forest (dark green), warm temperate evergreen deciduous forest (middle green), sclerophyllous forest (lime green), mixed mesophytic forest (red), and temperate mixed needle forest (brown).

SOURCE: Reprinted from *Palaeogeography, Palaeoclimatology and Palaeoecology*, Volume 247, Issues 3–4, Utescher, T. and V. Mosbrugger, "Eocene vegetation patterns reconstructed from plant diversity - a global perspective," Pages 243–271, Copyright 2007, with permission from Elsevier.

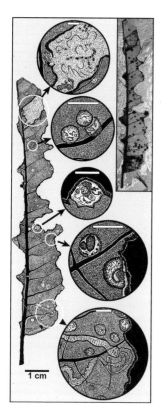

Figure 6.38 In the Eocene, insects attacked plants by mining, galling, and external feeding. Photographed specimen (inset upper right) is shown in detail on the left. From top to bottom: a blotch mine with insect feces (frass); three galls, one with exit holes along a secondary vein, one with the margin consumed by an external feeder, and two galls, one partially consumed; and two linear mines with early sap feeding. Much of the leaf edge has been consumed.

SOURCE: Wilf et al. 2005. "Richness of plant–insect associations in Eocene Patagonia: A legacy for South American biodiversity," *PNAS*, vol. 102 no. 25 8944–8948. Copyright 2005 National Academy of Sciences, U.S.A.

1 cm

A 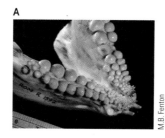 **B**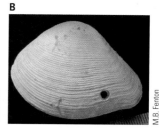

Figure 6.39 There are at least two ways to defeat armour. **A.** Fish like the jolthead porgy (*Calamus bajonado* (1258)), a teleost, just crush armoured shellfish. The lower mandible is well equipped with crushing teeth (scale in mm). **B.** The shell of a bivalve (*Eucrassatella speciosa* (543)) with a hole bored by a predatory naticid gastropod. The fossil is from the Lower Pliocene in Florida.

6.4f.2 Adaptive Radiation: Defeating Armour.

Adaptive radiation also occurs when some species are able to thwart the defenses of others. By the Eocene, there had been a radiation of small, powerful predators capable of crushing protective shells (Fig. 6.39A) These predators include teleosts, stomatopods, and decapod crustaceans. This development coincided with a marked reduction in the diversity and extent of distribution of sessile animals that could not re-attach to the substrate after being dislodged. The communities of benthic animals had also changed from those living on the surface of the sediment to species burrowing into the sediment.

Some predators such as gastropods attacked hard-shelled species by boring holes in their shells. Eocene fossils from Nigeria provide evidence of intraphylum predation in the form of bivalve shells (Fig. 6.39B) holed by a gastropod "labral tooth," an extension of the lip of the shell. Since the Late Cretaceous, labral teeth have evolved at least 58 times, half of them in gastropods in the family Muricidae. Forty-five clades with labral teeth have appeared since the Eocene, 30 originating in tropical waters, apparently at times of high plankton productivity as suggested by the large size of suspension- (filter-) feeding bivalves, barnacles, and gastropods. Eocene fossils from Nigeria indicate predation by two groups of gastropods with labral teeth: naticids that eat mainly bottom-dwelling prey, and muricids that more often take prey at the water's surface.

Defense against crushing attacks appears in gastropod molluscs capable of remodelling chambers in their shells and demonstrating a broad range of shell types (Fig. 6.40). Of particular note is the combination of short spires, narrow openings, strong external sculpture, and dental folds around the apertures. Many of these features, especially remodelling of the interior of the shell, occurred in shells lacking mother-of-pearl, the nacreous shell layer. Smooth and lustrous, **nacre** is widespread in some molluscs (monoplacophorans, some bivalves, gastropods, and cephalopods, see Box 6.3). Nacre is absent from the shells of more evolutionarily advanced prosobranch molluscs. These have cross-lamellar shells that are harder than nacreous shells, offering more protection against shell-crushing predators. Cross-lamellar shells are less energetically expensive to produce than nacreous ones. Changes in the mollusc community reflected an apparent arms race between shelled prey and their predators.

6.4f.3 Adaptive Radiation: Herbivory and Reefs.

Herbivory has often been the driving force in adaptive radiation. Herbivores are aquatic or terrestrial animals that eat parts of living multicellular, attached, photosynthetic plants (i.e., primary consumers). Herbivory is a derived trait that evolved in lineages of animals whose ancestors had eaten animal prey (small and large), as well as those that were scavengers

Figure 6.40 Variety of architecture of shells of living molluscs.
SOURCE: M.B. Fenton

BOX 6.3 Molecule Behind Animal Biology

Nacre

The iridescence of pearls and "mother of pearl" is due to nacre, a biocomposite mostly composed of aragonitic calcium carbonate in an organic matrix of polysaccharides and protein. Nacre is the strongest substance in mollusc shells and, phylogenetically, the oldest. The fracture toughness of nacre is about 3000 times higher than that of aragonitic calcium carbonate by itself. The tensile strength of nacre is 140–170 MPa (N.mm^{-2}). The equivalent value for bone is 130 MPa. Tensile strength is the maximum load nacre can support while being stretched without fracturing. Nacre has a stiffness measure (Young's modulus) of 60–70 GPa (gigaPascals), compared to 2.5 and 5 GPa for bone tissue from adult and young cattle, respectively.

Nacre is a natural ceramic with a microstructure arrangement of "brick and mortar" (Fig. 1), in which the bricks are flat polygonal crystals of aragonite and the organic matrix is the mortar. Its nanostructure is an important key to the strength of nacre, specifically mineral bridges between the crystals of aragonite.

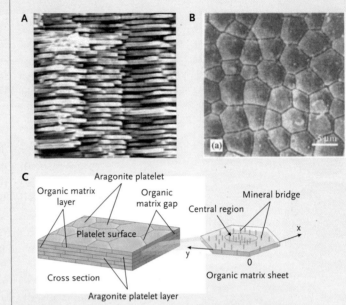

FIGURE 1

Different views of nacre. **A.** *Scanning electron micrograph of a cross-section of nacre showing the brick and mortar structure.* **B.** *Electron micrograph of nacre parallel to the surface of the shell, showing the aragonite platelet surface.* **C.** *Diagrammatic representation of the brick and mortar architecture and the mineral bridges.*

SOURCE: Reprinted from *Biomaterials*, 24, Song, F., A.K. Soh, and Y.L. Bai, "Structural and mechanical properties of the organic matrix layers of nacre," Pages 3623–2631, Copyright 2003, with permission from Elsevier.

---

and detritus feeders. In the Eocene, herbivorous fish had direct access to primary production on reefs, and this move down the trophic ladder resulted in an increase in biomass and diversity of fish. Earlier in Earth's history, invertebrate grazers had been the primary consumers, but compared to most of them, the fish were much more mobile, feeding and dispersing metabolic end products and feces over much larger areas.

Compared to predatory fish, herbivorous actinopterygian (ray-finned) fish had small, forceful jaws. Their lower jaws were the main site of force transmission from the jaw muscles and body to the jaws themselves (Fig. 6.41). In these fish of the Eocene, the lower jaw is short and functions as a single unit. Shorter jaws mean smaller gapes and more forceful bites, but access to a smaller size range of food. Moving down the trophic ladder meant access to larger pools of biomass. A comparison of fish faunas from Triassic to recent times indicates that by the Eocene, the basis for modern reef biotas had been established.

6.4f.4 Birds: Changing Patterns of Evolutionary Relationships. Fossils can change our views of evolutionary relationships. In the Eocene, birds in the fossil family Palaelodidae had a combination of features that could have given rise to grebes (Podicipediformes) and flamingoes (Phoenicopteriformes). The derived (specialized) skull features of flamingoes combined with the grebes' adaptations for hind limb propulsion, confirmed that the two are sister groups despite the striking differences between living representatives. Another example of an unexpected evolutionary relationship between birds involved Eocene fossils of the family Plotopteridae which had a mosaic of evolutionarily derived characteristics suggesting a close evolutionary relationship among penguins (Spheniscidae) and boobies, gannets, cormorants, and anhingas (Sulidea).

The radiation of one group may result in opportunities for others. Birds were undergoing an adaptive radiation during the Eocene, and the recent discovery of a fossil Eocene bird louse (*Megamenopon*

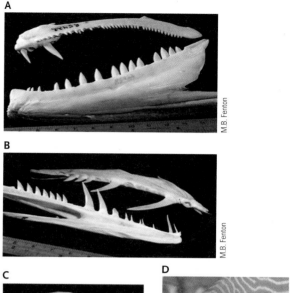

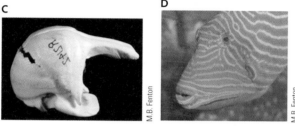

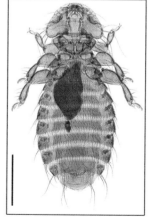

Figure 6.41 While many predatory fish have long narrow jaws and heterodont teeth (**A.** *Sphyraena barracuda* (1287); **B.** *Alepisaurus ferox* (1166)), herbivorous fish (**C. D.** parrotfish) have much shorter and stronger jaws.

rasnitsyni; order Phthiraptera, family Menoponidae; Fig. 6.42) demonstrates an ancient association between birds and arthropods. It is possible that the association between arthropods and feathers could extend back to feathered theropods.

Figure 6.42 A. Fossil *Megamenopon rasnitsyni* (911), an Eocene feather louse, ectoparasitic on birds, showing the complete exoskeleton. Scale bar = 2 mm. **B.** For comparison, an extant *Holomenopon brevithoracicum* (912) from a mute swan. There is evidence of blood in the crop of the fossil and the extant form. Scale bar = 0.5 mm.

SOURCE: Wappler, T., V.S. Smith, and R.S. Dalgleish, "Scratching an ancient itch: an Eocene bird louse fossil," *Biology Letters*, 2004, Vol. 271:S255–S258, by permission of the Royal Society.

6.4f.5 Adaptive Radiation: Whales. The Eocene was also a time of mammalian adaptive radiation. Whales (Cetacea) and bats (Chiroptera) are two groups of placental mammals lacking a Jurassic (Mesozoic) equivalent morphotype (Fig. 6.36). The story of *Indohyus*, the involucrum, and the Cetartiodactyla reminds us how important fossils can be in advancing our understanding of evolution. The Eocene fossil *Indohyus* from Pakistan (an ancestor of whales, Fig. 6.43) had a distinct involucrum, revealing that cetaceans and even-toed ungulates (artiodactyls) are a monophyletic lineage, the order Cetartiodactyla. The **involucrum** (Fig. 6.44) is a thickening of the medial wall of the tympanic bone, a feature shared between *Indohyus* and cetaceans (living and fossil).

The adaptive radiation of whales in the Eocene coincided with exploitation of a "new" ecological feeding opportunity (Fig. 6.45). The raccoon-sized Eocene *Indohyus* lived a hippopotamus-like existence, feeding on plants on shore and moving to the water to avoid predators. Isotopic evidence from *Indohyus* indicated a switch from a diet of vegetation to one of fish, providing an initial feeding opportunity for the ancestors of whales. Echolocation (see page 478) is one feature allowing the toothed whales (odontocetes) to effectively find and exploit fish. Other whales, the mysticetes, developed mass-feeding adaptations to exploit the huge biomass of plankton and the organisms that feed on them. The community of organisms exploited by mysticetes must have been highly productive with efficient energy transfer through short food chains. Diatoms may figure prominently in the evolution of these productive ecosystems, and climate warming also may have played an important role (Fig. 6.45). The Mesozoic presence of other very large planktivores (Fig. 6.29) suggests that adaptive responses to high energy availability have been a recurring theme in the evolutionary history of animals.

The Eocene fossil whale *Basilosaurus cetoides* retained functional pelvic limbs and foot bones

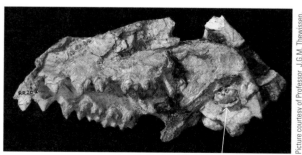

Figure 6.43 Skull of the Eocene *Indohyus* (an even-toed ungulate) which has a well-developed involucrum (white arrow). This feature indicates that whales and even-toed ungulates belong in a single order, Cetartiodactyla. The fossil is about 15 cm long.

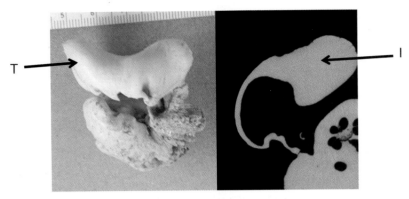

Figure 6.44 A. Lateral view of the boney part of the right ear of a beluga (*Delphinapterus leucas* (1752)) showing the lateral wall of the tympanic plate (T). **B.** A CT scan (bone in white) illustrates the thin wall of the tympanic plate and the thicker involucrum (I). Scale is in mm; dorsal is at top of images.

SOURCE: Image from J.G.M. Thewissen, NEOMED

modern cetaceans, unlike the head-first presentation of modern artiodactyls. Breech birth appears to be an adaptation for giving birth in water. The advantage of breech presentation in cetaceans is keeping the young in a position to obtain oxygen from its mother's blood until it is free of the birth canal and can surface to breathe. In modern artiodactyls, head-first presentation allows the young to breathe on its own through the process of birth.

The adaptive radiation of whales and perhaps marine reptiles brought with it accumulations of bones of these animals on the ocean floor. One group of annelids, the *Osedax* worms, has specialized on this opportunity.

STUDY BREAK

1. How are annelids and whales associated in the Eocene?
2. How did algal blooms change the fauna of the Eocene?

(Fig. 6.46) whose function remains unknown but which might have been involved in mating. Other Eocene fossil whales from Pakistan (genus *Maiacetus*) show evidence of tail-first (breech) birth, typical of

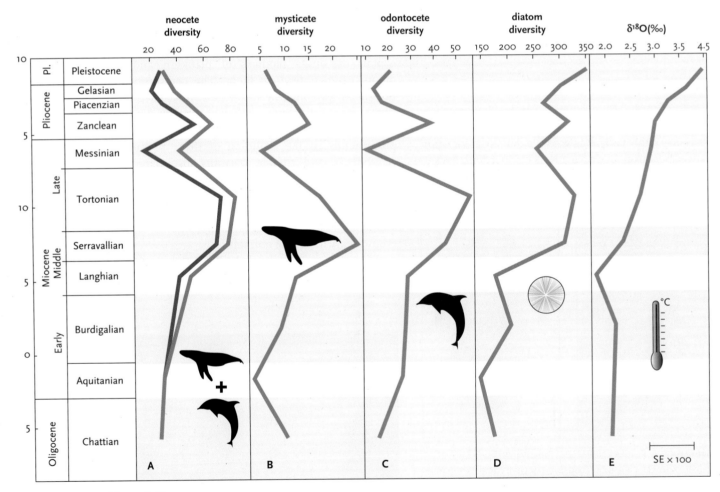

Figure 6.45

From the Oligocene to the present, changes in the diversity of whales (neocetes, mysticetes, odontocetes) and diatoms is shown relative to ocean temperatures (δ^{18} values). These data suggest that the diversity of diatoms has been an indicator of diversity in cetaceans.

SOURCE: From Marx, F.G. and M.D. Uhen. 2010. Climate, critters, and cetaceans: Cenozoic drivers of the evolution of modern whales. *Science*, 327:993–996. Reprinted with permission from AAAS.

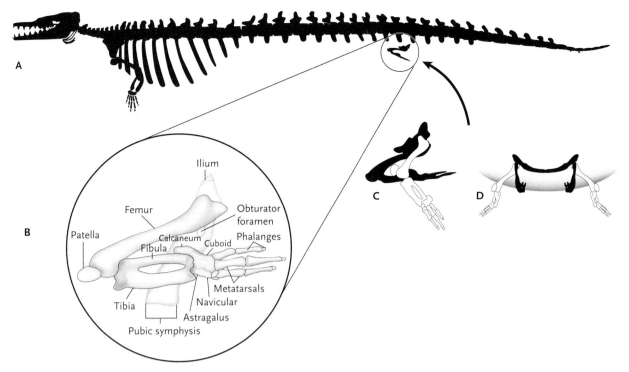

Figure 6.46 Unlike modern whales (Cetacea), the Eocene whale *Basilosaurus cetoides* from Egypt had a functional pelvic girdle and foot bones. **A.** The skeleton, **B.** skeletal details of the left hind limb and pelvis, **C.** the pelvic girdle in resting posture, and **D.** extended posture. The pelvic limbs may have served as guides during copulation.

SOURCE: Based on Gingerich, P.D., B.H. Smith and E.L. Simons. 1990. "Hind limbs of Eocene Basilosaurus: evidence of feet in whales'" *Science* 249:154–157. Gingerich, P.D., M. ul-Haq, W. von Koenigswald, W.J. Saunders, B.H. Smith and I.S. Zalmout. 2009. "New protocetid whale from the Middle Eocene of Pakistan: birth on land, precocial development, and sexual dimorphism," *PLoS ONE*, 4:e4366.

6.5 The Fossil Record

Adaptive radiation and opportunities for animals are two themes that are repeatedly seen in the fossil record. The radiation of one group opens opportunities for others, whether it's whales and then bone-eating worms, or ectoparasites of birds finding shelter in feathers and becoming food for the birds themselves. There is a detective lurking in the mind of any paleontologist, thriving on multidimensional jigsaw puzzles.

Scarcely a month goes by without a new fossil discovery opening the minds of biologists to some other reality. In August 2010, the description of cat-like teeth in Cretaceous fossils of crocodiles made it obvious that there was more to these predators than meets the eye. The combination of the fossil record and genetic tools continues to rewrite the history of life.

6.5a What Were Pivotal Changes for Moving to Land?

On Ellesmere Island in the Canadian Arctic, the discovery in 2006 of *Tiktaalik roseae* (1354) focused our attention on the anatomy of important changes in the pectoral appendage that had occurred by the Devonian. *Tiktaalik*, a lobe-finned (sarcopterygian) fish, could move its pectoral limbs (forelimbs) through a range of postures, including flexing the elbow and shoulder and extending

the distal part of the fin. A comparison of the pectoral girdles and forelimbs of stem genera of tetrapods (Fig. 6.47) highlights the anatomy underlying this degree of limb movement and the changes that occurred as vertebrates became more mobile on land. The move onto land allowed them to move as ponds of water dried up, and gave them access to terrestrial arthropods as food.

6.5b When Did Insects Start Mimicking Leaves?

In addition to exploiting plants as food, many species of insects use plants for concealment by resembling leaves or thorns. Middle Jurassic fossil lacewings from China are preserved in enough detail (Fig. 6.48) that we can see their astonishing resemblance to leaves. The fossil lacewings mimic the pinnate leaves of gymnosperms (Cycadales or Bennettitales) of the time. The evolutionary line that includes these fossils disappeared in the Early Cretaceous at the same time as the plants whose leaves they mimicked disappeared.

6.5c Cretaceous Mosquitoes

The females of many mosquitoes require a blood meal before they can reproduce. When a mosquito bites and feeds on several hosts, it has the potential to spread diseases such as malaria (see pages 471–472 in

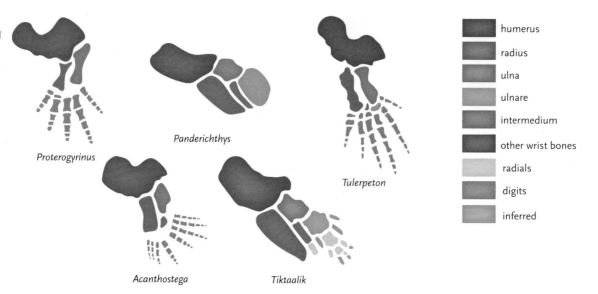

Figure 6.47

Variations in pectoral girdles and fins among tetrapod stem genera. Colour coding identifies the bones (or groups of bones) involved. Note the development of radials in *Tiktaalik*.

SOURCE: Based on "Tetrapods Answer," *Devonian Times*.

Proterogyrinus

Panderichthys

Tulerpeton

Acanthostega

Tiktaalik

| humerus |
| radius |
| ulna |
| ulnare |
| intermedium |
| other wrist bones |
| radials |
| digits |
| inferred |

Chapter 17 and page 521 in Chapter 18). Humans find chemicals that repel mosquitoes invaluable in protecting them from the nuisance and hazards of bites. Amphibians appear to be prime prey for mosquitoes by virtue of their thin moist skin and choice of damp habitats near water. The skins of at least three species of Australian frogs contain mosquito repellents. By applying the skin secretions of frogs to mice, researchers demonstrated that at least the mosquito *Culex annulirostris* were repelled. Frogs and one species of bird, crested auklets (*Aethia cristatella*), are now known to produce mosquito repellents. The history of these chemicals may shed more light on the history of mosquitoes.

Mosquito repellents in the skin of amphibians could indicate a long evolutionary relationship between mosquitoes and hosts. Known from Cretaceous amber from Myanmar, the first fossil mosquito, *Burmaculex antiquus*, lived 100–90 Ma. Its degree of specialization indicates that mosquitoes must be considerably older than previously expected. The characteristics of the single specimen, a female, suggest that it is the sister group of all other living and fossil Culicidae (the family that includes mosquitoes).

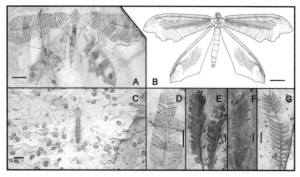

Figure 6.48 A., C. Fossil leaf-mimicking lacewings, **B.** reconstruction, **D.** the forewing, and **E., F.,** and **G.** potential models of pinnate-leafed plants.

SOURCE: Wang et al., "Ancient pinnate leaf mimesis among lacewings," *PNAS* September 14, 2010 vol. 107 no. 37: 16212–16215.

6.5d Animals Large and Small

We began the chapter with sauropod dinosaurs, the largest terrestrial animals ever known to have lived on Earth (Fig. 6.3). Our sampling has included other dinosaurs, as well as whales and other enormous plankton-feeding animals from ichthyosaurs to boney fish. We have not yet discussed the largest known flying animal, *Quetzalcoatlus northropi* (1448), a pterosaur that weighed about 85 kg and had a 12-m wing span. On each continent the emergence of large terrestrial mammals (often related to elephants and rhinos) occurred after the extinction of dinosaurs, suggesting an adaptive radiation to fill ecological niches. In the Carboniferous, the appearance of very large insects coincided with higher levels of atmospheric oxygen (see page 306).

As with living animals, for every gigantic fossil species there were hundreds of small species. Although the gigantic animals catch our attention, some of the most exciting fossils are small to microscopic.

6.5e Birds and Their Ancestors

It has been evident for some time that birds represent an evolutionary lineage that developed within the theropodan dinosaurs. The initial discovery of *Archaeopteryx* gave biologists and palaeontologists a glimpse of something that was "almost" a bird, and generated considerable discussion about two important questions: could *Archaeopteryx* fly? and was it a bird? By 1999, the fossil record provided a good picture of the skeletal changes involved in progressing from theropod to modern bird (Fig. 6.49). Nine important stages are:

1. hollowing of the long bones and loss of weight support as a role for pedal digit I
2. emergence of a rotary wrist joint associated with a grasping hand

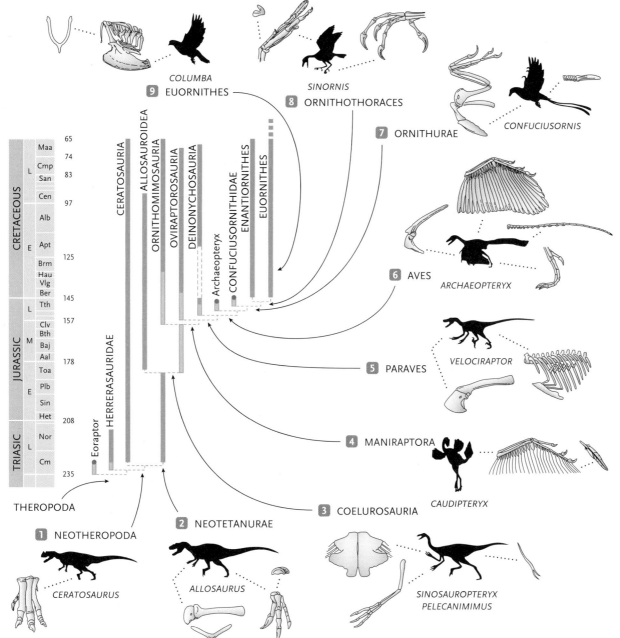

Figure 6.49
There are at least nine important stages in the development of a bird from a theropod. See text for details.

SOURCE: From Sereno, P.C., A.L. Beck, D.B. Duthell, H.C.E. Larsson, G.H. Lyon, B. Moussa, R.W. Sadleir, C.A. Sidor, D.J. Varricchio, G.P. Wilson and J.A. Wilson. 1999. Cretaceous sauropods from the Sahara and the uneven rate of skeletal evolution among dinosaurs. *Science*, 286:1342–1347. Reprinted with permission from AAAS.

3. expansion of coracoids and sternum to accommodate large pectoral muscles and emergence of feathers for insulation
4. presence of vaned feathers
5. shortened trunk and stiffened tail for balance and manoeuvrability
6. appearance of basic flight and perching behaviour (by end of Jurassic)

Three other adaptations are associated with powered flight:

7. deep thorax
8. triosseal canal for housing the main wing rotator muscles (supracoracoideus)
9. the elastic furculum (wishbone) and deeply keeled sternum

A general picture of the phylogeny of the lineage leading to birds (Fig. 6.49) identifies the Deinonychosauria as the sister group for birds. In 2011 the description of *Xiaotingia zhengi* (1535), a Late Jurassic theropod from China, upset this view of bird phylogeny. *Xiaotingia* was morphologically enough like *Archaeopteryx* to be classified in the same family (Archaeopterygidae). But *Xiaotingia* lacked an ossified sternum and uncinate processes as well as other features, meaning that the archaeopterygians (including *Archaeopteryx*) belong to the Deinonychosauria (Fig. 6.49). This dramatically changes our view of *Archaeopteryx* as the first bird. Many biologists and palaeontologists have used a variety of arguments to suggest that *Arachaeopteryx* could fly, while others did not, and still do not, agree.

There is more to birds than feathers, a keeled sternum, uncinate processes, or a furculum. The situation provides yet another example of additional fossils changing our view of the details of the evolutionary relationships among animals.

STUDY BREAK

1. What are the changes associated with moving onto land?
2. What is a bird?

FOSSILS **IN PERSPECTIVE**

The examples explored in this chapter set the stage for the rest of the material in the book. In some cases there are specific links between fossils we have presented and material that comes in later chapters. Please use the electronic library to find information about fossils we have not covered in this chapter. Our examples demonstrate how we can learn about the history of life on Earth, and how knowledge of living organisms can inform our interpretations of those we know only as fossils.

Questions

Self-Test Questions

1. How is the age of a fossil often best determined?
   a. By deposition of minerals such as silica or pyrite within microscopic pores in supporting structures (permineralization)
   b. By the replacement of structural materials such as collagen by harder minerals
   c. By comparison with other fossils found in similar strata of sedimentary rock
   d. By the variety of trace fossils recovered

2. Which of the following is correctly matched with a key distinguishing feature of that time period?
   a. Ediacaran Period; diversification of angiosperms, nectivorous insects, and planktivorous fish
   b. Cambrian Period; radiation linked to appearance of forms with protective coverings and forms with jaws well-suited to crushing shells
   c. Carboniferous Period; earliest evidence of complex multicellular animals
   d. Jurassic Period; diversification of life on land and radiation of amphibians

3. Which of the following is characteristic of an organism whose present-day body form appears unchanged since the Jurassic?
   a. Pachycormids: giant, bony, planktivorous fish
   b. Multiplacophora: plated molluscs (chitons)
   c. Horseshoe crabs
   d. Terrestrial vertebrates such as *Eucritta*

4. What environmental event likely contributed to the mass extinction that marks the Triassic–Jurassic boundary?
   a. An increase in atmospheric carbon dioxide concentration which generated 3°–4°C warming
   b. Extensive glaciation at high latitudes
   c. Rising levels of atmospheric oxygen and appearance of aerobic organisms
   d. Abrupt climate warming (5°–10°C) apparently linked to the release of gigatonnes of carbon

5. Which of the following illustrates an opportunity that opened up for one group through the adaptive radiation of other animals?
   a. Diversification of insects as food and radiation of many herbivores and nectarivores
   b. Diversification of angiosperms for mobile terrestrial amphibians
   c. Diversification of bony fish (pachycormids) and radiation of modern sharks and rays
   d. Diversification of jaws and teeth, and specializations in protective armour

Questions for Discussion

1. How do reconstructions of fossil organisms influence our view of evolution?

2. How do discoveries of new fossil taxa influence our view of phylogenies?

3. Does the fossil record for humans bring us closer to nature?

4. Use the electronic (or other) library to find the latest paper about one of the fossils we have discussed. How does the paper further our knowledge?

Figure 7.1 A six-wheeled Mars rover is adapted to rough terrain by being able to move with two wheels on one side and one on the other touching the ground—a mechanical example of a tripod gait.

SOURCE: Image by Maas Digital LLC for Cornell University and NASA/JPL

Figure 7.1 A six-wheeled Mars rover is adapted to rough terrain by being able to move with two wheels on one side and one on the other touching the ground—a mechanical example of a tripod gait.

SOURCE: Image by Maas Digital LLC for Cornell University and NASA/JPL

7 Locomotion

WHY IT MATTERS

We chose an extra-terrestrial exploration vehicle to lead this chapter because of the stability it achieves with its tripod gait. In this gait, two legs on one side and one leg on the opposite side of an animal are in contact with the ground at each step (Fig. 7.2). This arrangement is stable as long as the animal's centre of gravity is within the area defined by the tripod's legs. A stiff-legged tripod gait maintains stability, even if the terrain is not completely flat. Another advantage is the ease and simplicity of control. A simple six-legged mechanical device can walk on a level surface using only nine servo units (small motors that respond to a control system) (Fig. 7.3A). This parallels life because a cockroach in which all sensory inputs have been severed can still walk on a level surface, using only the motor neurons and connectives in its thoracic ganglia.

An insect leg is jointed and has proprioceptors that register the stress within the leg and the position of each joint. Sensory feedback from these elements is integrated and converted to a signal that modifies the output from the motor programs in the thoracic ganglia, allowing an insect to maintain a steady gait over rough terrain. Even these complexities are easily modelled through the incorporation of

Figure 7.2 **A.** Alternating tripod gait. The legs in contact with the ground at successive steps are marked. **B.** The alternating positions of the right legs of a beetle (*Heliocopris andersoni* (936)) as it walks. On the left, legs 1 and 3 are on the ground (red arrows) and leg 2 is elevated (blue arrow). On the right, legs 1 and 3 are raised (blue arrow) and leg 2 (red arrow) is on the ground.

SOURCE: Courtesy of Naas Rautenbach

sensors, joints, and springs into mechanical models. Simple mechanical robots that move on six legs are common (Fig. 7.3A). Notice that the legs of the model are not on the same front to back line, allowing the legs to swing without blocking each other. This again mimics the living world situation where the swing of the legs is arranged so that each leg can move freely, especially important for animals with many legs (Fig. 7.3B).

Running insects appear to abandon the tripod gait, particularly at high speeds. Cockroaches, which can reach speeds of 1.5 m·s⁻¹ (7.4 km·h⁻¹), run on their back legs and gain stability from the aerodynamic effect of the body which is raised and leaning forward. The fastest running insects, tiger beetles, reach 2.5 m·s⁻¹ (8.9 km·h⁻¹).

SETTING **THE SCENE**

Some form of motility is one of the fundamental characteristics of all animals. This may be expressed in animals that occupy a fixed position in their environment through an ability to pump water or produce water currents, or through movements of appendages to capture food and move it to the mouth. In free-moving animals, some form of locomotory mechanism is the norm. Animals may swim, using fins, appendages, or jets of water. In the water as well as on land, animals crawl, walk, run, and jump. The ability to take to the air and glide or fly has evolved in a number of different animal groups, using the modification of a variety of body parts as wings. The ways in which animals move provide a rich assemblage of examples of convergent and parallel evolution, repeatedly demonstrating how the medium affects movement (in water, in land, in air). Look for recurring themes among animals.

7.1 Introduction

While moving parts are characteristic of animals, not all animals move. **Sessile** animals, such as adult barnacles, move their limbs during feeding (see page 209) but live attached to a substratum. When the substratum is a whale or the hull of a ship, then adult barnacles depend on their host for transport. The larvae of sessile species (see Chapter 5) are typically mobile and responsible for dispersal.

Hitchhiking is common among animals. For example, in many species of anglerfish, adult males live attached to females (Fig. 14.16) and do not move on their own. In species of the land snail genus *Balea* (471), the distribution pattern of adults reflects transport by migrating birds and has led to different species of *Balea* being found in locations between Iceland and Tristan da Cunha. Analysis of the DNA sequences of cytochrome oxidase from snails from different locations has allowed reconstruction of the phylogeny of the different species. Fish such as remoras (Perciformes, Echeneidae, Fig. 7.4) use sucker-like disks on the top of their heads to attach themselves to predatory fish such as sharks. In this case, hitchhiking means being able to feed on the scraps when the animal transporting them feeds.

The locomotion of self-propelled animals is strongly influenced by size. The mass of an animal increases geometrically with length, but the strength of its skeleton increases arithmetically with cross-sectional area (see also page 27). Therefore when an animal doubles in length, its mass increases eight times while the strength of its skeletal elements increases only four times. Size also influences cost of locomotion, involving fuel use (measured by metabolic rate), the size of the animal (a reflection of the muscle mass), the distance covered, and the speed of travel. A comparison of energy costs, expressed as calories per gram per km (cal·g⁻¹·km⁻¹), reveals differences among flying, swimming, and running (Fig. 7.5).

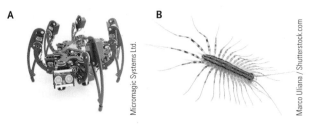

Figure 7.3 **A.** A hexapod mechanical robot. **B.** In centipedes such as *Scutigera* (702), the different lengths of the legs allow their swings to overlap without the legs interfering with each other.

A

B

Figure 7.4 A remora hitching a ride on a green turtle (*Chelonia mydas* (1421)). Inset shows the remora's face.

SOURCE: Geoff Spiby

When animals fly or swim, size also affects the ratio of inertial to viscous forces (expressed through **Reynolds number**, Re; see also pages 207–208). For very small animals, viscous forces dominate; for larger animals, inertial forces dominate. Flying animals range from 1 µg to 75 kg with Re of ~100 to 10^5, respectively. Insects have a range of 1 µg–20 g, bats 1.5 g–1.5 kg, birds 1.5 g–15 kg, and pterosaurs are thought to have ranged in mass from 4 g–75 kg.

When moving in air (flight) and in water, animals are subject to rotation around three axes: the transverse axis (**pitch**), median axis (**roll**), and vertical axis (**yaw**). The density of water supports virtually all the mass of the animals swimming in it, while animals moving on the ground or in the air must support their full mass. However, swimming animals must overcome the resistance posed by water's high viscosity and density.

STUDY BREAK

1. What are the advantages of a tripod gait?
2. Why are Reynolds numbers more important to small aquatic animals than to large swimming animals?

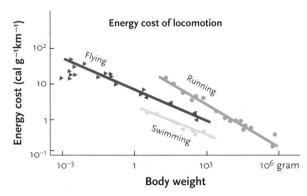

Figure 7.5 A comparison of the energy costs of flying, running, and swimming across a range of body sizes. The single open circle above the line for running is the cost of swimming for a duck. The cost of swimming for a bull sperm is over 10^3 cal·g^{-1}·km^{-1}.

SOURCE: From Schmidt-Nielsen, K. 1972. "Locomotion: energy cost of swimming, flying and running." *Science*, 177:222–228. Reprinted with permission from AAAS.

7.2 Skeletons and Locomotion

Typically, the movement of animals involves muscles, directly in motile animals and indirectly in sessile animals. Usually organized in pairs, muscles operate by contracting (discussed on pages 22–24). In this chapter we will see examples of various combinations that produce movement: muscle-to-muscle pairs or muscles paired with an elastic or flexible element.

Often, the origin of a muscle is the attachment that does not move and the insertion is the part that moves. For example, muscle contraction may move limbs or jaws, compress the body to expel water, or extend a proboscis (feeding apparatus). Where muscles are associated with limbs or jaws, they often attach by apodemes (in arthropods), ligaments, or tendons.

One role of animal skeletons is to transmit the force generated by muscular contraction and translate it into movement. This is a recurring pattern, whether it involves changes in the shape of a hydrostatic skeleton or the position of the legs of an insect, a spider, a bird, or a mammal.

In a locust, the joint between the femur and tibia (Fig. 7.6) is a simple hinge that moves in a single plane. While the femur and tibia appear to be separate tubular elements, they are covered by a continuous cuticle that is heavily sclerotized (hardened) over the tibia and femur, but unsclerotized (flexible) over the joint. Extensor and flexor muscles in the upper part of the femur pivot the

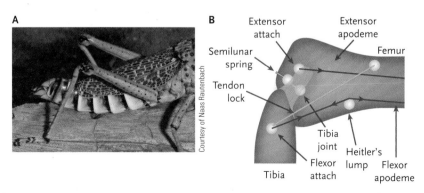

Figure 7.6 **A.** The hind leg of a locust (*Phymateus morbillosus* (903), Insecta, Orthoptera) showing the large femur and the femur-tibia joint. **B.** A model of the locations of the muscle apodemes and pivot points in the femur-tibia joint. Notice the spring element, an important component in the force required for a jump.

tibia about two articulation points with the femur. The muscles are connected to the tibia by **apodemes**. Notice the way in which the flexor apodeme passes over Heitler's lump to change the angle at which the muscle exerts force on the tibia. The extensor muscle is much larger than the flexor and generates about 15 N of force, compared to 0.7 N for the flexor. The extensor powers the jump of locusts and grasshoppers. Positional and slow movement control of arthropod joints is through the balance between flexor and extensor muscles (each has both excitation and inhibition innervation). In some examples, two extensor muscles may operate at different speeds, allowing one to act as a brake on the other.

A flea's jump illustrates the importance of connective tissue spring elements in the joints. Cat fleas (*Ctenocephalides felis* (1059), Siphonaptera) can jump over 100 times their own height, which is equivalent to a human jumping over a very tall building. While the flea's back legs are long and powerful compared with both its body size and its other legs, there is no outward indication of their extreme power (Fig. 7.7). The energy for the jump comes from the rapid release of energy stored in a pad of a rubber-like protein (**resilin**) associated with the joints. This spring is compressed slowly by the leg muscles as they flex and then expands very quickly to release the energy for the jump.

Joints in crustaceans and in many chelicerates operate like those described above. However, the femoral-patella and patella-tibia joints of spiders appear to work differently. In these animals, dissection reveals flexor muscles but no opposing extensor muscles. How, then, do spiders straighten their legs? The answer is that a spider leg is a hydraulic system. Fluid pressure, generated in the body of the spider, pumps blood into the leg and, with assistance from elastic elements in the joints, elevates pressure and straightens the leg. Salticid spiders that jump on prey demonstrate the fine control of this system.

Skeletal elements other than limbs also play important roles in locomotion. A running cheetah uses the articulations of its limbs as well as the extraordinary flexibility of its backbone to increase the length of its strides. Some articulations, such as those between adjacent vertebrae, permit flexibility in one plane, while others, such as joints between elements in a limb (Fig. 7.8), may limit movement to one plane (hinge joints) or more than one plane (ball and socket joints). In the raccoon (and other terrestrial vertebrates), the knees and elbows are hinge joints, and the hip and shoulder connections with femur and humerus, respectively, are ball and socket joints.

Variation in the degree of contact between bones can be considerable. While the lumbar vertebrae of hero shrews (*Scutisorex* spp. (1921)) show extreme specialization with a great deal of bone-to-bone contact, white toothed shrews (*Crocidura* spp. (1919)) do not. The differences in intervertebral contact are obvious when you compare animals with different life styles (Fig. 7.9). Compared to terrestrial mammals, contact between vertebrae of whales, mosasaurs, and other marine animals is limited. Intervertebral contact is consistently higher in terrestrial than in aquatic animals, reflecting the relative densities of air and water.

STUDY BREAK

1. What is resilin and what does it do?
2. How does spider locomotion differ from that of other arthropods?

7.3 Swimming

Two modes of swimming are common in animals; one involves the use of appendages such as fins and flippers (appendicular swimming), the other uses undulations of the body. Many invertebrates as well as skates and rays, frogs, turtles, penguins, manatees, and sea otters use flippers. Many worms, cartilaginous and

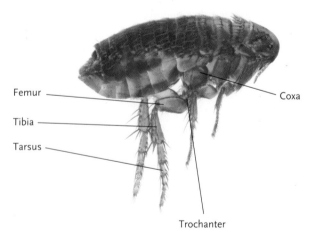

Figure 7.7 A flea from a cat (*Ctenocephalides felis* (1059), Insecta, Siphonaptera). Notice the size of the back leg compared to the front two pairs and the size of the body. Power for a flea's jump comes from muscle and resilin elements in the coxa and femur.
SOURCE: Cosmin Manci / Shutterstock.com

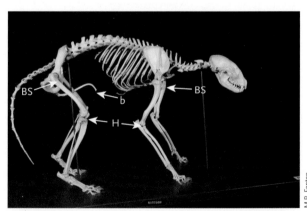

Figure 7.8 Skeleton of a male racoon (*Procyon lotor* (1728)) showing ball and socket (BS) joints, hinge joints (H), and the baculum (b) or penis bone.

BOX 7.1 Molecule Behind Animal Biology

Myosins

Myosins are a large super family of motor proteins that are involved with many aspects of movement. The original proteins were identified in association with muscle, which is where the "myo" prefix originates, but since that time myosin proteins have been identified in most eukaryotic cells in animals of all types including protozoans, as well as fungi and even plants! Although there are variations in the details of protein structure, all myosins are characterized by a globular motor domain that interacts with the filamentous protein **actin** to generate force and motion. These processes are powered by the hydrolysis of ATP, and ATPase enzymes are integral to the structure of the myosin globular "head". Each myosin class also has a distinct "tail" domain that serves in membrane binding, protein binding, and/or enzymatic activities, and targets each myosin to its particular subcellular location. The "neck" region, between head and tail, typically includes a binding site for calmodulin or other regulatory calcium-binding proteins (Fig. 1).

All myosin family members share these structural and enzymatic properties, but their physiological roles encompass a wide range of cellular motile functions including organelle transport, cell division, and muscle contraction. These specialized functions are likely due to structural and enzymatic variations that exist among the various classes of myosin, and individual proteins are assigned to different classes of myosin based upon their structure and cellular origin.

Myosin II is muscle myosin, also known as "conventional myosin" as it was the first to be characterized. It belongs to the class of myosins designated II because it contains two "heavy" chains of about 2000 amino acids in length, each of which has both a "head" and a "tail" domain. The myosin head interacts with the actin, whereas the tails are intertwined, along with four additional "light" chains of myosin, to form the thick filaments of muscle sarcomeres. Multiple myosin II molecules generate force in skeletal muscle through what is called a power stroke mechanism, fueled by the energy released from ATP hydrolysis. When the calcium-binding protein troponin moves the regulatory protein tropomyosin away from the myosin-binding site on actin, binding of ATP to the myosin head allows the formation of a "cross-bridge" between the head of the molecule and the associated actin filament (Fig. 2). There is a conformational change in the myosin molecule with the release of phosphate after ATP hydrolysis, while myosin is still tightly bound to actin, such that the myosin pulls against the actin filament. The binding of a new ATP molecule releases actin from the myosin, and subsequent hydrolysis allows for new binding of the myosin head farther along the actin filament,

with the phosphate release. In this manner, cross-bridge cycling allows the myosin molecule to pull the actin past in, working much in the way that a person might pull hand-over-hand on a rope during a tug of war. The heads of myosin proteins in a filament interact asynchronously with six surrounding actin filaments (Fig. 3), such that the combined effect of multiple cycles of power strokes by myosin heads causes the muscle to contract.

Invertebrate thick filaments are composed of an inner core of **paramyosin**, a rod-like molecule with a high α-helical content, that combines with a second paramyosin monomer to form a coiled coil. This central core is flanked by motor myosin proteins on the surface. Paramyosin is responsible for the "catch" mechanism that enables sustained contraction of muscles with very little energy expenditure, such that a bivalve can remain closed for extended periods. Paramyosin and myosin are the most abundant invertebrate thick filament proteins, although there are other motor proteins unique to invertebrate striated muscles. For example, *D. melanogaster* also possesses miniparamyosin, myosin rod protein and flightin, and *Caenorhabditis elegans* possess several forms of a protein called filagenin. The diversity of

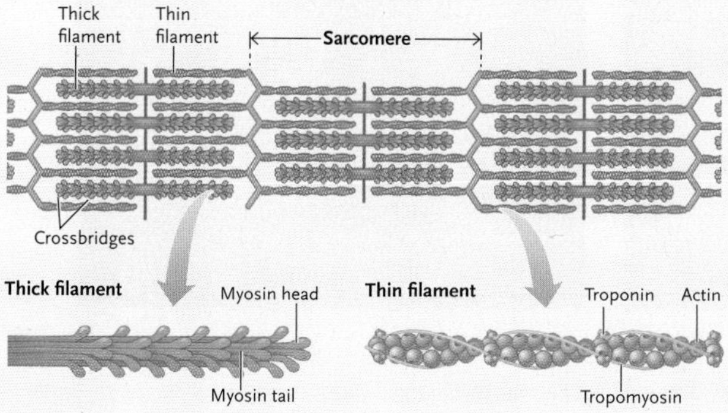

FIGURE 1 *Arrangement of actin and myosin proteins in the sarcomere, the functional unit of skeletal muscle.*

SOURCE: From RUSSELL/WOLFE/HERTZ/STARR. Biology, 1E. © 2010 Nelson Education Ltd. Reproduced by permission. www.cengage.com/permissions

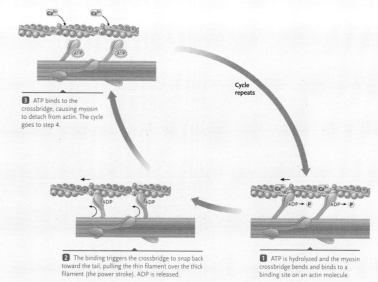

3 ATP binds to the crossbridge, causing myosin to detach from actin. The cycle goes to step 4.

Cycle repeats

2 The binding triggers the crossbridge to snap back toward the tail, pulling the thin filament over the thick filament (the power stroke). ADP is released.

1 ATP is hydrolyzed and the myosin crossbridge bends and binds to a binding site on an actin molecule.

FIGURE 2 *Interaction of myosin heads with actin to slide actin past myosin through crossbridge cycling.*

SOURCE: From RUSSELL/WOLFE/HERTZ/STARR. Biology, 1E. © 2010 Nelson Education Ltd. Reproduced by permission. www.cengage.com/permissions

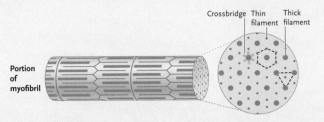

FIGURE 3 *Each myosin filament has myosin proteins which interact with six surrounding actin filaments, such that each actin filament is slid past myosin through the action of crossbridges extending from three myosin filaments.*

SOURCE: From RUSSELL/WOLFE/HERTZ/STARR. Biology, 1E. © 2010 Nelson Education Ltd. Reproduced by permission. www.cengage.com/permissions

thick filament components may account for the highly variable lengths and diameters of muscle thick filaments from different species.

Not all myosin proteins form complex assemblages, as shown by Myosin II filaments. Myosin I proteins, by comparison, are monomers, with a single head and tail region, located in enterocytes, the epithelial cells that line the luminal surface of the vertebrate small intestine. In these cells, the myosin-Ia protein localizes specifically with the brush border. Other animal myosins are specifically associated with the retina of the eye and the cochlea of the inner ear. In both animal and plant cells, intracellular movement of organelles and vesicles, and the complex movements of chromosomes during cell division are also directed by different classes of myosin proteins.

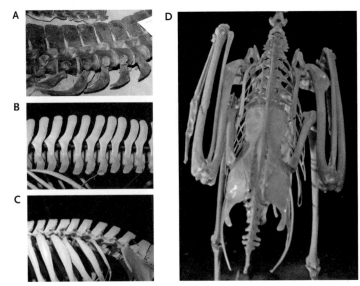

Figure 7.9 Variations in degree of contact among vertebrae: **A.** a marine reptile (a pleisiosaur), **B.** a whale, **C.** a terrestrial mammal (an armadillo) and **D.** a bird. The marine animals have minimal contact between vertebrae while those in the armadillo are extensive. The bird shows fusion of elements.

SOURCE: M.B. Fenton

bony fish, sea snakes, ichthyosaurs, and whales use undulations of the body that are often aided by an expanded tail fin (Fig. 7.10).

7.3a Flippers and Fins

Crustaceans show many examples of appendicular locomotion (Fig. 7.11). Calanoid copepods use long, setose antennae as locomotory paddles (Reynolds number comes into play here: the spacing of the setae and the rate of their movement through the water means that they operate as solid paddles). The posterior pair of legs of swimming crabs is typically expanded and flattened to be efficient paddles. Crayfish and lobsters use the uropods that form their tail fan for rapid backward escape-swimming. Among the gastropod molluscs, sea butterflies swim using an expanded wing-like foot. While most cephalopods are jet propelled, some squid and cuttlefish use fins for swimming.

The flippers of whales, dolphins, and porpoises show considerable interspecific variation. Flippers of

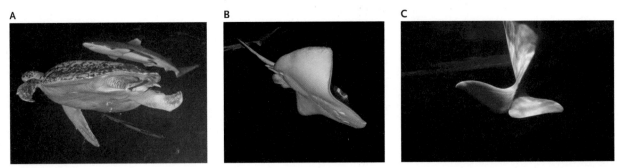

Figure 7.10 Various modes of propulsion by swimming animals. **A.** Propulsion by tail (caudal) fin (undulatory swimming) is typical of a blacktip reef shark (*Carcharhinus melanopterus* (1132)), which uses its pectoral and pelvic fins as stabilizers. The green turtle (*Chelonia mydas* (1421)) powers its appendicular swimming with its pectoral flippers, and uses its pelvic flippers for stabilization. **B.** A sting ray (*Dasyatis* sp. (1140)) powers swimming with enlarged pectoral "wings" (appendicular), and **C.** the tail fluke of a beluga whale (*Delphinapterus leucopterus* (1752)) illustrates undulatory swimming in a vertical plane rather than the more common lateral plane.

SOURCE: M.B. Fenton

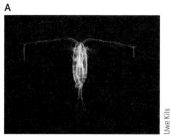

A

B

C

D

Uwe Kils

W. Scott / Shutterstock.com

Strong/Buzeta

© Jeff Rotman / Alamy

E

Bill Kennedy / Shutterstock.com

Figure 7.11 Invertebrates use a variety of structures in swimming (see text). **A.** *Calanus hyperboreus* (730) (Copepoda), **B.** *Callinectes sapidus* (811), **C.** lobsters (*Homarus americanus* (819), Decapoda), **D.** sea butterfly (*Clione limacina* (445), Gastropoda), and **E.** A cuttlefish (*Sepia officinalis* (518), Cephalopoda).

similar shape have similar hydrodynamic performance. Some flippers resemble the swept wings of modern aircraft and, like the aircraft wings, provide increased lift coefficients with increasing angles of attack, reflecting the onset of vortex dominated lift (see Section 7.7b).

7.3b Undulation

The mechanical basis of undulatory, sinusoidal, or serpentine (named for the wave-like motion of many snakes) motion is explained in Figure 7.12. Panels A, B, and C show a wave (marked by the circle) moving toward the back of an animal. Tangents have been drawn normal to the front and rear portions of the wave. At those points, the animal is producing a thrust (at right angles to the tangents) to the right or left and toward the posterior of the animal. Since action (in this case the thrust) and reaction must operate in opposite directions, the reactions at these points are shown in D. The force of the reactions (movement of the animal) can be divided into forward and lateral components (Fig. 7.12E). If the forces of the two parts of the wave are added together (F), the lateral components cancel each other out, and the forward components are joined.

Obvious examples of undulatory locomotion are most eels and many snakes. Perhaps less obvious is the lateral body bend and tail action of many fish and the vertical bend and fluke action in whales. Among the invertebrates, many annelids, nematodes, and the arrow worms (Phylum Chaetognatha) move by undulation.

7.3b.1 Nematoda. Many nematodes move using a very obvious sinusoidal series of waves (Fig. 7.13A). The surprising feature of nematode movement is that it does not depend on the opposed muscle groups (e.g., circular and longitudinal muscle arrangements) that are typical of most animal groups. A cross-section through a nematode (Fig. 7.13B) shows the presence of longitudinal muscles but no circular muscles.

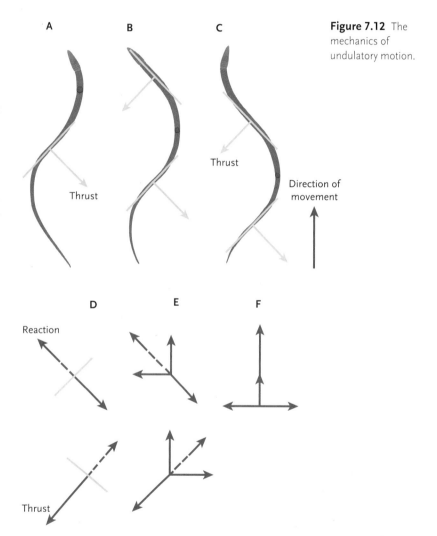

Figure 7.12 The mechanics of undulatory motion.

These longitudinal muscles are arranged as four cords (Fig. 7.13B, C), separated by lateral excretory canals and dorsal and ventral nerve cords. Notice in Figure 7.13C the arrangement of muscles and nerve cord; rather than nerves extending from the nerve cord to the muscle bundles, the muscle bundles have extensions that have synaptic connections with the nerve cord, an arrangement that is unique to the Nematoda.

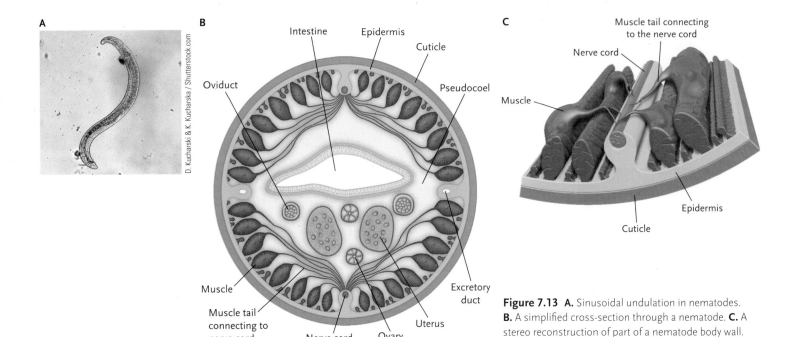

D. Kucharski & K. Kucharska / Shutterstock.com

Figure 7.13 A. Sinusoidal undulation in nematodes. **B.** A simplified cross-section through a nematode. **C.** A stereo reconstruction of part of a nematode body wall.

Nematodes also have a collagen-fibre-reinforced cuticle in which three layers of fibres are arranged at alternating angles, spiralling around the body. The spiral arrangement is important and is a feature of many animal body walls. If the fibres were arranged longitudinally and circumferentially (Fig. 7.14A), the body would be fixed at one length by the longitudinal fibres and one diameter by the circumferential bundles, making it unable to move. The spiral arrangement (Fig. 7.14B) allows the body to lengthen or to

fatten and frequently also adds a spring element to the system.

How do the nematode's longitudinal muscles generate sinusoidal movement? The answer lies in the fluid pressure in its body cavity. In some nematodes this may be as high as 30 kPa. When the longitudinal muscles in one region of the worm contract, they are opposed by the internal fluid pressure. Imagine a sausage-shaped balloon and what happens when you pinch one part of its wall. The balloon bends around the pinch, and, as soon as the wall is released, the internal pressure straightens the balloon. This is a model of what happens in a nematode when two adjacent fields of longitudinal muscle contract and produce a bend on that part of the worm. When these muscles relax, the internal pressure straightens the worm.

7.3b.2 Fish. The same mechanical principle of lateral forces cancelling each other as a partial wave is generated along the body explains the use of the tail in swimming by bony fish and sharks. In this situation, a rigid skeleton appears to be more effective than a flexible one. A swimming bony fish such as a scup (*Stenotomus chrysops* (1289)) moves at 80 cm·s⁻¹, powered by the contractions of red muscles on either side of the body. The muscles generate alternating lateral tail movements that propel the fish. The posterior muscles generate more force and work than the anterior ones (Fig. 7.15). The fish's bony skeleton provides origin and insertion points for the muscles to pull against so that their contractions move the tail.

Lacking an ossified skeleton, some cartilagenous fish (elasmobranchs) generate the force for swimming in a different way. Nurse sharks (*Negaprion brevirostris* (1134)) show dramatic changes in internal pressure

Figure 7.14
A. Longitudinal and circumferential fibre arrangements would inhibit movement.
B. A spiral arrangement of fibres allows elongation and contraction.

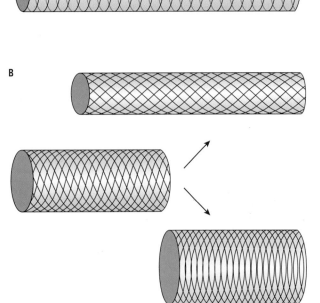

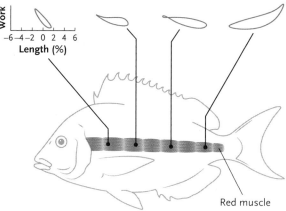

Figure 7.15 In a swimming scup, the patterns of force generation and work generated by red muscles along either side of the fish demonstrate that the anterior muscles produce more force (do more work) than the more posterior ones.

SOURCE: From Rome, L.C., D. Swank and D. Corda. 1993. "How fish power swimming." *Science*, 261:340–343. Reprinted with permission from AAAS.

between resting and swimming. Internal pressures range from 7–14 $N \cdot m^2$ and increase to 20–35 $N \cdot m^2$ when the shark is swimming and to 200 $N \cdot m^2$ when swimming fast. Layers of collagen, the white inner layer of shark skin, are arranged in helices and change from flexible to stiff when placed under tension. This is reminiscent of the tendons of mammals and the mutable (or catch) connective tissue of echinoderms. Changes in internal pressure could arise from changes in areas of compression relative to surface area. The net effect is efficient transfer of energy from muscle contraction to tail power during swimming. The data demonstrate that a skeleton with rigid elements is not essential to locomotion.

Sinuous waves characterize anguilliform locomotion (slipping-wave locomotion), used by swimming fish with eel-like bodies (Fig. 7.16). These waves occur along almost the entire length as each part of the body oscillates laterally across the direction of movement. As the waves pass from head to tail, they propel the animal forward. The general anguilliform pattern resembles the undulatory oscillations of snakes moving on the ground. Many other fish also use undu-

lation to generate forward propulsion, but the movements are restricted to smaller regions of their bodies, such as the tail. The difference is that snakes can push against the static ground, but when fish push against water as they generate forward movement they set up backward water movements. These may form vortices (discussed in relation to Fig 7.35) that can aid the fish in its swimming.

Fish such as sandlances (*Ammodytes hexapterus* (1251)) appear to swim rapidly into sand on the floor of a lake, stream, or aquarium. These fish use slipping wave locomotion when moving through water, but change to non-slipping locomotion as they burrow into the sand. In slipping wave locomotion, the water moves away from the animal when the animal pushes against it. In contrast, during non-slipping wave locomotion the animal pushes against a stationary substrate such as the ground or, in this case, sand. There are three phases in the burrowing behaviour of sand lances. First, the fish uses slipping-wave locomotion as it buries its head in the sand. Second, further movement into the sand is accomplished by more pronounced undulations of the body protruding from the sand. Third, the fish uses non-slipping locomotion when enough of the body is in the sand, combined with undulations of the body remaining above the sand.

Ropefish, *Erpetoichthys calabaricus* (1152), are amphibious, moving between land and water. In the water, slipping-wave locomotion involves short-wavelength, small-amplitude, high-frequency undulations that increase caudad along the body. On land, non-slipping locomotion involves long, slow, large-amplitude undulates that move along the length of the body (Fig. 7.17).

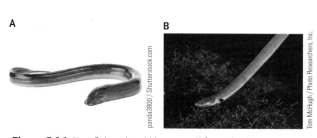

Figure 7.16 Two fish with eel-like (anguilliform) bodies. **A.** Pacific sandlance (*Ammodytes hexapterus* (1251)) and **B.** a rope fish (*Erpetoichthys calabaricus* (1152)).

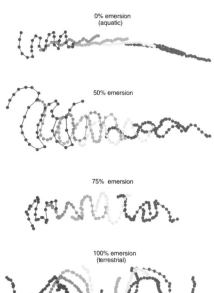

Figure 7.17 Midline traces of points along a ropefish's body (from 0 to 100% total length) documenting variation in the undulatory movements of a single individual from 0% emersed (totally immersed) to 100% terrestrial.

SOURCE: Reprinted with permission from Pace, C.M. and A.C. Gibb. 2011. "Locomotor behavior across an environmental transition in the ropefish *Erpetoichthys calabaricus*." *Journal of Experimental Biology*, 214:530–537. The Company of Biologists Ltd.

The change in wavelength and frequency seen as ropefish move between water and land exemplify a general principle. The same changes are seen in polychaete worms (e.g., *Nereis* sp. (321)) that move from swimming to crawling on the surface and in nematodes that move from a thin film of water, where they push against the surface tension of the film, to deeper water where they swim.

The division between appendicular and undulatory locomotion in fish is not complete—some fish combine the two modes. While some bony fish (e.g., piranha) use the caudal fin as the main source of propulsion and pectoral and pelvic fins for stabilization, others (e.g., a clownfish) use pectoral fins for propulsion and precise manoeuvring (Fig. 7.18).

7.3b.3 Snakes.
Some snakes use lateral undulations when moving in water and/or on land (Fig. 7.19). These undulations are generated by lateral flexion of the vertebrae. When using lateral undulation, snakes push the edges of ventral scales against the substrate, overcoming friction. In this mode of movement, all parts of the snake's body move at once, following a sinusoidal path.

Some snakes also use concertina locomotion and/or sidewinding. In these forms of terrestrial locomotion, one part of the snake's body is stationary and in contact with the substrate, while the rest of the body moves. Snakes often use concertina locomotion to move through small tunnels where wide undulations are impossible, and sidewinding is a common way of moving on surfaces that are smooth or tend to give way, like sand.

The diversity of snakes is illustrated by the arrangements of axial muscles in less specialized booid snakes compared to more specialized colubrid snakes. Among colubrid snake species, different patterns of tendons that interconnect muscles reflect

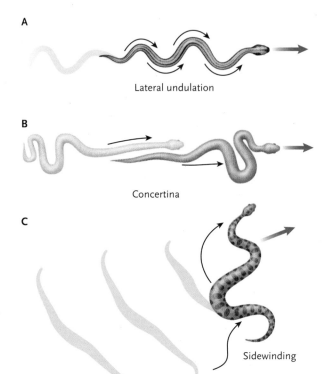

Figure 7.19 A. Many snakes move using lateral undulation, an alternating side-to-side flexion of the vertebral column. **B.** and **C.** A comparison of undulatory, concertina, and sidewinding locomotion in snakes. Large arrows indicate the overall direction of movement, while small arrows indicate the direction of movement of in different regions of the snakes.

diversity as do the percentage of the vertebral column spanned by segments of axial muscles. Further, snakes adjust their patterns of locomotion according to the situation in which they are operating. Some modes of locomotion are more expensive than others (Fig. 7.20).

Many snakes are arboreal, and the rat snake (*Pantherophis* sp. (1515)) from North America is a good example. When choosing a route from one perch to another in a laboratory setting, snakes usually chose routes with shorter gaps. In the horizontal plane, snakes usually went straight rather than taking a 90° turn, but when the gap was shorter if the snake turned, it chose that route. Biomechanics of movement, combined with snakes' ability to perceive appropriate routes, influenced their performance. In Jamaica, the endemic boa, *Epicrates subflavus* (1494), exploits introduced vines and lianas in edge habitats in its search for birds that nest in hollows. The vines increase the snakes' access, while their ability to detect heat (see page 438) makes them more efficient at finding birds' nests.

Modern sea snakes (Hydrophiidae) use their flattened tails to propel themselves through the water by contractions of lateral muscles. Marine hind-limbed

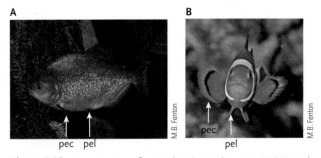

Figure 7.18 A comparison of a piranha (*Serrasalmus* sp. (1187)) and a clownfish (*Amphiprion ocellaris* (1253)) shows differences in the arrangement of pectoral and pelvic fins. **A.** In the pirhana, pectoral (pec) and pelvic (pel) fins are located ventrally, while **B.** the clown fish's pectoral fins are lateral and enlarged, and its pelvic fins ventral and more anterior. Both species power swimming by lateral movements of the tail (caudal) fin, but while clown fish use their pectoral fins for propulsion and for precise manoeuvring (stopping, turning), these fins in piranhas function more as stabilizers.

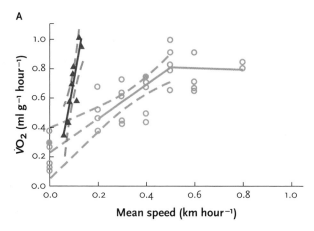

A

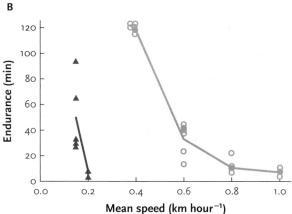

B

Figure 7.20 **A.** The cost of locomotion is shown in snakes (*Coluber constrictor* (1487)) moving by concertina (triangles) and undulatory (circles) locomotion. **B.** shows the speed and endurance.

SOURCE: From Walton, M., B.C. Jayne and A.F. Bennett. 1990. "The energetic cost of limbless locomotion." *Science*, 249:524–527. Reprinted with permission from AAAS.

snakes (e.g., *Eupodophis descouensi* (1496)) of the Cretaceous also had flattened tails, presumably for swimming, but also had hindlimbs (Fig. 7.21). The hindlimbs appear to have been regressive and not specialized for locomotion.

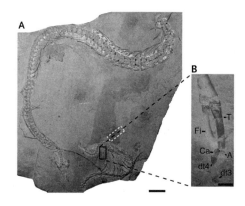

Figure 7.21 **A.** Fossil specimen of the hind-limbed snake *Eupodophis descouensi* (1496) in ventral view. Scale bar = 2 cm. **B.** Close-up of the visible right hindlimb, including tibia (T), fibula (Fi), calcaneum (C), astragalus (A), and two tarsi (dt3 and dt4). Scale bar = 1 mm.

SOURCE: Houssaye, A., F. Xu, L. Helfen, V. de Buffrénil, T. Baumbach and P. Tafforeau. 2011. Three-dimensional pelvis and limb anatomy of the Cenomanian hind-limbed snake Eupodophis descouensi (Squamata: Ophidia) revealed by synchrotron-radiation computed laminography. *J. Vert. Paleontology* 31:2–7, reprinted by permission of the publisher, Taylor & Francis Ltd, http://www.tandf.co.uk/journals.

STUDY BREAK

1. How can a feathery antenna act as a paddle?
2. How is the internal pressure in sharks important to their swimming?

7.4 Walking, Running, and Jumping

Animals that use their limbs to move on the ground (or rocks, tree trunks, etc.) show a great variety of postures and stances (Fig. 7.24). Some of the cursorial (running and walking) and scansorial (climbing) animals contact the surface with two appendages (bipeds: birds, humans), others with four (quadrupeds: chameleons, dogs, cats, and elephants), with six (many insects), with eight (spiders and scorpions), and also with many appendages (millipedes, Diplopoda;

BOX 7.2
Back to the Sea

Returning to the ocean is a recurring theme among vertebrates that had evolved from terrestrial ancestors. Many species of living and extinct reptiles, birds, and mammals breathe air but forage under water. The body forms of reptiles such as ichthyosaurs, and mammals such as cetaceans, resemble those of many fish. Although the body forms are similar, marine reptiles, birds, and mammals use lungs to breathe air, not gills for

extracting dissolved oxygen from water. Larger breath-holding diving animals can dive to greater depths and search for food more efficiently because they swim faster than smaller ones. They can stay underwater longer because they have larger stores of oxygen. Bony fish and sharks that feed or fed on krill do/did not reach the same gigantic size as baleen whales, although at least one species of ichthyosaur (*Shonisaurus* sp. (1440))

was 21 m long, almost the size of a blue whale (*Balaenoptera musculus* (1742)). Body shape in swimming animals affects the cost of swimming, specifically the drag generated as the body moves through the water.

It remains a mystery why diving birds and turtles do not reach the size of some of the extinct marine reptiles or living whales.

centipedes, Myriapoda). Some animals even use their tails when walking, running, or climbing. Animals such as sloths move quadrupedally while suspended below branches. Animals may walk, trot, or run with their entire feet making contact with the ground, with their toes making contact, or even with only their toenails contacting the ground.

7.4a Walking on Land

Walking locomotion is accomplished by a wide range of body forms. Vertebrate examples are shown in Figure 7.22. The range of invertebrate body forms and modes of walking (Fig. 7.23) also illustrate the diversity involved. Overall, animals show a range of locomotory abilities, as illustrated by mammals in Figure 7.24.

Hopping on two legs (Fig. 7.25), also known as richochetal or saltatory locomotion, is another recurring theme among terrestrial animals. In richochetal locomotion, both hind legs strike the ground simultaneously, and the anterior limbs may not touch the ground at all. Many species of grasshoppers, frogs, toads, and some mammals (kangaroos and several independent lineages of rodents) show this specialization. Richochetal animals often have proportionally large hindlimbs and are easy to recognize.

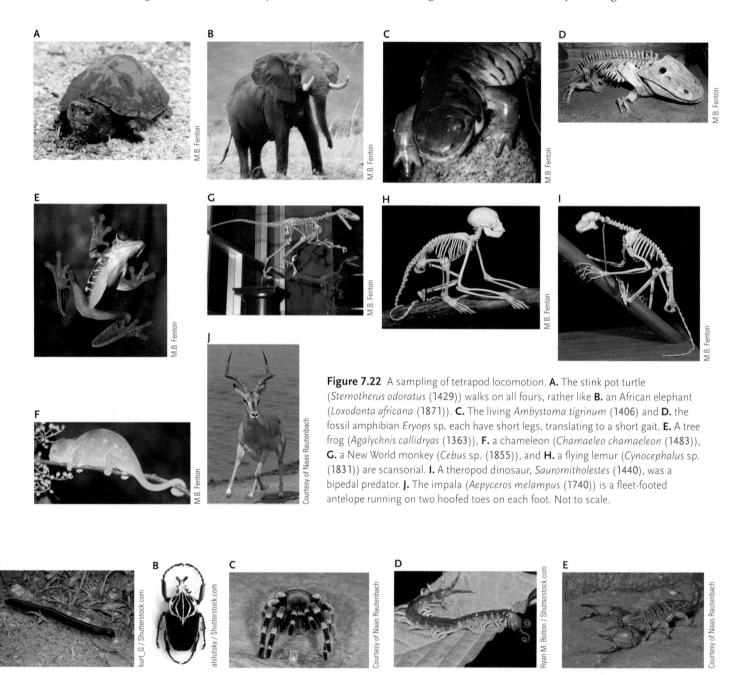

Figure 7.22 A sampling of tetrapod locomotion. **A.** The stink pot turtle (*Sternotherus odoratus* (1429)) walks on all fours, rather like **B.** an African elephant (*Loxodonta africana* (1871)). **C.** The living *Ambystoma tigrinum* (1406) and **D.** the fossil amphibian *Eryops* sp. each have short legs, translating to a short gait. **E.** A tree frog (*Agalychnis callidryas* (1363)), **F.** a chameleon (*Chamaeleo chamaeleon* (1483)), **G.** a New World monkey (*Cebus* sp. (1855)), and **H.** a flying lemur (*Cynocephalus* sp. (1831)) are scansorial. **I.** A theropod dinosaur, *Saurornitholestes* (1440), was a bipedal predator. **J.** The impala (*Aepyceros melampus* (1740)) is a fleet-footed antelope running on two hoofed toes on each foot. Not to scale.

Figure 7.23 Walking arthropods. **A.** Giant millipedes may reach 38 cm long. **B.** Goliath beetles (*Goliathus regius* (934)) can be 10 cm long and weigh up to 100 gm. **C.** A tarantula (*Brachypelma smithi* (663)) may have body length of 10 cm and a leg span of 15 cm. **D.** Centipedes (*Scolopendra gigantea* (700)) can be 30 cm long and prey on small rodents in the Peruvian Amazon area. **E.** A scorpion (*Opistophthalmus glabrifrons* (692) (*Scorpionae*)).

Upright	Non-climbing
	Scansorial (climbing)
	Mainly upright but can hang (quadripedal)
Not upright	Non-climbing
	Scansorial (climbing)
	Hang (quadripedal)

Figure 7.24 Hierarchical relationships among locomotor categories proposed for mammals. Some move in an upright posture (Type A) and show no ability to climb, while others (Type B) are mainly scansorial. Type C identifies animals that are mainly upright but can hang (quadripedal suspension). Others do not use an upright posture and are not scansorial (Type D), while Type E are not upright in posture but are scansorial. Type F are not upright and use quadripedal suspension.

SOURCE: Fujiwara, S., H. Endo and J.R. Hutchinson. 2011. "Topsy-turvy locomotion: biomechanical specializations of the elbow in suspended quadrupeds reflect inverted gravitational constraints." *Journal of Anatomy*, Volume 219, Issue 2, pages 176–191, August 2011. © 2011 The Authors. Journal of Anatomy © 2011 Anatomical Society of Great Britain and Ireland.

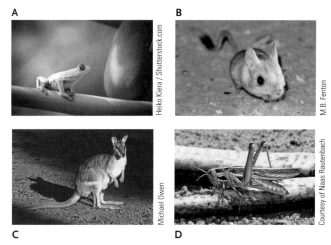

Figure 7.25 Four species that hop or jump using large hind legs. **A.** A frog (*Agalychnis callidryas* (1363)), **B.** a jerboa (*Jaculus jaculus* (1889)), **C.** a wallaby (*Macropus agilis* (1688)), and **D.** a green tree locust (*Cyrtacanthacris aeriginosa* (896)) (not to scale).

The skeleton of a frog (Fig 7.26A) shows specializations for richochetal locomotion in the hindlimbs and pelvic girdle, as well as the shortened thoracic skeleton, relatively small forelimbs, short neck, and skull with large openings (fenestrae) that apparently reduce its mass. The springhare's skeleton (Fig. 7.26B) also shows large hindlimbs with a strong attachment of the pelvic girdle to the spine (an expanded sacral region). The forelimbs are shortened and do not touch the ground when the springhare is hopping. The neck is shortened, partly by being recurved, and a long tufted tail serves as a counterbalance. The shortened neck reduces the danger of whiplash as richochetal animals move along. The details of the skeletal specializations differ considerably among the different evolutionary lineages of rodents and other animals that have adopted richochetal locomotion.

Elephants (*Loxodonta cyclotis* (1872), *Loxodonta africana* (1871) (Fig. 7.22), and *Elephas maximus* (1870)) are unique among mammals in that they use both forelimbs and hindlimbs in propulsive and in braking roles, making them analogous to four-wheel drive vehicles. Because of their mass, they are not light on their feet in the manner of richochetal or scansorial animals. However, the overall leverage on a walking elephant's foot—the ground reaction force at the foot per unit muscle force (effective mechanical advantage)—is about one third of the value predicted by their size alone; in fact it is smaller than that of horses (about 10% of their size). Elephants show a high level of compliance (allowing the limb joints to flex when their feet hit the ground), which minimizes ground reaction forces on their limbs and dampens the impacts associated with acceleration and braking. Increasing speed, however, results in a drastic reduction of effective mechanical advantage on the limbs. Other very large terrestrial animals, for example prehistoric mammals and reptiles, may have shown similar specializations.

As horses increase speed, they transition from a walk to a trot, a canter, and then a gallop. An inexperienced rider knows that staying on a horse can be increasingly difficult as it moves from walk to trot to

Figure 7.26 Skeletal specializations associated with richochetal (saltatory) locomotion as exemplified by **A.** a frog (*Rana* sp. (1380)) and **B.** a South African springhare (*Pedetes capensis* (1894)). **C.** A live springhare.

BOX 7.3 People Behind Animal Biology

Eadweard Muybridge

Eadweard Muybridge (born Edward Muggeridge) was the first to conduct an experimental analysis of animal locomotion. In the early 1870s Muybridge had achieved fame in San Francisco for a remarkable set of images of Yosemite. Producing these images had involved transporting a pile of 24 × 20" glass plates to the field where they were coated, just before use, with a sensitized wet colloidin emulsion. Every photographer was his own chemist and had his own processing secrets.

Leland Stanford, a horse owner and breeder, recruited Muybridge to provide an answer to the question, "Was there any time in a trotting horse's gait when all four legs were off of the ground?" Getting photographic evidence to answer this question was challenging at a time when photographic emulsions were so insensitive that an exposure was timed by taking the lens cap (or a hat!) off the camera lens for some seconds and then replacing it. Muybridge's approach was multifaceted. He worked to improve the sensitivity of his emulsions, brought in new high-quality lenses from Europe, and had a part of the horse track coloured white, with a white fence on the far side to improve the contrast between a dark-coloured horse and the background. Most importantly, he invented a shutter that only allowed a brief exposure as a horse passed a camera.

Muybridge set up a row of cameras, each being triggered as a horse passed in front of them. In early

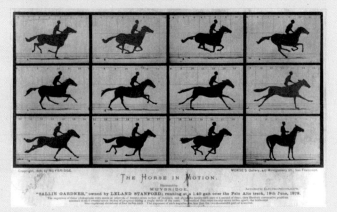

FIGURE 1
Sequential images showing the gait of a galloping horse.

SOURCE: The Horse in motion. Eadweard James Muybridge. "Sallie Gardner," owned by Leland Stanford; running at a 1:40 gait over the Palo Alto track, 19th June 1878.

experiments, triggering was done by the horse breaking threads stretched across the track. Later Muybridge worked with an electromagnetic release and a timer system. One of the products of this work is seen in Figure 1. In Muybridge's words,

> The negatives of these photographs were made at intervals of twenty-seven inches of distance, and about the twenty-fifth part of a second of time; they illustrate consecutive positions in each twenty-seven inches of progress during a single stride of the mare. The vertical lines were twenty-seven inches apart; the horizontal lines represent elevations of four inches each. The exposure of each negative was less than the two-thousandth part of a second.

Frames 2 and 3 in Figure 1 show the footfalls of a galloping horse with all four feet off of the ground. Muybridge also provided pictures of a number of different horses and horse types moving with different gaits and under different loads. He also developed an instrument that he called the zoöpraxiscope. This used a series of his images mounted around a disk that rotated in front of a bulb, allowing an apparently moving image to be projected onto a screen. Using the zoöpraxiscope, he toured the USA, Europe, and England (including presentations to the Linnean Society and the Royal Academy of Art), giving a series of public lectures. Muybridge later discussed his zoöpraxiscope with Thomas Edison; it is clear that Edison's development of moving pictures owed much to Muybridge's concepts and mechanisms.

gallop. Horses use each gait across a relatively narrow range of speeds, opening the question of what rules govern the changes in gait. There are two hypotheses. The first is that horses adjust their gait according to the cost of locomotion, the goal being to maintain low costs. The second is that musculoskeletal forces trigger the changes in gait to reduce the chances of skeletal injury. Data produced by adding weights (up to 23% of body mass) to three Shetland ponies walking, trotting, or galloping on a treadmill while measuring their oxygen consumption supported the second hypothesis (Fig. 7.27). Changes in gait resulted in a 14% reduction in ground reaction force, reducing the risk of injury.

Different ponies consistently changed gait at different points, illustrating the range of variation often found in a single species.

STUDY BREAK

1. What skeletal adaptations accompany richochetal locomotion in different groups of animals?
2. Explain the changes that accompany different gaits (walk, trot, canter, gallop) in horses. How do these changes correspond to those in humans from walk to speed walk to run?

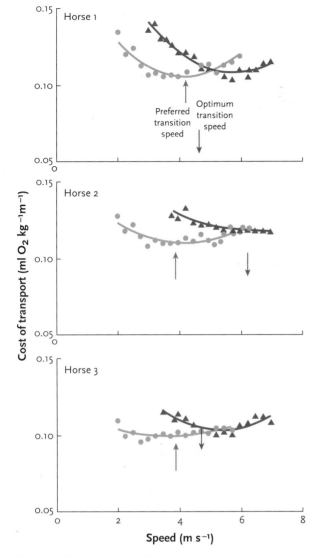

Preferred transition speed

Optimum transition speed

Figure 7.27 When three Shetland ponies were studied to determine at which points they changed from walk to trot (open circles) and from trot to gallop (closed triangles), it was clear that skeletomuscular forces stimulated the changes. Further, as measured by oxygen consumption, the changes in gait did not correspond to the energetically optimal speed.

SOURCE: From Farley, C.T. and C.R. Taylor. 1991. "A mechanical trigger for the trot-gallop transition in horses." *Science*, 253:306–308. Reprinted with permission from AAAS.

7.4b Walking on a Smooth Surface

Some animals, such as geckoes and many insects, can cling to and move effectively on very smooth surfaces; they can walk on the ceiling or on glass. Being able to adhere to a surface provides the traction needed to move on it.

Van der Waals forces are weak forces involved in molecular interactions other than those responsible for the formation of covalent bonds or ionic interactions. Van der Waals forces only act over very short distances (less than 0.5 nm), as illustrated by the way in which two very clean microscope slides seem to stick together, an adhesion that disappears if a few dust particles are trapped on their surfaces. Van der Waals forces are

fundamental to how animals attach and move on walls and ceilings.

If Van der Waals forces only operate at very short distances and between absolutely smooth surfaces, how does a textured and flexible animal appendage grip a surface, such as a painted ceiling, that has a textured roughness? The answer lies in the animal having very large numbers of very fine attachment endings, allowing the overall attachment great flexibility in accommodating surface roughness. Tokay geckoes (*Gekko gecko* (1498), Fig. 7.28A) have clumps of setae on their feet (Fig 7.28B), and each seta (Fig. 7.28C) ends in hundreds of finely divided, spatulate tips (Fig. 7.28D). These are the actual sites of attachment. When a gecko detaches its foot, the toes are hyperextended (Fig. 7.28E), breaking the attachment to the surface. The gecko's attachment mechanism is so strong that it can hang on a vertical wall attached by a single toe.

Many insect groups use a similar attachment mechanism with parts of their legs, or tarsi, covered in fine setae ("hairy"). Other groups have a highly flexible pad. Either morphology can conform to both smooth and rough surfaces (Fig. 7.29A). The important difference between insects and geckos is that insects secrete an attachment fluid at each step, this fluid being a mixture of water-soluble and lipid-soluble components (Fig. 7.29B). The importance of the attachment fluid was demonstrated by washing the attachment pads of a bug, *Rhodnius prolixus* (876), with an organic solvent; the procedure impaired the bug's attachment. It appears that capillary forces as well as Van der Waals force are important in insect attachment.

Insect attachment structures are located on different parts of the leg in different groups (e.g., Fig. 7.30) and have evolved independently a number of times. In some insects (e.g., the housefly *Musca domestica* (966)), the gait may be modified to provide better attachment. Houseflies walking on a horizontal surface use a tripod gait (see Why It Matters) but when

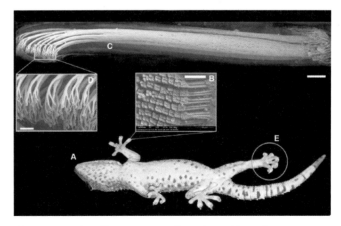

Figure 7.28 The adhesive system of a gecko (*Gekko gecko* (1498), Reptilia, Squamata).

SOURCE: Reproduced with permission from K. Autumn, A. Dittmore, D. Santos, M. Spenko and M. Cutkosky, "Frictional adhesion: a new angle on gecko attachment," *J Exp Biol* 209 (2006) 3569–3579. The Company of Biologists Ltd.

they are on an inverted surface, the gait is modified so four legs are in contact with the ground at each step.

Bats in the families Thyropteridae and Myzopodidae have discs on their wrists and ankles (Fig. 7.31) that allow them to adhere to and move on the smooth surfaces of leaves. While the discs of thyropterid bats (e.g., *Thyroptera tricolor* (1818)) work by suction, those of myzopodids (e.g., *Myzopoda aurita* (1802)) work by wet adhesion. The distinction between these two approaches is clear for two reasons. First, in wet adhesion by myzopodid bats, adhesion is stronger when the bat is pulled parallel to the surface than when it is pulled perpendicularly. Second, these bats can adhere to a perforated surface (which prevents the development of suction). Thyropterid bats that adhere by suction adhere well to a vertical flat surface, but not to a perforated surface. The same principles apply to other animals using wet adhesion such as some ants and tree frogs.

7.4c Walking on Water

Water striders (Hemiptera, Gerridae) are true bugs that live on the surface of water, with their mass supported by the surface tension (Fig. 7.32). The striders move freely over the water surface and feed on both live and dead insects trapped on the surface. The dimples formed under each foot may be over 4 mm deep before a leg actually breaks through the surface layer. The

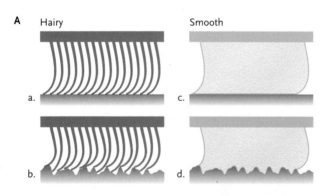

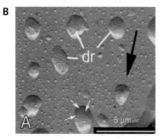

Figure 7.29 A. The way that hairy (a and b) and pad (c and d) attachment structures on insect legs and tarsi can conform to smooth (a and c) and rough surfaces (b and d). **B.** An electron micrograph of a carbon-platinum replica of the footprints of a fly (*Calliphora vicina* (957)). Notice that the main droplets (dr) contain microdroplets (white arrows) of an immiscible fluid.

SOURCE: A. Based on Gorb, Stanislav N. 2005. "Uncovering insect stickiness: Structure and properties of hairy attachment devices." *American Entomologist* 51(1): 31–35; B. Gorb, Stanislav N. 2005. "Uncovering insect stickiness: Structure and properties of hairy attachment devices." *American Entomologist* 51(1): 31–35.

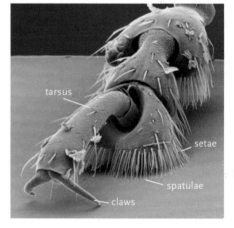

Figure 7.30 A false colour SEM of the tarsus of a beetle (*Gastrophysa viridula* (933), Coleoptera, Chrysomelidae) attached to a smooth surface.

SOURCE: Gorb, Stanislav N. 2005. "Uncovering insect stickiness: Structure and properties of hairy attachment devices." *American Entomologist* 51(1): 31–35.

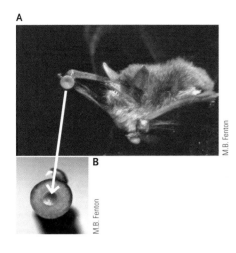

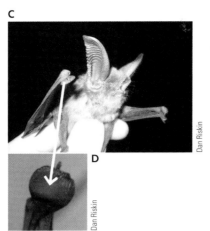

Figure 7.31 Adhesive disks have evolved in two families of bats (Thyropteridae, **A.** & **B.**; Myzopodidae, **C.** & **D.**). In each case the bats have disks on their wrists (A, C) and ankles. The discs look different (B, D) because bats in the genus *Thyroptera* use suction, and those in *Myzopoda* use wet adhesion. *Thyroptera* spp. occur in South and Central America, *Myzopoda* spp. occur in Madagascar. These bats roost on smooth, slippery surfaces.

depth and diameter of these dimples are greater than might be formed by the physical dimensions of the insect's leg. This discrepancy is explained by the physical structure of the leg, which is covered by an ordered array of microsetae that trap air and produce a highly hydrophobic surface.

When the water strider moves, the tips of the second pair of legs act as oars, sweeping backward and driving the dimple around the legs back while moving the insect forward. Successive sweeps leave a horseshoe-shaped series of vortices behind the insect.

Rove beetles (Coleoptera, Staphylinidae) in the genus *Stenus* (948) (Fig. 7.33) live on and near bodies of water and may be seen skimming on the water surface without moving their legs or wings. They do this by secreting a surfactant substance (identified as N-ethyl-3-(2-methylbutyl) piperidine, mixed with smaller amounts of 1,8-cineole, isopiperitenol and 6-Me-5-hepten-2-one) from a pair of pygidial glands located either side of the anus. The effect of lowering the surface tension behind the beetle is to drive the animal forward.

The surface tension reduction mechanism can be demonstrated in dead whirligig beetles (Coleoptera, Gyrinidae). When specimens of *Dineutus hornii* (932) were killed by freezing and then thawed, their pygidial glands became extended. When placed on a water surface, the dead beetles moved at speeds of up to $6 \text{ cm} \cdot \text{s}^{-1}$ for up to 20 s. When the dead beetles stopped moving, their pygidial glands were shown to be empty. In this example of what has been called "expansion swimming," the glands release the surfactant norsesquiterpene that may also function as an alarm pheromone. In life, a release of surfactant may give a whirligig beetle a speed boost to escape a predator.

Insects are light enough to be easily supported by the surface tension of water, but Jesus Christ lizards (basilisk lizard, *Basiliscus plumifrons* (1471), Reptilia, Squamata) (Fig. 7.34) weigh up to 200 g yet can run on water. Steven Vogel, a biophysicist, has calculated that the area of his size 9 sandals would support only 9.2 g. The splashes in Figure 7.34 are a result of the lizard's foot, held parallel to the surface, slapping the water. However, this provides a limited contribution to supporting the mass of the lizard. Most support comes from an air pocket that is formed under the foot as it strokes downward and backward while flexing the foot back. The foot is raised again before this air cavity collapses with the toes pulled together and the foot straightened so that it parallels the direction of movement (Fig. 7.35).

Figure 7.32 *Gerris remigis* (870) on a water surface. The short front legs do not rest on the water and are used for prey capture and handling. The long second pair of legs acts as oars in locomotion. Water striders may move as fast as $150 \text{ cm} \cdot \text{s}^{-1}$.

Figure 7.33 *Stenus comma* (948) (Coleoptera, Staphylinidae). In still water *Stenus* spp. use their legs to help swimming, supported by surfactant release. In fast water the legs are held against the body and propulsion depends only on surfactant release. In either case the beetles swim toward a dark object—a photonegative response that takes them back to the bank.

Figure 7.34 *Basiliscus plumifrons* (1471), the Jesus Christ lizard, running across the surface of a stream.

SOURCE: Bence Máté Photography, www.matebence.hu

Figure 7.35 Vortex formation during the "slap," "stroke," and "recovery" phases of a basilisk lizard step. Notice the changes in foot position at each phase of the step. The direction of travel is to the left.

SOURCE: Hsieh, S. T. (2004). "Running on water: Three-dimensional force generation by basilisk lizards." *PNAS* 101 (48): 16784–16788. Copyright 2004 National Academy of Sciences, U.S.A.

BOX 7.5
Gut-Propelled

Some caterpillars, such as the tobacco hornworm (Fig. 1), take advantage of the fact that their flexible, fluid-filled body and gut are essentially two separate hydrostatic compartments. The gut is anchored at the mouth and anus, but can slide freely inside the body of the animal. When the caterpillar lifts its hind feet and moves the posterior end of its body forward, the body slides over the gut, compressing it. Then the caterpillar lets go of the substratum with its anterior feet and the gut decompresses, pushing the anterior end of the caterpillar forward. These caterpillars can be considered gut-propelled. The sequence of movement is explained in Figure 2. Discovering the separate movement of the gut and body wall required a combined application of phase contrast synchrotron X-ray imaging and light microscopy to superimpose internal movement onto movement of the body wall.

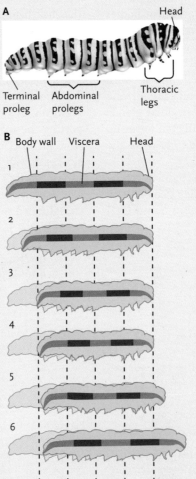

A · Head · Terminal proleg · Abdominal prolegs · Thoracic legs

B · Body wall · Viscera · Head

Figure 2
Body wall and visceral motion during caterpillar locomotion. **A.** *Anatomical sketch of a caterpillar.* **B.** *Example schematic of the motion of the body wall and the viscera of a caterpillar during a step. The body wall is represented in green and viscera are represented as a line of light and dark boxes within the animal. The body wall and viscera have been schematically divided into fifths by dashed lines so that the movement of the body wall can be clearly differentiated from visceral movement. In diagrams 2–6, the initial position of the caterpillar is shown in red. From rest (1), the posterior-most portion of the body wall compresses and moves forward, compressing and moving the viscera forward (2). Note that the viscera of the middle section have moved forward while the body wall of the middle section has simply compressed. The posterior of the animal continues to compress and move forward (3), until the terminal proleg is placed on the ground (4). This motion has compressed and moved the viscera forward. Then the middle and anterior portions of the body wall are moved forward, sliding along the viscera (5), until the viscera and body wall are back in alignment (6).*

SOURCE: Reprinted from *Current Biology* 20(16), Simon, M. A., Woods, W. A., Jr., Serebrenik, Y. V., Simon, S. M., van Griethuijsen, L. I., Socha, J. J., et al. "Visceral-locomotory pistoning in crawling caterpillars." Pages 1458–1463, Copyright 2010, with permission from Elsevier.

Steve Bower / Shutterstock.com

Figure 1
A caterpillar of the tobacco hornworm (Manduca sexta (1041), Lepidoptera).

The "slap" phase of the running movement provides a greater contribution to support in young lizards than in mature specimens. This explains the different water-running abilities of young and mature animals. Young lizards can generate support forces of more than double their body mass while adults produce barely enough upward force to support their body mass.

When terrestrial animals run, the hindlimb flexes while in contact with the ground, indicating that energy is being stored in the tendons and ligaments. When basilisk lizards run on water, the hindlimbs do not flex, indicating that the limb acts more like a piston than a spring, only producing force during its downward movement. Analysis of water movement behind basilisk lizard steps shows, as in water striders, the formation of vortices rotating outward from each side of the foot and trailing behind the step (Fig. 7.35).

STUDY BREAK

1. How do Van der Waals forces operate in the movement of an insect on a vertical surface?
2. Explain how a basilisk lizard's weight is supported as it runs on water.

7.5 Creeping

Animals without limbs face the challenge of finding a grip, or resistance, against which they can develop a propulsive force. The variety of ways that animals achieve this fall into two main categories: direct grip mechanisms (suckers, claws, glue, etc.) and undulation involving partial or complete waves of change in body form to develop force by pushing against the medium or substratum (see Section 7.3b).

Some animals use hooks or suckers to grip the substratum. Leeches (Annelida, Hirudinea) have anterior and posterior suckers. By attaching the posterior sucker, extending its body, attaching the front sucker, releasing the posterior one and drawing the posterior forward, the animal can advance. *Hydra* (135) uses a similar approach when it bends, grips the substratum with its tentacles, releases its basal attachment, and somersaults forward. Mammals such as canids (dogs and their relatives) use claws to gain purchase on surfaces. Most felids (cats) use retractable claws to hold prey or hold onto surfaces. The cheetah (*Acinonyx jubatus* (1705)) is the only cat with non-retractable claws, and its feet, toes, and claws are more like those of canids. Still other animals, such as gastropod molluscs (snails and slugs), secrete mucus to temporarily attach to the substratum.

As it crawls on a glass plate, the ventral surface of a slug foot shows a series of waves (Fig. 7.36) moving along the foot. Each wave has a region where the foot is lifted away from the substratum. In some species, locomotory waves may be direct (the wave moves from

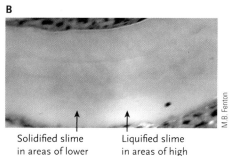

Solidified slime in areas of lower stress

Liquified slime in areas of high stress

Figure 7.36 A. A garden slug (*Limax maximus*) leaving an obvious mucus trail as it moves. **B.** A ventral view of the foot showing regions of liquid and gel mucus. Further details in text.

the back of the foot toward the anterior, the same direction as the animal is moving); in others it is retrograde (the wave moves from front to back along the foot, in the opposite direction to the animal's movement).

The morphology and mechanics associated with direct and retrograde waves are slightly different. Figure 7.37A shows the elements involved in a retrograde wave. Contraction of the dorso-ventral muscles

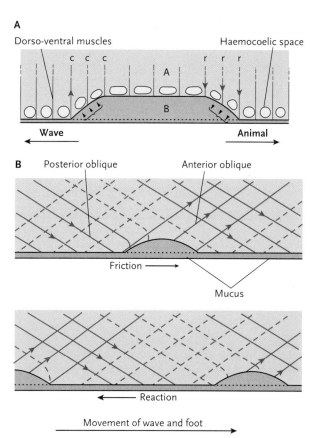

Figure 7.37 The morphological elements involved in **A.** a retrograde wave and **B.** a direct wave in the forward movement of a snail.

lifts the sole of the foot and distorts the hemocoelomic cavities; transverse muscle prevents the foot from expanding sideways so that this distortion results in the elongation of the raised portion of the foot, driving the animal forward. When the dorso-ventral muscles relax, the combination of suction under the raised portion of the foot and the elasticity of the hemocoelomic cavities draws the foot back down. Direct waves (Fig. 7.37B) are produced by anterior and posterior oblique muscles (which branch from a layer of longitudinal muscles). The anterior oblique muscles lift the foot and draw the lifted portion forward. The posterior oblique muscle draws the body of the snail forward, toward the anchored region behind the wave.

A further complication in looking at wave forms on snail feet is that in some species a wave may cross the entire foot (monotaxic) while in other species there is a division at the midline of the foot with direct or retrograde waves running out of phase on the two sides (ditaxic) (Fig. 7.38).

Snail mucus is unusual because it can be both a glue and a lubricant. When the snail is at rest, the 10–20 μm layer of mucus between the foot and the substratum is a thick, cross-linked gel. As the foot is pulled by the locomotory muscles, the mucus is placed under strain; at a critical strain level the molecular cross links in the gel separate, and the mucus becomes a fluid that lubricates the movement of the foot over the substratum. When the strain on the mucous layer is relaxed, the cross links are repaired, and the mucus returns to its gel state.

The metabolic cost of gastropod mucus production has been estimated to be 7–31 times the costs of locomotion. This means that any reduction in the need for mucus represents a significant metabolic benefit. Gastropod species have been shown to follow mucous trails of conspecifics in both marine and terrestrial environments. Trail-following snails can read the polarity of a trail, leading to feeding or protective aggregations in some species. Others, such as limpets, use mucous trails to return to a "home" spot. Trails may also provide some food as an aging trail will have been colonized by microorganisms that become food for the follower. More importantly, a snail following a trail uses the existing mucus as a base and reduces the amount of mucus it secretes as it moves.

Some snails (e.g., the moon snail, *Polinices duplicatus* (427)) exhibit a power gliding movement that uses muscular locomotion combined with pedal cilia. The movement does not involve muscular waves passing over the foot surface. This movement is unusual because moon snails are probably the largest animals using cilia for locomotion, and the locomotory cilia appear to be under neural control.

7.6 Burrowing

Burrowing—moving into a soft substratum—is a defensive strategy that appeared in the early Cambrian (see page 152). Animals that live most of their lives underground and burrow extensively are described as "fossorial". Today, burrowing is common across a variety of animals. In annelids such as earthworms (Fig. 7.39), the anterior segments of the worm (toward the bottom of the burrow) are swollen, compared to the posterior segments. At the moment that the photograph in Figure 7.39 was taken, the longitudinal muscles in the anterior segments had contracted around the hydrostatic skeleton in each segment, making those segments shorter and fatter. At the same time, muscle strands derived from the longitudinal muscle bundles

Figure 7.38 The feet of various snails and slugs, showing examples of direct, retrograde, monotaxic, and ditaxic wave forms.

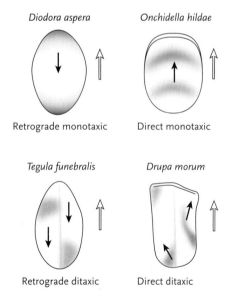

Diodora aspera

Retrograde monotaxic

Onchidella hildae

Direct monotaxic

Tegula funebralis

Retrograde ditaxic

Drupa morum

Direct ditaxic

Figure 7.39 Earthworm burrowing (for explanation see text).
SOURCE: DR JEREMY BURGESS/SCIENCE PHOTO LIBRARY

contract to extend the worm's setae, adding to its grip on the sides of the burrow. This fixes the worm in place and allows longitudinal muscle contraction to pull the rest of the worm down into the burrow while anchoring the posterior region in the burrow by forming a penetration anchor. Contraction of the circular muscle in the anterior segments thins and lengthens the worm, forcing the proboscis further into the soil. The anterior then anchors and the process repeats.

Some modern bivalve molluscs use the foot as a burrowing organ. This process of burrowing is illustrated in Figure 7.40. From the surface, a clam extends its foot down into the substratum (A and B) before pumping blood into the foot to expand its tip (C), forming a terminal anchor. The foot muscles then pull the rest of the animal into the substratum (D). The shell valves are sprung apart (D), forming a penetration anchor. If the burrow is to go deeper, the foot is extended further down to repeat the process. In many bivalves, additional muscles rock the shell relative to the foot, further assisting in moving the bulk of the shell down into the mud. Razor clams show a further refinement of this behaviour (Fig. 7.41). While razor clam musculature is not powerful enough for them to burrow more than a few centimeters, they burrow to about 70 cm by using shell movements to fluidize the surrounding sand. Moving in this fluidized medium allows rapid, and depth independent, burrowing into the substratum.

Some snakes also burrow. *Rhinophis drummondhayi* (1522) is a small burrowing snake from India that moves through the earth using an approach similar to the earthworm's (Fig. 7.42). Up to just past the heart, the axial muscles in this snake are thickened and comprise 60% of the cross-sectional area, compared to 50% of the cross-sectional area further back along the body. The anterior muscles are dark and rich in myoglobin, while the posterior muscles are translucent with no evidence of myoglobin. This snake anchors the rear end of its body (penetration anchor) and drives the head forward into the earth using contraction of the anterior muscles. It then wedges the head (terminal anchor) and draws the posterior part of the body ahead. This "freight-

Figure 7.41 A razor clam (*Phaxas pellucidus* (426)).

train" approach to burrowing allows the snake to move efficiently through loose, sandy soils and is a variant of concertina locomotion (see Section 7.3b.3).

Many other animals, from crickets to eels, caecilians to moles, golden moles, echidnas, and rodents are also fossorial, making and living in burrows. The

A

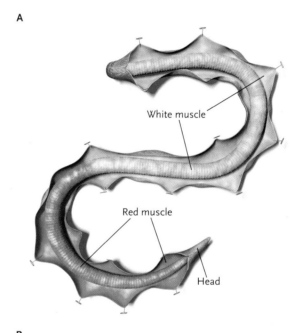

White muscle

Red muscle

Head

B

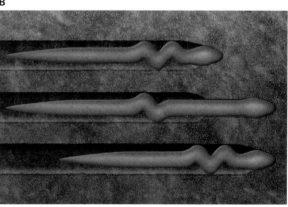

Figure 7.42 *Rhinophis drummondhayi* (1522), a snake from India, uses a "freight train" approach to burrowing. **A.** A partially dissected specimen shows the concentration of red muscle at the anterior end. **B.** The steps involved in burrowing.

SOURCE: Based on Gans, G., H.C. Dessauer, and D. Baie. 1978. Axial differences in the musculature of uropeltid snakes: the freight-train approach to burrowing. *Science*, 199:189–192.

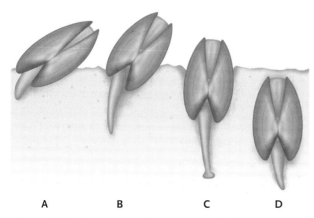

A	B	C	D

Figure 7.40 The sequence of movements in bivalve burrowing.

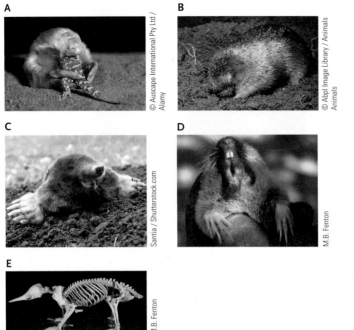

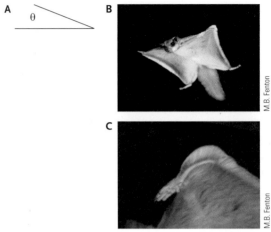

Figure 7.44 **A.** The flying squirrel (*Glaucomys sabrinus* (1686)) is a glider achieving an equilibrium angle of descent (θ) <45°. **B.** The posture during gliding illustrates the general importance of fore- and hindlimbs. **C.** The specific position and attitude of the forelimb can affect gliding performance.

Figure 7.43 A sampling of fossorial mammals. **A.** A marsupial mole (*Notoryctes* sp. (1695)), **B.** a golden mole (*Chrysochloris sp.* (1698)), and **C.** a mole (*Scalopus aquaticus* (1920)). These three species dig using specialized forelimbs. **D.** The gopher (*Geomys* sp. (1885)) burrows by breaking up the soil with its enlarged incisors and moving the soil out of the burrow by scratch-digging; note lightly built forelimbs. **E.** The echidna, or spiny anteater (*Tachyglossus aculeatus* (1678)), has enlarged forelimbs specialized for digging, allowing it to burrow and to tear open the nests of ants and termites on which it feeds.

appearance of fossorial species in different lineages of mammals is particularly impressive (Fig. 7.43). This sampling includes a monotreme (the echidna), a marsupial (the marsupial mole), a golden mole, a "true mole," and a pocket gopher.

STUDY BREAK

1. What roles does mucus play in the locomotion of different animals?
2. Explain the formation and roles of terminal and penetration anchors in different burrowing animals.

7.7 Moving in Air

Animals that move in air do so by **parachuting**, **gliding**, or **powered flying**. As the names imply, gliding and parachuting animals are capable of controlled descents, while those using powered flight generate lift and propulsion.

7.7a Gliders and Parachutists

Animal parachuters typically descend at angles >45°, while good gliders descend at angles of <45° (Fig. 7.44A). In gliding, gravitational potential energy is converted to aerodynamic work. During their descents, parachuters and gliders adjust their bodies to maximize the conver-

sion of energy. Gliders such as a northern flying squirrel (*Glaucomys sabrinus* (1686), Fig. 7.44B) control some aspects of their descent by manipulating the positions and attitudes of their appendages, for example, their forelimbs and wrists (Fig. 7.44C).

Gliding and parachuting have evolved independently in squid, bony fish, amphibians (Anura), reptiles (Squamata and Serpentes), birds, and mammals (both marsupials and placentals). Gliding or "flying" fish are found in oceans (Exocoetidae and Hemiramphidae) and fresh water (Gasteropelecidae), and use the combination of large pectoral fins, thin bodies, and heterocercal (shark-like) tails in which the vertebral column turns upward into the larger lobe of the tail) to launch themselves out of the water, usually when pursued by predators.

Dark-edged wing flying fish (*Cypselurus hiraii* (1172)) (Fig. 7.45) often spread both pectoral and pelvic fins when gliding. Pelvic fins increase the lift force generated by the fish. Spreading pectoral and pelvic fins simultaneously enhance longitudinal stability. Dark-edged wing flying fish approach the gliding performances of birds such as wood ducks. The fish reduce ground effect and drag by flying close to the water's surface.

Frogs in the families Rhacophoridae and Hylidae use enlarged webbing surfaces on their feet to parachute or glide. Some also flatten their bodies when airborne. When gliding, *Polypedates dennysi* (1377), a rhacophorid frog, uses two mechanisms for turning: banked turns in which the frog rolls into the turn, and crabbed turns involving a yaw into the turn. These frogs are about one-third as manoeuvrable when gliding as is a falcon (*Falco jugger* (1579)), one of the world's fastest and most manoeuvrable gliders.

Among living reptiles are examples of gliding/parachuting lizards (e.g., *Draco volans* (1492)) with flight surfaces supported by elongated ribs, geckos (*Ptycho-*

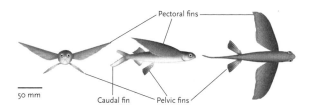

Pectoral fins

50 mm

Caudal fin   Pelvic fins

Figure 7.45 Three views of a gliding dark-edged wing flying fish.

SOURCE: Reproduced with permission from Park, H. and H. Choi. 2010. "Aerodynamic characteristics of flying fish in gliding flight," *Journal of Experimental Biology*, 213:3269–3279. The Company of Biologists Ltd.

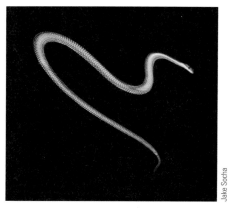

Jake Socha

Figure 7.46 A gliding snake, *Chrysopelea* sp. (1486), flattens its body and moves its head from side to side while gliding. It achieves a glide angle of 13°. In the air, the snake's body moves in travelling waves. The sinusoidal waves mean repeated shifts between leading and trailing edges of the gliding animal. This differs from many other animals that glide and fly in which leading and trailing edges are consistent.

zoon (1519)) with flight surfaces formed from flaps of skin and webbed feet, and snakes (*Chrysopelea* sp. (1486) from Borneo) that glide by flattening their bodies (Fig. 7.46).

At least two fossil reptiles appear to have been gliders. *Sharovipteryx mirabilis*, a small fossil reptile from the Upper Triassic of Central Asia, apparently had flight membranes supported by elongated hindlimbs. Further material is necessary to better understand the details of this animal. *Microraptor gui*, a dromeosaurid dinosaur of the Early Cretaceous in China, had four feathered wings (see also pages 160–161). Neither species appeared to have the morphological bone specializations that are associated with the powered flight of bats, birds, and pterosaurs.

Gliding is also a recurring theme among mammals, with living gliders among the marsupials (one genus in each of Pseudocheiridae and Petauridae), flying lemurs (two species of *Cynocephalus* (1831)), and Rodentia (~12 genera of Sciuridae and ~three genera of Anomaluridae). In each case the gliding membranes connect forelimbs and hindlimbs, but the details differ. *Volaticotherium antiquus*, a Middle Jurassic mammal of China, had a specialized gliding membrane covered with dense fur and supported by an elongated tail and limbs. The teeth suggest a diet of insects, but its relationship to other mammals remains unclear. *V. antiquus* has been placed in its own order, Volaticotheria.

In 2010 an eye-opening paper about the ability of some wingless ants to glide reminded us to keep an open mind on this topic. *Cephalotes atratus* (991) is an ant of the tree canopy in the Neotropics. When a worker ant falls from the canopy, it initially tumbles, but then rights itself, glides backward, and lands on a tree trunk. The ants appear to use their hind legs to effect the body movements necessary to adjust their trajectory and land on the tree trunk. This behaviour has been widely reported from arboreal ants, and demonstrates that controlled movement in air does not require a parachute, a gliding membrane, or wings.

7.7b Flapping Flight

Powered (flapping) flight has evolved independently in two animal phyla, Arthropoda (Insecta) and Chordata (Pterosauria, Aves, and Chiroptera). None of these animals evolved from an ancestral lineage capable of

flight, so powered flight provides clear examples of parallel (among Chordata) and convergent (between Arthropoda and Chordata) evolution.

Powered flight allows animals to exploit otherwise hard-to-access resources such as food or nest sites, as well as new situations and new habitats. Flight is metabolically expensive, in part limiting the maximum size of animals that can use it. Changes in power required over the range of airspeeds suggest that flying animals can adjust the cost of flight by regulating airspeed. The **minimum power speed** is the speed at which flight costs are minimal. The **maximum range speed** is the speed at which the animal can cover the maximum distance on a given amount of fuel.

For the most part, flying animals separate terrestrial locomotion (walking, climbing) from flying (most obviously birds and insects). In bats the wings often attach to the legs somewhere between toes to ankle, depending upon the species. In pterosaurs, there is a dichotomy of views, with some contending that the wings attached at the hip; others that the wings were more bat-like in their attachment. In any event, the morphology of the hip joints and hindlimbs of pterosaurs suggest that they were as bipedal as birds. The hindlimbs of bats, from size and hip joints to orientation of the knees (Fig. 7.47), are quite different from the situation in birds and pterosaurs. When bats walk on the ground, they use all four limbs and fold their wings up so that they are walking on their hindfeet and wrists. A few bat species, including some vampire bats, spend quite a bit of time on the ground and have robust hindlimbs.

The wings of birds, bats, and pterosaurs are all supported by modified forelimbs (Fig. 7.48). However, there are striking differences among them; for example the relative lengths of the humeri (upper arm bones)

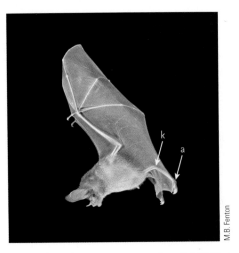

Figure 7.47 A flying bat (*Macrotus waterhousii* (1797)) showing the flexion of the knees (k) and ankles (a).

M.B. Fenton

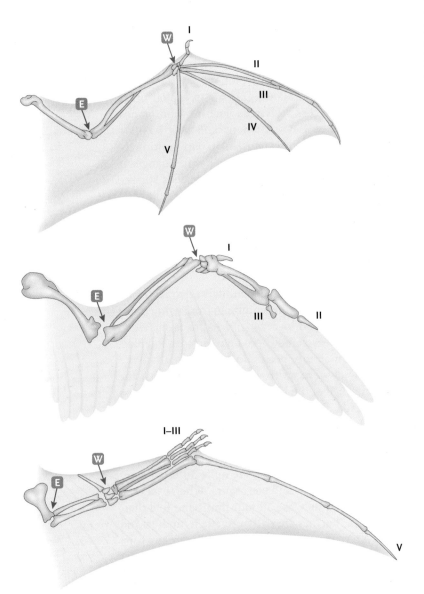

Figure 7.48 A comparison of the wing skeletons of a bat, a bird, and a pterosaur. Arrows identify elbows (E) and wrists (W). Numbers refer to digits I–V.

SOURCE: Re-drawn after Max Licht. From Fenton, M.B. 2001. *Bats*, revised edition. Facts On File, New York.

and the locations of the wrists (W). The wing membranes of bats are made of skin and supported by fingers II, III, IV, and V. Pterosaurs also had skin flight membranes, which were supported by a single elongated finger. In birds, feathers form the flight surface and the fingers are fused (Fig. 7.49).

Birds, bats, and pterosaurs also differ in the details of their breastbones (sterna), shoulder blades (scapulae), and other elements of their pectoral girdles (Fig. 7.49). Most birds have a prominent keel on the sternum while pterosaurs and bats have smaller keels on their manubria (a bone at the rostral end of the sternum). In birds and pterosaurs, the scapula is narrow and blade-like, while it is flat and broad in bats. The coracoids are large in birds and pterosaurs, while the clavicles are large in bats.

The diversity of details in the skeletons of the pectoral girdles and wings of birds, bats, and pterosaurs clearly demonstrates that one cannot identify any particular feature that leads to flight. In birds, feathers, a keeled sternum, a furculum (wishbone), and uncinate processes that brace the ribcage could be considered indicators of flight. A keel is absent from the sternum in many species of flightless birds, but the wishbone and uncinate processes are typically present. Penguins (Spheniscidae) "fly" through the water and, although flightless, retain keeled sterna. Flightlessness is relatively common in both birds and insects, but we know of no pterosaurs or bats that were or are flightless. We now know that other species of reptiles on the lineage leading to birds had feathers (see pages 160–161). In 2000, many people equated feathers with birds and flight, and thought none of the reptiles with feathers could fly. There is ongoing discussion about the flight abilities of pterosaurs and of primitive fossil birds. For example, one study reported that the rachises (shafts) (Fig. 7.50A) of the primary feathers of two fossil birds (*Confuciusornis* and *Archaeopteryx*) were much thinner and weaker than those of the primary feathers of modern birds and would not have been able to withstand the mechanical stress of flight (Fig. 7.50B). Other researchers disagree, citing features such as the size of the wings and areas for the attachment of wing muscles, measurements of rachises from other specimens and the fact that the large numbers of *Confuciusornis* skeletons found together suggest that they travelled in flocks.

We know most about flight in birds because they are conspicuous and many species are large enough to watch as they take off, fly, and land. Taking off and landing are both vital parts of flight, and while some birds launch themselves from branches or wires, others spring into the air from the ground or take a running start. Insects appear to launch themselves more like helicopters using wing beats to generate the lift they need to fly. By roosting upside down, many bats simplify take off by just letting go and spreading their wings. Pterosaurs may have used their flight

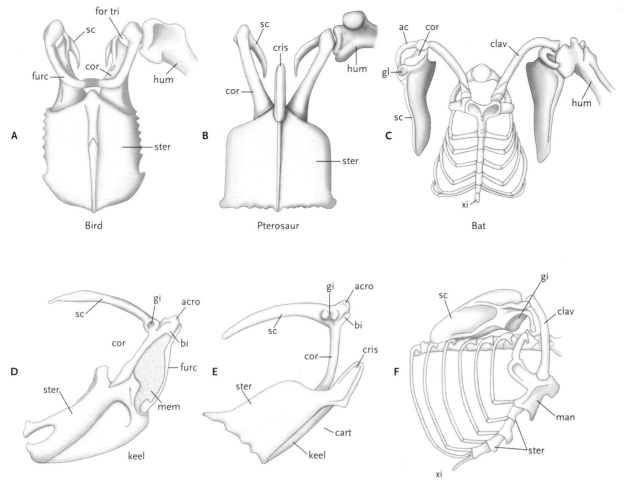

Figure 7.49 A comparison of the sterna and pectoral girdles of a bird (**A, D**), a pterosaur (**B, E**), and a bat (**C, F**), showing ventral (**A, B, C**) and right lateral (**D, E, F**) views. Labels: ac = acromion process; acro = acrocoracoid process; bi = biceps tubercle; cart = possible cartilagenous extension of sternal keel; clav = clavicle; cor = coracoid; cris = cristospine; for tri = foramen triosseum; furc = furculum; gl = glenoid fossa; hum = humerus; keel = sternal keel; man = manubrium; mem = membrane connecting sternum, coracoid, and clavicle; sc = scapula; ster = sternum; and xi = xiphoid process of sternum.

SOURCE: Padian, K. A, "Functional analysis of flying and walking in pterosaurs," *Paleobiology* by PALEONTOLOGICAL SOCIETY Copyright 1983 Reproduced with permission of PALEONTOLOGICAL SOCIETY, INC. - BIOONE in the format Republish in a textbook via Copyright Clearance Center.

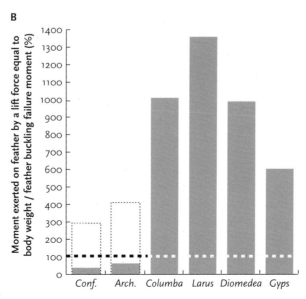

Figure. 7.50 The ability of the rachis (**A.**) of primary feathers to bend without buckling or breaking (moments of buckling failure and bending tensile failure) are compared among birds. **B.** The graph shows data from *Confuciusornis*, *Archaeopteryx*, a dove (*Columba* (1570)), a gull (*Larus* (1553)), an albatross (*Diomedea* (1633)) and a vulture (*Gyps* (1580)). Solid bars indicate buckling failure of the rachis, dotted line tensile failure, and the dashed line indicates a safety factor of 1).

SOURCE: B. From Nudds, R.L. and G.J. Dyke. 2010. "Narrow primary feather rachises in Confuciusornis and Archaeopteryx suggest poor flight ability." *Science*, 328:887–889. Reprinted with permission from AAAS.

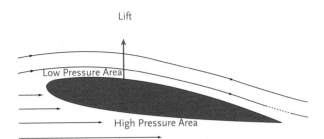

Figure 7.51 In cross-section, an airfoil (black) is shaped like an elongated teardrop. The black lines and arrowheads illustrate the direction of airflow. When air hits the leading edge of an airfoil, it separates such that some air flows over the upper surface and some air flows over the lower surface. Because the distance over the upper surface is longer, the air travels faster over the upper surface. Bernoulli's principle states that the pressure of air (or a fluid) decreases with increasing speed. Lift (red arrow) is produced because the air pressure is lower above the airfoil than below it.

muscles to launch themselves into the air, rather in the same way that vampire bats (*Desmodus rotundus* (1792)) use the leverage of long thumbs to take off from the ground, even after having taken on a full load of blood.

Powered flight requires a combination of lift and propulsion. Lift is generated as air moves over an **airfoil**, in this case a wing (Fig. 7.51). Increasing the angle of attack (the angle of the long axis of the airfoil relative to the air flow, Fig. 7.52) increases both lift and drag (forces that oppose the movement of the airfoil through the air), and at high angles of attack the animal (or airplane) stalls (Fig. 7.52). At this point the flying animal either reduces the angle of attack, dives to increase airflow (and lift), or it must land to avoid crashing. Feathers allow birds to use **slotting** to increase lift (Fig. 7.53), in this case because air flows faster over the upper surface of the wings and through the slots between feathers.

In flying animals, propulsion is provided by the movements of the wing. The primary structures used to generate forward momentum are the primary feathers of birds, the handwings of bats, and the wings of insects (Fig. 7.54). A wingbeat cycle consists of a **downstroke** (ventrad movement of the wings) and an **upstroke** that lifts the wings in preparation for another downstroke (Fig. 7.55). Power is generated on the downstroke, but the details of how the wings move and generate lift vary among flying animals. The movement of wings though air produces swirls of air called vortices (singular = vortex) that can help generate lift. The wingbeat cycle of birds is composed of a relatively simple, ventrally directed downstroke and a dorsally directed upstroke, but the wings of insects are flexible and rotate along their long axis through the wingbeat cycle. The wingbeat cycle of birds creates a single vortex that circulates around the wing to help generate lift. The rotational wingbeats of large insects create a swirling vortex that travels along the upper surface of the leading edge of the wing (a **leading edge vortex**) that enhances lift. This

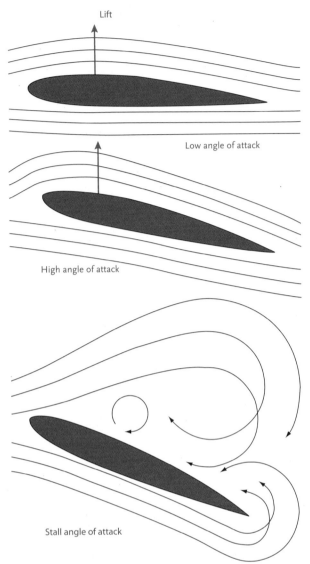

Figure 7.52 As the angle of attack increases, the airfoil produces more lift but also encounters more drag. At very high angles of attack, the airflow separates from the upper surface of the wing and swirls to form eddies. The pressure differential between the upper and lower surfaces (i.e., lift) disappears. At this point the animal (or airplane) stalls.

Figure 7.53 At the bottom of the downstroke of this Crowned Crane (*Balearica regulorum* (1585)), air flows over the wing and rapidly between the slots between the primary feathers (arrows) to increase lift.

Figure 7.54 **A.** Birds use primary feathers and **B.** bats use the handwings (arrows) to generate propulsion. Insects generate propulsion with their wings, which come in many different forms: **C.** beetles use their forewings as covers and fly with their hindwings, the fore- and hindwings of hemipterans are hooked to one another and function together, and **D.** in dragonflies the fore- and hindwings can move out of synch with one another.

same rotational movement and resulting leading edge vortex allows nectar-feeding bats and hummingbirds to hover (Fig 7.55). For very small insects, the viscosity of the air (due to very low Reynold's numbers, discussed in Section 7.1) weakens the leading edge vortex. These animals produce additional lift by rotating the wings very rapidly at the peak of each upstroke and downstroke and by interacting with the vortex generated by the previous wingbeat cycle.

Most insects have four wings, complicating the aerodynamic situation relative to birds, bats, and pterosaurs. Dragonflies and some other four-winged insects can move the forewings independently from the hindwings, allowing them to alter the timing of the wingbeat cycles of each pair. The performance of the forewings remains relatively constant, while the amount of lift generated by the hindwing varies by a factor of two. In some cases the motion of the hindwing precedes that of the forewing by about 25% of the stroke cycle. This phase shift accounts for climbing forward flight in locusts and dragonflies.

STUDY BREAK

1. How do different gliding animals control their glide path?
2. How do the skeletons of flightless birds differ from those of flying birds? What about penguins?

7.8 Jet Propulsion

Animal locomotion by jet propulsion involves a stream of water propelled away from the animal, and, according to the principle of conservation of momentum, the animal moves in the direction opposite to that of the water stream. Animal jet propulsion is typically pulsed, with a jet of water pushing an animal some distance, and the animal then glides as the water reservoir is refilled. This means that jet propulsion works best for relatively large animals, Reynolds numbers (Re) >6 (see Section 7.1).

Figure 7.56 shows the essential elements of pulsed jet propulsion. This is basically a two-stroke operation: a power stroke and a recovery stroke. Muscular contraction increases the pressure on a water reservoir and, after a short lag, a water jet produces thrust and accelerates the animal in the water. The power of the jet creates thrust and acceleration forces that are opposed by drag forces coming from both the animal's inertia and the hydrodynamic resistance of the medium. After the power stroke, the reservoir is refilled, a process associated with negative pressure in the reservoir. Several examples of jet-propelled animals follow.

7.8a Squid

Squid (e.g., *Loligo pealei* (505)) are fast swimming marine predators. Although most species of squid use fin movements to swim slowly, they use expulsion of water from the mantle cavity ("jet propulsion") to swim rapidly. Squid draw water into the mantle through spaces alongside the head (Fig. 7.57B & C), and then,

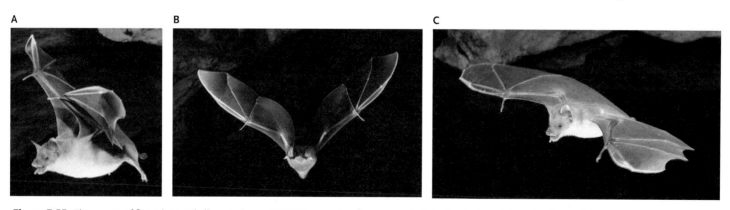

Figure 7.55 Three view of flying bats (*Phyllonycteris poeyi* (1808)) emerging from a cave in Cuba illustrate variations in wing position. In **A.** and **B.** the wings are at the top of the upstroke, in **C.** midway through the downstroke.

SOURCE: M.B. Fenton

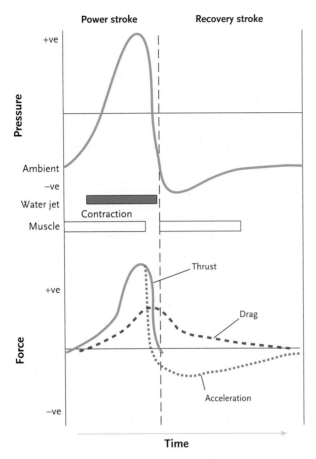

Figure 7.56 The essential elements of jet propulsion. See text for explanation.

when the body contracts around the mantle space, the resulting pressure in the funnel seals the edges of the funnel, closing inhalant openings and forcing water out through the small opening of the jet (Fig. 7.57D).

Muscles in the squid's body wall (Fig. 7.58) provide power for jet-propelled swimming. The flexible tunic lines the mantle cavity and the outer body wall. The tunic includes collagen fibres that prevent elongation of the body. Expansion of the mantle cavity, which draws water in, results from elastic contraction of collagen fibres embedded in the muscle (Fig. 7.58). These become thinner, which expands the v circumference of the body wall, partly assisted by water forced into the mantle as the squid moves. Contraction of circular muscles forces water out of the mantle cavity and stretches fibre layers 1 and 2.

There are three layers of squid circular muscle: a central layer of fast-twitch fibres sandwiched between inner and outer layers of slow-twitch muscle fibres (Fig. 7.58). The outer layers are rich in mitochondria and oxidative enzymes, while the middle layer has fewer mitochondria and a lower ratio of oxidative to glycolytic enzymes. This arrangement is analogous to the red (steady-state work) and white (burst activity) muscles of fish. In normal swimming, only the slow-twitch muscle fibres are active. The fast-twitch fibres contract in escape jetting, when more water is expelled with greater force than in normal swimming. The increased volume of water taken into the mantle occurs

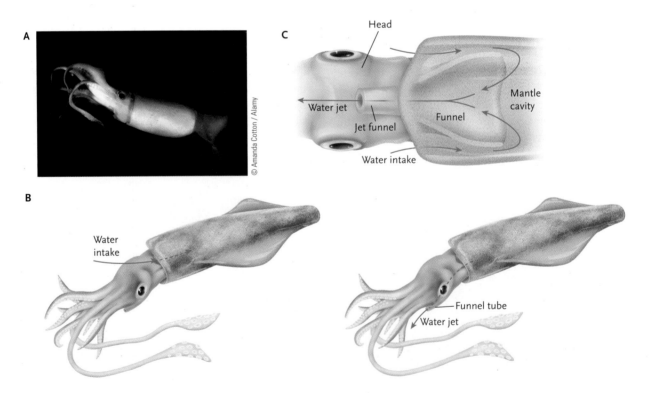

Figure 7.57 A. *Loligo vulgaris* (508). Notice the mantle opening just behind the eye. **B.** Water enters through the mantle openings and is forced out through the jet. **C.** Water pressure in the funnel closes the inhalant openings to the mantle cavity.

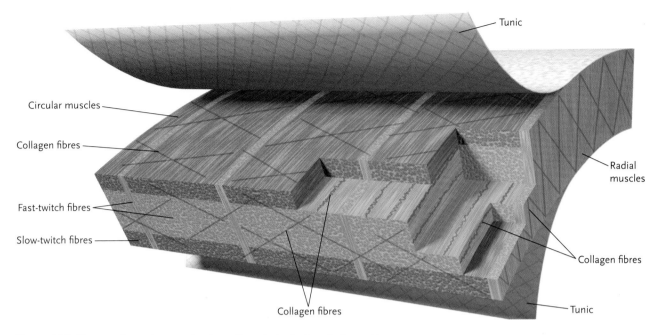

Figure 7.58 The muscles and connective tissue fibres in the body wall of a squid.

because of contraction of radial muscles thinning and expanding the body wall beyond the range used in normal swimming. As the escape jet is discharged, the pressure in the mantle cavity of *Illex illecebrosus* (502) may reach 30 kilopascals (kPa), compared with slow swimming when the pressure is about 15 kPa. (Normal human systolic blood pressure is also around 15 kPa.) Almost all of the water in the mantle cavity is discharged, and the mantle wall thickens by about 50%. In normal swimming, *Illex* produces 5–6 jet cycles in 10 s, a rate slightly accelerated (6–7 cycles in 10 s) in escape swimming.

We can compare the energetics and swimming speeds of squid and fish. At the same swimming speed (40 cm·s⁻¹), a 40-g *Loligo opalescens* (506) (California market squid) uses about 13 J·kg⁻¹·m⁻¹, a 400-g *Illex illecebrosus* (502) (short finned squid) uses 7.5 J·kg⁻¹·m⁻¹ and a 500-g salmon, *Oncorhynchus nerka* (1315), requires only 2 J·kg⁻¹·m⁻¹. A further comparison is with *Lolliguncula brevis* (507) (the brief squid) which swims slowly using its fins and uses a low-pressure jet to maintain its vertical position in the water column. At low speed, oxygen consumption by *L. brevis* is comparable to that of striped bass, *Morone saxatilis* (1276); mullet, *Mugil cephalus* (1234); and flounder, *Paralichthys* sp. (1306).

Some squid species briefly "fly" out of the water to escape predators. *Loligo pealei* (505) reaches speeds of 200 cm·s⁻¹ (7.4 km·hr⁻¹) in flight, and the larger *Dosidicus gigas* (494) (over 1 m long) reaches 26 km·hr⁻¹.

7.8b Scallops

Bivalve molluscs might not appear to be mobile, but several genera of scallops swim by jet propulsion. They close the shell valves and expel water through two channels close to the hinge (Fig. 7.59). The muscle familiar to those who love to eat scallops is not the adductor muscle, rather the catch muscle that holds the two valves tightly closed. The adductor muscle is a strap of striated muscle found to one side of the catch muscle. In scallop swimming, increased water pressure and outward water flow and power are correlated with the time at which the shell valves are being closed by the adductor muscle.

A contracting adductor muscle (Fig. 7.60) compresses the black triangular pad of the hinge, which acts as a spring and opens the valves when the adductor muscle relaxes. The triangular pad is made of an elastic protein, **abductin**, a cross-linked, amorphous protein with properties similar to **elastin** in mammals and **resilin** in insects. Abductin has a very high resilience (>90%), the amount of energy a material can release compared to the energy taken to compress it. Temperature affects the density of sea water and limits the

Figure 7.59 An artist's rendition of a scallop's swimming action.

A

B

Figure 7.60 **A.** A scallop (*Placopecten magellanicus* (538), Bivalvia) in which the mantle and other body parts have been removed to show the large catch muscle. Notice the black pad of abductin in the centre of the hinge. **B.** The adductor muscle being peeled away from the side of the catch muscle.

Figure 7.61 A larval *Aeshna cyanea* (850).

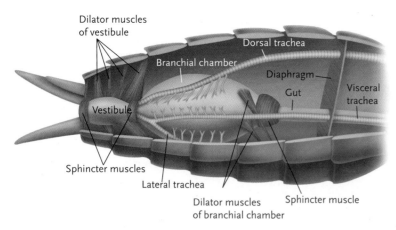

Figure 7.62 The anatomy of a dragonfly larval abdomen showing the branchial chamber and gills, and the tracheal supply to and within the branchial chamber.

SOURCE: Based on Hughes, G. M. and P. J. Mill. 1966. "Patterns of Ventilation in Dragonfly Larvae." *Journal of Experimental Biology* 44: 317–333.

amount of energy that can be stored in abductin. The Antarctic scallop, *Adamussium colbecki* (530), overcomes these problems by lightening the shell and changing the properties of abductin to increase resilience at colder temperatures. These scallops have fragile shells.

7.8c Dragonfly Larvae

The larvae of dragonflies (Insecta, Odonata, Anisoptera, Fig. 7.61) propel themselves using expulsion of water through the anus. The water is expelled from branchial chambers, specialized regions of the gut that house gills (sites of gaseous exchange). A muscular vestibule lies behind the branchial chamber. Three muscular sphincters control this system: one closes the anus, a second separates the vestibular chamber from the branchial chamber, and the third, at the anterior end of the branchial chamber, isolates it from the anterior parts of the gut (Fig. 7.62A).

During normal ventilation, water is rhythmically pumped through the anus into and out of the branchial chamber. Water is pumped out when dorsoventral muscles in the abdomen contract, compressing the abdomen and driving water out of the branchial chamber. Water is drawn in when the spring in the abdominal tergites and sternites restores their shape as the dorso-ventral muscles relax. The respiratory cycle involves small (<100 Pa) pressures on intake and 500 to 1000 Pa in the expulsion.

In jet-propelled swimming, outgoing pulses of water are about twice as fast as normal ventilation movements, and water expulsion pressures may reach 3000 Pa, enough to propel the larva at $0.5 \text{ m} \cdot \text{s}^{-1}$. A series of five to eight 200Pa pulses pair of pulses could propel the larva as far as 2 m.

7.8d Jellyfish

A jellyfish (Cnidaria, Scyphozoa) at the ocean surface does not look like a jet-propelled animal, but contraction of the muscles around the bell reduces the space under the bell, forcing water out and propelling the animal. Re-expansion of the bell is slower than contraction and depends on connective tissue elements around the bell. In larger jellyfish species, propulsion vortices in expelled water interact with each other and add power to the animal's propulsion (Fig. 7.63). The hydrodynamics of this interaction require that the height of the bell of the medusa be about a quarter the diameter of the bell—a finding that is supported by measurements of a large number of scyphozoan species. The recycling of the same body of water at each swimming stroke has an ecological significance in that it makes a significant contribution to bringing nutrients from deep water to the surface as the jellyfish swims upwards. In other medusoid classes (Hydrozoa and Cubozoa), the opening of the bell is partially closed by a **velum** (Hydrozoa) or **velarium** (Cubozoa), a shelf of tissue that projects inward from the lower edge. As water is expelled, the velum projects outward as a jet nozzle. As water refills the bell, the velum becomes an open funnel. Cubozoans (e.g., *Chironex fleckeri* (155), the sea wasp) are active swimmers moving in a controlled pattern back and forth along a coastline at $100–200 \text{ m} \cdot \text{hr}^{-1}$

Figure 7.63 A scyphozoan, *Mastigias papua* (151), swimming. Experiments in which green dye was injected under the bell have shown that a significant amount of the water expelled at each propulsive pulse is taken back into the bell and expelled again at the next contraction. These dye experiments also revealed the vortices that aid propulsion.

7.8e Salps

Designed for jet propulsion, salps (Urochordata, Thaliacea, Fig. 7.64) have a body plan with rings of muscles arranged around a tubular body and sphincters at anterior and posterior ends. In forward swimming, the anterior valve closes, the posterior valve opens, and the circular muscles contract, expelling up to 40% of the water in the jet chamber. The chamber is refilled by the elasticity of the body wall. Both anterior and posterior openings are relaxed during refilling, but most water enters anteriorly because the structure of the posterior valve appears to limit reverse flow. By reversing the sequence of anterior and posterior valve openings, a salp can swim backward.

Swimming speed depends on pulse rate (0.5 to 2.0 Hz) and variations in the contraction of the muscle bands. Large salps may swim at 8 cm·s^{-1} while continuous swimming speeds vary from 1.5 and 4.0 cm·s^{-1}. A salp jet is produced more slowly and through a larger aperture than the jets of most other animals. A jet develops thrust as the product of the mass of water ejected/second (m) and the velocity of the jet (u). The power needed to accelerate the fluid is the rate at which kinetic energy is supplied ($0.5 \times mu^2 \cdot s^{-1}$). Therefore it is more energy efficient to eject a large volume at low velocity than a small volume at higher speed. Further, the smaller the difference between jet speed and animal movement, the greater the efficiency of transfer of momentum from the jet, again demon-

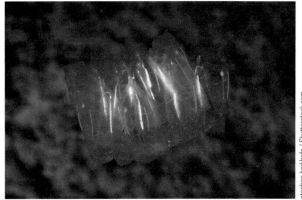

Figure 7.64 *Salpa cylindrica* (1117).

strating that a salp's large, slow jet is more efficient than the jets of other groups.

STUDY BREAK

1. Why do small animals not use jet propulsion as a means of locomotion?
2. In what ways does jet propulsion in squid and dragonfly larvae differ from that of cnidarian medusae?

LOCOMOTION **IN PERSPECTIVE**

This chapter has presented examples of the many ways in which animals move. When thinking about any single example it is important to consider how that example fits with generalizations about modes of locomotion. It is also important to consider the mechanical limitations that apply to any animal movement. These limitations may be environmental, for example the density of water relative to the size of the animal or the ambient temperature relative to the physiology of the animal. The limitations may also be imposed by the strength of the skeleton relative to muscle power and weight, the bulk of muscle that can be packed around a particular joint, or the strength and elasticity of tendons or apodemes. Any mode of animal locomotion demands energy. That energy comes from food, taken from the environment. The following chapters (8, 9, and 10) explore the ways in which food is acquired and processed by animals.

Questions

Self-Test Questions

1. Which of the following represents the skeletal–muscular interactions typically involved with locomotion?
   a. Animal skeletons transmit the force generated by muscular contraction and translate it into movement.
   b. Muscles always work in opposing pairs with skeletal elements to generate movement.
   c. All animal skeletal–muscular associations are structurally similar.
   d. Only the skeletal elements of limbs are important in locomotion.

2. Which of the following describes the mechanics of swimming?
   a. Appendages are always the primary means of propulsion in swimming animals.
   b. Nematode movements depend upon opposition of circular and longitudinal arrangements of muscle fibres.
   c. Flippers of similar shape have similar hydrodynamic performance and may provide lift as well as propulsion.
   d. Anguilliform (slipping wave) locomotion is used by all swimming fish to propel them through water.

3. Which of the following anatomical features is correctly matched with the mode of locomotion it favours?
   a. Secretion of surfactant to lower surface tension; cursorial
   b. Setae, microdroplets and discs on wings and ankles; flapping flight
   c. Changes in gait to reduce ground reaction force and reduce injury; adhesion to surfaces
   d. Short forelimbs, shortened thoracic and cervical skeleton, enlarged hindlimbs and expanded pelvic girdle; richochetal locomotion

4. Which of these factors is important in gastropod locomotion?
   a. Snail mucus is a powerful, permanent adhesive.
   b. The metabolic cost of producing gastropod mucus is great enough to favour trail-following behaviour by conspecifics.
   c. Direct locomotory waves propel the animal both forward and in reverse in all species.
   d. Contraction and relaxation of the same muscle groups are involved with forward and retrograde wave generation respectively.

5. Which of the following describes features of pulsed jet propulsion systems?
   a. An elastic protein, abductin, plays an important role in the recoil of the squid mantle.
   b. Contraction of rings of muscle around the tubular body wall with the anterior valve closed propels squid forward.
   c. Partial closure of the bell by a shelf of tissue, or velum, creates a jet nozzle to project the squid forward.
   d. Contraction of fast-twitch circular muscle fibres forces water out of the squid mantle cavity.

Questions for Discussion

1. Look back through this chapter and identify ten examples of animals that move in different ways. Then rank your examples in order of locomotory energy efficiency.

2. While many animals can use more than one locomotory mode, few are able to be highly efficient in multiple modes. Identify examples of animals that are able to perform well in more than one locomotory mode and explain how this plasticity is achieved.

3. In this chapter, the discussion of gliding and parachuting animals is restricted to these activities in air. What examples of gliding and parachuting in aquatic environments can you identify? What are the similarities and differences between these locomotory modes in the two media?

Figure 8.1 African Goshawk takes a snack.
SOURCE: Courtesy of Naas Rautenbach

STUDY PLAN

8   Feeding

WHY IT MATTERS

On 1 July 1992, the then Canadian Minister of Fisheries and Oceans John Crosbie was unhappy. Mr. Crosbie, a Newfoundlander himself, was confronted by an angry mob of fishermen in Bay Bulls, Newfoundland. They demanded to know why there were no fish to catch. Crosbie perhaps later regretted his response, "There's no need to abuse me. I didn't take the fish from the goddamn water." The next day he needed police protection when he announced a moratorium on the cod fishery.

Crosbie's problem, the lack of fish, was the culmination of a period of failed management and communication involving politicians, government scientists, fish packing companies, and fishers themselves. The result was a population of northern cod (*Gadus morhua* (1213)) at about 1% of the level it had been only decades before. The cod fishery had produced 200 000–300 000 tonnes a year up to the introduction of newer technology and larger boats in the 1960s, when the yield rose to about 800 000 tonnes in 1970 before crashing to under 50 000 tonnes in 1991 (Fig. 8.2). In 1992 a moratorium was imposed on the fishery.

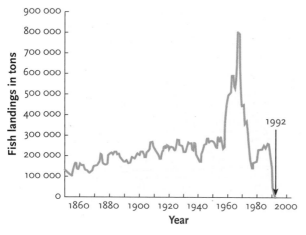

Figure 8.2 The history of commercial landings of Atlantic cod (*Gadus morhua* (1213)).

SOURCE: Millennium Ecosystem Assessment

Figure 8.3 A large (140 cm long and about 30 kg) cod taken during a Fisheries and Oceans research cruise in the late 1980s.

Capelin (*Mallotus villosus* (1240)) is a major food of cod. As cod catches declined, part of the fishery effort targeted capelin. This, along with the number of capelin taken as "by-catch" (animals caught and killed unintentionally) in the desperate effort to catch more of the diminishing cod population, led to a crash in the capelin population, limiting the food supply for the few surviving cod. Limited food supplies lead to smaller cod becoming sexually mature and result in fewer large fish in the population, another effect of overfishing. By the time of the major cod decline, few large fish (Fig. 8.3) survived to be caught.

Most of the fish eaten by North Americans—salmon, tuna, and pollock—are, like cod, piscivores: they eat other fish. Their prey feed on zooplankton, principally small crustaceans such as copepods. Copepods, in turn, eat mostly phytoplankton, the photosynthetic algae at the base of the marine food pyramid.

A food pyramid is a series of links between multiple trophic levels. Phytoplankton in water and plants on land are the producers—the first trophic level. Animals that eat phytoplankton (herbivores) are at trophic level 2. Carnivores hunting herbivores represent level 3, and larger carnivores level 4, etc. Charles Odum measured the flow of energy in a freshwater river system and showed how much energy is lost in the transitions from one trophic level to another (Fig. 8.4).

In the North Atlantic, a food chain based on a springtime phytoplankton bloom (level 1) feeds many species of small copepods (level 2), which are themselves food for shoals of young herring and capelin (level 3). These fish, in turn, support large cod, placing cod at level 4. When that cod is cooked and on your plate you are trophic level 5. However, your trophic level is not fixed. You might also eat a plate of oysters (level 2) putting you at level 3, or a tuna steak from a tuna that preyed on cod, putting you at level 6. If you are a vegetarian you are at trophic level 2.

The importance of recognizing trophic levels emerges from the cod story above. At market, the most valuable fish are generally large predators such as tuna and cod. As these fish are targeted by the fishery, their numbers decrease, and fishing effort moves to smaller individuals and other species. Consequently, the average trophic level of the total commercial fishery catch decreases. Fishers are now "fishing down the

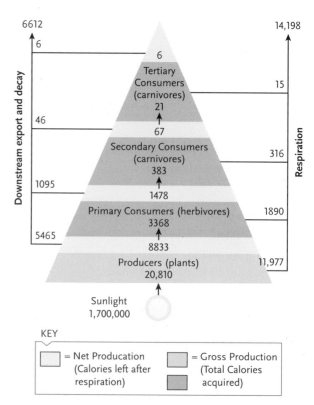

Figure 8.4 Energy flows and productivity in a food chain.

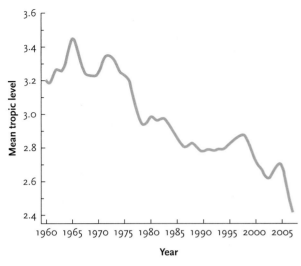

Figure 8.5 Trophic levels of commercial fishery landings in the Northeast Atlantic.

SOURCE: NOAA

food chain," a phenomenon that occurs at both regional and global levels (Fig. 8.5). Continuing to fish down the food chain can lead to a collapse in the fishery and damage to the whole ecosystem that may be irreversible.

SETTING **THE SCENE**

Feeding is one of the fundamental activities that define an animal. Food is the source of energy that powers other activities. It is also the source of the molecular building blocks that become tissues and organs as well as the materials that will be converted to reproductive output. Despite the enormous variety of animals it is possible to characterize feeding modes in a limited number of categories, and widely different morphologies are commonly adapted to feeding in similar ways. This chapter will explore examples of different feeding modes in examples from a wide selection of animal groups and morphologies.

8.1 Food Sources

Life requires energy, and animals acquire energy by feeding. Food is not randomly distributed in space or time—some foods are harder to gather or capture than others, and different types of foods offer different nutritional rewards. While food provides energy, animals must spend energy to acquire it. Energy is a precious resource, and natural selection has favoured the evolution of mechanisms that increase feeding efficiency. Animals exhibit a staggering array of morphological and behavioural adaptations to acquire specific food resources, adaptations that are different solutions to the cost/benefit game of finding, capturing, and consuming food.

Animal food falls into three broad categories: detritus, plants, and other animals, each presenting particular advantages and challenges. **Detritus** (Latin, *detere*, to wear away), the most abundant source of food, is organic debris formed by breakdown and decomposition of dead plants or animals or animal faecal material. Detritus occurs in soils, on the ground, suspended in water, and settled at the bottoms of oceans, lakes, rivers, and ponds. Its abundance makes detritus a good food choice because the cost of eating it can be small. A challenge to eating detritus is its small particle size, meaning that detritivores (detritus eaters) must be very efficient at gathering enough to obtain sufficient energy.

Plant resources are abundant in both terrestrial and aquatic environments. While unicellular plants, like detritus, may be abundant in fresh and marine water, feeding on unicells requires a mechanism to extract tiny particles from a large volume of water. Macroalgae, marine angiosperms and terrestrial plants are easier to find and are fixed, making them easy to capture. However, plants are low in protein, and many plants are protected by chemical and/or defensive structures (irritating hairs, spines, thorns, etc.).

Like plants, animals live in both terrestrial and aquatic environments. Animals contain more protein than plants (per gram) and therefore have higher nutritional quality as food. However, unlike plants, most animals are mobile and can escape or hide from predators.

This chapter reviews the mechanisms animals use to find, capture, and ingest foods. Evolutionary history, body size, and habitat influence both the strategies and the structures involved in feeding.

STUDY BREAK

1. How do the costs and benefits of foraging for and eating detritus, animals, and plants differ? What about the difference between fruit, leaves, and flowers?

8.2. Suspension Feeding Systems

8.2a. Background: Importance of Reynolds Number

Suspended small food particles (<1 mm in maximum dimensions) with limited locomotory ability are often plentiful in aquatic habitats, while airborne food particles in terrestrial habitats are rare. Consequently many more aquatic animals than terrestrial animals feed on small suspended food items; they are called **suspension feeders**. Some spiderlings (juveniles) of orb weaving spiders are an exception. These spiders periodically ingest their webs, deriving nourishment from protein-rich pollen grains captured by their sticky webs.

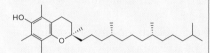

BOX 8.1 Molecule behind Biology
Alpha-Tocopherol

FIGURE 1 *Molecular structure of alpha-tocopherol.*

Alpha-tocopherol, better known as vitamin E, is one of four natural forms of tocopherol (*alpha, beta, gamma,* and *delta*). Synthetic tocopherols come in eight different stereoisomers, but the greatest vitamin activity occurs in the natural α-tocopherol $(R,R,R = CH_3)$ which is produced only by photosynthetic organisms (Fig. 1).

Alpha-tocopherol is a lipophilic antioxidant, an essential component in animal diets. Henry A. Mattill in the early 1920s demonstrated the impact of vitamin E on the fertility of males and females. He found that rats fed an artificial diet deficient in vitamin E were healthy but infertile. Their reproductive activity could be restored by feeding them lettuce. The name "tocopherol" is derived from the ancient Greek words *phero* = bring and *tocos* = childbirth. Despite these clear results, the *mechanism* by which these effects are achieved has long been the subject of research and debate.

Alpha-tocopherol is nature's most effective lipid-soluble, chain-breaking antioxidant, protecting cell membranes from peroxidative damage. When oxygen is used within cells, highly reactive free-radical products are produced that can damage cell structures and functions.

Most vitamins function as cofactors in metabolic pathways, as ligands for nuclear receptors, or in unique molecular roles in cellular activity. The antioxidant role of α-tocopherol may maintain the integrity of long-chain polyunsaturated fatty acids in the cell membranes, preserving their bioactivity.

Despite the huge literature on the antioxidant functions of α-tocopherol, some researchers maintain that, at physiological concentrations, it is incapable of protecting against oxidant-induced damage. Rather, the true importance of Vitamin E is as a signalling molecule that, by acting as a ligand to proteins, is involved in gene expression and signal transduction. A general role of α-tocopherol in signal processing is supported by experiments on snail neurons where α-tocopherol, but not a synthetic antioxidant, down-regulated acetylcholine receptors.

In insects, α-tocopherol may mimic the action of juvenile hormone.

Freshwater rotifers show a different signalling role for α-tocopherol: changing body form (phenotype). Zooplankton food chains in ponds and lakes depend on phytoplankton cells containing α-tocopherol. As these cells are eaten, some are broken, releasing α-tocopherol into the water where its concentration is an indicator of the amount of food available in the system. *Asplanchna sieboldi* (297), like many other rotifers, is polymorphic and reproduces asexually. At low food levels (i.e., low α-tocopherol) it occurs as a small form (saccate) (Fig. 2). As food levels increase, the population shifts to larger morphs, either "winged" (cruciform) or large bell-like (campanulate) forms. The phenotype of entire populations can be manipulated, in either direction, by changing the level of dissolved α-tocopherol in the water.

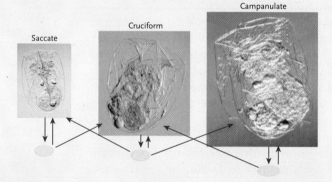

FIGURE 2
The different phenotypes in the asexual life cycle of Asplanchna sieboldi *(297).*
SOURCE: Gilbert, J. J. (1980). "Developmental polymorphism in the rotifer Asplanchna sieboldi." *Am Sci* 68(nov–dec): 636–648.

Suspended food in fresh and marine water includes

- organic debris ("particulate organic carbon")
- bacteria
- unicellular photosynthetic organisms (phytoplankton)
- small planktonic animals that cannot swim against prevailing currents (zooplankton)

The distribution of these food items is typically patchy in time and space. Because suspension feeders must separate these small food items from a large volume of fluid, they require a constant renewal of the supply of water that contains food particles past their feeding structures.

Suspension feeders are either passive or active. **Passive suspension feeders**, such as hydroids and corals (Cnidaria) and many species of brittle stars and sea cucumbers (Echinodermata), depend on ambient water currents to supply edible particles. **Active suspension feeders**, which must expend energy to generate water flow past their feeding structures, include sponges, ascidians, larvaceans, bryozoans, some poly-chaete annelids, and many species of bivalves and crustaceans. There are trade-offs in these categories of suspension feeding. Unlike active suspension feeders, passive feeders do not need to expend energy to achieve water flow, while active suspension feeders do. However, their dependence on ambient water currents limits the habitats available to passive suspension feeders.

An understanding of how suspension feeders capture edible material from water requires an understanding of how water behaves when encountering the animals' food capture devices. Viscous and inertial forces compete, governing the behaviour of moving fluids. **Viscous forces** act to keep fluid molecules together and flowing in smooth streamlines, called **laminar flow**. **Inertial forces** cause a fluid break up into uneven streamlines, called **turbulent flow**. The relative importance of inertial versus viscous forces in determining fluid behaviour is described by the **Reynolds number (Re)** (introduced on page 175, a dimensionless value.

$$\text{Reynolds Number} = \frac{\text{size} \times \text{velocity} \times \text{density}}{\text{viscosity}}$$

Reynolds number depends on object size, the velocity of the water flow or of the object through still water, and the density and viscosity of the water. Figure 8.6 illustrates how increasing the velocity of the water flowing past an object changes the behaviour of the water from laminar flow dominated by viscous forces to turbulent flow dominated by inertial forces. The same transition occurs when object sizes are changed but flow rate remains constant. With a small object or low flow, the Reynolds number is low, and fluid flow is dominated by viscosity: flow patterns are laminar and orderly, and a thick boundary layer of "adherent fluid" surrounds the object. When small hairy legs move slowly through water, they behave as solid paddles because the boundary layer around each hair overlaps the boundary layer of neighbouring hairs. Therefore, water flows around the periphery of the entire leg rather than between the individual hairs of the leg. Small hairy legs moving slowly cannot function as sieves. As objects increase in size or the flow velocity increases, so does the Re, and fluid movement is increasingly influenced by inertial forces. As inertial forces become dominant, the flow becomes turbulent, and the boundary layer of water that adheres to the surface of the object becomes thinner. A small object can also increase its Reynolds number and reduce the thickness of its boundary layer by increasing its speed of movement. Thus, small hairy legs can function as sieves when they are propelled rapidly enough through water or if they are held so that they intersect a rapidly moving water current.

It is difficult for us to have an intuitive feel for life at a low Re because we are very large organisms—when

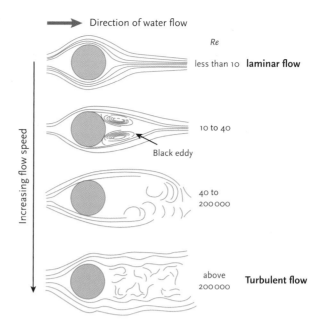

Figure 8.6 Changing behaviour of water as it flows past a cylindrical object at increasing velocity. At low velocity (and low Re), flow is dominated by viscous forces and is laminar. As velocity of flow increases, vortices form on the downstream side of the cylinder. Eventually the flow on the downstream side becomes completely turbulent due to the dominance of inertial over viscous forces.

SOURCE: Vogel, S. (1996). *Life in moving fluids* (2nd ed.). Princeton, N.J.: Princeton University Press.

we swim through water we glide between strokes and shed a turbulent wake because water flow around our bodies and appendages is dominated by inertial forces. The great majority of animals living and feeding in water are much smaller than humans, and their suspension feeding activities and small food capture devices must deal with the dominant influence of water viscosity.

Two extremes of Reynolds numbers for biological systems are epitomized by a large whale swimming at moderate speed and a bacterium swimming at 0.01 mm·s^{-1}. The Re of the whale is 300 000 000 whereas the Re of the bacterium is 0.00001. Inertial forces begin to significantly dominate over viscous forces when Re values exceed 10. Consider the importance of Re in the mechanisms of suspension feeding in the examples that follow.

8.2b. Barnacles

Acorn barnacles (Arthropoda, Crustacea, Cirripedia) are sessile, marine crustaceans that feed on suspended food particles by extending their appendages (**cirri**) through an opercular opening in the protective shell (Fig. 8.7A). Each cirrus bears side-branches called setae (Fig. 8.7B). Three pairs of longer cirri capture suspended food particles from the surrounding water, and three pairs of shorter cirri scrape the collected food from the long cirri and transfer it to the mouth.

A   **B**

Figure 8.7 **A.** Acorn barnacle (*Balanus glandula* (**748**)) sweeping its three pairs of long, setose cirri (suspension feeding appendages) through the water. One of the shorter pairs of cirri can be seen curving on the left side of the opening in the shell. **B.** Isolated cirri from *Balanus glandula*. Each cirrus consists of two stems (rami), both with filtering setae. The image shows the corresponding cirrus from two individuals of the same size. The cirrus on the left is from a barnacle on a protected shore and that on the right from a wave-swept shore. Shorter cirri on wave-swept shores may minimize damage to cirri when they extend from the shell.

SOURCE: Reprinted from *Zoology* 106, Marchinko, K.B. & Palmer, A.R., "Feeding in flow extremes: Dependence of cirrus form on wave-exposure in four barnacle species," Pages 127–141, Copyright 2003, with permission from Elsevier.

Barnacle feeding provides an example of suspension feeding using primarily a sieving mechanism. At low ambient current speeds, barnacles rhythmically sweep their cirral fan through the water to sieve out particles. The fan motion during each sweep is essential. This was demonstrated by an experiment in which a fan of barnacle cirri was held stationary in flowing water. The flow path of the water relative to the fan was visualized by releasing a stream of fluorescent dye into the flowing water. At low current speed, the boundary layer surrounding each setose cirrus overlaps that of the neighbouring cirri (because the Reynolds number is low) and the cirral fan behaves as a solid object within the path of the current. In this situation, the dye stream travels around the whole cirral fan rather than between individual cirri. Conversely, at higher current speed, the boundary layer is thinner, and the dye stream travels between neighbouring cirri. Thus, to sieve particles from slowly flowing water, the animal has to rake its cirri. The change in barnacle feeding behaviour as the ambient water flow speed changes illustrates the influence of Reynolds number on suspension feeding behaviour.

Repetitive sweeps of cirri through water cost energy. If water speeds are high enough to thin the boundary layer between cirri, then water can pass between the cirri and acorn barnacles need not sweep their fans. When water current speed exceeds a threshold level, acorn barnacles feed passively, spreading their cirri to intersect the current flow.

Feeding by gooseneck barnacles (*Lepas anatifera* (**749**)) further illustrates the influence of the Reynolds number in dictating the feeding behaviour of suspension feeders. Gooseneck barnacles are only capable of

Figure 8.8 Gooseneck barnacles (*Lepas anatifera* (**749**)) with cirri extended for feeding.

passive suspension feeding (Fig. 8.8); they cannot survive in quiet waters.

8.2c. Copepods

Among the most abundant organisms in the global ecosystem, calanoid copepods (Arthropoda, Crustacea) are mobile suspension feeders (Fig. 8.9). These tiny crustaceans occur in fresh and sea water in densities as high as $330\,000\ \mathrm{m}^{-3}$. Most copepods feed directly on phytoplankton and serve as a vital link in the food chain that supports almost all commercial fisheries.

High-speed cinematography shows capture of a diatom by a copepod (*Eucalanus pileatus* (**731**)) tethered in a glass chamber (Fig. 8.10). An ink stream shows patterns of water flow. Beating movements of three pairs of setose appendages create a continuous flow of water past the food capture appendages, the 2nd pair of maxillae. As an individual diatom approaches the copepod, these maxillae are flung apart at high speed, sucking the diatom into the newly created space. The high speed of movement increases the Reynolds number, allowing the diatom to be captured rather than pushed away. After capture, the diatom is transferred to other mouthpart appendages and pushed into the mouth. This is an active "scan and trap" mechanism to capture particles individually.

8.2d Ascidians

Barnacles and copepods actively transport water for suspension feeding by moving muscle-powered appendages. Many other active suspension feeders use beating cilia to transport water past their feeding structures. Ascidians, also known as tunicates or sea squirts (Chordata, Urochordata, Ascidiacea, Fig. 8.11),

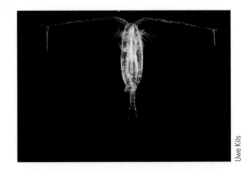

Figure 8.9 Calanoid copepod (*Calanus hyperborealis* (**729**)), one of many *Calanus* species that are extremely abundant in the world's oceans.

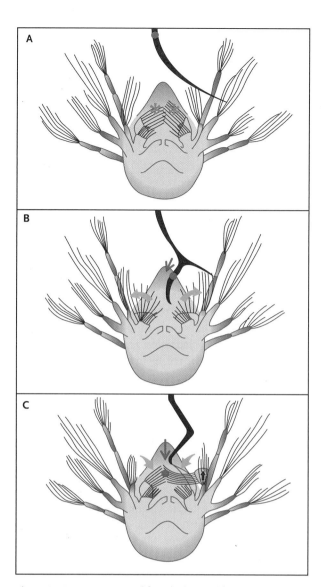

Figure 8.10 Diagrams traced from high-speed films of a *Eucalanus pileatus* (731) capturing an alga. The animal is upside down and is viewed from its anterior end. Black streaks are dye streams from a micropipette. Yellow arrows show movements of second maxillae, indicated by blue asterisks in **A**. The alga is represented by a red dot and its movements by red arrows. **A.** Water currents bypass the second maxillae until an alga comes near. **B.** Alga is captured by an outward fling and then **C.** an inward sweep of second maxillae.

SOURCE: Koehl & Strickler. "Copepod feeding currents," *Limnol. Oceanog.* 26(6):1062–73 (1981). Copyright 1981 by the Association for the Sciences of Limnology and Oceanography, Inc.

are sessile, suspension-feeding, invertebrate chordates that extract food from the water by sieving and by direct interception.

Typical ascidians have a sac-shaped body with a buccal (inhalant) siphon and an atrial (exhalent) siphon (Fig. 8.12A). Much of the interior is occupied by a large pharynx, which lies in a chamber, the atrium. Sea water is continuously circulated through the perforated pharynx, entering at the mouth (buccal siphon), flowing through the pharyngeal perforations to the atrium, and exiting through the atrial siphon. This flow of water is maintained by beating cilia on gill bars that border the pharyngeal perforations.

Figure 8.11 Lemon-tipped sea vase (*Ciona intestinalis* (1108)).

Strong/Buzeta

The actual filter is a fine net of mucus, secreted by a strip of glandular tissue, the **endostyle**, on the ventral side of the pharynx (Fig. 8.12B). As the mucous net is secreted, pharyngeal cilia transport it over the inner surface of the pharyngeal wall (Fig. 8.12B and C). Food particles are captured from the water flow by the mucous net. On the dorsal side of the pharynx, the food-laden mucus net is rolled into a string in the **dorsal lamina**. The string of mucus and captured food is then transported posteriorly in the dorsal lamina to the esophagus.

Mucous net filaments (10–25 nm diameter) surround openings about 0.5 μm wide and 1–2 μm long (Fig. 8.12D). The net captures 1–2 μm diameter particles with 100% efficiency and much smaller bacteria with about 70% efficiency. Sieving explains capture of larger particles by the net, but retention of smaller, bacteria-sized particles depends on direct interception: the adhesion of the particles to the mucous filaments.

Ascidian feeding is vulnerable to blockage if large particles become lodged in the incurrent siphon or interior of the pharynx. To prevent this, coarse particles are excluded from the pharynx by tentacles in the inhalant siphon (Fig. 8.12A). Large particles lodged in the pharynx can also be expelled by closing the exhalent siphon and abruptly contracting the walls of the pharynx and atrium, driving water and large particles back out the inhalant siphon. These forcible reversals of water flow also occur when an ascidian is disturbed and give rise to the name "sea squirts."

8.2e Larvaceans

Larvaceans (e.g., *Oikopleura* (1120), Chordata, Urochordata, Larvacea, Fig. 8.13) are motile and pelagic urochordates that secrete a mucoid "house" that is mainly a complex device for collecting food. It may also protect the animal. Muscular movements of the tail circulate water through the house, propelling the house through the water and bringing in suspended food. The house has coarse inlet filters that exclude large particles (>50 μm) and an outlet tube in which the tail of the animal beats. *Oikopleura* secretes a pair of fine filters that collect small particles. These are eaten, along with

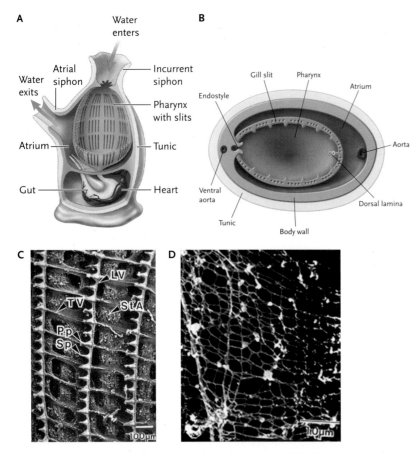

Figure 8.12 The anatomy of a generalized ascidian. **A.** Longitudinal section. **B.** Cross-section showing the perforated pharynx. Arrows show the movement of the mucous sheet over the gill bars, from the endostyle to the dorsal lamina. **C.** and **D.** Scanning electron micrographs showing the fine structure of the pharyngeal mucous net of the ascidian, *Ascidia paratropa*: **C.** the fine structure of the perforated pharynx; **D.** the structure of a part of the mucous net. Abbreviations: Pp and Sp = primary and secondary papillae which support the filter net; StA = stigmatal aperture; TV and LV = transverse and longitudinal blood vessels.

SOURCE: With kind permission from Springer Science+Business Media: Zoomorphology, "Functional morphology of the branchial basket of Ascidia paratropa (Tunicata, Ascidacea)," Volume 104, 1984, pp. 216–222, Pennachetti, C.A.

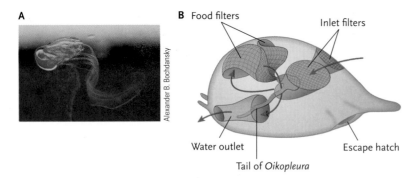

Figure 8.13 **A.** Adult larvacean (*Oikopleura* (1120)) out of its house. **B.** *Oikopleura* within its house (just below the centre of the picture).

the filters, when the filters are loaded with food. *Oikopleura* captures phytoplankton and bacterial cells, typically 0.5–10 µm in diameter.

When the coarse filters of the larvacean house are clogged with particles, the animal swims out of the house through an escape hatch. Before escaping, it secretes a new house as a compact layer covering its body surface. The new house is inflated into a functional food-capturing house once the animal is free of its old house. An individual *Oikopleura* may make and abandon from 4 to 16 houses daily.

Population densities of *Oikopleura* can be very high and their discarded mucoid houses, clogged with organic particles, are an important food source for other animals. As many as 10 000 m^{-3} *Oikopleura* houses have been recorded off the coast of British Columbia.

8.2f Echinoderms

Many echinoderms are passive suspension feeders, depending on water currents to bring suspended food particles. Their feeding behaviour typically involves trapping food on extended tube feet that are coated with sticky mucus. Particles are captured by direct interception: they adhere to the sticky mucus of the tube feet. These particles are collected into a bolus transported by the tube feet along each arm to the mouth.

The sea cucumber *Cucumaria frondosa* (1096) (Echinodermata, Holothuroidea) extends five branching tentacles (**buccal podia**, highly modified tube feet) to collect food particles from the water (Fig. 8.14). Papillae on the tentacles secrete a sticky mucus that traps food. While the *Cucumaria* tentacles collect particles, they do not select food by particle size or distinguish it from non-food. Mechanosensory feedback signals when a single tentacle is loaded with food. Then the tentacle is bent back into the mouth and the collected material transferred to the gut, reminiscent of licking food from your fingers!

The timing of tentacle extension by *Cucumaria* suggests that this process is regulated by the amount of food available in the water. Tentacle extension is also influenced by tidal cycles.

Figure 8.14 Suspension-feeding sea cucumber (*Cucumaria frondosa* (1096)) showing highly branched buccal tentacles (specialized tube feet) that capture suspended particles. Notice that one tentacle has been tucked inside the mouth for removal of trapped food particles.

8.2g Sponges

Sponges (Phylum Porifera) are sessile, suspension-feeding invertebrates that lack a gut. The body of a sponge consists of interconnected channels through which water is circulated. The channels are lined by distinctive cells called **choanocytes**, each with an apical flagellum surrounded by a collar of long microvilli-like extensions (Fig. 8.15A, B). The beating of choanocyte flagella creates the flow of water through the sponge's body. Choanocytes also capture particles from the water flow.

Sponges exhibit three grades of body complexity directly related to their need to maintain a continuous flow of water through their bodies as body size increases. **Asconoid** sponges have one internal chamber lined by choanocytes (Fig. 8.16). Water flows into this central chamber through holes (ostia) in the body wall and out through the **osculum**, a single large opening. Asconoid sponges are small because, as a body gets bigger, its volume increases by the power of three, but its surface area increases only by the power of two. Asconoid sponges cannot achieve large sizes because the choanocyte layer powering the water flow is a surface-area component while water flowing through the sponge is a volume component. At some limiting size, there is not enough power to move the increasing volume of water.

In **syconoid** sponges, the body wall forms outpockets, each lined by choanocytes. **Leuconoid** sponges have interconnected, choanocyte-lined chambers. In syconoid and leuconoid sponges, the surface area of choanocytes can generate the required water flow.

Old accounts suggested that water, driven by the flagellum of a choanocyte, passed through the collar of microvilli, which sieved food particles that were then phagocytized by the choanocyte. However, the Reynolds number for choanocyte microvilli is very small, and a beating flagellum cannot propel water through the narrow slits between neighbouring microvilli.

Experiments on syconoid sponges show that very few particles are caught on choanocyte collars. Most food particles are caught on the apical surface of choanocytes (outside the collar) and captured by elongate pseudopods extending from the surface of the choanocytes (Fig. 8.17).

Most sponge food items are 5 to 50 µm in diameter, small enough that they can be moved inside cells of sponges by phagocytosis or pinocytosis. Some particles can be phagocytosed by the choanocytes, while the microvilli collars on choanocytes trap and absorb the smallest particles.

Some sponges commonly host Cyanobacteria and even dinoflagellates (zooxanthellae) and sometimes incorporate green and red algae in association with the spongin fibres of the sponge's skeleton. These organisms are at least partially autotrophic. On the Great Barrier Reef off the coast of Australia, some sponges obtain more than half of their energy from the Cyanobacteria that live within them.

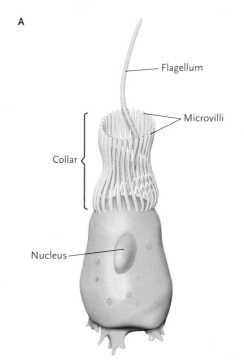

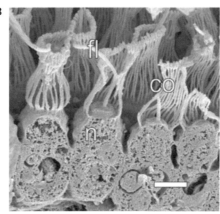

Figure 8.15 **A.** A choanocyte, the major food-collecting cell in members of Porifera. **B.** Scanning electron micrograph of choanocytes in the sponge, *Sycon coactum* (102). Note the flagellum (fl) surrounded by a collar (co) at the apical surface of each cell, n is the nucleus. Scale bar = 2 µm.

SOURCE: Figure 8.15.B from Leys, S. P., and D. I. Eerkes-Medrano. 2011. *Biol. Bull.* 211: 157–171. Reprinted with permission from the Marine Biological Laboratory, Woods Hole, MA.

Heterotrophic bacteria may account for 40% of sponge biomass. Although some of the bacteria may have been collected by filter-feeding, a large fraction resides in the mesohyl, for example of *Aplysina aerophoba* (112), and appears to be highly integrated into the sponge itself. In these sponges, treatment with antibiotics does not reduce the bacterial prevalence in the mesohyl, confirming the intracellular existence of these symbionts. Highly specific immunological cross-reactions indicate a long evolutionary history between sponge-specific bacteria and the sponges with which they are symbiotic. The data suggest that associations between sponges and bacteria date back to the Precambrian.

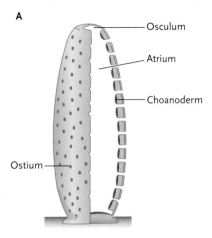

A

Osculum
Atrium
Choanoderm
Ostium

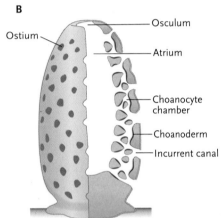

B

Osculum
Ostium
Atrium
Choanocyte chamber
Choanoderm
Incurrent canal

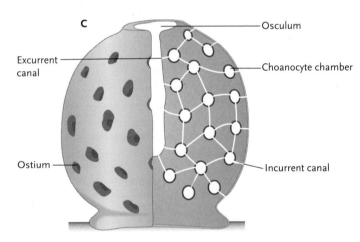

C

Osculum
Excurrent canal
Choanocyte chamber
Ostium
Incurrent canal

Figure 8.16 Sponge organization: **A.** asconoid, **B.** syconoid, and **C.** leuconoid. The red layer shows the locations of the choanocytes.

Like barnacles, some species of sponges use prevailing water currents to minimize the cost of transporting water during suspension feeding. Sponges with **oscula** (excurrent openings) at the top of otherwise low-growing bodies (Fig. 8.18A) have **ostia** (incurrent openings) located closer to the substratum. Adjacent to the substratum is a boundary layer with no movement of water, and water speed increases with distance from the substratum. Fast-flowing water

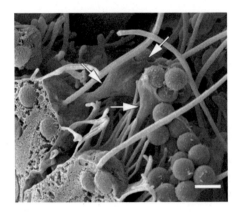

Figure 8.17 Scanning electron micrograph showing the apex of sponge choanocytes after being fed 1.0 μm latex beads. Note the cytoplasmic extensions (arrows) in contact with clusters of beads. Scale bar = 1 μm.

SOURCE: Figure 8.17 from Leys, S. P., and D. I. Eerkes-Medrano. 2011. *Biol. Bull.* 211: 157–171. Reprinted with permission from the Marine Biological Laboratory, Woods Hole, MA.

travelling past the oscula draws water from the sponge by viscous entrainment (Fig. 8.18B). The significance of this process for augmenting water flow through a sponge's body is demonstrated in the experiment in Figure 8.18C.

Carnivorous sponges such as *Asbestopluma hypogea* (113) capture prey by entanglement in filaments covered with hook-like spicules (Fig. 8.19). Prey are entangled by their setae or thin appendages. Captured prey is engulfed by sponge cells that aggregate around the victim's body. Digestion is mainly extracellular and involves a combination of autolysis and bacterial action. Fragments of prey are then phagocytosed and digested by choanocytes and bacteriocytes. The process from capture to digestion appears to take between 8 and 10 days for a large prey.

8.2h Bivalve Molluscs

Bivalves (e.g., clams, scallops, and oysters) use **ctenidia** (gills) to exchange gases with water (Figs. 8.20, 8.21, Section 11.3b) as well as for feeding purposes.

Ancestral bivalves were deposit feeders (see Section 8.4), and the few species of bivalves that retain this mode of feeding have long, ciliated palps that extend from the shell valves into surrounding sediment. There they collect particles for transport to the mouth. The ctenidia of these detritus-feeding bivalves are used only for gas exchange (**protobranch ctenidium**, Fig. 8.21A). Over the evolution of bivalves, opportunistic molluscs started to harvest organic particles suspended in the water brought in for gas exchange. The ctenidia gradually became elaborated to capture edible particles. Consequently, the ctenidial filaments of most living bivalves are elongated to provide an increased gill surface area for particle capture (Fig. 8.21). Increased gill filament length in the mantle cavity

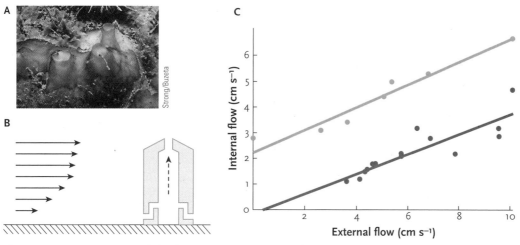

Figure 8.18 **A.** An encrusting sponge, *Halichondria panicea* (115), with oscula mounted on volcano-like elevations of the sponge body. **B.** A plastic model of a sponge with ostia (inhalent pores) close to the substrate and osculum (exhalent opening) high above the substrate. Water velocity (indicated by arrow length) is slow close to the substrate due to drag. As higher-speed water flows past the elevated osculum, it "sucks" water out of the osculum by viscous entrainment. **C.** An experimental demonstration of viscous entrainment for drawing water through a sponge when external water is flowing over the sponge. Measurements were made of the actual flow rate through a living sponge (filled circles) and a sponge that had been killed (open circles) when they were placed in different external current velocities.

SOURCE: B., C. R. McNeill Alexander. (1979). *The Invertebrates*, p. 110. Reprinted with the permission of Cambridge University Press.

requires that the filaments be folded into a V-shape, forming a **lamellibranch ctenidium**.

Tracts of cilia on the gill filaments of lamellibranch ctenidia are responsible for both water transport and capture and delivery of food particles to the mouth. Different types of lamellibranch ctenidia may use different mechanisms to capture particles with mucus appearing to play an important role. Particles trapped in mucus move along flow paths entrained by beating cilia. Eventually, particles are concentrated in food grooves, and the cilia in the groove transport the particles to the labial palps where edible and inedible particles are separated.

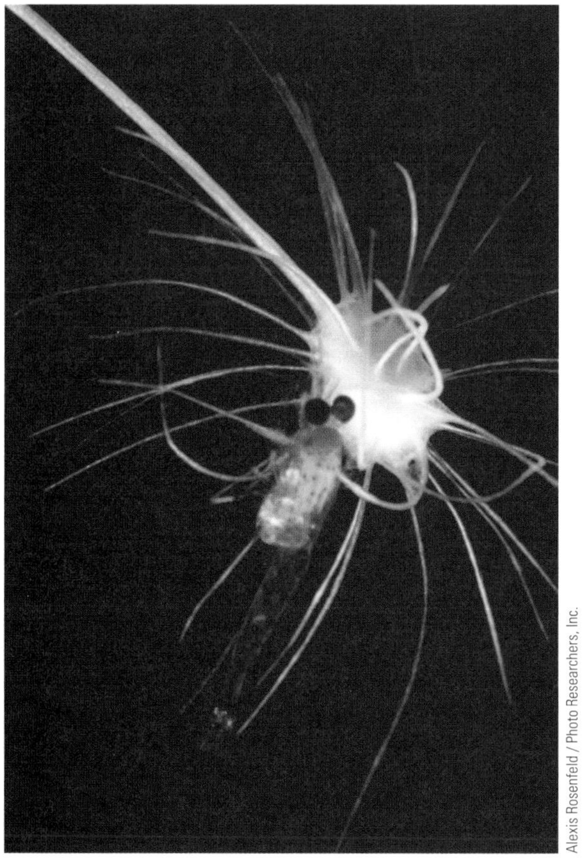

The efficiency of particle filtering by bivalves is impressive. A 4-g blue mussel (*Mytilus edulis* (536)) may clear phytoplankton from a litre of seawater every hour, and a scallop (*Aequipecten opercularis* (530)) can extract particles from as much as 7 L of water an hour.

8.2i Birds

Some aquatic birds filter food from water. A wide range of genera includes a variety of filtering behaviour, from accidental (phalaropes, *Phalaropus* (1558), Fig. 8.22A), to ram filtering (prions, *Pachyptila* (1634)), grasp pump filtering (ducks and geese, *Anser* (1656)), and through pump filtering causing distal inflow and proximal

Figure 8.19 A carnivorous sponge (*Asbestopluma hypogea* (113)) captures a crustacean (orange).

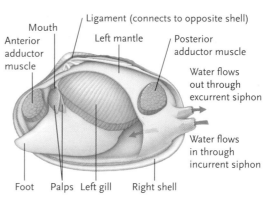

Figure 8.20 The anatomy of a generalized bivalve. The left shell valve and part of the underlying mantle fold have been removed to show the left ctenidium within the mantle cavity alongside the foot. A right ctenidium lies on the other side of the foot.

SOURCE: From RUSSELL/ WOLFE/HERTZ/STARR. *Biology*, 2E. © 2013 Nelson Education Ltd. Reproduced by permission. www.cengage.com/permissions

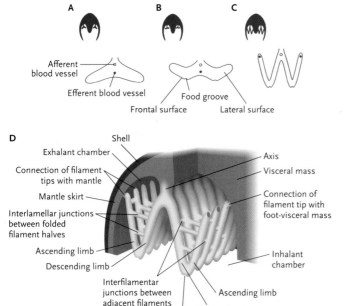

Figure 8.21 Bivalve ctenidia. **A.** Protobranch ctenidium with short gill filaments. **B.** Hypothetical intermediate showing lengthening of gill filaments and appearance of food groove. **C.** and **D.** Lamellibranch ctenidium with elongate and folded gill filaments to increase the surface area for particle capture. The delicate filaments are supported by junctions between neighbouring filaments and between the descending and ascending limb of each filament.

SOURCE: From RUPPERT/FOX/BARNES, *Invertebrate Zoology*, 7E. © 2004 Cengage Learning.

Figure 8.22 **A.** Wilson's Phalarope uses surface tension to draw drops of water containing small bits of food into the mouth, then extracts the food and spits out the water. **B.** Ducks and geese have lamellae along the edges of their beaks that strain small food particles out of the water. **C.** Lamellae are even further developed in flamingoes.

outflow (flamingos, *Phoenicopterus* (1627)). In *Phalaropus* (1558), prey are transported in a drop of water that runs rapidly along the open beak. Closing the beak and elevating the throat skin draws the water into the mouth. Ram filtering involves drawing water into the mouth as the bird swims and holding food particles in the mouth using recurved papillae when the water flows out. In some ducks and geese, the tongue draws water in at the tip of the beak and expels it to the side and back along the rim of the beak. Flamingos use a back-and-forth pump to draw water and particles in at the front and pass water out laterally through different sized lamellar meshes, filtering particles as water passes out.

STUDY BREAK

1. Explain why Reynolds numbers are of greater significance to small aquatic organisms than to larger ones.
2. What is viscous entrainment?
3. How do changes in water current affect the feeding of barnacles?

8.3 Ram and Suction Feeding

All vertebrates that feed in water are either **suction feeders** or **ram feeders**. Suction feeders move prey into the mouth by opening the jaws rapidly to create negative pressure. This pulls prey into the mouth, which is then closed to entrap it. The simple action of expanding the buccal cavity in water creates suction. Many aquatic vertebrates that are suction feeders are "sit and wait" predators. Rather than chasing their prey, these animals wait for a victim to wander too close. A good example is the mata-mata turtle (*Chelus fimbriatus* (1423), Fig. 8.23) of the Amazon and Orinoco river basins in South America. The mata-mata's brown, jagged outline camouflages it well in slow-moving water. It sits absolutely still and waits patiently for a small fish or insect to pass by. When prey is in range, the mata-mata opens its huge mouth very rapidly, creating a strong current that sweeps the meal into its mouth. This type of suction feeding is also common in fish such as anglerfish, stonefish, frogfish, and flounder.

Ram feeding requires movement of the whole body through the water with the mouth open. Ram feeders move food into their mouths by propelling their bodies forward and overtaking their prey. Fast-swimming predatory fish such as barracuda and some sharks are ram feeders, as are baleen whales. Fin whales (*Balaenoptera physalus* (1742), Fig. 8.24) and other baleen whales lunge with their mouths fully agape, taking in a huge volume of water and the prey it contains. Expansion of the buccal cavity allows an adult fin whale to engulf 60–82 m³, a volume exceeding that of its entire body. The cost of this approach is the

Figure 8.23 The jagged outlines of the mata-mata turtle help it blend into the background. When a prey item gets too close, the turtle opens its mouth rapidly and sucks it in.

drag generated by the open mouth, partly explaining why fin whales make short dives. Each lunge brings the whale almost to a halt, obliging it to accelerate again to lunging speed. The coordination of the extreme movements of the jaws and controlled filling of the mouth are monitored by two sensory organs. The vibrissae (whiskers) on the whale's chin have mechanoreceptors (page 428) that can sense the density of prey, enabling the whale to determine if it is dense enough to make it worth the high cost of opening its mouth. Mouth opening is monitored by a sensory organ that lies within the mandibular symphysis (joint between the two lower jaws) and contacts a fibrocartilage structure that is embedded in the floor

of the mouth. This organ also contains mechanoreceptors that sense the compression and shearing that occurs at the symphysis and the expansion of the mouth cavity.

Compared with suction feeders, ram feeders expend more energy pursuing prey because the animals must overcome drag to move through the water. Adaptations for reducing drag include streamlined body shapes and accommodations in fish that permit water to flow in through the mouth and out through the gills. Ram feeders cover larger distances than suction feeders, increasing the likelihood that they will encounter prey. Suction feeding, on the other hand, only works over relatively short distances. Large mouths and/or fast-opening jaws have evolved to increase the suction radius.

8.4 Deposit Feeding

A **detritivore** is a heterotrophic organism that eats decomposing organic material. Much of the food of a filter feeder may be detritus suspended in water. When the organic material is associated with the substratum, it becomes food for deposit feeders. This appears straightforward but consider the diet of an earthworm (*Lumbricus* sp. (357), Annelida, Clitellata, Oligochaeta), a well-known and ubiquitous detritivore (Fig. 8.25A). Certainly its food includes much decaying vegetable material, fitting the definition of a detritivore. At the

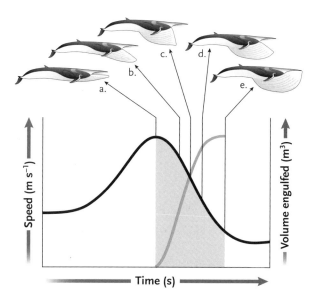

Figure 8.24 A fin whale feeding. Fin whales come to almost a complete stop (black line) after they open their mouths to feed (grey area under curve). The trade-off is that they take in a huge mouthful of water (green line) that contains a large amount of food.

SOURCE: Jeremy A. Goldbogen, Nicholas D. Pyenson, Robert E. Shadwick, "Big gulps require high drag for fin whale lunge feeding," *Mar Ecol Prog Ser* Vol. 349: 289–301, 2007. Inter-Research Science Center.

Figure 8.25 Common terrestrial detritivores. **A.** Earthworm (*Lumbricus* sp. (357)) and **B.** woodlouse (*Porcellio dilatatus* (807)).

same time it is eating many bacteria and fungi responsible for the breakdown of that organic matter (while the inorganic material of soil passes on through its body). This breadth of diet is inevitably true of all macrodetritivores.

Detritivores are important because they harvest organic carbon from the very lowest trophic level, making it immediately available to predators at higher levels. The "early bird" is the direct beneficiary of *Lumbricus* detritivory! Another familiar terrestrial detritivore is the woodlouse (*Porcellio* sp. (807), Crustacea, Isopoda, Fig. 8.25B), a terrestrial crustacean.

Some isopods are marine detritivores, for example, *Bathynomus giganteus* (803), a giant deep-sea isopod that may reach lengths of 35 cm (Fig. 8.26). Faced with the limited food availability of the deep sea environment, it is both a detritivore and a scavenger, and probably a predator on slow-moving prey.

Many polychaete species are detritivores in the marine environment. The echiurans (discussed in Section 8.6b) extend their proboscises to collect food. Others (e.g., *Amphitrite* (*Thelepus*) spp. (347), Fig. 3.47A) extend tentacles to collect detritus from the surrounding substratum. The worm is normally hidden in its tube with its gills at the mouth of the tube and highly extensile white tentacles extending across the substratum, collecting food particles and moving them back to the mouth via ciliated gutters that run the length of the tentacles.

STUDY BREAK

1. Why are ram feeders likely to be larger animals?
2. What are the advantages and disadvantages of detritivorous feeding?

8.5 Jaws and Teeth

8.5a Introduction

Many animals have **jaws**, specialized structures in the anterior part of the digestive system that capture or gather food and assist with the first stages of ingestion. The wide variety of jaws are made of different materials and take different forms. Teeth have similar hardness although different mineral forms and inclusions produce slight variations in hardness. The jaws of invertebrates have evolved in different ways in different groups. Invertebrate jaws are composed of a variety of

materials, commonly chitin and frequently with metallic or mineral components that increase hardness and strength. Ultimate strength is a better predictor of performance than hardness. Hardness is measured on Mohs scale, a nonlinear measure in which chitin (3) is about three times harder than fingernails (2.5). Apatite, the hard component of calcareous skeletons and vertebrate teeth, has a Mohs hardness of 5–6 (Fig. 8.27). Although chitin is relatively soft, it is pliable, resilient, and tough, a perfect material for arthropod exoskeletons and the mouthparts of many other invertebrates.

Teeth or tooth-like structures are widespread in animals (Fig. 8.28), usually associated with jaws and feeding. Teeth, plates, beaks, bills, and claws are used to crush and cut food. The structures of these features usually relate directly to function, analogous to the difference between a hammer and a chisel. Examine a sampling of teeth (Fig. 8.29), and use the caption to identify their function and owner.

8.5b Internal Jaws in Gnathostomulida, Rotifera, and Tardigrada

Pharyngeal jaws of animal phyla grouped in the Gnathifera may have been the first jaws to evolve. Chitinous hooks are common in Platyhelminthes, often around the mouth in monogeneans. Chitinous structures within the mouth opening, perhaps as an aid to food capture, may have subsequently developed into internal jaw mechanisms.

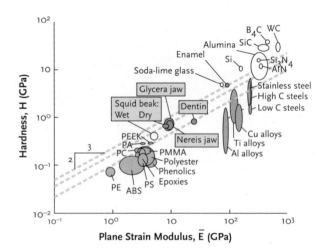

Figure 8.27 The hardness of some biological jaws relative to some synthetic materials, metals, and minerals. Notice that the range of hardness and strength of the biological materials (red) (shown by the lines indicating mean and range) are harder and stronger than synthetics (light blue) and may be as hard as a number of metallic compounds (grey), they are not capable of reaching the same strength (expressed as the plane strain modulus, a measure of the stiffness of a flat surface) as metals or harder and stronger compounds (yellow).

SOURCE: Reprinted from *Acta Biomaterials* 3(1), Miserez et al., "Jumbo squid beaks: Inspiration for design of robust organic composits," Pages 139–149, Copyright 2007, with permission from Elsevier.

Figure 8.26 *Bathynomus giganteus* (803), a giant deep-sea isopod.

Jany Sauvanet / Photo Researchers, Inc.

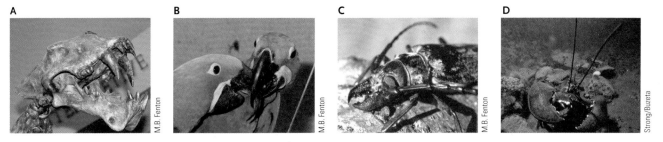

Figure 8.28 Teeth and tooth-like structures. Of these animals, only **A.** the sabre-tooth cat has teeth. **B.** Parrot beaks crack open hard fruits, **C.** long-horned beetle mouthparts impale their prey, and **D.** lobsters use their claws to crush and slice prey.

Gnathostomulida spp. (293), which have one of the simplest animal body plans, (most <1 mm long) eat bacteria and fungi and live in the spaces between grains of marine sand, particularly in low oxygen environments. Their gut is a simple sac with a single opening. Within the mouth opening, a basal plate appears to scrape food particles from the sand. To move food items back into the gut, a pair of jaws are brought together by pharyngeal muscles and reopened by cuticular springs (Fig. 8.30).

The body plans of rotifers (Phylum Rotifera) are more complex than those of gnathostomulids. Rotifers have a complete gut with differentiated regions, including a **mastax**, a muscular capsule in the pharynx that houses

the **trophi**, complex jaws. Rotifers are called "wheel animalcules" because of the appearance of beating cilia on the **corona** (Fig. 8.31A), an anterior structure that propels mobile rotifers and sets up feeding currents.

Asplanchna (297) eats other rotifers and small crustaceans. The sharp-pointed trophi of *Asplanchna* (Fig. 8.31B) are shaped for piercing and grasping. *Asplanchna*'s prey includes other rotifers. Populations of *Brachionus* (298) under predation pressure from *Asplanchna,* develop spines for protection.

Tardigrada are a sister group to Arthropoda. These tiny animals (<2 mm long) are found in marine, freshwater, and terrestrial environments and feed by piercing plant cells and sucking out the cytoplasm of

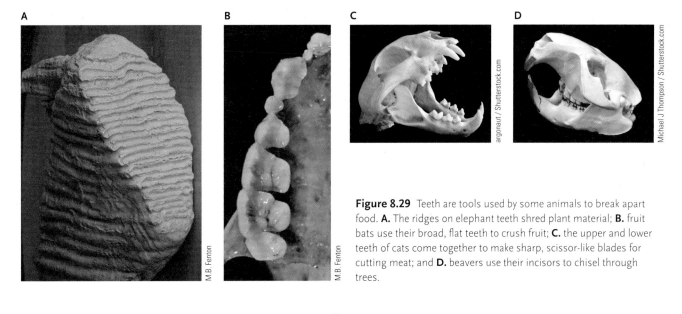

Figure 8.29 Teeth are tools used by some animals to break apart food. **A.** The ridges on elephant teeth shred plant material; **B.** fruit bats use their broad, flat teeth to crush fruit; **C.** the upper and lower teeth of cats come together to make sharp, scissor-like blades for cutting meat; and **D.** beavers use their incisors to chisel through trees.

Figure 8.30 **A.** *Gnathostomulida armata* (293). Note the position of the jaws close to the back of the head. **B.** The head, stained for muscle bundles, showing the jaws within the pharynx.

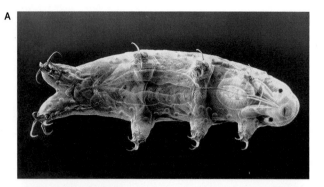

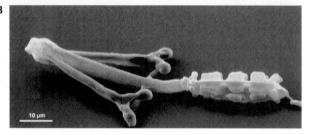

Figure 8.31 A. *Philodina* (305), a rotifer that feeds on detritus and small phytoplankton that are swept into the mouth by the cilia of the corona. Notice the mastax just inside the mouth. **B.** Isolated trophi from the mastax of *Asplanchna priodonta* (297). Scale bar = 10 μm.

Figure 8.32 A. *Halobiotus crispae* (619) (Tardigrada); notice the pharyngeal bulb and the stylet sacs. **B.** The isolated stylet apparatus.

SOURCE: A. Nadja Møbjerg. Cover of *Journal of Morphology* 270, 2009. This material is reproduced with permission of John Wiley & Sons, Inc. B. from *Zool. Anz.*, 240, Eibye-Jacobsen, J., "A New Method for Making SEM Preparations of the Tardigrade Buccopharyngeal Apparatus," Pages 309-319, Copyright 2001, with permission from Elsevier.

the cell. To achieve this, they have a muscular pharynx, a reinforced buccal tube (that prevents collapse when the pharynx is sucking food) and a complex stylet apparatus used to puncture the cell (Fig. 8.32). This mouth apparatus is perhaps more similar to those of the Gnathifera than to the jaws of arthropods.

8.5c Annelida Jaws

Many animals in the Class Polychaeta (Phylum Annelida) have an eversible pharynx (Fig. 8.33) that, when extended, reveals jaws at its tip that are used to attack and catch prey. It appears that Cambrian annelids did not have jaws, but, by the Ordovician, polychaete jaws are common fossils. These were only recently associated with annelids; before, disarticulated plates of polychaete jaws had been known as scolecodonts.

Polychaetes such as *Nereis* (321) have large, well-developed jaws with proteinaceous tips almost as hard as vertebrate teeth (Fig. 8.33). This hardness results from the incorporation of metallic zinc into the cross-linking of a histidine-rich protein matrix. Hardening of the relatively soft material of polychaete jaws has evolved in various ways. *Eunice* (332), a predatory polychaete, has calcite pads that harden the grinding faces of the jaws. *Glycera* (317), known as a bloodworm because its blood vessels show through its body wall, has a highly extensible proboscis with four very small jaws (Fig. 8.34) hardened by inclusion of atacamite, a copper compound ($Cu_2Cl(OH)_3$). *Glycera* jaws are hollow, and a duct transmits venom that immobilizes prey (other worms and small crustaceans). *Glycera* venom contains a **glycerotoxin**, a glycoprotein neurotoxin that induces a transient opening of Ca^{++} channels. Neurobiologists studying calcium channels use purified glycerotoxin as a valuable tool.

8.5d Molluscan Jaws

The Cephalopoda have a beak, probably a primitive molluscan feature lost in most modern molluscs. The beak (Fig. 8.35) is mounted in the buccal musculature and is used in attacking prey. The beak of a squid is two degrees of hardness greater at the tip than at the base. This gradient acts as a cushion to protect the squid from forces applied by the cutting action of its beak.

Unlike annelids and arthropods, Cephalopoda do not depend on a metal to give hardness to the beak tip.

NEL

BOX 8.2
Real Life CSI

In 2001, two men broke into a house and robbed a woman, leaving her tied to a chair. Forensic investigators found a leech that had recently fed (i.e., had blood in its gut) on the floor of the crime scene. The blood did not match the victim or any of the police personnel at the scene. The DNA sample from the blood was held in police records until eight years later, it was matched to DNA from a sample taken from a man arrested for an unrelated crime. Confronted with the evidence, the man confessed to the eight-year-old burglary.

Information from a personal communication from Dr. Sally Kelty, Tasmanian Institute of Law Enforcement Studies.

A.

B.

Figure 8.33 **A.** The head of a predatory polychaete, *Nereis virens* (323), showing the complex sensory palps and tentacles involved in prey detection (notice the eye, close to the base of the tentacles). **B.** *Nereis* everts its pharynx so that the jaws are in a position to capture prey. Notice the pointed denticles on the pharynx that help with gripping prey items.

A.   **B.**

 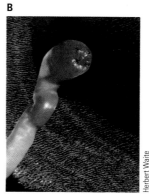

Michael Owen   Herbert Waite

Figure 8.34 **A.** *Glycera* (317). **B.** *Glycera* with its proboscis everted, showing its jaws.

The beak has a scaffold of chitin fibres that appear the same from the soft base to the hardened tip. The base of the beak is white and contains about 70% water (by weight), 25% chitin, and 5% protein. At the black hardened tip, the water content is 15–20%, protein 60%, and tanned pigment about 15%. The hardened part of the beak is rich in cross-linked, histidine-rich protein material. The damage that a giant squid (*Architeuthis dux* (491)) can do to an attacking sperm whale (*Physeter macrocephalus* (1766)) demonstrates the effectiveness of squid beaks.

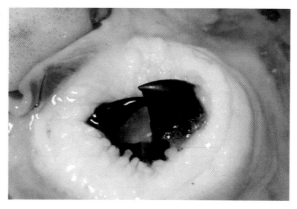

Figure 8.35 Beak of *Mesonychoteuthis hamiltoni* (509), the colossal squid.

SOURCE: Norman Heke/Museum of New Zealand Te Papa Tongarewa/ MA_I.088880

8.5e Jaws in the Echinoidea

The jaw apparatus of a sea urchin (Echinodermata, Echinoidea) and its relatives is called **Aristotle's lantern** (Fig. 8.36B). The name remembers Aristotle who originally described the structure. Translated, Aristotle's description (in *Historia Animalium*) reads: "The urchin has five hollow teeth inside, and in the middle of these teeth a fleshy substance serving the office of a tongue.... The mouth-apparatus of the urchin is continuous from one end to the other, but to outward appearance it is not so, but looks like a horn lantern with the panes of horn left out." The lantern is actually a framework of ossicles supporting five curved, sharply pointed teeth and the muscles controlling the movement of each tooth.

In operation, the teeth come together at a point (Fig. 8.36C) as though they were moving on the surface of a sphere, with tooth retraction and protraction following a curved path. The tooth tips operate like a five-jawed vice-grip, capable of enough force to crack a sand grain or crush a small calcareous pebble. Teeth are constructed from fibres of calcite (1 μm at the tip, broadening to 20 μm at the base), coated with an organic sheath and impregnated with $MgCO_3$. In several species of sea urchins, tooth hardness ranges from 4 to 5 on the Mohs scale, hard enough to scrape calcareous algae from rock surfaces. Teeth are continuously worn away and regenerated.

Strongylocentrotus (1091) and many other echinoids are primarily herbivores, rasping food from the surface of algae, or algae from the surface of the substratum. Some echinoid species are opportunistic omnivores, and groups of urchins may be seen cleaning the flesh off a fish carcass. Populations of *Strongylocentrotus* may be so dense that they completely clean the surface of sub-tidal rocky areas, eliminating newly settled growth and creating an "urchin barren." This situation is only reversed by outbreak of disease, sometimes caused by an amoeboid parasite, that kills large populations of *Strongylocentrotus*.

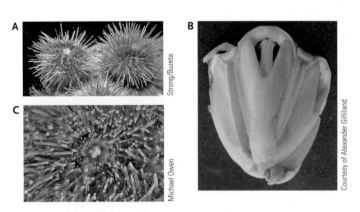

Figure 8.36 A. The green sea urchin, *Strongylocentrotus droebachiensis* (1091). **B.** Aristotle's lantern of a green sea urchin. The top part of the test (the outer skeleton) and the internal organs have been removed to leave the lantern in situ. **C.** Aboral view of *S. droebachiensus* showing the five jaws. Around the jaws, the minute pincer-like structures are pedicellaria, highly modified spines used in defense and cleaning.

8.5f Jaws of the Arthropoda

Not surprisingly, insects (class Insecta or Hexapoda) show the greatest diversity in mouthparts among animals. Insects range from being highly specialized feeders to broad generalists, and their mouthparts reflect this range. Insect mouthparts evolved as a series of segmented appendages in a primitive ancestor and specialized over evolutionary time to comprise the food-handling apparatus of modern arthropods.

When arthropods were thought to be polyphyletic, a key factor in that argument was the nature of the jaws—whether they originated from a limb tip or a limb base and used a pincer-like or a grinding movement. The discovery that insects are derived from a crustacean stem and that the gene *Dll* (*distal-less*) is expressed in all arthropod mandibles during development supports the view that Arthropoda are monophyletic.

The American cockroach (*Periplaneta americana* (860), a stowaway from Africa) has a basic insect jaw—it is an omnivore with unspecialized mouthparts (Fig. 8.37A). It has a pair of **mandibles** (jaws), each with a biting tip and a grinding surface at its inner base. Each mandible has two condyles (pivot points) at the mid-dorsal and mid-ventral region of the base. Muscles

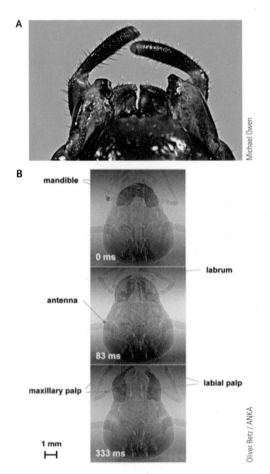

Figure 8.37 Jaws of *Periplaneta americana* (860). **A.** A ventral view of the mouthparts, showing the highly sclerotized faces of the mandibles. **B.** X-ray imaging shows the movement of the mandibles (coloured brown)

on either side open and close the jaws. Synchrotron-based, hard X-ray micro-imaging has allowed analysis of the movements of *Periplaneta* mandibles (Fig. 8.37B).

There are three broad categories of mouthpart adaptations in insects: piercing/sucking, raptorial, and chewing. Piercing/sucking mouthparts are best exemplified by the true bugs (Hemiptera) and their plant-feeding cousins the Homoptera. Here, mandibles (and sometimes other structures) are modified into stylets—thin sharp blades used to pierce prey and inject venom (in the case of predatory bugs) or to pierce the phloem of plants. In both cases the food (blood or sap) is sucked back through the stylets. Piercing/sucking mouthparts have evolved in a number of other insects, familiar examples being blood-feeding species such as some flies (mosquitoes and most horseflies, Fig. 8.38B), lice, and even some predatory moths, while most lepidopterans have sucking, but not piercing, mouthparts (Fig. 8.38C).

Raptorial mouthparts are present in many insect orders, but are probably most common in the predatory beetles (Coleoptera) where tiger beetles (Subfamily Cicindelidae) (Fig. 8.38A) are examples. Their mandibles are large, sharp, sword-like structures used to grasp and kill prey. Raptorial mouthparts are also common in wasps, ants, and lacewings (Neuroptera). These mouthparts are amazingly varied, from the lightning-fast snapping jaws of trap-jaw ants, to the razor-sharp, scissor-like jaws of leafcutter ants.

A modification from the raptorial theme, chewing mouthparts are shortened and hardened to function as tools for consuming or digging into plants. Insects such as beetles and carpenter bees use chewing mouthparts to burrow into solid hardwood. Caterpillars, the larvae of butterflies and moths, have chewing mouthparts and eat leaves.

While spiders (Chelicerata, Arachnida, Araneae) are typically carnivores, they do not take in solid food. Spiders either inject or "spit" enzyme-rich saliva (which is frequently venomous) to dissolve the tissues of their prey and then consume their food in liquid form (Fig. 8.39).

An East African jumping spider (*Evarcha culicivora* (668), Fig. 8.40) eats mosquitoes and, using visual and odour cues, preferentially chooses mosquitoes that have recently fed on vertebrate blood. This means that the spider indirectly eats vertebrate blood. *E. culicivora* is an unusual spider in that mate-choice is mutual. The surprising finding is that a spider of either sex making a mate choice demonstrates a strong preference for a mate that has recently eaten a blood-fed mosquito. So, at least for *E. culicivora*, what you eat may affect your love life!

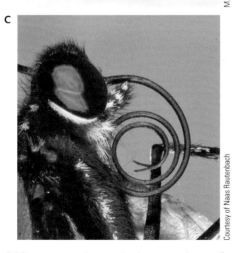

Figure 8.38 Insect mouthparts. **A.** The raptorial jaws of a predatory tiger beetle (*Cicindelida* sp. (929)); **B.** the mosquito's (*Aedes* sp. (953)) mouthparts modified for piercing and sucking, caught in the act; **C.** the crimson-rose butterfly (*Atrophaneura* (*Pachliopta*) *hector* (1026)) showing its long coiled proboscis that can be extended to suck nectar from a flower.

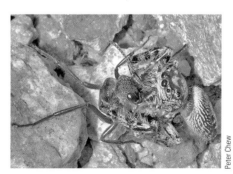

Figure 8.39 A jumping spider (*Opisthoncus parcedentatus* (679)) attacking a bull ant (*Myrmecia chrysogaster* (1001)). A droplet of spider venom is an enzyme-rich solution that, when injected, dissolves the tissues of the prey.

Figure 8.40 The East-African jumping spider, *Evarcha culicivora* (668), feeding on a mosquito.

8.5g Feeding in Primitive Fish

In vertebrates the **pharynx** is the passage from the mouth and nostrils to the oesophagus. All vertebrates have a muscular pharynx, supported by skeletal structures (cartilage or bone), that functions in feeding and respiration. The vertebrate pharynx develops from a series of **pharyngeal arches**, bars of skeletal elements and muscle covered by ectoderm on the side of the head (Fig. 8.41). In agnathans, vertebrates without jaws, all of the pharyngeal arches give rise to gills. In gnathostomes, vertebrates with jaws, the first pharyngeal arch gives rise to the jaws, and the second arch becomes the hyoid apparatus that braces the jaws of fish and supports the tongue in tetrapods. The remaining pharyngeal arches become gills, pharyngeal jaws or other muscles, ligaments, bones, and glands in the neck.

The only living vertebrates without jaws, lampreys and hagfish (Agnatha), rasp their food. **Ammocoetes**, larval lampreys, are filter feeders. In some species, adults do not feed, but spawn after metamorphosis and die. Species that live long adult lives are parasitic and usually feed on other fish. Lampreys use their round mouths lined with rows of keratinized "teeth" to attach and hold on to the bodies of fish. The keratinized plates on a tongue-like piston scrape a hole in their prey (Fig. 8.42) and while feeding, secrete **lamphredin** from a gland inside their mouths to break down the tissues. Some lampreys specialize on blood as food; others consume both flesh and blood.

Hagfish bite without jaws. They superficially resemble lampreys but feed in a different way. Hagfish evert keratinized dental plates and open the mouth to 180°. Once in contact with food, the plates are retracted into the oral cavity along with a large bite of food. A biting hagfish produces considerable force, showing that jaws are not essential to biting.

8.5h Fish Jaws and Teeth

In other vertebrates, jaws are composed of cartilage, bone, and sometimes teeth, and are used to capture and/or break down food. Some jaws are simple mechanisms with few moving parts; others are complex lever systems with many interacting components. Placoderms (Fig. 6.6) were the first jawed vertebrates and lived from the Late Silurian through the Devonian. At an estimated 10-m body length, *Dunkleosteus terrelli* (1130) was the largest known placoderm and had strong, bladed jaws. A biomechanical model of *Dunkleosteus* (Fig. 8.43) suggests that this predator was capable of fast and powerful bites. The model shows that two joints (one between head and thorax, the other between jaw and skull) were central to its bite by allowing contractions of four muscles to operate the jaws. To open the jaws, the epaxialis and coracomandibularis muscles contracted, and to close the adductors mandibulae and the cranial depressor muscles contracted. A 6-m-long *Dunkleosteus* would have bitten with a force of 4400 Newtons (N) at the tip of the jaws and 5300 N at the rear dental plates. For comparison, a bone-crushing spotted hyena (*Crocuta crocuta*) bites with a force of 2000 N. *Dunkleosteus* could probably eat any other aquatic species it encountered, armoured or not.

Elasmobranchs (Chondrichthyes), with shearing, bladed teeth and powerful bites, appeared in the Mesozoic. Many modern shark species are aquatic

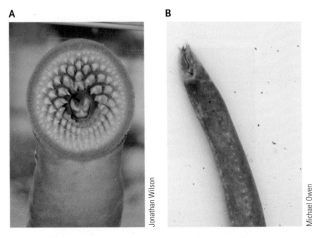

Figure 8.42 Lampreys and hagfish are jawless vertebrates. **A.** Lampreys are parasitic and attach to other animals with a round mouth ringed with teeth. **B.** Hagfish (Myxine glutinosa (1125)) are similar in appearance and sometimes feed on dead animals.

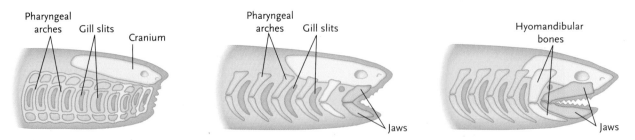

Figure 8.41 A. In jawless fish the pharyngeal arches are separated by gill slits. **B.** In early jawed fish, the arches contain more cartilage, and the first arch forms the jaws. **C.** In jawed fish, the jaws are supported by the hyomandibular bones, which are derived from the second pharyngeal arch.

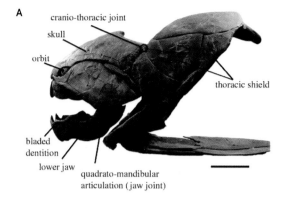

A

cranio-thoracic joint
skull
orbit
thoracic shield
bladed dentition
lower jaw
quadrato-mandibular articulation (jaw joint)

B

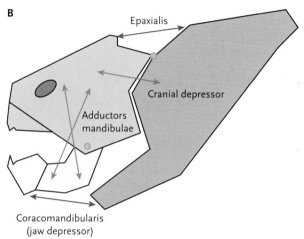

Epaxialis
Cranial depressor
Adductors mandibulae
Coracomandibularis (jaw depressor)

Figure 8.43 Model of the feeding apparatus in *Dunkleosteus* (1130). **A.** This extinct fish opened its mouth by raising the skull at the cranio-thoracic joint and simultaneously lowering the jaw at the quadrato-mandibular joint. **B.** The blue arrows illustrate the position and action of muscles that raised the head and lowered the jaw. The red arrows represent jaw-closing muscles.

SOURCE: Anderson and Westneat 3 (1): 76–79, *Biology Letters*. "Feeding mechanics and bite force modelling of the skull of Dunkleosteus terrelli, an ancient apex predator," 2007, The Royal Society.

predators with blade-like teeth. Among sharks, the white shark (*Carcharodon carcharias* (1133)) is notorious for its bite.

Powerful bites alone do not make a successful predator. The process of feeding also requires separating edible from inedible components of prey. Among elasmobranchs, structurally complex prey processing has evolved twice along different biomechanical pathways (Fig. 8.44). Skates and rays (batoids) use muscles of the lower jaw to manipulate food, especially armoured benthic prey such as bivalve molluscs. Ground and mackerel sharks accomplish prey manipulation with muscles of the upper jaw.

Many species of fish have two sets of jaws, oral jaws located in the mouth and pharyngeal jaws located in the pharynx (Fig. 8.45). Generalized percomorph fish use only the oral jaws to collect and process food. Many other ray-finned fish use oral jaws to capture and pharyngeal jaws (often bearing teeth) for prey manipulation. In more specialized percomorphs (cichlids and labrids), pharyngeal jaws and their associated musculature are derived and show three main features:

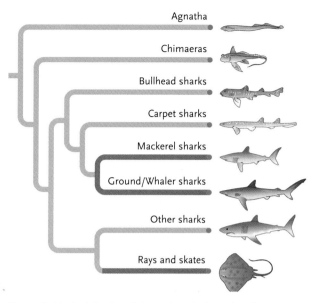

Agnatha
Chimaeras
Bullhead sharks
Carpet sharks
Mackerel sharks
Ground/Whaler sharks
Other sharks
Rays and skates

Figure 8.44 A phylogeny of elasmobranchs based on their primary food. The ability to eat prey that requires complex processing has evolved twice, once in sharks and once in rays. The agnatha are shown as an outgroup.

1. the left and right lower pharyngeal jaw elements are fused into one structure
2. the lower jaw is suspended in a muscular sling
3. the upper jaw elements articulate with the underside of the skull (the neurocranium)

Fossils show that the early radiations of fishes had characteristic changes in the jaw structure that allowed them to become primary consumers, increasing their access to food (see Chapter 6). Fundamental changes included differences in jaw bone lengths and biting behaviour.

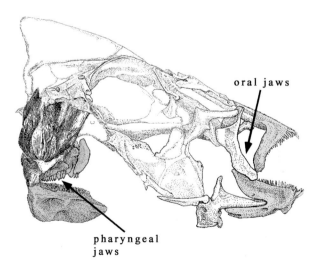

oral jaws

pharyngeal jaws

Figure 8.45 A schematic view of oral and pharyngeal jaws in a Lake Malawi cichlid (*Dimidiochromis compressiceps* (1263)) showing the relative locations of the two kinds of jaws.

SOURCE: Fraser, G.J. et al., 2009. "An ancient gene network is co-opted for teeth on old and new jaws," *PLoS biology*, 7(2), p.e31.

The appearance, jaw morphology, and teeth in carnivorous fish are obviously different between species that feed by biting or by suction feeding (Fig. 8.46). Some carnivores maximize the speed at which they open (suction feeders) or close (suction and ram feeders) their mouths. Other fish, from teleosts to chondrichthyans, can protrude their jaws to extend their reach, enhancing their efficiency in ram feeding on elusive prey (Fig. 8.47).

The specialized mouths of syngnathoid fish (pipefish, seahorses, seadragons, Fig. 8.48) are associated with **rapid pivot feeding**, which uses an explosively rapid dorsal rotation of the head, moving the mouth to approach and then capture prey. The evidence suggests that rapid pivot feeding does not involve (or require) suction. Prey are relatively stationary during the rapid pivot.

8.5i Jaws and Teeth in Reptiles, Birds, and Mammals

Some, if not many, of the elements in a vertebrate cranium may move considerably (**cranial kinesis**), and this movement can affect the power and nature of a

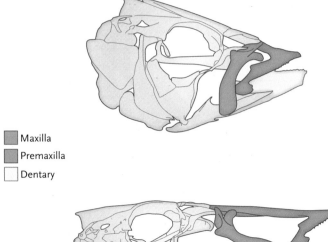

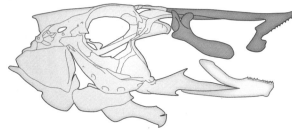

■ Maxilla
■ Premaxilla
□ Dentary

Figure 8.47 Extreme jaw protrusion in the Red Bay snook (*Petenia splendida* (1281)). This neotropical cichlid can dramatically extend its mouth forward from **A.** the resting position to **B.** fully open.

bite. *Dunkleosteus* shows cranial kinesis involving biting but not teeth. The concept of cranial kinesis is relatively unfamiliar in mammals because the only movement within the mammal skull is between the lower jaw (dentary) and the skull. Cranial kinesis is

Figure 8.46
Morphological differences between biting and suction feeding Lake Malawi cichlids. **A.** The suction feeder *Metriaclima zebra* (1274) has long narrow jaws with an outer row of widely spaced bicuspid teeth and a forward-facing mouth.
B. *Labeotropheus fuelleborni* (1269) is adapted for biting and has short, robust oral jaws with closely spaced tricuspid teeth and a downward-facing mouth.
Abbreviations:
LJ = lower jaw;
MX = maxilla; NCM = neurocranium;
PMX = premaxilla;
SUS = suspensory apparatus.
SOURCE: Albertson et al. 2003 Directional selection has shaped the oral jaws of late Malawi cichlid fishes. *PNAS* 100 (9):5252–5257. Copyright 2003 National Academy of Sciences, U.S.A.

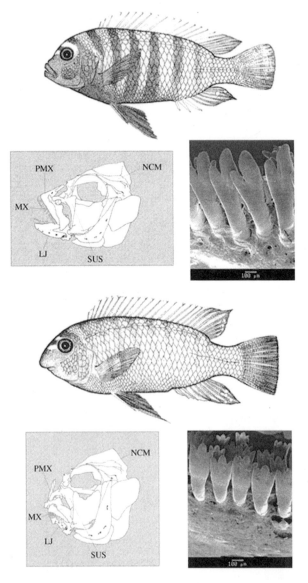

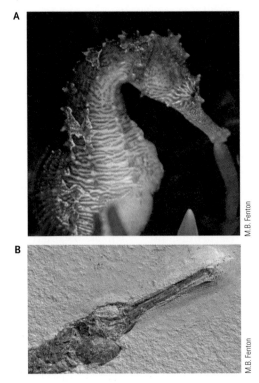

Figure 8.48 A. *Hippocampus erectus* (1341), a seahorse, and **B.** a fossil *Syngnathus acus* (1346) showing its specialized, elongated snout.

more evident in fish, but it also appears to have been important in some dinosaurs. Among living reptiles, snakes provide striking examples of cranial kinesis.

Eggs are obviously important to the individuals that produce them, but they are also nutritious food. Egg-eating predators include several snake species. *Dasypeltis* (1489) (Fig. 8.49), the main egg-eating snakes in Africa, are more proficient at this than egg-eating kingsnakes (*Lampropeltis getula* (1504)) in North America, apparently reflecting a pre-Miocene radiation of ground-nesting birds in Africa.

Crucial adaptations for egg-eating in *Dasypeltis* include mobile supratemporal bones that help the lower jaw extend further (Fig. 8.50A), a trachea with reduced rings, and an expandable region of connective tissue that allows the snake to breathe even with its mouth full of egg (Fig. 8.50B), and forward-facing, ventral extensions of vertebrae (Fig 8.50C) that help crack the eggs after they are swallowed.

Although several species of fossil birds had teeth, living species do not. However, their bills show a combination of strength and versatility (Fig. 8.51). The keratinous beaks of birds range from the thin and delicate insect-catching bills of goatsuckers (Caprimulgidae), to the crushing beaks of seed-eaters and the

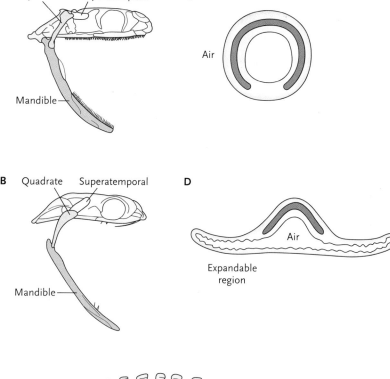

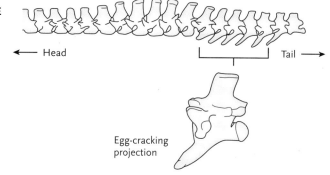

Figure 8.50 Both unspecialized (**A.**) and egg-eating snakes (**B.**) can open their jaws to 120°, but egg-eating snakes have mobile supratemporal bones (yellow) that allow the quadrate (blue) and mandible (pink) to open further. **C.** Most snakes have almost complete rings of cartilage (red) that help hold the trachea open. **D.** The tracheal rings are reduced in egg-eating snakes and replaced by tissue that can expand when eggs pass by. **E.** Egg-eating snakes also have vertebrae with forward-facing projections that help crack eggs.

SOURCE: *Biomechanics: An Approach to Vertebrate Biology*, by Carl Gans (Ann Arbor: The University of Michigan Press, 1980). Reprinted by permission of The University of Michigan Press.

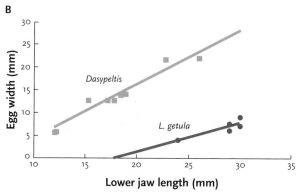

Figure 8.49 A. An African egg-eating snake (*Dasypeltis* sp. (1489)) ingesting an egg. **B.** The graph shows that African egg-eating snakes can eat much wider eggs (y-axis) than North American kingsnakes (*Lampropeltis getula* (1504)) even when their lower jaws are the same length (x-axis).

SOURCE: B. Gartner & Greene. 2008. "Adaptation in the African egg-eating snake: a comparative approach to a classic study in evolutionary functional morphology." *Journal of Zoology* 275(4) pg:368–374. © 2008 The Authors. Journal compilation © 2008 The Zoological Society of London.

cutting bills of birds of prey. Flower-visiting birds have exceptionally long bills and tongues, while woodpeckers have chisel-like beaks.

Mammalian jaws can appear relatively simple because they are composed of a few bones and show limited movement. Upper jaws are solidly fused to the cranium, and lower jaws articulate directly with the base of the cranium, usually (in living mammals) the **dentary-squamosal joint**. The lower jaw consists of a pair of dentary bones (left and right), homologous to the dentary bones of other gnathostomes. In reptiles, amphibians, and fish, the lower jaw includes bones in addition to the dentary. Because dentaries are the only mobile

Figure 8.51 A sampling of bird bills and their functions.

Generalist

Insect catching

Grain eating

Coniferous seed eating

Nectar feeding

Fruit eating

Chiseling

Dip netting

Surface skimming

Scything

Probing

Filter feeding

Aerial fishing

Pursuit fishing

Scavenging

Raptorial

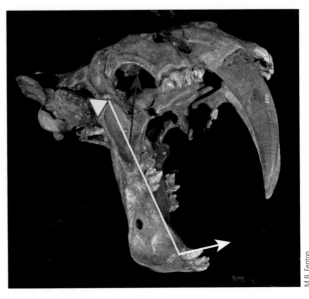

M.B. Fenton

Figure 8.52 Skull and mandible of *Smilodon* (1730), a sabre-toothed cat, illustrate the mammalian jaw, a class III lever. The yellow triangle is the fulcrum, the red arrow is the vector of the force generated by muscle contraction, and the white arrow is the vector of the force applied to the prey being bitten. To use its stabbing canine teeth effectively, *Smilodon* could open its mouth very wide.

jaw movements means more effective chewing (**mastication**) and quicker access to energy in ingested food. Mammals use their anterior (incisor) teeth to help ingest food into the mouth and then the cheek teeth (premolars and molars) to mechanically break up food by mastication. During this process, the posterior teeth repeatedly come together to break food into very small pieces. At the beginning of the masticatory cycle, the lower teeth are in contact with the upper teeth. The lower jaw then moves downward, and then outward, inward and upward to meet the upper jaw again. The lower teeth then move across the upper teeth, completing the cycle. Mastication increases the surface area of food exposed to digestive juices, speeding up the process of extracting nutrients.

In mammals, precise and complex movements of the jaw during mastication are linked to the complexity and developmental sequence of teeth. Most mammals are **diphyodont**, having only two sets of teeth, deciduous (baby) teeth and permanent (adult) teeth (Fig. 8.53). Some odontocete cetaceans have only adult teeth, and other mammals have no teeth at all. The upper and lower teeth have matching surfaces covered with sharp cusps (points), crests (blades), and basins. These features repeatedly shear, puncture, and crush food particles. Teeth must fit together precisely if they are to last for a mammal's lifetime. Mammals with extremely worn teeth cannot break down food efficiently, making it impossible for them to process enough food to meet their energetic needs.

elements in the mammalian jaw system, the jaws of mammals function as simple class III levers (Fig. 8.52).

Despite the simple lever mechanics of mammalian jaws, their movements are more complex and precise than those of other vertebrates. Precision of

BOX 8.3 Person Behind Biology

Anthony Herrel

Ecomorphology is the study of the relationships between the morphology of organisms and their ecological niches. To quantify these relationships, ecomorphologists measure performance, the ability of an organism to perform an ecologically relevant task that affects its fitness. The best-studied ecological tasks are the ability to escape predators (locomotor performance) and the abilities to feed and fight (biting performance). Dr. Anthony Herrel studies many kinds of performance, but is perhaps best known for his work on bite force (Fig. 1).

Most vertebrates bite into food to break it into smaller pieces before swallowing it. Bite force places limits on what an animal can eat because it cannot eat a food that it cannot break apart. Dr. Herrel demonstrated that selection for the ability to eat new foods can be very strong. By studying a population of Italian wall lizards (*Podarcis sicula* (1517)) that had been introduced to a new habitat 36 years before, he demonstrated the rapid evolution of higher bite force and features of the intestine that reflected a diet containing more plants than did the diet of the parent population. Differences in bite force can also be important in reducing competition for food within species. For example, while juvenile common anolis lizards (*Anolis lineatopus* (1467)) bite harder than expected for their size, they still cannot match the bite force of adult males and therefore do not compete for the same foods.

In sexually dimorphic lizards (males and females differ in size), bite force is also important in male–male competition for territories and mates. Dr. Herrel linked higher bite force to reproductive success in male tuataras (*Sphenodon punctatus* (1458)). He found that bite force was also linked to the body condition of the males: a male in good condition is heavier than expected for its body length. Since females often choose to mate with males that are in the best condition, bite force may be under sexual selection in these lizards.

Dr. Herrel's work has made significant contributions to how we think about feeding in the context of larger ecological and evolutionary processes, including competition, sexual selection, and adaptation to new ecological opportunities.

A

B

Anthony Herrel

Figure 1
A. Dr. Anthony Herrel uses a custom-made meter based on a piezoelectric crystal to measure bite force. B. When an animal such as this lizard bites the meter, a small electric charge is sent to an amplifier that returns a reading of force in Newtons.

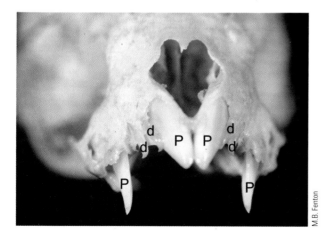

M.B. Fenton

Figure 8.53 Permanent (P) and deciduous (d) teeth of a vampire bat, *Desmodus rotundus* (1792). Deciduous teeth are smaller than permanent teeth and can be accommodated in the jaws of a young mammal.

STUDY BREAK

1. Describe the variety of forms found in insect mouthparts and relate these forms to the ways the insects feed.
2. Explain how a snake eats food items that are larger than its mouth.
3. How does a hagfish, without jaws, tear off food from its prey?

8.6 Tongues and Tentacles

8.6a Tongues as Muscular Hydrostats

Many animals have a muscle system associated with a skeletal support system (internal in chordates, external in arthropods). Animals without a skeleton

A

B

C

Figure 8.54 A. Like the elephant's trunk, **B.** blue-tongued skink tongues and **C.** octopus arms are examples of muscular hydrostats.

Courtesy of Naas Rautenbach

Tony Bowler / Shutterstock.com

Computer Earth / Shutterstock.com

are supported by fluid—a hydrostatic skeleton—that is enclosed in a connective tissue and muscle container. This allows changes in shape usually without changes in volume. Some body parts lack both rigid skeletal support and obvious hydrostatic skeleton. Consider the trunk of an elephant, the vertebrate tongue, and the flexible arms of squid and octopus, all of which are **muscular hydrostats**. Each is rich in muscle and connective tissue; all are highly mobile and capable of extension and retraction as well as changes in shape. None has an obvious hydrostatic skeleton because there is no fluid-filled compartment enclosed by muscles and connective tissue (Fig. 8.54).

Muscular hydrostats also use hydrostatic skeletons to change shape. However, unlike worms, sea cucumbers, or anemones, these hydrostatic skeletons are intracellular rather than extracellular. Intracellular hydrostatic skeletons maintain constant volume with no movement of fluid from one part of the body to another. This differs from movements such as the extension and swelling of a clam's foot as the animal burrows, movements that involve fluid movement into and out of the foot, rather like raising and lowering the blade of a bulldozer by pumping fluid from one cylinder to another.

Chameleons (*Chamaeleo* sp. (1485)) use their tongues to catch prey (Fig. 8.55). They project their tongues ballistically at least 1.5 body lengths, with accelerations of almost 500 m·s^{-2} and requiring 3000 W·kg^{-1} of energy. The chameleon's tongue is cylindrical and consists of a core composed of a carti-

laginous rod (entoglossal process) that supports the tongue surrounded by an accelerator muscle. The sliding-spring theory explains the action of the tongue. A layer of connective tissue that consists of at least 10 springs surrounds the entoglossal process. Each spring contains helical arrays of collagen fibres. Just before projecting the tongue, a combination of contraction of the accelerator muscle and hydrostatic shortening of the tongue loads the springs. The loaded springs try to expand radially (outward), which releases the energy stored in the collagen fibres. These springs increase the work rate of the accelerator by at least a factor of 10.

Temperature affects the elastic tissues that power ballistic movements of the chameleon's tongue less than muscle performance. This allows chameleons to keep their tongues loaded and operational, even during cool periods when temperatures fall to 15°C (Fig. 8.56).

Animals that collect nectar from tubular flowers must have long tongues to reach the flower's nectaries. There are many examples of co-evolution of flowers and pollinators. In addition to long muzzles and reduced teeth, some flower-visiting bats have extensible tongues that extend hydraulically when the bat pumps blood into the tongue. One nectar-feeding bat, *Anoura fistulata* (1781), can extend its tongue to 150% of its body length, more than twice as far as two other species of *Anoura* (*A. caudifer* (1780) and *A. geoffroyi* (1782)) that occur in the same location in the Ecuadorian Andes. The extraordinary tongue of *A. fistulata* is housed in a tube that extends back down the neck and into the thoracic cavity. *A. fistulata* feeds heavily on the nectar and pollen of one plant, *Centropogon nigricans*, whose flowers have corollas that are 8–9 cm long.

8.6b Tentacles

The use of **tentacles** for food capture is a feature of invertebrates that have a hydrostatic skeleton. Feeding tentacles take many forms and may be used in a variety

Figure 8.55 A chameleon extending its tongue ballistically to catch an insect.

Cathy Keifer / Shutterstock.com

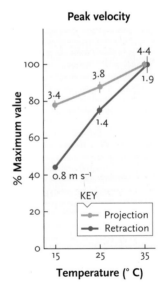

Figure 8.56 Maximum values of peak velocity of tongue projection and retraction at three different temperatures. Retraction, accomplished by contraction of the hyoglossus muscle, declines rapidly with temperature. Projection is less affected by temperature because it relies on elastic recoil rather than muscle power.

SOURCE: Anderson, C.V. & Deban, S.M., 2010. Ballistic tongue projection in chameleons maintains high performance at low temperature. *PNAS*, 107(12), pp.5495–5499.

Figure 1
A honeycomb moray eel (Gymnothorax favagineus (1158)).

Moray eels are predatory fish that inhabit crevices in reefs, coming out only to ambush smaller fish that get too close with a quick and deadly bite (Fig. 1). Like many other living ray-finned fish, moray eels have both oral and pharyngeal jaws. In most fish, the pharyngeal jaws are fixed in place within the pharynx. In moray eels, the pharyngeal jaws actually move forward from the pharynx into the oral cavity

and work alongside the oral jaws to seize and draw prey into the mouth and down the oesophagus (Fig. 2).

This previously unknown feeding mode wasn't discovered until 2007 when a young researcher, Dr. Rita Mehta, looked at high-speed video of a moray eel feeding on pieces of squid. She noticed that just after the eel's oral jaws bit into the squid, a second set of jaws came up from the throat and bit the squid again. These pharyngeal jaws pulled the squid down the throat while the oral jaws opened in preparation for

taking another bite. Dissections showed that the pharyngeal jaws have sharp, hook-like teeth that are very effective at grasping and holding prey. The double-jawed system of moray eels functions much like the ratcheting mechanism seen in snakes, where the upper and lower jaws move independently to slowly move prey from the oral cavity to the esophagus. The functionally convergent feeding mechanisms of moray and snakes systems may be associated with their independent evolution of long, limbless bodies.

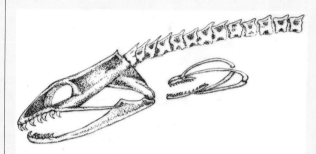

Figure 2
Lateral view of the skull, vertebral column, and pharyngeal jaws of a reticulate moray eel (Muraena retifera (1160)). When the mouth is closed, the pharyngeal jaws are relaxed and located in the pharynx (throat). When the mouth opens, muscles pull the pharyngeal jaws into the mouth to grasp prey. When the eel closes its mouth again, the pharyngeal jaws pull the prey backward into its pharynx.

SOURCE: Illustrated by Leith Miller

of modes of feeding, from detritivores to suspension feeders and predators.

The tentacles of sea anemones may be simple (e.g., *Urticina felina* (174)) or finely branched (e.g., *Metridium senile* (172), Fig. 8.57). In both cases, tentacles are extended by transfer of water from the body cavity (**coelenteron**) to the tentacles, and retracted by contractions of muscle fibres in the tentacle walls, around the mouth, and up and down the column of the animal. Unlike most hydrostatic skeletons, those of anemones (and other Cnidaria) do not maintain a constant volume. When they contract, water is expelled through the mouth, and the animal shrinks. Re-expansion is slow, depending on cilia at the edges of the mouth to move water back into the coelenteron. In an experiment, tentacles of *Condylactis* sp. (169) were cut off at their base. Isolated tentacles contracted but could be re-expanded by gentle injection of water. Injection of too much water caused the tentacle to rapidly contract—a demonstration of local muscle activity in the tentacle.

Urticina (174) is a predator, using its tentacles to capture small shrimp and fish. Prey are captured using **nematocysts** that discharge coiled, sticky, or venomous threads, discharged from a capsule within a nematocyte (Fig. 8.58). One of the fastest events in the biological world, the discharge of a nematocyst can occur in 700 ns, involving an acceleration of 5.4×10^6 G.

Discharge is initiated by mechanical contact with the **cnidocil** (trigger), and the force of the discharge is due in part to the release of spring energy from minicollagen molecules, and in part to a very high internal

Figure 8.57 A. Plumed anemone, *Metridium senile* (172); **B.** Dahlia anemone, *Urticina felina* (174); (both Cnidaria, Anthozoa).

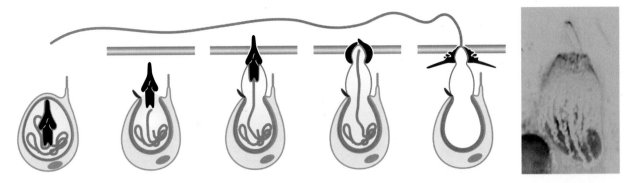

Figure 8.58 A. Schematic view of the discharge of a penetrating nematocyst. Nematocytes (blue; cell and vesicle membranes in dark blue) harbour one cyst (pink) with stylets (black) punching a hole into prey. **B.** Micrograph shows microtubules anchoring the capsule (tubulin antibody capsule wall, green, tubulin brown). The light blue projection from the nematocyte is the cnidocil, the trigger that initiates discharge.

SOURCE: Reprinted from *Current Biology*, 16(9), Nüchter, T., Benoit, M., Engel, U., Özbek, S., & Holstein, T. W., "Nanosecond-scale kinetics of nematocyst discharge," Pages R316–R318, Copyright 2006, with permission from Elsevier.

osmotic pressure. The pressure at the tip of the stylet as it hits the prey has been estimated at 7 GPa.

The sensitivity of the cnidocil is regulated by the nutritional state of the animal, with a well-fed animal having a very high discharge threshold. This regulation can be mimicked by varying the Ca^{++} levels outside the capsule. Low Ca^{++} levels slowed or inhibited capsule discharge. The venom injected from nematocysts varies with species but can include phospholipase A_2, hemolytic toxins (which create pores in lipid bilayers), and neurotoxic fractions that affect both Na^+ and K^+ channels in nerve membranes. No wonder some sea slugs extract unexploded nematocysts from their cnidarian prey and use them in their own defense!

Many suspension feeders have ciliated tentacles (Fig. 8.59). The cilia are partly responsible for the generation of feeding currents, which are usually aided by natural water flows. Cilia are also involved in the capture and transport of food particles. In many polychaete worms, the cilia work in conjunction with channels running along the tentacles to sort food from non-food before the stream reaches the mouth. In others (e.g., bryozoans) the muscles of tentacles may be used to flick non-food particles away from the zooid.

Many animals use an **extensible proboscis** to collect food. In the phylum Echiura (the name refers to the forked proboscis and comes from the Greek *echis* = viper and *ura* = tail) the proboscis simply extends when body fluid is forced into the proboscis and retracts when fluid moves from the proboscis back into the body cavity (Fig. 3.50).

The proboscis of the echiuran *Bonellia viridis* (**371**) is 15–17 cm long when contracted (about the same as the length of the worm) and can extend to about ten times this length (around 1.5 m). The contracted proboscis is coiled anterior to the worm, and its extension appears to be due to ventral cilia on the terminal ends of the forked tip. These cilia steadily draw the proboscis out at about 1 mm.s^{-1}, and it uncoils and becomes thinner as it extends. The proboscis can tie itself in a simple knot due to twisting and movement as it extends, but Echiurans can untie these knots. The extended proboscis has a dorsal gutter in which cilia move copious amounts of predator-repelling mucus that trap particles of detritus from the substratum. Proboscis retraction, by contraction of longitudinal muscles, takes about 1.5 s. If the proboscis is anchored, these muscles draw the worm forward.

Other animals have a proboscis that operates by turning inside out as it extends, called an eversible proboscis. Think of how a finger of a rubber glove may be tucked back inside the palm of the glove and then popped out to its extended position. In the phylum Nemertea (ribbon worms), the eversible proboscis evolved separately from the gut although later evolution brought the proboscis into close association with the mouth. In these worms, the proboscis is in a separate body cavity (the rhynchocoel). The proboscis is rapidly everted when contraction of the body

Wim van Egmond/Visuals Unlimited, Inc.

Figure 8.59 *Spirorbis borealis* (**343**) (Annelida, Polychaeta) attaches to any available surface and extends ciliated tentacles to feed on suspended particles.

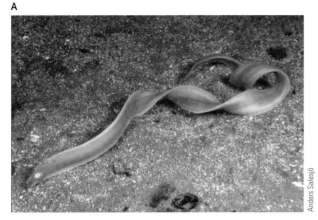

A

B

Proboscis
pore

Proboscis

Rhynchocoel

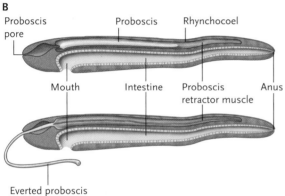

Mouth

Intestine

Proboscis
retractor muscle

Anus

Everted proboscis

Figure 8.60 A. *Cerebratulus marginatus* (563) (Nemertea);
B. proboscis extension in a nemertean.

SOURCE: B. From RUSSELL/WOLFE/HERTZ/STARR. *Biology*, 2E. © 2013
Nelson Education Ltd. Reproduced by permission. www.cengage.com/
permissions

musculature exerts pressure, via the body cavity, on the rhynchocoel. Retraction of the proboscis depends on a retractor muscle bundle. In some nemertines (e.g., *Cerebratulus* (563), Fig. 8.60), the tip of the everted proboscis is equipped with a stylet connected to a venom gland. The venom of *Cerebratulus* includes a potent polypeptide neurotoxin that binds to sodium channels and is lethal to crustaceans.

STUDY BREAK

1. Explain what is meant when a structure is described as a "muscular hydrostat."
2. What is a nematocyte, and how does a nematocyst work?

8.7 Other Structures Modified for Feeding

8.7a Raptorial Appendages

Raptorial appendages are used to capture prey, usually by impaling it on spines or similar structures. For example, although the phylum Chaetognatha (the arrow worms) includes only a few species, it is of great ecological significance because these mostly planktonic

worms are voracious predators on copepods and larval fish. Chaetognaths, like Nematoda, lack circular muscle (see page 179), and their darting movements are powered by longitudinal muscles operating against internal fluid pressure. Chaetognaths are sit-and-wait predators, detecting swimming copepods by their vibrations. They usually attack prey that comes within about 3 mm. To capture a copepod, a chaetognath darts rapidly forward and grasps it in the spines on either side of its head; hence the name *chaetognatha* = hairy jawed (Fig. 8.61). The grasping spines hold the prey against the anterior and posterior teeth of the jaws, which bite into the prey and may inject venom.

Different species of chaetognaths respond to different vibrational frequencies (30–150 Hz). Faster swimming copepods generate a greater disturbance in the water, perhaps accounting for some of the observed differences between copepod species abundance and numbers of copepods eaten by chaetognaths.

Significant quantities of tetrodotoxin (TTX) have been found in the heads of several species of chaetognath. TTX is the principal toxin in puffer fish and is a potent neurotoxin, binding irreversibly with sodium channels and blocking neural and muscular activity. *Vibrio alginolyticus*, a bacterium known to secrete TTX, occurs as a symbiont in the head region

A

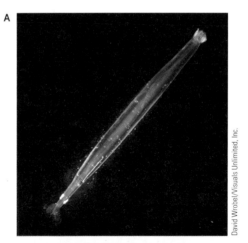

B

Figure 8.61 A. Arrow worm (*Sagitta* sp. (580), phylum Chaetognatha). **B.** Ventral view of the head of a chaetognath (*Sagitta marri* (581)) showing the grasping spines and the anterior teeth.

of chaetognaths. Injecting these bacteria into prey could deliver TTX to them. Copepods extracted from the gut of feeding chaetognaths appear to be paralyzed and enveloped in a mucus sheath. Prey extracted from the pharyngeal region are similarly wrapped but, when dissected free of the wrapping, retain some motility. Had they been injected with TTX, paralysis should be almost instantaneous. The role of TTX in chaetognath feeding remains open to question.

Many other animals have raptorial appendages, including praying mantises, mantis shrimp, and amblypygids. The teeth of some fish may also serve as impaling devices.

8.7b Crushing Appendages

Many predators, especially those feeding on armoured prey, use teeth or other structures (Fig. 8.28) to crush their victims. The first pair of walking legs of decapod crustaceans (e.g., American lobsters, *Homarus americanus* (819)) are **chelae**, pincer-like claws. In many decapod species, the two chelae differ in size and function. In *Homarus*, one chela is a broad crusher, the other a thinner and less powerful cutter (Fig 8.28D). The dactylopodite or terminal joint of a non-chelate leg is modified to become the moveable pincer, and is opened and closed by opposing muscles (Fig. 8.62).

A chela operates as a simple lever, pivoted around a fulcrum. The opener and closer muscles, whose size reflects the power needed by their functions, attach by **apodemes** to the inner (closer) and outer (opener) bases of the dactyl. Both opener and closer muscles are pinnate: muscle fibres are arranged at an angle to the axis of the apodeme. This arrangement has two advantages: 1. more muscle fibres can be packed into a given space and 2. muscle contraction does not result in the expansion in diameter of the bundle that would occur if the fibre arrangement were linear. Lack of expansion of the contracted muscle is important when the muscle is enclosed in the thick, inflexible carapace.

The base of the dactyl, even of the broader crusher claw, is significantly narrower than its length and the distance between pivot point and the closer apodeme even shorter than the dactyl base width. This means that there is significant loss of mechanical advantage when actual crushing force (250 N) is compared to muscle power. This force is about a third of the power generated by the closer muscle. The mechanical advantage of the slender cutting claw is only about 0.15× the muscle power. However, compared with other crustaceans, the lobster is very strong. Fiddler crab (*Uca pugilator* (842)) and crayfish (*Procambarus clarkii* (836)) claws develop 1–10 N of force. Some larger species of *Cancer* crabs may develop crushing forces of almost 500 N. Putting these numbers into context, 25 N is sufficient to break the shell of a small littorinid snail (common prey for shore crabs). Clearly a lobster has no difficulty in breaking the shell of a mussel or scallop, or the test (outer covering) of a sea urchin.

Mantis shrimp (stomatopods) use feeding appendages (Fig. 8.63) to smash shells and impale fish. Mantis can move their feeding appendages at 14–23 m·s⁻¹, making *Odontodactylus scyllarus* (786) the fastest appendicular striker among animals. Achieving such rapid strikes means releasing a large amount of energy in a short period of time, 2.7 ms. The striking power the mantis shrimp delivers (4.7×10^5 watts) is much higher than could be achieved by muscle contraction. *O. scyllarus* uses a compressive saddle-shaped spring as a source of energy for its strike and a latch mechanism to hold the tension for rapid release. The spring is located dorsally on the enlarged proximal segment (merus) of the striking appendage, and its strike

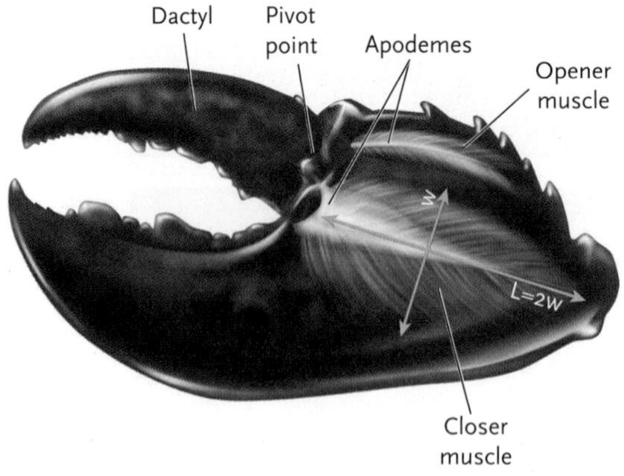

Figure 8.62 Mechanics of a lobster chela. The moveable element, the dactyl, pivots about a point on its base (P). Attached to two apodemes are the opener muscle (O) and the much larger closer muscle (C). The lower diagram shows that the closer muscle is about twice as long as wide and has its sarcomeres arranged in a pinnate pattern, angled at about 30°. This arrangement generates about twice the power than if the sarcomeres were parallel to the axis of contraction.

Beverly Speed / Shutterstock.com

Figure 8.63 The dactyl bulb (yellow arrow) of a peacock mantis shrimp (*Odontodactylus scyllarus* (786)), is used to smash open hard-shelled prey such as snails, crabs, and clams.

involves cavitation, a phenomenon occurring in fluids when areas of low pressure form vapour bubbles that collapse, releasing considerable energy as heat, light, and sound. Cavitation explains the loud "pop" accompanying a stomatopod strike. In *O. scyllarus*, cavitation occurs between the surface being struck and the heel of the dactyl, which becomes pitted and damaged over time. Regular moulting replaces the dactyl.

8.7c Traps and Tools

In addition to mandibles, primary mouthparts analogous to jaws, insects and other arthropods use other pairs of appendages as mouthparts Palps, for example, can look like miniature legs and usually serve sensory functions. Many aquatic insects have additional mouthparts modified into fine-toothed brushes and combs These manipulate silk produced in the salivary glands and used to form a variety of nets, either for shelter or for capturing prey. Some caddis flies (Trichoptera) build nets to capture prey (Fig. 8.64). The U-shaped nets are attached to the surface of a rock in a river current, with the wide mouth of the net taking in water and particles. The caddis fly larva lives at the narrow end and catches and eats anything edible that gets caught in the net. The larvae "knit" the nets using their complex mouthparts.

Adult antlions (Neuroptera: Myrmeleontidae) are short-lived and do not feed, but their larvae are ferocious predators. An antlion larva excavates a conical pit in sand (Fig. 8.65A) and lies partially concealed under the sand with only antennae and mandibles visible. Antlions use the inherent instability of heaped sand as a trap for prey. The angle of the cone is close to the point at which the sand slope is unstable.

When an insect such as an ant walks to the edge of the pit, it starts a small avalanche by dislodging sand grains. The insect slides down into the pit. The waiting antlion seizes the ant between its jaws, with the distal tooth of one mandible holding the "waist" while the other mandible probes to find a joint between sclerites, where it is inserted (Fig. 8.65B). Enteric fluid, perhaps

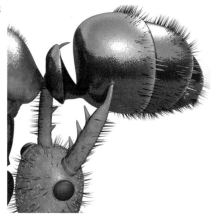

Figure 8.65 **A.** An antlion (*Cueta punctatissima* (1055)) buried and waiting to capture the ant tumbling into its pit. **B.** The jaws of the ant lion, *Morter obscurus* (1056), showing how the petiole of an ant prey is held by one jaw while the other jaw injects toxin into the abdomen.

including a poison, is injected into the prey through this mandible, along with digestive enzymes. Digested material is sucked up via the hollow mandible. The empty exoskeleton of the prey is later tossed out of the pit that is repaired in readiness for the next victim.

Antlion larvae preferentially build their pits in fine sand because these slopes are steeper before they become unstable. The larva uses its head and mandibles to excavate the pit. Some species retain smaller sand particles as they dig, discarding larger ones.

New Caledonian Crows (*Corvus moneduloides* (1595)) use sticks as tools to extract the larvae of woodboring beetles from their burrows. The crows' diet includes a wide range of prey species, including snails, lizards, carrion, nuts, and fruit as well as a variety of arthropods. Isotopic analysis ($\delta^{15}N$ and $\delta^{13}C$) reveals that the beetle larvae account for a great deal of the protein and lipids that the crows ingest. These findings demonstrate the advantages of tool-use, a topic explored in more detail beginning on page 492.

Northern blue jays (*Cyanocitta cristata* (1596)) raised in captivity have been found to tear strips of newspaper from the lining on the bottom of their cage and to use the strips to pull spilled food pellets back within their reach. The intensity of the birds' use of the newspaper tools reflected their level of hunger (time since their last meal).

Figure 8.64 Silk nets used by caddisfly larvae to capture prey.

8.7d Mollusc Radulae

Snails, slugs, and other molluscs use a **radula** as a feeding structure. The radula is secreted as a ribbon in a pocket off the oesophagus and is supported by the **odontophore**, a cartilaginous base, and bears longitudinal rows of chitinous teeth (Fig. 8.66 and 8.67). The radula is moved by groups of opposed muscles. The odontophore is protruded from the mouth by contraction of the odontophore protractor muscle. Alternating contractions of the radula protractor and radula retractor muscles move the radula around the odontophore. On retraction, the radula moves upwards and back into the mouth and the radula teeth scrape or cut the surface and direct food particles into the mouth.

In many species, the chitinous teeth of the radula are hardened by the inclusion of iron (in the form of magnetite) or silica. Even so, they are worn down by their scraping against hard substrata. Radula teeth are continuously replaced, and new teeth are secreted by **odontoblasts** in the radula sac. As many as five rows of teeth may be added to a radula in a day, appearing on a sub-radular epithelium that moves forward as the radula grows. At the other end of the radula the sub-radular membrane detaches and old teeth are shed.

Molluscan radulae show considerable diversity (Fig. 8.67). The radula of the common limpet (*Patella* (407)) is used to scrape the algal and bacterial film from intertidal rocks. Slipper limpets (*Crepidula* (421)) scrape food from the rock surface as well as filter feed in a manner similar to that of the bivalves. To this end, an expanded gill collects food particles, traps them in a mucus string, and moves the string to the buccal region where the radula helps manipulate it into the esophagus. The sea hare (*Aplysia* (440)) feeds on macroalgae such as the sea lettuce (*Ulva*), and its radula cuts the surface. The terrestrial wolf snail (*Euglandina* (476)) hunts other snails by sensing and following their mucus trails; its radula is designed for cutting and tearing flesh. Another

Figure 8.67 A. The radula of *Aplysia juliana* (442) is designed for cutting soft algae. **B.** The radula of *Crepidula fecunda* (420) is used both to scrape the substratum and to manipulate a string of food into the mouth.

SOURCE: B. O. R. Chaparro, R. J. Thompson, S. V. Pereda, "Feeding mechanisms in the gastropod Crepidula fecunda," *Mar Ecol Prog Series* 234:171–181 (2002). Inter-Research Science Center.

predatory snail, the oyster drill (*Urosalpinx* (431)), uses radula teeth, in conjunction with acid secretions from an accessory boring organ, to drill through the shell of another gastropod or bivalve mollusc (see Chapter 4).

The radula of the cone snail (Fig. 8.68) is specialized for delivering venom to mobile prey. Two harpoon-shaped teeth are formed side-by-side in the radula sac and passed forward so that the hollow canal behind the point can be loaded with venom secreted by a modified salivary gland. The animal can fire its harpoon in any direction. The harpoon venom includes potent neurotoxic peptides that act so quickly that a small fish is stunned before it can escape the snail's reach. Cone snail venom is potentially lethal to humans.

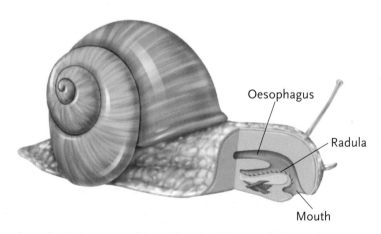

Figure 8.66 The position of the radula and radula sac inside the mouth of a snail.

Oesophagus

Radula

Mouth

A

B

C

Figure 8.68 A. California cone snail (*Conus californicus* (418)). Compare the size of the extended proboscis here to that in C. **B.** The harpoons from a cone snail radula. Notice the groove that carries the venom. **C.** *Conus geographus* with fish prey in its dilated proboscis.

8.7e Sea Star Stomach

The eversible cardiac stomach of a sea star (Echinodermata, Asteroidea) is neither a proboscis nor a tentacle. When the sea star encounters a bivalve mollusc, it crawls over the bivalve, grips the two shell valves with the suction cups of its podia, and exerts a steady tension until the mollusc opens 1–2 mm (Fig. 8.69). This gap is large enough for the star to evert its stomach into the mussel and release digestive enzymes, digesting prey tissue *in situ* and absorbing it. After feeding, sea stars may be seen with the stomach still everted, presumably for cleaning. Echinoderms are among the few animals that can regenerate a lost stomach, necessary if the stomach is torn off within the prey. Some species of sea stars, such as the crown-of-thorns star (*Acanthaster planci* (1070)), evert the cardiac stomach over an area of substratum to digest and absorb food from that area. *A. planci* feeds on living coral tissue. It settles on an area of coral and in 4–6 hours digests all of the polyps under the stomach. A single *A. planci* eats (and kills) 5–6 m^2 of coral a year, and an outbreak of these predators can seriously damage a reef.

STUDY BREAK

1. Explain the mechanics of a lobster's chela.
2. What is the ideal design for an antlion pit trap?
3. What is a radula, and how does it operate?

A

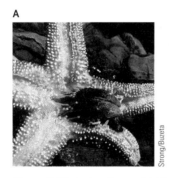

B

Figure 8.69 A. *Asterias* sp. (1071) showing tube feet pulling mussel shell valves apart. **B.** Everted stomach.

FEEDING **IN PERSPECTIVE**

This chapter has reviewed food types and the ways in which they are acquired by a wide variety of animals. While the detail of particular examples can be both fascinating and astonishing, it is important to recognize how particular examples can be viewed as generalizations about animal feeding modes. Once food is acquired, it must be digested; aspects of digestion are considered in Chapter 9. Feeding is a frequent component of parasitic, symbiotic, or commensal associations between animals; these aspects of feeding will be a large part of Chapter 10.

Questions

Self-Test Questions

1. Let's say you dine out on a fantastic meal that includes Atlantic salmon and shrimp cocktail. The salmon are piscivores that were feeding on cod that ate capelin that ate copepods that fed on phytoplankton. The shrimp feed on smaller, slow-moving benthic organisms, and may also be scavengers. Given this information, which of the following is true?

   a. You, the salmon and the shrimp may all be at the same trophic level.
   b. Eating the salmon puts you at trophic level 6, but for the shrimp you may be at trophic level 4.
   c. Even with the different food chain, eating the salmon with the shrimp puts you at trophic level 5, since your trophic level is fixed.
   d. The shrimp can only be at trophic level 2, so you could be at trophic level 3 for this part of your diet.

2. Why is the Reynolds number a useful tool in assessing factors important in suspension feeding?
   a. Knowing the ratio of inertial forces to viscous forces can provide an idea of the movement of small particles in water and the effectiveness of capture mechanisms.
   b. The movement of organisms through water is influenced primarily by inertial forces.
   c. Large organisms have a higher Reynolds number, reflecting the influence of water viscosity on the movement of the animals.
   d. In turbulent flow, where inertial forces are dominant, a small organism can increase its Reynolds number, which slows its speed of movement.

3. How does the efficiency of feeding compare between ram feeders and suction feeders?
   a. Ram feeders have an advantage because of the energy that must be expended to chase prey, which favours adaptations to reduce drag in water.
   b. The drag imposed by the open mouth of a suction feeder is reduced through sensory feedback, which controls extreme movements of the jaws and limits filling of the mouth.
   c. Suction feeding works only over a short range, but requires less energy and is aided by rapidly opening jaws to increase suction area.
   d. Suction feeding requires movement of the whole body through the water, which is energetically expensive and favours adaptations to streamline the body.

4. Which of the following invertebrate jaws are correctly matched to their taxonomic group?
   a. Phylum Rotifera; a beak composed of chitin fibres with histidine-rich proteins at the hardened tip
   b. Phylum Echinodermata; a framework of ossicles supporting five curved, sharply pointed teeth and associated muscles, known as Aristotle's lantern
   c. Phylum Annelida; paired mandibles, each with a biting tip and a grinding surface
   d. Phylum Arthropoda; muscular mastax housing complex jaws called trophi

5. Which of the following statements describes features of allvertebrate jaws?
   a. Have an upper jaw that is fused to the braincase
   b. Frequently involve cranial kinesis, in which the mandible moves freely relative to the braincase
   c. Are suspended to provide for mastication
   d. Are derived from the pharyngeal arches that support gills

Questions for Discussion

1. What is meant when an animal is described as a carnivore? What are the common adaptations of body form that make an animal a successful carnivore?

2. Non-selective suspension feeding yields poor quality food. If this is true, why have so many animals, from different groups, adopted this mode of feeding?

3. This chapter includes many examples of highly specialized modes of feeding. Select the three examples that you think show the greatest degree of specialization and explain the reasons for you choices.

4. Use the electronic library to find a recent paper that explores some aspect of animal feeding. Does the new information revolutionize any of the information in this chapter?

Figure 9.1
Puff adder, *Bitis arietans* (1472).

SOURCE: Courtesy of Naas Rautenbach

9 Digestion

WHY IT MATTERS

Snake bite is a story of digestion.

Imagine that you are one of a group of tourists enjoying a wilderness holiday in southern Africa. A member of your group encounters a snake while walking in open sandals on a nocturnal visit to the pit toilet. Your unfortunate companion steps on a puff adder (*Bitis arietans* (1472)) lying on one of the warm stones in the path, and is bitten on her foot. The tour guide knows just what to do: keep the victim as calm as possible, look for fang marks at the bite site, and keep the site of the bite below the level of the heart. The guide does not apply ice to the bite site or cut at the fang mark. The bitten foot is fitted with a splint and wrapped with an expansion bandage as tightly as one would a sprained ankle. This slows the spread of the venom through the victim's lymphatic system. The well-equipped guide has antivenom available, but arranges immediate transport to a local hospital so that the antivenom can be administered under medical supervision.

Puff adders produce toxic venom designed to immobilize and digest prey. The bites of puff adders have an impressive effect (Fig. 9.2), but in spite of the venom's toxicity, human deaths from puff adder bites are rare, usually resulting from secondary symptoms or bad

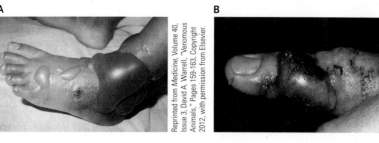

Reprinted from *Medicine*, Volume 40, Issue 3, David A. Warrell, "Venomous Animals," Pages 159-163, Copyright 2012, with permission from Elsevier.

Figure 9.2 A. Reaction to a puff adder (*Bitis arietans*) bite that injected venom in the ankle. **B.** The aftermath of a puff adder bite on a finger.

clinical management. Cardiotoxins (that slow heart rate) and hematoxins (that slow blood flow) are involved in immobilization of the prey, and are partly responsible for the severe pain associated with bites. Severe swelling, hemorrhages, and nausea are usually also experienced. Cytotoxins and protease enzymes in the venom begin digestion, which is why the guide did not cut at the bite site; this would increase exposure of tissues to the venom. These toxins also cause lysis of tissues that can lead to gangrene. Cytotoxins and protease enzymes are largely responsible for the missing fingers of people who have been bitten there by a puff adder (Fig. 9.2B).

In hindsight, we should expect to find a snake such as a puff adder resting on a warm stone at night, using its heat to remain active, promoting digestion of its previous meal, or allowing it to remain alert and actively hunting for its next one. Proximity to the pit toilet means ready access to prey: animals that come to feed on dung or on other animals that eat dung. The array of predators and prey around a latrine can be impressive. Many of the predators are venomous— spiders, scorpions, and snakes—and can use their venom in offense or defense. If you use a flashlight on your way to the outhouse at night, you can see and avoid animals you prefer to avoid. Wearing shoes or boots is also wise. A study of 5639 cases of snake bites on humans in India between 1999 and 2008 showed that 83% of snake bites were on the lower limb, most below the knee. The fact that 13% of these snake bites are on the hand or lower arm alerts us to the dangers of trying to catch or handle the snake. The best advice about snakes and other venomous animals is: **leave them alone**.

Knowing how poisonous animals use venom helps you reduce your risk of being bitten and know what symptoms to look for in someone who has been bitten. It also tells us something about the initial stages of digestive processes in venomous animals: these occur in the prey rather than in the predator!

SETTING **THE SCENE**

In Chapter 8 (Feeding) we sampled the range of specializations for getting food; now we will examine the diversity of approaches animals take

when digesting it. Digestion involves mechanical and chemical breakdown of food into its constituent components. In some cases, these processes are essential to reduce the chances of allergic reaction to chemicals in food. While humans often use cooking to effect this detoxification, other animals use other approaches. Whether the process involves ingesting grit to aid in mechanical digestion or paracellular absorption (material moving between cells), digestion is a window on the diversity of animals.

9.1 Fundamentals of Digestion

Obtaining food is the first step in gaining access to the **nutrients** (the components of food: materials that the cells will convert into energy and structural materials) that an animal requires. In the next steps, digestion, the food is broken down into smaller and smaller units, involving processes that combine mechanical and chemical processes of digestion. Smaller units are easier to move into, and distribute throughout, an animal's body. When the food is composed mainly of proteins, breaking them into their component amino acids also reduces the chance of allergic reactions to foreign proteins.

In many animals, digestion of complex molecules takes place simply in food vacuoles within cells (**intracellular digestion**). In sponges, for example, flagellated collar cells or **choanocytes** generate currents that help in the capture of small food particles by directing them into the mesohyl. Amoeboid cells called **archaeocytes** wander through this acellular layer by cytoplasmic streaming, and engulf food particles so that digestion is entirely intracellular.

In most other animals, digestion occurs in a gastrovascular cavity or tube (**extracellular digestion**). When animals ingest larger food items, sometimes entire other organisms, the site of digestion must be large enough to accommodate the food and store it so that digestion and absorption of nutrients can take place efficiently. Regional specialization of gastrovascular cavities and tubes is a recurring theme in animal physiology. However, some animals whose ancestors had digestive tracts with associated organs have lost these structures. This usually occurs among animals living in a milieu of food materials that do not require considerable additional digestion. Endoparasites are an obvious example, but there are, as we shall see, many others.

In general, feeding ranges from filter-feeding (also known as suspension feeding) to taking higher quality food in larger packages. Bacteria and other symbiotic organisms can also play a role in digestion, and we shall explore some of these interactions.

Complete loss of the digestive tract is a specialization found in animals that live in a soup of nutrients, even if a digestive tract had been present in the ancestral forms. Obvious examples are parasitic Platyhelminthes such as tapeworms, but some free-living animals also lack digestive tracts. For example, the protobranch bivalve mollusc (*Solemya borealis* (527)) from the northeastern Pacific Ocean has neither digestive system nor any provisions for secretion of digestive enzymes into the mantle cavity. Another *Solemya* (*S. parkinsoni* (528)), from the southern Pacific, has a vestigial digestive tract. *Solemya* often form dense populations around warm water outflow from pulp mills. Presumably these clams absorb dissolved organic molecules across their ctenidial lamellae. In spite of their gutless condition, these small bivalves live an energetic life style. Pogonophora is another non-parasitic group of free-living invertebrates that lack digestive tracts. These tube worms may be as much as 3 cm in diameter and live in areas around thermal vents with an abundance of food. Pogonophora also occur in association with chemosynthetic bacteria.

STUDY BREAK

1. Why are ingested materials broken down into smaller units?
2. What is intracellular digestion?
3. Why do some animals lack digestive systems?

9.2 Moving along the Digestive Tract

In animals where digestion occurs in the digestive tract but outside cells (extracellular digestion), the digestive system (gastrointestinal tract) typically shows regional specialization reflecting localization of different digestive operations. This regionalization is not evident in animals such as cnidarians and free-living Platyhelminthes (Fig. 9.3), but is obvious in many other animals (Fig. 9.4). In addition to regionalization of the tract, specialized organs (e.g., liver and pancreas) assist in the digestive process (Fig. 9.4C).

The digestive system not only serves in the digestion and assimilation of nutrients, but the system, including the associated organs, is also involved in sensing and signalling in the physiology of homeostasis. Chemical and mechanical signals about gut contents help control the rate of transit through the digestive tract and so affect the efficiency of digestion and absorption of nutrients. In conjunction with the islets of Langerhans (in the pancreas), blood vessels in the hepatic portal system, and visceral adipose tissue, the vertebrate gut produces neural and endocrine signals that affect controllers of energy balance. Signals from the digestive tract provide information about energy stores in terms of nutritional state, and this information is integrated in the hypothalamus and other parts of the central nervous system (Fig. 9.5). The importance of the digestive tract's role in the broader

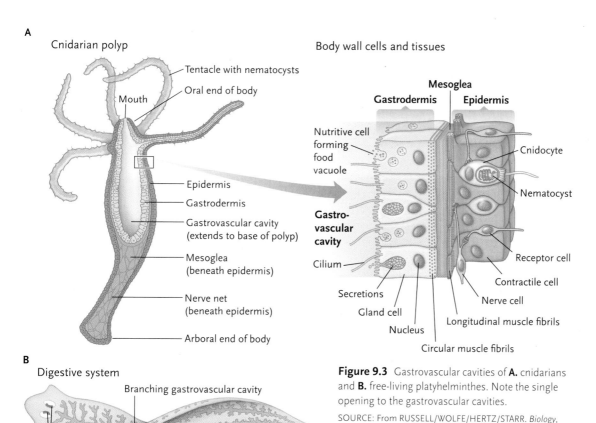

Figure 9.3 Gastrovascular cavities of **A.** cnidarians and **B.** free-living platyhelminthes. Note the single opening to the gastrovascular cavities.

SOURCE: From RUSSELL/WOLFE/HERTZ/STARR. *Biology*, 2E. © 2013 Nelson Education Ltd. Reproduced by permission. www.cengage.com/permissions

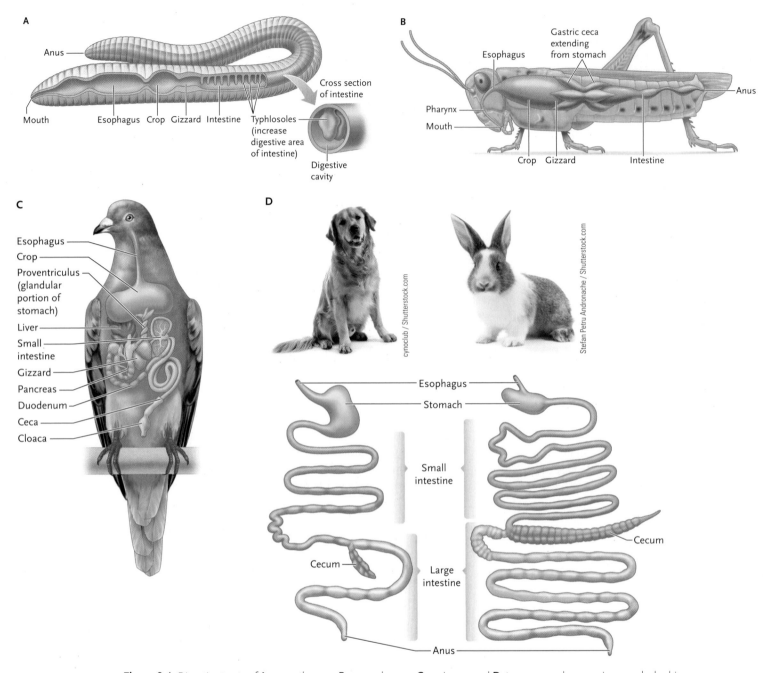

Figure 9.4 Digestive tracts of **A.** an earthworm, **B.** a grasshopper, **C.** a pigeon, and **D.** two mammals, a carnivore and a herbivore.
SOURCE: From RUSSELL/WOLFE/HERTZ/STARR. *Biology*, 2E. © 2013 Nelson Education Ltd. Reproduced by permission. www.cengage.com/permissions

realm of energy balance and behaviour identify it as a primary target in understanding and treating obesity. Some of the feedback loops that are involved are shown in Figure 9.6.

Leptin is a circulating hormone that provides information about energy stores to the brain. Leptin-deficient mammals are hyperphagic, consuming more food than required. Treatment with leptin has been shown to correct the hyperphagy, demonstrating the importance of this circulating hormone in eating behaviour.

In humans, an obese parent is an independent risk factor for obesity in childhood. The impact of maternal obesity has been well documented, but that of fathers has only recently emerged. When male laboratory rats are fed a diet high in fats, there is a resulting dysfunction in the programming of β-cells in the pancreas of their female offspring. In the fathers, chronic high-fat diets induced higher body mass, adiposity, and impaired glucose tolerance and insulin sensitivity. The daughters of these male rats showed early onset of impairment of glucose tolerance and insulin secretion

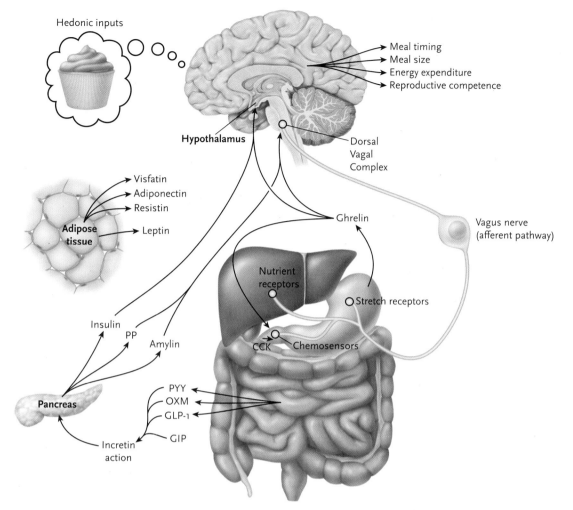

Figure 9.5 Long-term energy balance is achieved by several feedback systems. Leptin and insulin (produced by adipose tissue and the pancreas respectively) provide information about nutritional state, which is integrated at the hypothalamus. Neuronal feedback via the vagus nerve informs the nucleus of the tractus solitarius about gut distension and the chemical and hormonal milieu in the upper small intestine. Included in these steps are the DVC (dorsal vagal complex) and incretin hormones GLP-1, GIP, and OXM. Cholecystokinin (CCK) and insulin, released during absorption, act as satiety signals to decrease appetite. Peptide YY (PYY) is released from the lower intestinal tract and pancreatic polypeptide (PP) from the islets of Langerhans on the pancreas. Ghrelin is released by the stomach and stimulates appetite, while leptin inhibits appetite.

as reflected by the relative areas of β-cells in the pancreas. The effects are obvious in a detailed comparison (Table 9.1). Human interest in this topic relates to our understanding of the dysfunction in glucose management that arises with the development of late-onset diabetes.

Hormones not related to digestion also can influence feeding behaviour and the digestive tract. Normally, tadpoles of spadefoot toads (*Scaphiopus couchii* (1387)) eat detritus. In response to exposure to exogenous corticosterone (see also Section 16.4c) and novel diets (animal protein such as shrimp), tadpoles showed differences in body size and gut length (Fig. 9.7) compared to control animals. The work demonstrates how hormonal milieu and diet can result in changes in the morphology and food-processing ability of animals.

STUDY BREAK

1. Compare the digestive tracts of the animals in Figure 9.4. Give reasons for the similarities and differences.
2. What role do hormones play in digestion?

9.2a Receiving and Holding

Being able to collect food quickly may minimize an animal's exposure to predators. Many animals collect and swallow food without first processing it in the mouth, then store it in the **crop** for later consumption (Fig. 9.4A, B, C). Other animals, notably mammals, may collect food in cheek pouches (Fig. 9.8). In either situation, crops (or cheek pouches) are often thin-walled

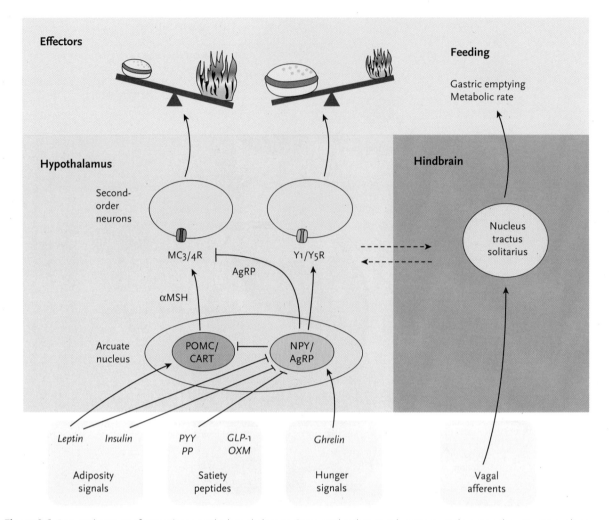

Figure 9.6 Potential actions of gut proteins on the hypothalamus. Gut peptides that signal nutrient uptake act on the arcurate nucleus and on second-order neurons in the hypothalamus. POMC (pro-opiomelanocortin) peptides (red) inhibit appetite as do cocaine- and amphetamine-stimulated transcript peptide (CART), through action on the melanocortin receptors (MC3 and MC4). Appetite-stimulating neurons and agouti-related peptide (AgRP) (green) in the arcuate nucleus act on Y receptors (Y1, Y5). Peripheral signals are integrated within the brain through exchanges between hypothalamus and structures in the hindbrain that receive input from the nucleus tractus solitarius (input from vagal afferent system). Other input to the hypothalamus comes from the cortex, amygdala, and brainstem nuclei.

SOURCE: From Badman, M.K. and J.S. Flier, "The gut and energy balance: visceral allies in the obesity wars," *Science* 25 March 2005: Vol. 307 no. 5717 pp. 1909–1914. Reprinted with permission from AAAS.

| Table 9.1 | Hormonal and metabolic parameters in father rats fed a high fat diet (HFD) and their daughters. Adiposity is the sum of white adipose tissue, glucose tolerance and insulin sensitivity are measures of homeostasis model assessment, and β-cell area is a percent of the area of pancreas. P-values compare rats on HFD and control diets. |

	Control (n = 8)	HFD (n = 9)	Probability
Fathers			
Body mass (g)	550±13	705±17	<0.0005
Adiposity (g)	21.17±0.75	64.1±4.84	<0.0005
Glucose tolerance (mM)	4.71±0.08	5.43±0.16	0.002
Insulin sensitivity	0.88±0.11	2.29±0.18	<0.0005
Female offspring			
Body mass (g)	253±8	260±5	0.92
Adiposity	6.76±0.32	7.87±0.89	0.28
β-cell area	0.72±0.06	0.58±0.05	0.09

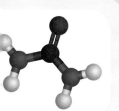

BOX 9.1 Molecule Behind Biology

Leptin: Regulating Appetite and Energy

Typically, animals eat until they have "had enough." But how do they know? When you are offered a second (or third!) helping of a meal, what prompts the reply that you "are full"? Researchers at the Jackson Laboratory in the early 1950s were prompted to ask this question when confronted with the emergence of a strain of mutant mice that were massively obese and voracious. Clearly, nothing was prompting their body to suggest that they had "had enough" (Fig. 1B)!

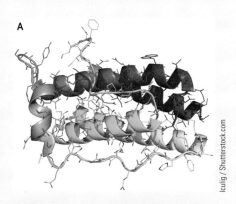

FIGURE 1

A. *The leptin molecule is a 4-helix bundle produced as a transcript of the ob gene.* **B.** *Mice with a normal copy of this gene (right) regulate food intake and body mass, whereas mice homozygous for the ob gene, which results in the synthesis of a defective leptin molecule, are voracious eaters that become massively obese.*

Four decades later, Jeffrey M. Friedman and colleagues at Rockefeller University linked the behaviour of obese mice to the role of the protein **leptin**. Although there are several strains of obese mice, they generally fall into two different categories: *ob/ob* mice with mutations in the gene that results in the production of the protein hormone leptin, and *db/db* mice with mutations in the receptor to which leptin binds. In both cases, the mice are homozygous recessive for the mutation, which means that they produce no functional proteins as either hormone or receptor.

In normal individuals, leptin is produced primarily by adipocytes, the cells of white adipose tissue. Consequently, the amount of leptin in circulation is proportional to the amount of white fat in the body. This small peptide (167 amino acids) can also be produced by brown adipose tissue, gastric chief cells in the fundic glands of the stomach, skeletal muscle, and cells in the placenta, mammary glands, and ovaries. Encoded by the obese gene (*ob*) expressed in these cells, leptin provides a mechanism for the body to assess nutritional status. As adipocytes increase their fat stores, the ob gene is activated to increase the amount of leptin synthesis.

Leptin enters the blood stream in proportion to the amount of body fat. It crosses the blood-brain barrier to enter the central nervous system where it binds to receptors at the hypothalamus in regions specifically involved with appetite, thermoregulation, and energy expenditure. Through its action at the hypothalamus, leptin has been shown to regulate body weight and metabolism with effects on appetite and metabolism. Specifically, leptin inhibits the synthesis of the powerful appetite stimulant **neuropeptide Y,** released by the hypothalamus to trigger feeding behaviour. Leptin's second major effect is to increase the amount of energy expenditure, both through increased metabolic rate (as shown by increased oxygen consumption) and by increasing body temperature. Both of these reduce adipose stores and the levels of circulating leptin. Injection of leptin into ob/ob mice significantly reduces both appetite and body mass, whereas similar injections show no effect in db/db mice, which lack functional leptin receptors at the hypothalamus.

Leptin also plays a role in reproductive behaviour, through actions on both luteinizing and follicle-stimulating hormones released by the anterior pituitary gland. Very low body fat in females, either through exercise (as in human female athletes) or through nutritional deprivation, is often correlated with cessation of menstruation. Low levels of circulating leptin in animals with low body fat appear to reduce the amount of gonadotropin-stimulating hormone from the hypothalamus, thus adversely affecting reproductive function. In young animals, these same conditions can slow the onset of puberty.

and expandable, maximizing their capacity. Many other animals cache food—storing it in the environment outside the body—including many species of spiders, birds, and mammals.

9.2b Mechanical Breakdown

Many animals break up food in their mouths by chewing. These instruments of breakdown may be teeth or radulae, beaks, or mouthparts (see page 218). Some mechanical breakdown of food is also achieved with appendages such as the talons of birds or claws of mammals. Some animals have a **gizzard**, which uses muscular action to grind food, often in combination with **gastroliths** (literally "stomach stones"), pieces of stone or other abrasive material the animal has swallowed. Gastroliths were first reported in 1832 when Geoffroy Saint-Hillaire found some in a fossil crocodile (Fig. 9.9). Since then they have been reported in many tetrapods, including other reptiles such as crocodiles and alligators, many species of birds, and mammals such as sea lions. In tetrapods, gastroliths are rounded

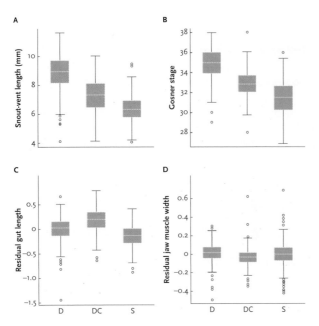

Figure 9.9 Cluster of gastroliths (arrows) in ceratopsian dinosaur *Psittacosaurus* (1443) from the Lower Cretaceous Ondai Sair Formation of Ussuk/Mongolia (AMNH 6253).

Figure 9.7 Tadpoles were given three food types: D (on the x-axis) is detritus, the normal diet; DC is detritus plus exogenous corticosterone; and S is shrimp, a novel diet. The impacts of these conditions include **A.** overall size (snout–vent length), **B.** developmental stage, **C.** gut length, and **D.** jaw muscle size.

SOURCE: Ledón-Rettig, C.C., D.W. Pfennig, and E.J. Crespi, "Diet and hormonal manipulation reveal cryptic genetic variation: implications for the evolution of novel feeding strategies," *Proceedings of the Royal Society B*, 2010, by permission of the Royal Society.

and polished stones, apparently reflecting grinding action in the gizzard or stomach. Gastroliths occur in crayfish as a pair of discs in the anterior part of the stomach. Here gastroliths also serve as a storage site for calcium carbonate during periods of moulting. The origin of gastroliths is presumed to have been the accidental ingestion of stones.

Presumed gastroliths have been found in birds, crocodiles, plesiosaurs, and dinosaurs such as prosauropods, psittacosaurs, and maniraptorans. In the case of fossil material, some researchers have questioned the tendency to label any rounded, polished stone associated with a skeleton as a gastrolith. The role of gas-

troliths in marine animals such as plesiosaurs, penguins, and eared seals has been questioned. These stones appear in the animals that "fly" through the water using the motion of their forelimbs. Gastroliths are not usually found in marine animals that use the tail for power swimming (cetaceans, ichthyosaurs, mosasaurs, and some pinnipeds (walruses and earless seals)). This suggests that gastroliths may function in buoyancy control in the "flying" marine animals. In crocodilians, gastroliths are used to adjust buoyancy and stability, and records of intact fragile prey in the stomach could argue against their use in grinding food. Gastroliths provide an interesting example of the impact of circumstantial evidence and the need for experimentation in relating structure to function.

STUDY BREAK

1. What are gastroliths and what do they do?
2. What advantage is provided by cheek pouches?

9.2c Chemical Breakdown

Although the study of organic chemistry includes many more complex concepts, the basic truth is that organic macromolecules—carbohydrates, lipids, proteins, and nucleic acids—are essentially long chains of repeating subunits (monomers) linked together to form the larger molecule (polymer). Chemical digestion involves the breakdown of ingested polymers to their component subunits. The end products of digestion are sugars, fatty acids, and amino acids, essential to fuel the organism's growth and maintenance, as well as provide vitamins and trace minerals. The smaller monomers are more readily transported across cell membranes from the digestive tract to the blood stream, and then into the cells, than their polymer families.

Digestion typically involves the breakdown of bonds between each pair of monomers through the addition of a water molecule, a reaction known as **hydrolysis**. Some digestive enzymes are proteins bound to membranes that link the energy released from breaking bonds to the transport of the monomers across the cell membranes, whereas other enzymes are

Figure 9.8 An eastern chipmunk (*Tamias striatus* 1907) with cheek pouches empty (left) or filled (centre and right).

SOURCE: M.B. Fenton

free within the digestive tract. Specific enzymes assist in the chemical breakdown of specific organic molecules. Most digestive enzymes are named for the substrate on which they act simply by adding the suffix -ase.

Proteases hydrolyze proteins by breaking the peptide bonds between amino acid monomers in the polypeptide chain. Also known as **peptidases**, proteases are produced in several types, classified by the specific amino acids at which they act or often by the pH at which they are most active. Most digestive proteases work best at low pHs, as acids are also very good at hydrolysis!

Typically, fat digestion begins with emulsification of large globules into many smaller ones. This process is exactly the same manner in which household detergents break up lipids (grease) into smaller droplets, and is purely a physical process, not a chemical one. Emulsification does not break the bonds between the monomers of fat molecules, but rather produces smaller droplets with a greater surface area relative to their volume, so that digestive enzymes are more effective. In vertebrates, bile (produced by the liver) acts as the body's emulsifying agent. **Lipases** are the specific enzymes that chemically break down long-chain fat molecules into smaller-chain fatty acids, which are a source of metabolic energy.

Carbohydrate digestion involves the breakdown of complex polysaccharides, such as starch and cellulose, to their monosaccharide monomers, simple sugars such as glucose and fructose. Carbohydrate digestion may begin in the oral cavity with the secretion of salivary amylase, but most of this process is completed farther down the digestive tract where it is coupled with the absorption of simple sugars. As humans, we typically consider **starch** or **glycogen** digestion when we think about carbohydrates, since these are the storage polysaccharides of plants and animals on which our enzymes function.

Of more significance to other animals, however, is **cellulose**, a structural component of all plants. Although cellulose is also a polysaccharide, the molecule is never branched, the way the starch and glycogen molecules are, so the glucose monomers are free to interact with adjacent molecules lying parallel to them. In plant cell walls, parallel cellulose molecules held together in this way are grouped into units called microfibrils. These cable-like microfibrils are a strong structural material for plant cell walls. This means that cellulose is insoluble and extremely resistant to chemical digestion. Many herbivores depend upon it as a major energy source, yet surprisingly, only two animals in the entire animal kingdom—both invertebrates— have been shown to be able to manufacture **cellulases**, enzymes that can digest cellulose. These enzymes are produced by some species of termites and some recently studied isopod crustaceans known as gribbles (see Section 9.3). Otherwise cellulases are produced only by fungi, bacteria, and protozoans. However, these include symbiotic microorganisms that live in the digestive tracts of host animals and produce cellulases to digest the cellulose of ingested plants. This

microbial process of breaking down cellulose anaerobically, known as **fermentation**, yields small organic molecules that are absorbed by the host to provide metabolic fuel. Carbon dioxide and methane (CH_4) are non-metabolized by-products of fermentation that are released by the expulsion of digestive gases. Parts of the digestive tracts of the hosts are specialized as fermentation chambers in which the symbiotic microorganisms eventually digest cellulose. This process is relatively slow, so the chambers can be large relative to the overall length of the digestive tract (see Box 9.3, Fig. 1B). This allows increased transit time in this region, increasing the opportunity for breakdown of the resistant cellulose molecules.

9.2d Absorption

The regions of the digestive tract where food components are absorbed are located distal to the mechanical preparation and chemical digestion regions. Most absorption occurs in the midgut of invertebrates and the small intestine of vertebrates. In all animals with gastrointestinal tracts, these regions are histologically differentiated along their length. The secretion of enzymes for digestion occurs at the proximal end of the midgut or small intestine, whereas absorption of the products of digestion takes place at the distal end. Absorption requires the movement of the breakdown products of digestion—fatty acids, simple sugars, amino acids, and small proteins (peptides)—from the lumen of the digestive tract, across endothelial cells lining the tract, and from there into blood and lymph. These movements may occur by passive or active transport, and the mode of transport is largely dictated by the chemical nature of the molecule. Fatty acids, being lipid soluble, readily cross the epithelial cell membrane, whereas simple sugars and amino acids are less soluble in the lipid membrane and require the action of membrane proteins for transport.

In mammals, the movement of a simple sugar across a cell membrane is mediated by a sodium glucose cotransporter (SGLT, a membrane-bound protein that is a member of a glucose transporter family) working with glucose transporter facilitator proteins of the family GLUT (glucose transporters) (Fig. 9.10A). As with all cells, intestinal epithelial cells have a lower intracellular concentration of sodium than the concentration outside the cell. Generally, SGLTs and GLUTs are integral membrane proteins that couple conformational changes that facilitate the diffusion of sodium down its gradient from the intestinal lumen across the epithelial cell membrane with the transport of glucose or related hexoses. The genes that encode these transporters have been identified and the functional properties of corresponding proteins described. Functional and morphological studies have revealed that GLUT sugar transporters also occur in invertebrates. Specifically, GLUT2–like transporters occur in the apical and basolateral cell membranes of the midgut in several

Figure 9.10 Glucose transport across mammalian epithelial cells occurs in conjunction with sodium at the apical membrane. **A.** The major mechanism for glucose entry into the cells is SGLT (see text). Glucose concentrated inside the cell moves outward through the basolateral membrane via GLUT, transporters, which do not rely on Na⁺. **B.** Amino acids are absorbed through a three-step process known as transcytosis.

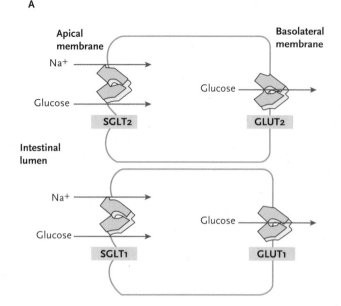

A

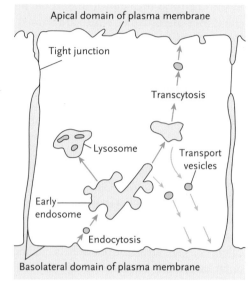

B

BOX 9.2 People Behind Biology

Alessio Fasano M.D.

What happens when the mechanisms required for moving the products of digestion from the intestinal lumen across the epithelial lining fail? When nutrients are not absorbed, animals exhibit signs of malnourishment, including weight loss and gastrointestinal symptoms that typically include chronic diarrhea (as unabsorbed nutrients draw water across the intestinal wall by osmosis) and abdominal bloating and pain. In humans, these signs of malabsorption are sometimes correlated with an intolerance for gluten, a protein found in wheat, rye, and barley. When this gluten-intolerance triggers an immune response, the affected individual's immune system destroys the villi that line the small intestine. These villi are critical for the surface area and mechanisms of absorption. This condition, known as **celiac disease**, is a genetic disorder that affects as many as 1 out of 130 North Americans.

The research of Dr. Alessio Fasano focuses on mucosal biology in humans, drawing together the disciplines of microbiology, molecular and cellular biology, and gastrointestinal physiology. In particular, Dr. Fasano and his team of researchers at the University of Maryland's Mucosal Biology Research Center have been exploring the role of bacterial toxins in the pathogenesis of diarrheal diseases. The structural characterization of one of these toxins (zonula occludens toxin from *Vibrio cholerae*) led his team to the discovery of a molecule they named **zonulin**. This molecule is involved in the regulation of the permeability of intercellular tight junctions of epithelial and endothelial barriers. By revealing more about the regulation of molecules trafficking between body compartments, this discovery has been a key factor in linking the pathogenesis of enteric disorders such as celiac disease to a particular molecular mechanism.

In people without celiac disease, zonulin selectively allows the passage of proteins between epithelial cells lining the gut cavity. This facilitates transport of only certain large molecules into circulation, while keeping harmful bacteria and toxins within the gut cavity and out of circulation. Based upon the research of Dr. Fasano's team, we now know that individuals with celiac disease have a greater number of zonulin proteins, allowing the passage of large proteins such as gluten to the extent that an immune response is triggered to this foreign plant protein. This reaction results in the production of anti-zonulin antibodies that attack the intestinal cells, ultimately disrupting the process of absorption and producing the enteric symptoms of celiac disease.

Dr. Fasano's research team has shown that zonulin is also a critical protein in the very selective permeability between the bloodstream and the brain, regulating movement across the blood-brain barrier. Their work has also linked zonulin to the pathogenesis of a series of other autoimmune diseases, including type 1 diabetes and multiple sclerosis.

This information has also been used to develop new vaccines to combat cholera and shigella, both diseases caused by bacterial toxins that disrupt the function of normal zonulin proteins in the process of absorption. Future research is targeted to the prevention and treatment of inflammatory and autoimmune diseases. The discovery of the genetic component of these disorders, through the production of a normal protein—zonulin—in abnormal amounts or in a dysfunctional state, has also opened the doors to research exploring long-term treatment through drugs, antigens, and gene delivery.

species of insects. It appears that the mammalian model for transepithelial transport of sugars applies equally well to invertebrate mechanisms.

The route followed by amino acids and small proteins across the epithelium involves endocytosis, a three-stage process (Fig. 9.10B):

1. vesicle-mediated internal transport moves these polar molecules into endothelial cells.
2. the vesicles move across the cells.
3. exocytosis, the release of vesicle contents into the extracellular milieu on the other side of the cells (opposite the lumen), occurs.

This overall mechanism of crossing an epithelial layer is called **transcytosis**, and it occurs at other sites in an animal's body where large polar molecules cross a cellular barrier.

The speed with which materials are transported across a cell layer is, in part, a function of the number of steps involved. Compared with active transport sites, **paracellular absorption** (molecules moving between the cells of an epithelial layer) is less selective but more direct, faster, and less energetically expensive. Fuelling hovering flight is an example where this mode of transport provides an advantage.

Digestive systems of birds and mammals tend to be specialized for higher feeding rates and have relatively longer intestines than other animals. But the mass of the digestive system and associated increases in flight costs may explain why the small intestines of small birds and bats are shorter than those of nonflying mammals of the same size. Small birds appear to compensate for shorter small intestines by the presence of extensive villi that increase the absorptive surface. Paracellular pathways, where molecules move between the cells of an epithelial layer, are used for the passive movement of water-soluble compounds (glucose and amino acids) by sieve-like absorption. These pathways may also account for shorter small intestines in bats. In large fruit bats (*Artibeus lituratus* (1784)), availability of carbohydrates through paracellular movement appears to be a function of the size of the molecule. In this species, 70% of total absorption of glucose is paracellular. These bats also depend upon carrier-mediated transcellular movement of the products of digestion. Paracellular absorption may explain the success of bats in the rapid absorption of the products of digestion in spite of their shorter small intestines to compensate for flight.

Animals such as hawk moths (Sphingidae), hummingbirds (Trochilidae), and nectar-feeding bats (Glossophaginae), all of which hover in front of flowers while feeding, quickly mobilize the sugars from the nectar to supply the energy demands of foraging. While hovering, the flight muscles account for >90% of a hummingbird's rate of oxygen consumption (its whole-body metabolic rate). It has been shown that sugar ingested from nectar is used immediately to fuel the cost of hovering flight. The presence of paracellular absorption in

birds and bats may account for the rapid uptake of recently ingested glucose. By comparison, humans lack the capacity for paracellular absorption and can fuel only 25–30% of energy costs associated with exercise from sugars ingested just before or during exercise (Fig. 9.11). Like hummingbirds, hawkmoths can fuel flight with either carbohydrates or fatty acids, while other insects may specialize on carbohydrates (e.g., honeybees, *Apis mellifera* (984)), amino acids such as proline (tsetse flies, *Glossina morsitans* (963)), or fat (some Lepidoptera, e.g., *Philosamia cynthia* (1046)). Still other animals use mixtures of different fuels or switch among fuel types according to the situation.

The relationship of ATP production to oxygen consumption, known as the P/O value, varies according to the metabolic substrate from which energy is

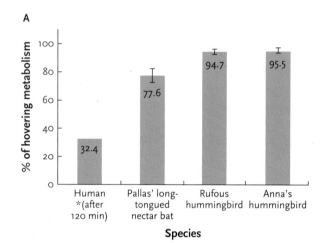

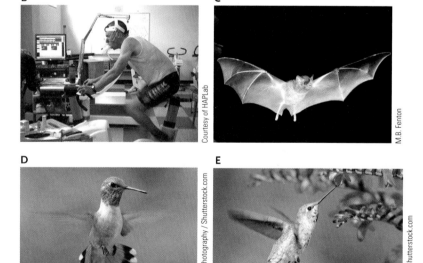

Figure 9.11 A. Sources of energy during exercise. **B.** Data for human (here shown exercising on an antique ergometer). Hovering flight data for **C.** Pallas' long-tongued nectar bat (*Glossophaga soricina* (1972)), **D.** Rufus Hummingbird (*Selasphorus rufus* (1546)), and **E.** Anna's Hummingbird (*Calypte anna* (1545)). Values are means ± standard error of the mean.

SOURCE: A. Reproduced with permission from Welch, K.C. Jr., L. G. Herrera M., and R. K. Suarez. 2008. "Dietary sugar as a direct fuel for flight in the nectarivorous bat Glossophaga soricina," *Journal of Experimental Biology*, 211:310–316. The Company of Biologists Ltd.

transferred to ATP. When the substrates are fatty acids, energy is transferred more readily and the P/O ratio is 15% lower than when glucose is the fuel. Associated with this difference are differences in the **RQ (respiratory quotient)**, the relationship between carbon dioxide produced relative to oxygen consumed. It is a useful measure of metabolic rate based upon CO_2 production. The RQ is only 0.71 when fatty acids are the fuel, reflecting the incomplete breakdown of these molecules, compared with 1.0 when the fuel is glucose. When body mass is plotted against metabolic power input for hovering (Fig. 9.12), there are clear differences among hawk moths, hummingbirds, and glossophagine bats, with the bats showing the least amount of power input, relative to body mass, required for flight.

Paracellular absorption has also been demonstrated in Egyptian fruit bats (*Rousettus aegyptiacus* (1815)), an Old World fruit bat (family Pteropodidae). We can expect rapid access to ingested sugars in the other fruit bats of the Old World (Africa, India, East Indies, Australia), including those specialized for feeding at flowers. It remains to be determined if other animals that exploit nectar as food have comparable adaptations for rapid access to the energy obtained from their food.

STUDY BREAK

1. What is paracellular absorption? Which animals use it? What advantage does it provide? At what cost?

9.2e Preparing to Evacuate

The invertebrate hindgut and the vertebrate large intestine largely function to reabsorb water and some water-soluble vitamins and minerals. Otherwise, this region of the digestive tract is simply preparing the undigested materials for evacuation from the body. These regions are characterized by both the absence of absorptive epithelia and the presence of many mucous-secreting glands.

The uses that other animals make of dung, particularly as food, indicate that although the animal that passed the feces may be finished with them, they may be useful to others. In coprophagous species, such as rabbits, hares, and pikas (Lagomorpha), eating their own droppings gives them access to the products of hind gut fermentation. In animals where the fermentation chamber is beyond the area in the gut where absorption takes place, coprophagy is a practical strategy. Coprophagic species produce two kinds of droppings: the soft mushy ones that they eat, and the hard pelletized ones they leave behind.

Feces accumulate in the distal part of the digestive tract before being passed. In insects, contact between Malpighian tubules and the hind gut allows for reabsorption of material, usually water, from feces before they are passed. The Malpighian tubules of blood-feeding insects (e.g. *Rhodnius* (875) and *Aedes* (953)) are particularly important in concentrating the contents of the digestive tract. The Malpighian tubules regulate hemolymph volume and composition, but the final composition of the feces is determined by the hindgut.

Feces are widely used by other animals. Humans use dung as fertilizer and fuel and, mixed with mud, as bricks or wall plaster. Some insects mimic the appearance of bird droppings (page 473), and some insects and other animals use feces to deter predators (page 480).

Sperm whales (*Physeter macrocephalus* (1766)) hunt prey at depth but defecate iron-rich feces near the ocean's surface (photic zone). It has been estimated that the 12 000 sperm whales of the Southern Ocean annually defecate 50 tonnes of iron into the photic zone, enhancing primary productivity and stimulating the export of 4×10^5 tonnes of carbon into the deep ocean.

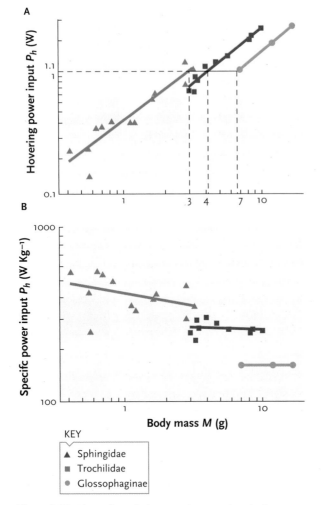

Figure 9.12 Plots of metabolic power input against body mass for hawkmoths (Sphingidae), hummingbirds (Trochilidae), and nectar-feeding bats (Glossophaginae). **A.** Data from individuals, **B.** data are mass-specific averages.

SOURCE: With kind permission from Springer Science+Business Media: Journal of Comparative Physiology B, "Energetic cost of hovering flight in nectar-feeding bats (Phyllostomidae: Glossphaginae) and its scaling in moths, birds and bats," Volume 169, 1999, pp. 38–48, Voigt, C.C., and Y. Winter.

Dung-feeding animals such as scarab beetles feed mainly on the bacteria in vertebrate dung. In addition to specializations for locating, transporting and burying dung, dung beetles usually lack mouthparts that would be useful in predation (Fig. 9.13A). In some locations the diversity of dung beetles can be high, including >80 species that show interesting differences in their approaches to using dung as food. Dung beetles such as *Agrilinus constans* (920) respond differently to the dung of different mammals such as cattle, sheep, horse, and wild boar, and their choices are also influenced by the presence of other insects in the dung. Normally, cow and sheep dung are the most attractive to this species. Some dung-feeding beetles have adopted carrion-feeding (necrophagy), using clypeal teeth (Fig. 9.13B) to cut carrion. Still others, such as *Deltochilum valgum* (931), kill and eat millipedes.

For reasons of hygiene, some animals expel feces with considerable force. Penguins provide a good example, achieving distances of 40 cm. Pressures associated with penguin defecation (Fig. 9.14) exceed the situation in humans. The pressure on rectal muscles of a human standing upright is ~20 mm Hg and these muscles cannot withstand pressures much over 50 mm Hg.

A

B

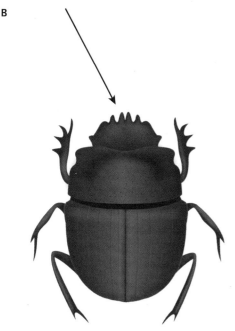

Figure 9.13 **A.** Dung beetle atop its ball of dung. **B.** Clypeal teeth (arrow) on a dung beetle.

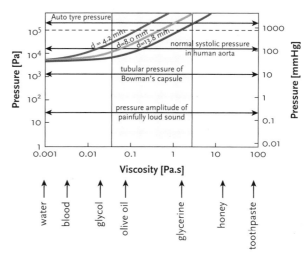

Figure 9.14 The pressure under which feces from three species of penguins is ejected from the cloaca. The y-axis depicts rectal pressure (in Pa) on the left axis and mm of Hg on the right. Viscosity is presented on the x-axis. Data are presented for Rockhoppers (4.2 mm Hg), Adelie (8.0 mm Hg), and Gentoo (13.8 mm Hg).

SOURCE: With kind permission from Springer Science+Business Media: *J Polar Biology*, "Pressures produced when penguins pooh - calculations on avian defaecation," Volume 27, 2003, pp. 56–58, Meyer-Rochow, V.B. and J. Gal.

STUDY BREAK

1. What is coprophagy, and what advantage does it provide to animals that use it?
2. How far behind a penguin should you stand?

9.2f Challenges of Digestion

In general, the mass of an animal's gut tissue increases with the mass of food ingested and processed. Gut tissue shows a high rate of turnover because it is abraded by food materials passing through and is under attack by microbes within the gut lumen as well as acid and enzymes active in digestion. Upkeep of the gut is energetically expensive, explaining why there is a relationship between gut mass and mass of food ingested. A corollary of this relationship is that an animal on a starvation diet develops a gut with a very thin lining, limiting its capacity to return to a normal feeding regime. One example of intraspecific variability comes from cattle. On pasture, cattle ingest more food (of lower digestibility) and have larger guts. At a feedlot where the diet is grain, which is of higher quality and more digestible, the gut mass is smaller as is the mass of food ingested.

Many plant secondary metabolites are toxic to animals, and plants use them to deter foraging herbivores. Some animals also sequester plant secondary metabolites for use in their own defense (see pages 477–478). Plant secondary metabolites may deter herbivory when they increase the costs of digestion needed to detoxify ingested plant material. Many insect herbivores specialize on a single type of plants (monophagous) whose secondary metabolites they are specialized to detoxify, making them efficient predators. Polyphagous insects eat a variety of plants and must maintain

BOX 9.3

Belching, Farting, and Climate Change

Greenhouse gases, methane (CH_4), carbon dioxide (CO_2), and nitrous oxide (NO_2) are produced by natural processes. Among these, CH_4 is a major concern because its warming potential over 100 years is >20 times greater than that of CO_2. The main removal system (sink) for methane is a chemical reaction with hydroxyl radicals (OH); this produces alkyl radicals (CH_3) and water. But this natural removal process can only deal with about 500×10^6 T a year.

The foregut fermentation digestion of ruminants (cows and sheep) is a significant natural source of methane. In Australia, for example, enteric methane accounts for 67% of total agricultural emissions, the equivalent of 66% of emissions produced by the transportation sector. By 2020, Australia could reduce its production of greenhouse gases by 16 mega-tonnes by removing 7 million cattle and 36 million sheep from its agricultural operations and replacing them with kangaroos. Why would this switch reduce methane emissions from livestock?

Kangaroos, cattle, and sheep are foregut fermenters in which the biota of the foregut decompose plant material to produce H_2, CO_2, and short-chain fatty acids (Fig. 1). The approaches to digestion are analogous and are a good example of convergent evolution. The key to the process is maintaining a low partial pressure of H_2, enabling re-oxidation of NADH. In cattle and sheep, slow-growing methanogens achieve this by reducing H_2 and producing methane. In kangaroos and wallabies, food moves more quickly through the forestomach, and acetogenesis is the dominant H_2-consuming reaction.

Selective breeding for cattle with reduced time that digesta (ingested food) remains in the tract (mean retention time, MRT) or in the rumen (ruminal retention time, RRT) is another approach to resolving production of CH_4 by ruminants (Fig. 2). Shorter MRT or RRT, coupled with changes in the biotic composition

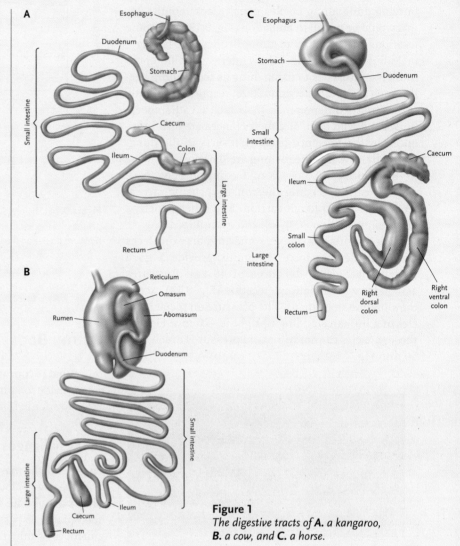

Figure 1
*The digestive tracts of **A.** a kangaroo, **B.** a cow, and **C.** a horse.*

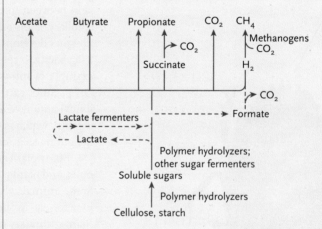

Figure 2
Rumen fermentation of plant polysaccharides including the production of methane (CH_4).

SOURCE: From Wolin, M.J., "Fermentation in the rumen and human large intestine," *Science* 25 September 1981: Vol. 213 no. 4515 pp. 1463–1468. Reprinted with permission from AAAS.

of the microorganisms in the rumen, could increase the number of long-chain volatile fatty acids that are produced, perhaps making cattle more like kangaroos in production of CH_4. It is also possible to administer higher unsaturated fatty acids or chloroform which could decrease CH_4 production, but these treatments do not improve animal performance and may cause other problems.

In other mammals there is genetic variation in methane production (Fig. 3).

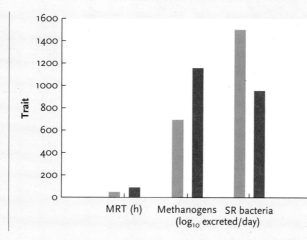

Figure 3
Variation among cattle in the time that digesta are retained in the tract (MRT, in hours) shown with total populations of gut methanogens and sulphide reducing (SR) bacteria. These data are from cows classified as methane emitters (green bars) versus non-methane emitters (red bars).

SOURCE: Hegarty, R.S. 2004. "Genotype differences and their impact on digestive tract function of ruminants: a review." *Australian Journal of Experimental Agriculture*, 44:459-467. SCIRO PUBLISHING. http://www.publish.csiro.au/nid/72/paper/EA02148.htm.

a larger array of detoxification systems. Vertebrate herbivores may show the same patterns.

In some parts of their winter range, Ruffed Grouse (*Bonasa umbellus* (1664)) selectively eat staminate flower buds of quaking aspen trees (*Populus tremuloides*). These birds feed more at some trees than at others, and their choices correlate with the presence of coniferyl benzoate (CB), a phenyl-propanoid ester that is concentrated in the aspen buds, giving them a bad taste and making them an irritant to other predators. The consumption of aspen flower buds by Ruffed Grouse is predictable from the content of coniferyl benzoate and crude proteins in the buds, and the birds will typically feed more from trees with lower CB levels in the flower buds. Ruffed Grouse have been shown to assimilate 24% less energy from their food when their diet was higher in coniferyl benzoate as opposed to low, and birds fed a diet with CB show a decline in body mass compared with controls (Fig. 9.15). With

increasing levels of CB in the diet, Ruffed Grouse excrete more glucuronic acid and ornithine as well as NH3 as a result of the detoxification activity of the birds' liver, required to metabolize CB. The increased production of metabolic acids and NH3 also affects the ability of Ruffed Grouse to maintain acid-base balance. These factors result in higher energetic costs to the birds on the high coniferyl benzoate diet, and may also lower the birds' capacity for consuming aspen buds, which partly explains the difference in energy assimilation between the two diets. But the ability to consume the flower buds of quaking aspen, despite the dietary challenges of CB levels, allows Ruffed Grouse to take advantage of a food source unavailable to others.

When an animal eats different food in winter and summer, is there a corresponding variation in its digestive tract? It is tempting to think that features of the body, such as the gastrointestinal tract, are consistent within a species. In some cases, the real picture is quite different. For example, studies of Phainopepla (*Phainopepla nitens* (1609)), small fruit-eating songbirds that occur in the Sonoran Desert, have shown that the birds' main food is berries of desert mistletoe (*Phoradendron californicum*) between October and December. During the rest of the year, the birds eat a more generalized diet, including a variety of fruits. Coincident with the change from a generalized to a specialized diet (and vice versa) is a change in the size of the gizzard, which is larger when the diet is more generalized (Fig. 9.16).

Migration can pose a special challenge to animals. While some migrating animals use stopovers to rest and replenish their energy supplies, others experience in-flight starvation because they either do not or cannot stop to rest and refuel. The stopover situation is illustrated by Western Sandpipers (*Calidris mauri* (1552)), which follow shorelines when migrating and have many opportunities to stop over. During migration, as compared with nesting, Western Sandpipers (Fig. 9.17) show striking

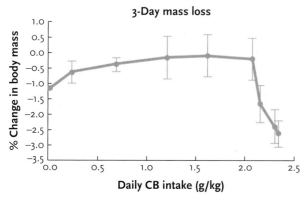

Figure 9.15 Percent change in body mass (±SE) over 4 days with different concentrations of ingested coniferyl benzoate (CB) in the diet of Ruffed Grouse (*Bonasa umbellus* (1664)).

SOURCE: Jakubas, W.J., W.H. Karasov and C. G. Gugliemo, "Ruffed grouse tolerance and biotransformation of the plant secondary metabolite coniferyl benzoate," *The Condor*, 95:625–640. © 1993, The Cooper Ornithological Society, University of California Press Journals.

Figure 9.16 Seasonal variation in the size of the gizzard of **A.** *Phainopepla.* **B.** The bird's diet in October, November, and December is much more specialized than it is during the rest of the year. Values are means ±95% confidence limits; sample sizes are shown within each bar.

A

Martha Marks / Shutterstock.com

B

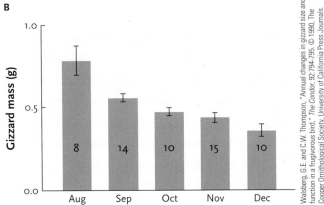

Walsberg, G.E. and C.W. Thompson, "Annual changes in gizzard size and function in a frugivorous bird," *The Condor,* 92:794-795. © 1990. The Cooper Ornithological Society, University of California Press Journals.

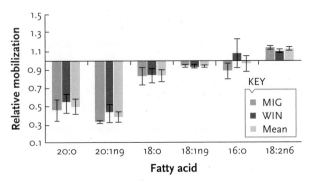

Figure 9.18 Mobilization of fatty acids from adipocytes of White-crowned Sparrows. More lipophilic fatty acids are mobilized first. Fatty acids on the x-axis are ordered from least to most mobilized, based upon structural properties and membrane characteristics. There were no significant differences between migratory and non-migratory birds for any fatty acid. Here MIG = migrant and WIN = winter, and the blue bars represent the means.

SOURCE: Reproduced with permission from Price, E.R., A. Krokfors, and C.G. Guglielmo. 2007. "Selective mobilization of fatty acids from adipose tissue in migratory birds," *Journal of Experimental Biology,* 211:29–34. The Company of Biologists Ltd.

changes in the sizes of small and large intestines, caecae, gizzard, liver, and pancreas, all indicating increased capacity for digestion. They also show changes in capacity for exercise, as indicated by the increased size of heart, lungs, and pectoral muscles during migration.

Animals that migrate across great expanses of inhospitable habitat (e.g., open ocean or large deserts such as the Sahara) do not have any chance to stop over. These long-distance migrants fuel part of their flight by metabolizing proteins. Bar-tailed Godwits (*Limosa lapponica* (1555)) start their 11 000 km non-stop flight from Alaska to New Zealand weighing about 485 g, which is reduced to ~213 g on arrival. By then, they have consumed all of their body fat as well as body components such as elements of the digestive tract and even flight muscles. The same is true of Garden Warblers (*Sylvia borin* (1617)) crossing the Sahara. These birds lose up to about 50% of the mass of kidney, heart, proventriculus, and small intestine during their northwards (spring) crossing of the Sahara. During these flights only the testes increase in mass, not surprisingly because the birds are going north to breed. When migrating Garden

Warblers can stop over, as they do on the migration from southern Africa to the south side of the Sahara, they do not show these reductions in the sizes of body parts. At some point, reductions in the size of an animal's digestive tract will limit its capacity to refuel before regenerating lost tissue.

During high-energy-demand periods such as migration, some birds selectively mobilize fatty acids from adipose tissue. In White-crowned Sparrows (*Zonotrichia leucophrys* (1622)), the mobilization is not a random process: the more lipophilic fatty acids are mobilized before the more hydrophilic ones (Fig. 9.18).

Flexibility in the structure of digestive systems can also involve variation among species in a group. Data on the lengths of small intestines in rodents suggest that these variations also reflect climatic conditions. Latitude and numbers of habitats occupied by species of rodents are significantly reflected in the different lengths of small intestines within the same species (Fig. 9.19). The broader range of tolerance allows species to occupy more habitats. These data demonstrate possible connections between digestive flexibility in mammals and local conditions, and are important for our understanding of the adaptive value of digestive flexibility in the distribution of animals.

STUDY BREAK

1. How is seasonal availability of food reflected in the function of the digestive tract?
2. What do plant secondary compounds do? Why do some animals specialize in eating plants with lots of secondary compounds?
3. How do the digestive systems of birds change during migration? Are all migrating birds the same? Why?

Figure 9.17 Western Sandpiper (*Calidris mauri* (1552)).

Lukich / Shutterstock.com

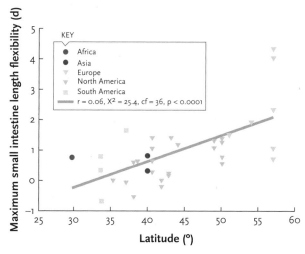

Figure 9.19 Relationships between maximum flexibility (or difference) in lengths of rodent small intestines and geographical latitude. The range of flexibility in length of small intestine reflects a broader range of tolerance in diet, which allows species to occupy more habitats.

SOURCE: Naya, D.E., F. Bozinovic and W.H. Karasov, "Latitudinal trends in digestive flexibility: testing the climate variability hypothesis with data on the intestinal length of rodents," *The American Naturalist*, 172:E122–E134. © 2008, The University of Chicago Press.

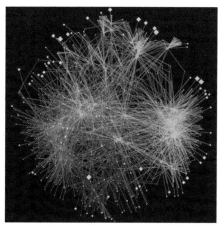

Figure 9.20 A network-based analysis of the fecal bacteria communities of 60 species of mammals. The data are colour-coded by diet: blue = omnivore, green = herbivore, and red = carnivore. The different colours indicate the variation in the fecal bacteria community associated with diet.

SOURCE: From Ley, R.E., M. Hamady, C. Lozupone, R.J. Turnbaugh, R.R. Ramey, J.S. Bircher, M.L. Schlegel, T.A. Tucker, M.D. Schrezel, R. Knight and J.I. Gordon, "Evolution of mammals and their gut microbes," *Science* 320, 2008:1647–1651. Reprinted with permission from AAAS.

9.3 Gut Biotas

Many animals, including humans, are **metagenomic**, in that they are composed of their own genome as well as those of their associated microbes. The animal gastrointestinal tract harbours a complex microbial network containing thousands of species and >100 times the number of genes than are present in the host's genome. The compositions of the microbial communities reflect the constant co-evolution of these organisms with their host environment. These endosymbiotic organisms include bacteria, protozoa, and fungi. About ~100 trillion individual endosymbionts live in a typical human gut, far greater than the population of microorganisms found elsewhere in the body. Most are confined to the hindgut regions. The endosymbionts are fundamental to digestion of almost all food. Some nutrients would not be available without their contribution.

Historically, classical microbiological techniques were used to study microbial colonization of animals, including the isolation and culture of individual bacterial species associated with the gut. More recently, metagenomic techniques have been used to characterize both the composition and the potential physiological effects of entire microbial communities without having to culture individual community members (Fig. 9.20).

The relationship between diet and gut microflora has been mainly studied in mammals. The presence of these microbes has meant that mammals have not had to evolve their own capacity of obtaining specific nutrients (such as vitamins and nutrients from cellulose) from food. The gut microbiome of mammals is influenced by both diet and phylogeny, meat-eaters showing the lowest diversity and herbivores the highest (Fig. 9.20).

By far the greatest diversity of microbes in the digestive systems of mammals lives in the distal regions of the tracts, which is of particular importance in the digestion of plant materials and subsequent absorption of the products of digestion, which often involves the retention of food in the gut for longer periods of time.

Microbial activity releases volatile fatty acids through digestion as well as vitamins and proteins that are essential to the host. During the digestion process, 40–60% of food protein from the cellulose is broken down by microbial organisms, which, in turn, serve as protein for the host. Microbes also upgrade proteins by converting them to microbial proteins through incorporation of NH3 derived by decomposition of urea from the hosts' saliva and by diffusion from its blood. This adds to the proteins available to the host for further digestion by host enzymes.

Endosymbiont bacteria in aphid guts aid in the synthesis of tryptophan, an amino acid involved in growth and nitrogen balance. The bacterium *Buchnera aphidicola* occurs in a variety of species of aphids and has a distinct evolutionary history. It is most closely related to *E. coli* and endosymbiosis in *Buchnera* genus began about 200–250 million years ago.

The bacteria *Aeromonas hydrophila* lives in the digestive tract of medicinal leeches (*Hirudo medicinalis* (367)), a blood-feeding annelid, where it produces enzymes that assist in digestion, produces vitamins, and prevents the growth of other bacteria by secreting an antibiotic. *Aeromonas hydrophila* also lives in the gut of *Desmodus rotundus* (1792), a blood-feeding bat. In blood-feeding arthropods and leeches, the symbionts are housed in a symbiotic organ called a bacteriome or mycetome. The symbionts are thought to provide the host with thiamine, pantothenic acid, pyridoxine, folic acid, and biotin—B-complex vitamins that are not common in the blood of vertebrates. Blood-feeding leeches have a pair of bulbous mycetomes that connect by separate ducts to the esophagus. The bacteria in these mycetomes are from a monophyletic group that also includes the endosymbiotic bacteria found in sap-feeding insects such as aphids.

In the evolution of mammals, the move to herbivory was a major breakthrough (80% of living mammals are herbivores), reminiscent of the colonization of coral reefs by fishes (see pages 165–166). It is remarkable that the diversification of the diet of humans, in part reflecting domestication of food species and cooking, is not obviously associated with a change in microbe fauna compared with other omnivorous primates.

It is often difficult for animals to extract energy from plant materials. Plant cell walls, particularly in woody materials, have low solubility and tend to form crystalline arrays with ether and ester linkages that make them very resistant to digestion by enzymatic activity. Cellulose and hemicelluloses are the main ingredients in plant cell walls. Cellulose is a linear homopolymer of β 1-4 linked glucose residues. Hemicellulose is a branched homopolymer of xylose units. Lignins are heterogeneous polymers of aromatic molecules. Pectin, another component of cell walls, is a heteropolymer with α 1-4 linked galacturonic acid.

The heterogeneous structure of lignin provides plants with structural rigidity and protects cellulose and hemicelluloses from degradation. To eat wood that has not yet rotted, animals must circumvent lignin. The Asian longhorned beetle (*Anoplophora glabripennis* (921), Fig. 9.21) is a cerambycid beetle that eats the inner wood of a variety of hardwood trees. The Pacific dampwood termite (*Zootermopsis angusticollis* (885)) eats the dead wood of coniferous trees. Both insects have the ability to alter the lignin polymer, specifically by side-chain oxidation that depolymerises it. A soft-rot fungus, called the *Fusarium solani/ Nectria haematococca* complex, lives in the gut community of the longhorned beetle and appears to catalyze the side-chain oxidation of lignin. *Z. angusticollis* lacks the soft-rot fungus, but still achieves side-chain oxidation and demethylation of lignin. In both insects, lignin degradation occurs in microbial ecosystems in the insects' guts.

Figure 9.21 Asian longhorned beetle (*Anoplophora glabripennis* (921)).

SOURCE: panda3800 / Shutterstock.com

Complete breakdown of wood involves enzymes that, in nature, are usually produced when multiple microorganisms act in concert. Three types of hydrolytic activities are involved in degradation of cellulose, while digestion of hemicellulose involves carbohydrases and esterases that break xylan backbones and decouple the side-chains involved in binding lignin. Glycoside hydrolases are the enzymes most often involved in degrading complex polysaccharides. There are at least 112 families of glycoside hydrolases; these show considerable variation in specificity to substrates, sites, and modes of cleaving substrates. Other enzymes may also be involved in degrading the carbohydrate esterases, polysaccharide lysases, and carbohydrate binding molecules in wood. The composition and organization of proteins that degrade polysaccharides are highly variable, another illustration of animal diversity.

Marine bivalves (Teredinidae) constitute one family of "ship worms," (wood borers) that are economically important because of the damage they do to wooden structures such as docks and the hulls of ships. The marine gamma proteobacterium *Teredinibacter turnerae* T7901 is an intracellular symbiont in the gills of teredinid ship worms. *T. turnerae* appears to provide the ship worms with cellulases and nitrogenases that are critical for digesting wood and supplementing the nitrogen-deficient diet of the host. This proteobacterium is very likely to be a facultative intracellular endosymbiont that was perhaps recently free-living. *Teredinibacter turnerae* has more glycoside hydrolase domains that are specific to cellulose, xylan, mannans, and rhamnogalactans—all major components of wood—compared with other known genomes of organisms producing the enzymes needed to digest cellulose (Fig. 9.22).

Termites are renowned for their capacity to digest wood, and their reputation is supported by many sorrowful homeowners faced with the need to make extensive repairs (Fig. 9.23). In the past, it was common to

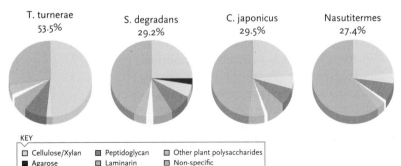

T. turnerae 53.5% S. degradans 29.2% C. japonicus 29.5% Nasutitermes 27.4%

KEY

☐ Cellulose/Xylan ■ Peptidoglycan ☐ Other plant polysaccharides
■ Agarose ☐ Laminarin ☐ Non-specific
☐ Chitin ☐ Pectin

Figure 9.22 Glycoside hydrolase domains in organisms with some capacity to digest wood, specifically the bacteria *Teredinibacter turnerae*, *Saccharophagus degradans*, *Cellvibrio japonicus*, and a sample from the midgut of *Nasutitermes* (886), a termite. The coloured pieces represent the proportion of eight different glycoside hydrolase domains.

SOURCE: Yang JC, Madupu R, Durkin AS, Ekborg NA, Pedamallu CS, et al. (2009) The Complete Genome of *Teredinibacter turnerae* T7901: An Intracellular Endosymbiont of Marine Wood-Boring Bivalves (Shipworms). PLoS ONE 4(7): e6085. doi:10.1371/journal.pone.0006085.

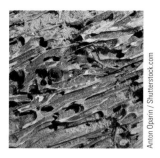

Anton Oparin / Shutterstock.com

M.B. Fenton

Figure 9.23 Termite damage and three termite soldiers (*Reticulitermes flavipes* (887)). The soldier termites guard the colony, while the workers do the damage.

attribute the wood-digesting accomplishments of termites to flagellates living in their digestive tract, but we now know that symbiotic bacteria produce the cellulose-digesting enzymes used by termites. The evidence for this conclusion is the effect of antibiotics on cellulase activity in the midgut of the termite *Nasutitermes takasogoensis* (886) (Fig. 9.24).

In some animals, notably some mammals, the microbial community in the digestive tract digests cellulose by fermentation, providing starch, soluble carbohydrates, proteins, and lipids to the host (see Section 9.2c).

Animals lacking endosymbionts may be the exception rather than the rule, and gribbles may be an example. Gribbles (Fig. 9.25) are isopod crustaceans (family Limnoriidae) that ingest wood and digest its cellulose components. The guts of gribbles are sterile,

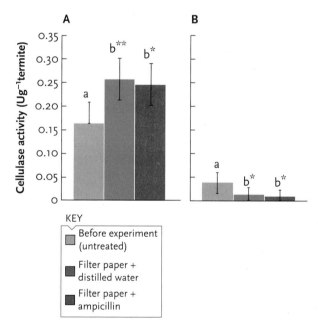

KEY

☐ Before experiment (untreated)

■ Filter paper + distilled water

■ Filter paper + ampicillin

Figure 9.24 Antibiotics affect cellulase activities in the termite *Nasutitermes takasogoensis* (886). Cellulase activity in **A.** a crude extract of the midgut and **B.** a crude extract of the hindgut. Bars show means of five extracts ± one standard deviation. In each plot, b denotes a statistically significant difference from a.

SOURCE: Tokuda, G., H. Watanabe, "Hidden cellulases in termites: revision of an old hypothesis," *Biology Letters*, 3:336–339, 2007, by permission of the Royal Society.

Simon Cragg/University of Portsmouth

Figure 9.25
Gribble (*Limnoria quadripunctata* (806)).

showing no evidence of endosymbionts. However, gribbles appear to produce cellulases without the need of endosymbionts. The transcriptome of *Limnoria quadripunctata* (806) is dominated by glycosyl hydrolase genes, and >20% of the expressed sequence tags (ESTs) are associated with genes encoding cellulases. This situation is rarely reported in animal genomes: glycosyl hydrolase genes are usually associated with wood-degrading fungi or protists symbiotic in the guts of termites. Hemocyanins may also be involved in the digestion of lignin by gribbles.

STUDY BREAK

1. Which animals digest cellulose? How do they do it?
2. Why do leeches and vampire bats appear to have similar symbionts?

9.4 More than Energy

Not all ingested materials are used as a source of energy. Chemicals or even structures in food can be used for other purposes. Ciliary tracts in the stomachs of some sea slugs (*Elysia chlorotica* (466), Mollusca) extract chloroplasts from their food. These are moved into the sea slug's skin and used as solar panels (see page 271). In other sea slugs (e.g., *Glaucus* spp. (453)), the same mechanisms extract unexploded nematocysts from ingested tentacles of Portuguese-man-of-war, which are moved to the slug's antennae and used in defense. In addition to these spectacular examples, many animals routinely extract alkaloids and other compounds from their food and use them in defense (e.g. cardiac glycosides from milkweed extracted by the caterpillars of Monarch butterflies (*Danaus plexippus* (1031)) and retained into adulthood. Such extractions are common among herbivores, but also extend to carnivores.

The skin glands of frogs and toads often contain chemicals that are toxic to predators. The list of toxins is impressive, including amines, peptides, proteins, and steroids. Most of these are produced by the amphibians themselves, setting them apart from alkaloids that are derived mainly from food. With the exception of steroidal samandarine alkaloids found in (and apparently produced by) European salamanders, the origin of water-soluble alkaloids (tetrodotoxins) remains unclear, although their production in amphibians may be associated with symbiotic micro-organisms. Lipid-soluble alkaloids of amphibians derive mainly from their prey. They are ingested and then moved to and sequestered in specialized skin glands. The ability to sequester lipid-soluble alkaloids has evolved in four distinct phylogenetic lineages (Fig. 9.26B), including poison-dart frogs (Dendrobatidae), *Melanophryniscus* (1372) (a genus of bufonid toad), a mantellid frog from Madagascar, and *Pseudophryne* (1378), a genus of Australian toadlets. Most recently, lipid-soluble alkaloids have been reported in a tiny eleutherodactylid frog from Cuba (Fig. 9.26A). The pumiliotoxins in these tiny frogs (*Eleutherodactylus iberia* (1370)) appear to have come from mites the frogs had eaten, but the source of indolizidines is not clear. Pumiliotoxins are also known from poison-dart frogs.

In Chapter 17 (Behaviour), we will see other examples of animals that sequester chemicals for use in defense.

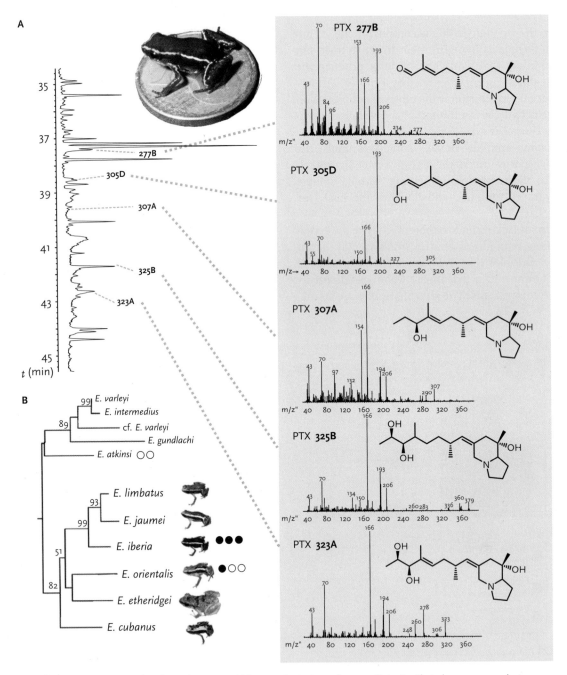

Figure 9.26 A. The diminutive Cuban frog (shown on a U.S. penny) produces five pumiliotoxins that show up as peaks in a gas chromatogram (in addition to PTX325B which is an artefact of processing). **B.** The phylogenetic relationships of the frog (*Eleutherodactylus iberia* (1370)) are shown. The number of black circles identifies the number of samples with alkaloids, the open circles, the number without.

SOURCE: Rodriguez, A., D. Poth, S. Schulz and M. Vences, "Discovery of skin alkaloids in a miniaturized eleutherodactylid frog from Cuba," *Biol. Lett.* published online 3 November 2010, by permission of the Royal Society.

1. What factors other than food do some sea slugs get from their diets?
2. How do some animals obtain chemicals they use in defense?

DIGESTION **IN PERSPECTIVE**

The importance of intestinal biotas in the digestion of food materials is a recurring theme among animals. The obvious examples of this are animals that use fermentation to break down cellulose (or other plant structural components), but the importance of gut biotas to most other animals is an interesting side to digestion. The occurrence of venom in animals is another example of assists not only to feeding (immobilization of prey) but also digestion. The extraordinary challenges facing animals that make long, non-stop migrations adds yet another dimension to studies of digestion and maintaining access to food.

Questions

Self-Test Questions

1. Which of these characteristics is included in regional specialization of the digestive tract?
   a. Histological differences between areas of storage, digestion, and absorption
   b. The presence of symbiotic bacteria that aid in chemical breakdown
   c. Gastroliths and other specialized structures which promote digestion
   d. Vacuoles where intracellular digestion occurs

2. Why are food items typically broken down into smaller pieces through mechanical processes?
   a. To enable the animal to use its mouthparts or gizzard
   b. To increase the surface area exposed to digestive enzymes
   c. To slow the rate of digestion and absorption
   d. Because smaller pieces are more readily stored within the digestive tract

3. What factors are typically involved in the control of the rate of digestion?
   a. Neural reflexes mediated by the Central Nervous System coordinate motility with gut contents.
   b. Chemical signals, including hormones, correlate sensory feedback about gut contents with the release of digestive enzymes and absorption of nutrients.
   c. Leptin and other circulating chemical signals that provide information about the state of digestion of gut contents.
   d. Only internal chemical signals are involved in the control of digestive processes.

4. Which of the following correctly pairs a mechanism of transport with molecules likely to be moved in such a manner?
   a. transcytosis; sugars
   b. paracellular absorption; fatty acids
   c. cotransport; glucose
   d. simple diffusion; amino acids

5. What role do endosymbiotic bacteria play in the digestive process?
   a. They break down the amino acid tryptophan, which is involved in nitrogen balance.
   b. They secrete antibiotics.
   c. They break down volatile fatty acids, vitamins, and proteins.
   d. They prevent foregut fermentation of cellulose in herbivores.

Questions for Discussion

1. What happens when omnivores turn to a more specialized diet? Consider this from the point of view of the omnivore that becomes a herbivore or a carnivore.

2. Pick an animal that appears to have a specialized diet. What adaptations does it show for digestion of its food? A good choice would be an animal that eats blood.

3. Why do studies of animals that hibernate offer interesting models for people studying eating disorders?

4. What mechanisms do animals use to endure long periods of fasting? What are the consequences of this behaviour?

5. Use the electronic library to find a recent scientific paper about the importance of digestion.

PHYLUM CHOANOFLAGELLATA

- A protistan phylum
- Viewed as the sister group to the Metazoa
- Collar flagellated cells

Reference Number	Genus, Species	Common Name	Page Reference
100	*Monosiga brevicollis*		41

PHYLUM PORIFERA

- The sponges
- Numerous well-differentiated somatic cell types but no organization into tissues, no nervous system
- Body structured around channels, in which flagellae drive a unidirectional flow of water, and chambers, lined by choanocytes (cells resembling choanoflagellates) that collect food particles

Class Calcarea

- Calcium carbonate (calcite) skeletal spicules
- Asconid, syconid, or leuconid body plan

Reference Number	Genus, Species	Common Name	Page Reference
101	*Sycon* sp.		50f
102	*Sycon coactum*		213f

Class Demospongiae

- Leuconid body plan
- Silica spicules
- Often with spongin fibres

Reference Number	Genus, Species	Common Name	Page Reference
111	*Agelas* sp.		50f
112	*Aplysina aerophoba*		213
113	*Asbestopluma hypogea*	a carnivorous sponge	214, 215f
114	*Chondrocladia gigantea*		17f
115	*Halichondria panicea*		215f
116	*Spongilla lacustris*	a freshwater sponge	366
117	*Tethya aurantia*	orange puffball sponge	272f

Class Hexactinellida

- Leuconid body plan
- 3- or 6-rayed silica spicules that are frequently fused to form a complex skeleton
- Cells are syncytial

Reference Number	Genus, Species	Common Name	Page Reference
125	*Acanthascus* sp.		50f
126	*Euplectella* sp.	Venus' flower basket	25f

PURPLE PAGES

PHYLUM CNIDARIA

- Hydroids, jellyfish, sea wasps, anemones, and corals
- Diploblastic animals (ectoderm and endoderm layers but no mesoderm)
- A nerve net but no ganglia or central nervous system
- Phylum name comes from *cnid*, meaning "nettle" and referring to the stinging nematocysts that are characteristic of members of the phylum

Class Hydrozoa

- Life cycle typically includes both polyp and medusa
- Acellular mesoglea
- No nematocysts in endodermal layer

Reference Number	Genus, Species	Common Name	Page Reference
130	*Aglantha*		52f
131	*Clava leptostyla*	club-headed hydroid	113, 113f
132	*Clytia*		35
133	*Clytia (Phialidium) gregarium*		402f
135	*Hydra* spp.		52, 52f, 116, 191, 386, 386f, 424, 448
136	*Hydractinia* spp.	snail fur	33, 33f, 281, 281f
137	*Muggiaea atlantica*		33
138	*Obelia*		52f
139	*Physalia* sp.	Portugese man-of-war	33, 33f
140	*Tubularia*	pink-hearted hydroid	52f

Class Scyphozoa

- The jellyfish; the name means "cup animals"
- Medusa stage dominant
- Polyp buds off multiple medusae

Reference Number	Genus, Species	Common Name	Page Reference
150	*Aurelia aurita*	moon jelly	51f, 108f
151	*Mastigias papua*	lagoon jelly	203f

Class Cubozoa

- Box jellyfish or sea wasps
- The medusa is cube-shaped.
- A simple polyp buds off a single medusa.
- Nematocysts produce an extremely potent venom, and some species can cause human death.

Reference Number	Genus, Species	Common Name	Page Reference
155	*Chironex fleckeri*	sea wasp	51f, 203
156	*Chiropsella bronzie*	box jellyfish	435f
157	*Tripedalia cystophora*	box jellyfish	432, 448

Class Anthozoa

- Name means "flower animals"
- Anemones, corals, sea pens, etc.
- Nematocysts on the numerous septae that divide the gastrocoel
- Ectodermally lined pharynx extends inward from the mouth

ORDER PENNATULACEA: SEA PENS

Reference Number	Genus, Species	Common Name	Page Reference
161	*Ptilosarcus gurneyi*	a sea pen	51f, 151f

ORDER ACTINARIA: SEA ANEMONES

Reference Number	Genus, Species	Common Name	Page Reference
165	*Actinoscyphia* sp.	Venus fly trap anemone	17f
166	*Aiptasia pallida*		270f
167	*Adamsia carciniopados*	cloak anemone	281
168	*Calliactis tricolor*	hermit anemone	281
169	*Condylactis* sp.	pink tip anemone	231
170	*Heteractis* sp.	magnificent or Ritteri anemone	275f
172	*Metridium senile*	plumose anemone	16f, 231, 231f
173	*Nematostella vectensis*	starlet sea anemone	386, 386f, 399, 399f
174	*Urticina felina*	dahlia anemone	16f, 26, 51f, 231, 231f

ORDER SCLERACTINIA: TRUE CORALS

Reference Number	Genus, Species	Common Name	Page Reference
180	*Acropora millepora*	staghorn coral	114, 180
181	*Acropora palmata*	elkhorn coral	98, 506
184	*Monastrea* sp.	star coral	51f
185	*Pocillopora* sp.	cauliflower coral	276
186	*Porites* sp.	finger coral	284

PHYLUM CTENOPHORA

- The sea gooseberries
- Swim through the action of plates (ctenes) of fused cilia
- Colloblasts (lasso cells) for prey capture

Reference Number	Genus, Species	Common Name	Page Reference
195	*Pleurobrachia*	sea gooseberry	52f
196	*Beröe*	melon jelly	52f

THE LOPHOTROCHOZOA

ACOELA AND CATENULIDA

The acoels and catenulids are not formally recognized as phyla but have been separated, as independent clades, from their historical placement in the Turbellaria.

Acoela

Reference Number	Genus, Species	Common Name	Page Reference
200	*Amphiscolops* sp.		54f
201	*Convolutriloba longifissura*		40
202	*Convoluta roscoffensis*		271, 272f
203	*Praeconvoluta castinea*		53, 54f

Catenulida

Reference Number	Genus, Species	Common Name	Page Reference
208	*Stenostomum unicolor*		54f

XENOTURBELLIDA

- Simple, ciliated bilaterians
- No brain (they do have a statocyst and a diffuse nerve net), gut, or excretory system
- No differentiated gonads
- Eggs and embryos described in follicles within the animal

Reference Number	Genus, Species	Common Name	Page Reference
210	*Xenoturbella*		47, 47f, 53, 70, 74

PHYLUM PLATYHELMINTHES

- The flatworms: free-living and parasitic forms, including tapeworms and flukes
- Triploblastic body plan with a single opening into a simple gut and mesenchyme cells packing the body
- No circulatory or respiratory systems
- Protonephridial excretory system present
- Nervous system has anterior ganglia and longitudinal nerve trunks
- Parasitic groups (Monogenea, Digenea, and Cestoda) are included within the Turbellaria in modern phylogenies (see Chapter 3); they are listed separately here for convenience

Class Turbellaria

- Mostly free-living flatworms
- There is no cuticle, and the monociliated epidermal cells (syncitial in a few species) are supported on the basement membrane
- Gut lined by phagocytic cells
- All are simultaneous hermaphrodites
- Can divide or bud asexually
- Mostly aquatic species (both marine and freshwater), a few species in damp terrestrial environments

ORDER MACROSTOMIDA

Reference Number	Genus, Species	Common Name	Page Reference
215	*Macrostomum*		8, 9f

ORDER POLYCLADIDA

Reference Number	Genus, Species	Common Name	Page Reference
221	*Prostheceraeus giesbrechtii*		55f
222	*Pseudobiceros gloriosus*		15, 15f
223	*Pseudoceros* sp.		30f, 55f

ORDER TRICLADIDA

Reference Number	Genus, Species	Common Name	Page Reference
230	*Dugesia*		54f, 448
231	*Microplana terrestris*		54f
232	*Planaria*		54f
233	*Polycelis auricularia*		432

ORDER UROSTOMIDA

Reference Number	Genus, Species	Common Name	Page Reference
235	*Urastoma cyprinae*		55f

Class Monogenea

- Mostly ectoparasites, commonly on fish
- Attach to the host using an anterior prophaptor and posterior opisthaptor with the latter usually the larger and more complex structure
- Direct life cycle with no intermediate host

Reference Number	Genus, Species	Common Name	Page Reference
242	*Discocotyle* sp.		287f
243	*Polystoma* spp.		279, 285, 285f
244	*Pseudacanthocotyla pacifica*		287f
245	*Sphyranura* sp.		287f

Class Cestoda

- The tapeworms
- Gut parasites of vertebrates
- Terminal attachment structure, the scolex, commonly viewed as anterior although this is now questioned (see Chapter 3)
- Scolex has hooks and/or suckers for attachment to the host gut wall
- Body divided into repeating units, proglottids, each of which contains a complete set of hermaphroditic reproductive organs
- Body covered by a syncytial tegument
- Life cycle involves one or more intermediate hosts

Reference Number	Genus, Species	Common Name	Page Reference
250	*Dibothriocephalus latus*		290
251	*Hymenolepis* spp.	rat tapeworm	282, 282f, 283, 284, 286
253	*Platybothrium tantalum*		287f
254	*Pseudanthobothrium hanseni*		283f
255	*Rhinebothrium* sp.		287f
256	*Taenia solium*	pork tapeworm	287f

Class Digenea

- The flukes
- Parasites of the gut and other internal organs of vertebrates
- Two attachment suckers, one around the mouth and the other (acetabulum) ventral
- Mouth opens to a muscular pharynx that usually leads to a two-branched gut
- Body covered by a syncytial tegument
- Most species protandrous hermaphrodites
- One or more intermediate hosts with first intermediate host commonly a gastropod mollusc
- Asexual multiplication of larval stages is common

Reference Number	Genus, Species	Common Name	Page Reference
260	*Euhaplorchis californiensis*	a trematode	126f
261	*Gigantobilharzia* sp.		288
262	*Himasthla*	a trematode	125, 126f
263	*Microbilharzia*		288
264	*Podocotyloides stenometra*		284
265	*Schistosoma* spp.		284, 288, 524f
266	*Trichobilharzia* sp.		288, 288f

PHYLUM GASTROTRICHA

- Freshwater and marine, bilaterally symmetrical acoelomates
- Ciliated epidermis, commonly a pair of posterior projections equipped with adhesive glands
- Through gut with separate pharynx and intestine
- Protonephridial excretory system
- Simultaneous hermaphrodites, some species are parthenogenetic and may have degenerate, or missing, male organs
- Eggs hatch as miniature adults
- Eutelic

Reference Number	Genus, Species	Common Name	Page Reference
270	*Chaetonotus*		55f

PHYLUM CYCLIOPHORA

- Commensal on the mouthparts of lobsters where there may be several hundred on a single seta
- Suspension feeders with food particles drawn into an anterior buccal funnel that may be replaced several times as the animal grows
- Sac-like body, posterior stalk, and an adhesive attachment disc
- Ciliated U-shaped gut with anus near the base of the buccal funnel
- Complex life cycle with sexual and asexual phases that can be coordinated with the moulting of the lobster host

Reference Number	Genus, Species	Common Name	Page Reference
275	*Symbion* spp.		58f, 279, 280f

PHYLUM ENTOPROCTA

- Also known as the Kamptozoa (Goblet worms)
- Sessile, solitary, or colonial and mostly marine
- Name Entoprocta comes from position of the anus inside the ring of ciliated tentacles that collect food into mucus and move it on the inner face of the tentacles to the mouth
- Reproduction may be by asexual budding
- Some species hermaphroditic, others dioecious
- Planktotrophic larva common

Reference Number	Genus, Species	Common Name	Page Reference
278	*Pedicellina*		64f

PHYLUM BRYOZOA

- The moss animals
- Also known as the Ectoprocta (the anus is outside the ring of tentacles)
- Marine and freshwater, typically sessile colonial animals
- Individual zooids enclosed in an exoskeleton that may be soft, chitinous, or mineralized
- Hollow, ciliated tentacles extended to collect food
- Colonies often polymorphous with specialized zooids for feeding, reproductive, cleaning, and defensive functions
- Varying levels of communication between zooids
- Typically hermaphroditic with sperm being released via pores in the tentacles, eggs retained and fertilized *in situ*
- Free-swimming larva

Reference Number	Genus, Species	Common Name	Page Reference
280	*Bugula* sp.		34f
281	*Bugula neritina*		269, 269f
282	*Celleporaria sherryae*		34f
283	*Cupuladria* sp.		35f
284	*Membranipora membranacea*		64f, 386
285	*Nematoflustra* sp.		35f
286	*Plumatella* sp.		64f
287	*Watersipora*		34f
288	*Zoobotryon*		35f

THE GNATHIFERA

A clade that includes the Micrognathozoa, the Gnathostomulida, and the Syndermata (Rotifera and Acanthocephala). The clade is defined by the similarity of the internal jaw structure of animals in these groups.

THE MICROGNATHOZOA

- Known from a single species from Greenland, now also known from the Antarctic
- Ciliated animal, less than 1 mm long,
- Complex pharyngeal jaw apparatus that includes three pairs of jaws
- Known only as females although protandry has been suggested
- Thin-walled eggs hatch rapidly, thick-walled eggs overwinter

Reference Number	Genus, Species	Common Name	Page Reference
290	Limnognathia maerski		57, 58f

THE GNATHOSTOMULIDA

- Small, ciliated acoelomates from the interstitial marine fauna
- Mouth just behind the head
- Jaws used to scrape bacteria from sand grains
- Cross-fertilizing hermaphrodites

Reference Number	Genus, Species	Common Name	Page Reference
293	Gnathostomulida spp.		56f, 219f
294	Rastrognathia sp.		56f

THE SYNDERMATA

The Syndermata is a clade created to include the Rotifer and Acanthocephala, both previously viewed as phyla.

ROTIFERA

- Aquatic animals, mostly in freshwater, named for the arrangement of cilia, forming a ring or corona around the mouth
- Pharyngeal jaws, known as trophi (see Fig. 8.31B)
- Syncitial epidermis, protonephridial excretory system
- Rotifers are dioecious although many species reproduce parthenogenetically

Class Monogonata

Reference Number	Genus, Species	Common Name	Page Reference
297	Asplanchna spp.		208, 208f, 219
298	Brachionus sp.		219
299	Collotheca ornata		57f
300	Epiphanes		57f
301	Floscularia sp.		57f
302	Sinatherina socialis		57f

Class Bdelloidea

Reference Number	Genus, Species	Common Name	Page Reference
305	*Philodina* sp.		56f, 220f

Class Seisonidae

Reference Number	Genus, Species	Common Name	Page Reference
307	*Seison africanus*		57f

ACANTHOCEPHALA

- Parasitic worms named from the eversible proboscis, armed with spines that provide attachment to the host
- No gut, food taken up from the host through body wall
- Found as parasites in vertebrate and some invertebrate hosts
- Complex life cycle with a crustacean as the usual first intermediate host
- Dioecious

Reference Number	Genus, Species	Common Name	Page Reference
310	*Corynosoma* sp.		287f
311	*Moniliformis dubius*		409
312	*Polymorphous* spp.		288, 288f, 289f
313	*Pomphorhynchus bulbocilli*		287f

PHYLUM ANNELIDA

Segmented worms (segmentation lost in some groups). Recent work on annelid phylogeny has invalidated the traditional divisions of the Annelida, as well as bringing groups previously viewed as phyla to positions in the Annelida. The arrangement used here recognizes current divisions of the phylum but does not assign these divisions to a taxonomic level.

Polychaeta: Errantia

- Errant means travelling (as in knights errant) and refers to swimming, burrowing, or crawling worms
- Well-differentiated head with sensory appendages
- Body segments all similar with parapodia, supported by aciculae and bearing setae. Mostly marine predators
- Dioecious, external fertilization
- Trochophore larva

Reference Number	Genus, Species	Common Name	Page Reference
314	*Abarenicola pacifica*		303f
315	*Alvinella* spp.		264, 266
316	*Aphrodita aculeata*	sea mouse	303
317	*Glycera* spp.	blood worm	61, 220, 221f
318	*Hermodice carunculata*	bearded fireworm	61, 62f
320	*Myrianida pachycera*		17
321	*Nereis* sp.		31, 31f, 147, 182, 220
322	*Nereis diversicolor*	clam worm	426f
323	*Nereis virens*	clam worm	61, 62f, 221f
324	*Platynereis dumerilii*		434
326	*Scolelepis* spp.		108f

Polychaeta: Sedentaria

- Non-motile polychaetes living in tubes or permanent burrows
- Head often with suspension feeding tentacles and sometimes gills
- Body frequently tagmatized (different regions specialized for different functions)
- Mostly marine suspension feeders
- Dioecious, external fertilization
- Trochophore larva

Reference Number	Genus, Species	Common Name	Page Reference
330	*Boccardia proboscidea*	shell worm	124, 125f
331	*Chaetopterus* sp.*	parchment worm	31, 32f, 61
332	*Eunice* spp.	sand worm	220
333	*Filogranella elatensis*		62f
334	*Hydroides dianthus*		114
335	*Hydroides elegans*		114, 115f
336	*Myxicola infundibulum*	slime fan worm	30f, 62f, 511, 511f
337	*Myzostoma**		61f
338	*Pectinaria koreni*	trumpet worm	62f
340	*Polydora* spp.		281
342	*Spirorbis*		339
343	*Spirorbis borealis*		232f
344	*Spirobranchus* sp.	Christmas tree worm	108f
345	*Thelepus (Amphitrite) crispus*	spaghetti worm	62f
346	*Thelepus (Amphitrite) ornata*		303f
347	*Thelepus (Amphitrite)* spp.		218, 338

*NOTE: The Chaetopteridae and Myzostomidae are now placed with the Sipunculida (below) as groups branching from the base of the annelid stem.

Clitellata: Oligochaeta

- Earthworms and their relatives
- Detritivores (few species predatory); terrestrial and freshwater with a few marine species. Setae on all segments except first (the prostomium)
- Some segments modified to form glandular clitellum
- Cross-fertilizing hermaphrodites
- Direct development

Reference Number	Genus, Species	Common Name	Page Reference
355	*Diplocardia floridana*		439
356	*Diplocardia mississippiensis*		439
357	*Lumbricus* spp.	earthworm	63f, 217, 217f
359	*Megascolex australis*		16, 16f
360	*Tubifex*		39

Clitellata: Hirudinea

- The leeches
- Segmented annelids, mostly in freshwater, some marine, few in wet terrestrial habitats. Fixed segmental pattern: 4 form head, 21 mid-body, 7 posterior sucker (32 in all)
- Cross-fertilizing hermaphrodites
- Direct development

Reference Number	Genus, Species	Common Name	Page Reference
366	*Haemopis marmorata*	horse leech	447
367	*Hirudo medicinalis*	medicinal leech	255, 506, 525

Echiura

- The spoon worms
- Formerly regarded as a phylum, now included in the Annelida
- Burrowing marine worms
- Highly extensile proboscis extends to collect food
- Gut coiled, anus posterior
- Some setae
- Dioecious, gametes released via metanephridia, external fertilization
- Trochophore larva

Reference Number	Genus, Species	Common Name	Page Reference
371	*Bonellia viridis*	green spoonworm	63f, 232, 394
372	*Echiurus* sp.	spoonworm	63f

Sipunculida

- The peanut worms
- Formerly regarded as a phylum, now included in the Annelida
- Marine, coelomate worms that show no segmentation in the adult
- Body divided into a posterior trunk and anterior introvert that can be completely inverted and retracted within the trunk
- Tentacles at tip of introvert collect food
- Coiled intestine loops forward to anus near the trunk–introvert junction
- Metanephridial excretory system
- Dioecious, release gametes for external fertilization
- Eggs hatch as trochophore larvae

Reference Number	Genus, Species	Common Name	Page Reference
375	*Sipunculus* sp.		62f
376	*Themiste* sp.		62f

Siboglinida

- The beard worms
- Formerly the Phylum Pogonophora (and in some classifications the Pogonophora was split to make the Siboglinidae and the Vestimentifera separate phyla)
- Benthic worms living in a chitin tube
- Four-part body with anterior cephalic lobe bearing cluster of thin tentacles, glandular forepart involved in tube secretion, elongate trunk and posterior opisthosoma that is segmented and bears setae
- No mouth, gut, or anus in adult
- Depend on a symbiotic relationship with bacteria and oxidation of sulphides for energy (see Chapter 10)

Reference Number	Genus, Species	Common Name	Page Reference
380	*Lamellibrachia* sp.		63f, 264, 265f
381	*Osedax* sp.	zombie worm	63f, 266
382	*Osedax rubiplumus*	zombie worm	266f
383	*Riftia* sp.	giant tubeworm	63f, 264, 303

Continued

Reference Number	Genus, Species	Common Name	Page Reference
384	*Riftia pachyptila*	giant tubeworm	263, 263f, 264f, 328, 328f
385	*Tevnia*		264

PHYLUM MOLLUSCA

- Snails, clams, squid, and their relatives
- Dorsal body wall secretes calcareous spicules or shell(s)
- A pocket off the foregut secretes a toothed ribbon, the radula (absent in bivalves)

Uncertain

Reference Number	Genus, Species	Common Name	Page Reference
388	†*Acaenoplax*	A Silurian fossil form	147, 147f

Class Neomeniiomorpha (the Solenogastres)

- Foot is reduced in the mid-ventral groove of the spicule-covered body
- Predators on hydroids and other attached fauna

Reference Number	Genus, Species	Common Name	Page Reference
390	*Alexandromenia crassa*		59f

Class Chaetodermomorpha (the Caudofoveata)

- Vermiform animals with no foot
- Gills in a posterior chamber

Reference Number	Genus, Species	Common Name	Page Reference
392	*Chaetoderma*		59f

Class Polyplacophora

- The chitons
- Shell is divided into 7 or 8 plates
- The divided shell gives flexibility to shape to a rock surface
- Browsers on the microalgal and bacterial film coating marine substrata

Reference Number	Genus, Species	Common Name	Page Reference
395	*Acanthopleura granulata*	a chiton	59f
396	*Katharina tunicata*	black Katy chiton	399f
397	†*Polysacos vickersianum*		157
398	*Sypharochiton pelliserpentis*	snake-skinned chiton	85

Class Monoplacophora

- Single-shelled, limpet-like animals
- Serial repetition of some organs and structures
- Well known in the fossil record
- Now known from deep oceans (hence *Neo* = new pilina, named after *Pilina*, a fossil from the Silurian period)

Reference Number	Genus, Species	Common Name	Page Reference
400	*Neopilina* sp.		61f

Class Gastropoda

- *Gastro* (stomach) *poda* (foot)
- Typical gastropods have a spiral shell and, as a result of torsion, the mantle chamber (enclosing the gills) is anterior
- Gastropoda is in a state of flux: old subclasses, the Prosobranchia, Opisthobranchia, and Pulmonata not now valid
- Prosobranchs divided into a number of clades
- Opisthobranchia and Pulmonata are recognized as being paraphyletic
- Old subclass names are retained here as a way of providing convenient groupings

PROSOBRANCHS

Patellogastropoda

- Limpets (simple conical shells)

Reference Number	Genus, Species	Common Name	Page Reference
405	*Lottia*		35
406	*Nacella concinna*	a limpet	84
407	*Patella* spp.	limpets	236

Vetigastropoda

- Keyhole limpets, slit shells, abalone, and turban snails
- Paired nephridia and bipectinate gills (the right gill may be reduced or absent)

Reference Number	Genus, Species	Common Name	Page Reference
410	*Calliostoma ligatum*	blue top snail	398f
411	*Haliotis rufescens*	red abalone	113, 113f
412	*Margarella antarctica*	a top snail	84
413	*Trochus niloticus*	a top shell	81
414	*Turbo brunneus*	dwarf turban snail	81

Caenogastropoda

- Most of the living gastropods
- Monopectinate gill, usually a pair of eyes at the base of the single pair of tentacles, and an operculum that closes the opening of the coiled shell.

Reference Number	Genus, Species	Common Name	Page Reference
415	*Buccinum* spp.	waved whelk	280
416	*Cerithidea californica*	California horn snail	125
418	*Conus californicus*	California cone snail	237f
419	*Conus geographus*		237f
420	*Crepidula fecunda*		111f, 236f
421	*Crepidula* spp.	slipper limpet	236
422	*Crepipatella* spp.	a slipper limpet	111
423	*Littorina* spp.	periwinkle	280
424	*Nucella lapillus*	dog whelk	78
425	*Nucella* spp.	dog whelk	280

Continued

Reference Number	Genus, Species	Common Name	Page Reference
427	*Polinices duplicatus*	moon snail	192, 280
429	*Strombus lentiginosus*	a conch	60f
430	*Thais tuberosa*	knobbed rock shell	81
431	*Urosalpinx* spp.	oyster drill	236

OPISTHOBRANCHS

- *Opistho* (behind) *branch* (gills): gills, when present, are behind the heart. Two pairs of tentacles.
- In many, the shell is reduced or lost in the adult stage
- Several independently evolved groups in the artificial group Opisthobranchia

Order Anaspidae

Reference Number	Genus, Species	Common Name	Page Reference
440	*Aplysia*	sea hare	236, 467, 495
441	*Aplysia californica*	sea hare	511f
442	*Aplysia juliana*		236f

Order Gymnosomata

Reference Number	Genus, Species	Common Name	Page Reference
445	*Clione limacina*	naked sea butterfly	179f

Order Nudibranchia

Reference Number	Genus, Species	Common Name	Page Reference
450	*Chromodoris* sp.		15, 16f
452	*Flabellina iodinea*	Spanish shawl nudibranch	302f
453	*Glaucus atlanticus*	sea swallow	257
454	*Glossodoris symmetricus*		15, 15f
455	*Hermissenda crassicornis*	a nudibranch	511f
456	*Nembrotha kubaryana*	a nudibranch	60f
457	*Onchidoris bilamellata*	rough mantled doris	113, 113f, 405
459	*Pteraeolidia ianthina*		17

Order Saccoglossidae

Reference Number	Genus, Species	Common Name	Page Reference
465	*Alderia willowi*	a sea slug	124, 124f
466	*Elysia chlorotica*	emerald sea slug	11, 12, 12f, 257, 271, 271f

PULMONATA

- Gas exchange is over a vascularized mantle (a "lung") rather than over the gills
- Common as land and fresh water snails

Reference Number	Genus, Species	Common Name	Page Reference
469	*Achatina glutinosa*	African snail	6of
470	*Australorbis (Biomphalaria) glabratus*		525
471	*Balea*	a land snail	174
472	*Biomphalaria* spp.		525
473	*Bulinus forskalii*		525
474	*Bulinus globosus*		439f, 525
475	*Cepaea nemoralis*	grove or brown-lipped snail	81, 81f
476	*Euglandina* spp.	wolf snail	236
477	*Helix aspersa*	garden snail	447
478	*Limax maximus*	leopard slug	431f, 447
479	*Mandarina*	a land snail	100
480	*Viviparus*	river snail	127

Class Cephalopoda

- Molluscs with a well-developed head and the foot modified to form tentacles

NAUTILOIDEA

- Chambered shell with the animal living in the outer (largest) chamber

Reference Number	Genus, Species	Common Name	Page Reference
485	*Nautilus pompilius*	chambered nautilus	432
486	*Nautilus scrobiculatus*		6of

COLEOIDEA

- Cuttlefish, octopus, and squid
- Shell reduced or absent

Reference Number	Genus, Species	Common Name	Page Reference
490	*Amphioctopus marginatus*	veined octopus	497, 497f
491	*Architeuthis dux*	giant squid	221
492	*Argonauta argo*	argonaut	324
494	*Dosidicus gigas*	Humboldt squid	201
495	*Enteroctopus dofleini*	giant Pacific octopus	18f, 134f
496	*Euprymna scolopes*	Hawaiian bobtailed squid	6of, 267
497	*Gonatus onyx*	clawed armhook squid	489
499	*Haliphron atlanticus*	giant octopus	324
500	*Hapalochlaena maculosa*	blue-ringed octopus	6of, 477f
501	*Histioteuthis heteropsis*	jewel squid	364f

Continued

Reference Number	Genus, Species	Common Name	Page Reference
502	*Illex illecebrosus*	short-finned squid	201, 426
503	*Loligo* spp.	squid	340, 426f, 511
504	*Loligo (Doryteuthis) bleekeri*	Bleeker's squid	87
505	*Loligo (Doryteuthis) pealei*	longfin squid	199, 201, 441
506	*Loligo opalescens*	California market squid	201, 340
507	*Lolliguncula brevis*	Atlantic brief squid	201
508	*Loligo vulgaris*	squid	200f
509	*Mesonychoteuthis hamiltoni*	colossal squid	221f
510	†*Nectocaris pteryx*		143, 143f, 145
511	*Octopus defilippi*		426f
512	*Octopus mototi*	Mototi octopus	477f
513	*Octopus salutii*		426f
514	*Octopus vulgaris*	common octopus	426f
515	*Ocythoe tuberculata*	football octopus	324
516	*Sepia apama*	Australian giant cuttle	60f, 410
517	*Sepia* spp.	cuttle	340, 426f
518	*Sepia officinalis*	cuttle	179f
519	*Sepiola* spp.		268
521	*Todarodes*	flying squid	426f
522	*Watasenia scintillans*	firefly squid	437

Class Scaphopoda

- The tooth shells
- Tapered shell, open at both ends
- Foot adapted for burrowing
- Tentacles capture food particles

Reference Number	Genus, Species	Common Name	Page Reference
525	*Dentalium* sp.	tooth shell	60f

Class Bivalvia

- "Bivalve" describes the two-part hinged shell that is typical of the group
- The foot of bivalves is modified for burrowing
- The gill area is expanded in most bivalves for collection of food particles

The Opponobranchia

- Named for the opposite arrangement of gill filaments along the gill axis
- Gills are not enlarged for feeding

ORDER SOLEMYOIDA

- A group of opponobranchs that depend on chemoautotrophy and have a reduced gut

Reference Number	Genus, Species	Common Name	Page Reference
526	*Solemya velum*		264
527	*Solemya borealis*	boreal awning clam	241
528	*Solemya parkinsoni*		241

The Autolamellibranchia

- All bivalves that have their gills enlarged for food collection

Subclass *Pteriomorpha*

- Epibenthic bivalves. Many attach to the substratum by byssal threads

Reference Number	Genus, Species	Common Name	Page Reference
529	*Adamussium colbecki*	Antarctic scallop	202
530	*Aequipecten opercularis*	queen scallop	215
531	*Argopecten gibbus*	Atlantic calico scallop	424f
532	*Bathymodiolus childressi*	a cold seep mussel	264, 265
534	*Crassostrea gigas*	Pacific oyster	517
535	*Crassostrea virginica*	American oyster	55f, 517
536	*Mytilus edulis*	blue mussel	215, 517
537	*Ostrea edulis*	mud or edible oyster	517
538	*Placopecten magellanicus*	Atlantic sea scallop	60f, 202f

Subclass *Heteroconchia*

ORDER UNIONIDA: FRESHWATER CLAMS

Reference Number	Genus, Species	Common Name	Page Reference
540	*Lampsilis reevesiana*	broken ray mussel	116f
541	*Lasmigona compressa*	creek heelsplitter mussel	116f

ORDER ARCHIHETERONDONTA

- Infaunal clams that lack siphons

Reference Number	Genus, Species	Common Name	Page Reference
543	†*Eucrassatella speciosa*	Gibb's clam	165f

ORDER EUHETERONDONTA

- A very large and variable group of bivalves with most shells lacking nacre

Reference Number	Genus, Species	Common Name	Page Reference
545	*Calyptogena magnifica*		165f
546	*Calyptogena soyoae*		405
547	*Driessina polymorpha*		527
548	*Lasaea subviridis*		390
549	*Phaxas pellucidus*	razor clam	193f
550	*Tridacna* sp.	giant clam	55f, 270

PHYLUM NEMERTEA

- The ribbon worms
- Mostly marine (few species in fresh water and damp terrestrial niches)
- Eversible proboscis in a separate body cavity: the rhynchocoel
- Through gut and simple blood circulatory system
- Once viewed as acoelomate, now recognized as coelomate
- Most dioecious and fertilization external

Reference Number	Genus, Species	Common Name	Page Reference
560	Amphiporus		338f
561	Cerebratulus lactuca	milky ribbonworm	63f
563	Cerebratulus marginatus		233, 233f
564	Micrura alaskensis		119f
565	Nipponnemertes pulcher		63f
566	Planktonemertes		338f
567	Tubulanus		338f

THE LOPHOPHORATA

PHYLUM BRACHIOPODA

- The lamp shells
- Bivalve shell (calcium carbonate or calcium phosphate) has dorsal and ventral valves (in bivalve molluscs the shell valves are left and right)
- Attached or burrowing marine coelomates
- Ciliated feeding tentacles arranged as a "U"-shaped lophophore
- Most species do not have an anus

Reference Number	Genus, Species	Common Name	Page Reference
570	Laqueus californianus		65f

PHYLUM PHORONIDA

- Unsegmented tube worms from low oxygen environments
- Live in chitin tube. Ciliated crown of tentacles forms a "U"-shaped lophophore
- U-shaped gut with anus outside lophophore
- Stomach in swollen region at the bottom of the tube
- Reproduce asexually by budding
- Some hermaphrodite, others dioecious

Reference Number	Genus, Species	Common Name	Page Reference
575	Phoronis	a horseshoe worm	310
576	Phoronis australis		65f
577	Phoronis vancouverensis		30f
578	Phoronopsis	a horseshoe worm	310

THE ECDYSOZOA

PHYLUM CHAETOGNATHA

- The arrow worms
- Phylogeny of the chaetognaths uncertain, placement among Ecdysozoa is tentative
- Elongate, predatory worms with lateral fins
- Hooked spines on the head give the phylum its name
- Cross-fertilizing hermaphrodites, eggs hatch as miniature adults

Reference Number	Genus, Species	Common Name	Page Reference
580	*Sagitta elegans*	an arrow worm	364
581	*Sagitta marri*		233f
582	*Sagitta* sp.	an arrow worm	233f

ECDYSOZOA: CYCLONEURALIA

PHYLUM NEMATODA

- The roundworms
- Parasitic or free-living and found in all environments
- May be over a million species
- Protective collagenous cuticle
- Longitudinal muscle operates against hydrostatic skeleton (no circular muscle)
- Most are dioecious
- Larva resembles miniature adult

Reference Number	Genus, Species	Common Name	Page Reference
585	*Caenorhabditis elegans*		66f, 108, 395, 432, 511
586	*Haemonchus contortus*		285, 285f
587	*Heterorhabditis bacteriophora*		268, 269f
588	*Meloidogyne incognita*		288
589	*Onchocerca volvulus*		520
590	*Phocanema decipiens*	cod worm	66f
591	*Trichinella* sp.		66f
592	*Trichuris muris*		5

PHYLUM NEMATOMORPHA

- The horsehair worms
- Adults long and thin, free-living and non-feeding
- Larvae parasitic in arthropods (marine species in crustaceans, freshwater and terrestrial species in Coleoptera and Orthoptera)
- No functional gut in adult or larva

Reference Number	Genus, Species	Common Name	Page Reference
596	*Nectonema*		66f
597	*Spinochordodes* sp.		66f

PHYLUM KINORHYNCHA

- Sometimes called mud dragons
- Marine in meiobenthos
- Head has mobile scalids
- Body divided into 11 segments
- Feed on unicellular algae or organic detritus
- Dioecious
- Hatch as miniature adult form

Reference Number	Genus, Species	Common Name	Page Reference
600	*Echinoderes* sp.		66f

PHYLUM PRIAPULIDA

- Known as the penis worms for their resemblance to a human penis
- Burrowing, predatory marine worms
- Body divided into trunk and extensile spiny proboscis
- Simple gut
- Dioecious

Reference Number	Genus, Species	Common Name	Page Reference
605	*Priapulus caudatus*	penis worm	66f
607	†*Treptichnus pedum*		149, 152

PHYLUM LORICIFERA

- Name comes from the lorica, a case, or sheath, surrounding the animal
- Minute marine animals from the interstitial fauna of mud and sand
- Scalids surround the head
- Dioecious, some species parthenogenetic
- Larva resembles adult

Reference Number	Genus, Species	Common Name	Page Reference
610	*Nanaloricus* sp.		67f
611	*Spinoloricus* sp.		67f

ECDYSOZOA: PANARTHROPODA

PHYLUM TARDIGRADA

- Water bears
- Four segments behind the head, each with a pair of lobe-like legs equipped with claws, pads, or suckers
- Pharyngeal stylets used to puncture plant cells to extract food
- Some species parthenogenetic, most dioecious and fertilization external
- Eggs hatch as miniature adults

Reference Number	Genus, Species	Common Name	Page Reference
616	*Cornechiniscus* sp.		67f
617	*Echiniscoides*		67f

Continued

Reference Number	Genus, Species	Common Name	Page Reference
618	*Echiniscus* sp.		139
619	*Halobiotus crispae*		220f
620	*Macrobiotus hufelandii*		139f
621	*Panagrolaimus* sp.		67f
622	*Paramacrobiotus richtersi*		139f

PHYLUM ONYCHOPHORA

- The velvet worms
- Caterpillar-like animals in damp terrestrial environments
- Metamerically segmented with a pair of lobe-like legs on each of 13–43 segments
- Head has sensory antennae and openings of slime glands used to trap prey
- Tracheal respiratory system lacks valves to close spiracles, allows water loss and restricts possible habitats
- Flexible body wall
- Dioecious and fertilization internal
- Eggs hatch as miniature adults (some species viviparous)

Reference Number	Genus, Species	Common Name	Page Reference
625	†*Hallucigenia*		153
626	*Peripatoides novaezealandiae*	velvet worm	128f
627	*Peripatopsis capensis*	velvet worm	305f
628	*Peripatus*	velvet worm	127

PHYLUM ARTHROPODA

- Literally "joint-footed" animals, usually translated as joint legged
- Segmented animals in all environments, with a chitinous exoskeleton that is moulted for growth

†Anomalocarida
- Large predators with a great claw from the Cambrian era

Reference Number	Genus, Species	Common Name	Page Reference
630	†*Anomalocaris*		145, 154, 155f
631	†*Opabinia*		153, 154f

†Marrelomorpha
- Fossil group from the Cambrian to Devonian, soft bodied

Reference Number	Genus, Species	Common Name	Page Reference
635	†*Mimetaster hexagonalis*		147

SUBPHYLUM CHELICERATA

- Horseshoe crabs, scorpions, harvestmen, spiders, ticks, mites, and their relatives
- No antennae, segment two has pair of chelicera, segment 3 a pair of pedipalps
- Typically a cephalothorax and abdomen (fused in the mites)

Class Pycnogonida

- The sea spiders
- Marine predators on hydroids and bryozoans
- Four pairs of legs with gut diverticulae extending into the legs
- Small body
- Chelifores, at base of proboscis, may not be homologous with chelicerae of rest of the subphylum.

Reference Number	Genus, Species	Common Name	Page Reference
641	*Nymphon* sp.	a sea spider	134f
642	*Pycnogonum*	a sea spider	69f

Class Merostomata

- Heavy protective cuticle
- Body divided into prosoma and opisthosoma with long caudal spike
- Pair of chelicera and five pairs of legs

ORDER XIPHOSURA

- The horseshoe crabs

Reference Number	Genus, Species	Common Name	Page Reference
645	*Limulus polyphemus*	horseshoe crab	69f, 307
646	†*Mesolimulus*	horseshoe crab	159

Class Arachnida

- Scorpions, spiders, mites, and their relatives
- Mostly terrestrial and carnivorous
- Have chelicera, pedipalps, and four pairs of legs
- Dioecious, internal fertilization by transfer of a spermatophore

ORDER ACARI: MITES AND TICKS

Reference Number	Genus, Species	Common Name	Page Reference
650	*Ixodes pacificus*	Western black-legged or deer tick	69f, 521
651	*Ixodes scapularis*	deer tick	521
652	*Neumania papillator*	a water mite	91

ORDER AMBLYPYGIDAE: WHIPSPIDERS

Reference Number	Genus, Species	Common Name	Page Reference
654	*Damon variegatus*	Tanzanian tailless whip scorpion	89f
655	*Phrynus marginemaculatus*	Florida whipspider	293, 294f

ORDER ARANEAE: SPIDERS

Reference Number	Genus, Species	Common Name	Page Reference
659	*Araneus diadematus*	garden or cross spider	69f, 500f
660	*Argiope argentata*	silver argiope	499
661	*Argyroneta aquatica*	diving-bell spider	293f, 309
662	*Bagheera kiplingi*		275
663	*Brachypelma smithi*	a tarantula	184f
664	*Caerostris darwini*	Darwin's bark spider	499
665	*Cyrtophora hirta*	Russian tent-web spider	499
666	*Deinopis ravidus*	net-casting spiders	500f
667	*Eurypelma californicum*	a tarantula	310
668	*Evarcha culcivora*	East African jumping spider	223, 223f, 472
669	*Frontinella pyramitela*	bowl and doily spider	500
670	*Habronattus pugillis*	a jumping spider	97, 97f
671	*Latrodectus hasselti*	redback spider	408
672	*Latrodectus mactans*	black widow spider	366, 366f
673	*Linyphia litigiosa*	Sierra dome spider	500
674	*Mastophora hutchinsoni*	bolas spider	500
675	*Misumenoides formosipes*	crab spider	499f
676	*Nephila antipodiana*	batik orb weaver	500
678	*Nephila* spp.		499f
679	*Opisthoncus parcedentatus*	a jumping spider	223f
680	*Phonognatha graffei*	leaf-curling spider	499
681	*Scytodes thoracica*	a spitting spider	501f
683	*Stegodyphus lineatus*	a wolf spider	500
684	*Tegenaria* sp.	house spider	309
685	*Uloborus glomosus*	feather-legged spider	310

ORDER SCORPIONES: SCORPIONS

Reference Number	Genus, Species	Common Name	Page Reference
692	*Opistophthalmus glabrifrons*		69f, 184f
693	*Parabuthus* sp.		134f
694	*Paruroctonus mesaensis*	giant dune scorpion	439, 439f, 446

SUBPHYLUM MYRIAPODA

- Centipedes, millipedes, pauropods, and symphylans
- Single pair of antennae, simple eyes, mouth ventral with labrum (upper lip), mandibles and maxillae forming lower lip
- Spiracular respiratory system

ORDER CHILOPODA: THE CENTIPEDES

Reference Number	Genus, Species	Common Name	Page Reference
700	*Scolopendra* spp.	red-headed centipede	70f, 184f
702	*Scutigera*	house centipede	174f

ORDER DIPLOPODA: THE MILLIPEDES

Reference Number	Genus, Species	Common Name	Page Reference
705	*Apheloria virginiensis*	Kentucky flat millipede	70f

THE PANCRUSTACEA

- Subphylum Crustacea now recognized as paraphyletic with Hexapoda
- This includes Insecta evolving from a crustacean stem

SUBPHYLUM CRUSTACEA

- Segmental pattern varies from all segments similar (Remipedia) to highly specialized segment groups seen in crabs
- Body divided into head, thorax, and abdomen
- Head and thorax may be fused (cephalothorax)
- Two pairs of antennae
- Appendages typically biramous
- Mostly aquatic but some terrestrial species

Class Remipedia

Reference Number	Genus, Species	Common Name	Page Reference
710	*Godzilliognomus frondosus*		32f
711	*Lasionectes entricoma*		70f

Class Branchiopoda

ORDER ANOSTRACA: FAIRY SHRIMP AND BRINE SHRIMP

Reference Number	Genus, Species	Common Name	Page Reference
715	*Artemia*	brine shrimp	140

ORDER CLADOCERA: WATER FLEAS

Reference Number	Genus, Species	Common Name	Page Reference
720	*Daphnia* spp.	water fleas	93, 389, 389f, 485

Class Maxillopoda

Subclass Branchiura

ORDER ARGULOIDA

Reference Number	Genus, Species	Common Name	Page Reference
722	Argulus japonica		68f

Subclass Copepoda

ORDER CALANOIDA

Reference Number	Genus, Species	Common Name	Page Reference
725	Acartia hudsonica		486
726	Boeckella propinqua		290
727	Calanus newmani		70f, 486
728	Calanus finmarchicus		486f
729	Calanus hyperborealis		210f
730	Calanus hyperboreus		179f
731	Eucalanus pileatus		210, 211f
732	Labidocera sp.		109f

ORDER HARPACTICOIDA

Reference Number	Genus, Species	Common Name	Page Reference
737	Tigriopus japonicus		408, 408f

ORDER SIPHONOSTOMATOIDA

Reference Number	Genus, Species	Common Name	Page Reference
740	Caligus elongatus	sea louse	281
741	Lepeophtheirus salmonis	sea louse	281
742	Lernaeocera branchialis		286

Subclass Pentastomida

- The tongue worms

Reference Number	Genus, Species	Common Name	Page Reference
745	Linguatula arctica		68f
746	Sebekia oxycephala		68f

Subclass Thecostraca

INFRACLASS CIRRIPEDIA

- The barnacles

Reference Number	Genus, Species	Common Name	Page Reference
748	*Balanus glandula*	acorn barnacle	210f
749	*Lepas anatifera*	goose barnacle	210, 210f
750	*Sacculina* spp.	a parasitic barnacle	113, 289, 289f, 489
751	*Semibalanus* spp.	acorn barnacles	70f, 77f

Class Malacostraca

Subclass Phyllocarida

ORDER LEPTOSTRACA

Reference Number	Genus, Species	Common Name	Page Reference
783	*Nebalia*		56

Subclass Hoplocarida

ORDER STOMATOPODA

Reference Number	Genus, Species	Common Name	Page Reference
785	*Hemisquilla californiensis*	mantis shrimp	18f
786	*Odontodactylus scyllarus*	mantis shrimp	234, 234f

Subclass Eumalacostraca

Superorder Peracarida

ORDER AMPHIPODA

Reference Number	Genus, Species	Common Name	Page Reference
789	Caprella sp.	skeleton shrimp	17
790	*Cystisoma* spp.		437
791	*Ericthonius punctatus*		89f
792	*Gammarus* spp.		70f, 288, 289f
793	*Gammarus duebeni*		394
795	*Phronima sedentaria*	pram bug	437

ORDER ISOPODA

Reference Number	Genus, Species	Common Name	Page Reference
803	*Bathynomus giganteus*		218, 218f
805	*Ligia*	sea roach	70f
806	*Limnoria quadripunctata*	gribble	257, 257f
807	*Porcellio dilatatus*	woodlouse	217f, 218

Superorder Eucarida
ORDER DECAPODA

Reference Number	Genus, Species	Common Name	Page Reference
810	*Birgus latro*	robber crab	375, 375f
811	*Callinectes sapidus*	blue crab	179f, 277f, 281, 339, 339f
813	*Carcinus maenas*	shore crab	290f, 374f
814	*Crangon crangon*	common shrimp	70f
815	*Discoplax (Cardisoma) hirtipes*	Christmas Island blue crab	375, 375f
816	*Gecarcinus quadratus*	red land crab	482
817	*Gecarcoidea natalis*	Christmas Island red crab	460, 460f
818	*Geograpsus grayi*	little nipper crab	375
819	*Homarus americanus*	American lobster	31, 32f, 179f, 234, 279, 486, 486f
820	*Hymenocera picta*	harlequin shrimp	18f
821	*Jasus lalandii*	Cape rock lobster	434f
822	*Metopaulias depressus*	bromeliad land crab	133, 134f
824	*Nephrops*	Norway lobster	58, 58f, 279
825	*Notostomus gibossus*	a shrimp	364
826	*Pagurus acadianus*	Acadian hermit crab	70f, 280, 281f
827	*Pagurus* spp.	hermit crabs	280
830	*Palinurus argus*	spiny lobster	447, 486f
831	*Palinurus gilchristi*	spiny lobster	486
832	*Palinurus ornatus*	ornate spiny lobster	486
835	*Potamocarcinus richmondi*	a freshwater crab	482
836	*Procambarus clarkii*	a crayfish	234
837	*Rimicaris* spp.	shrimp	264, 266
838	*Sergia lucens*	Sakura shrimp	486f
839	*Trapezia rufopunctata*	Coral crab	276f
840	*Uca annulipes*	fiddler crab	89f
841	*Uca capricornis*	fiddler crab	88
842	*Uca pugilator*	fiddler crab	234

SUBPHYLUM HEXAPODA

- 1.3 million species at current count; 6 million estimated
- Body divided into head thorax and abdomen
- Single pair of antennae
- Three thoracic segments each bearing a pair of legs
- One or two pairs of wings in many
- Mostly terrestrial and freshwater, few marine species

Class Entognatha

ORDER COLLEMBOLA: THE SPRINGTAILS

Reference Number	Genus, Species	Common Name	Page Reference
845	*Folsomia candida*	a springtail	277

Class Insecta
Infraclass Paleoptera

ORDER EPHEMEROPTERA: THE MAYFLIES

Reference Number	Genus, Species	Common Name	Page Reference
847	*Siphlonurus*	gray drake mayfly	117f

ORDER ODONATA: DRAGONFLIES AND DAMSELFLIES

Reference Number	Genus, Species	Common Name	Page Reference
850	*Aeshna cyanea*	blue darner	202f
851	*Anax imperator*	emporer dragonfly	343f
852	*Calopteryx maculata*	ebony jewelwing	408
853	*Diceratobasis macrogaster*	a damsel fly	133
854	*Libellula quadrimaculata*	four-spotted chaser	343

Infraclass Neoptera

Superorder Exopterygota

ORDER BLATTARIA: COCKROACHES

Reference Number	Genus, Species	Common Name	Page Reference
857	*Arenivaga investigata*	desert cockroach	439
859	*Leucophaea maderae*	Madeira cockroach	426f
860	*Periplaneta americana*	American cockroach	222, 222f, 374

ORDER HEMIPTERA: BUGS

Reference Number	Genus, Species	Common Name	Page Reference
865	*Acyrthosiphon pisum*	a pea aphid	3, 4f, 478
866	*Aphis glycines*	soybean aphid	389f
867	*Aradus*	a flat bug	437, 438f
868	*Cimex lectularius*	bed bug	510
869	*Geocoris* sp.	big-eyed bug	478
870	*Gerris remigis*	water strider	189f
871	*Hydrometra* sp.	water measurer	480f
872	*Icerya purchasi*	cottony cushion scale	262
873	*Macrosiphum liriodendron*	a pea aphid	3
874	*Magicicada*		462f
875	*Rhodnius*	kissing bug	250
876	*Rhodnius prolixus*	kissing bug	187, 355f, 437, 454, 467, 468
877	*Sitobion avenae*	a pea aphid	3
878	*Stiretrus anchorago*	predatory stink bug	480
879	*Triatoma infestans*	barber bug	447, 447f

ORDER ISOPTERA: TERMITES

Reference Number	Genus, Species	Common Name	Page Reference
885	*Zootermopsis angusticollis*	Pacific dampwood termite	256
886	*Nasutitermes takasogoensis*	termite	256f, 257
887	*Reticulitermes flavipes*	eastern subterranean termite	257f
888	*Reticulitermes speratus*	Japanese termite	492

ORDER MANTODEA: MANTIDS

Reference Number	Genus, Species	Common Name	Page Reference
890	*Pseudomantis albofimbriata*	false garden mantid	89

ORDER ORTHOPTERA: GRASSHOPPERS AND LOCUSTS

Reference Number	Genus, Species	Common Name	Page Reference
893	*Acheta domesticus*	house cricket	307
894	†*Archaboilus musicus*		159f
895	†*Cyphoderris buckelli*		159f
896	*Cyrtacanthacris aeriginosa*	green tree locust	185f
897	*Docidocercus gigliotosi*	a katydid	449f
898	*Gryllus campestris*	field cricket	487
899	*Laupala* spp.	Hawaiian crickets	96
900	*Lirometopum coronatum*	pitbull katydid	88, 88f
901	*Neoconocephalus ensiger*	sword-bearing conehead	449f
902	*Oecanthus* spp.	tree cricket	498
903	*Phymateus morbillosus*	milkweed locust	175f
904	*Schistocerca gregaria*	desert locust	510

ORDER PHTHIRAPTERA: LICE

Reference Number	Genus, Species	Common Name	Page Reference
910	*Columbicola* sp.	wing-feather louse	102
911	*Holomenopon brevithoracicum*		167f
912	†*Megamenopon rasnitsyni*		167f
913	*Physconelloides* sp.	body-feather louse	102

ORDER PLECOPTERA: STONEFLIES

Reference Number	Genus, Species	Common Name	Page Reference
915	*Perla marginata*	a stonefly	307

Infraclass Neoptera

Superorder Endopterygota

ORDER COLEOPTERA: BEETLES

Reference Number	Genus, Species	Common Name	Page Reference
920	Agrilinus constans	a dung beetle	251
921	Anoplophora glabripennis	Asian longhorned beetle	256, 256f
922	Astylus atromaculatus	spotted maize beetle	481f
923	Bolitotherus cornutus	forked fungus beetle	478, 478f
924	Brachinus spp.	bombardier beetles	449
925	Calleida viridipennis	green beetle	480
926	Cassida rubiginosa	thistle tortoise beetle	480
927	Ceruchus piceus	a stag beetle	348
928	Chiasognathus grantii	Chilean stag beetle	89f
929	Cicindelida sp.	tiger beetles	223f
930	Cycloneda sanguinea	ladybird beetle	480
931	Deltochilum valgum	a dung beetle	251
932	Dineutus hornii	a whirligig beetle	189
933	Gastrophysa viridula	green dock beetle	188f
934	Goliathus regius	goliath beetle	184f
935	Harmonia axyridis	a ladybird beetle	117f
936	Heliocopris andersoni	a dung beetle	174f
937	Hemisphaerota cyanea	tortoise beetle	480
938	Lasiorhynchus barbicornis	giraffe weevil	89f
941	Melanophila	a buprestid beetle	437, 438f
942	Merimna	a buprestid beetle	437
943	Onthophagus taurus	a dung beetle	121f
944	Photinus sp.	a firefly	307
945	Platynus decentis	a ground beetle	306
946	Rodolia cardinalis	vedalia beetle	262
947	Stenaptinus insignis	African bombardier beetle	449
948	Stenus comma	a rove beetle	189, 189f
949	Tribolium castaneum	flour beetle	284
950	Trypoxylus dichotomus	Japanese rhinoceros beetle	88

ORDER DIPTERA: FLIES

Reference Number	Genus, Species	Common Name	Page Reference
953	Aedes spp.	a mosquito	223, 250, 374, 472f, 520, 522
954	Anopheles spp.	mosquito	11, 472, 521, 522

Continued

Reference Number	Genus, Species	Common Name	Page Reference
955	*Aposthonia gurneyi*		512
956	*Blaesoxipha fletcheri*		278
957	*Calliphora vicina*		188f
958	*Climaciella brunnea*		18f
959	*Chironomus riparius*	harlequin fly	134f
960	*Drosophila* spp.	fruit fly	29, 41, 94, 108, 120, 306, 306f, 345, 395, 419, 425, 425f, 447, 472, 488, 511
961	*Eristalis tenax*	drone fly (a hoverfly)	449
962	*Eurosta solidaginis*	goldenrod gall fly	138f, 349
963	*Glossina morsitans*	tsetse fly	127, 249, 520, 520f
964	*Laphria* sp.	robber fly	88f
965	*Metriocnemus knabi*		278
966	*Musca domestica*	housefly	187, 437
967	*Pseudolynchia canariensis*	louse fly	280f
968	*Sarcophaga argyrostoma*	flesh fly	138f
971	*Teleopsis dalmanni*	a stalk-eyed fly	90f
973	*Toxomerus marginatus*	a hoverfly	449f
974	*Voriella uniseta*	a parasitic tachinid	262f
975	*Wyeomyia smithii*		278

ORDER HYMENOPTERA: ANTS, BEES, AND WASPS

Reference Number	Genus, Species	Common Name	Page Reference
980	*Acromyrmex*		1, 10, 277
981	*Ammophila* sp.	thread-waisted wasp	498, 498f
982	*Aphaenogaster* spp.	a myrmicine ant	498
983	*Apis adamsoni*	African bee	515
984	*Apis mellifera*	honey bee	21f, 133, 135f, 249, 434f, 477, 492, 515, 515f
985	*Apterostigma*		10f
986	*Atta*	a leaf-cutter ant	1, 10, 10f, 277
987	*Atta cephalotes*	a leaf-cutter ant	2f
990	*Camponotus pennsylvanicus*	carpenter ant	306
991	*Cephalotes atratus*	giant gliding ant	195
992	*Ceratosolen* spp.	fig wasps	101f, 275
993	*Copidosoma* spp.		290, 291f
994	*Crematogaster* spp.		275
995	*Dorymyrmex bicolor*	a dolichoderine ant	498
996	*Elisabethiella baijnathi*		274

Continued

Reference Number	Genus, Species	Common Name	Page Reference
997	*Glyptapanteles* sp.	a braconid wasp	291f, 473
998	*Kradibia chinensis*		274f
1000	*Monomorium pharaonis*	pharaoh ant	500
1001	*Myrmecia chrysogaster*	a bull ant	223f
1002	*Nasonia vitripennis*		290, 290f
1003	*Oecophylla* spp.		275
1004	*Phylotrypesis caricae*		275
1005	*Pleistodontes imperialis*		274f
1006	*Polistes humilis*	a paper wasp	18f, 133, 134f
1008	*Pseudomyrmex* sp.		272f, 273
1009	*Sphex* spp.	digger wasp	498
1010	*Synagris cornuta*	tusked wasp	88
1011	*Temalucha minuta*	a parasitic wasp	262f
1012	*Trichogramma brassicae*	a parasitic wasp	279, 280f
1013	*Trichogramma evanescens*	a parasitic wasp	262f
1014	*Wiebesia frustrata*		274f
1015	*Xanthopimpla rhopaloceros*	a parasitic wasp	262f

ORDER LEPIDOPTERA: BUTTERFLIES AND MOTHS

Reference Number	Genus, Species	Common Name	Page Reference
1026	*Atrophaneura hector*	crimson-rose butterfly	223f
1027	*Bicyclus anynana*	squinting bush brown butterfly	485
1028	*Bombyx mori*	silkmoth	122, 137, 512
1029	*Campaea perlata*	pale beauty moth	478
1030	*Cycnia tenera*	dogbane tiger moth	478, 479f
1031	*Danaus plexippus*	monarch butterfly	137, 257, 481, 481f, 483
1032	*Drepana arcuata*	masked birch caterpillar	475, 475f
1033	*Epiblema scudderiana*	a gall moth	349
1034	*Epiphyas postvittana*	light-brown apple moth	117, 261, 261f
1035	*Galleria* spp.	wax moth	121f, 269f
1036	*Gonometa rufobrunnea*		512
1037	*Haploa confusa*	confused haploa moth	478
1038	*Hyposmocoma* spp.	a moth	308, 308f
1039	*Hyposmocoma molluscivora*	a predatory moth larva	117, 117f
1040	*Lymantria dispar*	gypsy moth	137
1041	*Manduca sexta*	tobacco hornworm	190, 478
1042	*Paches loxus*	brilliant blue skipper	80, 80f
1043	*Papilio xuthus*	swallowtail butterfly	120f, 473, 475f
1044	*Pectinophora gossypiella*	pink bollworm	510

Continued

Reference Number	Genus, Species	Common Name	Page Reference
1045	*Pero ancetaria*	Hübner's pero moth	478
1046	*Philosamia cynthia*	Ailanthus silkmoth	249
1047	*Pieris brassicae*	cabbage white butterfly	117, 279, 280f
1048	*Spodoptera litura*	oriental leafworm moth	479, 479f
1049	*Thyrinteina leucocerae*	a geometrid moth	291, 291f, 473

ORDER NEUROPTERA: LACEWINGS

Reference Number	Genus, Species	Common Name	Page Reference
1055	*Cueta punctatissima*	antlion	235f
1056	*Morter obscurus*	antlion	235f

ORDER SIPHONAPTERA: FLEAS

Reference Number	Genus, Species	Common Name	Page Reference
1059	*Ctenocephalides felis*	cat flea	176, 176f

THE DEUTEROSTOMIA

PHYLUM ECHINODERMATA

- The spiny-skinned animals
- Typically pentaradial (5 axes) symmetry in adult, larvae bilaterally symmetrical
- Calcium carbonate skeleton of ossicles
- Water vascular system with tube feet
- Dioecious, fertilization external in most (some sea stars brood eggs under skin)
- Planktonic larvae, different forms in the different classes

Class Crinoidea

- The sea lilies
- Benthic, marine suspension feeders
- Body in three sections: stem, calyx, and arms
- In many the stem vestigial and the animal free-swimming
- "U"-shaped gut in calyx with anus on the edge of the disc
- Dioecious, fertilization external

Reference Number	Genus, Species	Common Name	Page Reference
1065	*Antedon* sp.	feather star	72f
1066	*Florometra serratissima*	a feather star	397f

Class Asteroidea

- The sea stars
- Typically five arms (some species may have up to 50)
- Arms not sharply set off from disc
- Gonads and digestive glands extend into arms
- Mouth on underside of body, anus on opposite surface

Reference Number	Genus, Species	Common Name	Page Reference
1069	*Acanthaster planci*	crown-of thorns sea star	237
1070	*Asterias* sp.	sea stars	237f
1071	*Asterias amurensis*	North Pacific sea star	467, 468f
1072	*Culcita novaeguineae*		276
1073	*Echinaster graminicola*		341
1074	*Henricia*	blood star	72f
1075	*Linckia multifora*		386f
1076	*†Taeniaster beneckei*		147
1077	*Xyloplax* spp.		264

Class Ophiuroidea

- The brittlestars
- Normally five-armed although arms may be extensively branched
- Arms set off sharply from disc
- Arms used for locomotion
- No gonad or digestive gland in arms
- No anus
- Mostly dioecious, a few protandric hermaphrodites
- Fertilization external

Reference Number	Genus, Species	Common Name	Page Reference
1081	*Ophiothrix*	a brittle star	72f

Class Echinoidea

- Sea urchins, heart urchins, and sand dollars
- Skeletal ossicles fused to form a globular or disc-like skeleton
- Pentaradial symmetry clear in arrangement of the tube feet and gonads
- Body surface covered by spines that may be long, some spines modified to form small-jawed structures known as pedicellaria
- Dioecious, fertilization external

Reference Number	Genus, Species	Common Name	Page Reference
1085	*Asterechinus elegans*		264
1086	*Dendraster*	biscuit urchin	113
1088	*Echinarachnius*	sand dollar	113
1089	*Lytechinus*		39f
1090	*Strongylocentrotus*	sea urchin	72f, 222, 380
1091	*Strongylocentrotus droebachiensis*	green sea urchin	222f
1092	*Strongylocentrotus purpuratus*	purple sea urchin	434, 435f

Class Holothuroidea

- Sea cucumbers
- Ossicles reduced and embedded beneath tough body wall
- Typically elongated on the oral–aboral axis

- Modified tube feet at anterior form suspension-feeding tentacles
- Internal respiratory trees open from cloaca inside the anus
- Typically dioecious (some protandric hermaphrodites), fertilization external.

Reference Number	Genus, Species	Common Name	Page Reference
1095	*Cucumaria*	sea cucumber	72f
1096	*Cucumaria frondosa*	orange-footed sea cucumber	212, 212f
1097	*Neothyonidium magnum*		30f
1098	*Synaptula*		35

PHYLUM HEMICHORDATA

- Acorn worms
- Burrowing marine worms
- Typically deposit feeders
- "Chordate" part of their name is inappropriate, coming from a misinterpretation of a dorsal gut diverticulum as a vestigial notochord
- Body has proboscis, collar, and an elongate trunk
- Mouth at junction of proboscis and collar
- Dioecious, fertilization external

Class Enteropneusta
- Typical acorn worms

Reference Number	Genus, Species	Common Name	Page Reference
1100	*Balanoglossus* spp.	acorn worm	108f
1101	*Saccoglossus*	acorn worm	72f

Class Pterobanchia
- Colonial and attached forms

Reference Number	Genus, Species	Common Name	Page Reference
1103	*Rhabdopleura*		72f

PHYLUM CHORDATA

- Chordates are characterized by four main features: notochord, dorsal hollow nerve cord, post-anal tail, and pharyngeal slits.

SUBPHYLUM UROCHORDATA (TUNICATA)
- Marine filter feeders that show chordate characteristics as larvae but lose them in adult forms (except the Class Larvacea).

Class Ascidacea
- Solitary and colonial urochordates
- Mostly filter feeders catching particles in an expanded pharyngeal basket
- External covering (tunic) contains cellulose

Reference Number	Genus, Species	Common Name	Page Reference
1105	*Boltenia*	sea potato	73f
1107	*Botryllus schlosseri*	gold star tunicate	73f, 115, 116f, 341
1108	*Ciona intestinalis*	lemon-tipped sea vase	211f
1109	*Halocynthia*	sea peach	73f
1110	*Lissoclinum patella*		271
1111	*Megalodicopia hians*		17f
1112	*Molgula pedunculata*		17f

Class Thaliacea
- Salps and their relatives
- Pelagic forms, often with complex life cycles

Reference Number	Genus, Species	Common Name	Page Reference
1115	*Pyrosoma*		33f
1116	*Salpa*		73f
1117	*Salpa cylindrica*	a salp	203f

Class Larvacea (Appendicularia)
- Larval characteristics retained in adult
- Secrete an external structure that acts as the food collector

Reference Number	Genus, Species	Common Name	Page Reference
1120	*Oikopleura*		73f, 211, 212f

SUBPHYLUM CEPHALOCHORDATA
- The lancets
- Have a notochord, pharyngeal gill slits, and hollow dorsal nerve cord

Reference Number	Genus, Species	Common Name	Page Reference
1122	*Branchiostoma*		72, 73f, 311
1123	*Branchiostoma lanceolatum*	amphioxus	312

SUBPHYLUM CRANIATA

SUPRACLASS AGNATHA
- Jawless vertebrates that lack paired fins

Class Myxinoidea
- Hag fish (only living representatives)
- No ossified backbone

Reference Number	Genus, Species	Common Name	Page Reference
1125	*Myxine glutinosa*	hagfish	224f

Class Placodermi

- Silurian and Devonian fish
- Head and thorax covered by heavy plates

ORDER ARTHRODIRIFORMES

Reference Number	Genus, Species	Common Name	Page Reference
1130	†*Dunkleosteus*		146f, 224, 225f

SUPRACLASS GNATHOSTOMATA

- Jawed fishes
- Vertebrates with jaws and paired fins

Class Chondrichthyes

- Cartilaginous fishes with jaws and paired fins

Subclass Elasmobranchii

ORDER SELACHII: TYPICAL SHARKS

Reference Number	Genus, Species	Common Name	Page Reference
1131	*Alopias vulpinus*	common thresher shark	343
1132	*Carcharhinus melanopterus*	black tip reef shark	178f
1133	*Carcharodon carcharias*	great white shark	225, 347
1134	*Negaprion brevirostris*	lemon shark	180
1135	*Sphyrna lewini*	hammerhead shark	287f

ORDER BATOIDEA: SKATES AND RAYS

Reference Number	Genus, Species	Common Name	Page Reference
1140	*Dasyatis* sp.	sting ray	178f
1142	*Potamotrygon* spp.	freshwater sting ray	363, 448
1143	*Raja binoculata*	big skate	287f
1144	*Raja radiata*	thorny skate	283f

Class Osteichthyes

- Bony fish with paired fins

Subclass Actinopterygii

- ray-finned fishes

INFRACLASS CHONDROSTEI

- Sturgeon, paddlefish
- Primitive ray-finned fish

Reference Number	Genus, Species	Common Name	Page Reference
1150	*Acipenser fulvescens*	sturgeon	531
1151	*Erpetoichthys*	reedfish or ropefish	315
1152	*Erpetoichthys calabaricus*	ropefish	181, 181f, 315
1153	*Polypterus*	Senegal or gray bichir	315

INFRACLASS HOLOSTEI: GARS, BOWFINS

Reference Number	Genus, Species	Common Name	Page Reference
1155	Amia calva	bowfin	315, 322

INFRACLASS TELEOSTEI: MOST BONY FISH

Reference Number	Genus, Species	Common Name	Page Reference
Family Anguilliformes			
1158	Gymnothorax favagineus	honeycomb moray eel	231f
1159	Gymnothorax moringa	spotted moray	276
1160	Muraena retifera	reticulate moray eel	231f
Family Aulopiformes			
1166	Alepisaurus ferox	lancetfish	167f
1167	Scopelarchus michaelsarsi	bigfin pearleye	436
Family Batrachoidiformes			
1170	Opsanus beta	Gulf toadfish	363
Family Beloniformes			
1172	Cypselurus hiraii	a flying fish	194
1173	Oryzias latipes	Japanese ricefish	517
Family Beryciformes			
1175	Anoplogaster cornuta	fangtooth	314, 314f
1176	Photoblepharon spp.	flashlight fish	268, 268f
Family Characiformes			
1182	Colossoma macropomum	tambaqui	315
1183	Hyphessobrycon pulchripinnis	lemon tetra	411, 411f
1184	Piabucina spp.	a pencilfish	315
1186	Serrasalmus niger	redeye piranha	303f
1187	Serrasalmus sp.	a piranha	182f
Family Clupeiformes			
1190	Clupea pallasii	Pacific herring	440
Family Cypriniformes			
1194	Carassius auratus	goldfish	295t, 440
1195	Catostomus commersoni	white sucker	532
1196	Ctenopharyngodon idella	grass carp	518, 518f
1197	Danio rerio	zebrafish	447
1198	Misgurnus spp.	oriental weatherfish	376, 376f, 377f
1199	Pimephales promelas	fathead minnow	403, 532
Family Cyprinodontiformes			
1203	Gambusia hubbsi	mosquito fish	91, 472
1204	Kryptolebias marmoratus	mangrove rivulus or killifish	372f, 373

Continued

Reference Number	Genus, Species	Common Name	Page Reference
	Family Cyprinodontiformes		
1206	*Phalloceros harpagos*		68f
1207	*Poecilia formosa*	Amazon molly	390, 390f
1208	*Poecilia latipinna*	sailfin molly	390
	Family Esociformes		
1210	*Dallia* spp.	Alaska blackfish	315
	Family Gadiformes		
1213	*Gadus morhua*	Atlantic cod	205, 206f
	Family Gasterosteiformes		
1217	*Gasterosteus aculeatus*	three-spined stickleback	486
	Family Gymnotiformes		
1220	*Borophryne apogon*	net devil anglerfish	392f
1221	*Caulophryne*	hairy anglerfish	392f
1222	*Electrophorus*	electric eel	315
1223	*Electrophorus electricus*	electric eel	445f
1224	*Haplophryne mollis*	soft leafvent angler	392f
1225	*Linophryne argyresca*	an anglerfish	392f
1226	*Melanocetus johnsonii*	humpback anglerfish	392f
1227	*Oneirodes* sp.	dreamer	268
1228	*Photocorynus spiniceps*	an anglerfish	392f
	Family Lophiiformes		
1230	*Diceratias pileatus*	anglerfish	268f
	Family Mormyridae		
1231	*Camplyomormyrus compressirostris*	elephant fish	476
1232	*Campylomormyrus rhynchophorus*	elephant fish	476
1233	*Campylomormyrus tamandua*	wormjawed mormyrid	476
	Family Mugiliformes		
1234	*Mugil cephalus*	striped mullet	201
	Family Myctophiformes		
1236	*Lampanyctus macdonaldi*	rakery lanternfish	436
	Family Osmeriformes		
1238	*Alepocephalus agassizii*	Agassiz' slickhead	432
1239	*Bathylychnops exilis*	javelin spookfish	436
1240	*Mallotus villosus*	capelin	206
1242	*Searsia koefoedi*	Koefoed's searsid	436
	Family Osteoglossiformes		
1245	*Arapaima gigas*	pirarucu or paiche	315
1247	*Osteoglossum* sp.	arawana	445f
	Family Perciformes		
1250	*Allothunnus fallai*	slender tuna	343

Continued

Reference Number	Genus, Species	Common Name	Page Reference
1251	*Ammodytes hexapterus*	Pacific sandlance	181, 181f
1252	*Amphiprion* sp.	clown fish	275, 395, 395f
1253	*Amphiprion ocellaris*	clown anemonefish	182f, 275f
1254	*Amphiprion percula*	orange clownfish	507, 508f
1255	*Anabas* spp.	climbing perch	315, 321
1256	*Astatotilapia burtoni*	Burton's mouthbrooder	491
1257	*Betta splendens*	Siamese fighting fish	315, 315f, 321
1258	*Calamus bajonado*	jolthead porgy	165f
1259	*Chaenocephalus aceratus*	blackfin icefish	348
1260	*Chaetodon* spp.	butterfly fish	284
1261	*Channa*	snakehead fish	315
1263	*Dimidiochromis compressiceps*	Lake Malawi cichlid	225f
1264	*Dormitator*	Pacific fat sleeper	315
1265	*Echiichthyes vipera*	lesser weeverfish	314
1266	*Gasterochisma melampus*	butterfly kingfish	343
1267	*Haemulon* spp.	grunts	506
1268	*Howella sherborni*	Sherborn's pelagic bass	436
1269	*Labeotropheus fuelleborni*	blue mbuna	226f
1270	*Labroides dimidiatus*	bluestreak cleaner wrasse	276f, 395, 395f, 491
1271	*Lepomis macrochirus*	bluegill sunfish	410, 410f, 481
1272	*Macropodus*	paradise fish	321
1273	*Meiacanthus atrodorsalis*	forktail blenny	481, 482f
1274	*Metriaclima zebra*	zebra mbuna	226f
1275	*Mnierpes*	rockskipper	315
1276	*Morone saxatilis*	striped bass	201
1278	*Pelvicachromis taeniatus*	striped kribensis	489, 489f
1279	*Periophthalmodon schlosseri*	giant mudskipper	376
1280	*Periophthalmus* spp.	mudskipper	314, 315, 315f, 376f
1281	*Petenia splendida*	red bay snook	226f
1282	*Plagiotremus laudandus*	bicolour fangblenny	481, 482f
1283	*Polydactylus sexfilis*	six-fingered threadfin	507
1284	*Pseudapocryptes*	elongate mudskipper	315
1287	*Sphyraena barracuda*	barracuda	167f
1288	*Stenotomus chrysops*	scup	180
1289	*Symphysodon* spp.	discus fish	135f
1290	*Tetrapturus* sp.	a marlin	343
1291	*Thunnus maccoyii*	Southern bluefin tuna	98, 333, 343
1292	*Thunnus thynnus*	Northern bluefin tuna	98, 333
1293	*Tilapia natalensis*	tilapia	518f
1294	*Toxotes chatareus*	archerfish	413, 414f
1296	*Trichopodus trichopterus*	blue gourami	321, 322f

Continued

Reference Number	Genus, Species	Common Name	Page Reference
Family Pleuronectiformes			
1305	*Paralichthys dentatus*	summer flounder	121, 123f
1306	*Paralichthys sp.*	a flounder	201
1307	*Pleuronectes platessa*	European plaice	112t
1308	*Pseudopleuronectes americanus*	winter flounder	314
1309	*Reinhardtius hippoglossoides*	Greenland halibut	121, 122f
Family Salmoniformes			
1314	*Coregonus clupeaformis*	lake whitefish	287f
1315	*Oncorhynchus nerka*	sockeye salmon	201
1316	*Salmo salar*	Atlantic salmon	105, 105f, 517
Family Siluriformes			
1323	*Agmus lyriformis*	banjo catfish	314
1324	*Ancistrus spp.*	an armoured catfish	315
1327	*Clarias batrachus*	walking catfish or magur	322
1329	*Heteropneustes fossilis*	singee	322
1330	*Hoplosternum littorale*	atipa	301, 315, 322
1331	*Hypostomus spp.*	an armoured catfish	315
1332	*Lithogenes wahari*	climbing catfish	322
Family Synbranchiformes			
1337	*Monopterus albus*	swamp eel	313
1338	*Monopterus cuchia*	mud eel	322, 322f
Family Syngnathiformes			
1340	*Hippocampus denise*	pygmy sea horse	73f
1341	*Hippocampus erectus*	sea horse	226f
1345	*Hippocampus spp.*	seahorses	83
1346	*Syngnathus acus*	pipefish	226f
1347	*Syngnathus scovelli*	Gulf pipefish	127, 129f
1348	*Syngnathus typhle*	broad-nosed pipefish	127

Subclass *Sarcopterygii*

- lobe-finned fishes

ORDER CROSSOPTERYGII: ANCESTORS OF LAND VERTEBRATES

Reference Number	Genus, Species	Common Name	Page Reference
1351	†*Pederpes*		156, 156f
1352	†*Acanthostega*		156f, 170f
1353	†*Eucritta melanolimnetes*		156
1354	†*Tiktaalik roseae*	tiktaalik	169, 170f

ORDER DIPNOI: LUNGFISHES

Reference Number	Genus, Species	Common Name	Page Reference
1357	*Protopterus dolloi*	slender African lungfish	314, 376, 376f

Class Amphibia

- Amphibians
- Tetrapods with no membrane around the egg

Subclass †Labrynthodontia

ORDER †TEMNOSPONDYLI

- Presumed ancestors of modern amphibians

Reference Number	Genus, Species	Common Name	Page Reference
1360	†Thoosuchus yakovlevi		156f

Subclass Lissamphibia

ORDER ANURA: THE FROGS AND TOADS

Reference Number	Genus, Species	Common Name	Page Reference
1363	Agalychnis callidryas	red-eyed tree frog	123f, 184f, 185, 439
1365	Barbourula kalimantanensis	Bornean flatheaded frog	317, 317f
1366	Bufo bufo	bullfrog	378, 449
1367	Bufo marinus	cane toad	442f
1368	Chiromantis xerampelina	grey foam-nest treefrog	378
1370	Eleutherodactylus iberia	Monte Iberia eleuth	258, 258f
1371	Leptodactylus bufonius	white-lipped frog	378
1372	Melanophryniscus	a toad	258
1373	Nectophrynoides occidentalis	Mount Nimba toad	129
1374	Odorrana tormota	concave-eared torrent frog	443
1375	Pelobates fuscus	a toad	351, 352f
1376	Phyllomedusa sauvagei	waxy monkey frog	378f
1377	Polypedates dennysi	Chinese gliding tree frog	194
1378	Pseudophryne	corroberee frog	258
1379	Pyxicephalus assperses	African bullfrog	377
1380	Rana sp.	a frog	123f, 185f
1381	Rana catesbeiana	bullfrog	156f, 317f
1382	Rana esculenta	edible frog	390
1383	Rana lessonae	pool frog	390
1384	Rana ridibunda	marsh frog	390
1385	Rana sylvatica	wood frog	88, 88f, 349, 349f
1386	Rheobatrachus silus	gastric brooding frog	129, 129f
1387	Scaphiopus couchii	spadefoot toad	243, 379f
1389	Spea multiplicata	spadefoot toad	125, 126f
1390	Telmatobius culeus	Titicaca water frog	317
1391	Tomopterna spp.	sand frog	73f, 378
1392	Xenopus		41
1393	Xenopus laevis	African clawed frog	378, 399

ORDER GYMNOPHIONA: THE CAECILIANS

Reference Number	Genus, Species	Common Name	Page Reference
1400	*Dermophis mexicanus*		16

ORDER CAUDATA: THE SALAMANDERS AND THEIR RELATIVES

Reference Number	Genus, Species	Common Name	Page Reference
1404	*Ambystoma maculatum*	spotted salamander	11, 12, 12f
1405	*Ambystoma mexicanum*	axolotl	156f, 302f
1406	*Ambystoma tigrinum*	tiger salamander	184f, 351
1407	*Amphibolurus muricatus*	Jacky dragon lizard	394, 394f
1409	*†Chunerpeton tianyiensis*		159
1410	*Hynobius keyserlingi*	Siberian salamander	349
1411	*Necturus maculosus*	common mudpuppy	378f
1412	*Plethodon cinereus*	red-backed salamander	123f

Class Reptilia

- Reptiles
- Fully terrestrial
- Lay amniote eggs

ORDER CHELONIA: TURTLES, TORTOISES

Reference Number	Genus, Species	Common Name	Page Reference
1420	*Caretta caretta*	loggerhead turtle	406f
1421	*Chelonia mydas*	green turtle	175f, 178f
1422	*Chelydra serpentina*	snapping turtle	87
1423	*Chelus fimbriatus*	mata-mata turtle	216
1424	*Chrysemys picta*	a painted turtle	349
1425	*Dermatemys mawii*	Central American river turtle	442f
1426	*Dermochelys coriacea*	leatherback turtle	346, 380f
1427	*Pseudemys*	a freshwater turtle	309
1428	*Pseudemys scripta elegans*	red-eared turtle	317f
1429	*Sternotherus odoratus*	stinkpot turtle	184f
1430	*Trachemys scripta*	pond slider turtle	8
1431	*Terrapene* sp.	a box turtle	349

ORDER CROCODILIA: CROCODILES, ALLIGATORS, CAIMANS, GAVIALS

Reference Number	Genus, Species	Common Name	Page Reference
1435	*Alligator mississippiensis*	American alligator	323, 323f
1436	*Caiman crocodilus*	spectacled caiman	317f
1437	*Crocodylus niloticus*	Nile crocodile	73f, 505

ORDER †ICHTHYOSAURIA: ICHTHYOSAURS

Reference Number	Genus, Species	Common Name	Page Reference
1440	†*Shonisaurus*	ichthyosaur	157, 183

ORDER †ORNITHISCHIA

- Dinosaurs with bird-like pelvis (e.g., sauropod)
- Includes coleurosaurs and lines leading to birds

Reference Number	Genus, Species	Common Name	Page Reference
1442	†*Ankylosaurus magniventris*	armoured dinosaur	89f
1443	†*Psittacosaurus*		163, 246f
1444	†*Saurornitholestes*		184f
1445	†*Tyrannosaurus rex*	t rex	160, 162f
1446	†*Triceratops* sp.	triceratops	344

†ORDER PTEROSAURIA: PTEROSAURS

Reference Number	Genus, Species	Common Name	Page Reference
1448	†*Quetzalcoatlus northropi*		170

ORDER †SAURISCHIA: LIZARD-HIPPED DINOSAURS

Reference Number	Genus, Species	Common Name	Page Reference
1450	†*Argentinosaurus*		142f
1451	†*Barosaurus lentus*		142f
1452	†*Brachiosaurus*		141, 142f
1453	†*Camarasaurus supremus*		162f
1455	†*Haplocheirus sollers*		160
1456	†*Shantungosaurus*	duck-billed dinosaur	142f

ORDER RHYNCHOCEPHALIA: THE TUATURA

Reference Number	Genus, Species	Common Name	Page Reference
1458	*Sphenodon punctatus*	tuatara	8, 229, 433, 443

ORDER SQUAMATA: SNAKES AND LIZARDS

Reference Number	Genus, Species	Common Name	Page Reference
1461	*Amblyrhynchus cristatus*	marine iguana	379f
1462	*Anguis fragilis*	slow-worm	8
1463	*Anolis carolinensis*	green anole	100f
1464	*Anolis cooki*		8

Continued

Reference Number	Genus, Species	Common Name	Page Reference
1465	Anolis gundlachi		100f
1466	Anolis krugi		100f
1467	Anolis lineatopus		229
1468	Anolis poncensis		100f
1469	Anolis spp.	West Indian anoles	100
1470	Aspidoscelis (Cnemidophorus) uniparens	grassland whiptail lizard	389, 390f
1471	Basiliscus plumifrons	Jesus Christ lizard	189, 190f
1472	Bitis arietans	puff adder	239, 239f
1473	Bitis caudalis	horned adder	480f
1474	Bothrops asper	fer de lance	439f
1475	Bradypodion spp.	dwarf chamelion	475
1476	Bradypodion atromontanum		476f
1477	Bradypodion caffrum		476f
1478	Bradypodion damaranum		476f
1479	Bradypodion gutturale		476f
1480	Bradypodion taeniabronchum		476f
1481	Bradypodion transvaalense		476f
1482	Carlia spp.	rainbow skink	499
1483	Chamaeleo chamaeleon	a chameleon	184f
1485	Chamaeleo sp.	chameleons	230
1486	Chrysopelea sp.	a flying snake	195, 195f
1487	Coluber constrictor	eastern racer snake	183f
1488	Crotalus oreganus	Pacific rattlesnake	479
1489	Dasypeltis sp.		227, 227f
1490	Diadophis punctatus		16, 16f
1491	Dispholidus typus	boomslang	335
1492	Draco volans	flying dragon	195
1494	Epicrates subflavus	Jamaican boa	182
1495	Erpeton tentaculatum	tentacled snake	427
1496	†Eupodophis desouensi		183, 183f
1498	Gekko gecko	a gekko	187, 187f
1499	Gekko ulikovskii	a gekko	8
1500	Hoplodactylus maculatus	a gekko	129
1501	Huia cavitympanum	hole-in-the-head frog	443
1502	Hydrophis	sea snakes	129
1504	Lampropeltis getula	kingsnake	227, 227f
1507	Leptotyphlops dulcis		16, 16f
1508	Mabuya heathi	a skink	129
1509	Micrurus tener	Texas coral snake	335

Continued

Reference Number	Genus, Species	Common Name	Page Reference
1510	*Naja mossambica*	Mozambique spitting cobra	413f
1511	*Naja nigricollis*	black spitting cobra	413
1512	*Naja pallida*	red spitting cobra	413
1513	*Naja siamensis*	Indochinese spitting cobra	413
1514	*Niveoscincus ocellatus*	spotted skink	129
1515	*Pantherophis guttatus*	rat snake	182
1516	*Pituophis melanoleucus*	gopher snake	480
1517	*Podarcis sicula*	wall lizard	229
1518	*Pseudemoia entrecasteauxii*	White's skink	129
1519	*Ptychozoon* sp.	a gekko	195
1520	*Python molurus bivittatus*	Burmese python	453, 453f
1521	*Python regius*	ball python	505
1522	*Rhinophis drummondhayi*	Drummond-Hay's snake	193, 193f
1525	*Thelotornis capensis*	twig snake	335
1526	*Tupinambis nigropunctatus*	tegu lizard	317f
1528	*Varanus komodoensis*	Komodo dragon	133, 442f

Reptilia or Aves?

Reference Number	Genus, Species	Common Name	Page Reference
1535	†*Xiaotingia zhengi*		171
1536	†*Archaeopteryx lithographica*		160

Class Aves: The Birds

Neognathae

Superorder Neoaves

ORDER APODIFORMES: HUMMINGBIRDS AND SWIFTS

Reference Number	Genus, Species	Common Name	Page Reference
1545	*Calypte anna*	Anna's Hummingbird	249f, 365
1546	*Selasphorus rufus*	Rufous Hummingbird	249f

ORDER BUCEROTIFORMES: HORNBILLS

Reference Number	Genus, Species	Common Name	Page Reference
1548	*Bucorvus leadbeateri*	Southern Ground Hornbill	276
1549	*Ceratogymna elata*	Yellow-Casqued Hornbill	490

ORDER CHARADRIIFORMES : SHOREBIRDS, PLOVERM AND PIPERS

Reference Number	Genus, Species	Common Name	Page Reference
1551	*Actophilornis africana*	African Jacana	276
1552	*Calidris mauri*	Western Sandpiper	253, 254f
1553	*Larus* sp.		197f
1554	*Limosa fedoa*	Marbled Godwit	483
1555	*Limosa lapponica*	Bar-tailed Godwit	254
1556	*Numenius phaeopus*	Hudsonian Curlew	319f
1557	*Numenius tahitiensis*	Bristle-thighed Curlew	497
1558	*Phalaropus* spp.	phalaropes	215, 216
1560	*Sterna paradisaea*	Arctic Tern	77, 483

ORDER CICONIIFORMES: HERONS, EGRETS, IBISES, AND SPOONBILLS

Reference Number	Genus, Species	Common Name	Page Reference
1565	*Ardea* spp.	heron	73f, 342
1566	*Bubulcus ibis*	Cattle Egret	279

ORDER COLUMBIFORMES: DOVES

Reference Number	Genus, Species	Common Name	Page Reference
1570	*Columba livia*	Rock Pigeon	197f, 528
1571	*Ectopistes migratorius*	Passenger Pigeon	509
1572	*†Raphus cucullatus*	Dodo	98, 99f
1573	*Zenaida macroura*	Mourning Dove	136

ORDER CORACIIFORMES: KINGFISHERS, BEE EATERS, MOTMOTS AND ROLLERS

Reference Number	Genus, Species	Common Name	Page Reference
1575	*Ceryle rudis*	Pied Kingfisher	414f

ORDER FALCONIFORMES: FALCONS

Reference Number	Genus, Species	Common Name	Page Reference
1578	*Aquila audax*	Wedge-tailed Eagle	432
1579	*Falco jugger*	Laggar Falcon	194
1580	*Gyps* spp.	a vulture	197f
1581	*Micronisus gabar*	Gabar Goshawk	87
1582	*Neophron percnopterus*	Egyptian Vulture	498
1583	*Stephanoaetus coronatus*	Crowned Eagle	490

ORDER GRUIFORMES: COOTS AND CRANES

Reference Number	Genus, Species	Common Name	Page Reference
1585	*Balearica regulorum*	Grey Crowned Crane	198f
1586	*Grus americana*	Whooping Crane	509

ORDER PASSERIFORMES: PASSERINE BIRDS

Reference Number	Genus, Species	Common Name	Page Reference
1589	*Acridotheres tristis*	Common Myna	528
1590	*Camarhynchus pallidus*	Woodpecker Finch	498
1591	*Catharus guttatus*	Hermit Thrush	484
1592	*Catharus ustulatus*	Swainson's Thrush	484
1593	*Cinnyris osea*	Palestine Sunbird	365
1595	*Corvus moneduloides*	New Caledonian Crow	235, 498
1596	*Cyanocitta cristata*	Blue Jay	235, 481
1597	*Dendroica caerulescens*	Black-throated Blue Warbler	485f
1598	*Dicrurus adsimilis*	Fork-tailed Drongo	490
1599	*Euplectes orix*	Red Bishop	469f
1600	*Ficedula albicollis*	Collared Flycatcher	485
1601	*Ficedula hypoleuca*	Pied Flycatcher	405, 533
1602	*Hirundo rustica*	Barn Swallow	90
1603	*Hylocichla mustelina*	Wood Thrush	483
1604	*Hylophylax naevioides*	Spotted Antbird	405
1605	*Lanius collaris*	Fiscal Shrike	476f
1606	*Loxia curvirostra*	Red Crossbill	473
1607	*Luscinia megarhynchos*	Nightingale	531
1608	*Passer domesticus*	House Sparrow	527, 528
1609	*Phainopepla nitens*	Phainopepla	253
1610	*Philetairus socius*	Sociable Weaver	86, 87f
1611	*Poecile atricapillus*	Black-capped Chickadee	494
1612	*Progne subis*	Purple Martin	410, 483
1613	*Pycnonotus xanthopygos*	Yellow-vented Bulbul	365
1615	*Setophaga* spp.	N. American Wood Warblers	96
1616	*Sturnus vulgaris*	European Starling	347f, 480, 527, 528
1617	*Sylvia borin*	Garden Warbler	254
1618	*Taeniopygia guttata*	Zebra finch	91
1619	*Tijuca atra*	Black-and-gold Cotinga	99, 99f
1620	†*Xenicus (Traversia) lyalli*	Stephens Island Wren	98
1621	*Troglodytes aedon*	House Wren	488
1622	*Zonotrichia leucophrys*	White-crowned Sparrow	254, 404, 404f

ORDER PHOENICOPTERIFORMES: FLAMINGOES

Reference Number	Genus, Species	Common Name	Page Reference
1627	*Phoenicopterus* spp.	flamingoes	216

ORDER PICIFORMES: WOODPECKERS AND TOUCANS

Reference Number	Genus, Species	Common Name	Page Reference
1629	*Indicator indicator*	Greater Honeyguide	516
1630	*Indicator minor*	Lesser Honeyguide	516, 516f

ORDER PROCELLARIIFORMES: SHEARWATERS AND PETRELS

Reference Number	Genus, Species	Common Name	Page Reference
1633	*Diomedea* spp.	Great Albatross	197f, 379f
1634	*Pachyptila* spp.	prions	215
1635	*Pterodroma cookii*	Cook's Petrel	528

ORDER SPHENISCIFORMES: PENGUINS

Reference Number	Genus, Species	Common Name	Page Reference
1640	*Aptenodytes patagonicus*	King Penguin	532
1642	*Pygoscelis adeliae*	Adelie Penguin	277

ORDER STRIGIFORMES: OWLS

Reference Number	Genus, Species	Common Name	Page Reference
1645	*Athene cunicularia*	Burrowing Owl	497, 498f

ORDER SULIFORMES: BOOBIES

Reference Number	Genus, Species	Common Name	Page Reference
1647	*Sula nebouxii*	Blue-footed Booby	344f

Galloanserae

ORDER ANSERIFORMES: DUCKS AND GEESE

Reference Number	Genus, Species	Common Name	Page Reference
1651	*Alopochen aegyptiacus*	Egyptian Goose	407f
1653	*Anas platyrhynchos domestica*	Pekin Duck	324
1654	*Anser anser*	Greylag Goose	324

Continued

Reference Number	Genus, Species	Common Name	Page Reference
1655	*Anser indicus*	Bar-headed Goose	324, 345, 345f
1656	*Anser* spp.	ducks & geese	215
1657	*Aythya affinis*	Lesser Scaup	288f
1659	*Histrionicus histrionicus*	Harlequin Duck	392f
1660	*Oxyura jamaicensis*	Ruddy Duck	527
1661	*Oxyura leucocephala*	White-headed Duck	527

ORDER GALLIFORMES: TURKEYS, QUAIL, MEGAPODS

Reference Number	Genus, Species	Common Name	Page Reference
1663	*Alectoria graeca*	Rock Partridge	510
1664	*Bonasa umbellus*	Ruffed Grouse	253, 253f
1665	*Coturnix coturnix japonica*	Japanese Quail	532
1666	*Gallus gallus*	Red junglefowl	514
1667	*Gallus gallus domesticus*	Domestic chicken	514
1668	*Gallus sonneratii*	Grey junglefowl	515
1669	*Meleagris gallopavo*	Turkey	503

Palaeognathae

ORDER APTERYGIFORMES: KIWIS

Reference Number	Genus, Species	Common Name	Page Reference
1672	*Apteryx australis*	Brown Kiwi	319f

ORDER STRUTHIONIFORMES: OSTRICHES

Reference Number	Genus, Species	Common Name	Page Reference
1674	*Struthio camelus*	Ostrich	442f, 505

Class Mammalia

Subclass Prototheria

ORDER MONOTREMATA: EGG-LAYING MAMMALS

Reference Number	Genus, Species	Common Name	Page Reference
1677	*Ornithorhynchus anatinus*	duck-billed platypus	162
1678	*Tachyglossus aculeatus*	short-beaked echidna (spiny anteater)	93f, 194f

Subclass Metatheria

- Marsupials: Young born at early stages of development (often before appearance of hind limbs), move to a pouch where they attach to a teat and complete development.

ORDER DASYUROMORPHIA: AUSTRALIAN CARNIVOROUS MARSUPIALS

Reference Number	Genus, Species	Common Name	Page Reference
1680	Sminthopsis douglasi	dunnart	321, 321f
1681	†Thylacinus cynocephalus	Tasmanian tiger	98, 99f

ORDER DIDELPHIMORPHIA: AMERICAN OPOSSUMS

Reference Number	Genus, Species	Common Name	Page Reference
1683	Didelphis marsupialis	common opossum	533
1684	Didelphis virginiana	Virginia opossum	163
1685	Marmosa mexicana	Mexican mouse opossum	442f
1686	Glancomys sabrinus	Northern flying squirrel	194

ORDER DIPROTODONTIA: KOALAS, WOMBATS, POSSUMS, WALLABIES, AND KANGAROOS

Reference Number	Genus, Species	Common Name	Page Reference
1688	Macropus agilis	wallaby	185f
1691	Trichosurus vulpecula	brush-tailed possum	130f

ORDER NOTORYCTEMORPHIA: MARSUPIAL MOLES

Reference Number	Genus, Species	Common Name	Page Reference
1695	Notoryctes spp.	a marsupial mole	194f
1696	Notoryctes typhlops	Southern marsupial mole	78, 78f

Subclass Eutheria: Placental Mammals

ORDER AFROSORICIDA: GOLDEN MOLES, TENRECS, AND WATER SHREWS

Reference Number	Genus, Species	Common Name	Page Reference
1698	Chrysochloris sp.	a golden mole	194f
1699	Chrysochloris asiatica	a golden mole	440f
1702	Eremitelpa granti	Grant's golden mole	440f
1703	Neamblysomus julianae	Juliana's golden mole	78, 78f

ORDER CARNIVORA: DOGS, CATS, HYAENAS, VIVERRIDS, MUSTELIDS, BEARS, AND SEALS

Reference Number	Genus, Species	Common Name	Page Reference
1705	*Acinonyx jubatus*	cheetah	12, 12f, 191, 387, 389f, 434f
1706	*Alopex lagopus*	Arctic fox	527
1707	*Callorhinus ursinus*	northern fur seal	406, 406f
1708	*Canis latrans*	coyote	505
1709	*Canis lupus*	gray wolf	514, 531
1710	*Crocuta crocuta*	spotted hyena	407
1711	*Cryptoprocta ferox*	fossa	487, 488f
1712	*Enhydra lutris*	sea otter	498
1713	*Felis catus*	domestic cat	510, 527
1714	*Gulo luscus*	wolverine	531
1715	*Helogale parvula*	dwarf mongoose	490
1716	*Herpestes auropunctatus*	mongoose	527
1717	*Herpestes javanicus*	small Asian mongoose	262
1718	*Lontra canadensis*	N. American river otter	503
1719	*Lutra annectens*	American otter	482
1720	*Lynx canadensis*	lynx	505
1721	*Mellivora capensis*	honey badger	517
1722	*Mirounga leonina*	southern elephant seal	347f
1723	*Mungos mungo*	banded mongoose	494
1724	*Mustela vison*	mink	505
1725	*Paguma larvata*	palm civet	525
1726	*Panthera leo*	lion	73f, 407
1727	*Panthera pardus*	leopard	490
1728	*Procyon lotor*	raccoon	176f, 320, 505
1729	*Pteronura brasiliensis*	giant otter	77f
1730	†*Smilodon* spp.	sabre-toothed cat	228f
1731	*Ursus maritimus*	polar bear	99

ORDER CETARTIODACTYLA: EVEN-TOED UNGULATES (ARTIODACTYLS: PIGS, DEER, ANTELOPES, CAMELS, LLAMAS, AND WHALES)

Reference Number	Genus, Species	Common Name	Page Reference
1740	*Aepyceros melampus*	impala	184f
1741	*Antilocapra americana*	pronghorn antelope	405
1742	*Balaenoptera* spp.	whale	183, 216
1743	*Bison bison*	American buffalo	509
1744	*Bos grunniens*	yak	345f

Continued

Reference Number	Genus, Species	Common Name	Page Reference
1746	*Bubalus arnee*	water buffalo	77
1747	*Camelus dromedarius*	Arabian camel	509
1748	*Capra aegagrus hircus*	goat	527
1749	*Capra pyrenaica*	Spanish ibex	88
1750	*Dama dama mesopotamica*	Persian fallow deer	77
1751	*Damaliscus lunatus*	topi	487
1752	*Delphinapterus leucas*	beluga	168f, 178f, 442f
1754	*Eubalaena glacialis*	N. Atlantic right whale	324
1755	*Gazella dorcas*	Dorcas gazelle	510
1756	*Gazella gazella*	mountain gazelle	510
1757	*Giraffa camelopardalis*	giraffe	333
1758	*Hippopotamus amphibious*	hippopotamus	276
1761	*Megaptera novaeangliae*	humpback whale	279, 280f
1762	*Orcinus orca*	killer whale	2, 4, 5f, 491
1763	*Ovis aries*	Soay sheep	488
1764	*Phacochoerus africanus*	warthog	276
1766	*Physeter macrocephalus*	sperm whale	221, 250, 324
1767	*Rangifer tarandus*	woodland caribou or reindeer	68f, 531
1768	*Sus scrofa*	pig	527
1769	*Syncerus caffer*	Cape buffalo	279, 280f
1771	*Tursiops aduncus*	Indian Ocean bottle nose dolphin	497, 497f
1772	*Tursiops truncatus*	bottle nose dolphin	531

ORDER CHIROPTERA: BATS

Reference Number	Genus, Species	Common Name	Page Reference
1780	*Anoura caudifer*	tailed tailless bat	230
1781	*Anoura fistulata*	tune-lipped nectar bat	230
1782	*Anoura geoffroyi*	Geoffroy's tailless bat	230
1783	*Antrozous pallidus*	pallid bat	494
1784	*Artibeus lituratus*	great fruit-eating bat	249
1785	*Barbastella barbastellus*	barbastelle bat	101, 478
1787	*Carollia perspicillata*	short-tailed fruit bat	94f, 131, 132f, 446f
1788	*Desmodus rotundeus*	vampire bat	198, 229f, 255, 334f, 364, 438
1789	*Dyacopterus spadiceus*	Dyak fruit bat	136
1790	*Epomophorus wahlbergi*	Wahlberg's epauletted bat	321
1791	*Eptesicus fuscus*	big brown bat	431, 446, 494
1792	*Glossophaga soricina*	Pallas' long-tongued nectar bat	249f, 350, 350f
1793	*Kerivoula hardwickii*	Hardwick's woolly bat	279

Continued

Reference Number	Genus, Species	Common Name	Page Reference
1794	Lasionycteris noctivagans	silver-haired bat	485f
1795	Lasiurus borealis	red bat	478
1796	Macrotus californicus	California leaf-nosed bat	364
1797	Macrotus waterhousii	Waterhouse's leaf-nosed bat	196f
1799	Molossus ater	black mastiff bat	132f
1800	Myotis lucifugus	little brown bat	471, 494
1801	Myotis myotis	mouse-eared bat	449
1802	Myzopoda aurita	Madagascar sucker-footed bat	188
1803	Noctilio leporinus	bulldog bat	442f, 502
1804	Nyctalus leisleri	Leisler's bat	478
1805	Otonycteris hemprichii	desert long-eared bat	442f
1806	Perimyotis subflavus	tricoloured bat	471
1808	Phyllonycteris poeyi	Poey's nectar-bat	199f
1810	Pteronotus	mustached bats	444f
1813	Rhinolophus	horseshoe bat	444f, 525
1814	Rhinolophus ferrumequinum	greater horseshoe bat	101f
1815	Rousettus aegyptiacus	Egyptian fruit bat	250, 449
1816	Sturnira lilium	yellow-shouldered bat	469f
1817	Tadarida brasiliensis	Mexican free-tailed bat	364, 510
1818	Thyroptera tricolor	Spix's disk-winged bat	188
1819	Trachops cirrhosus	fringe-lipped bat	494

ORDER CINGULATA: ARMADILLOS

Reference Number	Genus, Species	Common Name	Page Reference
1826	Dasypus hybridus	short-nosed amrmadillo	385
1827	Dasypus novemcinctus	nine-banded armadillo	93f, 385f

ORDER DERMOPTERA: GLIDING LEMURS FROM SOUTHEAST ASIA

Reference Number	Genus, Species	Common Name	Page Reference
1831	Cynocephalus sp.	flying lemur	184f, 195

ORDER †DOCODONTA

Reference Number	Genus, Species	Common Name	Page Reference
1833	†Castorocauda	Mesozoic beaver	162

ORDER LAGOMORPHA: HARES, RABBITS AND PIKAS

Reference Number	Genus, Species	Common Name	Page Reference
1835	*Ochotona princeps*	pika	345f

ORDER PERISSODACTYLA: ODD-TOED UNGULATES, HORSES, TAPIRS, AND RHINOS

Reference Number	Genus, Species	Common Name	Page Reference
1838	*Ceratotherium simum*	white rhinoceros	465
1839	*Diceros bicornis*	black rhinoceros	465, 509
1840	*Equus caballus*	horse	509
1841	*Equus ferus*	wild horse	98
1842	*Equus quagga*	quagga	98, 99f

ORDER PHOLIDOTA: SCALY ANTEATERS AND PANGOLINS

Reference Number	Genus, Species	Common Name	Page Reference
1845	*Manis temminckii*	pangolin	93f

ORDER PILOSA: ANTEATERS AND SLOTHS

Reference Number	Genus, Species	Common Name	Page Reference
1848	*Bradypus* spp.	three-toed sloth	272
1849	*Choloepus* sp.	two-toed sloth	272, 272f

ORDER PRIMATES: LEMURS, MONKEYS, APES, AND HUMANS

Reference Number	Genus, Species	Common Name	Page Reference
1851	*Alouatta* sp.	howler monkey	92
1852	†*Australopithecus afarensis*	Lucy	6, 7f, 75, 75f, 144
1854	*Callithrix jacchus*	common marmoset	532
1855	*Cebus* sp.	Capuchin monkey	184f
1856	*Cercopithecus diana*	Diana monkey	490
1857	*Gorilla gorilla*	gorilla	7f, 98, 496
1858	†*Homo erectus/ergaster*		504
1862	*Homo sapiens*	human	77f
1863	*Macaca mulatta*	rhesus monkey	403
1865	*Pan paniscus*	bonobo	77f, 473
1866	*Pan troglodytes*	chimpanzee	7f, 75f, 76, 473
1867	*Pongo pygmaeus*	orangutan	464f, 529

ORDER PROBOSCIDEA: ELEPHANTS

Reference Number	Genus, Species	Common Name	Page Reference
1870	*Elephas maximus*	Indian elephant	185, 497
1871	*Loxodonta africana*	African elephant	184f, 185, 465, 478f, 497
1872	*Loxodonta cyclotis*	African forest elephant	185

ORDER RODENTIA: RODENTS, RATS, MICE, AGOUTIS, PACAS, JUMPING MICE, BEAVERS

Reference Number	Genus, Species	Common Name	Page Reference
1876	*Bandicota bengalensis*	lesser bandicoot rat	510
1877	*Castor canadensis*	North American beaver	162, 371, 505
1878	†*Castoroides ohioensis*	giant beaver	149f
1880	*Dipodomys* spp.	kangaroo rat	371
1881	*Dipodomys merriami*	Merriam's kangaroo rat	380f
1882	*Dipodomys microps*	chisel-toothed kangaroo rat	380f, 381
1883	*Dipodomys spectabilis*	banner-tailed kangaroo rat	439
1884	*Erethizon dorsatum*	North American porcupine	93f
1885	*Geomys* sp.	gopher	194f
1886	*Heterocephalus glaber*	naked mole rat	446f, 492, 496
1887	*Hystrix indica*	Indian porcupine	510
1889	*Jaculus jaculus*	jerboa	185f
1890	*Mus domesticus*	house mouse	488
1891	*Mus musculus*	mouse	94f
1892	*Papagomys theodorverhoeveni*	giant tree rat	529
1893	*Paulamys naso*	Flores long-nosed rat	529
1894	*Pedetes capensis*	springhare	185f
1895	*Peromyscus* sp.	deer mouse	478
1896	*Peromyscus californicus*	deer mouse	473
1897	*Peromyscus leucopus*	white-footed deermouse	531
1898	*Peromyscus maniculatus*	a deer mouse	409
1899	*Peromyscus polionotus*	oldfield or beach mouse	409
1900	*Psammomys obesus*	fat sand rat	381
1901	*Rattus argentiventer*	rice-field rat	510
1902	*Rattus exulans*	Pacific rat	528
1903	*Rattus norvegicus*	brown rat	282, 488, 527
1904	*Spalax leucodon*	lesser mole rat	77f
1905	*Spelaeomys florensis*	Flores cave rat	529
1906	*Spermophilus* spp.	ground squirrel	479, 487
1907	*Tamias striatus*	chipmunk	246f
1909	*Tympanoctomys barrerae*	plains or red vischaca rat	381

ORDER SCANDENTIA: TREE SHREWS OF SOUTHEAST ASIA

Reference Number	Genus, Species	Common Name	Page Reference
1912	*Tupaia montana*	mountain tree shrew	534, 534f

ORDER SIRENIA: MANATEES AND SEA COWS

Reference Number	Genus, Species	Common Name	Page Reference
1915	*Hydrodamalis gigas*	Stellar's sea cow	98, 99f
1916	*Trichechus* spp.	manatee	323, 323f, 344

ORDER SORICOMORPHA: SHREWS AND MOLES

Reference Number	Genus, Species	Common Name	Page Reference
1918	*Condylura cristata*	star-nosed mole	426, 427f, 447
1919	*Crocidura* spp.	white-toothed shrews	176
1920	*Scalopus aquaticus*	common mole	194f
1921	*Scutisorex* spp.	hero shrew	176

ORDER †TRICONODONTIA

Reference Number	Genus, Species	Common Name	Page Reference
1925	*†Repenomamus giganticus*		163

ORDER †VOLATICOTHERIA

Reference Number	Genus, Species	Common Name	Page Reference
1928	*†Volaticotherium*		163

UNCERTAIN AFFINITY

Reference Number	Genus, Species	Common Name	Page Reference
1930	*†Fruitafossor*		163

Figure10.1
A. *Epiphyas postvittana* (1034), an adult light brown apple moth. **B.** The larval stage (caterpillar) that damages fruit.

SOURCE: David Williams. This material is © State of Victoria Department of Primary Industries. Reproduced by permission.

10 Symbiosis and Parasitism

WHY IT MATTERS

A hundred million dollars a year matters in almost any economic consideration. This is the estimated cost that would result from a full-blown invasion of light brown apple moth ("LBAM," *Epiphyas postvittana* (1034)) to the apple and soft fruit industries of California. LBAM was accidentally introduced from Australia and is now found in several California counties. LBAM attacks apple, pear, peach, apricot, nectarine, citrus, persimmon, cherry, almond, avocado, oak, willow, walnut, poplar, cottonwood, Monterey pine, and eucalyptus. It also feeds on common shrub and herbaceous hosts, including grape, kiwifruit, strawberry, various berries, corn, pepper, tomato, pumpkin, beans, cabbage, carrot, alfalfa, rose, camellia, pittosporum, jasmine, chrysanthemum, clover, lupine, and plantain. The variety of plant hosts used by LBAM is a significant problem in designing control measures, and it is all too easy for a nursery-bought plant to carry the moth into a commercial fruit-growing area.

LBAM infestations can be controlled by chemical pesticides, but biological control methods can be effective for LBAM control. A classical, if light-hearted, illustration of the principle of biological control is in the tale of the old lady who ate a spider to eat a fly, and then

Nina Fatouros

David Williams. This material is © State of Victoria Department of Primary Industries. Reproduced by permission.

David Williams. This material is © State of Victoria Department of Primary Industries. Reproduced by permission.

Figure 10.2 Species being considered for biological control of light brown apple moth. **A.** *Trichogramma evanescens* (1013), an egg parasite; (1011), **B.** The tachinid fly, *Voriella uniseta* (974), that parasitizes larvae at their 5th and 6th instar stage; and **C.** the ichneumon wasp, *Xanthopimpla rhopalocerus* (1015), that parasitizes pupae.

a bird to eat the spider, a cat to tackle the bird, and so on into absurdity. However, such a tale has unfortunate results in real life. Native species in Hawaii have been enormously damaged by the predatory behaviour of introduced mongoose (*Herpestes javanicus* (1717)) that were imported to control introduced rats that had, in the 1800s, become a pest in the sugar cane fields. Such negative experiences (including cane toads and rabbits in Australia) have led to great caution being exercised before a non-native species is introduced to an area to control a pest species. On the other hand, the dramatic success in the control of cottony-cushion scale (*Icerya purchasi* (872)), a serious pest of citrus plantations in California, by the vedalia beetle (*Rodolia cardinalis* (946)) (imported from Australia) demonstrates that biological control can be successful.

Potential biological control species for LBAM are parasitic wasps from different families that attack different stages (egg, early larval instars, late larvae, and pupae) in the LBAM life cycle. Species of *Trichogramma* (an egg parasite, Fig. 10.2A) are being mass reared and released to control LBAM in vineyards in both Australia and New Zealand (where LBAM has invaded). A number of species (e.g., Fig. 10.2B–D), known to attack different life-cycle stages of LBAM in its native Australia, are being investigated as potential biological control agents against LBAM in California.

SETTING **THE SCENE**

Parasites are typically inconspicuous. This makes it easy to not realize that parasitism is the most common mode of life in the animal kingdom. The separation of "parasitology" as a discipline also helps hide the fact that parasitism is simply a specialized mode of feeding. Closely associated with parasitism are relationships described as symbiotic (perhaps equivalent to perfect parasite adaptation) and commensalism (perhaps a stage in the evolution of a parasitic or symbiotic relationship). This chapter looks at the adaptations of symbiotic, commensal, and parasitic animals to their way of life.

10.1 Introduction

The term **symbiosis** (from the Greek *syn* = with, *biosis* = living) is used in two ways. The broad definition describes any close association between organisms in which there is a benefit to at least one of the partners (a definition that includes mutualistic, commensal, and parasitic relationships). A narrow definition restricts the term to long-lasting relationships in which both partners benefit (making symbiosis synonymous with **mutualism**). We use the narrower definition in this discussion, defining symbiosis as describing a relationship between two species that is long lasting (typically life-long), involves close physical and biochemical interactions, and provides benefit to both species involved.

Commensal relationships are an association of two species where one species benefits and the other receives neither benefit nor harm. Included in commensal relationships are **inquilism** (one species uses the other for housing), **phoresy** (one species uses the other as a means of transport), and **metabiosis** (one organism uses something created by an organism of another species after the death of that organism). This simple definition of metabiosis is extended by microbiologists, who use the term to describe a situation where one microorganism prepares a habitat for a successor.

Parasites (from the Greek *parasitos* = one who eats at another's table, coming from *para* = by and *sitos* = food) are species that benefit at the expense of another (the host). While medicine tends to restrict "parasite" to describing eukaryotic species and views harmful bacteria and viruses as pathogens, these prokaryotes are, by definition, parasites.

An interspecific interaction falling between parasitism and predation is termed **parasitoid**. Parasitoids spend a part of their life cycle associated with a host that is ultimately killed by the amount of nutrient consumed by the parasitoid.

STUDY BREAK

1. Are parasitoids more like typical predators or typical parasites?
2. What risks are involved in establishing a biological control program ?

10.2 Symbiosis

All living organisms are involved in some way with a symbiotic interaction with another species—even humans, even those with a sterile, North American lifestyle. Several hundred species of bacteria are normal symbionts associated with a healthy human, especially the gut bacterial flora. These bacteria are present in incredible numbers. For example, a human adult houses about 10^{12} bacteria on the skin, 10^{10} in the mouth, and 10^{14} in the gastrointestinal tract.

This section explores examples of symbiosis between different groups of organisms and the ways in which the symbiotic association benefits the partner species.

10.2a Animal–Bacterial Symbiosis

Chapter 9 introduced animal–bacterial symbiosis directly associated with the digestive system. While some of the associations to be discussed here are related to food and energy supply, there are many other examples of animal–bacterial symbiotic relationships.

10.2a.1 Chemoautotrophy. The biology of the sioglinid worm, *Riftia pachyptila* (384), and its association with sulphide-oxidizing bacteria (see Box 10.1), was the starting point for the identification of chemoautotrophic symbiotic relationships in animals that live near deep ocean hydrothermal vents and, later, in a broad range of other marine environments. Table 10.1 maps the

BOX 10.1 People Behind Animal Biology
Colleen Cavanaugh: Chemoautotrophy

Few young graduate student scientists enjoy a public "*Eureka* moment" that not only overturns current scientific thinking but also becomes the foundation for a highly successful career. Colleen Cavanaugh, a Harvard graduate student, did just that. She was attending a 1980 lecture by Meredith Jones in which he presented findings on *Riftia pachyptila* (384) (Annelida, Sioglinidae, Fig. 1), a sioglinid worm from the then recently discovered deep-sea hot-water vent sites. *Riftia* has no mouth or gut and was assumed to absorb nutrients through its tentacles and body wall. The trophosome (Fig. 2), a large organ in the body cavity that is 40–60% of the body mass, was then thought to be a detoxifying organ, protecting the worm from the high (and

potentially toxic) levels of sulphide and minerals in the vent water. When Jones mentioned that crystals of sulphur had been found in the trophosome, Cavanaugh had her inspiration: "It's perfectly clear that tubeworms have symbiotic sulphide-oxidizing bacteria," she said. Cavanaugh was familiar with sulphide-oxidizing bacteria from previous work in a salt-marsh environment. Jones was unconvinced by Cavanaugh's idea but did supply a sample of Riftia trophosome tissue for investigation.

Fluorescent reagent staining for DNA revealed 3.7×10^9 granules, 3–5 μm in diameter, per gram of trophosome tissue. These granules appeared to be either prokaryotic cells or organelles (e.g., mitochondria)

from eukaryotic tissue. Electron microscopy showed the granules to be prokaryotic cells with a cell wall resembling that of gram-negative bacteria. A positive test for lipopoly-saccharide (characteristic of the outer cell wall of gram-negative bacteria) supported this conclusion.

Important support for Cavanaugh's hypothesis came from the identification of enzymes involved in generating ATP through the oxidation of reduced sulphur compounds (thiosulfate sulphur-transferase (rhodanese), APS reductase, and ATP sulphurylase) in trophosome extracts. A simplified pathway (Fig. 3) shows the route by which sulphide from the vents is taken up by *Riftia* and oxidized in the bacterial cells in the trophosome to provide energy (the equivalent to energy from the sun in photosynthesis) for the synthesis of organic carbon molecules through the fixation of carbon dioxide. *Riftia* hemoglobin is a vital part of this process. Like other hemoglobins, it transports oxygen and carbon dioxide but is also the vehicle by which sulphide (at a separate binding site on the hemoglobin molecule) is carried from sea water to the bacterial symbionts without exposing other tissues to toxic sulphide molecules.

FIGURE 1
R. pachyptila (384). *The tentacle plumes extend from their tubes in a group of worms near a Pacific hydrothermal vent.*

SOURCE: NOAA Okeanos Explorer Program, Galapagos Rift Expedition 2011

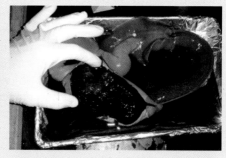

FIGURE 2
R. pachyptila (384). *A dissection of the tropho-some, the organ containing the symbiotic bacteria.*

SOURCE: Photo©Woods Hole Oceanographic Institution

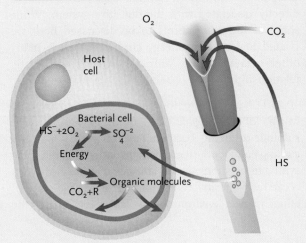

FIGURE 3

A simplified view of the pathways in chemoautosynthesis by Riftia pachyptila (384).

SOURCE: Based on TUNNICLIFFE, V. (1992) "Hydrothermal vent biota: who are they and how did they get there?" *American Scientist*, 80: 336–349; http://www.whoi.edu/oceanus/viewImage.do?id=36968

From her early leap to fame, Cavanaugh went on to an exceptional research career centred on the symbiotic associations between bacteria and marine animals using non-photosynthetic energy sources. She showed that use of sulphur oxidation is not confined to deep-sea vents but was also part of the biology of *Solemya velum* (526) (Mollusca, Gastropoda, Protobranchia), a bivalve living in sulphur-rich, shallow-water, eel-grass beds. She also found that *Bathymodiolus childressi* (532), a mussel from the Gulf of Mexico, uses bacterial symbionts that oxidize methane (as well as sulphide) as a source of energy for chemoautotrophy. She has investigated other bacteria–mollusc symbiotic associations (including the sulphur-oxidizing bacteria in the gill of the giant clam, *Calyptogena magnifica* (545)), as well as the genome of the bacteria involved in symbiotic associations and the routes by which bacteria move from one generation of host to the next. Cavanaugh's demonstration of chemoautotrophy has opened the door to the recognition of non-heterotrophic energy sources in a wide variety of marine animals.

habitats and organisms that are now recognized as depending on sulphur- or methane-oxidizing bacterial symbionts. Notice that in addition to hydrothermal vents and methane seeps, the habitat list includes whale skeletons (see pages 266–267) and wood falls; and shallow water, coral reef, and continental shelf sediments. More than 100 strains of bacteria have been isolated from chemoautotrophic animals.

Siboglinids (*Riftia* (383), *Lamellibrachia* (380), *Tevnia* (385), etc.), other polychaetes (*Alvinella* (315)), and bivalve molluscs (*Calyptogena* (545), and *Bathymodiolus* (532)) were the first vent fauna recognized as symbiont dependent. *Solemya* (526) (bivalve molluscs with either no gut or a reduced gut) was the first shallow-water species added. *Rimicaris* (837), a shrimp (Crustacea, De capoda), was discovered when mid-Atlantic vents were explored. As other sulphur-rich environments were investigated, other species from these groups were recognized and new groups added. Among these are gastropods, gutless nematodes and oligochaetes, a catenulid flatworm (Platyhelminthes, Turbellaria), a poriferan, and ciliates.

An important addition has been the recognition that the Echinodermata includes chemoautotrophic symbionts. Sulphur-metabolizing bacteria and enzymes have been identified in *Asterechinus elegans* (1085), an echinoid that feeds on sunken wood. The unusual asteroids in the wood-associated genus *Xyloplax* (1077) probably depend on similar symbionts.

Sea grasses are an important contributor to coastal productivity in many tropical and warm temperate regions. Seagrass meadows trap dead organic material into deposits that have low oxygen levels, despite the release of some oxygen from the roots of the grass. Bacterial decomposers, using the sulphate in sea water as an electron acceptor, produce sulphides that accumulate in the organic material and may then kill the sea grasses. The system is brought into balance by the presence of lucinid clams that have sulphide oxidizing bacteria as symbionts in their gills. The clams depend on oxygen from the sea grass roots and, through their bacterial symbionts, reduce sulphide levels to a point at which the sea grass survives, and example of an angiosperm–animal–bacterial three-way symbiotic system.

Riftia (383) may grow at a rate of 160 cm a year, an amazing figure for an animal living in the low-energy deep-sea environment. *Lamellibrachia* (380) (Annelida, Polychaeta, Siboglinidae, Fig. 10.3), on the other hand, is extremely slow growing. Worms reaching lengths of 2 m may be 170–250 years old. *Lamellibrachia*, a worm that depends on the oxidation of both sulphide and methane, is associated with methane seeps where sulphide levels in the surrounding water are very low. It obtains sulphide supplies by extending branching "roots" up to 1 m deep into the substratum, where the sulphide levels are 1.5×10^4 higher than in the water. At least four species of endosymbiotic bacteria are present in the trophosome of *Lamellibrachia*.

Riftia and *Lamellibrachia* obtain their symbionts from the water (described as "horizontally"; in vertical transmission the bacteria come directly from the parent). Both have larval stages that, for a brief period,

Table 10.1   Genera living in different marine habitats that have symbiotic relationships with chemoautotrophic bacteria.

Environment	Habitat Type	Phylum	Examples
Shallow water sediments	Marsh and intertidal	Annelida, Oligochaeta	*Tubificoides*
		Mollusca, Bivalvia	lucinid clams, *Solemya*
	Mangrove peat and sediments	Protista, Ciliophora	*Karyorelictea, Zoothamnium*
		Platyhelminthes	*Paracatenulida*
		Nematoda	*Astomonema*
	Sea grass sediments	Protista, Ciliophora	*Zoothamnium*
		Nematoda	*Stilbonematinae*
		Annelida, Oligochaeta	*Olavius*
		Mollusca, Bivalvia	lucinid clams, *Solemya*
	Coral reef sediments	Platyhelmithes	*Paracatenula*
		Nematoda	*Astomonema*
Continental slope sediments		Annelida, Polychaeta	*Siboglinum*
		Annelida, Oligochaeta	*Olavius*
		Mollusca, Bivalvia	*Thyasira, Acharax*
Deep sea cold seeps	Asphalt and petroleum seeps	Annelida, Polychaeta	*Escarpia, Lamellibrachia*
		Mollusca, Bivalvia	*Bathymodiolus*
	Gas seeps and mud volcanoes	Annelida	*Siboglinum*
		Mollusca, Bivalvia	*Bathymodiolus, Acharax, Thyasira, Calyptogena*
Deep sea whale and wood falls	Wood	Annelida, Polychaeta	*Sclerolinum*
		Mollusca, Bivalvia	*Adipicola, Idas*
	Whale bone	Annelida, Polychaeta	*Osedax, Escarpia*
		Mollusca, Bivalvia	*Idas, Adipicola, Vesicomya, Axinodon, Thyasira*
Deep sea hydrothermal vents		Annelida, Polychaeta	*Riftia, Tevnia, Ridgeia, Sclerolinum, Alvinella*
		Mollusca, Bivalvia	*Calyptogena, Bathymodiolus, Acharax*
		Mollusca, Gastropoda	*Alviniconcha*
		Arthropoda, Crustacea	*Rimicaris*

have a functional mouth through which bacteria are taken up.

Calyptogena (545) and *Bathymodiolus* (532) (Fig. 10.4), common bivalve molluscs at vent sites, acquire their bacteria by vertical transmission. In *Calyptogena*

Image courtesy of NOAA Okeanos Explorer Program, Galapagos Rift Expedition 2011

Figure 10.4
A. A bed of *Calyptogena* (Mollusca, Bivalvia, Vesicomyidae) clams near a vent. These clams may be 25 cm long (inset shows a clam with one valve removed to show the tissue colour due to a high concentration of hemoglobin). **B.** *Bathymodiolus* (Mollusca, Bivalvia, Mytilidae) is found near both hydrothermal vents and methane seep sites. The photo includes galatheid crabs and shrimp (*Alvinocaris muricola*). Alvinocarids are common around vents and feed on bacterial films and dead animal material.

NOAA

Figure 10.3 *Lamellibrachia* (380), a slow-growing tube worm at methane seeps.

Dr. Robert W. Embley—PMEL/NOAA

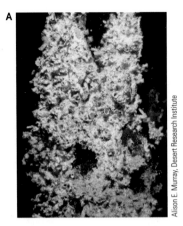

Figure 10.5 *Alvinella pompejana* (315). **A.** A dense colony of A. pompejana surrounds the "Biovent" hydrothermal chimney at 9 degrees North on the East Pacific Rise, image taken from the DSV Alvin in 2004prior to a major eruption the following year. **B.** Close up image of an individual A. pompejana specimen collected by DSC Alvin in 2004.

(Vesicomyidae), chemoautotrophy is sulphide based, and the bacteria are endobiotic in the gills. The red appearance of the gills and mantle of *Calyptogena* (Fig. 10.4A, Inset) is due to hemoglobin in the tissues. Yellow deposits of sulphur are sometimes seen in the gills. *Bathymodiolus* (Mytilidae) occurs at hydrothermal vents and methane seep sites. In this mussel the bacterial symbionts are both endo- and ectobionts, and *Bathymodiolus* can oxidize both sulphide and methane as energy sources.

Hydrothermal vents may be surrounded by dense colonies of the tubeworm *Alvinella pompejana* (315) (Fig. 10.5, named for the research submersible *Alvin*).

The distribution of its symbiont bacteria differs from that of the siboglinid worms. Dense cultures of bacteria are found associated with its tubes and under its cuticle but extracellularly to the ectoderm.

The decapod crustacean *Rimicaris* (837), dependent upon epibiont (bacteria living outside the tissues of other organisms) chemoautotrophic symbionts, is found in swarms around hydrothermal vents on the mid-Atlantic Ridge (Fig. 10.6). *Rimicaris* is mobile, unusual for an animal depending on vent sulphides. This raises the question of how the shrimp maintains its position close to vent chimneys. In its thorax, *Rimicaris* has two unusual organs that are infra-red light receptors (Fig. 10.6B), able to "see" the infra-red radiation emitted by vents.

In 2002 the ROV *Tiburon* discovered *Osedax* (381), a new genus of worms, living on the bones of a grey whale carcass. *Osedax* (the name means "bone eating"; Annelida, Siboglinidae) has no gut and lives with a crown of tentacles extended into the water (Fig. 10.7A). The base of the animal has branched roots that arise from around a posterior ovisac and extend through a hole bored, by the secretion of acid, into the decomposing bone. Unlike *Riftia* and other siboglinids, *Osedax* has no segmented posterior region and no trophosome (the organ in which the chemoautotrophic bacteria are found). The roots of *Osedax* house a culture of bacteria from the Order Oceanospirillales—heterotrophic, aerobic, rod-shaped bacteria that metabolize complex organic compounds. A number of *Osedax* species are now known, and the bacterial cultures in

Figure 10.6 A. *Rimicaris* (837) (Crustacea, Decapoda) is found in dense swarms on the mid-Atlantic Ridge vent sites. **B.** A group of *Rimicaris* showing the dorsal, two-lobed infrared receptor in the thorax.

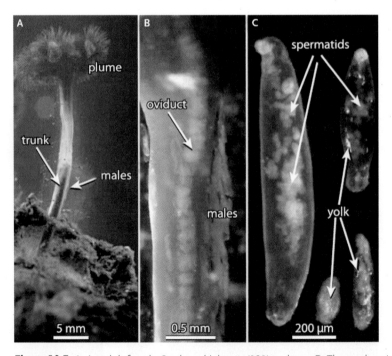

Figure 10.7 A. An adult female *Osedax rubiplumus* (382) on bone. **B.** The trunk region of a female with males beside her oviduct. **C.** Four isolated males showing size variation.

SOURCE: R . C. VRIJENHOEK, S . B. JOHNSON and G . W. ROUSE, "Bone-eating Osedax females and their 'harems' of dwarf males are recruited from a common larval pool," *Molecular Ecology* (2008) 17, 4535–4544. © 2008 The Authors. Journal compilation © 2008 Blackwell Publishing Ltd. John Wiley and Sons.

different species differ slightly. The evidence shows that these bacteria metabolize collagen from the skeletal remains, which is a source of protein for the host. There is no evidence in *Osedax* of the gene that codes for ribulose 1,5-bisphosphate carboxylase/oxygenase, a key enzyme in the Calvin-Benson cycle that characterizes chemoautotrophic siboglinid/bacterial symbiosis in *Riftia* and other siboglinids.

All of the plumed *Osedax* individuals collected from whale skeletons are female, about 10 cm long. A number of 0.2–1 mm long males are found in the gelatinous tube of each female (Fig. 10.9B, C). The males are paedomorphic, and their development is arrested at a metatrochophore (late larval) stage. The number of males found in the gelatinous tube of each female increases as the female grows, with more than a hundred in the tube of a mature female. DNA analysis shows that the males come from a wide pool of larvae and are not the offspring of either their female "host" or of neighbouring individuals.

The history of *Osedax* can be traced back to the appearance of whales in the fossil record (see Chapter 6). *Osedax* worms are known as trace fossils from the bones of Oligocene whales, but molecular clock estimates suggest that they first appeared in the Eocene or Cretaceous. A Cretaceous origin would imply that *Osedax* worms had also fed on the bones of marine reptiles such as mosasaurs or plesiosaurs.

Chemoautotrophy in arthropods may extend back to the late Cambrian and Ordovician. Trilobites in the Family Olenidae may have had a symbiotic relationship with chemoautotrophic symbionts (Fig. 10.8). The sulphur-rich nodules containing the trilobite fossils indicate an original living environment of the fossils, an environment that was oxygen poor and is marked today by the presence of pyrite. In a layered Upper Cambrian shale, olenids occur when pyrite is present but are absent (replaced by an ostracod) in non-pyritic layers.

Olenids appear to have poor development of musculature, indicating that they were not highly active animals. Most trilobites have a hard and well-developed hypostome anterior to the mouth, a structure associated with food handling. The olenid hypostome is atrophied, suggesting that they did not manipulate solid food. Olenids were dorso-ventrally flattened and had both more segments and expanded pleura compared to other groups. The ventral surfaces of the pleura may have provided a culture site for chemoautotrophic symbionts.

10.2a.2 Bioluminescence.
Bioluminescence (living light) is the release of energy in the form of light from a chemical reaction—the oxidation of a luciferin (pigment) molecule (sometimes involving co-factors such as Ca^{++} ions or ATP) to an excited state, catalyzed by the enzyme luciferase. When this excitation relaxes, energy is released as light, typically in the blue-green spectral range.

Examples of bioluminescence are found in almost all animal phyla, and 90% of deep-sea animals have some form of bioluminescence, associated with different roles such as communication, camouflage, and defense. Bioluminescent reactions can involve many different luciferins and luciferases, and may occur within an animal's tissues (e.g., the light organs of fireflies) or depend upon a host's symbiotic relationship with a bioluminescent bacterium.

In many marine animals the bioluminescent bacterium is *Vibrio fischeri* (or *Photobacterium fischeri*). The light enzyme in *V. fischeri*, luciferase, is controlled by the concentration of an autoinducer (N-β-ketocaproyl homoserine lactone) secreted into the environment by the bacterium. When bacterial cells reach high density and the inducer reaches a level above a threshold concentration (a process known as "quorum sensing"), the light is turned on.

A. Euprymna. The Hawaiian bobtailed squid, *Euprymna scolopes* (496) (Mollusca, Cephalopoda, Fig. 10.9), is nocturnally active and uses a ventral light organ to match ambient moonlight and obscure its silhouette to predators living below them. Light receptor molecules

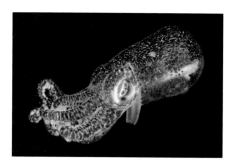

Figure 10.9 *Euprymna* (496), the Hawaiian bobtailed squid.
SOURCE: © David Fleetham / Alamy

Figure 10.8 *Wujiajiania sutherlanda*, a Cambrian trilobite that is thought to have been chemoautotrophic.
SOURCE: Brian Chatterton

associated with the light organ measure ambient light levels and allow light intensity matching.

Euprymna depends on horizontal transfer of bacteria from the environment to acquire its initial culture of *V. fischeri*, starting within seconds after they hatch. An embryonic light organ is connected to the mantle cavity by ducts in which cilia start to beat and drive water from the mantle cavity through the light organ. The water contains many species of bacteria, but in the next stage of the process, bacteria other than *V. fischeri* are inactivated, and *V. fischeri* cells appear to be selected through a receptor-ligand interaction within the light organ. Once the symbiont culture has been established, it appears to trigger the elimination of the parts of the embryonic organ that were important in establishing the relationship. The established culture also triggers development of the features of the adult light organ, including light shields and the lens that gives the bioluminescence a yellow colour.

Daily at dawn, *Euprymna* expels some 95% of the bacteria from its light organ as it stops swimming and burrows into the sand. The bacterial count in the organ rises steadily during the day, and before dusk, when the squid moves back to the water column, the light organ is again full of bacteria. This behaviour maintains an active and healthy culture in the light organ and seeds the environment with bacteria that can be picked up by other squid.

Squid in the genus *Sepiola* (519) live in cooler waters than *Euprymna*. While the light organs of *Sepiola* contain *Vibrio fischeri*, a different species, *V. logei*, is dominant in *Sepiola*. *V. logei* is more luminescent that *V. fischeri* at lower temperatures, an interesting example of the modification of a symbiotic relationship to match environmental conditions.

B. Flashlight Fish. Flashlight fish (*Photoblepharon* spp. (1176), Fig. 10.10) have large bioluminescent organs just below the eyes. The light emitted by these organs, generated by a culture of *V. fischeri*, is bright enough that a school of *Photoblepharon* over a reef may be seen from a considerable distance. By day and on bright moonlit nights, *Photoblepharon* hide in caves in the reef. The bacteria, estimated at 10^8–10^9 bacteria per light organ, are packed in tubules which come together just under the surface of the light organ.

Three patterns of light emission have been described for *Photoblepharon*:

1. The light is on almost all the time, but blinks off for a quarter of a second about three times a minute). This attracts the photophilic crustaceans that the fish eat.
2. The light blinks on for almost a second and then off for a similar time. This behaviour, observed in aquaria, seems to be a background rhythm in undisturbed fish. The blink rate is faster in daytime than at night.
3. In the "blink and run" scenario, the light is on for only 160 ms and off for almost 1000 ms. This is interpreted as a predator avoidance behaviour: the light confuses the predator and erratic swimming allows the flashlight fish to escape during the dark period.

Additional blink patterns are involved in intraspecific communication.

C. Anglerfish. The first dorsal fin ray of female deep-sea anglerfish (e.g., *Oneirodes* sp. (1227), Teleostei, Lophiformes, Fig. 10.11) is modified to form a lure. The lure includes a stalk, the illicium (line), and a globular luminescent bulb, the esca (bait). The esca contains a culture of bioluminescent bacteria (perhaps a species of *Vibrio*). Prey may be attracted to the bioluminescent esca because faecal pellets and organic debris (food for many deep-sea animals) may be weakly bioluminescent. The restriction of the bioluminescent lure to female fish reflects the dimorphism of anglerfish. Males are much smaller than females and, after a brief, free-swimming juvenile phase, attach to females, essentially as parasites, for the rest of their lives (see Fig. 14.16).

D. Insecta. Most bioluminescence in insects (e.g., fireflies that produce their own luciferin and luciferase) does not depend upon bacteria, but a bacteria symbiont makes some insect corpses glow in the dark. *Xenorhabdus luminescens* is a bioluminescent bacterium symbiotic in an insect pathogenic nematode, *Heterorhabditis bacteriophora* (587) (Fig. 10.12). The

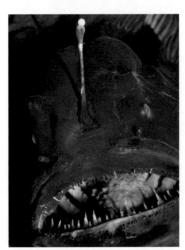

Figure 10.11 An anglerfish (*Diceratias pileatus* (1230)) with bioluminescent lure.

SOURCE: © Doug Perrine / Alamy

Figure 10.10 *Photoblepharon* (1176), the flashlight fish.

SOURCE: © WaterFrame / Alamy

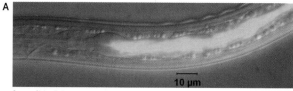

A

10 µm

Source: Ciche, T. The biology and genome of Heterorhabditis bacteriophora (February 20, 2007), *WormBook*, ed. The C. elegans Research Community, WormBook, doi/10.1895/wormbook.1.135.1, http://www.wormbook.org.

B

Marshall W. Johnson

Figure 10.12 A. *Heterorhabditis bacteriophora* (587) with symbiont *Xenorhabdus luminescens*. The bacteria have been stained with a green fluorescent protein. **B.** Luminescent larvae of the wax moth (*Galleria* sp. (1035)) infected with *H. bacteriophora*.

bacterium is cultured in pouches off the intestine of non-feeding stages of the nematode. When the nematode invades an insect prey, the bacteria are released into the host, where they appear to break down host tissues by antibiotic activity and preserve them as food for the nematode. The bioluminescence of the bacterium causes the insect corpse to glow in the dark, attracting other nematodes to feed. Nematodes that have fed and completed their development in an infected host then transfer the bacterium to a new host as their life cycle continues.

10.2a.3 Bacterial Symbionts as Defense Agents. Bryostatins are complex polyketides that are potent cytotoxins. These compounds have been collected from the free-swimming larvae of the bryozoan, *Bugula neritina* (281) (Fig. 10.13). The bryostatins protect the zooplanktonic larvae, which are vulnerable to predation by fish. The toxins are produced by a symbiont bacterium, *Endobugula sertula*, which is housed in depressions on the larval surface. Bryostatins show considerable potential as anticancer drugs and, despite a complex structure, have now been synthesized in the laboratory.

STUDY BREAK

1. How is symbiosis important to sea grass survival?
2. What is luciferin?

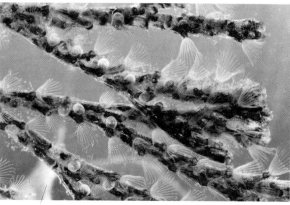

Figure 10.13
Bugula neritina (281) zooids.

SOURCE: Lovell and Libby Langstroth © California Academy of Sciences

10.2b Animal–Algal Symbiosis

While the Great Barrier Reef is a must visit for every tourist to Australian, few pause to realize that the reef is a product of an animal–alga symbiotic relationship. Without the symbiotic association of algal cells and coral polyps, shallow-water tropical reefs would not exist. The coral–algal association must be understood if we are to appreciate anthropogenic threats that currently confront the world's reefs (see Fig. 18.5 and Section 18.2).

10.2b.1 Zooxanthellae. Symbiotic algal cells in cnidarian polyps are known as zooxanthellae, an umbrella term for multiple species of the genus *Symbiodinium* (Fig. 10.14A) in which 160 strains (species?) have been identified. The zooxanthellae are endosymbionts, confined to the endodermal (gastrodermal) cells of the polyps. The algal cell is separated from the cytoplasm of the host cell by a symbiosome membrane, secreted by the host cell (Fig. 10.14B). This membrane regulates the metabolic interactions between the host and symbiont cells.

Zooxanthellae are fundamental to the life and success of reef-building corals. In the coral exchange with its symbionts (Fig. 10.14C), the coral host transports inorganic nitrogen, phosphate, and carbon (augmented by metabolic CO_2 from the host cells) from the surrounding sea water to the symbiont, along with acetate (used in fatty acid synthesis by *Symbiodinium*) and glycine from its own metabolism. In return the host receives a range of sugars, amino acids, and lipids from its symbiont.

Two elements in Figure 10.14C require some explanation: the role of "host release factors" and "MAAs". Host release factors come from the animal host and stimulate the photosynthetic productivity of the algal symbionts. If isolated *Symbiodinium* cells are grown in culture, they release a limited amount of glycerol to the culture medium. Glycerol production is increased several-fold if extracts of the coral cells are added to the culture. The stimulating effect of these host release factors is not species specific. If coral extracts are added to zooxanthellae isolated from giant

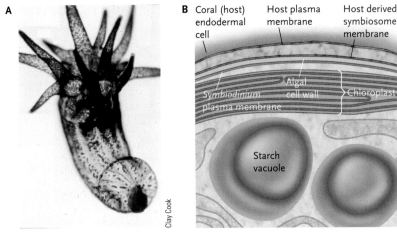

A

B Coral (host) endodermal cell — Host plasma membrane — Host derived symbiosome membrane

Symbiodinium plasma membrane — Algal cell wall — Chloroplast

Starch vacuole

Clay Cook

Figure 10.14

A. *Symbiodinium* sp., zooxanthellae (brown) in the endoderm of an anemone (*Aiptasia pallida* (166)); **B.** An electron micrograph showing the wall of a zooxanthellal cell, the host plasma membrane, and the symbiosome membrane that is secreted by the host. **C.** A simplified summary of the interactions between zooxanthellae and coral symbiont partners.

C

Environment

PO_4^- Inorganic carbon Inorganic nitrogen

Epithelial cells

Mesoglea

Lipids
Fatty acids
Acetate

MAAs

Host release factors

Krebs cycle intermediates
Sugars
O_2
Glycerol
Glycolic acid
Lipids
NH_4^+
Amino acids

Gastrodermal cell

Symbiodinium

Host derived symbiosome

Glycine

pH conditions needed for the different parts of the process are created and maintained. Work using pH-sensitive dyes shows that the pH regimes of the coral endodermal cytoplasm, the fluid between the symbiosome membrane and the zooxanthellal cells, and the zooxanthellae themselves are different, but details remain elusive.

Bleaching of coral (and other zooxanthellal symbionts, Fig. 10.15) has received much publicity. Bleaching is either the loss of colour from the zooxanthellae (*Symbiodinium* has chlorophyll a and c2 plus carotenoids that give the yellow-brown colour) or the expulsion of the algal cells from the host. Bleaching can be caused by increases in seawater temperature, high light levels, prolonged darkness, cold shock, metals (particularly copper and cadmium), and pathogens. While prolonged bleaching may kill the coral, in many bleaching episodes one strain of *Symbiodinium* is expelled and is replaced by another strain that is resistant to the environmental stimulus involved. Bleaching might then be regarded as a survival adaptation of the coral.

Giant clams, *Tridacna* sp. (550) (Mollusca, Bivalvia, Fig. 10.16), also have symbiotic zooxanthellae. The *Tridacna*–*Symbiodinium* relationship shares many similarities with corals and their zooxanthellae, including similar metabolic exchanges, calcification pathways, and bleaching. A major difference is that in corals the symbiont is intracellular, while in *Tridacna*

clams (*Tridacna* (550)) (or vice versa), the same stimulus effect occurs. MAAs are mycosporine-like amino acids—the arrow in the diagram indicates a link between MAA production of and UV radiation. MAAs are produced at high light levels and appear to protect the symbiont from damage by high levels of UV radiation. A possible addition to Figure 10.14C is the diel cycle of superoxide dismutase (SOD) levels. Under high light levels, photosynthesis rates in *Symbiodinium* are high, resulting in maximal oxygen production (a by-product of photosynthesis). Measurements on anemones with zooxanthellal symbionts have recorded O_2 levels at 2% of air saturation in the dark and 250% saturation in light. SOD protects the coral tissue from oxygen toxicity arising from free radicals.

Zooxanthellae are also involved in a process known as light-enhanced calcification, the deposition of calcium carbonate in the skeleton of reef building corals. Despite much experimental work the explanation of the calcification process is not complete. Calcium ion transport and inorganic carbon supplies are intrinsic to the process as is the enzyme carbonic anhydrase. The major problem is clarifying how the

Figure 10.15 Adjacent areas of bleached and healthy coral (*Acropora* sp.).

SOURCE: © Brandon Cole Marine Photography / Alamy

Figure 10.16 Giant clam (*Tridacna pineda* (550), Mollusca, Bivalvia). The green in the mantle edge is from the zooxanthellae in its tissue.

SOURCE: ligio / Shutterstock.com

the symbiont is extracellular and housed in fine tubules that are extensions of the stomach and digestive gland. These tubules permeate the mantle and hemolymph spaces, providing extensive surfaces for the exchange of symbiont-related metabolites.

10.2b.2 Elysia. The emerald green sea slug, *Elysia chlorotica* (466) (Mollusca, Gastropoda, Saccoglossidae) (Fig. 10.17) carries animal–algal symbiosis one step further. *Elysia* feeds on a filamentous green alga, *Vaucheria litorea*, and incorporates photosynthetic plastids into its digestive system cells. The plastids photosynthesize, providing the sea slug with organic carbon compounds for several months, even though the plastids are isolated from the algal nuclear genome. The plastid genome does not include the full complement of genes required for photosynthesis. Investigation of the adult *Elysia* genome shows that at least some of the genome of the alga (a gene involved

in oxygenic photosynthesis, *psbO*) has been incorporated into the genome of the host.

10.2b.3 Convoluta. At low tide, the beach at Roscoff, France, appears to be covered by dark green algae (Fig. 10.18.A). Closer inspection shows that the "algae" are huge numbers (up to a million/m²) of an acoel flatworm (Platyhelminthes), *Convoluta roscoffensis* (202). The flatworms (Fig. 10.18B) contain symbiotic green algae (*Platymonas* (*Tetraselmis*) *convolutae*) from the family Prasinophyceae (Chlorophyceae). Metabolic exchanges between *C. roscoffensis* and its algal symbionts are very different from those of corals and molluscs. Following ¹⁴C-labelled CO_2 incorporation, most of the label in the algae appears in mannitol (a disaccharide), while the sugars transferred to the flatworm host are glucose and fructose. Remarkably, the symbiotic exchange involves the symbiont obtaining nitrogen from uric acid produced by the metabolic activity of the host.

Convoluta roscoffensis is unable to synthesize long-chain saturated or unsaturated fatty acids and depends on its symbionts for these compounds. It follows that the long chain fatty acids in *C. roscoffensis* are plant-type ϖ-3 compounds, rather than typical animal ϖ-6 forms. The acoel can synthesize complex lipids (e.g., triglycerides) from the raw materials supplied by the symbionts. Similarly, *C. roscoffensis* cannot synthesize sterols *de novo*. The algal symbionts supply 2,4-methylene-cholesterol which *C. roscoffensis* can then convert to cholesterol, which does not occur in the algae.

10.2b.4 Ascidians. There is an obligate symbiosis between *Prochloron*, a cyanobacterium, and a number of species of tropical colonial ascidians. *Prochloron* is found inside the cloacal cavity and is associated with the tunic of its ascidian host, benefiting from its protected location. Further, the tunic of *Lissoclinum patella* (1110) contains UV-absorbing mycosporine-like substances that protect *Prochloron*, which is sensitive to high levels of UV light. The benefit to the host ascidian is less obvious. The location of the symbiont cells in the cloacal chamber makes it almost impossible for the host to acquire metabolites from its partner. Indirect evidence suggests that toxins, secreted by the *Prochloron* cells, may inhibit the growth of fouling organisms (bacteria and other algae) on the tunic, as well as offering protection against the feeding of commensal and parasitic crustaceans living in the cloacal cavity.

10.2b.5 Light Pipes in a Sponge. While fibre optic cables are considered modern technology, they may well have been "invented" 500 million years ago. Some sponges (Porifera, Demospongia and Hexactinellida, e.g., *Tethya aurantia* (a demosponge), Fig. 10.19) have long been known to house photosynthetic cyanobacteria and zooxanthellae. The ability of these cells to photosynthesize deep inside the sponge tissues, where light does not penetrate, is due to the sponges' siliceous

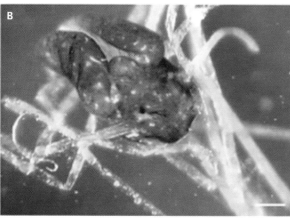

Figure 10.17 A. The emerald green sea slug, *Elysia chlorotica* (466). The green colour is due to plastids incorporated from its food, *Vaucheria litorea*. **B.** A juvenile *Elysia* feeding on *Vaucheria*. At this stage the sea slug has not yet incorporated algal plastids into its digestive epithelium.

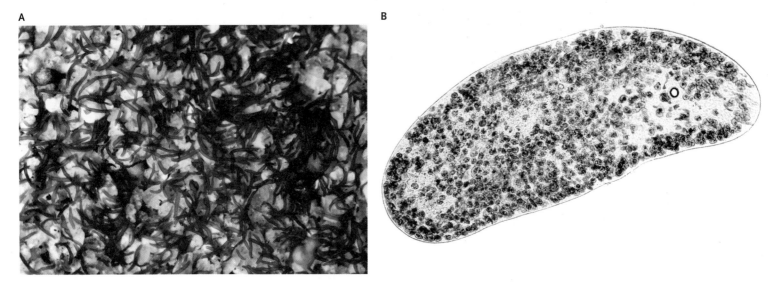

Figure 10.18 A. A mass of *Convoluta roscoffensis* (202) on the beach at Roscoff (France). **B.** Symbiont cells of *P. convolutae* in a young *C. roscoffensis* (adults may have 20 000–70 000 symbiont cells in each flatworm).

SOURCE: Arthur Hauck, Germany

spicules, which act as light pipes, carrying light energy deep into the sponge tissues.

10.2b.6 Sloth–Alga. An example of a mammalian–alga symbiosis is the association between a two-toed sloth (*Choloepus* sp. (1849), Fig. 10.20) and green algae. Two-toed sloth hair has grooves running lengthwise, while the hair of three-toed sloths (*Bradypus* spp. (1848)) has cracks running across the long axis. In each case, these irregularities provide anchorage for green algae, including diatoms, cyanobacteria, and most commonly, *Trichophilus welckeri* (a filamentous green alga). A minimum value in this association is that the sloth benefits from the green camouflage cover provided by the alga, and the alga benefits from shelter and exposure to light. It is possible that sloths absorb some nutrients from the alga and that the alga may protect the sloth from exposure to UV light.

10.2c Animal–Plant Symbiosis

10.2c.1 Ants and Acacia Trees. The symbiotic association between ants and acacia trees is a classic example of the coevolution of insects and flowering plants. The acacias provide the ants with room and board. Swollen, hollow thorns provide the ants with homes (domatia, Fig. 10.21), as well as extrafloral nectaries and Beltian bodies (lipid- and protein-rich buds on leaf tips) that provide food. In return, the ants defend the tree against attack by phytophagous insects and browsing

Figure 10.19 *Tethya aurantia* (117), an orange puffball sponge.

SOURCE: © Carver Mostardi / Alamy

Figure 10.20 *Choloepus* sp. (1849). The green colour is due to algal symbionts on the animal's fur.

SOURCE: Kjersti Joergensen / Shutterstock.com

megafauna (goats, giraffes, and even rhinoceros, although reports suggest that it takes a number of ant attacks to deter a browsing rhinoceros!).

Ants will also defend their home trees against other plants. The ground around an ant–acacia association is kept clear of other plants and any plants touching the *Acacia* are removed by the ants (Fig. 10.22).

Many acacias are insect pollinated. How do potential pollinators avoid the attacks of the guard ants? *Acacia hindsii* is guarded by the very aggressive *Pseudomyrmex veneficus*. *A. hindsii* pollen is released early in the morning, and pollinators visit the flowers throughout the morning without being attacked by *Pseudomyrmex* guards. In the mornings most of the guard activity is concentrated at the extrafloral nectaries on young leaves, while flowers develop on the previous year's growth. Additionally, flowers (but not flower buds or post-reproductive flowers) produce an ant repellent that keeps the guards away. The combined effect is to separate pollinators and guards in space and time.

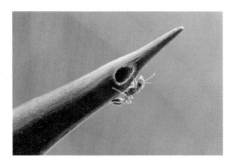

Figure 10.21 *Pseudomyrmex spinicola* (1008) by the entrance hole to its nest in an acacia thorn.

SOURCE: Alex Wild/Visuals Unlimited, Inc.

Figure 10.22 *Pseudomyrmex* sp. (1008) removing the tendril of a vine from its home *Acacia* tree.

SOURCE: Mark Moffett/Minden Pictures

BOX 10.2
Of Figs and Fig Wasps

The apparently simple association between an insect pollinator and a plant illustrates the complexities of symbiotic and parasitic associations in the natural world.

Figs are the fruit of any of about 850 species in the genus *Ficus*, one of the most highly speciated plant genera. Non-cultivated figs contain an amazingly complex biological microcosm. Central to this is the symbiotic relationship between a fig and a fig wasp (Hymenoptera, Chaliciae, Agaonidae, Fig. 1). Fig flowers are fertilized by the fig wasps, and each fig species has coevolved with a particular species of fig wasp. The closeness of the coevolution is supported by strong evidence for parallel cladogenesis where, in general, each section in the genus *Ficus* is pollinated by a particular genus, or group of genera, of Agaonida.

A fig is essentially an inside-out cluster of flowers with the male and female flowers enclosed within a protective synconium. At the opposite end to the synconium stalk is a small opening, the ostiole. Flower development within the fig is accompanied by the release of a chemical attractant that draws a female wasp to the ostiole, which it enters (Fig. 2). The ostiole is so tight that wasps frequently lose wings or antennae as they crawl in. In the synconium are three types of flowers:

Figure 1
Fig wasps are minute—this fig wasp is standing on the wing of a fruit fly that is roughly the size of a house fly.

SOURCE: Copyright Deeble and Stone

Figure 2
A female fig wasp entering the ostiole of a fig.

SOURCE: Mark Moffett/Minden Pictures

Figure 3
Male (light body, black head) and female Pleistodontes imperialis *(1004) in a synconium of* Ficus rubiginosa.

SOURCE: W.P. Armstrong

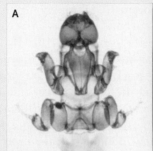
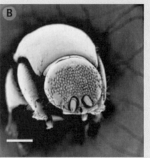

SOURCE: Republished with permission of Annual Reviews, Inc, from Weiblen, "How to be a fig wasp," *Ann Rev Entomol* (2002) 47: 299–330; permission conveyed through Copyright Clearance Center, Inc.

Figure 4
Male fig wasps. **A.** Wiebesia frustrata (1014) *appears four-legged as a result of the atrophy of the middle pair of legs.* **B.** Kradibia chinensis (998) *has vestigial eyes (scale bar = 0.1 mm).*

male flowers, and short-style and long-style female flowers. The fig wasp lays her eggs in the short-style flowers and the eggs develop there, forming a gall where the wasp larvae feed on the seed endosperm. As the female moves around in the synconium searching for short-style flowers, pollen from the fig in which she developed is brushed onto long-style flowers and fertilizes them.

Male fig wasps emerge before the females (Fig. 3). They are wingless, have vestigial eyes, and may have the middle pair of legs reduced or absent (Fig. 4). They search for galls containing females, open the gall and fertilize the females (the sex ratio is female biased). The males then bore an exit hole for the female through the wall of the synconium before dying. Pollen-bearing females escape through this hole and fly to another receptive fig tree of the right species. Newly emerged females of *Elisabethiella baijnathi* (**996**), for example, fly high and then glide downwind (fig wasps have large wings for their size) until they pass over an appropriate fig tree; they then turn and fly upwind to the tree. The life cycle of a generalized fig wasp is summarized in Figure 5.

Essential to a balanced relationship between figs and their pollinating fig wasp species is the length of the wasp's ovipositor and the relative length of the short- and long-style flowers in the synconium. A puzzle to understanding the coevolution of figs and fig wasps is the frequent presence of non-pollinating wasps (Subfamilies Torymidae (Fig. 6) and Eurytomidae) that have long ovipositors and lay eggs in the long-style fig flowers, creating galls in flowers that would normally produce seeds. These non-pollinating gall makers are but a small part of the complex fig–fig wasp community. Up to 20 species may be involved, even including parasites of the fig wasp.

Figure 5
Life cycle of a fig wasp (Agaonidae).

SOURCE: Reprinted with permission from Encyclopaedia Britannica, © 1999 Encyclopædia Britannica, Inc.

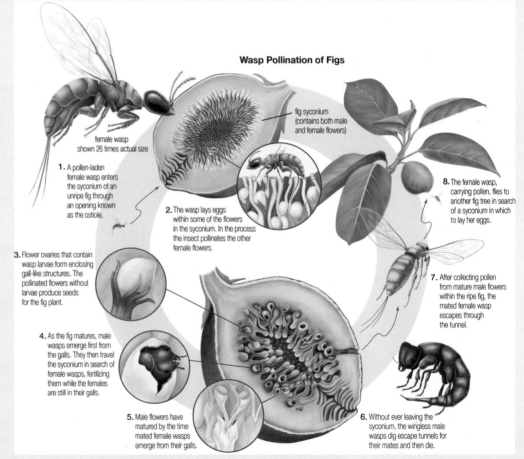

Wasp Pollination of Figs

female wasp shown 26 times actual size

fig syconium (contains both male and female flowers)

1. A pollen-laden female wasp enters the syconium of an unripe fig through an opening known as the ostiole.

2. The wasp lays eggs within some of the flowers in the syconium. In the process the insect pollinates the other female flowers.

3. Flower ovaries that contain wasp larvae form enclosing gall-like structures. The pollinated flowers without larvae produce seeds for the fig plant.

4. As the fig matures, male wasps emerge first from the galls. They then travel the syconium in search of female wasps, fertilizing them while the females are still in their galls.

5. Male flowers have matured by the time mated female wasps emerge from their galls.

6. Without ever leaving the syconium, the wingless male wasps dig escape tunnels for their mates and then die.

7. After collecting pollen from mature male flowers within the ripe fig, the mated female wasp escapes through the tunnel.

8. The female wasp, carrying pollen, flies to another fig tree in search of a syconium in which to lay her eggs.

Figure 6
A torymid wasp, one of a family of parasitic wasps that use their long ovipositors to lay eggs in the long-style flowers in fig synconia (this example is on an oak gall).

SOURCE: Tristram Brelstaff

Some of these parasites enter the synconium and attack their prey directly. Others, e.g., the chalcid *Phylotrypesis caricae* (1004), have long ovipositors and pierce the synconium to lay their eggs at their targets.

The complex fig–wasp symbiotic relationship extends yet further. Ants from the genera *Oecophylla* (1003) and *Crematogaster* (994) tend the associations between the fig wasps *Ceratosolen constrictus* (992) and *Ficus condensa* and between *C. fusciceps* and *F. racemosa*, by helping to control parasitic wasps from the genus *Apocrypta*. In the absence of ants the number of parasites increased; in the presence of ants (and if more ants were experimentally added), the number of parasites decreased.

Another complexity is the interaction between fig wasps and nematodes that, while parasites of the fig, use the fig wasp as an intermediate host and transport system. Nematode wasp associations are species-specific with one nematode species associated with particular fig and fig wasp species. Immature nematodes in the synconium are in the dispersal stage. They attach to newly emerged fig wasps and are carried to another fig. The nematodes burrow into the fig wasp host, feed on it (the wasp dies), emerge, and continue their life in the fig. In order to repeat the cycle, they time their own oviposition so that larval nematodes emerge in synchrony with the emergence of adult fig wasps.

There is an interesting aside to the story of ant–acacia symbiosis. The spider *Bagheera kiplingi* (662) lives on acacia in Central America (*Vachellia* sp.), where it builds a nest on leaf tips not normally patrolled by the resident ants (*Pseudomyrmex* sp.). Although almost all spiders are carnivorous, this spider is largely herbivorous, feeding on Beltian bodies and nectar (and some ant larvae).

STUDY BREAK

1. Giant clams and the nudibranch *Elysia* both have symbiotic associations with a green alga. How do these associations differ?
2. What benefits do *Acacia* spp. get from their symbiotic association with ants?
3. How might the fig–fig wasp symbiotic association have evolved?

10.2d Animal–Animal Symbiosis

10.2d.1 Anemones and Clownfish. Some 26 species of anemone fish (or clown fish, Fig. 10.23), 25 in the genus *Amphiprion* (1252), establish symbiotic relationships with 10 species of anemones (distributed among five genera and three families). Each fish lives between the tentacles of an anemone, which is the centre of the fish's territory. The fish rarely strays more than a few metres from its host or for more than a few minutes. The benefit of this association to the fish is presumed to be the protection offered by the anemone's nematocyst-laden tentacles (fish have also been reported nibbling on their host's tentacles). The benefit to the anemone is that clownfish attack and drive off butterfly fish (Chaetodontidae) that prey on anemones.

Juvenile *Amphiprion* raised in the presence of potential host *Heteractis* anemones recognize their host by olfactory cues and move into a host within minutes of this being possible. Fish raised apart from anemones take more time to recognize a host and take up to two days to move into the protection of an anemone.

Tests of isolated anemone toxins dissolved in water (so that the toxin was delivered to the gill membranes, rather than via nematocyst injection) show that some *Amphiprion* species are resistant to some anemone toxins. However, toxin immunity is not a requirement for a symbiotic association. Protection against anemone nematocyst discharge depends on the mucous layer covering the *Amphiprion* skin, which is thicker in anemone fish than in other fish. Components of the anemone mucus may also be added to the fish mucous layer.

10.2d.2 Coral Guard Crabs. Coral guard crabs (*Trapezius* sp. (839), Fig. 10.24) have a symbiotic relationship with a reef coral, typically a *Pocillopora* sp. (185) The crab

Figure 10.23 Clown fish (*Amphiprion ocellaris* (1253)) among the tentacles of *Heteractis* sp. (170).
SOURCE: Levent Konuk / Shutterstock.com

guards the reef against predators such as the crown-of-thorns sea star. *Trapezius* uses chemosensors that detect the approach of the sea star and attacks the predator by pinching its tube feet. In an experimental test, isolated colonies of two species of reef coral (*Acropora* and *Pocillopora*) were exposed to attacks by a predatory sea star (*Culcita novaeguineae* (1072)) in the presence and absence of coral crabs. There was no mortality in colonies protected by crabs. These protected colonies show faster growth and lower sensitivity to bleaching than unprotected colonies. *Trapezius* also removes sediment from the reef. In return, the crab benefits by being provided with living space and by eating mucus from the surface of the coral.

10.2d.3 Cleaners. Symbiotic cleaning associations are common, both within and between animal groups. Typically one organism (the cleaner) feeds on ectoparasites, dead skin, and debris from the surface of its partner (the client). The client benefits from the removal of the unwanted material. Cleaning associations typically suppress normal predatory behaviour, as shown in Figure 10.25 when a cleaner wrasse (*Labroides* sp. (1270)) enters the mouth of a spotted moray eel (*Gymnothorax moringa* (1159)).

A less familiar relationship involves ground hornbills (*Bucorvus leadbeateri* (1548)) and warthogs (*Phacochoerus africanus* (1764), Fig. 10.26). Warthogs approach the hornbill and lie down to have ectoparasites removed from their skin, particularly from around the head and neck region. Similarly, the African Jacana (lily trotter) (*Actophilornis africana* (1551)) cleans parasites from hippopotamus skin (*Hippopotamus amphibius* (1758), Fig. 10.27). However, this relationship edges towards ectoparasitism because jacanas also eat the flesh exposed by the removal of the parasite.

A

Figure 10.26 A ground hornbill and a warthog.

SOURCE: Courtesy of Naas Rautenbach

B

Figure 10.24 *Trapezia rufopunctata* (839). Most species of coral crabs are brightly coloured.

SOURCE: © WaterFrame / Alamy

A

Figure 10.27 A Jacana and a hippopotamus.

SOURCE: Courtesy of Naas Rautenbach

Figure 10.25 A cleaner wrasse feeding on detritus around the teeth of a spotted moray eel.

SOURCE: James A Dawson / Shutterstock.com

B

BOX 10.3 Molecule Behind Animal Biology
CHITIN

Chitin is a natural polymer of linked molecules of an amino sugar, N-acetyl glucosamine ($C_8H_{13}O_5N$, Fig. 1). Chitin is globally abundant, second only to cellulose in natural polymer biomass. Arthropods produce an estimated 1360 million tonnes/year in aquatic ecosystems alone. Chitin is not found in vertebrates or plants but is present in most invertebrates, fungi, and other microorganisms. Pure chitin is leathery and flexible, but it is hardened ("tanned") by the incorporation of sclerotized proteins and calcium carbonate. This creates both a tough protective layer for animals, as well as a material that resists enzymatic digestion.

Chitin is a vital part of arthropod exoskeletons and constitutes between 2 and 20% of the dry weight of an arthropod body. Before a moult an arthropod recovers much of the protein and minerals from its exoskeleton, leaving only a thin exuvium (Fig. 2). Much chitin is rapidly turned over. Small arthropods (e.g., marine copepods) may go through a rapid series of moults, with the chitin in each discarded exoskeleton becoming part of the marine food supply. About 90% of the chitin produced in the marine water column is broken down again within a week.

Most chitin degradation in nature is microbial, both by free-living forms and in association with animal guts. Symbiotic relationships occur between gut bacteria and animals as diverse as springtails, *Folsomia candida* (845) (Collembola), earthworms (Annelida), insectivorous bats, and numerous

FIGURE 2
*Maryland blue crab (*Callinectes sapidus *(811)) moulting, with shed exoskeleton shown above crab.*

SOURCE: Mary Hollinger, NODC biologist, NOAA

species of seabirds that feed almost entirely on crustaceans. The diet of Adelie penguins (*Pygoscelis adeliae*) (1642) includes significant amounts of chitin because up to 10% of the krill dry matter they ingest consists of chitin. Their gut microbes augment the activity of endogenous enzymes that are responsible for the host Penguin's degradation of chitin.

Why is chitin the selected molecule to accompany a chapter on parasites and symbionts? The answer lies in a major element in Neotropical terrestrial ecology, fungus farming by leaf-cutter ants (see Figs. 1.3 to 1.6). There are about a dozen genera of fungus-growing ants in the tribe Attini. Two of these, *Acromyrmex* (980) and *Atta* (986), cut fresh leaves and flowers as substrates for their fungus farms. More primitive genera collect dead organic material as a substrate. A single large colony of *Atta* may collect as much as 400 kg (dry weight) of vegetation each year, and leaf-cutters are estimated to take 12–17% of the leaf production of a neotropical rainforest; small wonder that the jaws of leaf-cutter ants become worn (pages 1–2).

All of this harvest becomes the substrate for a colony of fungus maintained by the ants. The fungus, aided by a community of bacteria, breaks down leaf material and uses it to support its own growth. The fungal hyphae, with their chitin walls and specialized swellings called

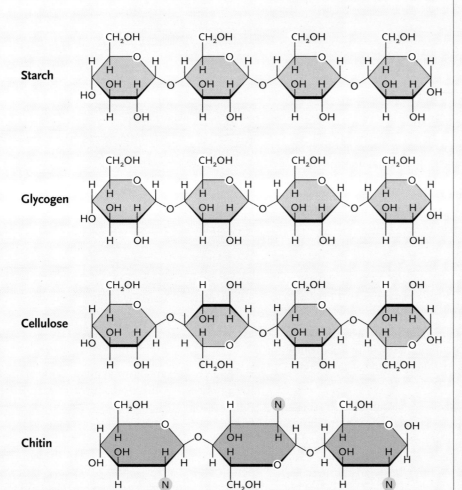

FIGURE 1
Molecular diagrams of polysaccharides. Starch and glycogen are storage polysaccharides, whereas cellulose and chitin are structural polymers.

gongylidia, become food for the ants. Ant workers may remove hyphae for nutritional reasons as well as trimming the fungus to stimulate its growth. Ant larvae are totally dependent on the protein and carbohydrate-rich gongylidia for their nutrient supply. Ant workers also eat gongylidia as well as fungal hyphae. About 10% of their diet is chitin. The complexity of the ant-fungus relationship (introduced in Chapter 1) is made more complex by invasive bacteria and fungi, particularly *Escovopsis*. Leaf-cutters have an association with a filamentous bacteria (*Pseudonocardia*) that grows on their cuticle and synthesizes antibiotics that attack *Escovopsis*. Ants also accumulate material cleaned from the fungus garden in **infrabuccal** pockets where pathogens are inactivated by antibiotic secretions from a bacterial flora in the pocket. Killed and compressed pathogens are then deposited as a pellet some distance from the fungal garden.

Chitin has very limited solubility (hence its value in exoskeletal and protective structures). A chitin derivative, chitosan, is synthesized by removing the acetyl groups from the chitin chain and is used in several biomedical applications. Chitin for industrial production of chitosan comes from shrimp and crab shells, a byproduct of the food canning industry. Insect chitin is difficult to recover because of its tight linkages with proteins. Chitosan is soluble and can be formed into threads by pressure extrusion through a small nozzle. Chitosan threads have important applications in the world of surgery where chitosan sutures are both biodegradable and immunologically inert, allowing internal sutures to be left in place to dissolve. Sheets of chitosan material are being tested in surgical repairs, for example, in rotator cuff surgery where the chitosan forms a scaffold on which natural tissues can regenerate. Chitosan microspheres, with different degrees of crosslinking between strands, are also used as drug containers, providing timed release of a drug.

A recent development is the combination of a layer of chitosan with fibroin extracted from silkworm silk (see pages 512–513). The fibroin is then treated with methanol to form an insoluble ß-sheet. This composite material, known as shrilk, has the strength of aluminum alloys but only half the density of such alloys. The properties of shrilk offer potential as a biodegradable plastic that could be used in bottles, packing materials, and diapers, as well as potential medical applications.

10.3 Commensalism

The definitions of commensalism, and of the various types of commensalism (phoresis, inquilism, and metabiosis) in Section 10.1 seem clear. However, a look at real world examples reveals that the complexity of biological relationships resists the limitations imposed by these artificial divisions.

10.3a Inquilism

Carnivorous plants have always fascinated biologists. In his book *Insectivorous Plants*, Charles Darwin wrote:

> *"The fact that a plant should secrete, when properly excited, a fluid containing an acid and ferment, closely analogous to the digestive fluid of an animal, was certainly a remarkable discovery."*

Even more remarkable is that some insects (mostly midge and mosquito larvae) have adopted the fluid in the pitcher plant leaf as home (rather than their death bed!) and, in some cases, their biology has become integrated with that of their host (an animal–plant symbiosis).

While some pitcher plants secrete proteolytic enzymes, others (such as *Sarracenia* spp. of North America, Fig. 10.28) depend on the microflora (bacte-

Figure 10.28
A pitcher plant, *Sarracenia purpurea*.
SOURCE: M.B. Fenton

rial and protist) and animal inhabitants of the fluid in the pitchers for the breakdown of trapped insects. Pitchers of *Sarracenia purpurea* attract insect prey when they first open (as the leaves age they lose their ability to attract prey) and may trap and digest insects as large a yellow jacket wasps (Fig. 10.29). Given this potency it is a surprise that other insects actually live in the pitcher fluid. *Sarracenia* is home to the larvae of three species of flies: *Blaesoxipha fletcheri* (956) (Sarcophagidae, a flesh fly), *Wyeomyia smithii* (975) (Culicidae, a mosquito), and *Metriocnemus knabi* (965)

Figure 10.29
Sarracenia purpurea traps and consumes insects as large as yellow jackets.

SOURCE: Nature's Images / Photo Researchers, Inc.

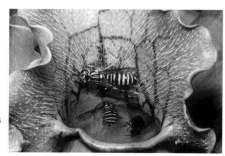

(Chironomidae, a midge). *B. fletcheri* larvae are buoyant and feed on freshly trapped insects at the pitcher fluid surface. Typically only one larval *B. fletcheri* lives in a pitcher. The larvae of *W. smithii* are free swimming and filter-feed on decomposing particulate matter, while those of *M. knabi* feed on the accumulated debris at the bottom of the pitcher. There is a complex interaction between the populations of these dipteran species in the niche environment of the pitcher plant.

The interaction between pitcher plants and their insect prey and denizens is complex symbiotic nutrient cycling between the prey animals, the inquiline resident animals in the pitchers (protists, rotifers, midges, and mosquitoes), and the plant cells. The greater the metabolic activity within the pitcher, the more CO_2 (from respiration) and NH_4 (from nitrogenous excretion) are made available to the plant, and, conversely, increased photosynthesis by the plant generates more oxygen for the inquilines. In some tropical environments, numbers of ants (common prey for pitcher plants) are low. Some *Nepenthes* spp. (Fig. 10.30D) secrete materials that attract tree shrews around the mouth of the pitcher. After feeding, the shrews defecate into the pitcher, their faeces providing an additional source of nitrogen for the plant. Another species, *Nepenthes rafflesiana*, has modified its pitchers to provide a roosting place for Hardwicke's woolly bats (*Kerivoula hardwickii* (1793)) (Fig. 10.30B & C), with the bat feces then providing a third of the plant's nitrogen needs.

10.3b Phoresy

Phoresy—hitchhiking—involves one species using another (the host) for transport without harming the host. Examples range from mites carried by insects (Fig. 10.31A&B) to barnacles on the pectoral fins of a humpback whale (*Megaptera novaeangliae* (1761)) (Fig. 10.31C). Cattle egrets (*Bubulcus ibis* (1566)) associate with a variety of herbivorous mammals, including domestic cattle and many wild species (e.g., Cape buffalo, *Syncerus caffer* (1769), Fig. 10.31D). The egrets benefit not only from being transported but also by feeding on insects stirred up from the grass by their host's movement. Egrets feeding in association with a host appear to get 1.5–2 times as much food in the same time period as egrets foraging without a host.

Members of the recently discovered Phylum Cycliophora are commensal on lobster mouthparts (*Symbion pandora* (275) on *Nephrops* (824), the Norway lobster, and *Symbion americanus* (Fig. 10.32) on *Homarus americanus* (819), the American lobster). *Symbion* has a complex life cycle in which the feeding funnel (at the apex of the animals) is replaced several times as the animal eats particles of food generated as the host breaks up its food. Asexual reproduction increases the population of the commensal on a host. *Symbion* is also in tune with the life cycle of its host. The commensals recognize the changes in the host that signal an impending moult and respond by triggering a sexually reproductive phase in their own life cycle. This produces free-swimming larvae that are able to colonize a new host, or recolonize the newly moulted lobster on which their parents lived. Sensitivity to host changes can be an important element in the life of a commensal or parasitic organism (another example, *Polystoma* (243), is discussed in Section 10.4b).

An example of phoresy with further consequence is the behaviour of a parasitic wasp, *Trichogramma brassicae* (1012). When cabbage white butterflies (*Pieris brassicae* (1047)) mate, the male deposits an anti-aphrodisiac pheromone (benzyl cyanide) on the female, making her less likely to mate with another male. The wasp detects this pheromone and is attracted to the newly mated female butterfly who transports the wasp

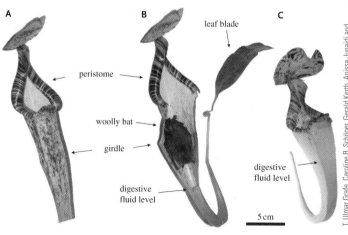

T. Ulmar Grafe, Caroline R. Schöner, Gerald Kerth, Anissa Junaidi and Michael G. Schöner, "A novel resource–service mutualism between bats and pitcher plants," *Biol. Lett.* 23 June 2011 vol. 7 no. 3 436-439, by permission of the Royal Society.

© Ch'ien C. Lee

Figure 10.30 Two pitcher plants from Borneo: **A.**, **B.**, & **C.** *Nepenthes rafflesiana* and **D.** *Nepenthes loweii*. These plants produce two kinds of pitchers, the typical ones that trap insects (A), and others specialized to house bats (B, C) or serve as latrines for tree shrews (D).

Figure 10.31 **A.** Louse fly, *Pseudolynchia canariensis* (967), carrying phoretic mites. **B.** A cluster of mites carried by the louse fly. **C.** A humpback whale (*Megaptera novaeangliae* (1761)), showing its pectoral fin, which is home to a population of phoretic barnacles. **D.** Cattle egrets associated with Cape buffalo (*Syncerus caffer* (1769)).

A

Alvesgaspar

B

Alvesgaspar

C

Michael Owen

D

Pål A. Olsvik

to the site where she lays eggs. The wasp then parasitizes the newly laid eggs (Fig. 10.33).

10.3c Metabiosis

An apparently simple example of metabiosis is a hermit crab's use of an empty shell originally secreted by a gastropod mollusc. This association may be ancient—tracks in a Cambrian sandstone originate from the ancestor of a hermit crab with part of its body enclosed by a dextrally coiled shell.

When a hermit crab encounters an empty shell, it carefully investigates its size, weight, and condition before moving from its old shell to the new one. Shell selection by hermit crabs is not random. Different species exhibit preferences for shells of different gastropods. For example, *Pagurus acadianus* (826) and *P. pubescens* (826) are sympatric species on the east coast of North America. Young *P. acadianus* choose shells of *Littorina* (423) over those of *Nucella* (425) (*Thais*), while *P. pubescens* enters shells of *Littorina* or *Nucella* in proportion to their abundance. *P. acadianus* grow larger than *P. pubescens*, and larger individuals select shells of the whelk, *Buccinum* (480), or moon snail, *Polinices* (427). As crabs grow and their shells become tight, they become more aggressive. Growing *P. acadianus* compete for the limited supply of *Buccinum* shells.

A

Nina Fatouros

Figure 10.32 A number of *Symbion americanus* (275) individuals attached to a seta (bristle) on the mouthparts of an American lobster.

SOURCE: © Peter Funch, Aarhus University

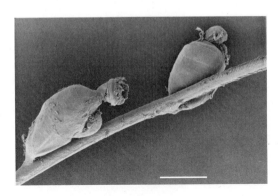

B

Nina E. Fatouros, *PNAS*, July 22, 2008; 105 (29). Copyright 2005 National Academy of Sciences, U.S.A.

Figure 10.33
A. *Trichogramma brassicae* (1012) on a female *Pieris brassicae* (1047). **B.** *T. brassicae* on an egg cluster of *P. brassicae*.

Pagurus acadianus and other hermit crab species frequently inhabit shells colonized by the hydroid, *Hydractinia symbiolongicarpus* (136) (sometimes known as "snail fur" and rarely found not associated with hermit crabs, Fig. 10.34). Some hermit crabs have a similar association with sea anemones, for instance, between *Pagurus pollicaris* (826) and the sea anemone *Calliactis tricolor* (168). The crab actually places anemones on its shells and arranges them to best balance the shell. The anemone benefits from scraps from crab feeding, and the anemone defends the crab from attacks by octopus—a clear example of symbiosis. The relationship between *Pagurus* spp. and *Hydractinia* is not as clear. Certainly the hydroid benefits from food scraps and the opportunity for reproductive exchange with another colony when hermit crabs gather. The crab benefits because the hydroid prevents settlement of fouling organisms (e.g., barnacles and bryozoans). However, shells covered with hydroids attract a polychaete, *Polydora* (340), that burrows into the shell, weakening it, and making the crab more vulnerable to attacks by blue crabs (*Callinectes sapidus* (811)).

Pagurus prideaux has a symbiotic relationship with the cloak anemone, *Adamsia carcinopados* (167), a species also found in association with other hermit crabs. In this association the crab places the anemone on its ventral surface, where the anemone has good access to food but does not defend the crab. The anemone pedal disc grows up and around the gastropod shell, becoming an extension of the shell and freeing the crab from the need to change to a new shell as it grows. This eliminates the vulnerable moment when the crab is out of its protective shell.

STUDY BREAK

1. Define phoresis, inquilism, and metabiosis.
2. What benefits do hermit crabs gain from associations with hydroids or anemones?

10.4 Parasitism

Parasites are almost ubiquitous in the biological world. The exceptions are in western civilization where medical and veterinary science fight an ongoing (and sometimes losing) battle to banish parasites from humans and domestic animals. Examples of human interactions with parasites are included in Chapter 18.

In the non-human world, life with parasites is the norm. Some 80 000 species of nematodes have been described, with estimates of 5–10 times as many species yet to be described, of which roughly half are parasitic. A nematologist once suggested that if one could remove the remainder of the biotic world and leave only the nematodes, then things would look pretty much the same. A tree trunk would be outlined by the nematodes living under its bark and its leafs sketched by the nematodes burrowing into them; a squirrel on a branch would appear as a parasite-drawn outline of skin and muscles and its gut (loaded with nematodes) would stand out in clear relief.

10.4a What Does It take to Be a Parasite?

Parasites may live on (ectoparasites) or in (endoparasites) the body of their host. The following are examples of some of the general features of a parasitic way of life.

10.4a.1 Lepeophtheirus salmonis (Ectoparasite).
In 2009, the global cost of infection by, and control of, sea lice (*Lepeophtheirus salmonis* (741) and *Caligus elongatus* (740), Crustacea, Copepoda, Caligidae) parasitic on salmon was estimated at US$480 million—emphasizing the importance of knowledge about this ectoparasite. An *L. salmonis* attaches to a host (ten species of salmonids) and feeds on the skin, mucus, and blood of the host. *L. salmonis* also attaches to sticklebacks, suggesting that these fish may act as temporary hosts to the parasite. Damage to the host (Fig. 10.35A) provides infection sites for pathogenic bacteria and fungi and seriously reduces the market value of aquaculture-farmed salmonids. *C. elongatus*

Figure 10.34 **A.** *Pagurus acadianus* (826) with *Hydractinia symbiolongicarpus* (136) on the whelk shell it is occupying; **B.** *Hydractinia symbiolongicarpus* with tall, tentacled gastrozoids and short, round gonozoids.

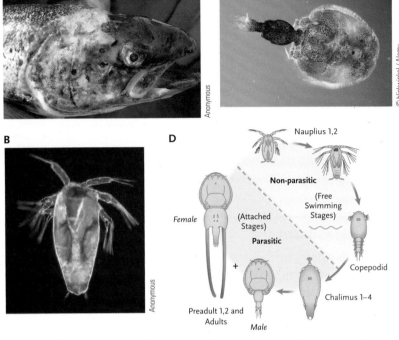

Figure 10.35 A. Damage to Atlantic salmon (*Salmo salar*) caused by severe sea lice infestation. **B.** Nauplius larva and **C.** adult salmon louse (Crustacea, Copepoda). **D.** Life cycle of sea lice.

attacks several species of fish in the North Atlantic, including salmonids.

The eggs of *L. salmonis* develop in an egg string attached to the female and hatch as free-swimming nauplius larvae (Fig. 10.35B). Egg development takes 5–40 days (depending on water temperature) and 70–300 nauplii hatch from each egg string. Females start to produce a new egg string within 24 hours of the detachment of the previous string. The nauplii follow a normal copepod development pattern through a **copepodid** stage. At this point, the free-swimming copepodid swims away from areas of low salinity and toward higher light levels, behaviours that take them to habitats of potential hosts. Once in the area of slow-moving fish, the copepodids sense the hydrodynamic changes in currents that accompany a fish's movement. This triggers the copepodid to swim in circles and spirals until it perceives a chemical signal. It senses a **kairomone** (an interspecific signal that benefits the receiver) released by the potential host. The sensitivity of *L. salmonis* to host scent in Y-tube experiments is obvious—the copepodids swim up the branch with water from a salmon aquarium. Physiologically, recordings from the antennal nerve demonstrate a response to water from salmonid aquaria.

When a copepodid lands on a potential host, it first grips with its maxillipeds and then stabs the host using its second antennae. It secretes an attachment filament that originates from material secreted through the filament duct, between the bases of the

second antennae and anchoring below the epithelium of the host. Once attached by the filament, the fish louse goes through four chalmonid stages before moulting to the adult form (Fig. 10.35C). At this stage, it is free to move around the surface of the host, using its hooked appendages to maintain a grip.

Four generalizations about life as an ectoparasite can be drawn from what is known of the biology of *L. salmonis*.

1. Transmission from one host to another must involve a free-living stage.
2. Successful transmission is a low-probability event so many infective larvae must be produced.
3. Locating the host requires an integrated complex of behavioural and physiological adaptations.
4. The parasite must have an effective means of attaching to the host.

10.4a.2 Hymenolepis diminuta (Endoparasite). Most rats (*Rattus norvegicus* (1903)) carry several species of parasites; the dominant one is the rat tapeworm, *Hymenolepis diminuta* (251) (Platyhelminthes, Cestoda, Fig. 10.36). A number of individual *H. diminuta* are visible—a normal situation for this tapeworm is 5–10 worms per host. Denser infestations reduce the fecundity of individual worms.

Figure 10.36 shows the holdfast (scolex) and a series of body divisions (proglottids) that form the tape of the body. The scolex is commonly regarded as anterior although this has been questioned because the holdfast of the Monogenea (opisthaptor), the sister group to the Cestoda, is posterior. Also, the arrangement of the testes and ovaries in cestodes would be the reverse of the norm for other platyhelminths if the scolex were anterior. The *hox* genes are present, but their arrangement (described as "scrambled" by one author) does not resolve the issue of polarity.

The scolex of *H. diminuta* (Fig. 10.37A) has four circular suckers (acetabula) that allow it to attach to the gut lining of its host. The scolex of its congeneric relative *H. microstoma* (Fig. 10.37B) shows a marked difference.

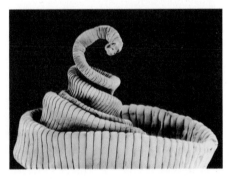

Figure 10.36 *Hymenolepis diminuta* (251) (Platyhelminthes, Cestoda).A false-colour micrograph of an individual worm showing the arrangement of proglottids as a ribbon formed from the scolex.
SOURCE: David Burder/Stone/Getty Images

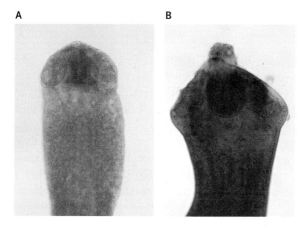

Figure 10.37 **A.** Scolex of *Hymenolepis diminuta*; **B.** scolex of *H. microstoma*.

SOURCE: Allen Shostak

H. microstoma has a central rostellum, equipped with curved hooks, that can be embedded in the host gut wall. In *H. diminuta* the rostellum is reduced, has no hooks, and surrounds an apical organ (a structure suggested to be neurosecretory). The absence of a hooked rostellum allows *H. diminuta* to detach from the host gut wall and change its position within the intestine, something it does in a rhythm associated with the host's feeding activity, triggered as a response to the pH conditions within the intestine. When parasitized rats are fed on a daily schedule, *H. diminuta* move to the anterior of the gut while the host has food available, reaching a forward point about 2 h after food is removed from the host. Over the next 18 h, the tapeworm gradually moves posteriorly in the intestine, only to move forward with the next daily feeding period.

Tapeworms have no gut, so they take up food in solution from the gut of the host. During this process, the tapeworm body surface expands into microtriches, minute extensions of the tegument similar to intestinal microvilli in their size and distribution (Fig. 10.38). Microtriches may help tapeworms attach to the gut wall because they cover the scolex as well as the proglottids. The presence of actin filaments in microtriches suggests that they move, a feature that would enormously enhance their effectiveness as an exchange surface with the rat's gut contents.

New proglottids are differentiated just distal to the scolex, and each develops a complete set of reproductive organs (Fig. 10.39). While there is a marked constriction between proglottids, this does not completely separate the individual proglottids, and the entire tape is linked by a continuous pair of excretory canals and lateral nerve trunks.

Hymenolepis (251) may self-fertilize (sperm from one proglottid transferring to eggs in another proglottid), and cross fertilization may occur between multiple worms when an infection involves several individuals. Once fertilized, eggs grow within the uterus. As the eggs develop, the proglottid enlarges to form a gravid proglottid (Fig. 10.39B), in which the

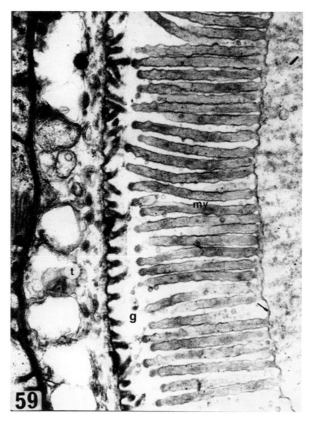

Figure 10.38 A section cut through part of a tapeworm (*Pseudanthobothrium hanseni* (254)) (t) and the spiral valve in the gut (g) of the host, the thorny skate (*Raja radiata* (1144)). Notice the dimensions and spacing of the microtriches on the tapeworm surface (about 0.3 μm long) and the microvilli on the lining of the gut (about 2.2 μm long).

SOURCE: A.E.Wilson, BSc Hons Thesis, UNB Biology Department, 1976. By permission of Mick Burt.

uterus is greatly enlarged and packed with eggs and other parts of the reproductive system disappear. Some eggs are released from gravid proglottids into the intestine and passed with the faeces, while others remain in the gravid proglottid, which breaks away from the tape and is shed in the faeces. *H. diminuta* produce up to 250 000 eggs a day (almost 10^8 eggs/year), and a worm may live for 2–3 years.

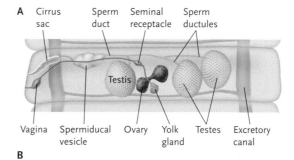

Figure 10.39 **A.** A mature proglottid of *Hymenolepis*; **B.** a gravid proglottid.

SOURCE: D. Kucharski & K. Kucharska / Shutterstock.com

The life cycle of *H. diminuta* (Fig. 10.40) requires that eggs be eaten by an intermediate host, typically the red flour beetle (*Tribolium castaneum* (949)). In the beetle the eggs develop to a cysticercoid stage which is transferred to the final host when a rat eats an infected beetle along with stored grains.

What generalizations about life as an endoparasite can be drawn from this summary of the biology of *Hymenolepis diminuta*?

- Transmission from one host to another usually involves an intermediate host.
- Successful transmission to an intermediate host is a low probability event so a very large number of infective stages must be produced.
- While the presence of a parasite may modify the behaviour of an intermediate host to make transmission to the final host more likely (see Section 10.4b), this remains a low probability event, further increasing the demands on reproductive output.
- The parasite must have an effective means of attaching to the host.
- The parasite must have a mechanism to take up food from the host.
- The parasite must be protected from the host's mechanical and immunological defenses (not mentioned in this summary of *Hymenolepis*).

STUDY BREAK

1. If a minor infestation of sea lice does not kill a salmon, why is it still a major problem to the industry?
2. Some species of cestodes are fixed in one place in the host intestine, other species are able to move within the intestine. What do you see as the advantages and disadvantages of each way of life?

10.4b Life Cycles and Reproductive Amplification

Animals in Class Digenea (Platyhelminthes) have complex life cycles involving two (and sometimes three) hosts. A typical digenean life cycle goes from an egg to a **miracidium larva** that infects a molluscan host and forms a **sporocyst**, from which a number of **redia** emerge (and redia may then produce daughter redia). The redia differentiate into a number of **cercaria** (and in some species **metacercaria**) which infect either a second intermediate host or the final host. Each larval stage, and individual larva, is a clone of the parental genome. The serial replication fits the requirement that a successful parasite must produce very large numbers of offspring to ensure at least one reaching a new final host. Each egg hatches to a sporocyst. If that sporocyst produces ten redia, each redia produces ten daughter redia, and then each redia releases ten cercaria, there would be 1000 copies of the genome. These animals often lay more than a thousand eggs, meaning that one parent can give rise to a million potentially infective cercaria. One digenean trematode, *Schistosoma* (265) (see Section 18.5c), meets our expectations of a parasite in that it has serious and negative effects on its host. Another interesting variant in a digenean life cycle is discussed in Chapter 5.

Another digenean, *Podocotyloides stenometra* (264), is a parasite of butterfly fish (*Chaetodon* spp. (1260), Fig. 10.41A) that feeds on corals. After *P. stenometra* eggs hatch, the miracidium larvae infect marine snails. Cercaria are released from the snails and infect a second intermediate host, a coral (*Porites* sp. (186)). The parasite triggers the development of pink swollen nodules containing metacercaria (Fig. 10.41B) on the surface of the coral, significantly reducing the coral's growth rate. Rather than avoiding the parasitized

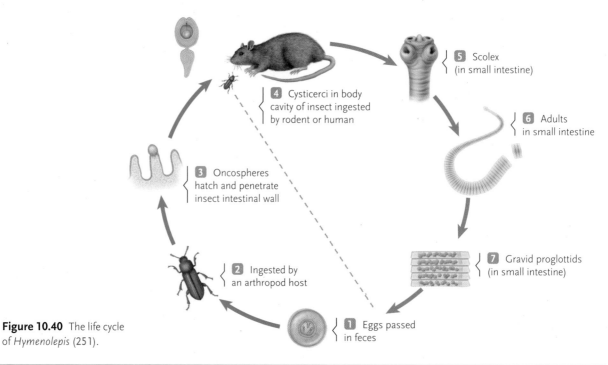

5 Scolex
(in small intestine)

4 Cysticerci in body cavity of insect ingested by rodent or human

6 Adults
in small intestine

3 Oncospheres hatch and penetrate insect intestinal wall

7 Gravid proglottids
(in small intestine)

2 Ingested by an arthropod host

1 Eggs passed
in feces

Figure 10.40 The life cycle of *Hymenolepis* (251).

Figure 10.41 A. *Chaetodon multicinctus*, a butterfly fish host of *Podocotyloides stenometra*. **B.** Pink nodules on *Porites* sp. are areas infected by cercaria of *Podocotyloides stenometra*. These nodules are a preferred food source for the butterfly fish.

nodules, *Chaetodon multicinctus* prefers the nodules as food because they offer significantly more tissue per bite than healthy *Porites*. It appears that the risk of trematode infection is relatively low and the infection does not have serious consequences to the fish, making the additional food a greater benefit than the risk of infection.

Haemonchus contortus (586) (Nematoda, Fig. 10.42) is a parasite in the abomasum (fourth stomach) of sheep and goats, where it feeds from the blood vessels supplying the abomasum. There may be as many as 5000 adult worms in the abomasum of a host. These worms use a lancet, formed from the dorsal wall of the buccal cavity, to pierce the blood vessels of the host, which may lose as much as 250 mL of blood a day as a result of these wounds. *Haemonchus* has a typical nematode life cycle. Eggs hatch as small nematodes, and growth involves a series of moults between larval stages (Fig. 10.43). If a female *Haemonchus* lays 10 000 eggs a day and half of these are female, in three or four generations (about three months) there are potentially 10^{12} worms.

Most members of the Class Monogenea (Platyhelminthes) are ectoparasites, many of them on fish. An exception is *Polystoma* spp. (243) (Fig. 10.44), found inside the bladder of frogs, where they feed on blood released when they damage capillaries in the bladder wall. The life cycle of *Polystoma* (Fig. 10.45), like other monogeneans, is simple. Parasite eggs are shed from the frog bladder into the water. The larvae (**gyrodactylids**, also known as **oncomiracidia**) attach to frog tadpole gills and transfer to the bladder as the tadpole metamorphoses to a frog, where they develop and reproduce.

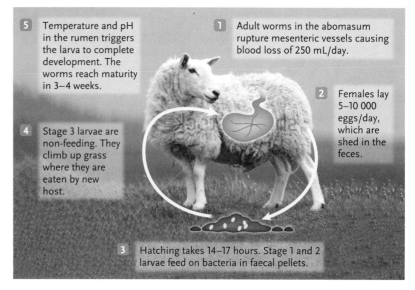

5 Temperature and pH in the rumen triggers the larva to complete development. The worms reach maturity in 3–4 weeks.

1 Adult worms in the abomasum rupture mesenteric vessels causing blood loss of 250 mL/day.

2 Females lay 5–10 000 eggs/day, which are shed in the feces.

4 Stage 3 larvae are non-feeding. They climb up grass where they are eaten by new host.

3 Hatching takes 14–17 hours. Stage 1 and 2 larvae feed on bacteria in faecal pellets.

Figure 10.43 The life cycle of *Haemonchus contortus*.

Two host-parasite interactions are superimposed on this simple cycle. The first is the timing of parasite reproduction, which is regulated by the gonadotropins released by the frog host as it controls its own reproductive timing. The second is the regulation of the development of the gyrodactylid larvae by the age of the tadpole on which it settles. If the tadpole is more than 14 days old, development is normal with the parasite moving to the frog bladder and developing there, followed by a period of 3–5 years before *Polystoma* reproduces in phase with its host. If the monogenean larva settles on a tadpole aged <14 days, it remains attached to the tadpole gills, develops rapidly to a neotenous adult, and reproduces in about three weeks,

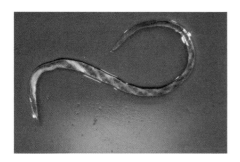

Figure 10.42 *H. contortus* (586) (the barber pole worm); females are 20–30 mm long, males are 10–20 mm long.
SOURCE: M. Blaxter, from an original photograph by E. Munn

Figure 10.44 *Polystoma galamensis* (243) (Monogenea). Notice the oral sucker (left end of the animal), the branched gut and the opisthaptor (posterior attachment organ) with its sucker + hook attachment structures.
SOURCE: Courtesy of Louis du Preez

CHAPTER 10 SYMBIOSIS AND PARASITISM

Figure 10.45 The life cycle of *Polystoma*.

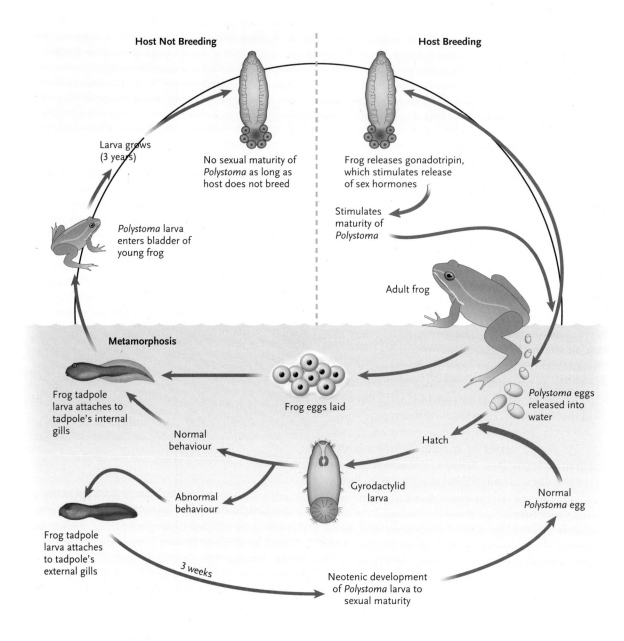

Host Not Breeding

Host Breeding

Larva grows (3 years)

No sexual maturity of *Polystoma* as long as host does not breed

Frog releases gonadotripin, which stimulates release of sex hormones

Stimulates maturity of *Polystoma*

Polystoma larva enters bladder of young frog

Adult frog

Metamorphosis

Polystoma eggs released into water

Frog tadpole larva attaches to tadpole's internal gills

Frog eggs laid

Hatch

Normal *Polystoma* egg

Normal behaviour

Gyrodactylid larva

Abnormal behaviour

Frog tadpole larva attaches to tadpole's external gills

3 weeks

Neotenic development of *Polystoma* larva to sexual maturity

producing larvae capable of infecting other tadpoles. The regulation of normal or neotenic development appears to be regulated by mucopolysaccharide and/or carbohydrate compounds released from the tadpole skin into the water.

10.4c Getting In and Holding On

The above examples suggest that a successful parasite is little more than a holdfast and a set of reproductive organs. This is almost true in the "mesoparasites": copepods (e.g., *Lernaeocera branchialis* (742)) in which the cephalothorax has been transformed to a holdfast that penetrates the tissues of a gadoid (cod and its relatives) fish host and drills into the aorta or bulbus arte-

riosus, leaving just the reproductive organs and egg sacs exposed.

The genus *Hymenolepis* (251) (see Section 10.4a.2) demonstrated that holdfasts can be structured for both temporary and long-lasting attachment to host tissues, and the salmon louse showed the hooked limbs used by some ectoparasites (Section 10.4a.1). Figures 10.46–10.48 illustrate the range of holdfast morphologies found in different parasite groups.

The monogeneans are commonly ectoparasites of fish and other aquatic animals. Attachment to a fast-moving fish places demands on the strength of an ecto-parasite's attachment. The Monogenea meet this demand by having both anterior (prohaptor) and poste-rior (opisthaptor) attachment structures (Fig. 10.47A–E).

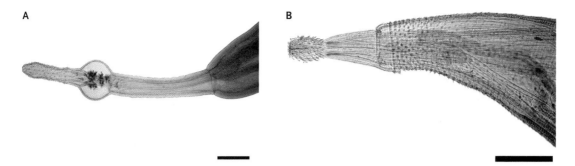

Figure 10.46 The thorny heads are the attachment structures from which the Acanthocephala (formerly a phylum, now viewed as a clade within the Rotifera) get their name. **A.** *Pomphorhynchus bulbocolli* (313) (its name means bulb neck), an acanthocephalan found in the gut of a number of fish species. **B.** *Corynosoma sp.* (310), a gut parasite in birds and mammals. Scale bars = 500 μm.

SOURCE: Allen Shostak

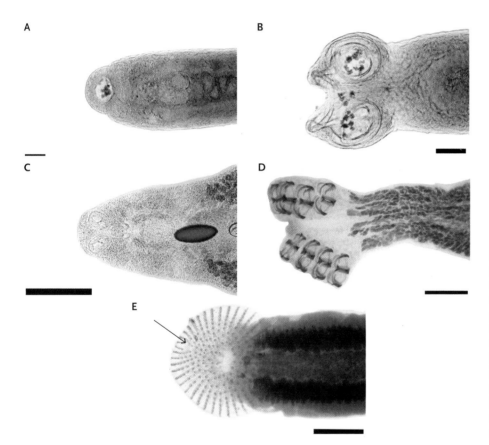

Figure 10.47 Monogenean attachment structures. **A.** The prohaptor of *Sphyranura* sp. (245) is a simple sucker while **B.** its opisthaptor has well-developed suckers and hooks. Scale bars = 100 μm. **C.** The prohaptor of *Discocotyle* sp. (242) (from lake whitefish, *Coregonus clupeaformis* (1314)) is a paired sucker while **D.** its opisthaptor has a complex array of hooks. Scale bars = 500 μm. **E.** Attachment structures may not be limited to the opisthaptor. *Pseudacanthocotyla pacifica* (244) (from a big skate, *Raja binoculata* (1143)) has a very small true opisthaptor (arrow) surrounded by a disk of hooks. Scale bar 500 μm).

SOURCE: Allen Shostak

Allen Shostak

Claire Healey

Claire Healey

Figure 10.48 A. The classic text image of a scolex is that of *Taenia solium* (256) (the pork tapeworm), which has a rostellum with hooks as well as four basal suckers. Other groups have quite different scolex structures that may have hooks but also large glandular suckers known as bothridia (or phyllidea). These muscular suckers, often ear-shaped, provide firm attachment to the host Examples include **B.** *Platybothrium tantulum* (253) from *Sphyrna lewini* (1135), a hammerhead shark and **C.** *Rhinebothrium* sp. (255) from a freshwater ray.

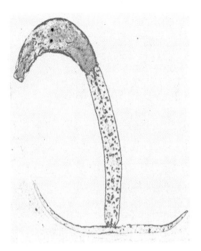

Figure 10.49 A cercaria larva of *Trichobilharzia* sp. (266), one of the freshwater digenean larvae that can cause swimmer's itch.

SOURCE: Cecilia Thors

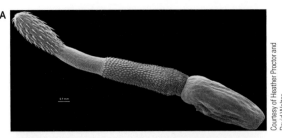

A

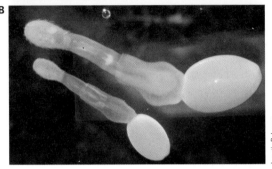

B

Courtesy of Heather Proctor and David Walter

Leontin Balanean

Figure 10.50 A. A reconstruction of an adult *Polymorphus marilus* (312) (Acanthocephala) from the gut of a lesser scaup (*Aythya affinis* (1657)). **B.** An everted cystacanth of *Polymorphus* dissected from its gammarid host.

Typically the opisthaptor is well developed and provides secure attachment while the less developed prohaptor allows temporary attachment while the parasite is moving on the surface of its host.

The attachment of tapeworms (Cestoda) is through the terminal scolex. Traditionally regarded as being anterior and now suggested to perhaps be posterior, scolices take many forms.

Endoparasites enter a host through the mouth in association with food, or by penetrating the skin as *Schistosoma* (265) does (see Section 18.5c). Other schistosomes that do not infect humans can be a problem in both fresh and salt water where they cause swimmer's itch in humans. Cercaria larvae (Fig. 10.49) of some schistosome genera that infect birds leave their molluscan intermediate hosts and, attracted up a temperature gradient to a warm-bodied host and perhaps to amino acids released from the host skin, attack the skin of humans and other non-host species. These cercaria use a sucker to attach to the skin, aided by mucus from a post-acetabular gland. Proteolytic enzymes from a pre-acetabular gland are the penetrating agents. The cercaria burrow through skin in about 30 minutes. In North America, *Trichobilharzia* (266) and *Gigantobilharzia* (261), common parasites of birds, are the major cause of swimmer's itch in fresh water. In coastal waters, *Microbilharzia* (263), a parasite of gulls, is the culprit.

The emerald sea slug (see Fig. 10.17) is an example of a symbiont incorporating genetic material from its partner. Most nematodes that parasitize plants have only one route into their hosts—through plant tissues. The root knot and cyst nematodes use stylets to penetrate plant cells and are among the most destructive of agricultural pests. Few animals have the enzymes needed to attack the components of plant cell walls, explaining the abundance of symbiotic associations in herbivores. However, root knot nematodes (e.g., *Meloidogyne incognita* (588)) have been shown to secrete at least six enzymes that degrade components of plant cell walls. The genome of *M. incognita* provides strong evidence that the enzymes were acquired from several different families of soil bacteria by lateral gene transfer.

10.4d Parasite Modification of Host Behaviour and Life Cycles

Many parasites manipulate the behaviour and physiology of their host to their own benefit. Adult *Polymorphus* (312) (Rotifera, Acanthocephala, Fig. 10.50) are parasites in waterfowl and some aquatic mammals. The intermediate host in their life cycle (Fig. 10.51) is typically an amphipod crustacean from the genus *Gammarus* (792).

The cystacanth larvae of some *Polymorphus* species are bright red and show through the amphipod exoskeleton (Fig. 10.51), perhaps making the amphipod host more visible to birds that eat them. Carotenoids

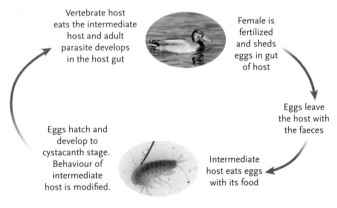

Vertebrate host eats the intermediate host and adult parasite develops in the host gut

Female is fertilized and sheds eggs in gut of host

Eggs leave the host with the faeces

Intermediate host eats eggs with its food

Eggs hatch and develop to cystacanth stage. Behaviour of intermediate host is modified.

Figure 10.51 The life cycle of *Polymorphus* sp. (312).

in the exoskeleton of a normal amphipod (Fig. 10.52A) give the animal a brown colour, protect it from UV light, and provide camouflage when the amphipod is on the bottom of the pond (as it normally is). When infected with a *Polymorphus* cystacanth, the gammarid loses its carotenoid pigments to the parasite, making the exoskeleton clear (Fig. 10.52B). In some conditions this allows the blue colour of its hemolymph to show through (as seen in the parasitized gammarid in the life cycle in Fig. 10.51). At the same time, the behaviour of the gammarid changes: it spends more time swimming at the surface (where it is more visible) or clinging to objects floating on the surface. This behavioural change makes it more likely that an infected gammarid will be eaten by a duck, thus transferring the parasite to its final host.

An extreme in the modification of the form and physiology of a parasite host is seen in the relationship between *Sacculina* (750) (Rhizocephala, meaning root-head) and its host. The Rhizocephala are barnacles (Crustacea, Cirripedia) that are parasitic on crustaceans (mainly decapods). A typical barnacle egg hatches to a nauplius larva (the first-stage larva of most crustaceans) that moults into a cypris larva, the stage that settles and metamorphoses to the adult barnacle form. In the Rhizocephala, the cypris settles on a crustacean host and moults to another larval stage, the kentrogon, which moults to form an elongate larva, the vermigon, that penetrates the host. Inside the host, the parasite develops as a branching series of root-like growths (known as the interna) that penetrate throughout the host (Figs. 10.53 and 10.54).

As the parasite matures, it produces an external growth that contains its female gonad (Fig. 10.55). This attracts a male cypris larva that moults to a **trichogon** larva that penetrates and parasitizes the female while producing sperm and fertilizing eggs (Fig. 10.54). The fertilized eggs are held in an external brood sac that is maintained by the host as if it were its own egg mass.

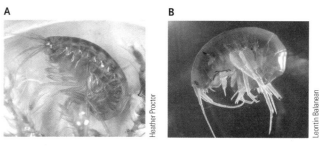

Figure 10.52 A. An unparasitized *Gammarus lacustris* (792), **B.** a gammarid infected by five *Polymorphus* (312) cystacanth larvae. These larvae have taken up the carotenoid pigment from their host, giving them the red colour and leaving the host exoskeleton transparent.

Figure 10.53 An illustration by the 19th-century German biologist, Haeckel, showing the externa of *Sacculina carcini* (750) (the pale oval under the tail) and its root system throughout the host body.

SOURCE: Crop of the 57th plate from Ernst Haeckel's *Kunstformen der Natur* (1904).

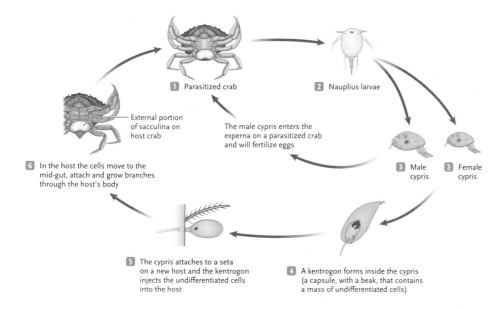

1 Parasitized crab

2 Nauplius larvae

3 Male cypris

3 Female cypris

External portion of sacculina on host crab

The male cypris enters the experna on a parasitized crab and will fertilize eggs

6 In the host the cells move to the mid-gut, attach and grow branches through the host's body

5 The cypris attaches to a seta on a new host and the kentrogon injects the undifferentiated cells into the host

4 A kentrogon forms inside the cypris (a capsule, with a beak, that contains a mass of undifferentiated cells)

Figure 10.54 The life cycle of *Sacculina* (750).

A

B

C

Hans Hillewaert

Michael Owen

Michael Owen

Figure 10.55 A. The externa of *Sacculina* (cream-yellow) between the triangular abdomen and the thorax of a male crab. **B.** Parasitic castration by *Sacculina* converts the appearance of the abdomen of a male *Carcinus maenas* (813) to look like **C.** that of a female.

As well as invading the host tissues and drawing nutrients from it, *Sacculina* has other subtle effects on its host. If the host is a male crab, the crab starts to take on the morphological appearance of a female with a broadening of the abdomen that would normally hold and protect the crab's egg mass. The wider abdomen now protects the externa of the parasite. At the same time, moulting is suppressed or completely blocked—protecting the externa from being removed with the moulted carapace. In some parasitized species when moulting does occur, it appears that the externa are shed but then regenerated and reinvaded by a male cypris that fertilizes the eggs produced by the newly developed externa.

Castration of a host by a parasite is not unique to the *Sacculina*–decapod host relationship. Parasitic castration of male copepods by protistan parasites has been reported, as has the feminization of male *Boeckella propinqua* (726) (a fresh water calanoid copepod) by the cysticercoid stage (the stage that infects the final host) of the tapeworm *Dibothriocephalus latus* (250).

STUDY BREAK

1. Why do parasites have such huge reproductive outputs?
2. In what ways could hosts control the life cycles of parasites?

10.5 Parasitoids

A parasite damages, but does not normally kill, its host. In contrast, a parasitoid typically kills its host (before the host is able to reproduce) in the process of its own development. Most parasitoids are insects, and a few are mites or nematodes. Typically the parasitoid coevolved with its host. The specificity of the relationship between parasitoids and their hosts has been exploited in many examples of biological control of insect pests (see "Why It Matters").

Like parasites, parasitoids face challenges in finding new hosts and completing their life cycle. These challenges have led to the evolution of many complex behaviours and specializations. Many parasitoids lay their eggs in a host egg, larva, or pupa, so it is clearly an advantage to not lay eggs in a host already parasitized by a conspecific. The chalcid wasp *Nasonia vitripennis* (1002) (Fig. 10.56) lays eggs in the pupae of muscid flies. A female *Nasonia* about to lay eggs will withdraw its ovipositor and seek a new host if the pupa already contains developing *Nasonia* larvae. Clearly the ovipositor has chemoreceptors that are able to identify these larvae.

A different challenge to an endoparasitoid is evading destruction by the host's immune system. Some parasitoid wasps inject a venom secretion with immunosuppressive molecules through the ovipositor. In others, the wasp has a symbiotic relationship with a virus from the family Polydnaviridae. The virus is injected into the host along with the parasitoid eggs and infects the host hemocytes, the source of most insect immune responses, as well as other tissues.

Like parasites, parasitoids maximize the probability of their offspring finding a new host by producing many larvae. Polyembryony (multiple embryos developing from a single fertilized egg) is one mechanism to achieve this. In wasps in the Family Encyrtidae, such as *Copidosoma* (993) (Fig. 10.57A), the egg divides holoblastically, and the cells continue division, without differentiation, to form a mass of cells known as a polygerm. As the polygerm grows, clusters of cells are grouped to form multiple embryos (as many as 2000 in some species), each of which becomes a developing larva. Some of these larvae (around 10%) develop precociously and form what are essentially mobile sets of jaws. These larvae, the soldiers, move through the host tissues killing larvae of other parasitic species, as well as clones of the same species. Soldier larvae may also have a role in manipulating the sex ratio of larvae that will develop fully.

Figure 10.56 *Nasonia* (1002) on a host pupa.

SOURCE: Hans Smid, http://bugsinthepicture.com

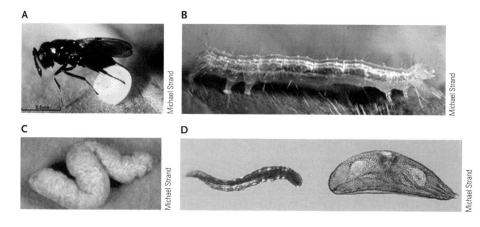

Figure 10.57 *Copidosoma* (993) life cycle. **A.** An adult oviposits in a lepidopteran host egg. **B.** There is clonal proliferation as the larvae develop within the host. **C.** The parasite larvae have killed the host and pupated. **D.** A soldier larva (left) and a normal larva (right).

Just as parasites may modify the behaviour of their hosts to its own advantage, so do many parasitoids. The parasitoid wasp *Glyptapanteles* sp. (Hymenoptera, Braconidae) lays its eggs in the first- or second-instar caterpillars of *Thyrinteina leucocerae* (1040) (Lepidoptera, Geometridae). Parasitized caterpillars develop to the fourth or fifth instar before about 80 wasp larvae emerge from their host and pupate. The host caterpillar then spins a silk covering over the parasitoid pupae and responds to disturbance with head swings that dislodge any intruder threatening the pupae (Fig. 10.58). Shortly after the adult wasps emerge from the pupae, the caterpillar host dies.

Figure 10.58 A caterpillar of *Thyrinteina leucocerae* (1040) guarding pupae of *Glyptapanteles* sp. (997).

SOURCE: Prof. José Lino-Neto

SYMBIOSIS AND PARASITISM IN PERSPECTIVE

This chapter has reviewed some of the morphological, physiological, behavioural, and life cycle adaptations associated with the relationships of parasitic and symbiotic animals. In some cases these adaptations are so extreme that early analyses placed parasites in separate taxonomic groups from their free-living relatives. Reviewing some of the samples of relationships illustrates how hard it can be to draw hard and fast lines. When does a commensal relationship become a parasitic one? How would you differentiate between a parasitoid and a predator? When looking over the detail presented in the chapter, it is important to also look for the generalities that emerge from the study of animal associations.

Questions

Self-Test Questions

1. Which of the following statements is describes the parasitic lifestyle?
   a. Parasitism is a very rare form of life in the animal kingdom.
   b. The study of parasitism is basically a study of specialized feeding systems.
   c. Parasitism is difficult to study because all parasites are small and inside other organisms in a symbiotic association.
   d. Each parasitic form may have associations with many different host species at the same time, which makes the study of parasitism challenging.

2. Which of the following pairs best illustrates a symbiotic association that benefits the partner species?
   a. hydrothermal vent fauna such as the tubeworm *Alvinella pompejana*; zooxanthellae
   b. bryostatins isolated from larval bryozoans; sulphide-oxidizing bacteria
   c. corals (cnidarian polyps); a filamentous green alga such as *Trichophilus welkeri*
   d. escae of deep-sea anglerfish; bioluminescent bacteria such as *Vibrio fischeri*

3. What is the symbiotic association between ants and acacia trees that makes it a great example of the coevolution of insects and flowering plants?
   a. Acacia flowers produce a repellent to keep ant guards away from potential pollinators.
   b. Ants defend the acacias from phytophagous insects, browsing megafauna, and even other plants.
   c. Acacias provide the ants with domatia for shelter, and food is available from extrafloral nectaries and Beltian bodies.
   d. All of the above show evidence of coevolution.

4. In comparing the general features of the lives of ectoparasites and endoparasites, it is apparent that:
   a. Only ectoparasites require a stage in the life cycle that is away from the host.
   b. Transmission of endoparasites is such a low probability event that many infective stages must be produced for successful transmission.
   c. Endoparasites require additional protection from the host's immunological defences.
   d. Ectoparasites must have enhanced features for attaching to a host.

5. Which of the following illustrates an adaptive response to the challenges faced by parasites in successful transmission to a new host?
   a. Infected coral develops pink swollen nodules full of metacercariae, preferred by Butterfly fish as they gain more tissue per bite than when feeding on healthy coral.
   b. *L. salmonis* copepodids are sensitive to kairomones released by potential host fish.
   c. Neotenic development by polystome larvae that land on young tadpoles, is triggered by secretions from tadpole skin.
   d. All of the above illustrate adaptive responses to the pressure of successful transmission.

Questions for Discussion

1. Many parasites have complex life cycles. Select different examples of such parasites and speculate on the way in which these life cycles may have evolved.

2. Why are symbionts particularly important to deep-sea animals?

3. Discuss the functions of bioluminescence.

4. Use the electronic library to find a recent paper that explores some aspect of the biology of parasitism and symbiosis. Does the new information revolutionize any of the information in this chapter?

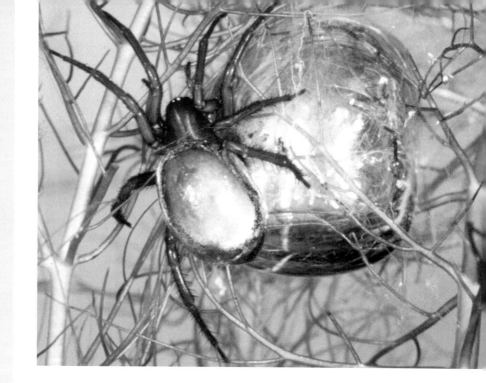

Figure 11.1 A spider (*Argyroneta aquatica* (661)) with a web-encased bubble of air (a diving bell).

SOURCE: Claude Nuridsany & Marie Perennou / Photo Researchers, Inc.

11 Gaseous Exchange

WHY IT MATTERS

Different species of terrestrial arthropods, such as the diving spider shown here, use bubbles of air when diving. This allows them to stay underwater for relatively long periods of time, like people using diving bells. The diving spider uses specialized silk to contain the air bubble (Fig. 11.2). This approach to diving depends upon a **plastron** (Fig. 11.3), a hydrophobic surface (hydrofuge hairs or cuticular projections) that creates an airspace on the spider's body surface. The plastron allows diving arthropods to take dissolved oxygen from the water, through the thin film of air that creates an incompressible exchange surface, into their gaseous exchange system.

Plastrons provide a large area of water-surface contact with the spiracle opening into book lungs and tracheae of a spider or the tracheae of an insect. Plastrons have evolved repeatedly and independently in terrestrial arthropods, and have also appeared in some arthropods living in habitats subject to flooding, where they are key to survival even for species that do not dive. The amblypygid *Phrynus marginemaculatus* (655) provides an excellent example. This denizen of the Florida Keys that regularly flood has a well-developed plastron (Fig. 11.3).

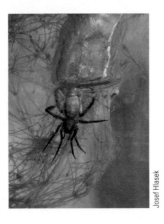

Figure 11.2 Specialized silk fibres form the diving bell of the water spider, *Argyroneta aquatica*.

Josef Hlasek

Diving bells for humans go back to around 384 BC when Aristotle described a prototype: "...one can allow divers to breathe by lowering a bronze tank into the water. Naturally the container is not filled with water but air, which constantly assists the submerged man." Some people used diving bells designed on this principle before the 17th century. The weighted chamber enclosed an air space in which a diver could breathe while working under water. When the enclosed air became unbreathable, the bell was raised to the surface to refresh the air. The development of air pumps in the 17th century allowed a major advance in dive bell technology: air could be continuously supplied to the underwater space. In 1619 Sir Edmund Halley patented a bell that incorporated ideas that (with the later addition of a non-return valve) would be used for the following 300 years.

Although insects do not come to mind when we think of marine invertebrates, more than 150 insect species are restricted to the intertidal zone. These include flies (Diptera) in the families Chironomidae, Tipulidae, Dolichopodidae, and Canaceidae, as well as some beetles (Coleoptera), bugs (Hemiptera), and at least one caddis fly (Trichoptera). Although the larval forms of many of these insects, particularly the early instars, effect gaseous exchange across the cuticle, some (especially species of tipulid flies) have plastron-bearing spiracular gills. The intertidal zone typically has well-aerated water but is alternately flooded and dry. In these situations the plastron can be used to extract oxygen from water or from air.

Plastrons also occur widely in the eggs of insects. The chorion (outer layer) of the eggs of many insects is a respiratory surface, penetrable to oxygen and carbon dioxide. Eggs of some species, particularly those subject to desiccation, are embedded in a hygroscopic "jelly." But the chorions of most insect eggs laid on land serve as plastrons. The surface of many of these eggs turns the chorion into a plastron, with a distinct architecture of fine fibers that forms a mesh layer capable of trapping air.

The features that make plastrons effective are similar to those used to achieve water repellency in many biological and industrial processes. Technically, hydrophobicity of a surface (water repellency) is enhanced by reducing surface energy through chemical modification of the surface. The modifications increase the angle of contact of a water drop toward 120°. When angles of contact exceed 150°, the surfaces become superhydrophobic. In addition to chemical modification, artificial superhydrophobic surfaces can be etched or machined. Increased surface roughness provides large geometric areas in relatively small projected areas (Fig. 11.3).

Engineers often mimic the surface features of animals and plants to achieve water repellency. We can learn from animals when it comes to developing water repellent surfaces, whether we are trying to keep snow from adhering to windows, reducing frictional drag on the hulls of ships, or trying to produce a stain-resistant textile.

SETTING **THE SCENE**

Gaseous exchange is the means by which animals acquire oxygen and rid themselves of carbon dioxide. The process includes exchange of gases with the respiratory medium, transport of gases in the circulatory system of the animal, and exchange at the tissues to supply mitochondria with oxygen and to remove carbon dioxide. This process is controlled by basic physical laws (e.g., Fick's Law). The situation for gaseous exchange, however, varies according to both the environmental setting and the animal itself. Animals living in water face different challenges from those living on land. Larger animals have greater demands for oxygen and produce more carbon dioxide. In this chapter we will explore the diversity of gaseous exchange in animals.

11.1 Basic Principles of Gaseous Exchange

Access to oxygen is a basic requirement of animal life to support the energy transfers involved in metabolic activity. However, oxygen is not uniformly available or accessible. There are huge differences in the availability of oxygen between air and water as well as across

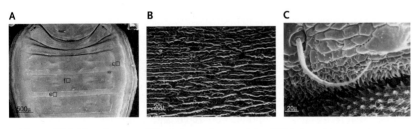

A   B   C

Figure 11.3 The plastron of *Phrynus marginemaculatus* (655) viewed by scanning electron microscopy. **A.** Ventral view of opisthosoma showing plastron with **B.** a series of transverse stripes. **C.** Opening to book lung.

SOURCE: Reprinted from *Journal of Insect Physiology*, 46, Hebets, E.A. and R.F. Chapman, "Surviving the flood: plastron respiration in the non-tracheate arthropod Phrynus marginemaculatus (Amblypygi: Arachnida)," Pages 13–19, Copyright 2000, with permission from Elsevier.

Table 11.1	Availability of Gases in Water and Air.		

The tremendous increase in the oxygen capacity of air compared to water results in a much more effective rate of diffusion and consequently a reduced energy requirement for gas exchange.

	Water	Air
Viscosity	100x	x
Density	1000y	y
Diffusion Rate	Low	High
O_2 mL $\cdot$ L^{-1}	0–10	100–130
CO_2 mL $\cdot$ L^{-1}	0–13	>100
O_2 extraction	<80%	25%
% of energy budget to run pump that drives breathing, whether air or water	20%	1–2%

SOURCE: From RUSSELL/WOLFE/HERTZ/STARR. Biology, 1E. © 2010 Nelson Education Ltd. Reproduced by permission. www.cengage.com/permissions

Table 11.2	The Impact of Temperature on Oxygen Availability for Goldfish (*Carassius auratus* (1194)).	

The inverse relationship between water temperature and gas saturation is opposite to the changes in metabolic rate exhibited by aquatic organisms. Warmer water becomes more readily saturated with oxygen and can therefore hold less of the dissolved gas, whereas the higher temperature favours higher metabolic activity and greater oxygen demand by the goldfish. The amount of oxygen is further reduced in salt water.

	5°C	35°C
O_2 available	9 mL $\cdot$ L^{-1}	5 mL $\cdot$ L^{-1}
Goldfish need	8 mL $\cdot$ kg $\cdot$ h^{-1}	225 mL $\cdot$ kg $\cdot$ h^{-1}
Ventilation rate	1.3 L $\cdot$ kg $\cdot$ h^{-1}	60 L $\cdot$ kg $\cdot$ h^{-1}

SOURCE: From RUSSELL/WOLFE/HERTZ/STARR. Biology, 1E. © 2010 Nelson Education Ltd. Reproduced by permission. www.cengage.com/permissions

temperatures (Tables 11.1, 11.2). Body size and shape, as well as metabolic rate, also affect access to and need for oxygen as well as the excretion of carbon dioxide. Body size and shape were introduced in Chapter 2 along with the progression from acellular to multicellular aggregations, and then to multicellular organisms exhibiting obligatory interdependence between cells. A brief summary of the **respiratory surfaces** (gaseous exchange surfaces) of animals in "major" phyla is presented in Table 11.3.

11.1a Simple Diffusion

Simple diffusion is basic to the movement of gases (and other substances) from one place to another. We can calculate the rate of simple diffusion using the equation for Fick's Law:

$$Q = \frac{DA \times (P_1 - P_2)}{L}$$

where:

Q = the rate of simple diffusion
D = the coefficient of diffusion
A = the area over which diffusion takes place
$(P_1 - P_2)$ = the concentration gradient (for gases, expressed as a difference in partial pressures, the relative contribution of a gas to the total pressure of a mixture)
L = the path length, the distance over which diffusion occurs.

The bulk or mass of an organism is generally proportional to its volume. When an organism is small in volume, it has a large relative surface area. The ratio of its surface area:volume is large. For example, a typical bacterium has a surface area:volume ratio of 6 000 000:1, while for a typical protist it is 60 000:1. At these sizes, the large relative area (A in Fick's Law) at the surface allows simple diffusion to supply enough oxygen and remove carbon dioxide. Bacteria and protists are small,

but some much larger animals also rely entirely on diffusion to acquire oxygen and expel carbon dioxide. For example, although some flatworms (Platyhelminthes) are several centimetres long, their flat shapes translate into a large surface area relative to their volume (Fig. 11.4). This, in turn, means there is a short path length (L) from the external environment to any internal cell, allowing basic diffusion to address the needs of gaseous exchange.

However, a round or spherical animal that is several cm in diameter has a much greater volume relative to its surface area, and so requires specialized structures to acquire oxygen and excrete carbon dioxide. Now, path length (L) puts most of the animal's cells out of the range over which passive diffusion is effective for gaseous exchange. We find specialized structures for gaseous exchange in those animals (adults or larvae) where diffusion does not meet this need. Clearly this is also influenced by metabolic rate—which drives the need for gas exchange—and the nature of the animals' outer coverings.

Fick's Law focuses our attention on how animals use **ventilation** (gas exchange with the environment) and **perfusion** (gas delivery within the body) to maximize diffusion. As they breathe, animals use simple diffusion to remove oxygen from, or add carbon dioxide to, the air or water adjacent to the respiratory surfaces. This lowers the concentration of oxygen in the air or water in the animal's immediate environment, and raises it on the other side of the respiratory surface. The concentrations of carbon dioxide on either side of the surface change in the other direction as the CO_2 produced by metabolism is lost to the environment.

For either gas, the concentration gradient $(P_1 - P_2)$ maximizes diffusion, as molecules move readily from an area of high concentration to lower concentration. By ventilation, animals move air or water past the respiratory surface, increasing P_1, while perfusion moves

Table 11.3 | **Respiratory Surfaces in Animal Phyla.**

Phylum	Body Surface	From Water			From Air			Respiratory Pigments			
		Gills	Book gills	Plastron	Lungs	Book lungs	Trachaea	None	Hemocyanin	Hemerythrin	Hemoglobin
Porifera	X							X			
Cnidaria	X							X			
Placozoa	X							X			
Ctenophora	X							X			
Platyhelminthes	X							X			
Nemertea	X										X
Gastrotricha	X							X			
Cycliophora	X							X			
Entoprocta	X							X			
Gnathostomulida	X							X			
Micrognathozoa	X							X			
Syndermata (Rotifera + Acanthocephala)	X							X			
Annelida	X	X						X		X	X*
Mollusca	X	X							X		X
Phoronida	X										X
Brachiopoda	X									X	
Dicyemida	X										X
Bryozoa	X							X			
Sipunculida										X	
Chaetognatha	X							X			
Kinorhyncha	X							X			
Loricifera	X							X			
Priapulida	X									X	
Nematoda	X										X
Nematomorpha	X							X			
Tardigrada	X							X			
Onychophora							X	X			
Arthropoda	X	X	X	X	X	X	X		X		X
Hemichordata	X							X			
Echinodermata	X										X
Chordata	X	X			X						X

* Chlorocruorin is a hemeprotein present in the blood of many annelids, particularly marine polychaetes.

Figure 11.4 A swimming flatworm showing extensive body surface. The large surface area:volume ratio is ideal for gas exchange by diffusion.

Jung Hsuan / Shutterstock.com

body fluids on the other side of the surface—most obvious in animals that transport gases in blood—and so lowers P_2. In this way, animals use Fick's Law to their advantage to maximize diffusion.

Animals must adjust ventilatory activity according to the internal concentration of carbon dioxide to ensure that appropriate levels of water or air reach the respiratory surfaces. In animals as diverse as mammals, insects, and pulmonate molluscs, specialized neuronal structures sense changes in the levels of carbon dioxide and pH. In mammals, these sensors are distributed throughout the medulla of the brain, and cells excited by carbon dioxide also connect to blood vessels. Close association of these sensors to cerebrospinal fluid and blood allows them to monitor levels of carbon dioxide (or related changes in pH) in these fluids, and cause changes in ventilation to maintain appropriate levels. In *Helix* species (Mollusca, Gastropoda, Pulmonata), similar sensors control the size of the opening of the mantle cavity, the **pneumostome**, to regulate ventilation and so adjust CO_2 levels and pH. In this way, changes in intracellular pH, of specialized receptor cells accurately reflect changes in ventilation. In the face of changes in extracellular pH, it is important that the mechanisms used by other cells to regulate internal pH be inactive in chemosensory cells, so that the response to ventilatory cues does not wane over time. The similarity of operation among mammals, insects, and terrestrial molluscs suggests an ancient lineage of sensors controlling ventilation across respiratory surfaces. Analogous structures that directly or indirectly govern the movement of oxygen and/or carbon dioxide must be part of any gaseous exchange system.

11.1b Pressure

The availability of oxygen in air varies considerably from sea level to the top of Mount Everest and beyond. At sea level, atmospheric pressure is 760 mm of mercury (mm Hg), which declines dramatically with increasing altitude (676.4 at 1000 m; 601.9 at 2000 m; 424.2 at 5000 m). At any altitude nitrogen (N_2) accounts for 78% of the volume of air, oxygen (O_2) 21%, and carbon dioxide (CO_2) and other gases, ~1%. The contribution of oxygen to 760 mm Hg (the **partial pressure** or PO_2) is calculated by taking 21% of 760, so PO_2 is 160 mm Hg (0.21×760). At 5000 m, the PO_2 is 89 mm Hg (0.21×424.2).

The values obtained from calculating the simple rate of diffusion (Q from Fick's Law), the effects of surface area:volume ratios on diffusion distance (L), or partial pressures and the resultant concentration gradients, strongly influence life on Earth by affecting animals' access to oxygen. The impact of pressure will be completely obvious to the person acclimated to living at sea level (PO_2 = 160 mm Hg) who, when transported to 2000 m altitude, experiences fatigue, nausea, and dizziness caused by relative lack of oxygen. Even though the atmosphere still contains 21% O_2, the lower PO_2 (0.21×601.9 = 126 mm Hg) reduces the concentration gradient and the subsequent rate of diffusion.

The other side of pressure is what happens underwater. At sea level, air pressure is 760 mm Hg, but below the surface things change quickly. At 10 m below the surface, the water pressure is 1520 mm Hg, 2280 mm Hg at 20 m, 3040 mm Hg at 30 m. These realities are very important for any animal that dives to or lives at depth. We will return to these issues when we consider the cases of animals that fly high or dive deep.

11.1c Respiratory Water Loss

Water vapour lost to the atmosphere during "breathing" is a basic cost of living for terrestrial animals and plants. One model predicts that water loss is proportional to gas exchange with an exponent of 1 (Fig. 11.5). The amount of water lost depends upon many factors, including surface temperature of the respiratory surface, the gas consumed, the steepness of concentration gradients for gas and vapour, and whether the mode of water transport is convective or diffusive. Achieving higher levels of oxygen uptake at the cost of increasing water loss is not a viable trade-off for many animals, particularly those living in dry terrestrial situations.

STUDY BREAK

1. What does Fick's Law describe? How are the components of this law relevant to anatomical adaptations for gaseous exchange?
2. How do the partial pressures of O_2 and CO_2 compare at sea level? Do the partial pressures of these gases vary with altitude? What mechanisms allow animals to maximize the differences in partial pressure relevant for gaseous exchange?

11.2 Sites for Gaseous Exchange

Animals too large to rely on simple diffusion for gaseous exchange have a specialized respiratory surface for that purpose. Fick's Law plays a major role in determining the size and shape of the respiratory surfaces and their suitability for gaseous exchange. Animal respiratory surfaces include skin, gills, lungs, tracheoles, mantle cavities, and others (see Table 11.3). Respiratory surfaces can be very large (increasing A in Fick's Law), for example, a 20-kg sea bass has more than 9 m^2 of gill lamellae! The respiratory surface is typically in close association with the animal's circulatory system, minimizing the diffusion distance (L) and creating a close relationship between ventilation and perfusion. When oxygen is being extracted from water, its uptake by blood is enhanced by **countercurrent exchange** (see Box 11.2). This occurs when the water flows in one direction and the blood in the opposite direction, maximizing the concentration gradient ($P_1 - P_2$).

Sites for gaseous exchange often change over the life cycle of an animal, and even according to the availability of oxygen in its environment. Although gaseous exchange is achieved using a number of very different respiratory surfaces, the constraints identified in Fick's Law always apply, as we shall see. There are many examples of parallel and convergent evolution in respiratory surfaces.

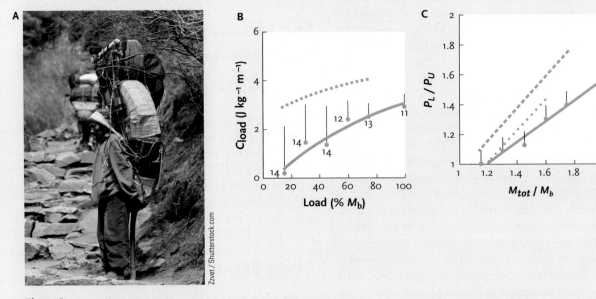

A

B

C

Figure 1

A. *Nepalese porters use a head strap to support a basket containing their load, and rely on a T-shaped stick to support the basket during rest periods.* **B.** *The metabolic cost of carrying 1 kg over 1 m is presented as C_{load} $(J \cdot Kg^{-1} \cdot m^{-1})$. The solid line and symbols show data for Nepalese porters, the dashed line the data for fit European hikers.* **C.** *Power increase $(P_L/P_U; power loaded/power unloaded)$ is plotted against relative size of load (M_{tot}/M_b) compared to body mass of subject. In economical terms, Nepalese porters carrying heavy, head-supported loads had smaller load costs than either African women carrying loads on their heads (dotted line) or fit European hikers (dashed line).*

SOURCE: From Guillaume J. Bastien, Bénédicte Schepens, Patrick A. Willems, Norman C. Heglund. 2005. "Energetics of Load Carrying in Nepalese Porters," *Science* 17 June 2005: Vol. 308 no. 5729. p. 1755. Reprinted with permission from AAAS.

How much do you weigh? Imagine carrying a loaded backpack weighing over 150% of your body mass about 100 km along a trail, involving ascents of about 8000 m and descents of about 6300 m. In the Himalayas in Nepal, porters carrying loads of up to 183% of their body mass take 7–9 days to cover the 100 km from Kathamandu to Namche (Fig. 1).

Nepalese porters walk slowly for many hours each day, taking frequent rests, 15 s of walking punctuated by 45-s rests. The details of how the Nepalese porters achieve this feat remain unclear.

Figure 11.5 Water loss scaled to gas exchange (O_2 for animals). Solid lines are derived from a fitted linear model. Data, as silhouettes, are shown for insects, bird eggs, birds, and mammals.

SOURCE: Woods, H.A. and J.N. Smith. 2010. Universal model for water costs of gas exchange by animals and plants. PNAS, 107:8469-8474.

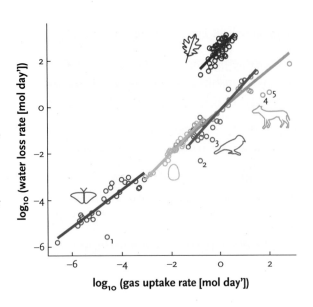

Metabolic rate has a major influence on the demand for oxygen, impacting both ventilation and perfusion. A typical resting primate has a heart rate of 200 beats/min and 8 $cm^2 \cdot g^{-1}$ of lung surface area, whereas a flying bat may have a heart rate of ~1200 beats/min and 100 $cm^2 \cdot g^{-1}$ of lung surface area.

Oxygen and carbon dioxide must be in solution in order to cross the epithelium of the cells that form the respiratory surface. For aquatic animals that extract dissolved oxygen from water, this is already the case. For terrestrial animals, it means that a thin film of water must cover the respiratory surface, whether the lungs of a vertebrate, the book lungs of a spider, or the tracheae of an insect.

Carbon dioxide produced by animals may be in solution, but more often occurs as bicarbonate (HCO_3^-). Typically, animals have little capacity to store or

BOX 11.2
Countercurrent Exchange: Maximizing the Concentration Gradient

In countercurrent exchange, blood leaving the capillaries has the same O_2 content as fully oxygenated water entering the gills.

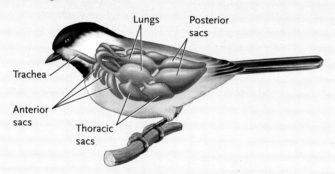

Figure 1

Countercurrent exchange: oxygen from the water diffuses into the blood, raising its oxygen content over the entire blood vessel/water flow interface. The percentages indicate the degree of oxygenation of water (blue) and blood (red).

SOURCE: From RUSSELL/WOLFE/HERTZ/STARR. *Biology*, 2E. © 2013 Nelson Education Ltd. Reproduced by permission. www.cengage.com/permissions

A. Lungs and air sacs of a bird

Figure 2

*The countercurrent exchange system in birds is the most efficient system of gas exchange in vertebrates. **A.** Lungs and air sacs of a bird. **B.** The unidirectional air flow is produced by a system of air sacs in conjunction with the lungs, such that air flows in one path through the entire system. With each aspect of the respiratory cycle, blood flows in the opposite direction of the air flow, so that the gradient for exchange is maintained throughout.*

SOURCE: From RUSSELL/WOLFE/HERTZ/STARR. *Biology*, 2E. © 2013 Nelson Education Ltd. Reproduced by permission. www.cengage.com/permissions

B. Crosscurrent exchange

Cycle 1

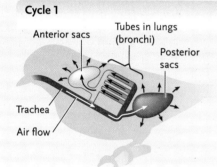

1 During the first inhalation, most of the oxygen flows directly to the posterior air sacs. The anterior air sacs also expand but do not receive any of the newly inhaled oxygen.

2 During the following exhalation, both anterior and posterior air sacs contract. Oxygen from the posterior sacs flows into the gas-exchanging tubes (bronchi) of the lungs.

Cycle 2

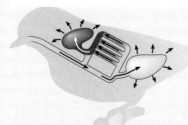

1 During the next inhalation, air from the lung (now oxygenated) moves into the anterior air sacs.

2 In the second exhalation, air from anterior sacs is expelled to the outside through the trachea.

Animals with external gills have the luxury of continuous water flow over their respiratory surfaces. This provides an almost limitless gradient for exchange between blood and water. Animals with internal gills, where the access to the respiratory medium is somewhat more limited, rely upon other physical mechanisms to make the most of the concentration gradient. Countercurrent exchange relies upon a basic engineering principle, useful in transferring energy or molecules across a gradient by enhancing the gradient. When the fluids between which the exchange is occurring travel across the exchange surface in opposite directions (Fig. 1), the gradient for exchange is maintained across the entire length of the surface. For fish gills, this means that gaseous exchange occurs readily across the gill surface at any point, because there is always more oxygen in the water than in the blood in the neighbouring vessels.

The arrangement of the flow across the tissues and in the vessels is such that fully oxygenated water first passes over a gill at the point where the blood flowing beneath it in the opposite direction is almost fully oxygenated. The oxygen in the water, however, is at a higher concentration than in the blood, so a gradient exists and the gas diffuses from the water into the blood, raising the concentration of O_2 in the blood almost to the level of the fully oxygenated water (Fig. 1). At the opposite end of the gill surface, much of the O_2 has been removed from the water, but the blood flowing in the gill filament, which has just returned from the venous circulation of the animal, is deoxygenated, and so contains even less O_2. As a result, there is still a gradient, and O_2 diffuses from water to blood. This relationship exists at every point along the gill filament: the water is always more highly oxygenated than the blood in the adjacent capillaries, and O_2 diffuses from water to blood (Fig. 1).

While most prevalent in fish, countercurrent exchange also makes the most of the opposing flow of air in the parabronchial system of birds. This is assisted by flow through the air sacs and lungs (Fig. 11.29) and the blood in the vessels that support the respiratory system (Fig. 2).

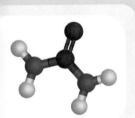

BOX 11.3 Molecule Behind Animal Biology

Respiratory Pigments: Hemoglobin and Hemocyanin

Hemocyanin and hemoglobin (Fig. 1) are the two respiratory pigments commonly found in the blood and/or tissues of animals. Both are allosteric proteins whose shape is altered when they bind to oxygen. These tetrameric (quaternary) structures or their protein subunits (globins) have been found in virtually all living animals. Globins are more consistent in form in vertebrates (where they are tetramers and monomers) compared to the situation in

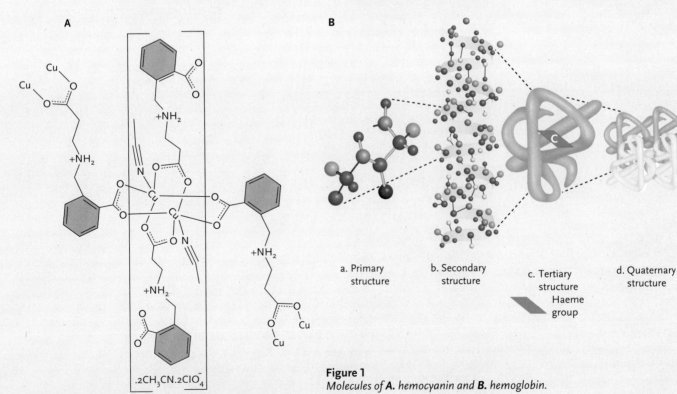

Figure 1
*Molecules of **A**. hemocyanin and **B**. hemoglobin.*

other animals. Vertebrates may have up to four types of globins distributed in the body: hemoglobins in blood, myoglobins in muscles, neuroglobins in nervous tissue, and cytoglobin in nuclei.

The affinity of hemoglobin or hemocyanin for oxygen is affected by pH. Specifically, the hemoglobin and hemocyanin of many animals loses its affinity for oxygen as the concentration of carbon dioxide increases and pH decreases. Known as the Bohr effect (Fig. 12.8), the change in affinity ensures that oxygen is released when blood pH is lower and the concentration of carbon dioxide increases, as they do in tissues where oxygen has been consumed (Fig. 2).

Many fish have two kinds of hemoglobin, anodic hemoglobin (He^{An}) and cathodic hemoglobin (He^{Ca}). He^{An} shows the typical Bohr effect, releasing oxygen as pH and concentration of carbon dioxide increase. He^{Ca} shows an anti-Bohr effect, namely increasing affinity for oxygen with rising pH and concentration of carbon dioxide. In the catfish *Hoplosternum littorale* (1330), a bimodal breather, He^{Ca} appears to serve a role in regulating metabolism. Hemoglobin with an anti-Bohr effect is known from some other bimodally breathing fish, as well as in some frog tadpoles and salamanders.

Hemocyanins show many similarities in operation to hemoglobins and may have as much

diversity in detail of structure and function. Ancestral globin genes had appeared by at least 1.9 billion years ago as oxygen started to accumulate in air and in water. The original function of globins may have been to scavenge oxygen, as well as to bind toxic carbon monoxide and nitrous oxide, to remove it from tissues.

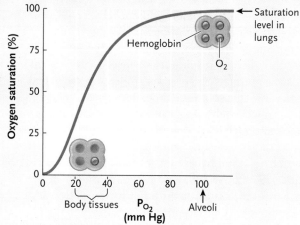

A Hemoglobin saturation level in lungs

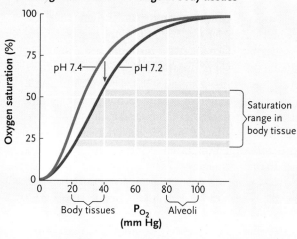

B Hemoglobin saturation range in body tissues

Figure 2
*Hemoglobin saturation in **A**. lungs and **B**. body tissues.*

SOURCE: From RUSSELL/WOLFE/HERTZ/STARR. *Biology*, 2E. © 2013 Nelson Education Ltd. Reproduced by permission. www.cengage.com/permissions

transport carbon dioxide as a gas. The H$^+$ ions released in the production of bicarbonate may increase the buffering capacity of blood or body fluids involved in the business of gaseous exchange. Because oxygen does not dissolve readily in body fluids, it is typically carried by specialized proteins known as **respiratory pigments**. The affinity of respiratory pigments to oxygen is responsible for distributing oxygen and increasing its concentration in the circulatory fluid well above the fluid's capacity to carry oxygen in solution. Hemoglobin and hemocyanin are two common respiratory pigments (see Box 11.3). Variations on these pigments, hemerythrin and chlorocruorin, are used as respiratory

pigments by some invertebrates. Insects are noteworthy in that their circulatory system plays a minimal role in transporting oxygen.

11.2a Surface Layers and Skin

The simplest animals use simple diffusion for gaseous exchange (provided the surface is wet so gases can be in solution), again driven by the principles of Fick's law. Although there are different designs of sponges, all members of the Phylum Porifera rely on a unidirectional flow of water that passes directly over the cells in a predictable path called an **aquiferous system**. The same current that drives filter feeding (see Figs. 8.16 and 8.18) functions for gaseous exchange. This dependence on simple diffusion limits the size of sponges. However, it has favoured designs that improve the surface area:volume ratio of larger sponges by either alternating surface inpockets with interior outpockets or restricting the flow of water to minute capillary-like vessels that enter and leave small chambers (see Fig. 8.16).

Simple diffusion is also used for gaseous exchange in Cnidaria, where the tentacles and general body wall serve as exchange surfaces. The circulation of water over the body surface by ciliated epidermal cells maintains the concentration gradient between the cells and the gases dissolved in water.

Aquatic or endoparasitic animals are more likely to use the skin as a respiratory surface, because many terrestrial animals have waterproof coverings that severely limit gaseous exchange. Both aquatic and terrestrial annelids rely mainly on the skin as a respiratory surface, although many polychaetes have gills on some or all of their parapodia. Very small crustaceans, such as water fleas and barnacles, take advantage of their small surface area:volume ratio by making use of the thin, permeable cuticle on the inner surface of the carapace as their only surface for gas exchange.

11.2b Gills

Gills extract dissolved oxygen from water. They are branched and folded evaginations of the body that provide a large surface area as a respiratory surface. The epithelial cell layer of the areas of gaseous exchange is typically very thin—a single-cell layer—minimizing L in the equation for Fick's Law. Although the gills of chordates, arthropods, annelids, and molluscs all serve as respiratory surfaces, they are each derived from different ancestral structures, so they are analogous rather than homologous.

At the macroscopic level we can distinguish between internal and external gills. Animals such as nudibranch molluscs and some salamanders (Fig. 11.6) have **external gills**, as do a variety of other molluscs, some annelids, and horseshoe crabs, as well as the larvae of animals such as aquatic insects, fish, and amphibians.

Echinoderm gills are hollow, thin evaginations (or invaginations) of the body wall that are ventilated with seawater externally and perfused with coelomic fluid internally. Tube feet, also involved with locomotion, serve as the gills for the water vascular system of all echinoderms, although this group shows many variations on this theme (see Section 11.3f).

Most animals with gills, however, house them in chambers of the body, protecting them from dangers such as abrasion, particulate matter, and desiccation. **Internal gills** occur in arthropods such as crustaceans, in many molluscs, and in vertebrates (cartilaginous and bony fish). Chambers housing internal gills must be ventilated by a current of water to maintain the concentration gradient of oxygen described in the equation for Fick's Law. In some cases chambers are ventilated by the actions of cilia (e.g., bivalve molluscs), directly by muscle contraction (the mantle chambers of cephalopods or the buccal cavity of tadpoles), or indirectly by the action of muscles on structures such as setae associated with the base of chelipeds in crabs (see Section 11.3d). In adult bony fish, the gill cover (one on each side of the head) is called an operculum (Fig. 11.7). Crustacean gills are external to the exoskeleton, but similarly enclosed within the carapace in most species. In adult elasmobranchs, spiracles and gill slits provide separate openings to the gill chamber.

11.2c Lungs and Tracheae

Lungs are internal structures for gaseous exchange. They range from small thin-walled sacs with relatively little folding or pocketing (as in amphibians and pulmonate land snails) to large structures with many infoldings and pockets (reptiles, birds, and mammals) that increase the surface area for exchange. In terrestrial arachnids, book lungs serve the same function, as do mantle cavities of pulmonate snails or the labyrinth organs of some air-breathing fish. Internal lungs require ductwork—a **trachea**—to transport gases from the surface of the animal to the site of gaseous exchange.

The **tracheae** of arachnids, onychophorans, and terrestrial insects serve a similar function. These tubular invaginations of the epidermis extend inward from

A

Joe Belanger / Shutterstock.com

B

Andrew Burgess / Shutterstock.com

Figure 11.6 External gills. **A.** The nudibranch *Flabellina iodinea* (452), and **B.** axolotl (*Ambystoma mexicanum*) (1405).

Figure 11.7 The operculum or external gill covering in a piranha (*Serrasalmus niger* (1186)).

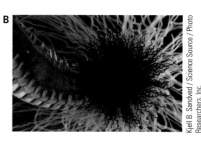

Figure 11.8 Polychaete gills. **A.** Modified parapodia of the Pacific lugworm (*Abarenicola pacifica* (314)) and **B.** specialized gills of the spaghetti worm (*Amphitrite ornata* (346)).

openings on the body surface called **spiracles**. Rather than ending at large respiratory surfaces such as lungs, however, the tracheae of arthropods branch into ever-smaller tubes, ultimately delivering oxygen directly to individual cells. In this way, gaseous exchange occurs without direct involvement of the circulatory system, in contrast to the mechanisms found in other groups that ensure close contact between respiratory systems that work in air and the blood or other body fluids, matching ventilation with perfusion.

STUDY BREAK

1. Compare the operation of lungs, gills, and skin as respiratory surfaces. How does each of these systems reflect the principles described by Fick's Law?

11.3 Respiratory Surfaces by Phylum

Most "primitive" animals take advantage of their great surface area:volume ratio by exchanging gases with the environment across their surfaces by simple diffusion. This is characteristic of the sponges (Phylum Porifera), members of the Phyla Cnidaria and Ctenophora, flatworms (Phylum Platyhelminthes), and ribbon worms (Phylum Nemertea). While simple diffusion remains predominant, with increasing body size and complexity we see more elaborate mechanisms to ensure gaseous exchange with all tissues. These respiratory mechanisms work in conjunction with more complex systems of internal circulation (Chapter 12) to match ventilation with perfusion.

11.3a Phylum Annelida

The skin (sometimes just specialized regions) is the usual respiratory surface in annelids, although in larger polychaetes, specialized gills serve as respiratory surfaces. These gills are unprotected outgrowths of the body surface and are most common in tube-dwelling species. Polychaete gills are often modified parts of parapodia (Fig. 11.8A), specifically the dorsal cirrus. But some tube-dwelling polychaetes have long thread-like gills, each attached to the base of a notopodium (Fig. 11.8B). These are ciliated and contractile so they pulsate rhythmically. In other polychaetes, the scale worms, respiratory surfaces are located on the dorsum and shielded by **elytra**, specialized plates on short stalks. Ventilation mechanisms maintain the concentration gradient. This is achieved in the sea mouse *Aphrodita aculeata* (316) by beating cilia, whereas tube-dwelling polychaetes ventilate their gills by muscular contractions of the parapodia or even peristaltic movements of the animal's body in the tube.

Hemoglobin, the common respiratory pigment in annelids, may be found in coelomic fluid, blood, muscle, and nerve cells. Although hemoglobin is packaged in coelomocytes in the coelom, it is also often dissolved in blood plasma and other body fluids.

Pogonophorans (beard worms) are tube-dwelling polychaetes with pronounced regional specializations. In some pogonophorans (e.g., *Riftia* sp. (383)), hemoglobin gives the crown of tentacles their red colour. Symbiotic bacteria pack the mid-gut and oxidize sulphur-containing compounds. Hemoglobin binds and delivers both oxygen and sulphur-containing compounds to the bacteria (see page 328).

11.3b Phylum Mollusca

Gills in the mantle cavity are the usual respiratory surfaces in molluscs, but the diversity of species in this group means that there are many variations on, and exceptions to, this theme. The generalized gill structure (Fig. 11.9A) is **bipectinate**, with two rows of leaf-like gill filaments along a central axis, usually staggered to maximize exposure to water flowing through the mantle cavity, as in cephalopods. Many molluscs have reduced **monopectinate** gills, with filaments on only one side of the central axis. Each filament has an upstream **frontal margin** on one side and a downstream **abfrontal margin** on the opposite side, arranged so the gills divide the mantle cavity into ventral **inhalant** and dorsal **exhalent** chambers. Powerful **lateral cilia** generate the respiratory current that moves water

A Internal gills: clam

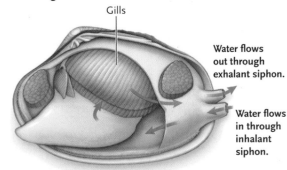

Gills

Water flows out through exhalant siphon.

Water flows in through inhalant siphon.

B Internal gills: cuttlefish

Water flows in around edges of mantle.

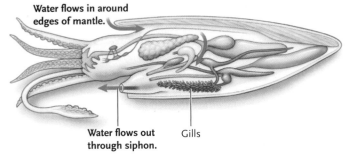

Water flows out through siphon.

Gills

Figure 11.9 The diversity of mollusc gills. **A.** Paired bipectinate gills of a bivalve and **B.** a cephalopod.

through the mantle cavity. Two other types of cilia in the gills of molluscs move mucus and particles, helping to keep the gill lamellae clean. **Osphradia** are sensory organs that monitor water coming into the mantle cavity, apparently to detect sediment or other contaminants. Detection of undesirable chemicals in the incurrent flow stops the beating of the lateral cilia and the incurrent water flow over the gills.

Gastropoda is the largest and most diverse molluscan class, including representatives from marine and freshwater habitats and the only terrestrial molluscs. As such, some gastropod species (prosobranchs) rely mainly on gills in the mantle cavity as respiratory surfaces, while in many families (pulmonates) the mantle cavity is converted into a lung-like structure for breathing in air. Still others (opisthobranchs) retain their gills but have, in many species, lost their shells (Fig. 11.6A). Torsion of the body mass of pulmonates during development rotates the visceral mass by 90–180°, and the elongation of a portion of the mantle beside the head forms a **siphon** on the left side of the animal near the head. This structure evolved repeatedly among aquatic gastropods, allowing them to choose the source of water drawn into the mantle cavity, away from the excurrent flow near the anus. This structure also has an important sensory function, especially when associated with chemoreceptors that allow the snail to test the water ahead of it or to follow a chemical trail.

In most cephalopod molluscs, a pair of bipectinate nonciliated gills (Fig. 11.9B) serve as the respiratory surface with some additional oxygen uptake occurring across the body surface. Some cephalopods (nautiloids) have four gills, while others (coleoids) have two. Muscular contraction of the mantle draws water into the mantle cavity, which turns dorsally and directs the water anteriorly and out across the gills. Large folded gill surfaces appear to compensate for the general absence of countercurrent exchange. The coupling of ventilation to "jetting" locomotion, which uses the thick circular musculature of the mantle wall to expel water from the mantle cavity, likely increases the ventilation of the gills and the efficiency of extracting oxygen from water, even in the absence of countercurrent exchange. Circulatory adaptations also increase the blood flow across the gills (Chapter 12); this is important for efficient and rapid gas exchange particularly for organisms that are swift, active, agile carnivores.

In bivalve molluscs, gills and the inner surface of the mantle cavity serve as respiratory surfaces. Structural adaptations to increase the surface area of the respiratory membrane allow for efficient gas exchange, despite the lack of countercurrent flow. This is most notable in the lamellibranchs, where the disproportionately large gill surface is also important in filter-feeding (see page 215).

In scaphopods (tusk or tooth molluscs), there are no gills in the mantle cavity. The cavity itself is ventilated by cilia arranged in dense rows (**preanal ciliary ridges**). The lining of the mantle is presumed to serve as the respiratory surface in these animals.

Respiratory pigments are important in the molluscs, although there is considerable diversity between forms. Hemocyanin, the most common pigment, is dissolved in blood plasma and used to transport oxygen from gills or lungs throughout the body. The freshwater pulmonate snails use hemoglobin for this purpose, and both forms of respiratory pigments are found in gastropod molluscs. Most bivalves lack respiratory pigments and rely on dissolved oxygen, but a few have hemoglobin and a smaller number, hemocyanin.

11.3c Phylum Onychophora

The living onychophorans are terrestrial and have well-developed tracheal systems that deliver oxygen directly to tissues. The trachea occur in tufts, which are invaginations scattered over the body surface, predominantly on the dorsal side. Trachea serve only the tissues close to the spiracle, which opens to the surface. These structures appear to exceed the levels of demand for oxygen and production of carbon dioxide in onychophorans, and may reflect ancestral conditions where the demand for oxygen was higher. Oncychophorans do not have respiratory pigments.

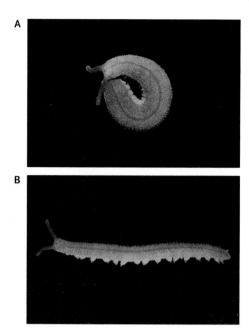

Figure 11.10 A. The curled and **B.** elongated body positions in a juvenile onychophoran (*Peripatopsis capensis* (627)).

SOURCE: Reproduced with permission from Clusella-Trullas, S., and S.L. Chown. 2008. "Investigating onychophoran gas exchange and water balance as a means to inform current controversies in arthropod physiology," *Journal of Experimental Biology*, 211:3139–3146. The Company of Biologists Ltd.

Available data indicate that the onychophorans studied to date experience high respiratory water loss relative to the situation in insects. The difference between the two groups is that onycophorans, unlike insects (see Section 11.3d), cannot close their spiracles. Curling behaviour reported in onychophorans (Fig. 11.10) may allow these animals to conserve energy and respiratory water loss.

11.3d Phylum Arthropoda

11.3d.1 Subphylum Crustacea. Most species of Crustacea are marine, but some live in fresh water and others are terrestrial. Aquatic crustaceans typically use gills as sites for gaseous exchange. These gills are found near the base of the appendages on structures called epipodites, and are housed in the branchial chambers. Specialized **gill bailers** ventilate the gill cavity. Hemolymph (blood) in the lamellae takes up oxygen dissolved in water in the same way as fish gills do. Sediment can accumulate in the protected branchial chambers and the gill lamellae are vulnerable to organisms that may grow on them and smother the respiratory surfaces. Aquatic crustaceans have developed several mechanisms for cleaning their gills and removing these organisms. In some species, grooming and/or cleaning chelipeds are used to brush the gills, while in others, gills are brushed by setae on the base of the thoracic legs. Setae involved in gill cleaning have similar structures across different species and appendages.

Terrestrial isopods (Oniscoidea) have respiratory surfaces across a thin cuticle (analogous to lungs);

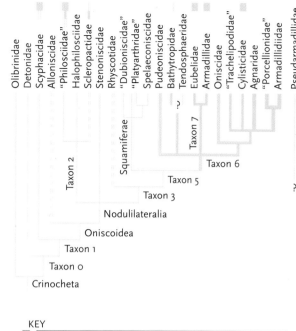

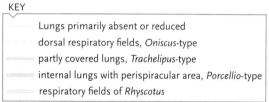

KEY

Lungs primarily absent or reduced
dorsal respiratory fields, *Oniscus*-type
partly covered lungs, *Trachelipus*-type
internal lungs with perispiracular area, *Porcellio*-type
respiratory fields of *Rhyscotus*

Figure 11.11 A morphology based phylogeny of Oniscoidea (isopods) showing the occurrence of pleopodal lungs (respiratory surfaces). Names of families shown in quotation marks identify units that are probably not monophyletic. Symbols above the names identify evolution or reduction of lungs within the taxon.

SOURCE: Schmidt, C. and J.W. Wagele, "Morphology and evolution of respiratory structures in the pleopod exopodites of terrestrial arthropods (Crustacea, Isopoda Oniscidea)," *Acta Zoologica*, 82:315–330. Copyright © 2002, John Wiley and Sons.

these have developed at least six times within this group (Fig. 11.11). These respiratory surfaces are located on the abdominal appendages, the **pleopods**, which typically lie flat against the underside of the abdomen. They are strongly wrinkled (= increased surface area) and are often enclosed in chambers that open via a spiracle to the outside.

11.3d.2 Tracheae of Insects (Subphylum Insecta). Insects typically use tracheae for gaseous exchange. These air-conducting blind tubes are invaginations of the outer epidermis consisting of a single layer of epithelial cells and a covering cuticle (Fig. 11.12). Air enters the tracheae through spiracles at the body surface. The tracheae branch repeatedly, opening into smaller and smaller vessels; the smallest are called **tracheoles**. Ultimately every cell in an insect's body is in contact with a tracheole which, at their ends, are <1 µm in diameter. The cells of large flight muscles, which often have the highest metabolic demand of any insect tissues, may be penetrated by the tips of tracheoles which are fluid-filled and the sites of gaseous

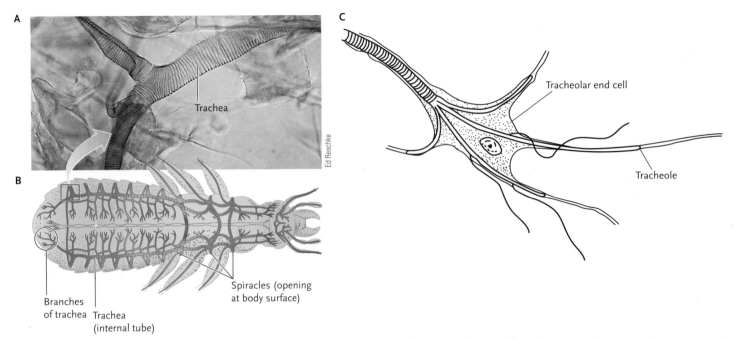

Figure 11.12 Tracheal systems of insects. **A.** Chitinous rings reinforce the tracheae and keep them from collapsing. **B.** Diagram of larval insect showing tracheal system in place. **C.** End of trachea showing tracheoles.

SOURCE: From RUSSELL/WOLFE/HERTZ/STARR. *Biology*, 2E. © 2013 Nelson Education Ltd. Reproduced by permission. www.cengage.com/permissions

exchange. This decreases the diffusion distance to the metabolically active muscle cells.

The insect system of gaseous exchange is extremely efficient. Flies (Diptera) flown to total exhaustion of energy show no evidence of lactic acid buildup or other products of anaerobic respiration. This raises the question about whether or not the tracheal system would have served the larger insects in the Carboniferous efficiently. These fossil insects include dragon flies (Odonata) with wing spans of 70 cm and mayflies (Ephemeroptera) with wing spans of 45 cm. Oxygen concentrations in the atmosphere of the Carboniferous were as high as 35%, compared to 21% today. Today, larger insects have disproportionately larger tracheal volumes than do smaller insects. Over their evolutionary history, notably in the Carboniferous, insects may have been able to reduce their investment in tracheae because of higher atmospheric oxygen concentrations, which would increase the efficiency of gas exchange through the increased concentration gradient. Reared at higher concentrations of oxygen, fruit flies (*Drosophila* (960)) reduce the diameter and overall volume of their tracheae relative to their body size. This could mediate the effects of differences in oxygen distribution through the tracheae, such as those arising from restricted diameters in long tracheae serving the brain and legs of large beetles. The higher oxygen concentrations in the Carboniferous may explain the occurrence of "gigantic" insects.

Insect tracheae can be open or closed (at the spiracle). Spiracular muscles control the size of spiracular openings and adjust the movement of air into and out of the tracheae (Fig. 11.13). In fruit flies (*Drosophila*), the area of open spiracles matches metabolic requirements by maintaining constant pressure for carbon dioxide and oxygen within the tracheae. Controlling the size of spiracles also reduces respirometric water loss. Previous suggestions that air moves along tracheae by passive diffusion have given way to the proposal that active processes are also involved. Changes in internal pressure, perhaps associated with muscle contraction in the abdomen, could change the volume of tracheae and actively draw or push air along the tubes. In a variety of insects (e.g., ground beetles, *Platynus decentis* (945); carpenter ants, *Camponotus pennsylvanicus* (990); and house crickets, *Acheta domesticus* (893)), rapid cycles of tracheal

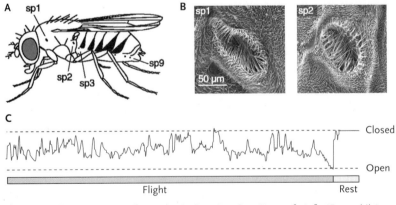

Figure 11.13 A. Locations of spiracles (sp1, sp2, sp3, sp9) on a fruit fly (*Drosophila*). **B.** A thick sclerite and protective hairs border each spiracle, and a narrow, flexible lid covers it. The elasticity of cuticular structures opens the spiracles, and the lids are held closed by spiracular muscles. **C.** Pattern of opening and closing of spiracles in flight and at rest in an individual *Drosophila mimica* (960).

SOURCE: From Lehmann, F., "Matching spiracle opening to metabolic need during flight in Drosophila," *Science* 30 November 2001: Vol. 294 no. 5548 pp. 1926–1929. Reprinted with permission from AAAS.

BOX 11.4

BOX 11.4
Oxygen and Flashing Fireflies

Fireflies (*Photinus* spp. (944)) produce rapid flashes of light by limiting the supply of oxygen to photocytes, light-emitting cells of the lantern organ. Bioluminesence requires enough energy to produce an excited single-state molecule that generates a single photon as it relaxes to its ground state. Two chemical reactions allow fireflies to produce bioluminescence.

$$ATP + luciferase + luciferin \rightarrow$$
$$Luciferase\text{-}luciferin\text{-}AMP + PP_i$$

$$Luciferase\text{-}luciferin\text{-}AMP + O_2 \rightarrow$$
$$Luciferase + oxyluciferin + CO_2 + AMP$$
$$+ light$$

It follows that bioluminescence could be controlled by controlling the availability of oxygen.

Scientists discovered how this mechanism works by using three ingenious steps, two involving adjusting access to oxygen, the third using electrodes to measure the delay between stimulation and peak light emission.

- Rapidly changing the supply of oxygen to the photocytes from very low to high (= normobaric hypoxia) turned on light emission.
- Onset of light emission could be adjusted using gas mixtures with differ-ent gas-phase diffusion coefficients for oxygen.
- The delay between stimulation of the lantern and peak light emission varied when the gas-phase diffusion of oxygen varied.

The mechanism of control appears to be the modulation of fluid levels in the tracheoles that supply the photocytes, in turn affecting the diffusion of oxygen across the respiratory surface to the photocytes. It remains to be determined if at least some of the other organisms that use bioluminescence use the same mechanisms to turn it on and off.

---

compression and expansion in the head and thorax have been visualized with a synchrotron beam. The magnitude of these changes exceeds that which had been proposed by active mechanisms. It is clear insects have a method analogous to the diaphragm or diaphragm-like structures that ventilate the lungs of vertebrates. While delivering adequate oxygen for metabolic processes is an obvious role in any system of gas exchange, limiting availability of oxygen in fireflies is also important for the process of bioluminescence (Box 11.4).

Because the circulatory system plays a minimal role in transporting oxygen in insects, most insects have no respiratory pigments. However, the surprising discovery of a hemocyanin pigment in the hemolymph of nymph and adult stoneflies (*Perla marginata* (915)) changed our view of gas transport in insects. In these stoneflies, the hemocyanin shows a moderate affinity for oxygen and so increases the oxygen content of the hemolymph. This discovery reinforces the suggestion that the loss of respiratory pigments is a derived condition in insects. The general absence of respiratory pigments in insects, particularly more evolutionarily advanced ones, underlines the primacy of the tracheae in gaseous exchange and transport. Indeed, hemocytes—insect blood cells—are the only cells in an insect's body not penetrated by tracheoles. During times of anoxia, some hemocytes in caterpillars change in structure and accumulate on tracheal tufts near the last (8th) pair of abdominal spiracles. Hemocytes also adhere to tracheae in the compartment at the tip of the abdomen (the tokus). Spending time on the tracheal tufts or in the tokus restores the normal structure of the hemocytes. This phenomenon is widespread in the larvae of Lepidoptera (moths and butterflies) and may be an example of a lung analogue in these insects.

11.3d.3 Subphylum Chelicerata. The ancestral chelicerate arthropods were aquatic and the first known fossils are eurypterids and horseshoe crabs from the Ordovician. Terrestrial chelicerates—Scorpiones—occurred in the Silurian and the first mites and spiders appear in the Devonian. By the Carboniferous, all extant orders of arthropods had appeared.

Chelicerate arthropods—spiders (Tetrapulmonata), horseshoe crabs (Xiphosura), eurypterids (Euripteryda) and scorpions (Scorpiones)—use book gills or book lungs for gaseous exchange. Horseshoe crabs (*Limulus polyphemus* (645)) have book gills and ventilate them with rhythmic movements of gill appendages (opisothosomal appendages, Figure 11.15C). Three classes of oxygen sensors in the gill nerve provide neural information that regulates the rate of ventilatory movements. Some of these units increase their activity in the presence of oxygen while the activity of others is depressed. The tactile sensiqtivity of other mechanosensitive units is oxygen-dependent.

Book gills, external structures on opisthosomal appendages (Fig. 11.15C), were the ancestral sites of gaseous exchange (in water), and they are homologous with book lungs. Specifically, the embryology indicates that book lungs and book gills develop as lamellae projecting from opisthosomal limb buds (Fig. 11.15). Book lungs consist of hemolymph-filled lamellae that project into an enclosed air space (Fig. 11.16). The book lungs of chelicerates are generally considered to represent an early stage of breathing oxygen in air. The primitive

BOX 11.5
Amphibious Caterpillars

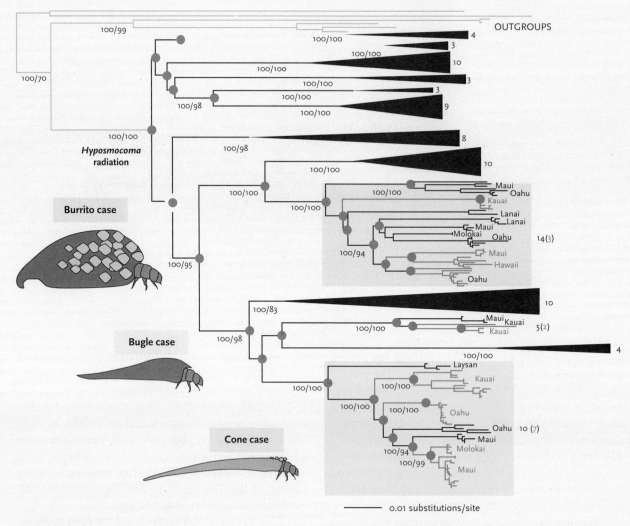

Figure 1

A molecular phylogeny of three lineages of amphibious pyralid moth larvae (Hyposmocoma (1038)) in Hawaii. Species per clade are shown as numbers to the right of each lineage. Branch lengths are drawn to scale. A Bayesian probability of at least 95% and nonparametric bootstrap values of at least 70% are presented under each corresponding node and clade.

SOURCE: Rubinoff, D., and P. Schmitz. 2010. Multiple aquatic invasions by an endemic, terrestrial Hawaiian moth radiation. PNAS, March 30, 2010 vol. 107 no. 13 5903–5906.

Cool, fast-flowing mountain streams are typically rich in dissolved oxygen, partly explaining the lungless frog of Borneo (see above). Other departures from typical approaches to gaseous exchange have been reported in caterpillars (*Hyposmocoma* spp. (1038), family Pyralidae) from Hawaii (Figure 1). These remarkable caterpillars are amphibious, faring as well in rushing mountain streams as on adjoining dry land. When in water, the caterpillars do not use gills, plastrons, or air bubbles, they appear to rely upon direct diffusion of oxygen across the hydrophilic skin along their abdomens. The same surfaces operate when the caterpillars are on land. Several species of these moths, endemic to Hawaii, have independently evolved amphibious behaviour, providing a striking example of parallel evolution.

Hyposmocoma caterpillars use special cases as camouflage and for protection whether on land or in water. Three different case types have been described as "burrito," "bugle," and "cone".

Hyposmocoma amphibious caterpillars have evolved at least three times, although most of the 350 species of *Hyposmocoma* are strictly terrestrial. The amphibious life style has evolved independently on five volcanic islands and first appeared about 6.3 Ma. Hawaii is the most isolated archipelago on Earth and its mountains are among the wettest places on the planet with repeated periods of heavy rains, rushing water, and flooding.

This example, along with others presented in this chapter (and throughout the book), illustrate a recurrent theme of adoption of novel solutions to common problems which, in turn, provide astonishing examples of the diversity of life.

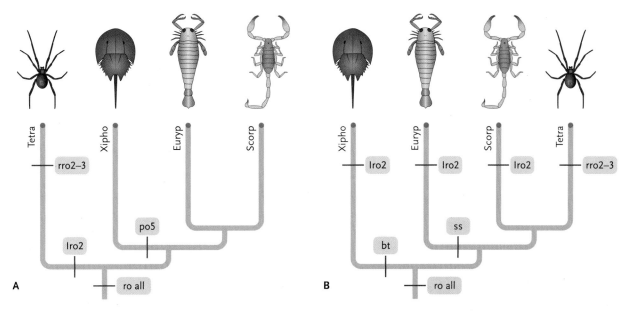

Figure 11.14 Compared here are **A.** a hypothetical and **B.** a traditional phylogeny of chelicerates based on the distribution of respiratory organs. Abbreviations: TETRA = tetrapulmonates, XIPHO = Xiphosura, EURYP = Eurypterida, SCOR = Scorpionida, rro 2–3 = restriction of respiratory organs to opisthosomal segments 2–3, iro 2 = loss of respiratory organs on opisthosomal segment 2, bt = basitarsus, po5 = postabdomen of five segments, ss = slit sensilla, and ro all = respiratory organs on all opisthosomal segments (pleisomorphic state).

SOURCE: Based on Dunlop, J.A. 1998. "The origins of tetrapulmonate book lungs and their significance in chelicerate phylogeny," pp. 9–16 in Proceedings of the 17th European Colloquium of Arachnology, P.A. Selden (editor).

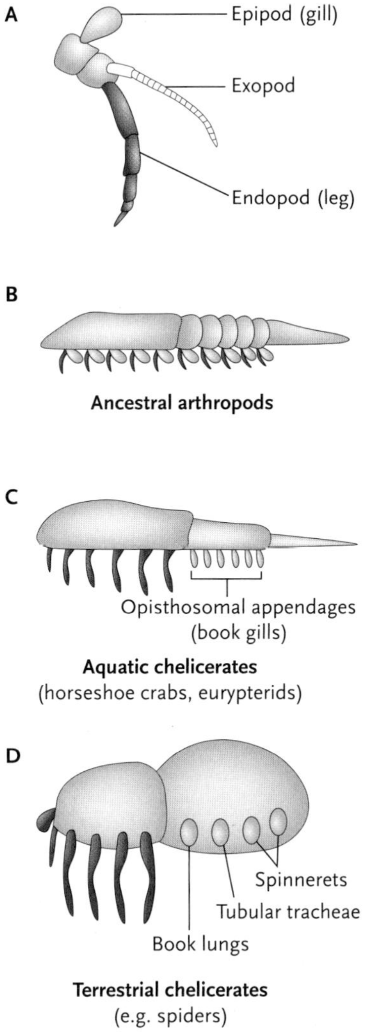

Figure 11.15 A. The typical branched arthropod appendage has a leg (endopod), a gill (epipod), and an exopod. **B.** Ancestral arthropods had limbs with distinct ventral and dorsal branches on most segments of their trunks. **C.** Chelicerates have two body regions: prosoma (anterior) and opisthosoma (posterior). Aquatic chelicerates have legs on the prosoma and book gills on the opisthoma. **D.** In spiders, like other terrestrial chelicerates, opisthosomal appendages have been transformed into book lungs, internal respiratory surfaces. Spinnerets are other opisthomal appendages used to spin webs.

SOURCE: Reprinted from *Current Biology*, 12, Daman, W.G.M., T. Sarkidaki, M. Averof, "Diverse adaptations of an ancestral gill: a common evolutionary origin for wings, breathing organs and spinnerets," Pages 1711–1716, Copyright 2002, with permission from Elsevier.

A
Epipod (gill)
Exopod
Endopod (leg)

B
Ancestral arthropods

C
Opisthosomal appendages (book gills)
Aquatic chelicerates
(horseshoe crabs, eurypterids)

D
Spinnerets
Tubular tracheae
Book lungs
Terrestrial chelicerates
(e.g. spiders)

condition in the Tetrapulmonata (Trigonotarbida, Araneae, Amblypygi, and Uropygi) is two pairs (four) of book lungs (Fig. 11.16).

The presence of book lungs and tracheal systems in some terrestrial arachnids raises questions about the role the two systems play in gaseous exchange. At times of high oxygen demand, the book lungs of the house spider (*Tegenaria* sp. (684)) must provide about twice as much oxygen per gram body mass as would be expected of an average vertebrate lung. Scaled to body mass, the diffusing capacity of the book lungs of this 100-mg spider is similar to that of a much larger turtle (*Pseudemys* (1427)) on a gram–body weight basis.

In spiders, oxygen demand is particularly high during moulting, and the efficiency of the tracheal system may reflect these periods of peak demand. The tracheal system also reduces the specific weight of a spider, which could minimize the chances of injury during a fall. It could also provide a better distribution of air throughout the body of water spiders (e.g., *Argyroneta aquatica* (661)), enhancing buoyancy. In any case, the relative impermeability of chitin to oxygen makes the 0.2 μm thickness of the respiratory surface the greatest barrier to diffusion of oxygen in spiders.

Two types of tracheae occur in conjunction with book lungs: **sieve tracheae** that are modified book lungs and **tube tracheae** that are analogous to the tracheae of insects and onychophorans. Sieve tracheae occur mainly in smaller arachnids, while tube tracheae occur in several groups. Tube tracheae are best developed in desert-dwelling solifugids, which have large,

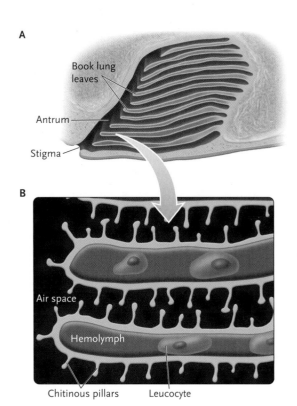

A

Book lung
leaves

Antrum

Stigma

B

Air space

Hemolymph

Chitinous pillars   Leucocyte

C

Book lung

Secondary
tracheae

Atrium

Primary
tracheae

Figure 11.16 Anatomy of the book lung of a spider. **A.** Main components (stigma, atrium, and leaves) and **B.** a close-up view of the leaves of the book lung. **C.** Book lungs and tracheal system of a female *Uloborus glomosus* (685).

well-ventilated tracheal systems. In some water mites, tube tracheae lack open stigmata and are blind tubes, separated from the surrounding water by areas of thin cuticle. In these mites and in other arachnids, a plastron (see above) enables them to use a thin layer of air as a physical gill.

Are the books lungs of arachnids ventilated? At least in a tarantula (*Eurypelma californica* (667)) weighing about 20 g, book lungs appear to operate by diffusion because there is little evidence of changes in lung volume. The low resting metabolic rates of terrestrial arachnids may explain the absence of ventilation. To put the book lung data in perspective, the tidal volume of air change in the lungs (0.25 mL) for a 20 g mammal is 16 250 times that in a 20 g tarantula. The difference in maximal oxygen consumption is 144 times higher in the mammal than in the tarantula.

11.3e Phylum Phoronida

The two genera of Phoronida (*Phoronis* (575) and *Phoronopsis* (578)) are sessile, benthic wormlike animals living in chitinous tubes that may be buried in sand or attached to objects in shallow water. These animals use hemocytes containing hemoglobin to transport oxygen from the respiratory surface (the lophophore). Enclosing hemoglobin in cells is unusual among invertebrates, but the small hemoglobin molecules in the hemocytes of phoronids have a high binding capacity for oxygen because of their large surface area. In this regard the oxygen capacity of phoronids is equivalent to that of most vertebrates. More commonly, invertebrates have large molecules of hemoglobin in blood plasma. The phoronids' approach to gaseous exchange has allowed

them to colonize the anoxic and hypoxic environments in which they are found and may explain why these animals do not ventilate the tubes in which they live.

11.3f Phylum Echinodermata

Echinoderms generally rely on thin-walled external processes, the tube feet, as gas exchange surfaces, but the surface area:volume ratio of these large animals favours the development of many types of specialized gills to supplement the gaseous exchange capability of the body surface. The tube feet of the water vascular system (WVS)—gill surfaces found in all echinoderms—provide surface area for simple diffusion. In many groups, additional surfaces provide gas exchange, complemented by circulatory mechanisms that enhance the ventilation/perfusion relationship. **Papulae** or **dermal gills** in Asteroidea are ciliated evaginations of the epidermis and peritoneum in which a countercurrent exchange is created between the coelomic fluid and the overlying water supply. Papulae are similar to the 10 invaginations of the body wall in ophiuroids, called **bursae**, which open to the outside through ciliated slits that help circulate water past the respiratory surface. In some species, additional musculature of the bursae helps pump water through the spaces, enhancing the diffusion gradient for gases.

Sea urchins (Echinoidea) possess five pairs of branched evaginations on the periostomium that have long been viewed as "gills" and the primary organs of gas exchange. These structures may also be involved in a mechanism to regulate pressure changes in the lantern complex during feeding, as well as an important source of oxygen to the lantern musculature. Sand dollars bear highly modified podia on the aboral **petu-loids** (the five ambulacral regions of the fused skeleton), which are named for their resemblance to the petals of a flower. These petaloid tube feet are actually broad flat gills over which the ciliary water current flows in the opposite direction to the flow of WVS coelomic fluid, creating a favourable gradient for gaseous exchange.

Most Holothuroidea (sea cucumbers) are suspension-feeders, so it's not surprising that feeding and gaseous exchange occur across the same surfaces. In addition to gas exchange across the external surfaces, highly branched outgrowths of the rectal region of the digestive tract form **respiratory trees** into which water is pumped, via the anus, for additional gas exchange. These "water lungs" arise as diverticula from the wall of the cloaca, and their blind-ended projection into the posterior portion of the coelom is similar to the mammalian lung configuration in the anterior body cavity. Respiratory trees are ventilated by muscular pumping of the cloaca and by contraction of the respiratory evaginations themselves, and serve as the primary surface for gaseous exchange in large-bodied species of echinoderms with thick body walls.

Coelomic fluid, not blood, is the typical transport medium for gases in echinoderms. Respiratory pigments are present in some forms, including hemoglobin-containing cells in the WVS and coelom of holothurians. Some large-bodied, sedentary burrowing species have been found to have hemoglobin-containing cells in the gut wall. Burrowing species of ophiurids also have hemoglobin in coelomocytes of the WVS.

STUDY BREAK

1. What groups of organisms show adaptations for gas exchange in both aquatic and terrestrial habitats? Choose three different animal groups and compare the adaptations between groups.
2. Describe the process by which countercurrent exchange enhances gas exchange. In what specific anatomical applications do you find this mechanism?

11.3g Phylum Chordata

Although we will see many different forms of respiratory organs in representatives of the Phylum Chordata, the only respiratory pigment in chordates is hemoglobin, which occurs in all but the tunicates.

11.3g.1 Urochordates and Cephalochordates. Tunicates (also known as Urochordata) have extensive branchial baskets that function in both gaseous exchange and filter feeding (Fig. 11.17). As such, these organisms lack specialized respiratory surfaces, as they also make use of their extensive surface area for direct diffusion. There is no evidence of respiratory pigments in tunicates, although the circulatory system is well developed. Again, large surface area relative to volume

A
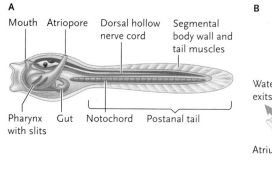
Mouth Atriopore Dorsal hollow nerve cord Segmental body wall and tail muscles

Pharynx with slits Gut Notochord Postanal tail

B
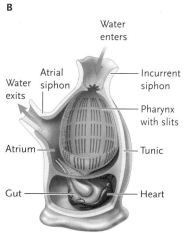
Water enters

Atrial siphon

Water exits

Atrium

Gut

Incurrent siphon

Pharynx with slits

Tunic

Heart

Figure 11.17 A. Tadpole-like larva and **B.** sessile filter-feeding adult urochordate. **C.** Internal anatomy of *Branchiostoma* (1122), a cephalochordate.

SOURCE: From RUSSELL/WOLFE/ HERTZ/STARR. *Biology*, 2E. © 2013 Nelson Education Ltd. Reproduced by permission. www.cengage.com/ permissions

C
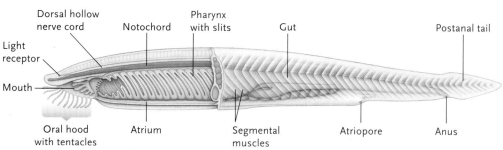

Dorsal hollow nerve cord Notochord Pharynx with slits Gut Postanal tail

Light receptor

Mouth

Oral hood with tentacles Atrium Segmental muscles Atriopore Anus

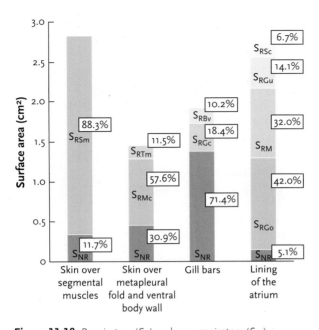

Figure 11.18 Respiratory (S_{Rx}) and non-respiratory (S_{NR}) surfaces of the four body regions in the amphioxis, where x is one of the following subscripts: Bv= blood vessel, Go = gonocoel, Gc = gill coelom, = Gu, gut coelom, M = muscles (segmental and transverse), Mc = metapleural coelom, Sc = subchordal coelom, Sm = segmental muscle, Tm = transverse muscle. Each of the coeloms indicated represent separate body cavities in amphioxis in which the respective anatomical structures are housed. Values are given for one hypothetical amphioxus of body mass 217 g.

SOURCE: Reproduced with permission from Schmitz, A., M. Gemmel and S.F. Perry. 2000. Morphometric partitioning of respiratory surfaces in amphioxus (Branchiostoma lanceolatum Pallas). *Journal of Experimental Biology*, 203:3381–3390. The Company of Biologists Ltd.

and low metabolic rates probably explain the lack of specializations for transporting oxygen and carbon dioxide. These situations for gaseous exchange appear to prevail in the three classes of tunicates: Ascidiacea, Thaliacea, and Appendicularia.

The branchial baskets of cephalochordates and urochordates differ substantially from the gills of vertebrates, primarily because these structures are well adapted for filter feeding. These organisms have ciliated gill bars—unknown in vertebrates—where the gills are often the main respiratory surface. The debate about the importance of the branchial baskets in gaseous exchange is obvious in the traditional use of the term "pharyngeal bars" rather than "gill bars."

Anatomical diffusing factors, the physical factors affecting gas exchange, were calculated (for cephalochordates such as amphioxus, *Branchiostoma lanceolatum* (1123)) as a function of the respiratory surface area relative to the body mass, with attention to the barrier thickness (diffusion distance in Fick's Law). Comparisons of respiratory and non-respiratory portions of different internal and external surfaces revealed that the lining of the atrium accounts for over 80% of gaseous exchange (Fig. 11.17 and 11.18). Other structures, such as the skin over the segmental muscles, skin over metapleural fold, and the gill bars

(tissues between gill slits), are much less important (~4% for gill bars). This strongly suggests that the "gills" of cephalochordates are used mainly in filter feeding rather than in gaseous exchange.

Further, the coelom of amphioxus may provide the main route for circulating oxygen within the body because the diffusing capacity of its blood for oxygen is only about 1% of total capacity. In amphioxus, the muscles alone account for 23% of the oxygen diffusing capacity, suggesting that they are self-sufficient for oxygen.

11.3g.2 Subphylum Vertebrata.

A. Fish. The gills of the three living lines of vertebrate fish are anatomically complex structures with tissues that perform several different functions. Gills are composed of several gill arches anchored to the pharynx which support complex arrangements of epithelial (external), and circulatory and neural (internal) components (Fig. 11.19).

The gills of lamprey (an agnathan) consist of one **hemibranch** and six pairs of holobranchs located in pouches on each side (Fig. 11.19B) for a total of seven paired gill pouches. A **holobranch** is a set of cranial and caudal hemibranchs. In lamprey with a parasitic life style, ventilation of the gill pouches is effected by contractions of the muscles surrounding the gill pouches to create a water flow ("tidal ventilation") when the animal is attached to a host and unable to effect a flow of water in through the mouth. Lamprey can reverse the flow of water through the gill pouches, presumably to clean the gill filaments and facilitate detachment from prey. The gills of larval lamprey (ammocoetes) are structurally similar to those of adults but differ in ventilation patterns because they are also used in filter-feeding.

Gills of hagfish (Fig. 11.19A) are also housed in pouches that differ from those of lamprey. Hagfish have 5–14 pairs of gill pouches, each with a discrete incurrent and excurrent opening that, depending upon species, may involve a common opening to the outside. Unlike all other vertebrates, hagfish gills lack well developed arches or supporting skeletal features. The gill filaments of hagfish are considered comparable to those of other fish, including lampreys, cartilaginous and bony fish. Water enters through a nasal opening, passes to the incurrent ducts, crosses the gill filaments, and exits via excurrent ducts. The pharyngeocutaneous duct in hagfish connects the pharynx to the environment, providing an alternative path for ventilating gill pouches and probably allowing ventilation during feeding.

In elasmobranch and teleost fish, gills are supported by bony or cartilaginous rods (gill rays) that radiate laterally from the internal base of each gill arch (Fig. 11.19C, D). The interbranchial septae are formed from connective tissues between gill rays. Septa support hemibranchs, rows of fleshy gill filaments

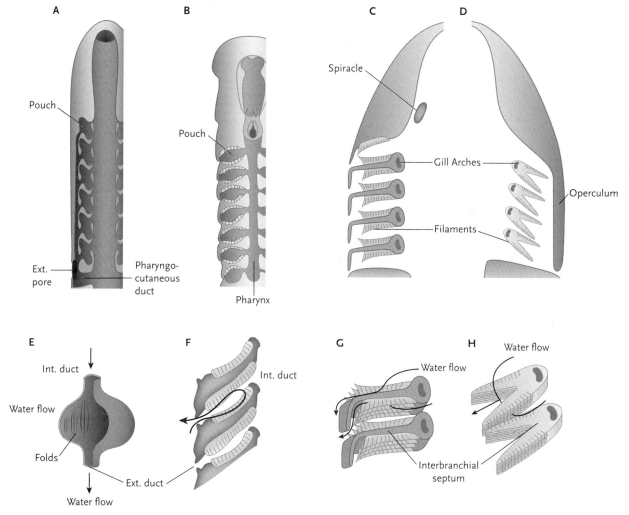

Figure 11.19 The generalized arrangements of gills and associated structures in hagfish (**A, E**), lampreys (**B, F**), elasmobranchs (**C, G**), and teleosts (**D, H**).

SOURCE: Based on Evans, D.H., P.H. Piermarini, and K.P. Choe. 2005. "Multifunctional fish gill: dominant site of gas exchange, osmoregulation, acid-base regulation, and excretion of nitrogenous waste," *Physiological Review*, 85:97–177.

running parallel to the gill rays on the cranial and caudal sides of each gill arch. Elasmobranch fish typically have four pairs of holobranchs in the branchial chamber and a pair of hemibranchs on the caudal side of the first gill arch. In elasmobranchs, gill slits are formed where each interbranchial septum extends from the base to the skin. In these fish, water enters through the mouth or spiracle (a cranial opening to the pharynx), passes across the gill filaments, and follows the interbranchial septa to exit through gill slits.

There are four pairs of holobranchs in teleost fish, where interbranchial septa are reduced in size compared to elasmobranchs. Reduced septa translate into gill filaments that move more freely. Teleosts lack spiracles, so water usually enters through the mouth, passes over the filaments, and exits by following the inner wall of the operculum. In some teleosts the operculum is modified to provide incurrent and excurrent openings.

Countercurrent exchange is fundamentally important to effective gaseous exchange in fish. Deoxygenated blood flowing from the heart flows in the opposite direction to the water moving through the fish's branchial cavities. This maximizes the gradients of oxygen and carbon dioxide across the gill epithelium, a combination of diffusion and perfusion (Fig. 11.20). The impact of countercurrent exchange is demonstrated by the situation in larval swamp eels (*Monopterus albus* (1337)). The eels' pectoral fins move oxygenated water toward the tail along the fish's body, while blood flows in the opposite direction under the fish's skin. Experimental manipulation of the direction of water flow demonstrates that there is much better uptake of dissolved oxygen during countercurrent rather than concurrent flow (Fig. 11.21). The same principle applies when gills are the respiratory surface—the common condition in fish.

In teleost fish, the usual pattern of ventilation of the gills involves a buccal force pump and an opercular suction pump. The two pumps work out of phase, ensuring an almost constant flow of water across the gill lamellae (Fig. 11.20). Specifically, the buccal force pump forces water through the gills when the mouth closes, while the opercular suction pump draws water through the sieve formed by the gills when the opercula

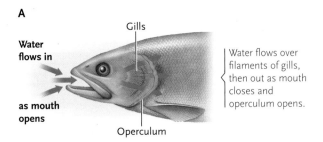

A

Gills

Water flows in

as mouth opens

Operculum

Water flows over filaments of gills, then out as mouth closes and operculum opens.

B The flow of water around the gill filaments

C Countercurrent flow in fish gills, in which the blood and water move in opposite directions

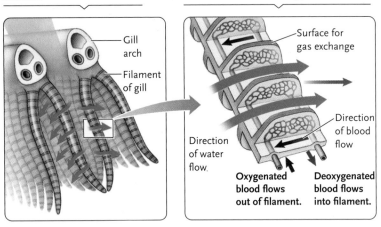

Gill arch

Filament of gill

Surface for gas exchange

Direction of blood flow

Direction of water flow.

Oxygenated blood flows out of filament.

Deoxygenated blood flows into filament.

Figure 11.20 A. Water flows in through a fish's open mouth, across its gills, and out through the operculum when the mouth is closed. **B.** Water flows around the gill filaments and **C.** across each filament against the flow of blood (countercurrent exchange). Countercurrent exchange enhances the gradient as oxygen diffuses from the water into the blood. (See Box 11.2).

SOURCE: From RUSSELL/WOLFE/HERTZ/STARR. *Biology*, 2E. © 2013 Nelson Education Ltd. Reproduced by permission. www.cengage.com/permissions

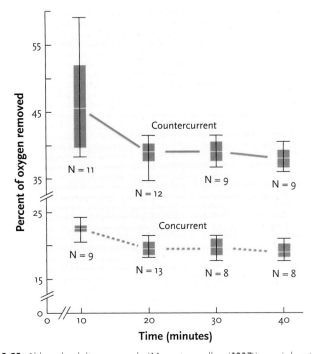

Figure 11.21 Although adult swamp eels (*Monopterus albus* (**1337**)) are air-breathers, their larvae use countercurrent exchange across the skin to obtain oxygen. Movements of the pectoral fins promote water flow from front to back, while blood flows from back to front. When the water flow is experimentally reversed (concurrent flow), less oxygen is extracted.

SOURCE: From Liem, K.F. 1981. "Larvae of an air-breathing fish as countercurrent flow device in hypoxic environments," *Science*, 211:1177–1179. Reprinted with permission from AAAS.

Dante Fenolio / Photo Researchers, Inc.

Figure 11.22 This fangtooth (*Anoplogaster cornuta* (**1175**)) swallows large prey whole (see also Fig. 8.50).

open. The close proximity of the tips of gill filaments of adjacent branchial arches maximizes water flow over a large surface area of gill lamellae by minimizing flow between gill arches. Gill ray adductor muscles can part gill filaments to expel foreign bodies that might lodge in them.

This general pattern of gill ventilation is not universal. Some fast-swimming pelagic fish do not ventilate their gills with muscular pumping, relying instead on their speed through the water to generate water flow across the gills with a ramming effect. For example, in scombrid fish such as mackerel and tuna, the ventilator stream is directed from buccal to opercular cavities. Other fish, such as a fangtooth (*Anoplogaster cornuta* (**1175**)) have a different problem. These fish swallow very large prey (almost their own body size) whole, and while swallowing must maintain water flow across the gills (Fig. 11.22). When swallowing large prey, *A. cornuta* distends its opercula, spreads its gill arches well apart, and uses its pectoral fins to direct water flow across the gills. This reverses the usual direction of water flow, reminiscent of the tidal ventilation in lampreys.

Other fish, such as banjo catfish (*Agmus lyriformis* (**1323**)) and winter flounder (*Pseudopleuronectes americanus* (**1308**)), expel jets of water through opercular valves to assist in locomotion. When burrowing in the sand, lesser weever fish (*Trachinus vipera* (**1265**)) use a combination of downward directed jets of water and fin movements. The arrangement of teeth and papillae in this fish filters sand from the water drawn into the buccal cavity.

The ability of some living fish to exchange respiratory gases in both air and water (**bimodal respiration**) is a recurring theme among living fish (agnathans, elasmobranchs, and teleosts) and molluscs. Although it is common for biologists to focus on the movement of vertebrates from water onto land with the associated changes in breathing, many lineages of teleost fish are successful because they can use bimodal respiration to make use of both aquatic and terrestrial habitats. Some fish, such as mudskippers (*Periophthalmus* sp. (**1280**), Fig. 11.23B), are often active on land where they may operate for several hours. Others, such as African lungfish (*Protopterus* (**1357**)), may spend months encased in cocoons, aestivating in dried mud. Some fish breathe

BOX 11.6
The Tambaqui, a Tasty Fat-Lipped Air-Breathing Fish

Under "normal" conditions, the tambaqui (*Colossoma macropomum* (1182), Actinopterygii) uses its gills to extract dissolved oxygen from water. When faced with hypoxic conditions, a dermal swelling on the fish's lower lip directs well-aerated surface water across the gills. This aquatic surface respiration (ASR) involves skimming the surface to exploit the oxygen-rich layer and allows tambaqui to continue to operate aerobically.

ASR appears to be triggered by stimulation of oxygen chemoreceptors innervated by cranial nerve V, while other mechanisms trigger the swelling of the lip that facilitates this behaviour. Both orobranchial and branchial chemoreceptors are involved in mediating cardiac responses to hypoxic conditions.

Tambaqui can reach 30 kg in size. While the young filter phytoplankton from water, adults are seed-eaters. These fish from the Amazon are often exhibited in aquaria but perhaps more important, they are prized as food, whether grilled or stewed. Tambaqui are readily farmed (see also aquaculture, page 517), because their ability to withstand hypoxic conditions is a key to their value in aquaculture. Their rapid rates of growth and the large sizes they can attain combine to make them promising as a source of animal protein. At a density of 50 fish.m³ in cages, the yield was 45.8 kg.m³. Wild-caught tambaqui can be more lucrative than farmed ones because shoppers prefer fish larger than 1.5 kg.

Naturally, tasty large fish are under pressure from harvesting, but local people in the Silves Region of Central Brazil have established zones to protect tambaqui and arapaima (*Arapaima gigas* (1245), up to 200 kg). The success of this project depends upon a combination of nursery grounds where fishing is forbidden, lakes where only local people can fish, and other lakes where fishing is open access. Lake Purema was declared a sanctuary in the 1980s, and arapaima there are doing better, meaning that more fish are reaching large sizes and their population is increasing.

Sanctuaries—combined with aquaculture that partly depends upon the tambaqui's fat lips—offer the prospect of continued harvesting as well as conservation.

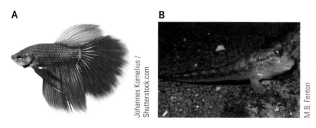

Figure 11.23 Air-breathing fish. **A.** Bettas (*Betta splendens* (1257)), **B.** mudskippers (*Periophthalmus* spp. (1280)).

air when there is not enough dissolved oxygen in the water to sustain them (e.g., armoured catfish *Ancistrus* spp. (1324) and *Hypostomus* spp. (1331)). Others operate bimodally more or less continuously, including some that are obligatory because they cannot survive without breathing in air (e.g., *Anabas* spp. (1255) and *Arapaima* spp. (1245)). Others that do this (e.g., *Piabucina* spp. (1184) and *Hoplosternum* spp. (1330)) are non-obligatory air breathers. (see also Section 11.4)

In living fish, air-breathing organs fall into one of three categories:

1. Respiratory surfaces are derived from the digestive tract (three genera of lung fish as well as *Erpetoichthys* (1152) and *Polypterus* (1153)), the pneumatic duct (*Dallia* spp. (1210)), the stomach (*Ancistrus* spp.), or the intestine (*Hoplosternum* spp. (1330)).
2. Respiratory surfaces are in the head region, including the buccal, pharyngeal, and branchial areas (Synbranchidae, *Electrophorus* (1222), *Hypostomus* (1331), *Betta splendens* (1257)), opercular surfaces

(*Periophthalmus* (1280), *Pseudapocryptes* (1284)), and pouches formed adjacent to the pharynx (*Channa* (1261)) which may be filled with diverticulae (Clariidae and anabatoids). Labyrinth organs, well-vascularized surfaces for gas exchange formed from an expansion of the epibranchial cavity of the first pharyngeal arch, may also be associated with sensory structures functions (see Section 11.4a).

3. Respiratory surfaces are existing structures such as gills or skin (Gobiesocidae, *Dormitator* (1264), *Mnierpes* (1275)). Some fish use air and water breathing respiratory surfaces simultaneously.

Variations in the arrangement of lung-like structures associated with the digestive tract in vertebrates (Fig. 11.24) reflect their relative positions and different functions. The ancestral condition for actinopterygian and sarcopterygian fish is the presence of a lung which may originate as an unpaired dorsal, lateral, or ventral outgrowth from the transition between the pharynx and esophagus. In the line leading to tetrapods and terrestrial existence, the lung becomes the respiratory surface. In actinopterygians, the lung was gradually transformed, first to a respiratory gas bladder connected to the digestive tract (physostomous) and then to an unconnected swim bladder (physoclistous). Among early actinopterygians, respiratory gas bladders occur in some living polypterids (*Polypterus* (1153), *Erpetoichthys* (1312)), gars, Ginglymodi, Halecostomi, and *Amia* (315) as well as in some basal euteleosts, the ostariophysi and protacanthopterygii.

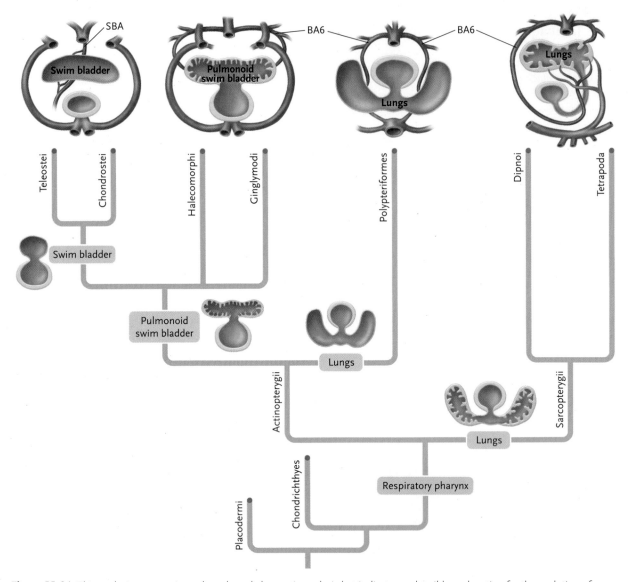

Figure 11.24 This evolutionary tree is not based on phylogenetic analysis but indicates a plausible explanation for the evolution of swim bladders (SB) which may be paired (PSB), and lungs (L). Lungs have appeared separately from the respiratory pharynx (RP), as have swim bladder arteries (SBA), derived from the vessels of the sixth branchial arch (BA6).

SOURCE: Based on *Respiratory Physiology & Neurobiology*, 144, Perry, S.F. and M. Sander, "Reconstructing the evolution of the respiratory apparatus in tetrapods," Pages 125–139, 2004.

B. Amphibia. In ancestral tetrapods, a buccal pump derived from the one used in gill ventilation inflated the lungs (Fig. 11.25). The buccal pump works by first drawing air into the buccal chamber. Then, after the mouth and/or nostrils are closed, air is pushed through the pharyngeal-esophageal connection into the lungs. Exhalation involves contraction of the abdominal wall muscles, resulting in movement of air from the lungs, through the pharyngeal-esophageal connection, and out via the mouth and/or nostrils.

Figure 11.25 The steps involved in moving air into and out of the lungs of a frog.

SOURCE: From RUSSELL/WOLFE/HERTZ/STARR. *Biology*, 2E. © 2013 Nelson Education Ltd. Reproduced by permission. www.cengage.com/permissions

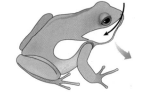  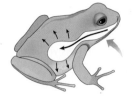 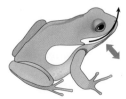

1 The frog lowers the floor of the mouth and inhales through its nostrils.

2 Air in the lungs is exhaled when the glottis opens due to elastic recoil of the lungs and body wall.

3 The frog closes its nostrils and elevates the floor of the mouth, forcing air into the lungs.

4 Rhythmic movements flush the mouth cavity with fresh air for the next cycle.

A

B

Figure 11.26 Amphibian skeletons. A. *Eryops*, a fossil amphibian, and B. a modern frog. Note the difference in the presence of ribs and the general arrangement of the skeleton. While *Eryops* clearly walked on four legs, the frog is specialized for hopping (see also Chapter 7).

A

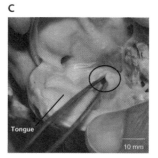

C

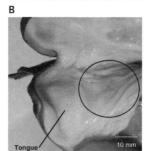

B

Figure 11.27 A. The lungless frog *Barbourula kalimantanensis* (1365). **B.** a typical frog, a bullfrog, *Rana catesbeiana* (1381). **C.** A view of its open mouth compared with that of. The frog from Borneo lacks a glottis (circled area) which is obvious in the bullfrog. The tongues are identified in each view. The glottis prevents solid and liquid food from entering the air passages below it and influences and partly controls the production of vocalizations.

SOURCE: Reprinted from *Current Biology*, 18, Bickford, D., D. Iskandar, and A. Barlian, "A lungless frog discovered in Borneo," Pages R1–R2, Copyright 2008, with permission from Elsevier.

The earliest land vertebrates, amphibians (stegocephalians), had well-developed ribs (Fig. 11.26A), probably indicating that ventilation of lungs by the actions of a buccal pump had been replaced by aspiration-type ventilation. In contrast, ribs are reduced in living amphibians (Fig. 11.26B) such as frogs and salamanders, indicating that these animals use a buccal pump to ventilate their lungs. In modern amphibians, lungs are fixed in place in the dorsum of the body cavity. Frogs and salamanders typically have short, one-chambered (unicameral) lungs with about 1 cm² of respiratory surface per gram of body mass. A typical medium-sized frog has 15–20 cm² of lungs. Typical anurans have lungs that are better suited for sound production than for breathing, and they do not directly inform us about the early evolution of lungs in tetrapods.

Among living amphibians, lungs vary in size from quite large to small, and some species (a caecilian, a frog, and several salamanders) have no lungs. Compared with frogs living at sea level, the Andean frog *Telmatobius culeus* (1390) that lives in Lake Titicaca (3812 m above sea level) has poorly developed lungs. But the blood of *T. culeus* has a high affinity for oxygen and a high capacity for transporting it. In hypoxic water, these frogs show "bobbing" behaviour apparently to use their highly vascularized skin as a respiratory surface. The next step in the reduction of lungs is clear in Bornean lungless frogs (*Barbourula kalimantanensis* (1365), Fig. 11.27) that live in fast-flowing, cool (14–17°C), mountain streams where dissolved oxygen is readily available. The large surface area of skin and low metabolic rate allow these frogs to exchange gases without the buoyancy problems of lungs.

Variations in lung size of amphibians reflect the importance of other respiratory surfaces, mainly skin, although many species of adult salamanders (urodeles) have external gills (Fig. 11.6B).

C. Reptiles. While basal unicameral lungs are still a single chamber, like the lungs of amphibians, variations on these lungs in reptiles can effectively increase the surface area of the respiratory surface (Fig. 11.28). The lungs in some species of lizards and snakes have interconnected folds that increase surface area. The most developed forms, multicameral lungs, have deep and narrow **faveoli**. These occur in large varanid

A

B

C

Figure 11.28 Complexity of reptilian lungs. **A.** The tegu lizard (*Tupinambis nigropunctatus* (1526)) has unicameral lungs. Multicameral lungs are found in **B.** turtle (*Pseudemys scripta elegans* (1428)) and **C.** crocodile (*Caiman crocodilus* (1436)).

SOURCE: Based on Dunker HR. "Functional morphology of the respiratory system and coelomic subdivisions in reptiles, birds and mammals," *Vehr Dtsch Zool Ges* 1978: 99–132, 1978.

lizards, such as goannas and monitors, and pythons. Turtles, tortoises, monitor lizards, and crocodiles also have multiple lung chambers developed in three rows with at least four chambers per row. These are most developed in large sea turtles. The posterior margin of the thoracic cavity is defined by a fibrous oblique process in most reptiles, although crocodilians have a muscular diaphragm (see Section 11.4b).

Some reptiles, notably small lizards, have lung extensions that are not respiratory surfaces, but which allow a dramatic increase in the volume of the body cavity. During defensive behaviour, maximum inhalation inflates these dilatations, providing access to additional oxygen and, in some cases, wedging the animal into a crevice. These dilatations are most developed in snakes, especially in pythons where they can translate into a several fold increase in tidal volume (air volume/breath). But they also occur in turtles, where the diversity of form reflects, in part, life style as well as shell size and shape.

D. Birds. The ability of a gaseous exchange system to deliver oxygen is governed by V (process of ventilation) and Q (process of perfusion—the movement of oxygen into blood). Greater oxygen demand requires more complex lungs with smaller gas exchange units. The alveolar lungs of mammals and the parabronchial lungs of birds (next section) are the ultimate gas exchangers in vertebrates. Subdividing the lung into many gas exchange units gives greater surface areas (relative to volume) and increases the capacity for diffusing oxygen. But matching V and Q is an important consideration, and V/Q heterogeneity is a measure of the balance between ventilation and perfusion.

The multicameral reptilian lung with three rows of secondary bronchi (Fig. 11.28A) is the basis of the parabronchial lung of birds, where **parabronchi** are the respiratory surfaces. Rather than terminating at the blind-ended alveoli of mammals, the conducting passages that branch off avian trachea continue to branch repeatedly to eventually form numerous tiny, one-way parabronchial tubes that allow air to flow through the lungs. Small, blind-ended **air capillaries** open off the walls of each parabronchus, and gas exchange with the blood occurs at the walls of the air and blood capillaries (Fig. 11.29). Avascular air sacs are nonrespiratory dilatatory structures that are thought to serve as bellows to direct air flow through the lungs. The trachea of birds is divided into two primary bronchi (or mesobronchi) that do not directly enter the lung, but instead extend to the posterior air sacs, with numerous branches (latero-, ventro- and dorsobronchi) leading to the parabronchi of the lung (Fig. 11.29).

If we follow a single breath of air, its passage through the bird's air sacs and lungs involves two complete cycles of inspiration and expiration. Inspiration occurs with inflation of the air sacs, drawing air from the trachea along the two primary bronchi where it is

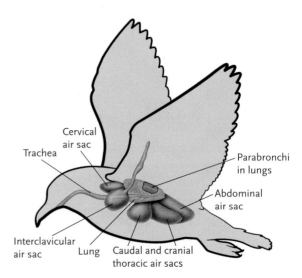

Figure 11.29 The lung and air sac system of a bird. The trachea branches into two mesobronchi that lead to three ventrobronchi that connect to the cervical, interclavicular, cranial and caudal thoracic, and abdominal air sacs. Along the way, numerous secondary bronchi lead to well-vascularized parabronchi in the lungs, to create a one-way flow of air back to the more anterior air sacs upon expiration. See also Box 11.2.

divided. Some air passes along the secondary bronchi to the parabronchi of the lungs, and the rest fills the posterior thoracic and abdominal air sacs (Fig. 11.29). The first expiration, caused by compression of these air sacs, directs air flow out into the parabronchi of the lungs, displacing the volume of air already there, which exits via the trachea. As the second inspiration begins, the entering air from the divided primary bronchi refills the posterior air sacs, and air from the secondary bronchi flows through the lungs, pushing the remainder of the air in the parabronchi out temporarily into anterior air sacs (Fig. 11.29). With the second expiration, air from these anterior air sacs exits along with air from the lungs, replaced by air from the posterior air sacs which flows into the parabronchi. This flow-through pattern of air flow and the use of air sacs lead to nearly continuous ventilation of parabronchi during both inspiration and exhalation, increasing the efficiency of gas exchange. This is further enhanced by the arrangement of capillaries (Fig. 11.30) that may establish a countercurrent exchange (Box 11.2).

The vertebral column of birds, combined with the broad sternal plate and incompressible thoracic skeleton, allows limited movements of the sternum (Fig. 11.31A). Wing movements are separated from movements associated with breathing, but both actions are achieved by intercostal muscles that insert on uncinate processes (Fig. 11.31B). The pleural cavity houses the lungs and is located dorsally in the thoracic cage. Contractions of the intercostal muscles are responsible for ventilation, and while the volume of the pleural cavity remains relatively constant, the volumes of air sacs change dramatically during ventilation. Although

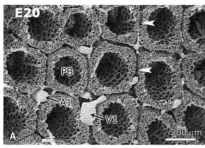

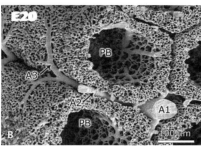

Figure 11.30 Scanning electron micrographs of intravascular casts showing the arrangement of the parabronchial (PB) blood vessels in the chick embryo at 20 days. **A.** The PB capillary meshwork forms hexagonal tubes that surround the PB lumen. Interparabronchial septa (arrowheads), which contain the interparabronchial artery (A1) and the interparabronchial vein (V1), separate adjacent parabronchi. **B.** At higher magnification, branches of the PB arterial system vessels are evident. The interparabronchial artery (A1) runs parallel to the long axes of the PB and gives rise to the orthogonal PB arteries (A2), which then form PB arterioles (A3). These arterioles give rise to the dense meshwork of capillaries that surrounds the air capillaries in the PB wall.

SOURCE: A. N. Makanya, V. Djonov. 2009. "Parabronchial angioarchitecture in developing and adult chickens." *J Appl Physiol* 106:1959-1969. Copyright © 2009, The American Physiological Society.

the air sacs are connected to the lungs, they are tucked in among the viscera and extend into the cores of most large bones adjacent to the thoracic cavity. The bird respiratory system works with minimal pressure gradients.

E. Mammals. Mammalian broncho-alveolar lungs reflect their reptilian ancestry, but show developments beyond the basic multicameral reptilian condition. The bronchial system in mammatls leads to passages of decreasing size, culminating in branched respiratory bronchioli and ducts terminating in

A 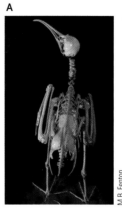 **B**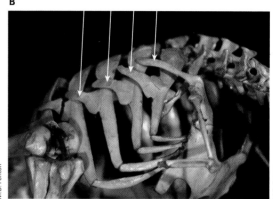

Figure 11.31 A. Dorsal view of the skeleton of a Hudsonian Curlew (*Numenius phaeopus* (1556)) and **B.** lateral view of the thoracic skeleton of a Kiwi (*Apteryx australis* (1672)) showing four uncinate processes (arrows), one on each rib. Intercostal muscles insert on the uncinate processes and are not unique to birds that can fly. The curlew can fly; the kiwi is flightless.

BOX 11.7 People Behind Animal Biology

Colleen G. Farmer

Colleen Farmer is a physiologist working in the Department of Biology at the University of Utah. She completed her B.A. in Physics at the University of Idaho in 1987 and her Ph.D. in Physiology at Brown University (Rhode Island) in 1991. She and members of her laboratory focus their research on the evolution of gaseous exchange and circulatory systems.

In 2010 she presented two scenarios for the origin of the patterns of the lungs of modern mammals and birds. Mammals have alveolar lungs that Farmer proposed evolved in the Palaeozoic when levels of atmospheric oxygen were higher than they are at present. Higher levels of atmospheric oxygen relaxed selection for thin blood-gas barriers within the lungs.

The alveolar lungs of mammals are ventilated by tidal movements of air into and out of the alveoli.

In birds and archosaurs, air flow through the lungs is unidirectional and gaseous exchange occurs at the parabronchi. Farmer's theory is that unidirectional flow appeared first in ancestral archosaurs, animals that were ectothermic (cold-blooded). Unidirectional flow of air allowed the blood to circulate oxygen and carbon dioxide during periods of apnea such as those experienced during diving, torpor, or estivation.

Farmer and her colleagues have also suggested that lungs appeared in fish that lived in near-shore marine environments where the supply of dissolved oxygen is usually

plentiful (i.e., hypoxia is rare). They proposed that lungs provided oxygen directly to the heart because the fish involved lacked a coronary circulation.

She and her colleagues also proposed that endothermy (warm-bloodedness) in birds and mammals provides a major advantage in reproduction. Specifically, warm-blooded bird and mammals can achieve faster rates of growth in developing embryos, which leads to faster development and more rapid attainment of sexual maturity.

Professor Farmer and her colleagues have repeatedly designed and conducted experiments that test the predictions arising from their innovative theories.

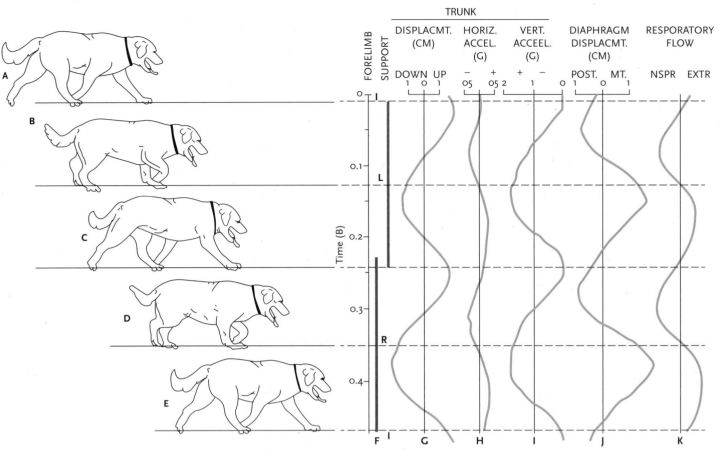

	TRUNK					
FORELIMB SUPPORT	DISPLACMT. (CM)	HORIZ. ACCEL. (G)	VERT. ACCEEL. (G)	DIAPHRAGM DISPLACMT. (CM)	RESPIRATORY FLOW
	DOWN UP	− +	+ −	POST. MT.	NSPR EXTR

Figure 11.32 A.–E. show the stride cycle of a trotting dog, illustrating changes in the trunk and diaphragm, and a pneumotechnographic record of inspiration and expiration.

SOURCE: From Bramble, D.M. and F.A. Jenkins. 1993. "Mammalian locomotor-respiratory integration – implications for diaphragmatic and pulmonary design," *Science*, 262:235–240. Reprinted with permission from AAAS.

blind-ended sacs called alveoli. While many species of small mammals have single-branched bronchioli, larger species have double branches ensuring sufficient ventilation of each alveolus. The alveoli of mammals have exponentially reduced diameters compared to the conducting tubes, resulting in exponentially increased respiratory surface area. Unlike the lungs of reptiles and birds that are anchored in the pleural cavity, the lungs of mammals move freely in the pleural cavity, which is separated from the peritoneal cavity by the muscular diaphragm. The diaphragm and intercostal muscles account for forceful inspiration and also maintain a pressure gradient between pleural and peritoneal cavities. Expiration is largely passive, involving simple relaxation of the inspiratory muscles, but during exercise expiration is strongly supported by contractions of muscles of the abdominal wall (Fig. 11.32).

During running, the movement of the mammalian vertebral column, caused by contractions of the trunk and abdominal wall musculature, generates high intra-abdominal peaks in pressure. The lumbar region of the spine differs distinctly from the thoracic region (Fig. 11.33), and the musculature of both regions

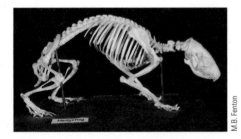

Figure 11.33 Mammalian skeleton (hedgehog) showing regional specialization of vertebral column into cervical, thoracic, lumbar, sacral, and caudal vertebrae.

generates ventilation of the lungs. In mammals, unlike in birds, locomotory and respiratory movements are closely coupled.

Mammalian lungs are relatively homogeneous across taxa, compared to those of other vertebrates. The mammalian respiratory passageway begins with the trachea, which branches into primary bronchi and then bronchioles, which divide to produce smaller and smaller branches until they finally terminate in the alveoli. Since only alveoli are vascularized, ventilation occurs throughout, but <20% of lung volume is

BOX 11.8
BOX 11.8
Wings and Gaseous Exchange

Many biologists support the view that insect wings evolved from gill-like appendages already present in the aquatic ancestors of crustaceans and insects. This idea is partly derived from *pdm* (*nubbin*) and *apterus*, two genes that have wing-specific functions in insects and for which there are homologues in crustaceans.

This hypothesis does not include the prediction that insect wings also serve as respiratory surfaces. In fact, as we have seen, respiratory surfaces functioning in air must have a combination of a wet surface and a thin membrane across which oxygen and/or carbon dioxide can move.

Water loss during gaseous exchange is an important factor for terrestrial animals (Section 11.1c).

Therefore the suggestion that bat wings also function as respiratory surfaces is surprising. Support for this proposal comes from measuring oxygen uptake and carbon dioxide production by Wahlberg's fruit bats (*Epomophorus wahlbergi* (1790)) roosting in a chamber. The rate of oxygen consumption and carbon dioxide release across wings was 10% of the total intake by lightly anesthetized bats. Body epidermis is thicker (61±3 μm) than wing epidermis (9.8±0.7 μm). The folded wing membranes of bats can be highly convoluted compared to body skin, perhaps representing a specialization as a respiratory surface.

The lack of data about oxygen consumption and carbon dioxide production during flight is an important shortcoming for the proposal that bats use their wings as respiratory surfaces. Furthermore, it is not clear whether the wing membrane is as convoluted in flight. Equally important is the amount of water a bat might lose across the wing membrane if it served as a respiratory surface. The suggestion remains intriguing as does its possible application to insects or birds, or even pterosaurs.

---

dedicated to gaseous exchange. In birds, ventilation and gaseous exchange are completely separated, and air sacs account for ~90% of the volume of the respiratory system, but ventilation through the countercurrent arrangement of parabronchi and continuous air flow is much more efficient.

The discovery that a newborn Julia Creek dunnart (*Sminthopsis douglasi* (1680), a marsupial mouse) breathes through its skin appears startling given the highly derived respiratory system of mammals. At birth, these marsupial mice are 4 mm long and weigh 17 mg the size of a grain of rice. The skin also serves as a respiratory surface in other very small newborn mammals (e.g., some rodents). Here, the moist surface and large surface area relative to volume make the skin a feasible respiratory surface. In these mammals the alveoli are not fully developed at birth, and until they are, the skin suffices. Other mammals, particularly precocial ones that are up and running shortly after birth, are born with fully functional alveoli.

STUDY BREAK

1. Construct a table comparing gaseous exchange in five phyla of animals. Be sure that your comparison includes respiratory surface, respiratory pigments, and other relevant details. What happens if you extend the comparison to all of the phyla treated above?
2. What is "bimodal" respiration? Which animals use it? Why is it adaptive?

11.4 Some Specific Examples

11.4a Air-Breathing Fish

The labyrinth organs of anabantid fish (*Anabas* (1255), *Macropodus* (1272), *Betta* (1257), *Trichopodus* (1296)) are derived from epibranchial regions of the first and second branchial arches and are located in the opercular cavity. The well-vascularized plate-like organs extend dorsally and fill the suprabranchial chamber, and all blood leaving the first and second gill arches enters the folds of this accessory respiratory organ. Gouramis (*Trichopodus trichopterus* (1296)) use a combination of gills and labyrinth organs for gaseous exchange. The water-to-blood distances across the gills of anabantoid fish are 15–29 μm compared to <0.3 μm between air and blood in the labyrinth organs. In gouramis, the labyrinth organs account for 40% of the oxygen uptake and 15% of carbon dioxide eliminated. The circulatory pattern in gouramis (Fig. 11.35) reflects

A

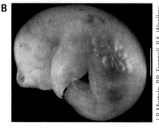
B

Figure 11.34
A. Adult dunnart (*Sminthopsis douglasi* (1680)), a marsupial mouse. **B.** One-day-old embryo.

Janelle Lugge / Shutterstock.com

J.P. Mortola, P.B. Frappell, P.A. Woolley; original publication: Nature 397, 161, 1999.

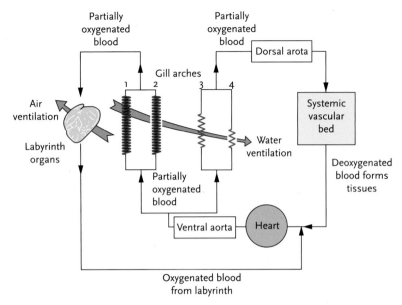

Figure 11.35 Circulation diagram of a gourami (*Trichopodus trichopterus* (1296)), a bimodal, air-breathing fish. Blood flow from the first two gill arches continues dorsally to exchange gases in the well-vascularized labyrinth organs when the fish is breathing air. The blood returns via the jugular veins to the heart, so blood in the ventral aorta is partially oxygenated before it enters the gills.

SOURCE: Reproduced with permission from Burggren, W.W. 1979. "Bimodal gas exchange during variation in environmental oxygen and carbon dioxide in the air breathing fish Trichogaster trichopterus," *Journal of Experimental Biology*, 82:197–213. The Company of Biologists Ltd.

their bimodal gaseous exchange strategy and generally resembles the situation in climbing perch, with both gills and labyrinth organs functioning in oxygen uptake and elimination of carbon dioxide.

Catfish (order Siluriformes) show a wide range of systems for extracting oxygen from air. These respiratory surfaces in two (magur, *Clarias batrachus* (1327), and singee, *Heteropneustes fossilis* (1329)) are also derived from the same basic structures as gills, but are different from labyrinth organs. In *Saccobranchus* the lung-like air sacs are large, often extending through one third of the body. The respiratory surfaces in this fish are secondary lamellae where the epithelial cells on the surface have numerous microvilli. Magur are also obligate air breathers that gulp air and consume more oxygen from air (51.4%) than from water (41.6%). In *Clarias*, respiratory surfaces for obtaining oxygen from air include the respiratory membrane in the paired suprabranchial chambers and the respiratory trees associated with the fused dendritic plates formed by the secondary lamellae on the second and fourth primary gill lamellae. Air from the pharynx enters the dorsal suprabranchial chambers via the incurrent aperture and leaves via the excurrent aperture that opens into the opercular chamber.

Tamoatas, a catfish (*Hoplosternum littorale* (1330)) from the Amazon, is also a bimodal breather, using its gills to extract oxygen from water and a thin-walled part of its intestines to obtain oxygen from air. The "lung" part of the intestine does not contain food. Young are placed in floating nests of dead weed in which the

levels of dissolved oxygen are higher than those in open water.

These three species of catfish reflect the diversity within this group and the repeated evolution of at least bimodal breathing, if not more dependence upon oxygen obtained from air. Some catfish take this a step further and regularly leave the water, for example, the South American *Lithogenes wahari* (1332) that climbs around on rocky surfaces. Although we have information about the specializations of its pelvic fins and musculature for climbing, we still lack data about its capacity for bimodal breathing.

As adults, mud eels (*Monopterus cuchia* (1338)) are obligatory air-breathers (family Synbranchidae) (see also Fig. 11.21). The gaseous exchange surfaces for extracting oxygen from air are paired lateral extensions of the pharyngeal cavity (buccal cavity, buccopharynx, hypopharynx and, from the oesophagus, the air sac). Respiratory islets (Fig. 11.36) on the inner mucosal linings are the sites of oxygen uptake. The posterior parts of the sacs appear to serve as air reservoirs because they lack the islets. Bowfins (*Amia calva* (155)) provide another example, relying primarily on their gills for oxygen extraction and excretion of carbon dioxide. But these fish can augment their oxygen uptake by gulping air and moving it to their swim bladder (Fig. 11.24) where another respiratory surface is located.

11.4b Diaphragms and Ventilation

The mammalian diaphragm, which separates the pleural and visceral cavities, consists of opposing costal and crural sheets of muscle; the former expands the dimensions of the lower rib cage, the latter contracts them. Actions of the diaphragm allow mammals very efficient aspiration ventilation, increasing their capacity for diffusion of oxygen. In rats, cutting the nerve that innervates the diaphragm (phrenic nerve)

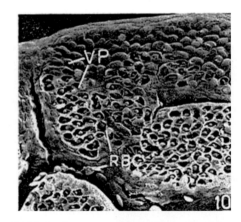

Figure 11.36 An electron micrograph of respiratory islets in the hypopharynx of a swamp mud eel (*Monopterus cuchia* (1338)). Also shown are some red blood cells (RBC) and ruptured papillae (VP). The surface of the respiratory membrane is wrinkled.

SOURCE: Munshi, J.S.D. 1961. "The accessory respiratory organs of Clarias batrachus (Linn.)," *Journal of Morphology*, 109:115–139, John Wiley and Sons. Copyright © 1961 Wiley-Liss, Inc.

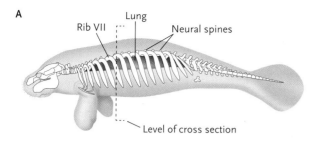

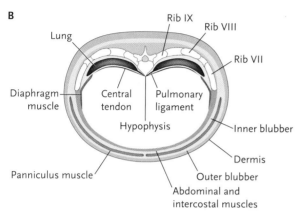

Figure 11.37 The lung and diaphragm of a manatee (*Trichechus trichechus* (1916)). **A.** The left lateral view shows the elongate and flattened nature of the lung in the dorsal pleural cavity. **B.** In cross section it is clear that the diaphragm stretches almost horizontally from the hypophysis at the midline to the ribs at mid shaft.

SOURCE: Rommel, S. and J.E. Reynolds III. 2000. "Diaphragm structure and function in the Florida manatee (Trichechus manatus latirostris)," *Anatomical Record*, 259:41–51, John Wiley and Sons. Copyright © 2000 Wiley-Liss, Inc.

reduces their capacity for activity. This demonstrates that, at least in this situation, high metabolic rates (fueled by high levels of oxygen), translate into higher levels of activity rather than increased thermoregulatory capacity.

The diaphragms of manatees (*Trichechus manatus* (1916), large aquatic mammals) are strikingly different from the normal pattern (Figs. 11.37). In manatees, the diaphragm is dorsal to the heart and does not attach to the sternum. The attachments are by tendons extending to hypapophyses, bony projections from vertebrae. There are two functional hemidiaphragms. These adaptations are presumed to change the volume of the pleural cavity and, in concert with abdominal wall muscles, compress gases in the abdominal cavity, both of which would allow better control of buoyancy.

In crocodylians, fan-shaped muscles of the diaphragm originate on the ventral posterolateral margin of the ischia, the cranial margin of the pubis and the ventral ribs, called the gastralia. These muscles insert on the pericardium and on the membrane enclosing the liver. Their contraction creates negative pressure in the pleural cavity, drawing air into the lungs. This demonstrates that the mammalian diaphragm is not the only way to effect negative-pressure ventilation. In turtles, muscles actually move the liver and other abdominal organs posteriorly, to allow the lungs to expand into the available space to effect this negative-pressure ventilation!

The gaseous exchange surface in crocodylians is the parabronchus as it is in birds. Furthermore, the cervical ventral bronchus in American alligators (*Alligator mississippiensis* (1435)) is analogous to the avian ventral bronchus that connects to the cervical air sacs, causing unidirectional air flow across the respiratory surface. Although air sacs are absent in crocodylians, the arrangement of the cervical ventral bronchus still results in unidirectional air flow (Fig. 11.38).

11.4c Atmospheric Gases and Body Size

We have seen above how higher volumes of oxygen (~35% vs 21%) in the atmosphere could explain the gigantic insects of the Carboniferous. It is evident that levels of atmospheric oxygen have varied over time, and the high during the Carboniferous was followed by a decline. Over the last 205 million years (from the Triassic-Jurassic boundary), carbon isotopic data indicate further variation in atmospheric levels of oxygen, with relatively rapid increases at the beginnings of the Jurassic and the Eocene, coinciding with an increase in the average size of placental mammals.

A gigantic anaconda-like snake (*Titanoboa cerrejonensis*, 13 m long, weighing >1100 kg) that lived in middle-to-late Palaeocene rainforests in what is now Colombia, supports the view of a greenhouse climate similar to that of today. Of particular importance then, as now, were high concentrations of carbon dioxide that resulted in general warming of the climate and sharp latitudinal gradients in temperature. It is not clear if a huge fossil frog (*Beelzebufo ampinga*) from the late Cretaceous of Madagascar was also a reflection of warmer climate and/or availability of oxygen.

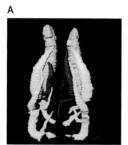

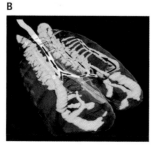

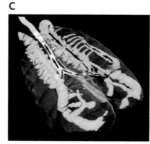

Figure 11.38 A. An oblique dorsal view of the thoracic cavity of an American alligator (*Alligator mississippiensis* (1435)) showing the dorsobronchus in blue (d), ventrobronchus in green (v), and a small pair of medial paracardiac bronchi in red. The pattern of air flow on **B.** inspiration and **C.** expiration are also shown.

SOURCE: Farmer, C.G. and K. Sanders, "Unidirectional airflow in the lungs of alligators," *Science* 15 January 2010: Vol. 327 no. 5963 pp. 338–340. Reprinted with permission from AAAS.

11.4d Flying High

In hypoxic conditions (reduced oxygen availability), animals typically increase ventilation rates because of the action of arterial chemoreceptors. Changing the breathing rate (ventilation) is one way to operate at high altitude.

Populations of humans living at high altitude use different adaptations allowing them to operate in hypoxic conditions (reduced availability of oxygen). Quechuans from the Andean plateau in South America show a reduced ventilation response under hypoxic conditions compared to people living at low altitude, however, they also have higher concentrations of hemoglobin in their blood. People living at high altitudes in Tibet show the opposite pattern, an enhanced hypoxic ventilation response (increased rate and depth of breathing), and they synthesize nitric oxide, a vasodilator, more than people from elsewhere. In contrast, high-altitude Ethiopians appear to show none of these adaptations. Variations in human responses suggest that animals also do not always use the same approach to solving the problem of operating at high altitude under hypoxic conditions.

Flying at high altitude requires matching the demand for oxygen with the supply in the face of reduced availability (see Section 11.1b). Twice a year, bar-headed geese (*Anser indicus* (1655)) fly over the Himalayas as they migrate from summer to winter ranges. These geese repeatedly fly above the highest peaks in the Himalayas at altitudes of 8000 m, where the oxygen level can be one-third that at sea level. Flying bar-headed geese use 10 to 20 times more oxygen per minute than they do at rest. One single amino acid point mutation in the α_A polypeptide chain means that the hemoglobin of bar-headed geese has a higher affinity for oxygen and facilitates oxygen download during hypoxia.

Flying bar-headed geese also breathe substantially more than other birds (e.g., greylag geese, *Anser anser* (1654), and pekin ducks, *Anser platyrhynchos* (1653)) because of enhanced tidal volume (volume per breath), which accounts for the flight performance rather than the increased breathing rate. These results demonstrate that changes in control of ventilation enhance oxygen loading into blood.

11.4e Diving Deep

Animals take in air not just for breathing; some use air for buoyancy. Pelagic animals must solve the problem of counteracting their body weight in water and maintain their vertical position in the water column. This is done by achieving neutral buoyancy. For example, pelagic octopods known as Argonauts are free-swimming, open-ocean cephalopods that have evolved from benthic stock. Female argonauts (e.g., *Argonauta argo* (492)) are much larger than males (eight times longer, 600 times heavier) and, unlike the males, occupy a brittle shell (a "paper nautilus"). Previously the female argonaut's shell was thought to be a place only to store and brood eggs. Female argonauts gulp air at the surface, trap it in the apex of their shells, use their flanged arms to seal off the captured air, and then forcefully dive to depths where compression of the stored gas provides buoyancy to counteract body mass. This behaviour allows females to adjust their buoyancy to compensate for the weight of developing eggs also in the shell.

Female argonauts use compressed air to achieve neutral buoyancy and reduce the costs of holding their position in a vertical column of water. But after getting the air (gulping) and diving deeply enough to reach the zone of neutral buoyancy requires energy expenditure. The situation in argonauts parallels the chambered shells of nautiluses, as well as the gas bladders that have evolved in other octopods such as *Ocythoe tuberculata* (515) and *Haliphron atlanticus* (499). In other cases, such as the lungless frog from Borneo, being lungless avoids problems related to buoyancy. Lungless frogs appear to breathe across their skins, while argonauts use the gills in their mantle cavities.

Diving tetrapods such as marine reptiles, birds, and mammals take in air to obtain the oxygen they need to fuel their operations. "Right whales" got their name from whalers because their carcasses floated, making it relatively easy to retrieve them after they had been killed. Like the female argonaut that is positively buoyant, a right whale (e.g., *Eubalaena glacialis* (1754)) exerts considerable energy during a dive, especially at the beginning, to overcome its positive buoyancy. As it dives, its lungs collapse, decreasing the volume and density of the animal. Buoyancy of whales is not much affected by density because, like sea water, animal tissues have low compressibility. There is seasonal variation in tissue density according to times of feeding and fasting, or, in the case of the female argonauts, the presence of eggs.

Diving whales may use swimming styles ranging from steady propulsive movements of the tail (fluking), to alternating between stroking and gliding. Differences in buoyancy forces coincide with different patterns of locomotion. While diving sperm whales (*Physeter macrocephalus* (1766)) stroked steadily during descents, they used stroke and glide swimming during ascents when positive buoyancy aided their movement.

STUDY BREAK

1. What gaseous exchange problems are faced by diving whales and birds flying at high altitudes? How do the principles of Fick's Law apply to these problems?
2. How are buoyancy and gaseous exchange related?
3. How has availability of oxygen (partial pressure of oxygen) varied over time? How has this affected animal life?

GASEOUS EXCHANGE **IN PERSPECTIVE**

It should be obvious that gaseous exchange is ubiquitous in animals and involves a great variety of specializations, providing many examples of parallel and convergent evolution. But basic laws (e.g., Fick's Law) govern how gaseous exchange systems work. The move from breathing oxygen dissolved in water (often via some form of gill) to breathing atmospheric oxygen increased the activity potential of animals and reduced their costs of operation. The recurrent appearance of structures designed to acquire atmospheric oxygen across the diversity of animals emphasizes this reality.

Questions

Self-Test Questions

1. Which of the following statements best illustrates the relationships quantified by Fick's Law?
   a. As the size of an animal increases, its relative surface area decreases.
   b. The distance over which diffusion takes place (L) is proportional to the area (A) available for exchange.
   c. Ventilation and perfusion are major factors in determining the rate of diffusion by maximizing the differences in partial pressure.
   d. The coefficient of diffusion will decrease relative to the surface area available for exchange.

2. Which of the following is characteristic of the respiratory surfaces of animals?
   a. Surfaces for gas exchange are moist.
   b. The anatomical surface area for exchange is maximized through numerous layers (such as gills) or convoluted surfaces (such as skin, tracheoles and lungs).
   c. The concentration gradient for exchange may be maximized through mechanisms such as countercurrent exchange or linking ventilation to extensive perfusion.
   d. All of the above are true for animal respiratory surfaces.

3. Which of these statements illustrates the importance of the role of respiratory pigments in animals?
   a. They are only found in animals with high metabolic activity, such as molluscs, arthropods, and vertebrates.
   b. They are primarily adaptations for terrestrial animals, to aid in the extraction of oxygen from air.
   c. They are allosteric proteins that bind oxygen under favourable conditions and then change shape to deliver oxygen in areas of high metabolic activity.
   d. They are particularly prevalent in insects.

4. Which of the following anatomical structures is correctly paired with the organism in which it maximized the respiratory surface of the animal?
   a. sieve trachea of book lungs; spiders
   b. parabronchi and air sacs; insects
   c. tube feet of the water vascular system; annelids
   d. mono- and bipectinate gills; elasmobranchs

5. To what does bimodal respiration refer?
   a. The presence of different gas exchange surfaces on a single organism
   b. Changes in the pattern of gas exchange in an organism under different physiological conditions
   c. The ability of some organisms to exchange gases in both aquatic and terrestrial habitats
   d. The double circulatory pattern of amniotes that supplies both the respiratory surfaces and the body

Questions for Discussion

1. What is the difference between hydrophobic and hydrophilic? How are these phenomena relevant to gaseous exchange?

2. What is a plastron? What role does it play in gaseous exchange? What advantages does it confer over animals that lack plastrons?

3. Fick's Law shows us that the rate of diffusion is dependent upon the area over which diffusion occurs, as well as the concentration gradient across the surface of exchange. How has natural selection maximized the factors that favour diffusion in a) polychaetes, b) echinoderm gills (tube feet), c) tracheae of terrestrial invertebrates, and d) mammals.

4. How are the demands for gaseous exchange different for aquatic and terrestrial animals? Choose a couple of examples from each environment and compare the anatomical features that meet the requirements for effective exchange between environment and tissues.

5. What mechanisms of ventilation facilitate gaseous exchange in different organisms?

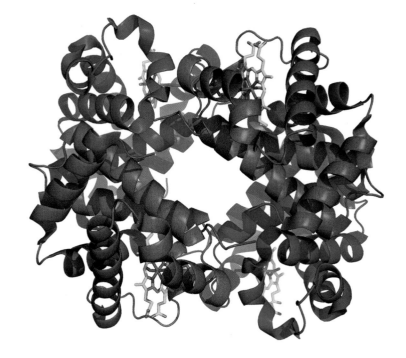

Figure 12.1 Many animals use hemoglobin to bind oxygen and distribute it throughout the body through the circulatory system. This model shows the quaternary structure of the human hemoglobin molecule. It consists of two α subunits (red) and two β subunits (blue). The four structures in green are the heme groups, which contain iron and serve as the oxygen-binding sites.

SOURCE: By Richard Wheeler (Zephyris) 2007

12 Circulation

WHY IT MATTERS

Probably one of the first things you learned in science is that your blood is red because it contains hemoglobin. This protein carries oxygen in the blood, and without bound oxygen the blood colour shifts to blue. Later you may have learned that your hemoglobin has four polypeptide chains (two alpha and two beta), each with a heme group that can bind a single oxygen atom (Fig. 12.1).

Even in humans, hemoglobin takes more than one form. A fetus receives the oxygen in its blood through the umbilical vein in the placenta. When maternal blood reaches the placenta, much of the oxygen has already been used. To overcome this and ensure an oxygen concentration gradient that favours the fetus, fetal hemoglobin has a significantly higher oxygen binding affinity than adult hemoglobin. Furthermore, fetal hemoglobin is synthesized in the yolk sac and has different globin chains. We should expect a wide variety of hemoglobins with different properties among the animals that use hemoglobin to carry oxygen.

Hemoglobin occurs in the blood of many invertebrate groups, including bivalve and gastropod molluscs, some nemerteans, parasitic nematodes, priapulids, brachiopods, and holothurians, as well as some crustaceans and the larvae of chironomid flies.

The symbiotic association of *Riftia pachyptila* (384) (Annelida, Siboglinida) and chemoautotrophic bacteria was introduced on page 263 (Fig. 12.2). Chemoautotrophy requires bacterial autotrophs to have a supply of oxygen and sulphide. *Riftia* has three different hemoglobins, two in the vascular system and the third in the coelomic fluid. Sulphide inhibits oxygen binding in typical hemoglobins. But in *Riftia*, hemoglobins can transport both oxygen and sulphide molecules, bound at different sites on the globin. Many annelids live in sulphide-rich environments, and their hemoglobins, like those of *Riftia*, have sulphide-binding free cysteine residues. The sulphide-binding property of ancestral annelids may have been lost when annelids colonized more aerobic environments. While *Riftia* hemoglobins have sulphide-binding free cysteine groups, their role in sulphide transport has been questioned. *Riftia*'s coelomic hemoglobin has 12 Zn^+ ions at its poles, and sulphide transport appears to occur via chelation of sulphide to these zinc ions. Here the cysteine residues may be responsible for preventing sulphide toxicity that occurs when sulphide binds to the iron in cytochrome enzymes and blocks cellular respiration.

Species in the phylum Phoronida have a parallel adaptation to that of vertebrates where hemoglobin is contained in cells rather than free in solution (an arrangement seen in some other invertebrate phyla). Phoronids (Fig. 3.53B) live with a crown of tentacles, the lophophore, extended into oxygenated sea water. Their bodies are encased in a tube situated in anaerobic substrata. Measurements of oxygen consumption by phoronids show that hemoglobin transport of oxygen, from lophophore to internal tissues, is essential to maintain aerobic metabolism. In the absence of an oxygen supply, for example when the worm withdraws into its tube, levels of oxygen in hemoglobin maintain normal metabolism for 15 minutes. Phoronid hemoglobin differs from that of vertebrates because it does not bind with sulphide. Therefore the high sulphide levels in the phoronid anaerobic environment do not impact oxygen transport.

While *Riftia* and phoronids illustrate extreme conditions, hemoglobin plays a vital role in the adaptation of many animal groups to the demands of their environments.

SETTING **THE SCENE**

The circulatory system is largely responsible for supplying all cells in an animal's body with essential materials (oxygen, nutrients, hormones, etc.) and collecting wastes (metabolic end products, whether carbon dioxide or nitrogenous wastes). The size and features of the circulatory system vary with the size of the animal and the rate at which it consumes energy and produces metabolic wastes. We will explore the diversity of circulatory systems and their components in this chapter and identify links to other components of animals' bodies.

12.1 Circulation: Basic Principles

Circulation is about delivering materials to and collecting other materials from different sites within the body. In Chapter 11 we saw that simple diffusion accounts for movement of oxygen and other substances through the bodies of very small animals. Most animals, however, are too large to rely solely on diffusion; they augment diffusion with circulatory systems. The problem of delivering materials (such as oxygen, nutrients, and hormones) and collecting wastes has been solved in several ways in the animal kingdom. In sponges, the aquifers, while part of the external environment, carry oxygen to the sponge surfaces and carry away wastes. In the Cnidaria (Fig. 2.17A), fluid in the coelenteron acts as a circulatory system. In the Ctenophora, polyclad and turbellarian Platyhelminthes, the branched gut acts as a transport system, reaching all areas of the body (look at the gut branches in the photograph of a transparent polyclad (Figure 3.19A). Most other animals—both invertebrate and vertebrate—have a specialized circulatory system located between the ectoderm and endoderm and isolated from the gastrovascular cavity. The circulatory systems of animals vary considerably in both their components and functions.

A circulatory system consists of a fluid (sometimes called "blood"), vessels (tubes, e.g., arteries and veins), and pump(s) (or heart(s)). Although **"open" circulatory systems** have traditionally been considered different from **"closed" circulatory systems**, this distinction is not as clear as once thought (Figs. 12.3, 12.4). In an open system, tissues are directly bathed in blood that circulates in sinuses and lacunae (small cavities). There is no separation between the blood and the tissues by an endothelial layer. In closed systems, the blood is always contained within vessels.

Figure 12.2 Giant tube worms (*Riftia pachyptila* (384)), molluscs, and white crabs at a deep sea geothermal vent in the Pacific Ocean.

Photo courtesy of C. Van Dover

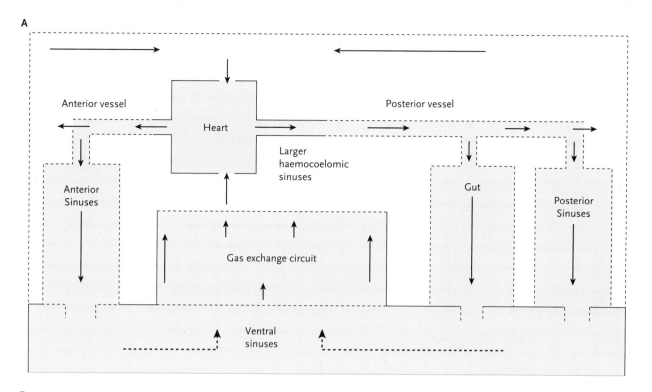

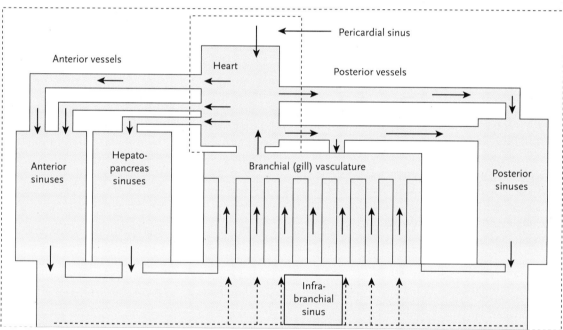

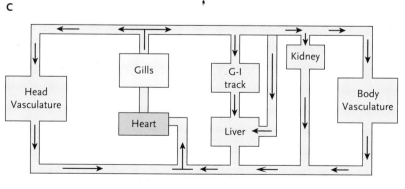

Figure 12.3 **A.** A typical "open" circulatory system such as that of annelids.
B. A "complex incompletely closed" system like that of a cephalopod mollusc.
C. A typical "closed" system like that of a fish. Solid lines are vessels or a muscular pump, while dashed lines are sinuses or vessels lacking a clear lining.

SOURCE: Reiber and McGaw. 2009. "A Review of the "Open" and "Closed" Circulatory Systems: New Terminology for Complex Invertebrate Circulatory Systems in Light of Current Findings," *International Journal of Zoology*, Volume 2009. Copyright © 2009 Carl L. Reiber and Iain J. McGaw.

A

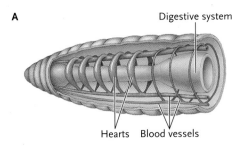

Digestive system

Hearts   Blood vessels

B

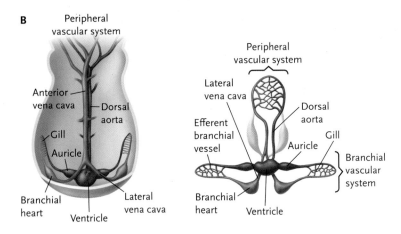

Peripheral
vascular system

Anterior
vena cava — Dorsal
aorta
Gill
Auricle
Branchial
heart   Lateral
vena cava
Ventricle

Peripheral
vascular system

Lateral
vena cava
Efferent
branchial
vessel   Dorsal
aorta
Gill
Auricle
Branchial
vascular
system
Branchial
heart   Ventricle

Figure 12.4 A. The "open" circulatory system of an annelid and **B.** the "incompletely closed" system of a cephalopod mollusc.

SOURCE: Reiber and McGaw. 2009. "A Review of the "Open" and "Closed" Circulatory Systems: New Terminology for Complex Invertebrate Circulatory Systems in Light of Current Findings," *International Journal of Zoology*, Volume 2009. Copyright © 2009 Carl L. Reiber and Iain J. McGaw.

A key to understanding the differences in circulatory systems is a combination of the size of the animal (surface area:volume ratio, see page 295) and its level of activity. Size and the surface area for exchange determine the capacity for diffusion between the surrounding medium and the cells of the body. Level of activity reflects the demand for oxygen and food as well as the rate at which the animal produces carbon dioxide and other metabolic wastes. Therefore we expect active animals such as birds, mammals, and cephalopod molluscs to have circulatory systems with an extensive surface area capable of delivering and collecting materials as rapidly as required (Fig. 12.5). Less active animals have lower demands for oxygen and food and lower waste-production rates, so the demands on their circulatory systems are lower. Open circulatory systems are frequently also associated with non-circulatory functions, for example, blood is pumped into the foot of a clam to expand the foot and provide an anchor as the clam burrows.

In the vertebrate closed system, the complex walls of vessels (consisting of epithelial, connective, and muscular tissues) effectively isolate blood from the tissues through which the vessels pass. This lining limits the diffusion between blood and tissues to areas where numerous vessels consist of endothelial tissues only (e.g., gills, lungs, muscles, etc.), maximizing the effects of concentration gradients there.

Figure 12.5 Researchers injecting a wake-up drug into a large vein in the ear of an immobilized elephant. Elephants are active animals with an efficient and fast-flowing circulatory system, so this drug will take effect quickly. The researchers will have to move quickly to avoid being trampled by a groggy elephant.

Circulatory systems operate at rates required by the organism. The rate of transport is determined by variables identified in the equation for Fick's Law:

$$\mathrm{d}S/\mathrm{d}t = -DA\,(\mathrm{d}C/\mathrm{d}x)$$

This equation shows that the rate of transport ($\mathrm{d}S/\mathrm{d}t$) depends on the diffusion coefficient (D), the area through which diffusion occurs (A), and the concentration gradient ($\mathrm{d}C/\mathrm{d}x$) at the site(s) where diffusion occurs. The gradient was described as $P_1 - P_2$ on page 295 because diffusion of gases is based on differences in partial pressures. In all organisms, A, the surface area, is the major factor limiting the rate of transport. In circulatory systems, A is minimized by delivering circulatory fluids with their important contents to multiple areas with thin enough linings and sufficient area to permit efficient diffusion.

Study Break

1. Is diffusion central to the operation of most circulatory systems? Explain.
2. How do the "circulatory" systems of sponges and cnidarians differ from those of annelids and molluscs?

12.2 Patterns of Circulation

The principles of fluid pressure and velocity are universal, but the circulatory systems of animals take many different forms. The simplest systems, such as those of sponges (Fig. 8.16A), rely on a unidirectional flow of water to bring nutrients into and take waste out of the body. However, in most groups of animals, the circulatory system is completely contained within the body, and the circulating fluid is pressurized by a pump (a muscular heart or vessel) that drives blood to the body region where gas and nutrient exchanges occur. The circulating fluid then returns to the pump to be recirculated.

12.2a Hemodynamics: Pressure and Flow

Like roadways, circulatory systems consist of large and small vessels (roads). It is interesting to compare the movement of blood in a circulatory system with the movement of traffic on roads. Efficient, long-distance movement of goods is best achieved with large diameter vessels (major highways), while local deliveries occur where traffic moves more slowly along shorter and narrower roads (streets and country roads). Capillaries, the smallest vessels, bring blood close to tissues at a slow rate, facilitating diffusion (driveways). Small vessels tend to be short.

Poiseuille's Law describes the relationship between the rate of fluid flow, pressure, and the lengths and radii of circular pipes. The equation for Poiseuille's Law is:

$$Q = \pi \mathrm{d} P r^4 (8 L \mu)^{-1}$$

This equation allows calculation of volumetric flow rate (Q) through a pipe with a radius of r. Important factors include the difference in the pressure (dP) at the two ends of the pipe, where r is the radius, L is the length and μ is the viscosity of the fluid. This equation demonstrates that the radius (actually r4) is the primary determinant of flow rate. This equation applies to fluid flow when viscosity is not a major factor (i.e., Reynolds number is ~2000) and flow is laminar, with minimal friction between the wall of the vessels and the stream through the vessel. The endothelial walls of blood vessels help to reduce friction and limit turbulent blood flow. Poiseuille's Law applies to the velocity and pressure of fluid flow in all animal circulatory systems (Fig. 12.6).

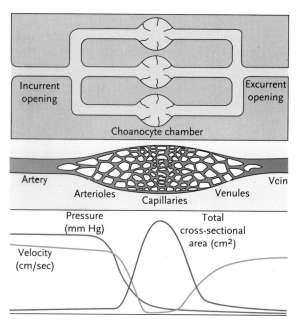

Figure 12.6 The basic structure of circulatory systems and their patterns of fluid velocity and pressure are similar in animals as distantly related as Porifera (top) and vertebrates (middle).

SOURCE: Based on Ruppert, *Invertebrate Zoology* 7th ed., 2003, Cengage Learning, based upon data from Reiswig, 1975, and Purves et al., *Life: The Science of Biology*, 4th Edition, by Sinauer Associates (www.sinauer.com) and WH Freeman.

STUDY BREAK

1. Compare and contrast Fick's and Poiseuille's laws. What are the applications of each?
2. What are the implications of vessel size (length and diameter) in a circulatory system?

12.3 Components of Circulatory Systems

The circulatory systems of animals have three main components: circulating fluids containing cells that deliver oxygen, a series of tubes (arteries, veins, and sometimes capillaries) that transport the fluids, and a pump or series of pumps that keep the fluids moving. In this section, we review the many different variations on these components and how they are put together.

12.3a Circulating Fluids

Several kinds of fluids are found in animal bodies. **Tissue fluid** is present in animals lacking a circulatory system. **Coelomic fluid** in coelomates circulates in the general body or perivisceral coelom. **Lymph** in vertebrates circulates into spaces in tissues and is collected in ducts. It may be mixed with blood. **Blood** is the fluid in the circulatory system. In some invertebrates, the blood is also known as **hemolymph**, particularly in groups that have an open circulatory system (Fig. 12.3A).

More than half of the volume of vertebrate blood is **plasma**, a watery fluid that transports carbon dioxide, nutrients, glucose, clotting factors, and wastes. Plasma moves through the endothelial linings of capillaries into surrounding tissues. There it fills spaces between cells and is called **interstitial fluid**. This fluid drains to lymphatic vessels where it is called **lymph**. The lymphatic system drains back into the circulatory system where it is again called plasma.

Blood or hemolymph [usually] contains oxygen-binding proteins (hemoglobin, chlorocruorin, hemerythrin, or hemocyanin) which transport oxygen. These proteins are also called **respiratory pigments**. In vertebrates, hemoglobin is contained inside **erythrocytes** (red blood cells). In invertebrates some oxygen-binding proteins are suspended in fluid while others may be carried intracellularly. Oxygen-binding proteins are less common in insects where most gaseous transport and exchange are through the tracheal system (see pages 305–306).

Enclosing hemoglobin in erythrocytes appears to be a way of reducing the viscosity of blood. Hemoglobin is a relatively small protein (mass = ~64 kDa) compared to hemocyanins (average mass >1500 kDa), so free-floating hemoglobin would be prone to filtration by the kidneys and spleen. More often found free-floating in the blood, hemocyanin can achieve greater oxygen-carrying capacity of the blood through increased density over hemoglobin. However, hemocyanin density is likely limited by the effects on blood viscosity and the increased energy expenditure that would be needed to circulate the blood.

In most cases, oxygen in physical solution is only a small fraction of the total oxygen content, so the oxygen capacity of blood increases in proportion to the concentration of oxygen-carrying pigment. Consequently, to compare bloods of different haemoglobin (or other pigment) content, we use the term **percent saturation**, expressing O_2 content as a percentage of O_2 capacity. **Oxygen dissociation curves** (Fig. 12.7) describe the relationship between percent saturation and the partial pressure of oxygen (P_{O_2}).

The oxygen dissociation curve of **myoglobin** (a respiratory pigment that stores O_2 in vertebrate muscle) is hyperbolic, whereas hemoglobin and invertebrate hemocyanin oxygen dissociation curves are sigmoid (Fig. 12.7). This difference in the shape of the curves is correlated with the structure of the molecules: myoglobin (and lamprey hemoglobin) has a single heme group, whereas other hemoglobins and hemocyanin have several (usually four) heme groups. The binding of oxygen to the first heme group in these molecules facilitates the oxygenation of subsequent heme groups owing to conformational changes in the protein globin. This is known as **subunit cooperativity**, and the steep portion of the dissociation curve corresponds to oxygen levels at which at least one heme group is already occupied by an oxygen molecule, thus increasing the affinity of the remaining heme groups for oxygen.

Hemoglobin molecules that have high oxygen affinities are saturated at low partial pressure of oxygen, whereas hemoglobins with low oxygen affinities are completely saturated only at relatively high partial pressures of oxygen. This difference in oxygen affinity is not due to differences in the heme group; rather it is related to differences in the properties of the protein globin. The affinity is expressed in terms of the $\mathbf{P_{50}}$, the partial pressure at which the hemoglobin is 50% saturated with oxygen.

An important property of respiratory pigments is that they combine reversibly with O_2 over the range of partial pressures normally encountered in the animal. At low P_{O_2}, only a small amount of O_2 binds to the respiratory pigment; at high P_{O_2}, however, a large amount of O_2 is bound. This property allows the respiratory pigment to act as an oxygen carrier, loading at the respiratory surface (a region of high P_{O_2}) and unloading at metabolically active tissues (a region of low P_{O_2}). In some animals, the predominant role of a respiratory pigment may be to serve as an oxygen reservoir, releasing O_2 to the tissues only when O_2 is relatively unavailable from the environment. For example, seals and other diving mammals have high levels of myoglobin in their muscles. Myoglobin acts as an oxygen store, releasing O_2 only during periods when oxygen levels in the muscle decrease, as they do during an extended dive.

The affinity for oxygen of respiratory pigments such as hemoglobin and hemocyanin varies according to local conditions of pH and temperature (Fig. 12.8). The affinity of respiratory pigments is labile and is reduced by an increase in P_{CO_2} (which decreases pH) or increased temperature. The term **Bohr effect** is used to describe the effect of pH on the hemoglobin–oxygen affinity, as triggered by increases in $[H^+]$ on the pigment molecule. These factors—lower pH and increased temperature—serve to facilitate the unloading of oxygen at metabolically active tissues. The oxygen dissociation curve for myoglobin (Fig. 12.7), unlike that of hemoglobin, is relatively insensitive to changes in pH.

Mammalian erythrocytes contain high levels of 2,3-diphosphoglycerate (DPG). DPG binds to deoxyhemoglobin and reduces oxygen affinity, facilitating

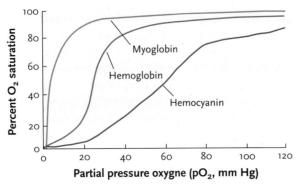

Figure 12.7 Characteristic oxygen dissociation curves for myoglobin, hemoglobin, and hemocyanin. Note the hyperbolic shape of the myoglobin curve and the sigmoid shape of the hemoglobin and hemocyanin curves.

SOURCE: Based on Morris and Bridges 1988. "Interactive Effects of Temperature and L-Lactate on the Binding of Oxygen by the Hemocyanin of Two Arctic Boreal Crabs, Hyas araneus and Hyas coarctatus," *Physiological Zoology* 62 (1): 62–82.

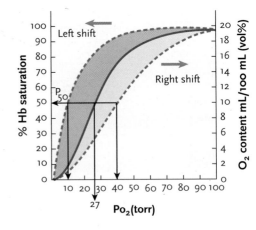

Figure 12.8 The effect of changes in pH on blood–oxygen dissociation curves (the Bohr effect). The P_{50} increases so that oxygen is unloaded from hemoglobin more readily under the conditions found in active tissues, which include lower pH, higher temperature, and an increase in organic phosphate ligands such as ATP or 2,3-diphosphoglycerate (DPG).

SOURCE: Based on Des Jardins (2013) *Cardiopulmonary Anatomy & Physiology: Essentials of Respiratory Care*, 6th Edition, Delmar.

oxygen delivery. Increases in DPG levels accompany reductions in blood O_2 levels, increases in pH, and reduction in hemoglobin concentrations in the blood. Low blood O_2 levels, such as those encountered at higher altitudes, trigger an increase in DPG production, so that oxygen is more readily delivered to tissues. Prolonged exposure to lower environmental oxygen pressures also triggers an increase in red blood cell production to facilitate oxygen uptake.

A decrease in pH results in an increase in oxygen affinity in hemocyanins from several gastropods and the horseshoe crab *Limulus* (645). This greater oxygen affinity is referred to as a **reverse Bohr effect**, and may serve to facilitate oxygen uptake during periods of low oxygen availability, when reductions in blood pH are maintained in these animals.

Another way to achieve high concentrations is to increase the numbers of erythrocytes. But more erythrocytes make blood more viscous and reduce its flow. Tuna have high concentrations of hemoglobin in their blood because of high average concentrations in their erythrocytes. In northern bluefin tuna (*Thunnus thynnus* (1292)) and southern bluefin tuna (*Thunnus maccoyii* (1291)), oxygen binding to hemoglobin is virtually independent of temperature. In *T. maccoyi*, the temperature of the tuna's blood can differ between gills and tissues by up to 20°C. In both species, the absence of impact of temperature on the hemoglobin–oxygen affinity ensures that oxygen is not downloaded at heat exchangers.

Erythrocytes of most vertebrates are typical eukaryotic cells with nuclei and other organelles. In mammals, however, the erythrocytes lose their nuclei and organelles during development, presumably to allow them to carry more hemoglobin. **Leukocytes** (white blood cells, Fig 12.9) are much less numerous than erythrocytes and are an important part of the immune system, attacking potential invaders including fungi, bacteria, viruses, and even tumour cells. Blood also contains **platelets**, cell fragments that play a key role in blood clotting (Fig. 12.9). Platelets are activated by

contact with collagen, which is released when the lining of a blood vessel is injured. Activated platelets send out long projections (pseudopodia) that adhere to the exposed collagen and to one another, forming a plug over the damaged area.

A good supply of oxygen and the ability to form blood clots are both critical features of vertebrate circulatory systems. It's no small wonder that blood is one target of some snake venoms!

STUDY BREAK

1. How does the structure of respiratory pigments such as hemoglobin and hemocyanin facilitate binding of oxygen, as shown in oxygen dissociation graphs?
2. What factors favour the unloading of oxygen by hemoglobin? Why is this adaptive?

12.3b Vessels: Arteries, Veins, and Capillaries

Closed circulatory systems are defined by the presence of arteries, veins, and capillaries, all of which are lined with epithelium. **Arteries** carry blood away from the heart, subdivide into smaller **arterioles** that ultimately connect with **capillaries**. Blood flows through numerous capillaries at very low pressure. It then flows into small veins (**venules**) that feed into larger **veins** and hence to the heart (Fig. 12.10). The diffusion of gasses, nutrients, hormones, wastes, and other molecules occurs only across the walls of capillaries, making them one of the most important parts of the circulatory system. Because all tissues must be close enough to capillaries for diffusion to occur, there are vast networks of **capillary beds** throughout the body and more capillaries than any other type of vessel in the vertebrate body.

Arteries are thick-walled, and their lining includes a layer of smooth muscle and elastic fibres, allowing them to accommodate changes in pressure that occur across a heartbeat cycle. The tone of the smooth muscle in blood vessels is regulated by the autonomic nervous system in response to pressure sensors in the arterial walls. Arterial tone can be relaxed to allow increased blood flow or tensed to restrict blood flow. Therefore, when you bend down to touch your toes, you do not blow your brains out because the blood pressure at the base of your brain remains constant. The same is true for a giraffe (*Giraffa camelopardalis* (1757)) bending down to get a drink of water (Fig. 12.11).

Veins have an extremely thin muscle layer because blood enters veins at considerably lower pressure. Larger veins have valves that prevent backflow, ensuring one way flow of blood back to the heart. Veins are capacitance vessels, able to serve as reservoirs during times of low metabolic need.

Erythrocyte (red blood cell)

Leukocyte (white blood cell)

Platelets

Figure 12.9 This colour-enhanced scanning electron micrograph of human blood illustrates flattened, enucleated erythrocytes (red), less numerous leucocytes (yellow) which are part of the immune system, and platelets (pink) that function in blood clotting.

SOURCE: Science Source/Photo Researchers, Inc.

BOX 12.1 Molecule Behind Animal Biology
Draculin

When bitten, stung, cut, or scraped, your body mobilizes platelets and proteins in your blood plasma to approach the site of the wound and close it off to prevent infection and further loss of blood. This process is called coagulation or, more commonly, **clotting**. Without clotting you could bleed to death after even the most minor of incisions. Clotting begins when platelets immediately start closing off the wound, while proteins in your plasma known as **coagulation factors** engage in a chain of complex chemical reactions with one another to eventually form the protein **fibrin**. Fibrin significantly strengthens the platelet clot and prevents it from easily reopening.

This same basic process of blood clotting is highly conserved through-out mammals. Unfortunately for some animals, this shared mechanism has negative consequences. Blood-feeding vampire bats (Fig. 1A) take advantage of the highly predictable nature of clotting by generating anti-coagulants that allow them to target a wide range of potential hosts. Vampire bat saliva contains several compounds that disrupt the clotting process. One very important molecule in this concoction is known as **draculin** (for obvious reasons). Draculin, a protein, attaches to some of the intermediate proteins in the fibrin-forming cascade and prevents two of these coagulation factors from activating the next steps in the process. This prevents fibrin from forming, and keeps blood flowing. So after a vampire bat bites, removing a 5 mm diameter divot of skin from a sleeping animal (Fig. 1A), draculin (and other compounds in its saliva) prevents the host's blood from clotting. With a constant flow of blood, the vampire can feed to its heart's content. Eventually the bat has a full stomach and leaves its host. The draculin molecules in the host's blood are degraded, bleeding stops in minutes to hours, and the wound begins to heal. All of this typically happens without the host animal even knowing!

The striking anticoagulant properties of draculin have attracted notice from the medical community. Strokes are often caused or exacer-bated by clots that form and are carried through the body by the bloodstream. When a clot lands within blood vessels in the brain or near the heart, the results can be catastrophic. Drugs with draculin as the active ingre-dient target these types of clots, and are currently being tested in Phase II clinical trials. The anticoagulant properties of draculin-based drugs are being compared against traditional anticoagulant drugs such as tissue plasminogen activator (tPA). Draculin apparently allows clots to be treated up to 12 h after the clot is first formed, as opposed to the approximately three-hour treatment window of tPA. If draculin holds up better than tPA in the clinical trials, it will be another important example of a powerful medical treatment derived from natural molecules.

Copyright Jim Clare, naturepl.com

M.B. Fenton

Figure 1
A. *Common vampire bat,* Desmodus rotundus (1788), *feeding on a cow. The bat feeds by licking the blood from a flow that is maintained by anticoagulants in its saliva.* **B.** *Close-up of the face of* D. rotundus, *showing the exceptionally sharp teeth used to make the initial incision in the host.*

12.3c Pumps and Circulation

The circulatory systems of most animals include a pump that circulates fluid around the body. The pump may be a central heart with more than one chamber and valves controlling blood flows. Many invertebrates, however, have **tubular hearts**, tubes with contractile muscle in their walls. In insects, hearts are usually **pulsating vessels** that propel blood by peristalsis. Some animals have **ampullar hearts**, accessory propulsive systems boosting flow to particular peripheral net-works, for example the **branchial hearts** of cephalopod molluscs (Fig. 12.4B).

In many arthropods, the hemolymph is partly con-tained in vessels but also bathes the tissues. Spiders are a good example (Fig. 12.12). The heart is suspended in a pericardial sinus. As the heart contracts (systole), blood is pumped into open-ended vessels that lead toward the brain, organs, and limbs. Blood enters the hemocoel (body cavity) surrounding these structures, allowing gas and nutrient exchange to take place. Blood flows back to the pericardial sinus where it is drawn into the heart through **ostia** (openings) as it relaxes (diastole).

The vertebrate heart pumps blood through a closed circulatory system (Figure 12.3C). Derived from

BOX 12.2
Venoms

Animals from several phyla (including Cnidaria, Annelida, Mollusca, Arthropoda, and Chordata) produce venom, which is used in offense and/ or in defense. The venoms of two African colubrid snakes, boomslangs (*Dispholidus typus* (1491)) and bird snakes (*Thelotornis capensis* (1525)) target the blood and cause coagulation—effectively immobilizing their prey. Metalloproteinases in the venom of boomslangs cause diffuse intravascular coagulation by consuming fibrinogen, causing hemorrhages into the muscle and brain.

More typically, venoms are a mixture of proteins that interrupt normal processes in the target organism. In some other venomous snakes, one active ingredient in the venom lyses red blood cells (Fig. 9.2), causing a different form of disruption of the circulatory system. Other venoms frequently have neurotoxins that may act on synapses affecting the heart and/or diaphragm. Protease enzymes also are common in venoms. These cause degradation of tissue and, when delivered as a defensive bite, can lead to gangrene.

When injected, venoms cause pain ranging from minor to extreme. A coral snake from Texas, *Micrurus tener* (1509), produces venom that causes intense and unremitting pain. This venom activates acid-sensing ion channels on nociceptors. The venom delivered by some hymenopterans and scorpions can also cause distracting pain, making them effective in defense. Venoms that are not particularly deadly to humans can cause sufficient interruption of normal activities to put the victims at risk. For example, people may be stung while swimming. Whether from cnidarians, molluscs, sting rays, or sea snakes, these stings may cause the victims to drown.

---

mesoderm, the vertebrate heart first appears as a pair of parallel endocardial tubes. Each tube receives inflow from the developing gut and passes outflow to an aorta. The endocardial tubes quickly fuse to form a single heart tube. The end that receives inflow ultimately forms one or two atria, while one or two ventricles form on the outflow end. Different patterns of folding of the heart tube and the development of fusions and partitions within the tube result in the varieties of hearts seen among vertebrates (Fig. 12.13).

The hearts of cartilaginous and boney fish are modified tubes consisting of a single atrium and a single ventricle (Fig. 12.13A). Deoxygenated blood in veins from the body enters the atrium, which contracts to push blood into the ventricle. The ventricle then contracts to push blood into arteries that carry it to the gills where gas exchange occurs. Backflow of blood from the atrium to the veins and the ventricle to the atrium is prevented by valves, flap-like structures composed of tough connective tissue. After passing through the gills, blood then travels to the rest of the body through arteries, arterioles, and capillary beds and travels back to the heart through veins. This simple loop of blood flow through the body is called a **single circuit**.

Gills were replaced by lungs as sites of gaseous exchange as vertebrates moved onto the land. This transition was accompanied by changes in the form of the heart and circulatory system (Fig. 12.13B) such that the **pulmonary circuit** (heart to lungs and back) became separated from the circulation going to the rest of the body (**systemic circuit**). Amphibian and most reptile hearts have two atria and one ventricle. The right atrium receives deoxygenated blood through

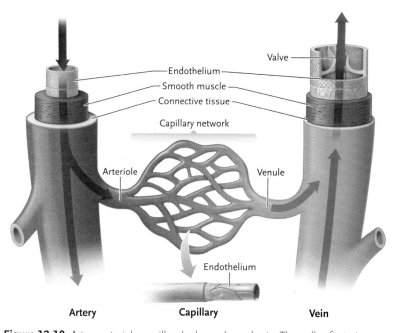

Figure 12.10 Artery, arteriole, capillary bed, venule, and vein. The walls of arteries contain a thick layer of smooth muscle, while veins are thin-walled structures. All blood vessels in vertebrates are lined with a layer of epithelial cells (endothelium).

SOURCE: From RUSSELL/WOLFE/HERTZ/STARR. *Biology*, 1E. © 2010 Nelson Education Ltd. Reproduced by permission. www.cengage.com/permissions

Figure 12.11 As a giraffe bends down to drink, it changes the challenge to its circulatory system. When standing, its brain is 162 cm above the heart. To maintain 90 mm Hg pressure in the brain, its cardiac output must be 208 mm Hg. The giraffe's cardiac output is about the same as that of a cow, but the aorta, the main blood vessel leaving the heart, is more resilient and maintains pressure more effectively than that of the cow (or a human).

Figure 12.12 The circulatory system of a spider.

SOURCE: From RUSSELL/ WOLFE/HERTZ/STARR. *Biology*, 1E. © 2010 Nelson Education Ltd. Reproduced by permission. www.cengage.com/ permissions

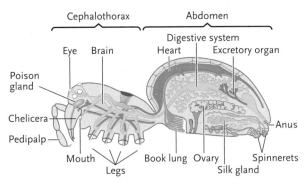

veins from the body and contracts to push it to the ventricle. At almost the same time the left atrium receives oxygenated blood through veins from the lungs and contracts to push it into the ventricle. The two atria contract at slightly different times, which helps keep oxygenated and deoxygenated blood moving in separate streams. Nevertheless, some mixing does occur. When the ventricle contracts, blood is pushed into the arteries that lead to both the lungs and the rest of the body. This slight mixing of oxygenated and de-oxygenated blood in a single ventricle is called an **incomplete circuit**.

Mammals, birds, and crocodylians have a **closed circuit** in which the pulmonary circuit is completely separated from the systemic circuit (Fig. 12.13C). A septum develops in the ventricle, dividing it in two. Now de-oxygenated blood from the body travels through the right atrium and ventricle to the lungs, and then back to the left atrium and ventricle, making a complete pulmonary circuit. Functionally, this results in a more richly oxygenated blood supply being sent to the body tissues. This is consistent with the evolution of endothermy and the relatively high metabolic rates of mammals and birds.

As body size increases, systemic pressure must also increase to provide enough power to circulate blood through a larger body. This could pose a problem for animals with single and incomplete circuits in which a single ventricle pressurizes blood that goes to both the lungs and body. Pressure high enough to support systemic circulation through a large body may be too high for effective gas transfer over the capillaries in the lungs. This is a primary reason why many researchers think that dinosaurs had four-chambered hearts—especially those with very long necks (Fig. 6.1). This point of view is further supported by the presence of four-chambered hearts in birds, as well as similarities in the gaseous exchange systems of birds and living archosaurs (crocodylians).

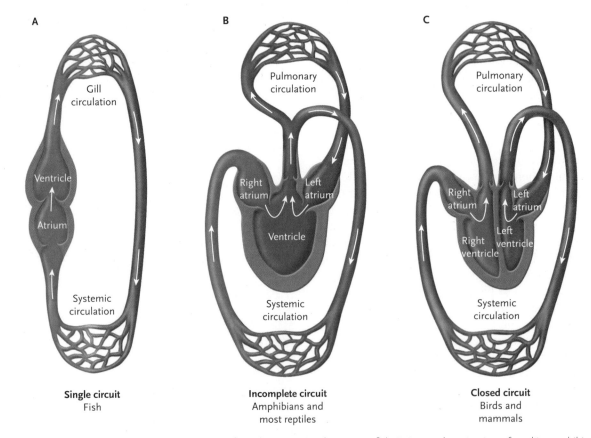

Figure 12.13 The hearts and circulatory systems of vertebrates. **A.** Single circuit in fish. **B.** Incomplete circuit, as found in amphibians and most reptiles. **C.** Closed circuit, as found in birds and mammals.

BOX 12.3 People Behind Animal Biology

Ibn al-Nafis

The modern understanding of the circulatory system began in earnest with the pioneering dissections and experimental work of a British physician, William Harvey, in the early 1600s. His research culminated in the 1628 publication of his famous work, *On the Motion of the Heart and Blood in Animals*. Harvey was the first to compare circulation in a wide range of species, bringing about a broad understanding of circulatory systems in animals. Prior to Harvey, most medical understanding of the circulatory system dated back over 12 centuries, to the scholar, Galen, in ancient Rome. As an example of the reach of his ideas—though bloodletting had been practiced for centuries before him—Galen was the first physician to prescribe bloodletting in treatments, a practice that would continue to the late 19th century.

Between the 1200 years separating Galen and Harvey, a Syrian-Egyptian scholar, Ibn al-Nafis, made fundamental discoveries about pulmonary circulation and blood flow. Although unnoticed by his Western contemporaries, Ibn al-Nafis demonstrated his understanding of the movement of blood and of pulmonary circulation in

the early 13th century (Fig. 1). He noted that the heart was always a "forward-pumping" organ and that blood always flowed in a circuit. (Galen thought blood moved via ebb-and-flow). Crucially, Ibn al-Nafis also understood that the heart had complete separation between its right and left sides. Before Galen's time, anatomists knew that the heart had two distinct ventricles. But the functions of the two ventricles were not well understood. One of Galen's theories held that the right ventricle only pumped a small amount of blood to the lungs to nourish them. He

thought that the rest of the blood was passed through to the left ventricle by invisible pores in the intraventricular septum. Ibn al-Nafis clearly stated that there were no pores and recognized that blood flows in a complete circuit to and from the lungs (the pulmonary circuit), omitting the need for intraseptal pores in the heart.

The discoveries of al-Nafis are even more surprising, considering that Muslim law in the 13th–14th centuries prohibited dissections. Although his writing hints that he may have performed some dissections, we cannot be sure how he arrived at his ideas. We can be sure that if the Western world had not overlooked his ideas, we might have stopped the practice of bloodletting long before the dawn of the 20th century!

FIGURE 1

Reproduction of a portion of Commentary on the Anatomy of the Canon of Avicenna, by Ibn al-Nafis. This portion of the manuscript describes the pulmonary circulation.

© The Art Gallery Collection / Alamy

STUDY BREAK

1. What is a pulmonary circuit?
2. How are the hearts of vertebrates different from those of molluscs?

12.4 A Sampling of Invertebrate Circulatory Systems

12.4a Nemertea

The Nemertea have perhaps the simplest and most primitive circulatory system. The most basic consists of a single central vessel linked to two lateral vessels (Fig. 12.14A). There is no heart, but the blood vessels are contractile and lined by a mix of endothelial, epitheliomuscular, and ciliated cells. Movements of the animal compress regions of the blood vessels, driving the blood along them. Unidirectional flow is main-

tained by non-return flap-like valves within the vessels. Blood flows forward in the central vessel and toward the back in the lateral vessels. There are more complex blood systems in some larger nemerteans (Figure 12.14B and C) to meet the demands of supply oxygen to all parts of the body.

Nemerteans are among the phyla with hemoglobin in their blood. This pigment is concentrated in cells circulating in the blood system. Experimental exposure of nemerteans to carbon monoxide reveals two important facts:

- hemoglobin in nemertean blood does carry oxygen (their metabolism is greatly slowed after CO treatment)
- these worms are not totally dependent on their hemoglobin for oxygen and are able to maintain slow aerobic metabolism in the presence of CO

Nemertean blood vessels are **coelomic**, completely surrounded by mesoderm. Recognition of this

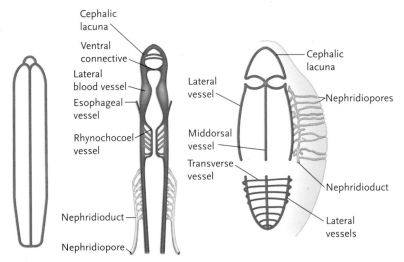

Figure 12.14 Circulatory systems of members of the Phylum Nemertea.
A. *Planktonemertes* (566), a simple system. B. The system of *Tubulanus* (567) is more complex (notice the close association with the nephridial system). C. The system in *Amphiporus* (560) has a series of transverse vessels. (Note: In B & C only parts of the circulatory system are shown.)

SOURCE: Based on Brusca and Brusca Invertebrates, 2003, Sinauer, and Gibson, R. (1998). "Epilogue - one hundred years of nemertean research: Bürger (1895) to the present," *Hydrobiologia*, 368, 301–310.

coelomic structure is one reason why the nemerteans have been moved away from their original placement as acoelomates. Nemerteans have an extensible proboscis (see page 232), and some gas exchange takes place across the wall of the proboscis and into the fluid in the coelomic rhynchocoel (proboscis cavity) (Fig. 8.60B), as well as through the gut wall from water taken into the gut lumen.

12.4b Annelida

Small annelids may have no specialized circulatory system and rely on the coelomic fluid for transport of oxygen, nutrients, and wastes. Large annelids, in all groups, have evolved a complex system of vessels. The common pattern shows contractile dorsal and ventral main vessels with blood pumped forward in the dorsal vessel and to the posterior in the ventral vessel. The directions of blood flow are controlled by valves in the main vessels and where the segmental vessels join the dorsal and ventral vessels. In each segment, the dorsal and ventral vessels are linked by vessels that surround the gut and that carry blood from the dorsal vessel to the ventral, and by vessels that run close to the body wall (or that extend into a gas exchange complex in the parapodia of polychaetes) and carry oxygenated blood back to the dorsal vessel (Fig. 12.15).

In earthworms, blood flows are augmented by the development of 5 pairs of accessory hearts (or pseudo-hearts) in segments 7–11. With this arrangement, there is never a complete separation of oxygenated and deoxygenated blood since each segment is receiving a supply of oxygenated blood through the dorsal vessel and deoxygenated blood through the ventral vessel. Some tube-dwelling polychaetes (e.g., *Thelepus* (*Amphitrite*) spp. (347), Fig. 3.47A) have developed a cluster of anterior gills. In each segment the dorsal vessel feeds an afferent vessel that enters the gill. Efferent vessels run from each gill to vessels surrounding the gut and to the ventral vessel. This morphology allows full separation of oxygenated and deoxygenated blood.

In the annelids, four pigments—hemoglobin, erythrocruorin, chlorocruorin, and hemerythrin—are

Figure 12.15
Blood circulation in an errant polychaete. Arrows indicate direction of blood flow.

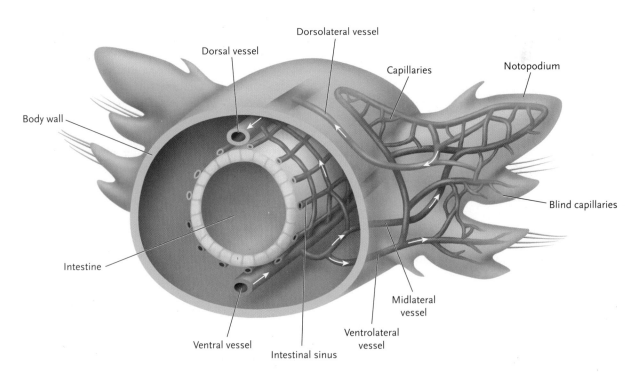

found with no apparent distributional pattern in different species. Relating a particular oxygen-transporting pigment to phylogeny is a challenge, although it is suggested that a form of hemoglobin was the first respiratory pigment and that others have evolved from that form.

Chlorocruorin, found in Serpulida and the Sabellidae (Fig. 3.48), is related to the high-molecular-weight hemoglobins although it has a significantly lower oxygen affinity than hemoglobin. It takes its name from its colour, green-yellow in dilute solution. The distribution of chlorocruorin is variable in the serpulids and sabellids. In some species it occurs with hemoglobins; in others chlorocluorin is absent. Three species in the serpulid genus *Spirorbis* (342) (Fig. 8.59) that live in similar habitats provide an example: one species has chlorocruorin, one has hemoglobin, and the third has neither pigment.

Hemerythrin, after the Greek word for blood, does not include iron atoms. Hemerythrin is found in sipunculids and some other annelids, and also in brachiopods and priapulids. Deoxygenated hemerythrin is colourless while oxygenated hemerythrin has a violet-pink colour.

Erythrocruorin has a high ($1-3 \times 10^6$) molecular weight. One molecule of erythrocruorin may carry 50 to 100 molecules of oxygen, making it a fairly low efficiency respiratory pigment when compared to vertebrate haemoglobin carrying one oxygen molecule on each 17 000 Dalton subunit. Erythrocruorin is found in a number of different annelid groups (including earthworms) and also occurs in some snails, arthropods (insects and crustaceans), and nematodes.

Some annelids may use the properties of more than one oxygen carrier. *Thelepus* (*Amphitrite*) (Fig. 3.47A) has a low-molecular-weight hemoglobin in its coelomic fluid and high-molecular-weight erythrocruorin in the vascular system.

12.4c Decapod Crustaceans

Decapod crustaceans were long thought to have relatively simple and completely open circulatory systems. This was partly because their circulatory systems are composed of many hard-to-find vessels that are often difficult to see with a dissecting microscope, much less the naked eye. A team of researchers perfected a method of illustrating the crab circulatory system called corrosion casting, and used it to map the circulatory systems of blue crabs (*Callinectes sapidus* (811)) and other crabs of the family Canceridae (Fig. 12.16).

Crabs have a series of arteries that lead from a single-chambered heart through one-way valves. These arteries carry hemolymph to different regions of the body, where they branch into arterioles and then into very finely divided capillary-like structures. Some capillaries are blind-ended, but many branch into very complex capillary beds. So far these are clear characteristics

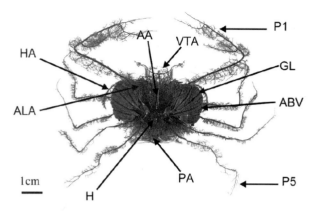

Figure 12.16 A corrosion cast of the circulatory system of a blue crab, *Callinectes sapidus* (811), was made by injecting resin into the blood vessels downstream of the heart. The resin is allowed to harden, and then the tissue that overlies the hardened resin is dissolved chemically. The end result is a detailed cast of the entire circulatory system. The labelled vessels are: AA = anterior aorta; VTA = ventral thoracic arteries; P1, P5 = pereiopod arteries; ABV = one of the afferent branchial veins, which lies between the gill lamellae (GL); PA = posterior aorta; H = heart and pericardial sinus; ALA = anterolateral arteries; and HA = one of the hepatic arteries.

SOURCE: McGaw, I. J., & Reiber, C. L. 2002. "Cardiovascular system of the blue crab Callinectes sapidus," *Journal of morphology*, 251, 1–21, John Wiley and Sons. Copyright © 2002 Wiley-Liss, Inc.

of closed circulatory systems. The anatomy of the venous system is more difficult to characterize. After supplying the crabs' organs and tissues, hemolymph drains to a series of vessels that are located between organs and/or muscles. Recent evidence from corrosion casts indicates that these sinuses are not just pools of hemolymph, as was once believed, but are actually complex networks of capillary-like structures. The sinuses are surrounded by connective tissue, and the walls of the capillary-like structures are formed by the outer layer of the organs they drain. This means that hemolymph passes through a series of sinuses before reaching the gills and then the pericardial sinus that returns hemolymph to the heart.

Technically, crabs have open circulatory systems because they lack veins that are lined with a layer of endothelium. With respect to function, however, the crab circulatory system performs as efficiently as an open system. One of the benefits of closed systems is that they can produce relatively high blood pressures that can be changed very rapidly. Decapods achieve this by having a functionally closed system in which heart rates are controlled by both the CNS and a pericardial endocrine gland that secretes neurohormones. Another benefit of most closed systems is that fluid flow to different parts of the body is regulated by contraction and relaxation of the smooth muscle within the walls of arteries. Crab arteries lack smooth muscle, but the flow of hemolymph to different parts of the body is controlled effectively by contracting the one-way valves that lead from the heart to each of the arteries. Although these are different mechanisms, the results are the same—rapid changes in circulation to different regions of the body. Far from being "primitive," crab circulatory systems are complex. The existence of well-developed circulatory systems has been linked with the impressive ecological and physiological diversification of crustaceans as a group.

12.4d Cephalopoda

Octopus, squid, and cuttlefish are active predators, many reaching large sizes and exhibiting behaviours as complex as those of most vertebrates. Despite these measures of success, the cephalopod molluscs operate close to physiological limitations. Many of those limitations are imposed by the anatomy and physiology of their circulatory system.

The anatomy of cephalopod circulatory systems reflects a spectrum of activity levels from slow-swimming nautiloids, more active cuttlefish and octopus, to highly active squid. Across this spectrum the circulation system design is based on the same pattern (Fig. 12.17). Parts of the cephalopod circulatory system show remarkable parallels to the vertebrate system. Unlike most invertebrate systems, cephalopods (except nautiloids) have a closed circulatory system with arteries and veins that connect through capillary beds. Like vertebrates, the blood vessels have an endothelium lining. The closed system coincides with a significantly lower blood volume compared with that of other molluscs. Squid blood is about 6% of body weight, compared with 55% in bivalves and 33–66% in gastropods, or about 5% in salmon and 9% in humans.

From the ventricle of a three-chambered heart, blood is pumped through the peripheral circulatory vessels and flows through paired vena cavae to accessory (branchial) vessels of the ctenidia. Efferent branchial vessels carry oxygenated blood from the ctenidia

to a pair of atria that feed the ventricle of the systemic heart. The major cephalopod arteries and veins are contractile. The circulation of venous blood into the ctenidia, and the flow of water in the mantle over the outer face of the ctenidia, combine to form an efficient countercurrent exchange system. In *Octopus* (511) and *Sepia* (517) swimming in oxygenated seawater, blood oxygen saturation is almost 100%. *Octopus* and *Loligo* (503) extract about 85% of the available oxygen, compared to about 33% in salmon.

The level of depletion of blood oxygen in tissues means that as a squid becomes more active, the rate of blood circulation must increase to make more oxygen available (Fig. 12.18). This is accomplished by increasing both heart rate and heart stroke volume. Estimates for *Loligo opalescens* (506) suggest a tenfold reduction in circulation time, from 34 s to 3.4 s, the accelerated blood flow being driven by an increase in the power output of the heart, from 4.2 W.s^{-1} to 41.1 W.s^{-1}, as the squid goes from rest to swimming at 0.6 m.s^{-1}. At the other end of the system, water flow through the mantle and over the ctenidia increases with swimming speed as the pulses used for jet propulsion increase in volume and speed. Blood being in the ctenidia for a shorter time means that the extraction of oxygen from the water to the blood is reduced from about 10% of the O$_2$ in the water to around 5%. This decline does not prevent the blood from becoming fully oxygenated as it passes through the gills.

As in other molluscs, hemocyanin is the oxygen-carrying pigment in cephalopods. The pigment is blue when bound to oxygen, the blue a result of the copper atoms that play a role similar to that played by iron in hemoglobin (Fig. 12.19). While hemocyanin also occurs in many arthropods, the molecular architecture of the protein in the two groups is quite different. Hemocyanins bind oxygen non-cooperatively between the two copper atoms. Hemoglobin undergoes a

Figure 12.17
A simplified view of the circulatory system of an octopus.

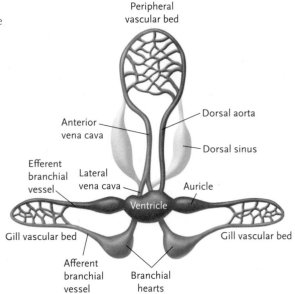

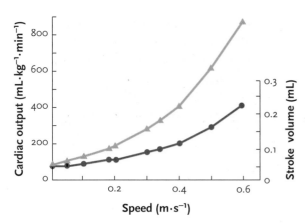

Figure 12.18 Cardiac output (▲) and cardiac stroke volume (●) for a swimming squid.

SOURCE: SHADWICK, R. E., R. K. O'DOR, et al. (1990). "Respiratory and cardiac function during exercise in squid." *Can J Zool* 68: 792–798. © 2008 Canadian Science Publishing or its licensors. Reproduced with permission.

Figure 12.19

The oxygen-binding site on a hemocyanin molecule is between the two copper atoms that are linked to a series of histidine molecules.

SOURCE: Based on "Bioinorganic chemistry," Access Science.

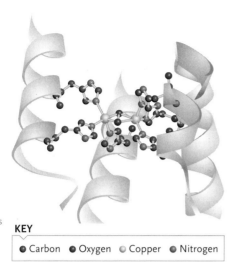

KEY

● Carbon ● Oxygen ○ Copper ● Nitrogen

conformational change when it binds oxygen (cooperative binding) and is able to carry four times the amount of oxygen carried by the same amount of hemocyanin (for equal blood volumes).

Squid must pump eight times as much blood to go half as fast as a fish. While the circulatory system of cephalopods is complex and very efficient, it operates close to the limiting constraints of its components. This is clearly a triumph of engineering over design.

12.4e Echinodermata

Species in the phylum Echinodermata have a water vascular system (WVS) consisting of coelomic canals (ring canal, radial canal, lateral canals, and stone canal) that connect to the **madreporite**, a porous ossicle (light coloured and shown by arrow in Figure 12.20A). The WVS consists of **podia** (tube feet, Fig. 12.20B) used for locomotion, feeding, and gaseous exchange. Podia are extended, moved, and retracted by muscles, and fluid in each tube foot can be isolated from the radial canal by a valve. In a typical echinoderm about 99% of the water in the WVS is in the podia. Echinoid (sea urchin) tube feet are long and include a membrane that separates the flow into each foot from that flowing out, creating a counter-current system for gaseous exchange with seawater. In asteroids (sea stars), thin, blister-like extensions of the WVS system (known as papulae) are the major sites of oxygen exchange with the WVS, although some oxygen

exchange occurs across the podia walls. The podia and WVS also play a part in oxygen and nutrient transport in the ophiuroids and crinoids.

The transport function in holothurians (sea cucumbers) has been taken over by the hemal system, which may include a functionally integrated heart, axial canal, and axial hemal vessel. The hemal system is closely associated with the respiratory trees (internal, highly branched, invaginations of the WVS). In some holothurians, hemoglobin is used as an oxygen carrier in the hemal system, while in others, the hemal system probably has no role in gaseous exchange, and hemal fluid movement is slow. In these groups, the functions of the hemal system are probably restricted to internal (endocrine) chemical communication.

The fluid in the WVS of echinoderms is rather like sea water but also includes coelomocytes, proteins, and a higher concentration of K^+ than in seawater. Using a fluorescent, high-molecular-weight tracer, researchers have demonstrated that the madreporite of a sea star (*Echinaster graminicola* (1073)) admits a steady trickle of seawater into the WVS (Fig. 12.20). "Trickle" is the correct term as it takes about 2.4 days for 1 mL of seawater to enter through the madreporite. Fluid exchange through the madreporite balances the pressure and volume of the WVS with the surrounding environment and supplies replacement fluid for both the water vascular system and the perivisceral coelom.

12.4f Urochordata

The circulatory system of Urochordata is unusual in several ways. The heart is tubular, ventral (see Fig. 8.12A), and enclosed in a pericardial cavity. From the heart, a ventral aorta runs close to the endostyle to a capillary bed that branches over the pharyngeal basket. These capillaries connect to a dorsal aorta that then supplies blood to the viscera and other organs before the blood returns to the heart. One unusual feature of urochordate circulation is the reversibility of the blood flow direction. Periodically the heart reverses the pumping direction, controlled by a pacemaker at each end of the heart. In the colonial ascidian *Botryllus* (1107) (Fig. 3.75B), these reversals occur every 2–3 minutes.

There are two main theories about the reasons for circulatory reversal. First, if it takes more energy to pump in one direction than the other, it evens out the workload on the heart. Second, and more likely, it evens out the distribution of nutrients and oxygen to the organs that are arranged in series along the circulatory path (Fig. 12.21).

The tunic (outer covering) of ascidians is separated from the body wall by a fluid-filled space (Fig. 8.12A). This space is vascularized, meaning that blood vessels actually extend out through the body wall. These extensions carry nutrients to the tunic and are the transport system involved in the continual breakdown and rebuilding of the tunic as the animal grows

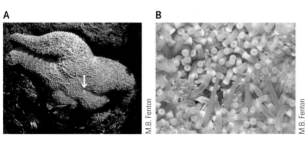

A B

M.B. Fenton

Figure 12.20 A. Dorsal view of a sea star showing the madreporite (arrow) and **B.** a ventral view of sea star showing podia (tube feet).

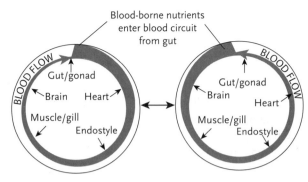

Figure 12.21 The blood system of a urochordate carries blood to the organs arranged in series. By reversing the flow, the sequence in which oxygen and nutrients reach different organs is alternated, ensuring an even supply to all parts of the body.

SOURCE: From RUPPERT/FOX/BARNES, *Invertebrate Zoology*, 7E. © 2004 Cengage Learning.

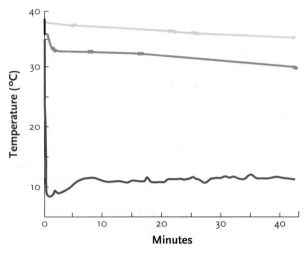

Figure 12.22 Change in body temperature (blue line), temperature deep in the foot (orange line), and footpad surface temperature (red line) in an Arctic fox whose foot was immersed in a mixture of ethanol, ethylene glycol, and water cooled to −38°. After an adjustment period, the temperature of the pad remained relatively constant, and the body and foot temperature decline only slightly.

SOURCE: From Henshaw et al 1972. "Peripheral Thermoregulation: Foot Temperature in Two Arctic Canines," *Science*, Vol. 175, No. 4025, pp. 988–990. Reprinted with permission from AAAS.

(it is never moulted). Blood cells, including vanadocytes, are found in the fluid between body wall and tunic. Vanadocytes are one of six to nine types of ascidian blood cells, and, as their name suggests, they contain significant levels of vanadium. The metal element vanadium is toxic because of its effects on the sodium-potassium pump, muscle contraction, and ciliary motility. Vanadium occurs in ascidians at levels a million to ten million times high than in seawater. Vanadium can be used in the manufacture of armour plating, but harvesting ascidians for vanadium is not commercially viable. Originally, vanadium was thought to be involved with oxygen transport, rather like iron in haemoglobin, an idea now disproved. More likely, the toxic nature of vanadium, combined with the level of sulphuric acid required to keep it in solution in vanadocytes, acts as a deterrent to any animal that might attempt to eat an ascidian. Niobium and tantalum are also concentrated by some ascidians.

STUDY BREAK

1. What do hemoglobin and hemocyanin have in common?
2. How do the circulatory systems of cephalopods differ from those of vertebrates?

12.5 Regional Heterothermy

Animals capable of **regional heterothermy** can maintain different temperatures in different parts of the body. This is most often accomplished by specializations of the circulatory system such as **countercurrent heat exchange** (see Box 11.2). A classic experiment demonstrated how Arctic foxes use countercurrent heat exchange to solve the problem of heat loss through their feet (Fig. 12.22). Wolves and wading birds use the same mechanism to keep their feet warm (Fig. 12.23).

Larger arteries within the fox's foot or heron's leg are surrounded by many small veins. The blood in the arteries loses heat to the venous blood, ensuring that the blood in the foot is colder than that in the body core. In addition, many small vascular bundles, consisting of a venule surrounded by a capillary network, lie under the skin of the fox's footpad. This system provides just enough warmth to keep the pads from freezing.

Many aquatic vertebrates also use regional heterothermy to retain heat. Penguins, for example, keep the core of the body warm during long dives in cold water. The primary site of penguin countercurrent heat exchange is between branches of the main wing artery and the surrounding veins. Again the cool venous blood is warmed by arterial blood as it returns to the body core. Diving penguins also appear to minimize blood flow to the extremities to retain heat in the body core. Fossil evidence suggests that the countercurrent exchange system in penguins evolved when they lived in more temperate climates. Then, regional heterothermy would

Figure 12.23 Wading birds such as this Goliath Heron (*Ardea goliath* (1565)) have a heat-saving countercurrent exchange system in their legs very similar to that in the Arctic fox.

have allowed penguins to hunt in deeper, colder waters, making their ability to live in harsh Antarctic environments an exaptation (see Section 6.3a).

Some fish provide another example of regional warming, this time of brains and eyes. Tuna, butterfly mackerel, and mackerel (Pisces, Scombridae) hunt in colder waters, so they need to keep their brains and eyes warm enough to function properly. The Scombridae exhibit at least three different ways to keep the eyes and brain warm. Tuna (*Thunnus* spp. (1291)) have high metabolic rates and generate power with red aerobic muscles. A combination of low thermal conductance and countercurrent heat exchangers in the brain, muscles, and viscera minimize heat loss to water. Butterfly mackerel (*Gasterochisma melampus* (1266)) use a modified extraocular muscle, the lateral rectus eye muscle, to generate heat. Slender tunas (*Allothunnus fallai* (1250)) use four fused extraocular muscles in each eye as eye and brain heaters.

The use of a modified extraocular muscle to maintain brain temperatures evolved convergently in billfish (Istiophoridae), for example the marlin (*Tetrapturus* spp. (1290)). In this case, the superior rectus eye muscles generate heat and, using countercurrent heat exchange, keep heat within the brain.

Common thresher sharks (*Alopias vulpinus* (1131), Alopiidae) also show evidence of endothermy. Most fishes have a bundle of red, highly aerobic muscle (RM) near the lateral surface of their bodies that powers continuous swimming over long distances. Typically, blood is carried to the RM by a few large branches of the centrally-located aorta. In common thresher sharks, the RM lies next to the vertebrae and is supplied by branches of a lateral artery that lies just under the skin along the side of the body. Hence, this arterial blood is relatively cool as it travels to the RM. However, because the RM is an aerobic, heat-producing muscle, the venous blood draining from it has been warmed by muscle activity. Small branches of these arteries and veins run side by side, forming a countercurrent exchange system that maintains higher temperatures in the centre of the body than at the body surface. Interestingly, two other species of thresher sharks lack this specialization for endothermy.

The circulatory systems of some dragonflies also allow them to regulate their body temperatures internally (Fig 12.24A). At low ambient temperatures, "fliers" vibrate their wing muscles to warm the thorax and head before taking flight. Fliers spend much of the day in the air patrolling their territories, foraging, and seeking potential mates. To avoid overheating during the hottest parts of the day, the warm hemolymph in the thorax and head is shunted through the dorsal vessel to the abdomen, which serves as a thermal window that gives off heat. This ability to thermoregulate by shunting hemolymph has evolved multiple times in dragonflies, suggesting that it is under positive selection. Flight is energetically expensive (see Section 7.7b), and fliers have long wings, low wing loading, large energy reserves, and they glide to increase flight efficiency.

Unlike fliers, perching dragonflies cannot regulate their temperature internally (Fig 12.24B). These species make only short, intermittent flights and control body temperature by orienting their bodies toward or away from the sun and by moving short distances between warm and cool microhabitats. The differences in the circulatory systems of fliers and perchers affect their flight behaviour, which, in turn, is closely linked to their mating strategies. This is a good reminder of how intensely anatomy, physiology, and behaviour are integrated.

Each breeding season, most birds develop a brood patch, an area of warm, bare, and well-vascularized skin on their breasts. The patch forms in response to hormonal and environmental cues in whichever sex incubates the eggs, and is used to keep the eggs warm. For each clutch of eggs, feathers over the patch are shed, and a dense network of arterioles, capillaries, and venules grows just under the skin. These vessels are resorbed, and the feathers grow back after the young are fledged, only to form again when the next clutch is laid. Some birds do not develop a brood patch but use their feet to keep their eggs warm. For example, the feet of an incubating Blue-footed Booby (Pelecaniformes) become more densely vascularized just prior to brooding, and the boobies use their feet to warm the eggs (Fig. 12.25).

Regional heterothermy is also used to keep certain body parts cool. Mammalian body temperature is often

A

federico stevanin / Shutterstock.com

B

wiw / Shutterstock.com

Figure 12.24 A. The emperor dragonfly (*Anax imperator* (851)) is a strong flier that is capable of internal thermoregulation, while **B.** the four-spotted chaser (*Libellula quadrimaculata* (854)) is an ectotherm that spends most of its time perching.

Figure 12.25
A Blue-footed
Booby (*Sula
nebouxii* (1647))
incubating eggs
with its feet.

too high to allow the production of fertile sperm. Therefore, testes are suspended in a scrotum outside the body cavity. The main artery to each testicle is surrounded by a large plexus of veins that cool the arterial blood before it reaches the testis. When the testes are too cold, they are drawn upward toward the body by contraction of a muscle in the scrotum. However, external testes in a scrotum of a male marine mammal would keep the testicles too cool and generate hydrodynamic drag by disrupting the streamlined body form. Since the core body temperature of marine mammals is too high to produce fertile sperm, the testicles are cooled by a countercurrent exchange mechanism using cooled venous blood from the fins and flukes to cool the arterial blood to the testes (Fig. 12.26).

A countercurrent exchange system that keeps the brain cool has evolved in at least two groups of vertebrates. Some species of order Cetartiodactyla (two-toed, hooved animals) live in very hot climates where over-heating can be a problem. When they enter the base of the skull, the carotid arteries of these animals divide into many small, thin-walled branches that pass through a large venous sinus (the cavernous sinus). This sinus drains blood from the eyes (cooled by contact with air) and the nasal cavities (cooled by evaporative cooling). As the arteries pass through the cavernous sinus, their heat is transferred to the venous blood, thus cooling the arterial blood before it reaches brain tissue.

By studying the oxygen isotope composition of fossil bone, paleontologists have suggested that the vascularized horn cores of frilled dinosaurs (*Triceratops* (1446), Ornithischia) regulated brain temperature, while the frill itself functioned more generally to cool body temperature in much the same way as the ears of elephants do.

STUDY BREAK

1. What is regional heterothermy?
2. How do animals use regional heterothermy to deal with heat and cold?

12.6 Altitude and Diving

Animals that live at or spend much of their time at high elevations have to cope with low atmospheric oxygen levels. More precisely, the *percentage* of oxygen is constant regardless of altitude, but the partial pressure of oxygen decreases with increasing altitude (Section 11.1). This low atmospheric pressure of oxygen causes a corresponding decrease in hemoglobin's ability to take up oxygen. Without any counteracting responses to low oxygen, these animals would encounter **hypoxia**—the body is deprived of adequate oxygen supply. Mammals that live at low elevations, including most humans, respond to moving rapidly from sea-level to high

A

Lateral view

Flank

B

Ventral view

Figure 12.26 A. Cool venous blood from the flanks of manatees (*Trichechus manatus* (1916)) drains to an internal plexus of veins (inguinal venous plexus) that cools the arterial blood traveling to their testes (**B.**). The inguinal venous plexus drains to the two inferior venae cavae, which lead to the heart.

SOURCE: Based on Rommel, S.A., D.A. Pabst, W. A. McLellan. 2001. "Functional morphology of venous structures associated with the male and female reproductive systems in Florida manatees," *The Anatomical Record* 264:339–347.

Blood from flank

Inguinal venous plexus

Body wall

Testis

Venae cavae

A

B

PHOTO COURTESY U.S. GEOLOGICAL SURVEY

© blickwinkel / Alamy

C

Galyna Andrushko / Shutterstock.com

Figure 12.27 A. Bar-headed Geese (*Anser indicus* (1655)), **B.** pikas (*Ochotona princeps* (1835)), and **C.** yaks (*Bos grunniens* (1744)) are adapted to high elevations.

altitude by increasing pressure in the pulmonary circuit (pulmonary hypertension), which can increase the amount of oxygen transferred to the blood. Over time this can lead to an increase in the size the right ventricle of the heart and a thickening of the pulmonary arteries. The extra load on the heart can lead to heart failure.

Certain animals have evolved a variety of responses to combat the effects of high altitude. One example is the Bar-headed Goose (*Anser indicus* (1655), Fig 12.27A). Twice yearly, Bar-headed Geese migrate over the Himalayas, flying at altitudes of about 8000 m, en route to and from their breeding grounds in central Asia and their overwintering grounds in south Asia. Other species that move between these same destinations typically migrate *around* the high mountains. Although this detour adds hundreds of extra kilometres, these detouring species avoid the extreme hypoxic conditions of such high elevations.

How does the Bar-headed Goose cope with such extreme altitudes? One way is by increasing the surface area of several organs that allow oxygen to diffuse more quickly. Compared with closely related, low-altitude geese species, Bar-headed Geese have much larger lungs. This increased lung capacity likely allows greater diffusivity of oxygen into the blood. The bird's heart ventricles also have a relatively high capillary density. This provides a greater surface area for oxygen to diffuse from the bloodstream into the heart muscle cells. These two factors increase oxygen diffusion, which helps to prevent negative effects of hypoxia.

Bar-headed Geese migrate at incredibly high altitudes, but many other animals live their whole lives at high elevations. Pikas (order Lagomorpha) typically live above 3000 m elevation (Fig. 12.27B). They have more erythrocytes and denser capillary beds than mammals that live at sea level, and these features are associated with the expression of certain proteins. In pikas and yaks (order Artiodactyla, Figure 12.27C), the amount of a protein called hypoxia-inducible factor Iα

in the lung, spleen, and kidneys increases with altitude. It is up-regulated when oxygen is low and works to influence cell metabolism, the growth of new blood vessels, and red-blood-cell production—each of which can increase the delivery of oxygen to the body.

While invertebrates are more tolerant of hypoxia than vertebrates, high-altitude species avoid lethal hypoxia by lowering metabolic rates through freeze-avoidance or freeze-tolerance (see Section 12.7), or by avoiding microhabitats with particularly low oxygen concentrations. For example, species within a community of springtails (Collembola) and mites (Family Oribatidae) from a 1200-m plateau in Norway vary in their tolerance to hypoxia. Three species of springtails and mites survived anoxia in the laboratory for three months at 0° C, but one springtail species survived only six days. It is likely that this species avoids anoxic environments, especially being encased in ice, by moving to the spaces between snow grains or vegetation. Carabid beetles from the highlands of Ecuador also use behaviour to minimize their exposure to the low temperatures that occur at night and the low humidity that is common during the day. These beetles are active only during certain times of the night when conditions are best. Otherwise they hide under rocks and in leaf litter, where temperature and humidity are more constant.

The combination of cold, low oxygen pressure, and low air density interact to significantly affect insect locomotion. Fruit flies (*Drosophila melanogaster* (960)) that were exposed to both low temperatures and low pressures had significantly reduced flight performance, measured by how high they could fly inside a closed cylinder. However, the combination of low temperature and pressure had a much stronger effect than either did alone. Even though the flies could fly at low pressures and temperatures, they became less and less motivated to fly as temperatures and pressures decreased. This is presumably a response to a combination of factors that increase the cost of flight, and may provide a clue to the mystery of why so many high-altitude insects either lost their wings or have very small wings that are often useless in flight.

Short-term hypoxia also influences development and behaviour in insects. *Drosophila* raised under hypoxic conditions have wider trachea that also have more branches. Flying adult insects compensate for the loss of lift at low air densities by increasing the amplitude of their wing beats. Over evolutionary time, it is likely that adaptations to high-altitude living evolved in at least some invertebrates. For example, many, but not all, insects that live at high elevations are small, and there are several explanations for why that might be. Small bodies may reduce wing loading (the ratio of insect weight to wing area), which allows them to generate the greater lift that is needed to fly in less dense air. Small body size could also simply be a response to the limited food resources available in high-altitude habitats.

Animals that live or travel at high altitudes have to deal with a low partial pressure of oxygen, but air-breathing animals that dive have to contend with the lack of oxygen altogether. Air-breathing vertebrates are breath-hold divers. They simply inhale (penguins, whales, leatherback turtles) or exhale (seals) a large breath of air at the water's surface and then dive. Whatever oxygen is stored within their bodies must last until they come up for air. These animals all have an **aerobic dive limit** (ADL), the point at which oxygen is depleted and their muscles begin to generate energy anaerobically. This leads to a build-up of lactic acid, which further diminishes muscle function. Marine tetrapods use several different mechanisms to extend the ADL and thus have more time to spend foraging underwater.

Diving birds and mammals exhibit a well-developed **diving reflex** that reduces the amount of oxygen used by the body. This diving reflex, elicited by immersing the face in cold water, includes a dramatic drop in heart rate and the constriction of vessels in the extremities. These responses decrease the amount of oxygen needed to produce energy to contract the heart, a very large muscle, and direct blood flow to essential organs, especially the heart and brain. The basics of the diving reflex are present in all tetrapods, even humans. Professional free divers do breath-hold dives to more than 200 m, and part of their training is aimed at enhancing their diving reflex.

Oxygen is stored in the blood, respiratory system, and muscles, and diving birds and mammals distribute it in different ways in order to extend the ADL (Fig. 12.28). Diving birds such as penguins store oxygen in their respiratory systems, and some gas exchange occurs throughout the dive. The presence of similar P_{O_2} values in atrial and venous blood during dives indicates that oxygen is also stored in blood, perhaps by arterio-venous shunts that circumvent non-essential organs and tissues. Diving mammals store a great deal of oxygen in their blood, rather than in the respiratory system. Seals even have relatively large spleens that store oxygenated red blood cells under normal conditions and release them during dives. Deep diving mammals store less oxygen in their lungs and more oxygen in their blood than do mammals that forage in shallower waters. Measurements of blood gasses also demonstrate that diving mammals and birds are more tolerant of hypoxia than their fully terrestrial relatives.

Whales, dolphins (order Cetacea), seals (order Carnivora), and leatherback turtles (*Dermochelys coriacea* (1426)) use an energy-saving locomotor strategy during diving. They glide to the bottom of the dive, level off, and continue gliding at the deepest depth. They only use their tails or flippers to propel themselves back to the surface. Essentially, they save energy by "turning the engines off" during descent. This same sequence of locomotor activity followed by downward

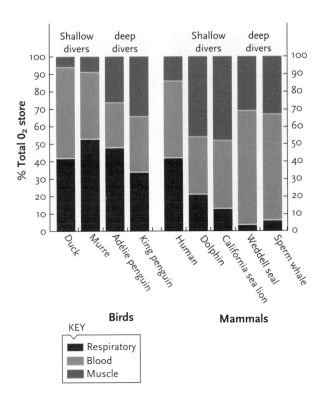

Figure 12.28 The proportion of oxygen sequestered in the respiratory system (red), blood (green), and muscle tissue (blue) of shallow- and deep-diving birds and mammals. Humans are included for comparison.

SOURCE: Reproduced with permission from Paul J. Ponganis In pursuit of Irving and Scholander. 2011. "A review of oxygen store management in seals and penguins," *Journal of Experimental Biology* 214, 3325–3339. The Company of Biologists Ltd.

gliding is also seen in sharks, sea turtles, and even flying birds, suggesting it is a common mechanism of saving energy while moving through fluids (Fig. 12.29).

Sharks are neutrally buoyant, but whales and seals have thick layers of blubber that provide buoyancy (the ability to float). How can these animals glide downward through the water column? As seals and whales descend, the increasing pressure squeezes their bodies. The alveoli in their lungs collapse, and the air is pushed into a cartilage-reinforced bronchial tree. The compression of the alveoli reduces the volume of the animal but does not change its mass. Therefore, the animal becomes denser. When it reaches the depth at which its density exceeds its buoyancy, it is able to glide downward.

The alveoli are the sites of gas exchange, which ceases when the alveoli collapse. Although this means the animal no longer gets oxygen from inspired air, it also means that nitrogen (and other inert gases) do not accumulate in the blood and dissolve into the body's tissues as pressure increases (Henry's Law). If that happens and the animal surfaces rapidly, potentially deadly nitrogen bubbles could form anywhere in the body. Scuba divers call this getting "the bends." It was long thought that diving mammals never get the bends, but bubbles have been found in tissues of whales and dolphins that have stranded on beaches.

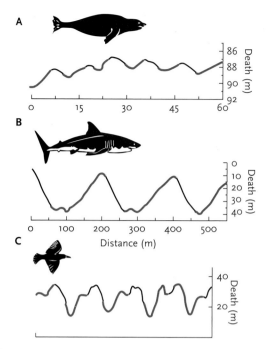

Figure 12.29 **A.** Southern elephant seals (*Mirounga leonina* (1722)), **B.** white sharks (*Carcharodon carcharias* (1133)), and **C.** European starlings (*Sturnus vulgaris* (1616)) all use intermittent locomotion (red portion of lines) and gliding (black portion of lines) as a way of saving energy while moving forwardv. Data for the seal demonstrate locomotion at depth during breath-hold diving.

SOURCE: Based on Gleiss et al. 2010. "Intermittent locomotion in flying and swimming animals," *Nature Communications* 2, Article number: 352, and Rayner, J. M. V., Viscardi, P. W., Ward, S. & Speakman, J. R. "Aerodynamics and Energetics of Intermittent Flight in Birds," *Am. Zool.* 41, 188–204 (2001).

STUDY BREAK

1. How does living at high altitude challenge the circulatory system?
2. What is the aerobic dive limit? How do mammals and birds differ in how they extend this limit?
3. What role does behaviour play in mitigating the effects of high altitude and diving?

12.7 Antifreeze

As we hear on the news every winter, cold can be lethal. Cold lowers metabolic rates and interferes with enzyme function. As well, the formation of ice within cells can debilitate crucial proteins and irreparably damage cell membranes and intracellular machinery. Outside of cells, ice can destroy thin-walled structures such as capillaries. This damage disrupts metabolic functions, leads to ischemia (loss of blood supply to tissues and subsequent oxygen starvation), and incapacitates vital functions such as heart and brain activity. Yet many animals live where ambient temperatures dip well below the freezing point of their bodily fluids. How do they do this?

Most animals avert the threat of freezing simply by avoiding it (Fig. 12.30). They may migrate to more temperate climates or, if they do stay around, live in or take refuge under water. Terrestrial animals that live in regions with seasonal subzero temperatures can avoid freezing by taking shelter. They may burrow deeply into the soil where temperatures are stable and above 0° C, retreat to well-insulated nests, dens, or crevices, or live under the snowpack where temperatures hover near freezing. However, some animals, such as those that overwinter above ground or in supercooled water, can't avoid the cold. Some of them avoid freezing by supercooling their bodily fluids, which allows them to remain liquid at temperatures below their freezing points. Other animals tolerate freezing and allow ice to form in their extracellular fluids while protecting the liquid state of their cells' contents.

Freeze-avoiding and freeze-tolerant animals harness similar physiological mechanisms and use them in different ways to deal with the problem of freezing (Fig. 12.30). For example, one way to stave off freezing is to increase the concentration of low molecular-weight solutes (antifreeze) in circulating fluids. Increasing the number of solute particles relative to the number of solvent particles lowers the solution's freezing point but not the melting point. This is a colligative property of all solutions, called **freezing point depression**. In addition to making antifreeze,

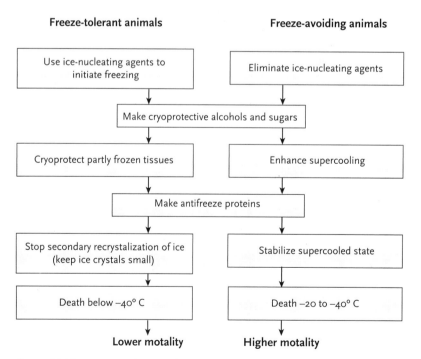

Figure 12.30 Freeze-avoiding and freeze-tolerant animals take advantage of the same mechanisms either to avoid the formation of ice or to permit its growth only in extracellular fluids.

SOURCE: Based on J. S. Bale 2002. "Distributions and abundance Insects and low temperatures: from molecular biology to distributions and abundance," *Philosophical Transactions of the Royal Society of London*, B 357, 849–862.

some animals lose water content as a means of concentrating antifreeze molecules. In order for antifreeze to work, the molecules must be small enough to be present in high concentrations, as well as nontoxic. Antifreeze has evolved convergently many times and is usually generated through the breakdown of stored glycogen. Insects produce many different antifreeze molecules but glycerol is the most common. Some insects use a combination of several antifreeze molecules, which may avoid the toxic effects of any one compound. Vertebrates produce glucose or glycerol for their antifreeze. Antifreeze concentrations usually rise as winter approaches and lower again in the spring.

Another mechanism used to manage freezing temperatures involves directly controlling ice formation. Antifreeze can supercool the fluids within animals' bodies, but if those fluids come into contact with an ice crystal, then the fluids can convert to ice almost instantaneously, causing massive damage and killing the animal. The offending ice crystal can come from outside the animal or form inside around any small particle that can serve as nucleator (seed) for ice crystal formation. Most animals are protected from external ice by choosing to overwinter in dry places or having a protective outer covering. Insects, for example, are covered with a hydrophobic exoskeleton, and some protect themselves further by spinning a cocoon. Even so, the price of failing to protect again instantaneous freezing is high.

Another way to control the formation of ice in circulatory fluids is to add antifreeze proteins (AFPs) to these fluids. These AFPs work in a very different way from glycogen-derived antifreeze molecules: they attach to nascent ice crystals that form in the body's fluids and arrest their growth. In this way, they lower the freezing point of the fluid. Like antifreeze, AFPs have evolved convergently many times in insects and salt water fish, and different AFPs interact with ice crystals in different, but equally effective ways. An advantage of AFPs is that they work well at very low concentrations. A drawback, however, is that rapid cooling can outpace AFP activity and expose animals to the danger of instantaneous freezing. Not surprisingly, AFPs are common in animals that live in relatively stable environments.

Icefish (suborder Notothenioidei) are freeze-avoiders that survive in seawater that is at or near their freezing point largely by having high concentrations of AFPs in their blood serum (Fig. 12.31). One of the dangers of living on the edge of one's freezing point is that once ice crystals form, they spread rapidly through the body. Icefish AFPs bind to ice crystals that form within the fish and prevent them from growing and spreading. Interestingly, juvenile icefish have lower levels of circulating AFPs then their adult counterparts. In the relatively low latitude *Chaenocephalus aceratus* (1259) for example, adult serum antifreeze activity is 2.7 times that of the youngest juveniles. At

Figure 12.31 An Antarctic icefish (suborder Notothenioidei).

first glance, it seems like juveniles should be especially vulnerable to freezing. However, these young icefish have physical characteristics that allow them to survive in ice-laden conditions. Their continuous, intact integument forms a tight seal over their bodies, and they have a very low gill surface area. Both of these features are physical barriers to ice entry, allowing juveniles to "make do" until they produce enough AFPs to survive as adults.

Another way to control the formation of ice crystals is to regulate the molecules, called ice nucleators, that can serve as nuclei for the formation of ice crystals within circulating fluids. These are typically either **exogenous** (originating outside of the body) ice-nucleating particles in the gut, or **endogenous** proteins that are suspended in either hemolymph or blood. Simply removing these molecules, as many insects do, can be an extremely effective way to avoid freezing. For example, one species of solitary bee (*Megachile rotundata* (999)) lowers its supercooling temperature by 20°C by emptying its gut contents. Stag beetle larvae (*Ceruchus piceus* (927)) lower their supercooling point by 13°C simply by getting rid of ice nucleators. Like many insects, they stop feeding to clear ice-nucleating particles and bacteria from their guts, but they also remove ice-nucleating proteins (INPs) from their hemolymph. These proteins aren't needed in the winter because the larvae's metabolic rate is extremely low, and they are replaced when they are needed again in the spring.

The challenges for all freeze-tolerant animals include:

- to control where and how quickly ice forms
- to regulate how much cells are desiccated (dried out) and deformed (which can destroy cell membranes)
- to tolerate the anoxia that occurs when circulation ceases
- to activate and control thawing

Many freeze-tolerant animals add ice-nucleating proteins in order to control where and how fast ice crystals form. The production of these proteins varies seasonally. As the weather gets colder, the proteins

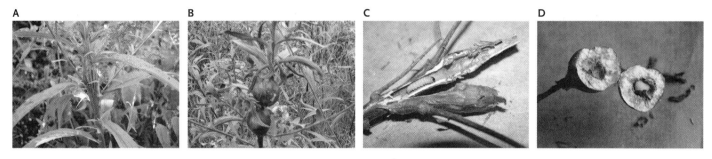

Figure 12.32 Larvae of gall moths (Epiblema scudderiana, A) and gall flies (Eurosta solidaginis, B) make galls on the stems of golden-rod plants. Each gall contains a single larvae (gall moth, C; gall fly D) that overwinters there and emerges in the spring.

SOURCE: Kenneth Storey

accumulate in hemolymph, blood, and interstitial fluid outside of cells, and begin the process of freezing just a few degrees below zero. By regulating the type, number, and location of ice-nucleating proteins, animals can slow the freezing process and keep ice from forming inside cells. As ice forms, it pulls water out of the extracellular fluid and, because of the resulting osmotic gradient, out of the cells. This results in freezing point depression, a good thing, but can also dehydrate cells so much that they are damaged as they shrink. High concentrations of solutes such as glycerol or glucose (antifreeze) prevent excessive dehydration. Ice-nucleating proteins are very common in freeze-tolerant insects, some frogs, and hatchling painted turtles (*Chrysemys picta* (1424)).

Although some animals use only one of the anti-freezing mechanisms we have mentioned, most use a combination of tricks to avoid or tolerate freezing. A good example of this are the freeze-tolerant gall flies (*Eurosta solidaginis* (961), Order Diptera) and freeze-avoiding gall moths (*Epiblema scudderiana* (1033), Order Lepidoptera), two animals that occupy very similar niches but survive the cold using a very different combination of adaptations. Both gall flies and gall moths lay eggs on goldenrod plants in the spring (Fig 12.32). When the larvae hatch, they burrow into the plant and begin to eat it. The larvae trick the plant into forming a gall around them and filling it with nutritious cells by producing secretions that mimic plant hormones. This solves two problems: finding high-quality food and a dry place to overwinter.

The gall moth is a freeze-avoider and does so primarily by producing glycerol, but it also makes AFPs. The production of glycerol and AFPs is triggered by the first subzero temperatures and increases as the weather becomes colder. By late autumn the larvae are super-cooled, on average, to 38°C and contain 19% more glycerol than they do in the autumn. Glycerol and AFP levels fall as temperatures warm in the spring.

Unlike the gall moth, the gall fly tolerates freezing. The gall fly makes ice-nucleating proteins that begin the slow formation of extracellular ice crystals just below freezing, in conjunction with antifreeze (glycerol and sorbitol). As is most often the case, the concentra-

tions of these molecules are lowest in the summer and highest in the winter.

Freeze-tolerance is rare among vertebrates. Several species of frogs, Siberian newts (*Hynobius keyserlingi* (1410)), painted turtles (*Chrysemys picta* (1424)), and box turtles (*Terrapene* sp. (1431)) are capable of freezing for days or weeks at a time. This allows them to survive the winter in shallow burrows (turtles and salamanders) or under leaf litter (frogs). A handful of other snakes, lizards, and turtles can survive short periods of freezing, but not long enough to suggest that freeze tolerance is ecologically meaningful in these species. It is more likely that true freeze tolerance evolved from ancestors with this rudimentary capability.

Wood frogs (*Rana sylvatica* (1385)) are perhaps the best-studied freeze-tolerant vertebrates (Fig 12.33). They live in North American as far north as the tree line. Unlike most frogs, which spend the winter at the bottom of lakes and ponds, wood frogs overwinter above ground under the leaf litter. Wood frogs avoid the damage caused by freezing by combining several strategies to control where and how quickly ice forms. The more slowly the ice forms, the more time an animal has to deploy cryoprotective strategies. Unlike animals that eliminate ice-nucleating proteins to avoid freezing, wood frogs maintain high levels of ice-nucleating proteins within their interstitial fluids. Ice first forms in the periphery of the body and progresses to the fluid surrounding the organs before entering the

Figure 12.33 This frozen wood frog (*Rana sylvatica* (1385)) will thaw and return to life as usual.

SOURCE: Kenneth Storey

BOX 12.4
Collecting Nectar with a Hemodynamic Mop

When visiting flowers to drink nectar, some bats hover over the flowers and lap nectar with their tongues. Nectar resources are limited in the wild, so some species have developed extremely long tongues to gather nectar (Fig. 1). Until recently it was not clear just how bats could extend their tongues. Like all mammalian tongues, the tongues of nectar-feeding bats are composed of orthogonally arranged muscle fibres. Nectar-feeding bats also have enlarged veins in their tongues (Fig. 2), leading to the suggestion that muscle contraction and resulting changes in blood distribution account for their exceptional tongue elongation. Colour high-speed video film of Pallas' long-tongued bats (*Glossophaga soricina* (**1793**)) lapping sugar water shows that blood flow through the lingual blood vessels not only increases tongue length but also increases its surface area (Fig. 3). Keratinized lingual papillae lie along the edges of the tips of the bat's tongue. Careful examination of the tongue's internal anatomy, not illustrated here, reveals a network of enlarged blood vessels associated with these papillae. During nectar-feeding, these blood vessels become engorged with blood, erecting the lingual papillae and increasing the surface area of the tongue (Fig. 3). Other mammals, such as cats and dogs, lap liquid by contracting the skeletal muscles within the tongue. These nectar-feeding bats, on the other hand, have developed a specialized hemodynamic feeding mechanism to collect a large amount of nectar with each lick.

Figure 1
A nectar-feeding bat, Glossophaga soricina *(1792), gathering nectar with its tongue.*

Caroline Harper

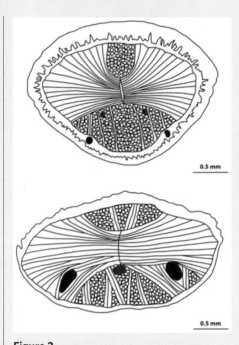

0.5 mm

0.5 mm

Figure 2
Cross-sections through the tongue of **A.** *an insectivorous bat and* **B.** *a nectarivorous bat. Lingual arteries are highlighted in red and lingual veins in blue.*

SOURCE : Redrawn by Caroline Harper, from "Muscular and Vascular Adaptations for Nectar-Feeding in the Glossophagine Bats Monophyllus and Glossophaga" by Griffiths appearing in *Journal of Mammalogy* 59.2, 1978, Allen Press Publishing Services.

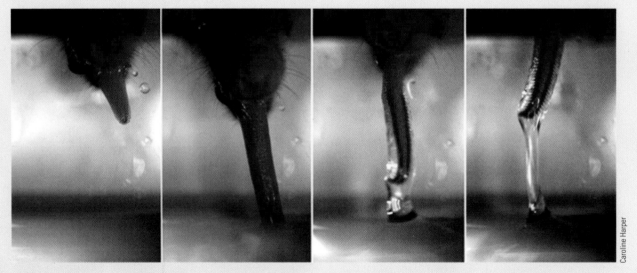

Caroline Harper

Figure 3
Frames from a colour high-speed video of a nectar-feeding bat collecting sugar water. Notice that the blood vessels along the lateral margins of the tongue are engorged with blood and the elongated lingual papillae are erect.

organs themselves. The formation of ice crystals triggers the rapid production of glucose, an antifreeze, by the liver which is distributed to the body by the circulatory system.

As the body freezes, circulation to the frozen areas is lost from the periphery inward, leaving the highest levels of glucose in the liver, followed by the heart, brain, and other organs. Ultimately, the entire body is infused with a very high concentration of glucose, which serves as antifreeze and helps to regulate cell shrinkage. Compared to closely related frogs that cannot tolerate freezing, wood frogs move more glucose across cell membranes by virtue of higher glucose transport rates. Thawing happens in reverse of freezing, from the inside out. The organs with the highest levels of glucose melt first, which quickly restores heart function. Wood frogs are also more tolerant of ischemia than relatives that cannot survive freezing.

Freeze tolerance has evolved at least three times among frogs. In each case, ice formation is controlled by a cryoprotectant of low-molecular-weight carbohydrates (glucose and/or glycerol) that is produced in the liver and rapidly distributed through the blood stream when ice crystals begin to form. Dehydration also triggers the accumulation of glucose, even in frogs that do not tolerate freezing. This suggests that dehydration tolerance may be a pre-adaptation to freeze tolerance. This fits well with the fact that anurans are among the most dehydration-resistant vertebrates.

Freeze tolerance in all animals requires a very complex set of physiological adaptations; imperfections in the process can prove deadly. In order for freeze tolerance to have evolved, it must offer a substantial benefit in terms of individual fitness. Freeze-tolerant animals typically emerge from overwintering earlier in the spring than do animals that avoid freezing. This allows them early access to places to mate and lay eggs, and their offspring get a head start on growth before the next winter season. It may also help them to avoid predators, which may be absent or less active early in the spring. Freeze tolerance also opens habitats that may not be available to freeze-avoiding species. Indeed, freeze-tolerant species often have larger ranges.

STUDY BREAK

1. What are freeze avoidance and freeze tolerance? What mechanisms do they share, and in what ways do they differ?

12.8 Development and Plasticity

In many animals the morphology and function of the circulatory system changes dramatically during development. Animals that undergo metamorphosis may exhibit some of the most profound changes, often because these animals live in very different habitats before and after metamorphosis. Anurans (frogs and toads) and some salamanders are good examples. Aquatic larvae typically have functional gills and take a relatively high percentage of their oxygen from the water, while terrestrial adults usually lose the gills and gain more oxygen from the air. For example, the larval tiger salamander (*Ambystoma tigrinum* (1406)) takes roughly 45% of its oxygen from the air, but the percentage increases to over 65% in a fully metamorphosed adult.

The change from gills to lungs requires a significant re-organization of the circulatory system. In frogs (e.g., *Rana pipiens* (1379), the leopard frog) and toads (e.g., *Pelobates fuscus* (1375)), the situation is similar. Early in their development, the larvae have four aortic arches that carry blood to the gills. Arch VI gives rise to the pulmonary artery. Arches V and VI arise from a common trunk (Fig. 12.34A). After metamorphosis, the gills have been lost completely. The only remaining branch of arch III is the external carotid artery, which provides blood to the head and neck (Fig. 12.34B). Arch IV becomes the main (systemic) route to the body. All deoxygenated blood passes into the lungs via the pulmonary artery, and to the skin via the cutaneous artery. The absence of gills forces the animals to rely primarily on lung ventilation; only a small amount of gas is exchanged through the skin. Similar changes happen among all amphibians as gill-breathing larvae metamorphose into lung-breathing adults.

Mammals shift from placental to pulmonary circulation at birth, a process that involves several central cardiovascular readjustments. Fetal blood is oxygenated at the placenta, where it also eliminates metabolic wastes and acquires nutrients by diffusion between fetal and maternal blood. The concentration gradient for oxygen is assisted by the structure of fetal hemoglobin, which has a higher affinity for oxygen than maternal (adult) hemoglobin. Oxygen- and nutrient-rich blood leaves the placenta via the umbilical vein, and heads to the liver on its way to the heart. In following this path, only about half of the blood from the placenta actually passes through the liver sinuses, and the remainder bypasses the liver to the inferior vena cava through the **ductus venosus** (Fig. 12.35). In the inferior vena cava, blood from the ductus venosus joins blood returning from the lower trunk and extremities, and this combined stream is in turn joined by blood from the liver through the hepatic veins. It is interesting to note that the streams of blood tend to maintain their identity in the inferior vena cava, and they are divided into two streams of unequal size by the edge of the interatrial septum (crista dividens). The larger stream, which is mainly nutrient- and oxygen-rich blood from the umbilical vein, is shunted directly to the left atrium through the interatrial septum via the **foramen ovale**, literally an "oval window" between the right and left atria. The other stream passes into the right atrium, where it is joined by venous return from the upper parts of the body, arriving through the superior vena cava. Having entered

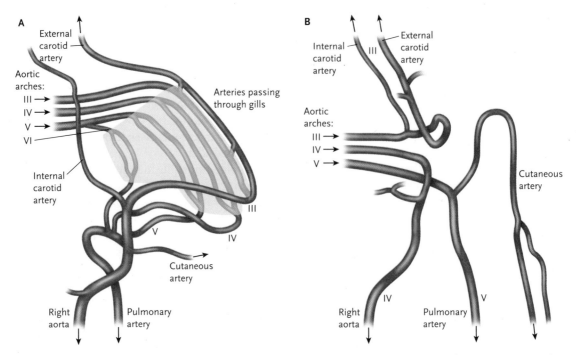

Figure 12.34 Thoracic circulation in the common spadefoot toad (*Pelobates fuscus* (1375)). **A.** Late larval, pre-metamorphic circulation. Aortic arches III–VI take blood to and from the gills (grey region), and the pulmonary artery (derived from arch VI) is small. **B.** Post-metamorphic circulation. The gills and associated arteries are lost. Arch III gives rise to the external carotid artery; arch IV, the right aorta; and arch V, the pulmonary artery, which is now large, and the cutaneous artery.

SOURCE: Based on Kolesova et al. 2007. "The evolution of amphibian metamorphosis: insights based on the transformation of the aortic arches of Pelobates fuscus (Anura)," *Journal of Anatomy*, 210:379–393.

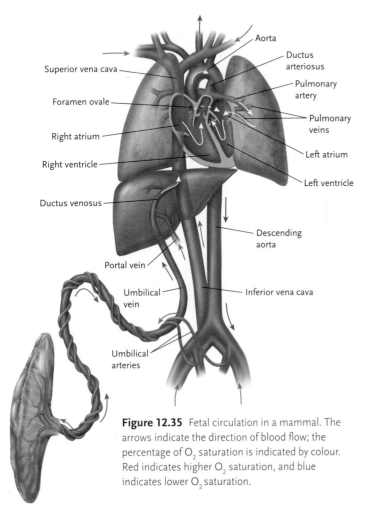

Figure 12.35 Fetal circulation in a mammal. The arrows indicate the direction of blood flow; the percentage of O_2 saturation is indicated by colour. Red indicates higher O_2 saturation, and blue indicates lower O_2 saturation.

the left atrium, the return from the umbilical vein is then pumped from the left ventricle to the systemic circulation through the aorta.

The lungs of the mammalian fetus are collapsed, presenting a high resistance to blood flow. The pulmonary artery is connected to the aortic arch by a short, large-diameter blood vessel, the **ductus arteriosus**. Most of the blood ejected by the right ventricle is thereby returned to the systemic circuit via the ductus arteriosus, bypassing the pulmonary circuit.

At birth, the lungs are inflated, reducing the resistance to flow in the pulmonary circuit. As a result, blood ejected from the right ventricle passes into the pulmonary vessels, creating an increased venous return to the left atrium from the pulmonary veins. This increased pressure in the left atrium closes the flap valve covering the foramen ovale, which then closes permanently, preventing blood from moving through. At the same time, the placental circulation disappears, and the resistance to flow increases dramatically in the systemic circuit. The ductus venosus closes and is distinguishable in later life as the ligamentum venosum. With the decrease in pulmonary vascular resistance, blood in the pulmonary artery passes more readily to the lungs, bypassing the ductus arteriosus, which begins to constrict and eventually closes. This closure of the ductus arteriosus appears to be initiated by the high P_{O_2} of the arterial blood passing through it; pulmonary ventilation with O_2 closes the ductus, whereas ventilation with air low in O_2 keeps the shunt open. Whether O_2 acts directly on the ductus or mediates the release of a vasoconstrictor substance is unknown.

STUDY BREAK

1. Why does development often require major re-arrangements of the circulatory system?
2. What physiological shift drives the major change in the configuration of the circulatory system from larval/fetal stages to the adult stages in terrestrial vertebrates?

CIRCULATION **IN PERSPECTIVE**

The circulatory systems of animals provide many clear examples of functional solutions to problems faced by living organisms. Circulatory systems provide repeated demonstrations of convergent and parallel evolution. For biologists accustomed to the circulatory systems of vertebrates, those of invertebrates open many windows onto the diversity of life.

Questions

Self-Test Questions

1. What would be a functional advantage of an open circulatory system that is not characteristic of a closed circulatory system?
   a. Fluids in an open circulatory system are delivered under lower pressure.
   b. Fluids in an open circulatory system bathe the tissues and organs directly, reducing the diffusion distance.
   c. In an open circulatory system, concentration gradients will be greater in tissues that are more metabolically active.
   d. In an open circulatory system, the coefficient of diffusion will decrease relative to the surface area available for exchange.

2. Which of the following statements best describes the dynamics of blood flow within vessels?
   a. Distribution of blood-borne materials is best achieved through small vessels that perfuse tissues.
   b. Maintaining high blood pressure is important for efficient exchange between vessels and tissues.
   c. Small diameter vessels lower blood pressure and flow, and the arrangement of numerous small vessels in active tissues increases the efficiency of exchange.
   d. Vessels of longer length need to be more compliant to allow blood to maintain pressure over a longer distance.

3. Which of the following correctly pairs a body fluid with its location and function?
   a. Tissue fluid is protein rich and circulates between cells.
   b. Coelomic fluid circulates around the visceral organs.
   c. Plasma is a minor component of vertebrate body fluids.
   d. Hemolymph is the term applied the circulatory fluid within vessels.

4. Which of these statements best describes the functional importance of a heart?
   a. A heart is a contractile organ that puts the circulatory fluid under pressure to maintain circulation.
   b. The pressure generated by the heart keeps blood moving in a forward circulation.
   c. Hearts that support single-circuit circulation generate higher pressures than when the circulation is divided between pulmonary and systemic circuits.
   d. Hearts are characteristic of closed circulatory systems where the fluid pressure must be maintained throughout the circuit.

5. How are cephalopod circulatory systems similar to those of terrestrial vertebrates?
   a. Both cephalopods and vertebrates have relatively high blood volumes, characteristic of organisms with closed circulatory systems.
   b. Efficient countercurrent exchange systems are found in both systems to dissipate body heat.
   c. Oxygenated blood returns from efferent vessels to paired atria that feed the ventricle of a systemic heart.
   d. The arrangement of heart and major vessels for distribution and return maintains constant cardiac output under all conditions.

Questions for Discussion

1. How have the circulatory systems of animals evolved to support different kinds of locomotion? What about feeding strategies? What about temperature tolerance?

2. Circulatory systems often change as animals pass through different life stages. In what ways do different life stages make different demands of the circulatory system?

3. How do the circulatory and respiratory systems work together in nemerteans? echinoderms? cephalopods? mammals?

4. Use the electronic library to find a recent paper that discusses an anatomical aspect of circulation in a non-human animal. What is new or different about the discovery that is discussed?

5. Use the electronic library to find a recent paper that discusses a physiological or molecular aspect of circulation in a non-human animal. What is new or different about the discovery that is discussed?

Figure 13.1 A fifth-instar kissing bug (*Rhodnius prolixus* (876)), before and after feeding. The fed insect is urinating.

SOURCE: Ian Orchard

13 Excretion

WHY IT MATTERS

You've drifted off to sleep in your budget-friendly hostel in southern Mexico, oblivious to the excited scrabbling that's begun within a crevice of the adobe walls. A kissing bug, *Rhodnius prolixus* (Arthropoda, Insecta), has detected a whiff of carbon dioxide and is frantically waving its antennae to locate the source—you! Rhodney eats blood but has not had a meal in several months—it is hungry. After finding you in the darkened room, it crawls onto your face and extends its beak, a head appendage specialized for piercing and sucking (Fig. 13.1). It pierces your skin with the beak and begins drinking your blood. Occasionally it secretes anticoagulants from its salivary glands to ensure an uninterrupted flow of unclotted blood into its gut.

This blood meal amounts to ten times Rhodney's unfed body weight! This is necessary because most of your blood is water and salts; the protein that Rhodney craves is very dilute. So it adds insult to injury while gorging on its blood meal by simultaneously urinating the water and salts from your blood onto your skin. This excretory act allows it to consume even more blood. In fact, Rhodney can excrete its original body weight in fluid every 40 min during and immediately after a meal.

This nasty scenario is made possible by the extraordinary osmoregulatory and excretory abilities of kissing bugs, a family of hemipteran insects. When *Rhodnius* ingests blood, the load of water and salts in the blood is selectively transported across the wall of the midgut and added to the hemolymph within the insect's body cavity. Although this separates the water and salts from the protein in the meal, it could upset the osmotic balance of the insect's body because the osmolarity of vertebrate blood is less than that of *Rhodnius* hemolymph. To prevent this upset, the water and salts are rapidly extracted from the hemolymph by Malpighian tubules that empty into the junction between the midgut and hindgut. The fluid is then passed out the anus as highly dilute urine. Hormones control the entire process.

After *Rhodnius* has dealt with the immediate crisis of massive fluid intake, it must deal with a second potentially fatal crisis arising from its protein meal. The end product of protein catabolism is ammonia, which is toxic unless highly diluted. After *Rhodnius* has unloaded the water from its meal and crawled back to its hiding place, it digests the protein and must rid its body of ammonia without the luxury of plentiful water to flush it out. For this purpose, the kissing bug converts ammonia into non-toxic uric acid, which can be excreted as a dry solid.

A further potential insult to you, the source of the blood meal, is a possible infection with *Trypanosoma cruzi*, a protozoan parasite that disperses to new human hosts by exploiting the blood-feeding and urination habits of *Rhodnius*. Urine from a *Rhodnius* infected with *T. cruzi* contains stages of this trypanosome's life cycle that infect humans. When you scratch the itchy wound created by the *Rhodnius* bite, the insect's urine is worked into the wound, and your body may be invaded by the causative agent of Chagas disease. Charles Darwin, who suffered chronic ill health during much of his life, may have contracted Chagas disease when he was bitten by a kissing bug (or similar blood-sucking insect) while travelling in Peru. He recorded this incident in his diary.

The purpose of this chapter is to provide the background for understanding differences in osmoregulatory and excretory systems in relation to inherited body plans, physiological frameworks, habitats, and ontogeny. Examples will demonstrate how the strong links among these parameters have generated the diversity of animal solutions to the dual but interrelated problems of osmoregulation and excretion.

SETTING **THE SCENE**

Animals must rid themselves of wastes (carbon dioxide, nitrogenous substances) produced during metabolism. Many also eliminate undigested food materials (as feces or frass). Accumulations of nitrogenous wastes (as urea or uric acid) can be used to osmotic advantage, while ammonia can be toxic and typically is produced by animals living in water. We will explore some of the many variations on excretion that occur in the animal kingdom.

13.1. Excretion and Osmoregulation: Basic Principles

At the outset we must clarify the distinction between different types of bodily wastes that result from digestion and metabolism of ingested food. As humans, we are all familiar with **feces** and **urine**, two obvious kinds of wastes regularly voided from our bodies. Carbon dioxide is a third and perhaps less obvious metabolic waste that we exhale from our lungs during breathing. Carbon dioxide is an end-product of aerobic respiration, a biochemical process that transfers chemical energy for use by cells. Feces constitute the indigestible remains of a meal that are eliminated from the digestive tract, a process termed defecation. We do not deal with gas exchange or defecation in this chapter (see Chapters 9 and 11).

Urine, at least for many animals, consists of nitrogenous wastes resulting from protein catabolism, together with excess solutes and a variable amount of water. Urine is therefore a waste product resulting from two physiological processes: excretion, which is the elimination of nitrogenous wastes, and **osmoregulation**, which is the maintenance of an appropriate balance of salts and water within the body. In terrestrial vertebrates, both excretion and osmoregulation are performed mainly by the kidneys (although the process of excretion begins in the liver), whereas in fish the gills are also important. Separation of defecation and excretion/osmoregulation into two discrete products holds for some animals, but in others the waste products may be either more consolidated or more distributed. For example, in insects and vertebrates with cloacas, feces consist of indigestible material as well as excretory products, because the excretory organs open into the lower intestine or hind gut. At theother extreme, aquatic crustaceans release wastes and excess salts from multiple sites. Marine crustaceans release both nitrogenous waste (ammonia) and excess salt from the gills, as do many teleost fish, but excess salts are also released from so-called antennary or coxal glands. Crustacean feces are expelled through the anus.

Excretion and osmoregulation are integral aspects of **homeostasis**, the business of maintaining an internal milieu conducive to the biochemical processes necessary for life. Although the processes of osmoregulation and excretion have different goals and outcomes, they are often interconnected by biochemical pathways and physiological mechanisms, and sometimes co-occur within the same anatomical structure.

1. What is the difference between excretion and osmoregulation?
2. What is homeostasis?

13.2 Osmoregulation

Osmoregulation occurs at two levels: the level of individual cells inside an organism and the level of the whole organism within its environment, be it terrestrial or aquatic. At both levels, the physiology of osmoregulation involves the processes of diffusion and osmosis across cell membranes which involve various passive and active transporters (proteins embedded in the cell membranes). The cell membrane is therefore of central importance for osmoregulation. Like all biological membranes, the cell membrane is a phospholipid bilayer with proteins decorating the external surface of the bilayer, embedded within the bilayer, and spanning the width of the bilayer (Fig. 13.2).

Biological membranes are semi-permeable because only some molecules can readily pass through them. Small, non-polar molecules can travel rapidly through most biological membranes, but membranes become progressively less permeable to solutes of increasing size. Furthermore, because the interior of cell membranes consists of long chains of non-polar fatty acids, non-polar solutes are miscible within the phospholipid bilayer and therefore pass through biological membranes much more readily than do polar solutes having a positive or negative charge.

Rapid movement of water through biological membranes has long been a mystery because water molecules, although small in size, carry a charge. The polar water molecules are not miscible with the non-polar lipid bilayer of membranes and should not pass through membranes rapidly. The research that solved this mystery won a Nobel Prize in Chemistry in 2003. Peter Agre demonstrated that water moves through cell membranes primarily via aquaporins, proteins embedded in membranes that form channels for the passage of water (Fig. 13.3). With only a few known exceptions, **aquaporins** are ubiquitous in the membranes of both plant and animal cells.

13.2a Diffusion and Osmosis

Diffusion is the process whereby molecules that initially have a clumped distribution spread out to attain a fully dispersed, equally spaced distribution. This equalization process occurs on both sides of an intervening biological membrane if the membrane is permeable to the solute. If we place small, membrane-permeable solutes inside a bag of permeable membrane immersed in a beaker of water, the solutes will redistribute themselves by diffusion until the solute concentration is the same both inside and outside the bag.

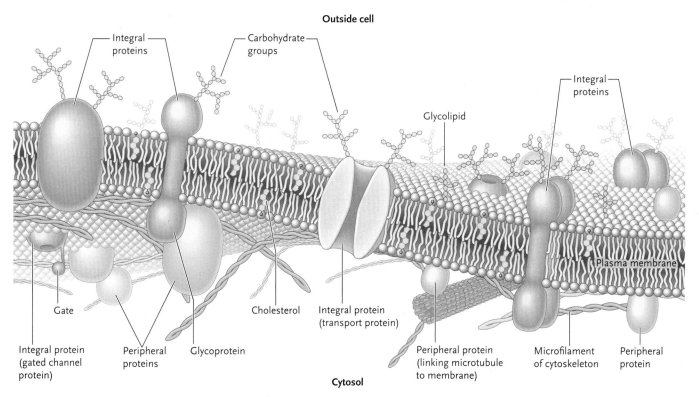

Figure 13.2 A cell membrane: a bilayer of phospholipids and associated proteins.

SOURCE: From RUSSELL/WOLFE/HERTZ/STARR. *Biology*, 2E. © 2013 Nelson Education Ltd. Reproduced by permission. www.cengage.com/permissions

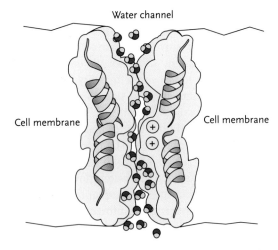

Water channel

Cell membrane

Cell membrane

Figure 13.3 Model of aquaporin embedded in a cell membrane. The aquaporin forms a channel for water molecules to move through the membrane.

SOURCE: From RUSSELL/WOLFE/HERTZ/STARR. *Biology*, 2E. © 2013 Nelson Education Ltd. Reproduced by permission. www.cengage.com/permissions

Osmosis is a special case of diffusion that refers to movement of water molecules through a biological membrane. Essentially, water moves through membranes from the side with a low solute concentration to the side with a higher solute concentration. A dramatic demonstration of osmosis would occur if the solutes in our membrane bag were too large to diffuse across the membrane. Water in the beaker would move into the membrane bag to the point of bursting the bag.

In real life, cells and the body fluids surrounding them contain a large number of different types of solutes, including small inorganic molecules and organic molecules of various sizes, all of which may or may not carry a charge (positive or negative). **Osmolarity** is the total concentration of all solutes in a solution. It is important to understand that the movement of each type of solute through the membrane depends not only on the difference in solute concentration on either side of the membrane, but also on the charge (if any) carried by the solute. Physical laws dictate that ions should distribute so as to establish neutrality inside the cell and between the cell and its immediate environment. Charged (polar) solutes move passively across membranes according to the **electrochemical gradient**, the sum of both the concentration differential and the electric charge differential on either side of the membrane. Each solute moves across the membrane according to the electrochemical gradient acting on that specific molecule.

13.2b Cellular Osmoregulation by Animal Cells: The Na⁺/K⁺ Pump

Although many solutes become distributed on either side of a cell membrane according to the electrochemical gradient acting on that solute, some cellular solutes cannot pass through cell membranes because they are too large and/or too highly charged. This creates a

universal problem for all cells in an aqueous setting. Even if all the permeable solutes diffuse toward electrochemical equilibrium across the cell membrane, the impermeable solutes that remain within the cell (large organic molecules, most of which have a negative charge) result in a higher osmolarity inside the cell. Therefore, water will flood into the cell by osmosis and the cell will eventually burst.

The problem of membrane-impermeable solutes within living cells (e.g., nucleic acids, proteins, carbohydrates) had to be solved very early in the history of life. Two different processes emerged. Bacteria, fungi, algae and plants solved the flooding problem by secreting a tension-resisting wall around their cells. Bacteria use peptidoglycans for their extracellular wall, fungi use chitin, and plants and algae use cellulose. The cell wall prevents these cells from bursting because osmotic pressure is counteracted by hydrostatic pressure. As a result, the turgid cells of bacteria, plants, algae, and fungi have only limited ability to change their shape and cannot do so rapidly.

Because animal cells lack a rigid cell wall, they often change shape drastically and rapidly, for example, the amoeboid crawling movements of fibroblasts and the rapid shortening of muscle cells. Animal cells are seemingly oblivious to the fatal tide of water that should be flooding inside by osmosis, due to the membrane-impermeable solutes within their cytoplasm. How do they do it?

The answer to the enigma lies in the fact that cell membranes consist of more than a naked phospholipid bilayer. Rather, a great variety of proteins are embedded within the bilayer (Fig. 13.2). One of these proteins is aquaporin, which provides a channel for water molecules to move through the membrane. In addition to channel proteins such as aquaporin, other transporters are integral membrane proteins that bind to solutes and then undergo a change in shape (conformational change) allowing transport of solutes through the membrane. There are two basic types of transporters: passive and active. Passive transporters carry solutes through membranes in the direction dictated by the electrochemical gradient acting on the solute. Passive transporters facilitate an increased rate of solute diffusion, but they do not carry solutes against their electrochemical gradient, and they do not require cellular energy. By contrast, active transporters carry solutes against a prevailing electrochemical gradient, and the process requires input of cellular energy. The sodium/potassium pump, often called the Na⁺/K⁺-ATPase pump, is the most important active transporter in animal cell membranes. The name reflects the fact that solute transport is coupled with ATP dephosphorylation. During each exchange cycle, this pump exports three sodium ions and simultaneously imports two potassium ions (Fig. 13.4). Eureka! This is the solution to the core enigma of animal cell osmoregulation. By actively generating a net efflux of positively charged

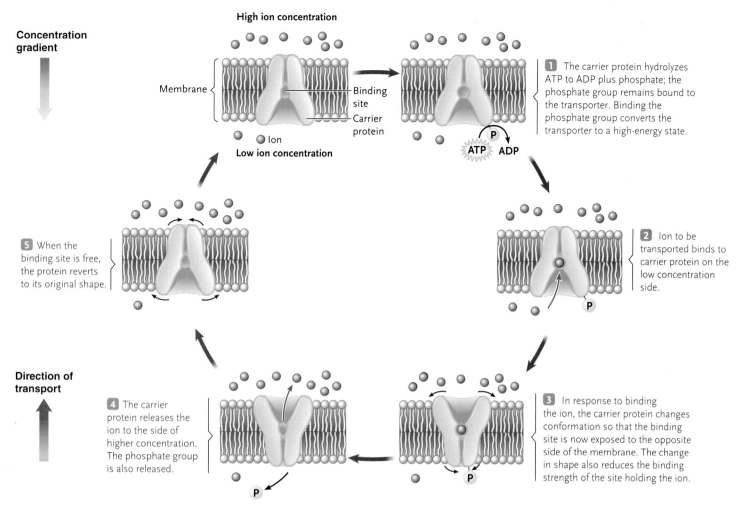

Concentration gradient

High ion concentration

Membrane

Binding site

Carrier protein

Ion

Low ion concentration

1 The carrier protein hydrolyzes ATP to ADP plus phosphate; the phosphate group remains bound to the transporter. Binding the phosphate group converts the transporter to a high-energy state.

ATP   ADP

2 Ion to be transported binds to carrier protein on the low concentration side.

P

5 When the binding site is free, the protein reverts to its original shape.

Direction of transport

4 The carrier protein releases the ion to the side of higher concentration. The phosphate group is also released.

P

3 In response to binding the ion, the carrier protein changes conformation so that the binding site is now exposed to the opposite side of the membrane. The change in shape also reduces the binding strength of the site holding the ion.

P

Figure 13.4 A model of the sodium/potassium pump embedded in a cell membrane. During each transport cycle, three ions of sodium are exported from the cell, and two ions of potassium are imported into the cell. The energy required for this exchange of ions across the cell membrane is provided by dephosphorylation of ATP.

SOURCE: From RUSSELL/WOLFE/HERTZ/STARR. *Biology*, 2E. © 2013 Nelson Education Ltd. Reproduced by permission. www.cengage.com/permissions

ions (three sodium ions exported but only two potassium ions imported), animal cells effectively create a population of cations (positively charged ions) exterior to the cell that balances the population of non-permeable anions (negatively charged ions) confined to the interior of the cell. After the membrane-permeable ions achieve equilibrium on either side of the membrane, there is no osmotic imbalance of solutes on either side of the membrane and therefore no net flux of water into animal cells.

The Na^+/K^+ ATPase membrane pump is unique to animal cells, and all cells in all animals have them. Remarkably, these pumps consume 5–40% of cellular ATP turnover, depending on the type of cell. Powering them requires a lot of energy. Importantly, although Na^+/K^+ pumps allow osmotic equilibrium between a cell and its environment, the pumps also maintain a separation of oppositely charged ions across the cell membrane—inside negative and outside positive. This charge separation, known as the **membrane potential**, is a form of energy that drives a variety of other membrane transporters and exchangers.

An important example of a membrane transporter that depends on the Na^+ gradient maintained across cell membranes by the Na^+/K^+ pump is the family **of Na^+-dependent co-transporters.** These co-transporters allow cells to take up nutrients such as glucose and amino acids and to manipulate the movement of ions. Na^+-dependent co-transporters have binding sites on the extracellular surface for both Na^+ and a second molecule, such as glucose or an amino acid. The co-transporters facilitate the inward diffusion of Na^+ along its strong, inward electrochemical gradient (maintained by the Na^+/K^+ pump) and couple this inward movement of Na^+ to the inward movement of the second molecule. The coupled importation is effected by a change in the shape of the co-transporter. Co-transporters do not dephosphorylate ATP directly, but they depend on the energy-consuming Na^+/K^+ pump to maintain the high external concentration of Na^+. This Na^+ gradient is also important in a process known as counter transport or ion exchange.

Of particular interest is the mechanism by which cells regulate intracellular pH, which relies upon a similar transport mechanism as described for co-transport. In this case, sodium:proton exchangers bind to extracellular Na^+ and facilitate its inward movement along its electrochemical gradient, while simultaneously carrying protons (H^+) out of the cell. As described later in this chapter, the Na^+/K^+ pump and secondary transporters that depend on it are critically important for many aspects of excretion and osmoregulation. Specifically, transporting epithelia use these membrane-bound transporters to manipulate the movement of solutes and metabolic wastes in and out of cells.

13.2c Different Habitats, Different Osmoregulatory Challenges

This discussion has dealt mainly with the osmoregulatory challenges of individual cells within the internal environment of a multicellular animal. However, whole animals occupy different habitats that present large additional problems for osmoregulation. In terrestrial habitats, the risk of desiccation is universal, and mechanisms that conserve and retain water are of primary importance. Aquatic habitats present three very different types of situations. First, many marine invertebrates are osmoconformers, meaning that the osmolarity of their body fluids matches that of their environment. These marine inhabitants live in an isosmotic environment. Second, many marine fish maintain a blood and tissue osmolarity that is lower than that of seawater; they therefore live in a hyperosmotic environment. These fish must deal with a constant diffusional influx of salts into their bodies, as well as an osmotic efflux of water to the outside. Third, all animals living in fresh water must deal with a hypoosmotic environment; they require mechanisms to prevent diffusional loss of salt ions and to export water that enters their bodies continuously by osmosis.

Osmoregulation at both the microscale (cells) and macroscale (whole organisms) involves controlling the movement of water and solutes, particularly ions. The process of excretion also involves the control of solute movement, but in the case of excretion the solutes are nitrogenous wastes resulting from protein catabolism. Cell membranes and their associated pumps, channels, and transporters are of central importance for both osmoregulation and excretion of nitrogenous wastes at both the microscale and macroscale.

STUDY BREAK

1. Define osmosis and diffusion.
2. Why does rapid movement of water through cell membranes require a channel protein embedded in the membrane?
3. What is the difference between passive and active transport?
4. Explain how the Na^+/K^+ pump prevents a catastrophic influx of water into animal cells.

13.3 Excretory Products

Excretory products are nitrogen-containing waste compounds resulting from digestion and catabolism of ingested proteins. Ammonia, urea, and uric acid (Fig. 13.5) are the most common nitrogenous waste products of animals; guanine is a less common form of nitrogenous waste. The primary factor that determines which of these three nitrogenous waste products is/are released by an animal is the availability of water in its environment.

13.3a Ammonia

Ammonia (Fig. 13.5A) occurs in two forms within animals. NH_3 is a gas that readily dissolves in water and can move through cell membranes by diffusion. NH_4^+, the ammonium ion, also dissolves readily in water, but its charge reduces its rate of diffusion across cell membranes. By convention, the term "ammonia" refers to total ammonia, whereas the chemical formulae are given when either the gaseous or ionic form needs to be specified. **Ammonotelic** animals produce ammonia as a principal nitrogenous waste.

Ammonia is the primary end product of protein catabolism. Ingested protein is hydrolyzed within the gut or within food vacuoles of epithelial cells lining the gut (page 240) to release free amino acids. This pool of free amino acids can be used to build proteins as needed by the animal. Amino acids in excess of what the animal needs for protein synthesis can be channeled into gluconeogenic pathways (pathways that synthesize glucose from non-carbohydrate precursors) or the citric acid cycle, but first the amine group(s) must be removed from the carbon skeleton of each amino acid. Amine groups can neither be metabolized for energy nor stored. Transdeamination is a two-step process occurring in the liver of vertebrates that removes the amine groups from excess amino acids, producing ammonia (Box 13.1). Ammonia is toxic above a threshold concentration and must be removed from the body.

Why is ammonia toxic? The answer to this question is complex. Ammonia can poison cells at many different levels, not all of which are relevant to all animals. Probably the most fundamental reason for ammonia toxicity comes from its ability to compete for

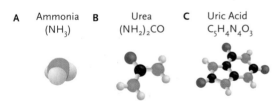

Figure 13.5 A space-filling model of **A.** ammonia as well as models of **B.** urea, and **C.** uric acid. Blue = nitrogen; black = carbon; red = oxygen; white = hydrogen.

BOX 13.1
Catabolism of Amino Acids

Catabolism of amino acids to remove the amino groups is a two-step process involving transamination (transfer of the amino group to another molecule) and deamination (removal of the amino group). Most transaminases share a common substrate and product (glutamate and oxoglutarate) with glutamate dehydrogenase. This allows the catabolism of many different amino acids to be channeled through glutamate, which is deaminated by glutamate dehydrogenase to produce ammonia.

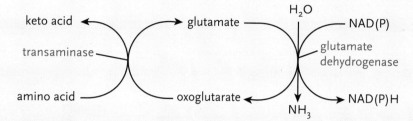

Figure 1

Most transaminases share a common substrate and product (glutamate and oxoglutarate) with glutamate dehydrogenase. This allows the catabolism of many different amino acids to be channeled through glutamate, which is deaminated by glutamate dehydrogenase to produce ammonia.

access to potassium ion (K$^+$) channels and for K$^+$ binding sites on membrane transporters. One of the most important sites in this regard is the Na$^+$/K$^+$ pump that maintains the membrane potential for all animal cells. Ammonia poisoning may cause mitochondrial dysfunction and production of damaging free radicals within cells. Excess ammonia may also interfere with glutamate signaling and cycling systems in the nervous system. This has dire consequences for vertebrates because glutamate is the major excitatory neurotransmitter within the vertebrate central nervous system. For vertebrates, ammonia poisoning produces convulsions, coma, and ultimately death.

The toxicity of ammonia depends on its concentration; most animals can tolerate low levels of ammonia. In habitats where water is plentiful, ammonia can be rapidly diluted and flushed from the body before it reaches a dangerous concentration. Excretion of nitrogenous waste in the form of ammonia carries the lowest metabolic price tag. Organisms living in aquatic habitats often exploit this cheap solution to the problem of disposing of excess amine generated by protein catabolism. These **ammonotelic** animals include most aquatic invertebrates, teleost fish, and fully aquatic amphibians.

Even for aquatic animals, however, diffusion alone may be too slow to remove ammonia from the deep tissues of structurally complex animals that have relatively high metabolic rates. Therefore, teleosts must have a mechanism to safely transport ammonia from deep tissues to peripheral sites (gills) where ammonia can be released to the environment. Teleosts typically convert ammonia to glutamine for safe, temporary storage and transport (Box 13.2). This conversion has been observed in other aquatic organisms, including mosquitoes and crustaceans. Glutamine synthetase is upregulated in the liver and brain of some teleosts after they ingest a meal, presumably anticipating the need to detoxify an incoming load of ammonia as the meal is digested and the amino acids catabolized.

Although glutamine is one option for detoxifying ammonia, organisms living where water must be conserved spend metabolic energy to convert ammonia to a non-toxic form that can be excreted without a large amount of water. Water-conserving forms of nitrogenous waste are typically urea and uric acid.

13.3b Urea

Many terrestrial and semi-terrestrial animals convert ammonia to urea (Fig. 13.5B). These **ureotelic** animals include mammals and some adult amphibians and turtles. Urea carries two amine groups and is less toxic than ammonia. It can therefore be concentrated to a greater degree than ammonia, which reduces water loss during excretion of excess nitrogen. The multistep processing pathway that generates urea from ammonia involves a cassette of enzymes, collectively known as the ornithine-urea cycle (Box 13.3). These enzymes are typically expressed within liver cells, with some of the enzymes localized within liver mitochondria and others within the cytoplasm.

The phylogenetic distribution of the genes encoding CPSI and III (see Box 13.3) shows a fundamental dichotomy among the major groups of vertebrates (Fig. 13.6). This distribution is consistent with the hypothesis that lungfishes, rather than coelacanths, are the closest extant relatives of tetrapods, because tetrapods and lungfish have the gene for CPSI, whereas other vertebrates have the gene for CPSIII.

BOX 13.2

Conversion of Ammonia to Glutamine

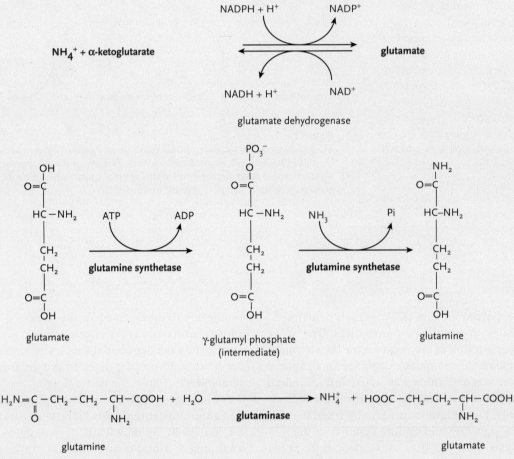

Figure 1
Conversion of ammonia to glutamine.

Ammonia can be temporarily stored or transported in the form of glutamine (non-toxic), and then converted back to ammonia for excretion. First, ammonia (NH_4^+) is added to α-ketoglutarate by the enzyme glutamate dehydrogenase to form glutamate (Fig. 1). A subsequent two-step reaction catalyzed by glutamine synthetase adds a second molecule of NH_4^+ to glutamate to form glutamine. The third equation shows how ammonia can be released from glutamine in a reaction catalyzed by glutaminase.

By converting ammonia to the less toxic urea, the ornithine-urea cycle allows terrestrial and semi-terrestrial animals to concentrate nitrogenous waste to a greater degree, thereby conserving water during the process of excretion. **Ureotelism** refers to animals that produce urea. The explanation for ureotelism in certain groups of fish, notably cartilaginous fish (elasmobranchs [sharks, skates, and rays] and holocephalans) and coelacanths, is less obvious. Indeed, marine cartilaginous fish and coelacanths actually retain high levels of urea in their blood and tissues. Urea is retained in these organisms to function as an osmolyte, a solute that increases osmolarity. Osmolytes can help reduce the osmotic difference between the tissues of marine animals and their hyperosmotic environment. The gill

epithelia of cartilaginous fish have low permeability to urea, preventing it from leaking into the environment. Their kidneys also conserve urea. So, these fish deal with the osmotic challenge of a salty environment by increasing the osmolarity of their body fluids using retained urea as the osmolyte.

The strategy of retaining high levels of urea to deal with life in a salty environment is not without complications. Unfortunately, urea is not entirely non-toxic. At sufficiently high concentrations, such as the concentration in the blood and tissues of cartilaginous fish, urea can disrupt or denature the secondary structure of proteins, including enzymes. To counteract these adverse effects, cartilaginous fish also have high blood and tissue levels of trimethylamine oxide, a compound that

BOX 13.3
The Ornithine–Urea Cycle

The initial steps of the series of biochemical transformations that convert ammonia to urea occur within mitochondria, while subsequent reactions occur within the cytoplasm. Ammonia (NH_3) enters the ornithine–urea cycle via two routes, depending on the species. In tetrapods, NH_3 is combined with bicarbonate ion (HCO_3^-) under catalysis by carbamoyl phosphate synthetase I (CPSI) to produce carbamoyl phosphate. Alternatively, in teleosts and invertebrates, ammonia is converted to glutamine (not shown), which in turn is combined with HCO_3^- under catalysis by CPSIII to form carbamoyl phosphate. Carbamoyl phosphate reacts with ornithine to form citrulline, which leaves the mitochondrion.

Several enzyme-catalyzed steps within the cytoplasm generate ornithine and urea. Ornithine enters back into a mitochondrion. An initial step in the production of urea from ammonia involves production of carbamoyl phosphate in a reaction catalyzed by the enzyme carbamoyl phosphate synthetase. However, there are two different types of carbamoyl phosphate synthetase. As shown in Figure 1, carbamoyl phosphate synthetase I (CPSI), combines ammonia with bicarbonate ion (HCO_3^-) to produce carbamoyl phosphate, whereas carbamoyl phosphate synthetase III (CPSIII) combines glutamine (generated from ammonia as illustrated in Box 13.2) with bicarbonate ion to generate carbamoyl phosphate.

Most transaminases share a common substrate and product (glutamate and oxoglutarate) with glutamate dehydrogenase. This allows the catabolism of many different amino acids to be channeled through glutamate, which is deaminated by glutamate dehydrogenase to produce ammonia. Glutamate has a central role in amino acid catabolism.

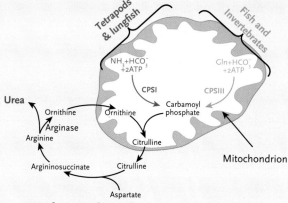

Figure 1
Ornithine-urea cycle.

SOURCE: Haskins et al. "Inversion of allosteric effect of arginine on N-acetylglutamate synthase, a molecular marker for evolution of tetrapods," *BMC Biochemistry* 2008 9:24.

CPSI & III - two forms of carbamoyl phosphate synthetase
NH_3 - ammonia
HCO_3 - bicarbonate ion
Gln   - glutamine

---

stabilizes and protects proteins from urea denaturation. The ratio of urea to trimethylamine oxide is 2:1.

In marine elasmobranchs, levels of urea are 750–1300 mg per 100 mL of blood. The high concentrations of urea make the body fluids of these elasmobranchs even more concentrated than sea water so that water flows into these fish by osmosis and must be flushed out by the kidneys. Such high levels of urea would make life in fresh water impossible, because the body would be flooded by water. Not surprisingly, freshwater skates (*Potamotrygon* spp. (1142)) that live 3000 km up the Amazon have only 2–3 mg of urea per 100 mL of body fluids. We can only surmise that pleurocanths, freshwater sharks known from the late Devonian to the Triassic, had levels of urea similar to those of contemporary freshwater skates and stingrays, rather than levels similar to modern saltwater elasmobranchs.

A third role for urea that operates under special circumstances is the nitrogen cycling that occurs between the liver and gut of ruminants such as cows, bison, deer, and giraffes. The diet of these herbivores is low in nitrogen, so conservation of nitrogen is perhaps a larger issue for herbivores than elimination of excess nitrogen. The digestive tract of these animals includes two specialized regions, the rumen and reticulum, which are populated by a diverse fauna of bacteria and unicellular eukaryotes that assist with digestion of plant material (see Chapter 9). Urea manufactured in the liver of ruminants is carried to the rumen and reticulum, where urease of bacterial origin degrades urea to release ammonia. This ammonia supports microbial protein synthesis, some of which is transported back to the host.

A final novel function for urea production has been reported for the Gulf toadfish (*Opsanus beta* (1170)). Although urea production is highly unusual among teleost fish, the Gulf toadfish excretes its nitrogenous wastes as both ammonia and urea, switching in pulses from all urea to all ammonia. Cyclical production of urea has been proposed as a defensive mechanism that may help these fish avoid predators that detect their prey by the scent of ammonia.

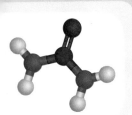

BOX 13.4 Molecule Behind Animal Biology

Ammonia (NH₃)

The average person can detect the odour of NH_3 at 17 parts per million (ppm), and regular exposure to 25 ppm is not considered dangerous. However, a 30-min exposure to 500 ppm affects the upper respiratory tract and causes the eyes to tear, and levels >5000 ppm can be fatal to humans. Our ability to detect even small traces of NH_3 is probably an important safety mechanism.

Anyone who has ventured inside a bat cave in the topics or subtropics may have immediately noticed the prevailing odour of NH_3, which is generated by bacterial urease acting on bat urine. Inside caves that house even small populations (~50 individuals) of vampire bats (*Desmodus rotundus* (1788)), the odour of NH_3 is usually noticeable. But inside caves housing tens of thousands or even millions of bats, it can be overwhelming. California leaf-nosed bats (*Macrotus californicus* (1796)) can withstand exposure to 3000 ppm for up to 9 h, and Mexican free-tailed bats (*Tadarida brasiliensis* (1817)) can withstand 5000 ppm for more than 4 days. Production of copious amounts of mucous within respiratory airways appears to be at least partly responsible for this astonishing tolerance to NH_3.

Although ammonia is usually thought of in the context of a toxic metabolic waste that must be excreted, we have seen how other metabolic end products, such as urea, can serve useful functions in some situations. Surprisingly, ammonia can also be useful, notably among pelagic organisms such as squid (Mollusca, Cephalopoda, Fig. 1). Ammonia is readily available for carnivorous animals such as predatory squid, which roam throughout the world's oceans in search of prey. Some species of squid that live at meso- to bathypelagic depths, replace sodium ions with lighter ammonium ions within their bodily fluids to achieve neutral or near neutral buoyancy to reduce the energetic cost of swimming. However, because the atomic mass of NH_4^+ is not a great deal less than that of Na^+, whole body buoyancy requires that the ammoniacal fluid constitute a large proportion of the total body volume. In some squid species, a large volume of ammoniacal fluid is contained within an enlarged coelomic compartment; in others the fluid is relegated to vacuoles within cells dispersed throughout the muscular arms and mantle. The absence of NH_4^+ in the blood of these squid suggest that the animals have effectively removed it from general circulation.

Other pelagic animals also use ammonia to enhance buoyancy. For example, the deep-sea caridean shrimp *Notostomus gibossus* (825) (Arthropoda, Crustacea), which lives off the coast of Hawaii, is positively buoyant because it has a greatly expanded carapace that houses a large volume of low-density fluid. The reduced density of this fluid is achieved by replacing nearly 90% of the Na^+ in the carapace fluid with trimethylamine (Me_3NH^+) and NH_4^+. These shrimp are neutrally buoyant, but sink when the fluid within the carapace chamber is removed.

Chaetognaths, or arrow worms, are another group of pelagic, predatory organisms. The chaetognath *Sagitta elegans* (580) has greatly enlarged gut cells that each contain a large vacuole filled with a fluid high in NH_4^+, presumably to confer buoyancy.

Finally, within the highly complex and intriguing realm of symbioses between animals and other organisms, ammonia provides a case where a toxin to one organism is an essential nutrient to another. Coral reefs (Cnidaria, Anthozoa) are huge ecosystems within the nutrient-poor marine waters of tropical and subtropical latitudes. Marine organisms associated with coral reefs represent an estimated 25% of global marine biodiversity. Primary productivity within a coral reef is accomplished predominantly by unicellular, photosynthetic symbionts living within the corals. The symbionts provide their coral hosts with organic carbon generated by photosynthesis, and the algae in this mutualistic relationship receive needed nitrogen from ammonia released by their host coral, which retains its ability to prey on other animals. Thus, a toxic waste product for one organism is an essential nutrient for another.

FIGURE 1
The squid Histioteuthis heteropsis *(501), a species that achieves buoyancy by replacing sodium ions with ammonium ions.*

© Richard E. Young

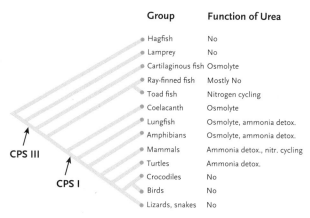

Group	Function of Urea
Hagfish	No
Lamprey	No
Cartilaginous fish	Osmolyte
Ray-finned fish	Mostly No
Toad fish	Nitrogen cycling
Coelacanth	Osmolyte
Lungfish	Osmolyte, ammonia detox.
Amphibians	Osmolyte, ammonia detox.
Mammals	Ammonia detox., nitr. cycling
Turtles	Ammonia detox.
Crocodiles	No
Birds	No
Lizards, snakes	No

CPS III

CPS I

Figure 13.6 Phylogenetic distribution of carbamoyl phosphate synthetase I and III (CPSI and CPSIII) of the ornithine-urea cycle of vertebrates. Functions of urea in those groups that synthesize urea is indicated. Sister-group relationships among amniote tetrapods are controversial.

Figure 13.7 Bird feces showing white flecks (crystals) of uric acid.

13.3c Uric Acid

The conversion of ammonia to uric acid (Fig. 13.5C) is biochemically complex and metabolically expensive. Each mole of uric acid produced from ammonia requires eight moles of ATP. Nevertheless, the advantage of uric acid as a nitrogenous waste product comes from its tendency to crystallize, particularly under acidic conditions. Once crystallized, uric acid is no longer in solution, and thus the osmotic work to reabsorb water from this nitrogenous waste is greatly reduced. Reptiles and birds, as well as many insects, are examples of **uricotelic** animals that convert ammonia to uric acid to minimize water loss during excretion. It is the white paste so characteristic of bird droppings (Fig. 13.7).

In humans and higher primates, uric acid is generated not from ammonia but by catabolism of the adenine- and guanine-based purines of nucleic acids. In its soluble form, uric acid has beneficial properties as a scavenger of damaging reactive oxygen species within cells and tissues. Normally, the amount of uric acid produced in humans is small and is eventually excreted by the kidneys. However, gout is a painful condition that occurs when uric acid, dissolved in blood, precipitates as uric acid crystals that become lodged in the joints.

Although birds are typically uricotelic, several species are occasionally ammonotelic. During periods of cold temperature, Anna's Hummingbirds (*Calypte anna* (1545)) must increase their intake of nectar, a diet low in protein and high in water content, to support thermogenesis. At these times, ammonia constitutes more than 50% of their excreted nitrogenous waste, presumably because it is less expensive to produce than uric acid and can be tolerated when fluid intake is high. Other birds having a diet of low protein but high water content, notably the nectar-feeding Palestine Sunbirds (*Cinnyris osea* (1593)) and the fruit-

eating Yellow-vented Bulbuls (*Pycnonotus xanthopygos* (1613)), also exhibit a higher proportion of ammonia in their excreta relative to most other birds. However, in these cases the explanation may be more complex. Birds do not have a bladder or urethra; the ureters discharge directly into the cloaca at the end of the hindgut. Research has shown that the concentration of uric acid within the ureters of these birds is much higher than in excreta. This observation led to two possible interpretations: 1) uric acid is converted to ammonia by bacteria within the hindgut or 2) uric acid may be actively reabsorbed by the hindgut, leaving mostly ammonia within the excreta. The value of reabsorbing uric acid may derive from its role as an antioxidant, an efficient scavenger of reactive oxygen species. Humans and some species of birds lack the ability to synthesize ascorbic acid, also a potent antioxidant, and may therefore benefit from retention of uric acid.

13.3d Guanine

Guanine, like uric acid, is a purine that forms crystals with low solubility in water. Although insects, the largest group of terrestrial arthropods, excrete nitrogenous waste as uric acid to conserve water, arachnids, the second largest group of terrestrial arthropods (e.g., spiders, scorpions, mites, and ticks), conserve water by converting their nitrogenous waste into guanine. There is no obvious functional reason for this difference in excretory products among terrestrial arthropods, but on the practical side, it inspired a proposal to diagnose infestations of tiny, allergenic dust mites by testing mattresses for the presence of guanine.

Production of guanine by spiders is an interesting example of evolutionary opportunism—a material that serves a particular function (excretion without water loss in this case) has been co-opted to perform an additional but completely unrelated function.

Figure 13.8 Scanning electron micrographs of different forms of guanine crystals. **A.** Small, cuboidal crystals of guanine reflect light as a matt white surface. **B.** Large, flat, overlapping crystals appear as a shiny, mirror-like surface. Scale bars: A = 5 μm; B = 10 μm.

SOURCE: Oxford, G.S. 1998. "Guanine as a colorant in spiders: development, genetics, phylogenetics and ecology." Selden PA (ed) 1998. *Proceedings of the 17th European Colloquium of Arachnology.* Edinburgh 1997. British Arachnological Society.

Some species of spiders place a layer of guanine crystals beneath the epidermis that secretes the exoskeleton. If the guanine crystals are small, cube-shaped, and jumbled together in various orientations, the crystal layer scatters reflected light in various directions and appears white (Fig. 13.8A). This white layer allows red and yellow pigments within the thin, overlying epidermis to be visible, because the colour of the pigment is not swamped by the dark underlying tissue of the spider's digestive organs. This explains why the location of subepidermal uric acid crystals in female black widow spiders (*Latrodectus mactans* (672)) exactly corresponds with the location of the red pigment having an hour glass shape within the overlying epidermis (Fig. 13.9). Alternatively, if the subepidermal guanine crystals have the form of large, flat, overlapping crystals (Fig. 13.8B), rather than small cubes jumbled together, the layer has mirror-like reflective properties and gives the spider's body a shiny appearance. A shiny spider sitting in a web may be ignored as a mere drop of water by potential prey and predators.

Figure 13.9 Female black widow spider (*Latrodectus mactans* (672)) showing the distinctive hourglass-shaped patch of red pigment on the ventral surface of the opisthosoma.

Melinda Fawver / Shutterstock.com

STUDY BREAK

1. Compare ammonia, urea, uric acid, and guanine with respect to toxicity, cost of biosynthesis, and solubility at normal physiological pH.
2. Why can urea function as an osmolyte? Is uric acid used in this way? Why?

13.4 Excretory Systems

13.4a Contractile Vacuole Systems

In high school, you probably watched a contractile vacuoles at work in an *Amoeba* or *Paramecium* on a microscope slide (Fig. 13.10). The contractile vacuole undergoes a period of expansion followed by a sudden contraction as fluid is expelled from the vacuole to the exterior of the cell. The rate of filling and contraction of a contractile vacuole can be changed by altering the concentration of solutes within the medium. The vacuoles contract more rapidly in distilled water than they do in a weak saline solution, presumably because the osmotic gradient into the cell is greater in distilled water.

Widespread in protozoans (single-celled animals) and members of the phylum Porifera (sponges), particularly those living in fresh water, contractile vacuoles eliminate excess water from cells. *Amoeba proteus* has one contractile vacuole at any point in time, species of *Paramecium* have two contractile vacuole systems, and each pinacocyte of a freshwater sponge (*Spongilla lacustris* (116)) contains at least 40 contractile vacuoles with a 5–30 min cycle of filling and discharge.

The exact mechanism by which contractile vacuoles concentrate and export water out of cells is still largely unknown. Proton pumps (active pumping of hydrogen ions across a membrane) must be part of the mechanism because they have been identified in contractile vacuole systems of many protozoans. The contractile vacuole systems of amoeboid stages of *Dictyostelium discoideum* (a slime mould) and of *Paramecium* consist of an array of tubules connected with the central vacuole (Fig. 13.10). The membranes of the tubules and central vacuole are studded with the head domains of vacuolar-type proton pumps (known as pegs).

13.4b Nephridia

A nephridium is a combination excretory and osmoregulatory organ that incorporates two steps to remove nitrogenous waste and excess salts and water from bodily fluids. The initial step is filtration of extracellular body fluid to produce a primary urine. The second step is modification of the filtrate (primary urine) by selective reabsorption of solutes. The final product is released from the body.

The first step is more accurately called *ultra*filtration because only water and very small molecules are allowed through the filter. The force for filtration is provided by a pressure gradient, but this pressure gradient is generated quite differently in the two basic types of nephridial systems. In protonephridia, ciliary beating generates the pressure gradient. In metanephridia, contraction of muscles associated with blood vessels generates the pressure gradient. The vertebrate kidney is a complex of many, very sophisticated metanephridia-like units called nephrons. Nevertheless, nephrons of the vertebrate kidney incorporate the two basic steps of pressure-generated filtration and selective reabsorption of solutes from the initial filtrate. Secretion-type excretory organs, such as the Malpighian tubules of insects and arachnids, and the excretory role performed by the gills of aquatic crustaceans and fish, work by a different mechanism.

13.4c Protonephridia

Protonephridia (Fig. 13.11) are found among invertebrates that lack a true coelomic compartment and a muscularized blood vascular system. This group includes platyhelminths, rotifers, and many invertebrate larval forms. A protonephridium is a slender, tubular invagination of epidermal epithelium into the interior of an organism. As shown in Figure 13.11, the wall of the tubule is formed by duct cells, with the end of the tubule capped by a terminal cell. In the simplest case, the terminal cell is cup-shaped with perforations through the sides of the cup. A motile cilium, or bundle of cilia, arises from the cup-shaped terminal cell and extends into the lumen of the protonephridial tubule. Protonephridia are epithelial invaginations, so the entire tubule, including the perforations through the terminal cell, is encased by a basal lamina consisting of a meshwork of collagen and other fibrillar proteins.

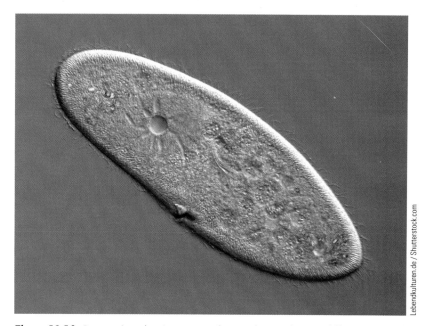

Figure 13.10 *Paramecium* showing contractile vacuole complexes at different stages of filling.

The terminal cell accomplishes the initial filtration step of the excretion and osmoregulatory process. The beating cilium (or bundle of cilia) of the terminal cell propels fluid down the protonephridial duct toward the **nephridiopore** that opens to the

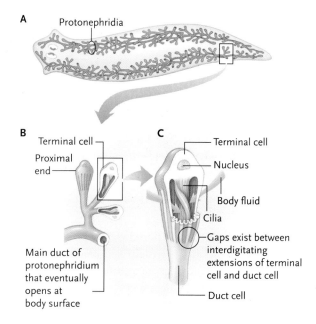

Figure 13.11 Protonephridia. **A.** Adult flatworm showing branched network of protonephridial ducts. **B.** Detail of a cluster of side-branches from the main duct. Each branch is capped with a terminal cell. **C.** Detail of a terminal cell (also called a flame cell) showing the bundle of cilia arising from the terminal cell and the gaps between the interdigitating extensions from the terminal cell and duct cell that allow body fluid to enter the duct.

SOURCE: From RUSSELL/WOLFE/HERTZ/STARR. *Biology*, 2E. © 2013 Nelson Education Ltd. Reproduced by permission. www.cengage.com/permissions

exterior. This fluid movement down the duct creates negative pressure within the duct so that body fluid surrounding the terminal cell (i.e., fluid between internal body cells or within a pseudocoel) is drawn through the perforations in the terminal cell and into the lumen of the protonephridial duct. However, the perforations of the terminal cell are too large to act as the ultrafilter. Instead, ultrafiltration is accomplished by the fibrous mesh of basal lamina that covers the perforations. This basal lamina mesh is so fine that only water and small solutes can pass through the filter into the protonephridial duct. The basal lamina covering the perforations of the terminal cell is a size-selective filter.

Some molecules that pass through the filter, such as ammonia, excess water, and various ions, need to be excreted. Other small molecules that pass through the filter, such as glucose, amino acids, and various essential ions, must be reclaimed. The reclamation process is the role of the protonephridial duct cells. As the primary urine is propelled down the protonephridial duct by the beating cilium (or ciliary bundle) of the terminal cell, molecules such as glucose and various ions are actively absorbed by the duct cells. It is likely that Na$^+$-dependent co-transporters (see Section 13.2b), embedded in the apical membrane of the duct cells, are part of this process of selective absorption. As a result, the final urine that exits from the nephridiopore consists only of nitrogenous wastes, excess ions, and water.

The preceding describes the simplest type of protonephridium, which is found only among larvae of some invertebrates. In most adult animals with protonephridia, such as flatworms and rotifers, the protonephridial duct is highly branched, with the end of each branch capped by a terminal cell. The terminal cells of protonephridia systems are often called "flame cells" because the rapidly beating ciliary bundle reminded early microscopists of a flickering candle flame.

13.4d Metanephridia

Metanephridial systems are found among animals that have a true coelom and a pressurized blood vascular system. This group includes most annelids (including sipunculids), molluscs, and brachiopods. The **metanephridia** of annelid worms provide a convenient and simple model to describe the anatomy and functioning of a basic metanephridial system. As shown in Figure 13.12, these animals have a muscularized dorsal blood vessel suspended within a dorsal mesentery that separates spacious left and right coelomic compartments within each segment of the body. Each segment also contains a pair of metanephridia. A metanephridium is a tubular invagination of epidermal epithelium into the interior of the organism. However, unlike a protonephridium, the metanephridial tubule

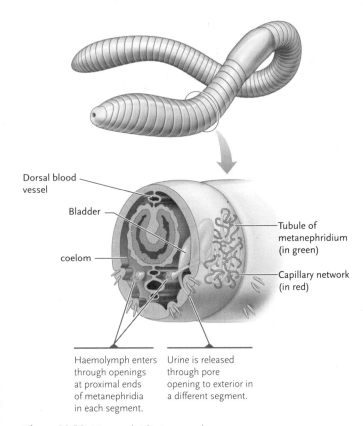

Dorsal blood vessel

Bladder

coelom

Tubule of metanephridium (in green)

Capillary network (in red)

Haemolymph enters through openings at proximal ends of metanephridia in each segment.

Urine is released through pore opening to exterior in a different segment.

Figure 13.12 Metanephridia in an earthworm.

SOURCE: From RUSSELL/WOLFE/HERTZ/STARR. *Biology*, 1E. © 2010 Nelson Education Ltd. Reproduced by permission. www.cengage.com/permissions

has a funnel-shaped opening into a coelomic compartment. The filtration step of the annelid excretory process occurs across the wall of the dorsal blood vessel, not by the metanephridium itself.

The wall of the dorsal blood vessel is formed by coelomic epithelium and is therefore surrounded by basal lamina. Some of these coelomic cells (podocytes, Fig. 13.13) form extremely flat sheets that are perforated. When the muscles of the vessel wall contract, the increase in blood pressure forces fluid through the perforations between the podocytes so that it enters the coelomic compartment, which already contains coelomic fluid. Blood is therefore filtered through the basal lamina that extends over the gaps formed by the podocytes. Thus, basal laminae are the ultrafilters in both protonephridial and metanephridial systems. Podocytes are also located along the lining of blood vessels in molluscs and in the glomerulus of the vertebrate nephron. In both cases, the podocytes form a filtration surface.

In the annelid, coelomic fluid—containing the filtrate from the dorsal blood vessel—enters the open funnel of the metanephridial ducts and travels down the duct to the nephridial pore (Fig. 13.12). Along the way, the fluid is modified as essential solutes and much of the water are selectively transported across the wall of the metanephridial duct, while excess solutes, wastes such as ammonia, and excess water are released

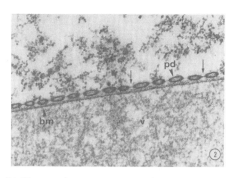

Figure 13.13 Cytoplasmic processes of podocytes (pd) and underlying basal lamina (bl) bordering a blood vessel (v). Arrows indicate gaps between cytoplasmic processes of podocytes.

SOURCE: Reprinted from *Tissue and Cell*, 27, Hansen, U., "New aspects of the possible sites of ultrfiltration in annelids (Oligochaeta)," Pages 73–78, Copyright 1995, with permission from Elsevier.

into the environment from the nephridial pore. A network of fine blood vessels is associated with the metanephridial duct, presumably so that reclaimed solutes and water can be delivered to the blood vascular system rather than to the coelomic compartment.

STUDY BREAK

1. Where are the sites of ultrafiltration in protonephridial and metanephridial systems?
2. Why is it important that the epithelial tubules of both protonephridia and metanephridia are transporting epithelia?

13.4e The Vertebrate Kidney: Focus on Mammals

Although the kidneys of vertebrates are much more complex than the metanephridial systems of annelids and other invertebrates, the functional unit of the vertebrate kidney, known as a **nephron**, accomplishes excretion and osmoregulation using the same basic design principles as the metanephridium. In both cases, a blood vascular system provides a pressure gradient for initial filtration of body fluid, and the filtrate is subsequently processed by a combination of active and passive transport across the wall of an epithelial tubule. The processing reclaims essential solutes and water from the filtrate while eliminating nitrogenous and other wastes and excess salt and water. Additional wastes that are too large to pass through the initial filter are added to the filtrate within the nephron by secretion across the wall of the tubule.

The vertebrate kidney consists of many nephrons. In hagfishes and larval lamprey, the structure of the nephron shows striking similarities to an annelid metanephridium, in that the nephron opens into the coelom as a ciliated funnel called a coelomostome. Tangled knots of capillaries lie adjacent to the coelomostomes, and blood filtrate from these presumably enters the coelom and gets wafted into the coelomostome. Like the metanephridium of annelids, this coelomic fluid is processed as it travels down the nephron, and reclaimed solutes are transferred to blood capillaries associated with the nephric tubule.

At some point during vertebrate evolution, the blood vessels became intimately associated with the coelomostome. In most modern vertebrates the coelomostome has become a sealed epithelial cup, called Bowman's capsule, that closely invests the knot of blood vessels, now called a glomerulus. Because the walls of these entities consist of podocytes, the glomerulus and Bowman's capsule accomplish filtration of blood. Blood pressure generated by the heart forces water and small molecules through the mesh of basal laminae covering the pores in the glomerular capillaries and the adjacent Bowman's capsule. This ultrafiltrate then enters the nephric tubule to begin a long and convoluted passage through the kidney. During its passage, the filtrate passes through sequential regions of qualitatively different epithelia forming the wall of the tubular nephron. Collectively, these epithelial regions use a combination of active and passive transport processes to ultimately reclaim glucose, amino acids, essential ions, and water from the filtrate, while allowing most of the nitrogenous waste and an appropriate amount of salts and water to exit from the kidney as urine.

The kidneys of fish, amphibians, and most reptiles cannot generate urine with a higher osmolarity than the tissues and blood of the animal. Birds can generate urine that is only slightly hyperosmotic to the body tissues. The ability to generate highly concentrated urine is truly an innovation of the mammalian kidney (Fig. 13.14). A great deal of research on the mammalian kidney, particularly that of humans, has provided detailed information about how filtrate processing within the nephron of these animals is accomplished. Remarkably, human kidneys process 150–200 L of initial filtrate from blood each day! Obviously a great deal of that water volume must be reabsorbed across the wall of the nephron and returned to the blood. The water reabsorption mechanism involves using salt ions absorbed from the filtrate to set up an osmotic pressure gradient between the interstitial fluid of the kidney and the fluid within the nephron. For that purpose, the shape of each nephron and the regional differences in epithelial transport functions along the length of the nephron are both critically important. As illustrated in Fig. 13.14C, beyond Bowman's capsule, the nephron of the mammalian kidney is subdivided into five main regions: the proximal convoluted tubule, descending limb (segment) of the loop of Henle, ascending limb (segment) of the loop of Henle, the distal convoluted tubule, and the collecting duct. Some regions of the nephron are located within the cortex (peripheral region) of the kidney, but the loop of Henle lies within the kidney's medulla (inner region). The collecting

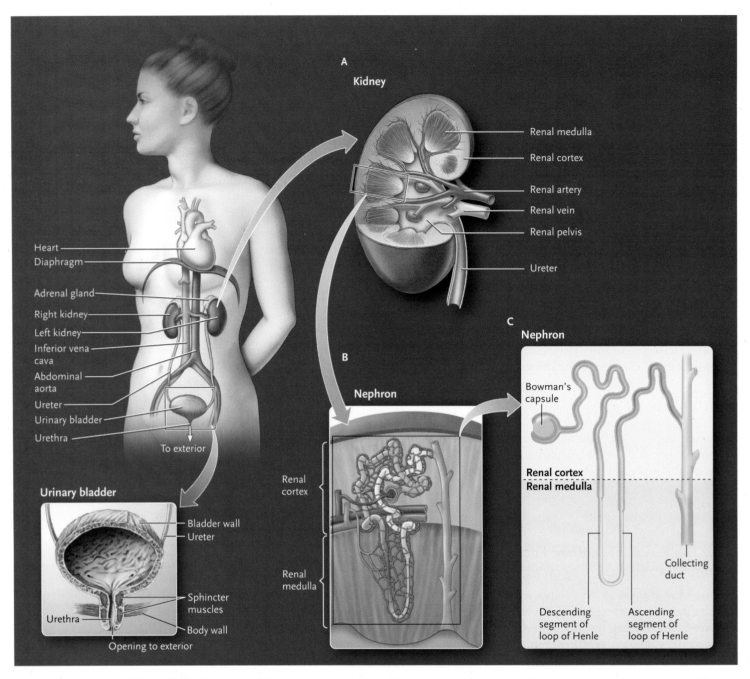

Figure 13.14 Mammalian kidney and nephron. **A.** Entire kidney, **B.** section through a portion of the cortex and medulla of the kidney, **C.** detail of an individual nephron showing its major regions.

SOURCE: From RUSSELL/WOLFE/HERTZ/STARR. *Biology*, 2E. © 2013 Nelson Education Ltd. Reproduced by permission. www.cengage.com/permissions

ducts of all the nephrons drain into the ureters, which carry urine from the kidney to the urinary bladder.

As shown in Figure 13.15, during its passage from the Bowman's capsule to the collecting duct, the filtrate goes through several stages of salt and water reabsorption. The first stage occurs in the proximal tubule. For this purpose, Na^+ and Cl^- ions are actively pumped across the epithelium, and water follows by osmosis. In addition, Na^+-dependent co-transporters embedded in the apical membrane of epithelial cells within this region reclaim valuable nutrients that were small enough to pass through the filtration step.

After the proximal tubule, the nephron extends into the medulla of the kidney as the descending limb of the loop of Henle, a region of nephric epithelium that bears many aquaporin channels but no ion channels or transporters. This region is highly permeable to water but not to salt. Furthermore, the osmolarity of interstitial fluid within the medulla progressively increases toward the innermost depths of the medulla. As a result, osmosis progressively draws water out of the filtrate into the interstitial fluid of the medulla as the descending limb penetrates deeper into the increasingly saline environment. Due to this export of water, the tubule fluid at the bottom of the loop is up to four times more concentrated than the fluid in the proximal tubule.

The explanation for the high osmolarity within the interstitial fluid of the medulla is partially provided by a drastic qualitative change in the epithelium of the nephron along the ascending limb of the loop of Henle relative to the descending limb. In the ascending limb, the epithelium has no aquaporin channels, yet the cells do have channels and transporters to actively pump Na^+ and Cl^- ions out of the filtrate (Fig. 13.15). It is these ions that set up the salinity gradient within the interstitial fluid of the medulla. As a result, when the filtrate reaches the top of the ascending limb of the loop of Henle, its osmolarity has returned to a level similar to that of the filtrate in the proximal tubule, but the *volume* of the filtrate has been drastically reduced. This mechanism of volume reduction of the filtrate helps explain the relationship between the lengths of loops of Henle and the availability of environmental water in different groups of mammals. The loop of Henle is relatively short in freshwater animals such as beavers (*Castor canadensis* (1877)), where water is abundant, but considerably longer in desert rodents such as kangaroo rats (*Dipodomys* spp. (1880)) and porpoises (Cetacea: Phocoenidae) that must be better adapted to retain water. The ocean can be thought of as an aquatic desert for marine animals that maintain their body tissues at a lower osmolarity than seawater.

After the loop of Henle, the next two regions of the nephron are the distal convoluted tubule and the collecting duct. Fine tuning of the filtrate's pH and ionic constituents occur within the distal convoluted tubule. The trip through the collecting duct is fundamentally important for allowing mammals to produce hyperosmotic urine (i.e., urine with a higher solute concentration than the blood). The collecting duct carries the filtrate once again down the steep osmolarity gradient of the medulla. Aquaporin channels within the wall of the collecting tubule allow water to flow out of the duct fluid by osmosis. Furthermore, the all-important osmolarity gradient within the medulla is enhanced by additional transport of Na^+ and Cl^- ions across the wall of the collecting duct and also by transport of some of the urea from the duct fluid to the adjacent interstitial fluid. As previously described for elasmobranch fish, urea can act as an osmolyte, and its presence in the inner medulla contributes substantially to the high osmolarity of this region of the kidney. Nevertheless, most urea within the filtrate is conducted to the urethra for eventual discharge from the body.

This account has described how water is reclaimed from the original filtrate, but how does it return to the blood? We need to go back to the glomerulus sitting within Bowman's capsule—the point where fluid was extracted from blood by pressure filtration. The capillaries of the glomerulus coalesce as an efferent arteriole that exits Bowman's capsule. The efferent arteriole then

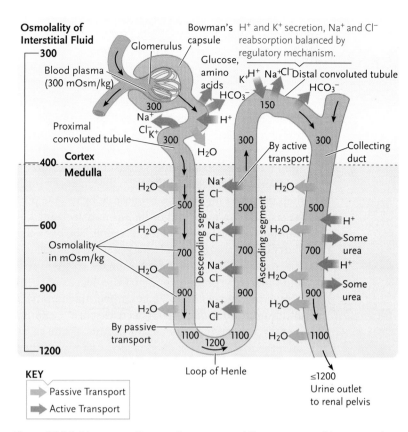

Figure 13.15 Movement of ions and water across different regions of the mammalian nephron (see text for details).

SOURCE: From RUSSELL/WOLFE/HERTZ/STARR. *Biology*, 2E. © 2013 Nelson Education Ltd. Reproduced by permission. www.cengage.com/permissions

subdivides into an interconnected network of capillaries that surround and intertwine with the entire nephron delivering oxygen and food (Fig. 13.14B). Water and ions that are transported from the filtrate within the nephron into the interstitial fluid of the kidney are subsequently taken up by this network of capillaries, although the design is such that an osmolarity gradient remains within the medulla of the kidney.

Although we will not describe the kidneys of nonmammalian vertebrates in any detail, a few points are worth mentioning. Although fish have never become truly terrestrial, the osmoregulatory challenges of life in fresh water are quite different from those of life in salt water. Freshwater fish generate a copious amount of dilute urine because their bodies receive a continuous osmotic influx of water. Marine fish live in the aquatic equivalent of a desert because water continually escapes to their hyperosmotic environment. Marine fish have responded by minimizing the filtration step of kidney function to reduce loss of water through the kidneys. Indeed the kidneys of some marine fish, as well as some amphibians and reptiles that are exposed to dry terrestrial conditions, are aglomerular. The nephrons lack a glomerulus altogether and are reduced to secretory-type excretory organs (much like the Malpighian tubules of insects). Fish gills play a large role for both excretion of ammonia and maintenance of proper ion balance.

Birds are the only non-mammalian group where the nephrons of the kidney have a loop of Henle for concentrating urine. Nevertheless, the bird kidney can produce urine that is only weakly hyperosmotic relative to body tissue. Production of nitrogenous wastes in the form of uric acid, which is secreted across the wall of the nephric tubule, helps to eliminate the need to flush copious water through the bird kidney. Nevertheless, birds further minimize water loss in urine by directing urine delivered from the ureters to an outpocketing of the cloaca (birds lack a urinary bladder). Fluid within this outpocketing is then propelled into the lower intestine by reverse peristalsis, and the water within the urine is reclaimed by transport across the wall of the intestine. The paste-like slurry of uric acid is then discharged with fecal waste.

As a final note, various hormones in mammals and other vertebrates can adjust the degree of urine osmolarity and volume. For example, in mammals, antidiuretic hormone (ADH) decreases urine volume, thereby conserving water. The hormone causes an increase in the number of aquaporin channels within the distal convoluted tubule and collecting tubule so that more water is reabsorbed from the filtrate.

STUDY BREAK

1. What are the similarities and differences between an annelid metanephridium and a nephron of a vertebrate kidney?
2. What is a major difference between the descending and ascending limbs of the loop of Henle with respect to types of molecules transported across the tubule wall in these two regions?
3. What is the functional significance of the network of blood vessels that surrounds the tubule of the nephron?

13.4f Gills: Fish and Crustaceans

For aquatic animals with well-ventilated gills, most nitrogenous wastes are excreted at the gills and flushed by the very large volumes of water that are transported across the gill surface in order for the animal to acquire oxygen (see page 302). This hyperventilation means that in the process of getting oxygen, these animals also can shed CO_2 and NH_3. Although it has long been known that NH_4^+ does not readily penetrate the membranes of gill epithelial cells (the non-polar lipids of the phospholipid bilayer impede passage of these charged ions), it was generally believed that NH_3 readily diffused through membranes. However, the discovery in the early 1990s of possible membrane channels for NH_3 suggested that lipid membranes may not be as permeable to NH_3 as previously presumed. Rather, NH_3 may move through lipid membranes by travelling through channels formed by members of the Rhesus (Rh) family of proteins. We usually think of Rh proteins in relation to blood groups

and immune reactions, but a subgroup of this protein family (Rhag, Rhbg, and Rhcg) has been implicated in the passage of ammonia through cell membranes in both vertebrates and invertebrates (Fig. 13.16). Although Rh proteins have been identified only in eubacteria, unicellular eukaryotes, and animals, they are distant homologues of a family of ammonia transporters in bacteria, yeast, and plants, where they function for the importation of ammonia as a nitrogen source.

In salt-water fish, Rhbg channels embedded in the basolateral membrane of gill epithelial cells allow NH_3 to move into the cytoplasm, whereas Rhcg channels embedded in the apical membrane allow NH_3 to move from cytoplasm into the boundary layer on the external surface of the gill. The boundary layer is acidified so as to convert NH_3 to NH_4^+, a charged form of ammonia that cannot easily diffuse back into gill epithelial cells. Instead, NH_4^+ is carried away in the flow of water that is pumped through the gills. Continuous conversion of NH_3 to NH_4^+ maintains a steep concentration gradient of NH_3 across the gill epithelium. Acidification of the gill boundary layer results from a continuous supply of H^+ to this region. The H^+ is supplied via two routes. First, CO_2 diffuses across the gill epithelium and is hydrated by carbonic anhydrase embedded in the apical cell membrane, generating HCO_3^- and H^+. Second, carbonic anhydrase within the cytoplasm can hydrate cytoplasmic CO_2, and the resulting H^+ is pumped across the apical cell membrane by a hydrogen ion (proton) pump, a process that requires energy from ATP dephosphorylation.

In freshwater fish, NH_3 traverses the basal cell membrane of gill epithelial cells via Rhbg channels, and crosses the apical membrane of these cells via Rhcg channels (Fig. 13.17). This movement occurs along the concentration gradient for NH_3 between blood and water. Nevertheless, additional cellular processes maintain the steepness of the gradient for NH_3 across the gills and thus promote a high rate of ammonia release from the gill surface. Acidification of

NKA (red)
Rhcg1 (green)
DAPI (blue)

25 μm

Figure 13.16 Immunohistochemistry of membrane proteins in the gill of mangrove killifish (*Kryptolebias marmoratus* (1204)) from brackish water. Red label = an antibody against the Rhesus glycoprotein, Rhcg; green label = an antibody against the sodium/potassium pump; blue = a marker for cell nuclei.

SOURCE: Reproduced with permission from Wright, P.A. & Wood, C.M. 2009. "A new paradigm for ammonia excretion in aquatic animals: role of Rhesus (Rh) glycoproteins," *J. Exp. Biol.* 212: 2303–2312. The Company of Biologists.

Patricia Wright

As an undergraduate student at McMaster University in Hamilton, Ontario, Canada, Patricia Wright researched ammonia transport across the gills in rainbow trout. Working under the supervision of Professor Chris Wood, she completed her undergraduate program and then studied for her Ph.D. at the University of British Columbia with Professor Dave Randall as her supervisor. She continued working on gas exchange and ion transport in fish. As a postdoctoral fellow at the University of Ottawa (working with Professors Tom Moon and Steve Perry), she changed her research focus to glucose metabolism, and still later as a postdoctoral fellow at the National Institutes of Health (with Dr. Mark Knepper) she studied metabolism in the mammalian kidney. Today in the Department of Integrative Biology at the University of Guelph, her research focuses on osmoregulation and respiration in aquatic animals.

One aspect of her work considers the challenges faced by amphibious fish—animals that move between aquatic and terrestrial situations. Her research has increased our understanding of the biology of mangrove killifish (*Kryptolebias marmoratus* (1204) Fig. 13.16), which lives in brackish water. The students working under her supervision also study the development of salmonid fish, noting that the prolonged period of development within an egg capsule exposes the fish to potentially high levels of ammonia. It follows that her research also focuses on the mechanisms by which fish detoxify ammonia, and she is one of the leaders in expanding our appreciation of the importance of Rh proteins in ammonia excretion.

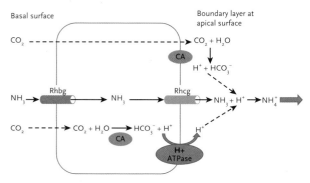

Figure 13.17 Model for the mechanism of ammonia excretion through gill epithelial cells of freshwater teleosts. CA = carbonic anhydrase.

SOURCE: Reproduced with permission from Wright, P.A. & Wood, C.M. 2009. "A new paradigm for ammonia excretion in aquatic animals: role of Rhesus (Rh) glycoproteins," *Journal of Experimental Biology* 212: 2303–2312. The Company of Biologists.

the external boundary layer of freshwater fish gills is important for this process. Boundary layer acidification (requiring an increase in hydrogen ions within the boundary layer) converts NH_3 to NH_4^+, so that the concentration of NH_3 within the boundary layer remains low and the diffusion gradient for ammonia across the gill remains steep. As illustrated in Fig. 13.17, acidification of the external boundary layer of the gill involves carbonic anhydrases and energy-requiring hydrogen ion pumps embedded in the apical membrane of gill epithelial cells. The NH_4^+ that results from these processes is continuously carried away in the gill ventilation current.

In saltwater fish, acidification of the boundary layer of water is less important and less feasible because of the buffering capacity of seawater. Furthermore, in these fish the gill epithelia lack apical proton (H^+) pumps. It is not clear how marine fish maintain a high partial pressure gradient of NH_3 across the gill epithelium.

Decapod crustaceans (Fig. 13.18A) are invertebrates with particularly well-developed gills. The gills are delicate branching structures arising from the base of the legs and are held within specialized gill chambers on either side of the thorax (Fig. 13.18B). Specialized extensions (gill bailers), arising from a pair of feeding appendages, project into the gill chambers of decapods crabs. These gill bailers pump water through the gill chambers. As in fish, the gills of crustaceans are important sites for release of ammonia.

Some crab species burrow into organically rich sediments and remain there for prolonged periods. Ammonia concentration within these sediments may rise to a level similar to that within the hemolymph of the crabs. How crabs continue to excrete ammonia under these conditions is a question begging further research.

13.4g Malpighian Tubules

Insects are generally described as uricotelic because they excrete nitrogenous waste as uric acid to facilitate water conservation. The excretory and osmoregulatory organs of insects are the Malpighian tubules (Fig. 13.19) and the hindgut. The blind distal ends of these epithelial tubules float within the hemolymph of the body cavity (a hemocoel), while the proximal ends empty into the junction between the midgut and hindgut. Unlike nephridial systems, the insect excretory system does not involve an initial filtration step for the extraction of wastes and excess salt ions and water from body fluids. Instead, ions and wastes are actively transported from the hemolymph into the lumen of

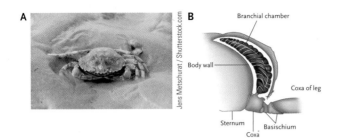

Figure 13.18 A. Brachyuran crab, *Carcinus maenas* (813). **B.** Sketch of a section through one side of a crab thorax showing a gill extending from the base of a leg and contained within a gill chamber formed by lateral overhangs of the carapace.

SOURCE: B. Redrawn after Richard Fox

each Malpighian tubule. The transporting epithelia of Malpighian tubules import Na^+, K^+, and Cl^- ions from the hemolymph into the lumen of the tubules, and water within the hemocoel follows by osmosis. This process involves the action of an H^+-transporting vacuolar ATPase (V-ATPase) on the luminal membranes of tubular cells. The V-ATPase, together with H^+-countertransport, constitutes a common ion transport mechanism that can move Na^+, K^+, or both ions from the tubular cells into the lumen of the Malpighian tubules. In addition, urate, the soluble form of uric acid, is imported from hemolymph into the lumen of the Malpighian tubules. Acidification of the lumen of the Malpighian tubule causes the urate to precipitate as crystalline uric acid. The uric acid, water, and ions within the Malpighian tubules are discharged into the

anterior end of the hindgut. Most of the water is subsequently reabsorbed through the wall of the rectum. As a result, the feces of insects are dry pellets consisting of indigestible material from the gut and uric acid crystals.

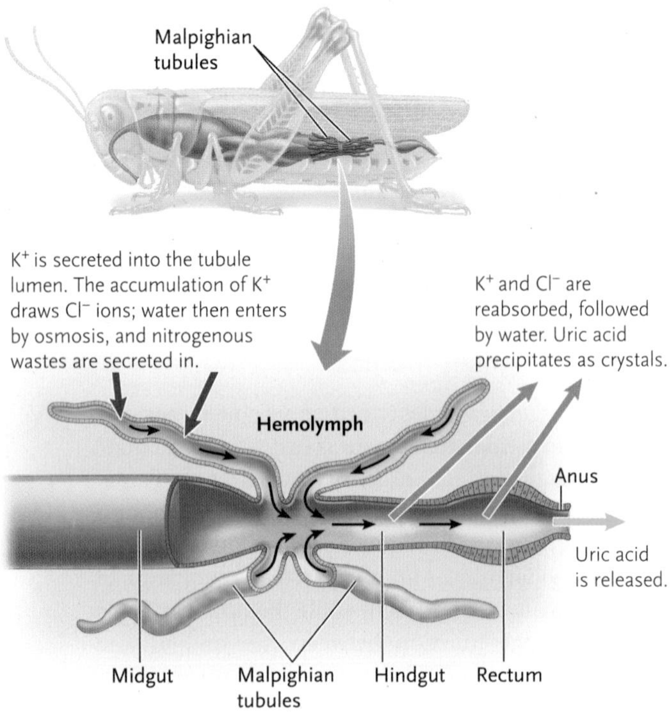

The foregoing account is the traditional description for the workings of a generalized Malpighian tubule/hindgut excretory system in insects. In reality, however, some insects do not excrete uric acid. For example, the American cockroach, *Periplaneta americana* (860), and the mosquito, *Aedes aegypti* (953), excrete ammonia. In the case of the cockroach, ammonia in the feces volatilizes after defecation.

Recent research has indicated that some insects have evolved resistance against various plant defensive chemicals (such as nicotine, caffeine, and morphine) because their Malpighian tubules have acquired novel transport functions. When ingested by an insect, these compounds move from the gut into the hemocoel. The Malpighian tubules then remove these chemicals from the hemolymph, and they are ultimately excreted.

STUDY BREAK

1. Name six types of organs that function for excretion and/or osmoregulation in different groups of animals. Compare the systems you have chosen with respect to where they occur and how they operate.
2. What are podocytes? Where do they occur, and what role do they play in excretion?

13.5 Environmental Settings

Section 13.4 outlined the basic framework of organs and physiological capacities that accomplish excretion and osmoregulation in different groups of animals. However, the history of life has witnessed the successful invasion of a host of novel habitats by both vertebrates and invertebrates, all of which have challenged the capabilities of their ancestral body plans and their physiological capabilities. This section gives examples of how inherited frameworks for excretion and osmoregulation have been embellished or altered to allow survival in different environmental settings. The diversity of these systems is often underestimated.

13.5a Land Crabs

The extraordinary diversification of arthropods has included multiple invasions of both fresh water and terrestrial habitats. Although insects are the ultimate success story on dry land, crustaceans provide an arguably more interesting story about escaping from the sea because living species include transitional forms ranging from partial to full independence from the

K^+ is secreted into the tubule lumen. The accumulation of K^+ draws Cl^- ions; water then enters by osmosis, and nitrogenous wastes are secreted in.

K^+ and Cl^- are reabsorbed, followed by water. Uric acid precipitates as crystals.

Figure 13.19 Malpighian tubules.

SOURCE: From RUSSELL/WOLFE/HERTZ/STARR. *Biology*, 2E. © 2013 Nelson Education Ltd. Reproduced by permission. www.cengage.com/permissions

ocean. By studying crustaceans, biologists can begin to understand how the raw materials of ancestral body plan and physiology have evolved to cope with a radically different habitat.

Christmas Island is a tiny remote island in the Indian Ocean that is a microcosm of terrestrial adaptations by land crabs (Fig. 13.20). Some of the adaptations relating to excretion and osmoregulation are described in this section.

Marine decapod crabs (Arthropoda, Crustacea) excrete ammonia and excess salt ions from their gills, housed within gill chambers on either side of the thorax (Fig. 13.18B). In addition, the antennary glands, which are interpreted as a type of metanephridium, assist in the maintenance of water and salt balance. Recall that aquatic organisms can afford to excrete nitrogenous waste as ammonia because abundant environmental water can dilute this toxin to a tolerable concentration. Many marine crustaceans are isosmotic relative to seawater so the ocean is not a desiccating environment for them as it is for marine teleosts. Water, however, is a luxury on dry land. Certainly a major reason for insect success on land is their excretion of nitrogenous wastes as dry uric acid, thereby conserving water. Remarkably, however, most land crabs on Christmas Island and elsewhere have clung to their ancestral habit of excreting nitrogenous waste from the gills as ammonia. The exception is the coconut (or robber) crab, *Birgus latro* (810). This crab is the largest terrestrial crustacean on the planet and the only known crab that excretes nitrogenous waste in the feces as uric acid.

How have the other land crabs of Christmas Island been able to cope with the toxicity of ammonia in an environment where water is available only occasionally? The Christmas Island blue crab (*Discoplax (Cardisoma) hirtipes* (815)) detoxifies and stores ammonia as glutamine (see Box 13.2). When water becomes available, these crabs convert glutamine back to ammonia and release the ammonia from the gills into water pumped through the gill chambers. The little nipper crab (*Geograpsus grayi* (818)), another land crab from Christmas Island, does not rely on a chance encounter with water to thoroughly flush the gills. Instead, these active carnivorous crabs excrete ammonia into a reservoir of acidified gill chamber fluid where it is volatilized and released as NH₃ gas. It is interesting to note

that other smaller terrestrial crustaceans, isopods (commonly known as woodlice or sowbugs), also use this method of volatilization to excrete ammonia as NH_3 gas.

Although the threat of desiccation and the toxicity of concentrated ammonia are primary challenges faced by land crabs that forage far from the sea, a second challenge comes from the fact that the environmental water they encounter is fresh water. Like all crabs, the urine released from the antennary glands of Christmas Island land crabs is isosmotic to body fluid (hemolymph). During a rainstorm, how do these crabs prevent loss of salts and influx of too much water? The solution comes from an ingenious mechanism to reprocess the urine. The isosmotic urine released from the antennary glands is directed into the gill chambers, where the transporting epithelium of the gill reclaims essential salt ions. The final urine released from the gill chambers may be either hypoosmotic or hyperosmotic, depending on environmental conditions (too much or too little freshwater).

13.5b Fish in Fresh and Salt Water

In Section 13.4f, we mentioned the different osmoregulatory challenges of freshwater teleost fish, which live in a hypoosmotic medium where water constantly enters the fish and ions diffuse outward, versus marine teleost fish, which live in a hyperosmotic medium where water escapes from the fish and salt ions move inward. Freshwater teleosts get rid of excess water by flushing it through their kidneys, resulting in copious hypoosmotic urine. Diffusive loss of salt ions is balanced by active uptake of ions across gill epithelial cells. Marine teleost fish have responded to the need to conserve water by reducing the filtration step of kidney function; some marine fish have lost the glomeruli completely. However, this is not sufficient to meet the dual challenges of desiccation and salt influx faced by these marine fish. To replenish osmotically lost water, marine teleosts drink sea water and distill this within their bodies by actively transporting divalent ions into the nephric tubules and actively excreting monovalent ions from gill epithelial cells.

Marine cartilaginous fish and coelacanths have responded to the physiological challenge of life in saline oceanic water in an entirely different way. As mentioned in Section 13.3b, these fish reduce the osmotic differential between their body tissues and seawater by using urea as an osmolyte. Indeed the blood of cartilaginous fish is slightly hyperosmotic to sea water, so excess water that enters through osmosis must be expelled by the kidneys. However, although urea minimizes the osmotic imbalance between sea water and the tissues of marine cartilaginous fish and coelacanths, it does not change the fact that salt ions are much more concentrated in sea water than in the tissues of these fish—sodium, chloride, and other ions

A **B**

© WaterFrame / Alamy

© FLPA / Alamy

Figure 13.20 Two Christmas Island land crabs: **A.** blue crab (*Discoplax (Cardisoma) hirtipes* (815)); **B.** coconut (robber) crab (*Birgus latro* (810)).

continuously diffuse into their bodies along the electrochemical gradient acting on each ion. Cartilaginous fish deal with the influx of ions from their saltwater habitat by active secretion of monovalent ions by the rectal gland, which dumps brine into the lower intestine for release with the feces.

13.5c Air-Breathing Fish

Fish have occasionally colonized habitats where oxygen in water is severely depleted, such as intertidal mudflats, swamps, or seasonally flooded plains alongside rivers. A few species within 49 families of fish have responded to the challenge of oxygen depletion in water by acquiring the ability to breathe atmospheric oxygen. Although air is a richer source of oxygen than water (even well-aerated water), the delicate gill lamellae of fish collapse when out of water, which greatly reduces the surface area for gas exchange. As described on pages 321–322, air-breathing fish have often evolved alternative body surfaces to exchange oxygen and carbon dioxide with air. Nevertheless, other mechanisms have been recruited to resolve the problem of build-up of toxic ammonia during the periods between immersion, when the gill surfaces are not being flushed with water.

Ammonia build-up to toxic levels is potentially acute for the giant mudskipper (*Periophthalmodon schlosseri* (1279), Fig. 13.21B), because this inhabitant of intertidal mudflats is both active and carnivorous. Studies have found that toxic ammonia concentrations are avoided in this species by intermittent suppression of proteolysis and amino acid catabolism. In addition, some amino acids can be shunted through a pathway that allows partial catabolism. This pathway generates α-ketoglutarate (a substrate for the tricarboxylic acid cycle), but without the production of ammonia as normally occurs. Instead of the deamination of glutamate by glutamate dehydrogenase to produce α-ketoglutarate and ammonia (see Box 13.1), glutamate instead reacts with pyruvate to form α-ketoglutarate and alanine.

Like the giant mudskipper, the slender African lungfish (*Protopterus dolloi* (1357), Fig. 13.21C) is also able to reduce ammonia production when out of water by down-regulating amino acid catabolism, but this fish also detoxifies ammonia by converting it to urea. As described previously, urea is a less toxic molecule than ammonia and can be tolerated at a much higher concentration, but most fish lack the enzymes to make this conversion.

The oriental weatherloach (*Misgurnus anguillicaudatus* (1198), Fig. 13.21A) exhibits a formidable arsenal of strategies to deal with the problem of ammonia build-up when water is not available to flush its gills. This species inhabits the muddy margins of rivers, lakes, ponds, and swamps where it burrows into the mud but keeps its mouth exposed to the surface. It gulps air and respires through a specialized region of its anterior digestive tract. When out of water, this species is able to down-regulate amino acid catabolism and partially catabolize amino acids, as in the mudskipper, but it can also detoxify ammonia by converting it to glutamine, and, remarkably, it can release gaseous ammonia from its skin and digestive tract (Fig. 13.22). For this purpose, both the skin and the gut tissues become alkaline to encourage the conversion of NH_4^+ to the gaseous NH_3. In addition, the oriental weatherloach can withstand remarkably high concentrations of ammonia. At present, the mechanism for this tolerance is entirely unknown.

13.5d Amphibians

Amphibians are traditionally viewed as animals that require a moist habitat as adults because their skin is vulnerable to evaporative water loss. They are also

Figure 13.21 Air-breathing fish: **A.** weather loach (*Misgurnus fossilis* (1198)), **B.** mudskipper (*Periophthalmus barbarus* (1280)), and **C.** slender African lungfish (*Protopterus dolloi* (1357)).

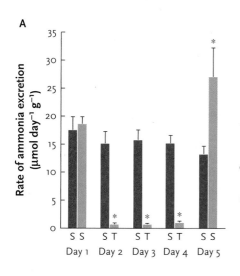

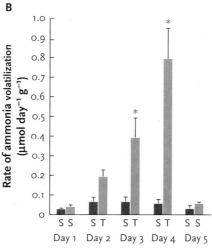

Figure 13.22 Effects of aerial exposure on A. ammonia excretion rate (µmol day^{-1}g^{-1}) and B. NH$_3$ volatilization rate (µmol day^{-1}g^{-1}) of the oriental weatherloach (*Misgurnus anguillicaudatus* (1198)), an air-breathing species of fish. Open columns represent control (no fish). Filled columns represent the experimental condition. S = submerged; T = aerial exposure. Values are means ±S.E.M. (n = 4). *Significantly different from the corresponding control condition, P<0.05.

SOURCE: Reproduced with permission from T. K. N. Tsui, D. J. Randall, S. F. Chew, Y. Jin, J. M. Wilson and Y. K. 2002. "Accumulation of ammonia in the body and NH3 volatilization from alkaline regions of the body surface during ammonia loading and exposure to air in the weather loach Misgurnus anguillicaudatus," *Journal of Experimental Biology* 205, 651–659. The Company of Biologists.

viewed as animals that require a body of fresh water for laying eggs, because their larval stage is aquatic. Although these views describe the situation for many amphibians, the generalizations mask the incredible range of habitats and life history patterns that appear among species in this group. For example, some species of amphibians are fully aquatic during both larval and adult stages, whereas others can survive remarkably dry and hot terrestrial environments, at least during the adult stage. Some species of amphibians can tolerate freezing, and many species, particularly among the caecilians, deposit their eggs in terrestrial nests rather than in water. These differences have major implications for both excretion and osmoregulation, not only because of differences in water availability but also because most adult amphibians are predators and therefore have a diet high in protein.

Excretory products and habitat type for the adults of 19 species of amphibians are listed in Table 13.1. (Two of these amphibians are shown in Fig. 13.23.) Four of the listed species are ammonotelic, because greater than 50% of their nitrogenous waste is excreted as ammonia. Nine others produce 10% or less of their nitrogenous waste as ammonia, relying instead on urea or uric acid. Habitat is a strong predictor of the type of excretory product, with aquatic species being ammonotelic and terrestrial species being ureotelic or even uricotelic. This correlation is entirely consistent with expectation: ammonia excretion is a metabolically inexpensive way to eliminate nitrogenous waste, but is possible only if sufficient environmental water is available to dilute this potentially toxic compound. In dehydrating environments, ammonia must be converted to a less toxic compound that can be excreted in a more concentrated form.

Nevertheless, we cannot conclude that the osmoregulatory challenges faced by adult amphibians living in dry habitats are resolved solely by a switch in excretory product. If water readily evaporates from the skin of amphibians, then water conserved by converting ammonia to urea or uric acid will soon be lost. To address this problem, amphibians living in dry habitats (such as the waxy monkey leaf tree frog (*Phyllomedusa sauvagei*, Fig. 13.23) and other species of tree frogs) secrete waxy lipids from cutaneous glands and use their limbs to wipe this waterproofing layer over the surface of their body.

Still other amphibians avoid evaporative water loss from the skin by retreating to cool, shaded crevices or by burrowing. Some burrowing species live underground in cocoons for part of the year. *Pyxicephalus adspersus* (1379), the African bullfrog, is a good example. These large (up to 1 kg) predators pass long periods of drought in cocoons made from sloughed layers of keratinized stratum corneum (the external layer of the skin). While residing in their underground cocoons, bullfrogs enter a state of dormancy. They are inactive and do not feed, and their rates of metabolism and breathing drop from active levels. Many frogs that become dormant within cocoons during dry periods enter this period of metabolic quiescence with a "water flask" in the form of a water-filled outpocketing of the cloaca (a sort of urinary bladder). Water can be reabsorbed across the wall of the bladder to keep the body hydrated for up to 100 days of dormancy.

Not all frogs manufacture cocoons when they enter a period of dormancy within underground burrows. In this circumstance, the water permeability of the skin, which allows frogs to absorb water when plentiful, becomes a liability when soil begins to dry. The spadefoot toad has been shown to increase the concentration of urea within the blood plasma during prolonged periods of dormancy within underground burrows. The concentration in the plasma depends on the dryness of the soil (Fig. 13.24). Accumulation of plasma urea should promote osmotic movement of water into the frog's body.

In addition to habitat differences among species of adult amphibians, many amphibians experience profoundly different habitats as they progress through their life cycle. The aquatic larvae of the European

Table 13.1

Table 13.1   Types of Nitrogen Excreted by Amphibians.

Order and species	Habitat	% Nitrogen excreted as:		
		NH3	Urea	Urate
Gymnophiona (caecilians)				
Chthonerpeton indistinctum	Soil	17	83	
Caudata (newts & salamanders)				
Necturus maculosus	Freshwater	89	11	
Ambystoma mexicanum	Freshwater	62	34	
Triturus vulgaris	Terrestrial	18	82	
Salamandra salamandra	Terrestrial	5	95	
Anura (frogs & toads)				
Pipa pipa	Freshwater	93	7	
Senupux laevis	Freshwater	62	35	0.2
Rana pipiens	Terrestrial	6	87	0.3
Bufo bufo	Terrestrial	15	85	
Bufo woodhousei	Terrestrial	23	77	
Scaphiopus chouchi	Terrestrial	19	81	
Ceratophys ornate	Terrestrial	7	81	
Hyla regilla	Arboreal	1	98	0.3
Hyperolius nautus	Arboreal	4	96	0.1
Racophorus leucomystax	Arboreal	17	83	
Chiromantis xerampelina*	Arboreal	1-8	20-35	60-75
Agalychinis annae	Arboreal	10	79	11
Phyllomedusa hypochonrialis	Arboreal	5	75	20
Phyllogedusa sauvagei	Arboreal	7	15	78

*Analysis based on % dry weight, not % Nitrogen

SOURCE: Loveridge, J.P. 1993. "Nitrogenous excretion in the Amphibia," pp. 135–144. *Society of Experimental Biology Seminar Series 52: New insights in vertebrate kidney function* (J.A. Brown, R.J. Balment and J.C. Rankin, editors). Reprinted with the permission of Cambridge University Press.

A E. R. Degginger / Photo Researchers, Inc.

B Michael Klenetsky / Shutterstock.com

Figure 13.23 Amphibians. **A.** Common mudpuppy (*Necturus maculosus* (1411)), a species of salamander that lives entirely in freshwater and excretes nitrogen mostly as ammonia; note the external gills. **B.** Waxy monkey leaf frog (*Phyllomedusa sauvagei* (1376)), an arboreal (terrestrial) frog that has waxy, waterproofing skin secretions and excretes nitrogen mostly as uric acid.

bullfrog (*Bufo bufo* (1366)) and the South African sand frog (*Tomopterna* (*Rana*) *delalandii* (1391)) excrete most of their nitrogenous waste as ammonia, but both species switch to excreting urea during metamorphosis to the terrestrial adult stage. Another African frog, *Chiromantis xerampelina* (1368), mates during the rainy season and places the eggs in foam nests over temporary bodies of water. Upon hatching, larvae drop into the water where they excrete mainly ammonia, switching to urea later in development. Newly metamorphosed froglets of this species climb into trees during the dry season and feed on insects. The dry environment and high protein diet of this final life history stage correlates with a switch to uricotelism. Meanwhile, the African clawed frog, *Xenopus laevis* (1393), is aquatic during both larval and adult stages and is ammonotelic during both these stages. Ammonia is released from both the cloaca and the skin. The picture changes again for amphibians that deposit eggs in terrestrial nests. For example, larvae of the white lipped frog, *Leptodactylus bufonius* (1371), which hatch in foam nests deposited in burrows, are ureotelic rather than ammonotelic.

Elsewhere in this chapter we have mentioned that various forms of nitrogenous excretory products can serve additional functions for organisms. Amphibians provide at least two examples, one being the use of urea as an osmolyte as previously described. A second novel function is urea as cryoprotection. In the wood frog,

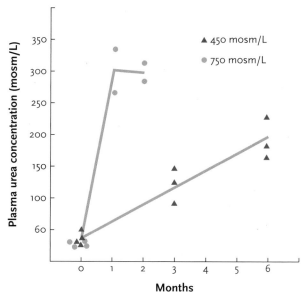

Figure 13.24 Plasma urea concentrations of spadefoot toads (*Scaphiopus couchii* (1387)) kept in soils with different water potentials as a function of number of months in the ground. The higher value (750 mOsm/L, mOsm = milliosmole) indicates a lower water availability in the soil. Toads in the drier soil accumulated urea more rapidly in the plasma and reached a higher total urea concentration.

SOURCE: Republished with permission of AMERICAN SOCIETY OF ICHTHYOLOGISTS AND HERPETOLOGISTS - BIOONE from McClanahan, L. 1972. Changes in body fluids of burrowed spadefoot toads as a function of soil water potential. *Copeia*, Vol 1972, no. 2 pp. 209–216, permission conveyed through Copyright Clearance Center, Inc.

Rana sylvatica, a northern hibernating species, an increase in both glucose and urea within body fluids protects tissues from freezing damage by lowering the freezing point for interstitial fluid.

13.5e Marine Reptiles and Birds

The ocean is home for a vast diversity of marine fish, but also for a number of tetrapods that have returned to the sea from more recent ancestors that lived on land. Although all marine vertebrates maintain their body fluids at a lower osmotic concentration than seawater, only mammals can produce urine more concentrated than seawater and relatively few mammals can produce urine with concentrations of Na^+ and Cl^- equal to the concentration of these ions in seawater. For non-mammalian marine vertebrates, maintaining a stable, hypoosmotic internal environment without access to fresh drinking water requires devices other than the kidneys (extrarenal) to eliminate excess salt. Rectal glands of elasmobranchs secrete salt, and the same is true for the gills of teleosts. Marine birds and other reptiles have various types of salt-secreting glands, usually located on the head. These include nasal glands of marine iguanas and marine birds (Fig. 13.25), glands beneath the tongue of sea snakes and marine crocodiles, and glands located near the eye of marine turtles. These salt-secreting glands quickly excrete salt

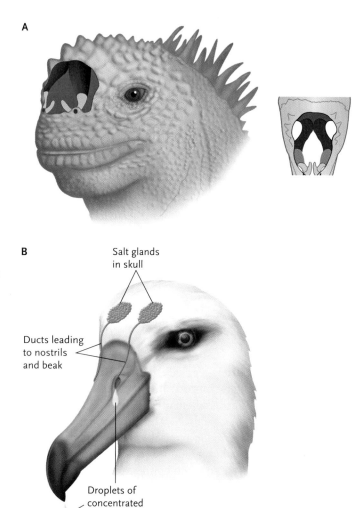

Figure 13.25 Salt glands: **A.** the black areas on the drawing of a Galapagos marine iguana (*Amblyrhynchus cristatus* (1461)) and **B.** in an albatross (*Diomedea* spp. (1633)).

SOURCE: A. Based on WA Dunson, "Electrolyte excretion by the salt gland of the Galapagos marine iguana," *American Journal of Physiology*, Consolidated, Apr 1, 1969; B. From RUSSELL/WOLFE/HERTZ/STARR. *Biology*, 1E. © 2010 Nelson Education Ltd. Reproduced by permission. www.cengage.com/permissions

and allow marine reptiles (including birds) to drink seawater even though their kidneys cannot produce a concentrated urine (Fig. 13.26).

Salt-secreting glands of marine reptiles and birds consist of branched epithelial tubules surrounded by blood vessels. Salt secretion is brought about by the combined action of four intrinsic membrane proteins residing within the membrane of the transporting epithelial cells of the tubules (Fig. 13.27). The entire mechanism is powered by Na^+/K^+ pumps (ATPases), and the high demand for ATP by these pumps explains the dense packing of mitochondria within the tubule cells of the salt gland. In addition, the extra area of cell membrane required to accommodate all the membrane-localized pumps, co-transporters, and channels is provided by extensive folding of the cell membrane of tubule cells. The Na^+/K^+ pump generates a gradient of Na^+ across the cell membrane. $Na^+/2Cl^-/$

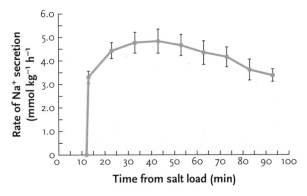

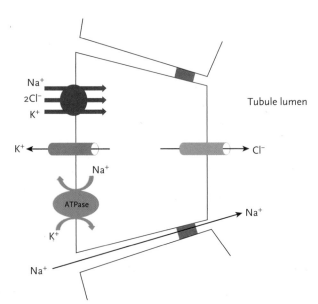

Figure 13.26 Functioning of salt glands in hatchlings of the leatherback sea turtle (*Dermochelys coriacea* (1426)). Researchers collected fluid released from the salt glands at regular time intervals after injecting the animal with a salt load. Data points show mean mass-specific rate of Na$^+$ secretion from 12 turtles ±1 standard error of the mean.

SOURCE: Reproduced with permission from Reina, R.D., T.D. Jones and J.R. Spotila. 2002. "Salt and water regulation by the leatherback sea turtle Dermochelys coriaecea," *Journal of Experimental Biology*, 205:1853–1860. The Company of Biologists.

Figure 13.27 The ion channels and passive and active membrane transporters that allow secretion of sodium chloride by salt-secreting glands of marine vertebrates.

SOURCE: Based on Shuttleworth T.J. and Hildebrandt, J.-P. 1999. "Vertebrate salt glands: short- and long-term regulation of function," *Journal of Experimental Zoology* 283: 689–701.

K$^+$ co-transporters, embedded in the basolateral membrane, facilitate inward diffusion of Na$^+$ along its electrochemical gradient and couple this with inward movement of K$^+$ and Cl$^-$. The Cl$^-$ subsequently exits through Cl$^-$ channels in the apical cell membrane. The electrochemical gradient of chloride—a negatively charged ion—across the epithelium leads to passive efflux of positively charged sodium ions through junctions between the tubule cells.

13.5f Food

Halophytic (salt-loving) plants grow in soils exposed to salty groundwater and can concentrate salt and move it from the soil into the plant. Increased osmotic

pressures in halophyte tissues result from the accumulation of salts and the resulting osmotic gradient allows the plants to draw in water. Worldwide, halophytic plants such as saltbush (*Atriplex*) occupy terrain that is otherwise inaccessible to plants.

Salts, especially sodium chloride, are concentrated within cells at the outer tissues of leaves. These cells eventually burst, depositing crystalline salt on the leaves' surfaces. Crystalline salt is a reflective coating that casts some shade on the underlying leaf tissue but it also

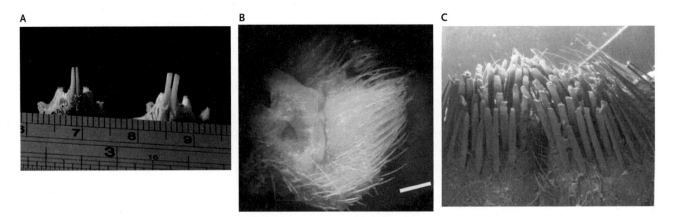

Figure 13.28 Rodent adaptations for eating saltbush. **A.** Anterior view of the lower incisors of two congeneric desert rodents. The left image shows the teeth of saltbush specialist *Dipodomys microps* (1882). These teeth are broader and more chisel-shaped than the incisors of the seed eater, *Dipodomys merriami* (1881), on the right. The chisel-like lower incisors of *D. microps* are used to scrape the salt-filled exterior tissues from saltbrush before eating the inner tissues. **B.** Specialized hair bundles of *Tympanoctomys barrerae* (1090), another species of saltbush-eating rodent. The hair bundles (arrow) are located on each side of the small round mouth, just posterior to the upper incisors. These hair bundles vibrate to strip the salty superficial layer from the saltbush. **C.** Higher magnification of the stiff hairs composing the specialized bundles.

SOURCE: Mares, M.A., R.A. Ojeda, C.E. borghi, C.E. Borghi, S.M. Giannoni, G. B. Diaz, and J.K. Braun. "How desert rodents overcome halophytic plant defenses," *Bioscience* 47:699–704, University of California Press Journals. Copyright © 1997, American Institute of Biological Sciences. .

makes these plants unpalatable to many herbivores. Yet, the succulent green saltbush leaves are an important food source for some herbivores, which eat saltbush leaves despite the salt crust. Three species of rodents from different families eat saltbush: chisel-toothed kangaroo rats (*Dipodomys microps* (1882), Heteromyidae), fat sand rats (*Psammomys obesus* (1900), Muridae), and red viscacha rats (*Tympanoctomys barrerae* (1909), Octodontidae). The chisel-toothed kangaroo rat and the fat sand rat use chisel-shaped lower incisors to strip away the outer leaf tissues laden with salts before eating the inner parts of the leaf (Fig. 13.28A). Red viscacha rats have stiff hairs on either side of the mouth. These bristle brushes work with the lower incisors to strip the salt-filled exterior layers from the leaves (Fig. 13.28B). It turns out that the bristle brush approach is more efficient (1–2 s to remove the salt layer) than the chisel teeth (15–20 s). The kidneys of fat sand rats are specialized to remove high salt loads from the blood and to produce extremely concentrated urine.

STUDY BREAK

1. What do land crabs, adult amphibians, and air-breathing fish have in common with respect to osmoregulation and excretion? How does each of these groups deal with the challenges?
2. What shared problem is faced by rodents that eat saltbush and oceanic birds? How does each of these groups deal with the problem?

EXCRETION **IN PERSPECTIVE**

Excretion takes us from sodium and potassium pumps to osmoregulation and homeostasis. It also involves looking at specialized systems within animals, and provides numerous examples of using natural and potentially toxic products to advantage. Excretion provides clear connections to diet and digestion, to gaseous exchange and to behaviour.

Questions

Self-Test Questions

1. How do animal cells survive the osmotic imbalance created by high concentrations of membrane-impermeable solutes, such as nucleic acids, proteins and carbohydrates, within living cells?
   a. Animal cells have reinforced walls that counteract osmotic pressure with hydrostatic pressure.
   b. The cells of animals lack aquaporins, so polar water molecules cannot freely cross the cell membrane.
   c. Na^+/K^+-ATPase maintains a high concentration of cations outside the cell, which balances the high intracellular concentration of anions, reducing the osmotic gradient.
   d. Active transport of water across the membrane uses cellular energy to maintain osmotic equilibrium.

2. Which of these statements best describes osmoregulation at the macroscale?
   a. Terrestrial animals employ mechanisms to conserve water to counteract the constant threat of desiccation.
   b. Osmoconformers adjust the osmolarity of their body fluids to maintain an osmotic gradient directed inward to retain water.
   c. In freshwater aquatic habitats, animals must contend with diffusional influx of salts into their bodies, as well as an osmotic efflux of water to the outside.
   d. Marine invertebrates require mechanisms to prevent diffusional loss of salt ions and to export water that enters their bodies continuously by osmosis.

3. How do animals control levels of nitrogenous wastes?
   a. Amino acids in excess of what an animal needs for protein synthesis can be channelled into pathways to synthesize glucose for metabolic energy.
   b. Ureotelic organisms convert ammonia to the less toxic urea, which can diffuse across membranes into aquatic habitats.
   c. When it cannot be readily released by diffusion, ammonotelic organisms may convert ammonia to glutamine for safe, temporary storage and transport.
   d. The production of uric acid and guanine allows sequestering of nitrogenous wastes in readily soluble forms that can be excreted.

4. Which functional characteristics are common to all types of nephridia?
   a. Muscular constriction of blood vessels generates a pressure gradient for ultrafiltration of body fluids.
   b. Terminal cells of protonephridia control reabsorption of filtered metabolites.
   c. Podocytes are specialized cells located along the lining of blood vessels in annelids and molluscs and in the glomerulus of the vertebrate nephron, which is important in reabsorption of filtered water and ions.
   d. Pressure-generated ultrafiltration of body fluids across the basal lamina is followed by selective reabsorption of needed solutes and water prior to excretion.

5. Which of these examples best illustrates a successful strategy employed by animals to solve the osmoregulatory challenges of life in a terrestrial environment?
   a. Many terrestrial crustaceans are isosmotic relative to seawater.
   b. Some terrestrial crustaceans store nitrogenous waste by converting ammonia to glutamine until water is available.
   c. Many terrestrial animals have increased surface area to maximize the glomerular filtration rate.
   d. Many species of air-breathing fish resolve the problem of buildup of toxic ammonia by increasing proteolysis and amino acid metabolism.

Questions for Discussion

1. How are excretory products used as osmolytes? What implications does this have for animals that move between marine and freshwater environments?
2. How are excretory products used in communication (see also page 480)? Does this apply equally to terrestrial and aquatic animals?
3. What is the evolutionary significance of the differences in the kidney structures of birds and mammals?
4. Using the electronic library, find a recently published study that provides a novel perspective on excretion and/or excretory products.

Figure 14.1 A syllid polychaete (Annelida) undergoing asexual reproduction (see text). Some members of this family generate gamete-filled reproductive stolons from the posterior end of the body; these swim to the surface and release gametes.

SOURCE: Greg Rouse

14 Reproduction

WHY IT MATTERS

A science writer from Australia recently posted a Blog entry entitled: "My genitals just grew eyes and swam away." Contrary to what you might think, this attention-grabbing title was not the lead-in to a pornographic fantasy. It was an entertaining article about the bizarre reproductive practices of a group of polychaete worms (Annelida) belonging to the family Syllidae. Diversity of reproductive strategies among animals as a whole should not be surprising, considering that reproduction is the ultimate goal of all animals. And they must reach this goal within the context of very different body plans and environmental settings. Nevertheless, the spectacular diversity and seemingly extreme measures that the syllid worms use to accomplish reproduction emphasize the extraordinary outcomes that can result from powerful selection on reproductive biology.

In general, the gametes of polychaetes mature within the spacious coleomic compartments of the adult's body segments and are spawned into the surrounding seawater by exiting through gonoducts or ducts of the metanephridia (excretory organs), or by rupture through the body wall. External fertilization in seawater results in embryos that develop into swimming larvae that eventually settle onto a benthic substrate and undergo metamorphosis to the juvenile stage. The ancestral reproductive strategy among syllids is a variation on this basic polychaete pattern. As spawning season approaches, both sexes of these worms develop enlarged eyes, their locomotory appendages (parapodia) change to become more suitable as swimming paddles, and the segments become swollen with gametes. This sexual transformation is known as epitoky. Under appropriate lunar cues and only after sunset, these otherwise bottom-dwelling worms swim up toward the ocean surface in a behaviour known as swarming. During swarming, males and females release their gametes in a mass orgy close to the ocean's surface. After swarming and spawning, the worms return to life on the bottom.

A more derived type of syllid reproduction involves a more complex morphological transformation, in that only the posterior end of the body, called a stolon or epitoke, undergoes the sexual transformation in morphology. Furthermore, the transformation of the stolon involves not only a change to paddle-shaped parapodia and filling of segments with gametes, but the stolon also develops a pseudohead that bears large eyes but no mouth. An environmental cue triggers the stolon to detach from the remainder of the worm, whereupon it becomes a swimming bag of gametes with eyes that engages in a gamete-releasing swimming dance with other stolons. This reproductive strategy allows the original worm to remain on the bottom, safe from pelagic predators, while only the short-lived stolon engages in the risky reproductive behaviour. Some species of syllids produce not just one stolon, but many stolons are budded from the main body of the worm, either as a sequential series of stolons (Fig. 14.1) or as multiple clusters—all eventually breaking away to swim off to the water surface for a sexual frenzy of gamete release with stolons from other individuals.

In yet another group of syllids, gametes are not released indiscriminately from the swimming stolons. Instead, a male stolon swims rapidly around a female stolon as he releases a strand of mucous laden with sperm. After the female becomes wrapped in the sperm-containing mucus string, she slowly releases her eggs. The eggs are fertilized by the sperm,' and she then retains the fertilized eggs on her body, thereby providing the eggs with protection as the embryos develop.

Although syllids are definitely bizarre in their reproductive practices, they nevertheless must accomplish processes that are basic to the reproductive requirements of all animals: gender specification, gametogenesis, fertilization, and sexual behaviour. This chapter will describe the critically important requirements of reproduction and the various ways that different animals achieve this central goal of all life.

SETTING THE SCENE

Arguably, the drive to reproduce underlies virtually everything in biology, from structure and function to behaviour and genetic control. Aspects of reproduction are central to the process of speciation (and evolution), as well as to understanding the links between ecology, evolution, and fitness. We will consider sexual and asexual reproduction, and aspects of gender determination, both genetic and environmental. The production of gametes is described as well as the processes around fertilization. The timing of reproduction is considered from the standpoints of environmental cues, synchronization of spawning and mating, promoting the survival of offspring, and resolving timing conflicts around reproduction. We also examine some aspects of sexual selection and how it reflects and influences reproduction in animals.

14.1 Reproduction: The Drive to Reproduce

Reproduction is the creation of new individuals from pre-existing individuals. Only "successful organisms" reproduce, meaning organisms that have survived the ruthless filter of a selective environment. The phrase "survival of the fittest" could be rephrased as "reproduction of the fittest," because reproduction is the mechanism by which genes that helped produce a successful organism are propagated to create additional organisms. Seen in this light, it follows that all aspects of an organism's morphology, physiology, and behaviour are gene-directed phenotypic traits that interact and cooperate to ultimately facilitate successful reproduction by the whole organism. Achieving reproduction, which is essentially the transmission of genes to progeny (with or without the combination of genes from two parent organisms), is the ultimate criterion of success in the game of life.

STUDY BREAK

1. Why bother to reproduce?

14.2 Reproducing With and Without Sex

Despite the tremendous diversity of animal reproductive parameters, the traditional approach has been to pigeon-hole all this diversity into only two basic reproductive modes: sexual and asexual reproduction (Fig. 14.2). **Sexual reproduction** is defined as the creation of new individuals, known as **progeny** or **offspring**, from genes of two parent individuals; two genomes combine to generate a novel genome in each

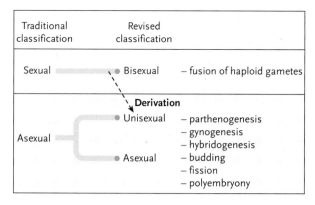

Figure 14.2 Classification schemes for modes of reproduction.

Traditional classification	Revised classification	
Sexual	Bisexual	– fusion of haploid gametes
	Derivation	
Asexual	Unisexual	– parthenogenesis – gynogenesis – hybridogenesis
	Asexual	– budding – fission – polyembryony

offspring. Alternatively, **asexual reproduction** is the production of progeny, new individuals, without a mingling of genes from separate parents.

Alas, nature abhors being pigeon-holed into rigid categories. Evolution usually proceeds through transitional forms that often blur the boundaries between categories. Furthermore, evolution within different lineages may follow very different routes to reach similar endpoints. Not surprisingly, there are inevitable exceptions and grey zones for almost every categorization scheme, and categories of reproductive strategies are no exception. In particular, cases of animal reproduction that do not involve the combination of genes from separate parents (which have traditionally been lumped together as asexual reproduction) include situations that have arisen from different starting points and proceed by different mechanisms. For this reason, it is appropriate to revise the traditional "asexual" category by subdividing it into two categories. Under this revised scheme (Fig. 14.2), the term **bisexual reproduction** refers to the familiar case of two sexes producing haploid gametes by the process of meiosis, with the haploid gametes fusing during fertilization to form a diploid **zygote** (fertilized ovum). The term **asexual reproduction** is confined to those cases where new individuals are produced from diploid somatic cells of a pre-existing individual through processes such as budding and fission. **Unisexual reproduction** is a form of reproduction that is clearly derived from bisexual ancestors, but no longer delivers genes from two parents into offspring (or at least not into the offspring of offspring—see the description of hybridogenesis in Section 14.2c). During unisexual reproduction, ova (egg cells) are produced within a female gonad, but only female genes are passed on during reproduction. These three modes of reproduction are described in further detail below.

STUDY BREAK

1. How does sexual reproduction differ from asexual reproduction?

14.2a Asexual Reproduction

Most people are aware that sexual mating does not always result in reproduction, but the notion of reproduction without sex is foreign to us—with the exception of twinning. **Polyembryony** occurs when an embryo developing from a single fertilized egg separates into two or more embryos. For example, monozygotic (identical) twins occur when an embryo developing from a single fertilized egg divides to form two genetically identical offspring, so the event is not independent of bisexual reproduction. The southern long-nosed armadillo (*Dasypus hybridus* (1826), Fig. 14.3) holds the record for twinning virtuosity among mammals, producing up to 12 offspring from a single fertilized egg.

Polyembryony also occurs routinely within the life cycle of some species of parasites, where it functions to amplify the number of offspring produced (see page 290). A good example is provided by trematode platyhelminths, in which two of the larval stages exhibit polyembryony. By greatly increasing the number of individuals generated from a single fertilization event, polyembryony improves the chance that at least one copy of a genome will complete the complex life cycle of these parasitic organisms.

Although reproduction without sexual mingling of genes is rare among mammals, well over half of the 30+ phyla of multicellular animals include at least some species that routinely create new individuals without sexual partners. This type of reproduction is common among some phyla of invertebrates, but it also occurs occasionally in a few species of fish, amphibians, and reptiles. Nevertheless, all described cases of vertebrate reproduction without the contribution of genes from two parents are examples of

Figure 14.3 Female armadillos produce a single egg, which, after fertilization, develops into several (4–12) genetically identical young. This makes these armadillos polyembryonic and thus unique among vertebrates. This is a nine-banded armadillo (*Dasypus novemcinctus* (1827).

unisexual reproduction (see Section 14.2c). We must look to the invertebrate phyla and to the invertebrate chordates within the subphylum Urochordata for examples of reproduction that does not involve sex and cannot be traced to an origin from bisexual ancestors.

The most common forms of asexual reproduction are budding and fission. **Budding** occurs when a new individual develops from somatic cells of a pre-existing animal's body—usually beginning as a protuberance of proliferating cells that eventually differentiate into all components of an entire organism. Buds may detach from the parent's body and assume independent lives. For example, cnidarians belonging to the genus *Hydra* (135) periodically grow a bud from the side of the body column. The bud differentiates a set of feeding tentacles and eventually drops off the parent to become a completely separate organism (Fig. 14.4A). Budding also occurs in some species of fan worms, a family of suspension-feeding polychaete annelids that live in proteinaceous tubes that they secrete (Fig. 14.4B).

In many cases, an asexual bud does not completely separate from the pre-existing individual, and the bud goes on to produce additional buds that also fail to completely separate. The result is the formation of a **colony** of interconnected individuals, each individual being genetically identical to the original founder individual. Colonies are formed by many species of cnidarians and urochordates (Chordata, Urochordata) and all species within the phylum Bryozoa ("moss animalcules," Fig.14.5). Colonies are particularly common among invertebrates that live permanently attached to a substrate, particularly in places such as the intertidal

Figure 14.5 Colony of a bryozoan (*Membranipora membranacea* (284)). New zooids bud from pre-existing zooids along the periphery of the colony.

and shallow subtidal zones of marine coastlines where competition for living space can be intense. A colony allows the genome of a single founder individual to occupy more living space and access more resources. Furthermore, although predators may eat part or even most of a colony, the genome of the individual that originally established the colony persists, provided that a few colony members survive the attack.

Fission is a form of asexual reproduction in which an animal separates into two or more pieces, with each piece subsequently regenerating any missing part(s). Fission by the starlet sea anemone, *Nematostella vectensis* (173), is preceded by one or more constrictions of the body column where the animal's body will completely cleave apart (Fig. 14.6A). After fission, the fragment with the oral tentacles is able to feed and remains fully functional, while each of the other fragments quickly regenerates new tentacles and soon also becomes capable of feeding.

Asexual reproduction by fission also occurs in some species of sea stars. A single arm detaches from a sea star body and proceeds to regenerate a full set of arms (Fig. 14.6B). Sea stars that are regenerating arms are called "comets," because of their superficial similarity to celestial comets.

A curious but highly functional case of asexual reproduction occurs in many freshwater sponges living in temperate latitudes. As winter approaches, clusters of a type of sponge stem cell accumulate

Figure 14.4 Asexual budding. **A.** Stages in the asexual budding of *Hydra* (135) (phylum Cnidaria). **B.** An asexual bud growing on a sabellid polychaete.

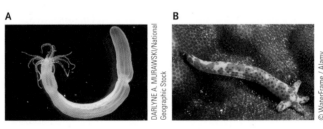

Figure 14.6 A. The sea anemone *Nematostella vectensis* (173) (Cnidaria) with several transverse fission planes. **B.** Comet stage of the sea star *Linckia multifora* (1075) (Echinodermata); four new arms are regenerating from a single arm released from a pre-existing individual.

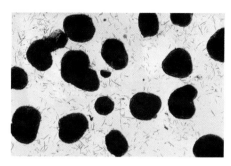

Figure 14.7 Sponge gemmule (phylum Porifera).

SOURCE: Carolina Biological Supply Co/Visuals Unlimited, Inc.

within the body of the sponge, and a tough protein-aceous capsule with embedded spicules is secreted around each cellular cluster. These are called **gemmules** or overwintering bodies (Fig. 14.7). With the onset of winter, the sponge dies, and the gemmules are released from its disintegrating body. Unlike the parent sponge, gemmules can survive freezing winter temperatures. When spring returns, the gemmule capsule ruptures, and the cells flow out to form a new sponge body.

14.2b Bisexual Reproduction

Most species of multicellular animals generate offspring through bisexual reproduction, even if many of these same species are also capable of asexual or unisexual reproduction. Bisexual reproduction involves fusion between a male and female gamete (sperm and ova, respectively), which are produced by the process of meiosis. Meiosis reduces a diploid set of chromosomes to four sets of haploid chromosomes. Fusion of a sperm and an ovum (fertilization) re-establishes the diploid state. Without meiosis, the number of nuclear chromosomes would double with each new generation of fertilized ova—hardly a tenable situation.

Sexual reproduction combines genetic material from two adult animals to produce offspring that have a genome derived from, but different from, the genome of the individual parents. Two reasons explain this difference. The most obvious reason is that the offspring contains a combined subset of the two parents' chromosomes. The less obvious second reason emerges from the process of haploid gamete formation, because the process of meiosis can produce a novel mix of gene alleles on individual chromosomes.

As illustrated in Fig. 14.8, something remarkable can happen during the transient tetrad stage of meiosis: interconnected chromatids can exchange equivalent pieces, a process known as **chromosomal crossover** (a form of genetic recombination). Crossover during meiosis produces hybrids of the parent's homologous chromosomal pairs. The individual chromosomes of each tetrad then become segregated during the two rounds of cytoplasmic divisions to produce four haploid cells, the future gametes.

Genetic differences between offspring and parents and between sibling offspring is the raw material for selection. Collectively, large populations of sexually reproducing animals have a large amount of genetic variation within their common gene pool, because the occasional gene mutations that arise, particularly neutral and advantageous mutations, will gradually become distributed throughout the population by chromosomal crossover and interbreeding. The high degree of genetic heterozygosity of large interbreeding populations can provide the genetic variability that may allow these populations to respond to changes in environmental parameters, such as those caused by climate change or the arrival of invasive species.

The cheetah (*Acinonyx jubatus* (1705)) is an example of a species with limited genetic heterozygosity (Fig. 14.9). This may be evidence of a time in its history when the gene pool of the species was limited—a phenomenon known as a population or genetic bottleneck. In the case of cheetahs, lack of genetic variability may make the species more vulnerable to extinction in the face of environmental change.

14.2c Unisexual Reproduction

Both asexual and unisexual reproduction are forms of propagation that do not involve the combination of genetic material from two parents. However, unisexual reproduction is derived from an ancestral, bisexual reproductive mode. The phenomenon has arisen independently many times among animals and may be accomplished by a variety of different cytological mechanisms, all involving a modification of meiosis during the maturation of ova. The progeny of unisexual species originate as ova within a female's ovaries, and the ova are provisioned with yolk protein, mRNA transcripts, and other materials necessary for the development of the future embryo. Unisexual reproduction is therefore very different from the processes of asexual propagation via somatic tissue, as occurs during budding and fission.

Figure 14.10 illustrates three common forms of unisexual reproduction—parthenogenesis, gynogenesis, and hybridogenesis—and compares them to bisexual reproduction. We will first describe these three forms of unisexual reproduction and the cytological mechanisms involved, and then we will provide specific examples of each situation.

Parthenogenesis is development of eggs without any requirement for sperm. The eggs are diploid and contain genes from the mother only. **Gynogenesis** also involves the production of diploid eggs, but these eggs require fusion with a sperm to activate development. Nevertheless, the sperm chromosomes do not combine with the egg chromosomes. Typically, the obliging males that provide sperm for activation of gynogenetic eggs come from one of the two "parental" species that originally hybridized to produce the gynogenetic lineage.

A Meiosis I

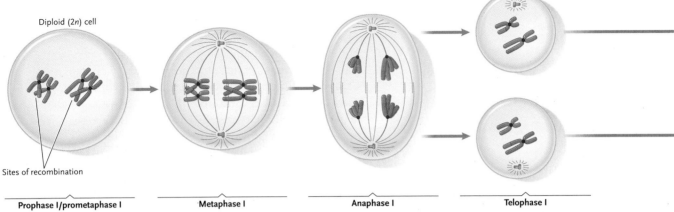

Diploid (2n) cell

Sites of recombination

Prophase I/prometaphase I	Metaphase I	Anaphase I	Telophase I
Duplicated chromosomes condense. Homologous chromosomes pair and exchange segments by recombination. Chromosomes attach to spindle in homologous pairs.	Each maternal chromosome (a pair of sister chromatids) and its paternal homologue align randomly at the spindle midpoint.	Homologous chromosomes, each as a pair of sister chromatids, separate and move to opposite poles.	Two haploid (n) nuclei form.

B Mitosis

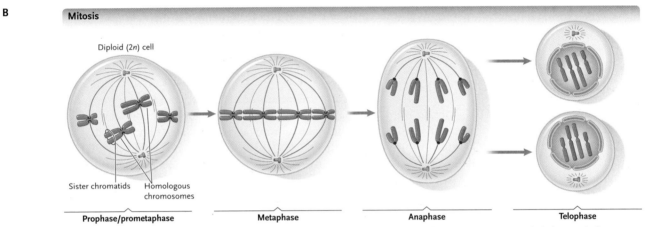

Diploid (2n) cell

Sister chromatids   Homologous chromosomes

Prophase/prometaphase	Metaphase	Anaphase	Telophase
Duplicated chromosomes (sister chromatids) become visible and attach to developing spindle.	Duplicate chromosomes line up individually at the spindle midpoint.	Sister chromatids of each chromosome move to opposite spindle poles.	Two diploid (2n) nuclei form. Cytokinesis produces two daughter diploid cells.

Meiosis II

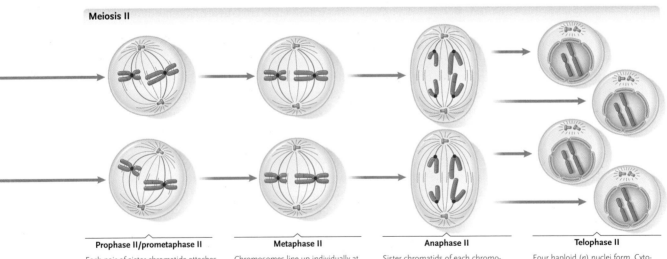

Prophase II/prometaphase II	Metaphase II	Anaphase II	Telophase II
Each pair of sister chromatids attaches to newly formed spindle.	Chromosomes line up individually at the spindle midpoint.	Sister chromatids of each chromosome move to opposite poles.	Four haploid (n) nuclei form. Cytokinesis produces four haploid cells.

Figure 14.8 Comparison of mitosis and meiosis. The starting cell in each case is shown with two sets of homologous chromosomes; the different homologues are distinguished by colour. **A.** Mitosis. **B.** Meiosis; the two stages distinguished as Meiosis I and II. See also page 80.

SOURCE: From RUSSELL/WOLFE/HERTZ/STARR. *Biology*, 2E. © 2013 Nelson Education Ltd. Reproduced by permission. www.cengage.com/permissions

Figure 14.9 Cheetah (*Acinonyx jubatus* (1705)), a species that shows very little genetic heterozygosity.

Hybridogenesis is an extremely interesting case that has been described as a type of "sexual parasitism." Unlike the other two forms of unisexual reproduction, ova produced by hybridogenetic species are haploid. These ova are fertilized by a male from one of the parental species that produced the hybrid lineage. The sperm activates the egg and becomes incorporated into the genome of the fertilized ovum. However, when offspring resulting from these fertilizations become adults and produce their own ova, the haploid condition is achieved not by meiosis but by jettisoning the chromosomes received from the father! Therefore, male genes are used for somatic processes by offspring of hybridogenetic forms, but the male genes are not allowed to enter the next generation of eggs.

As mentioned previously, both parthenogenesis and gynogenesis involve production of diploid eggs. Diploid eggs can result from a variety of possible mechanisms. The most common mechanism among

invertebrates is **apomixis**, the total arrest of meiosis during egg production. The second option is **automixis**, in which meiosis occurs but with appropriate modifications or embellishments that result in a diploid final condition. In some cases, chromosomes in early oocytes undergo a pre-meiotic duplication to produce a tetraploid condition, so that the reductive divisions of meiosis restore a diploid condition. Alternatively, the ovum completes meiosis and then fuses with one of the other haploid cellular products of oogenic meiosis—the so-called **polar bodies** (see Section 14.4c)—to produce a diploid chromosome number.

In summary, unisexual reproduction can involve:

1. parthenogenesis, in which diploid ova are produced by apomixis or automixis
2. gynogenesis, in which diploid ova may also be produced by either apomixis or automixis
3. hybridogenesis.

Apomictic parthenogenesis is the most common form of unisexual reproduction among invertebrates, such as water fleas (*Daphnia pulex* (720), Fig. 14.11A) and many species of aphids (Fig. 14.11B) and rotifers (which reproduce sexually at other times of the year). The latter are small, mostly freshwater-inhabiting invertebrates. Although males may appear seasonally within populations of some of these groups, males are completely unknown for all species of bdelloid rotifers.

An invertebrate with no males is one thing, but surely this situation could never exist for a vertebrate! Remarkably, parthenogenesis does occur naturally in a number of reptiles and fish species. A well-studied example is the desert grassland whiptail lizard (*Aspidoscelis* (formerly *Cnemidophorus*) *uniparens* (1470), Fig. 14.12), a species produced by hybridization between two closely related, bisexual whiptail lizards. The cytological mechanism underlying parthenogenesis in this lizard is a type of automixis. The chromosomes duplicate before the oocytes undergo meiosis so the resulting ovum is diploid.

Remarkably, although males are not needed to either activate or contribute genetic material to mature ova of desert whiptail lizards, females that have finished laying a clutch of eggs will perform the stereotypical

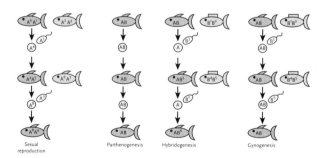

Figure 14.10 Reproductive modes. Capital letters depict genomes, different letters indicate different species, and numbers (0–9) differentiate individual genomes derived from recombination (crossover). In bisexual reproduction, both mating partners produce haploid germ cells that are unique due to recombination and result in highly variable offspring. In parthenogenesis, females produce diploid oocytes that are genetically identical to the mother. In hybridogenesis, the female genome passes through generations unchanged while the male genome is exchanged every generation. In gynogenesis, females produce diploid oocytes that develop into all female offspring. However, they need sperm from a closely related bisexual species to trigger the onset of embryonic development. The male genetic material does not contribute to the offsprings' genotypes.

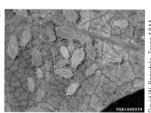

Figure 14.11 Two examples of invertebrates that reproduce by parthenogenesis. **A.** Water flea (*Daphnia pulex* (720)); note parthenogenetic eggs (arrow) being brooded within her body. **B.** Soybean aphid (*Aphis glycines* (866)).

Figure 14.12 Desert whiptail lizard (*Aspidoscelis uniparens* (1470)); a vertebrate that lacks males and reproduces by parthenogenesis.

"male" mating behaviour toward females that have not yet laid eggs. The behaviour appears to activate release of reproductive hormones within the mounted females. These hormones promote ovulation so that the process of females mounting other females results in production of more egg clutches during the reproductive season. The advantage of this pseudocopulation for the individual performing the mounting behaviour is obvious when you realize that this species of obligate parthegenotes consists of a single genome throughout their range. Behaviours that promote increased production of offspring by any one individual serve to propagate the genome of all individuals.

The best studied example of apomictic gynogenesis is a fish, the Amazon molly (*Poecilia formosa* (1207)), another case of a unisexual lineage arising from hybridization between two closely related bisexual species (Fig. 14.13). The diploid eggs of the Amazon molly are produced by arrest of meiosis (apomixis), but they cannot develop without activation by sperm from the sailfin molly (*P. latipinna* (1208)), which is one of the parental species that produced the Amazon molly. The Amazon molly's dependence on the sailfin molly for sperm means that the Amazon molly cannot stray outside the distributional boundaries of the sailfin molly.

An interesting comparison to the Amazon molly is the apomictic gynogenesis exhibited by at least one species within the bivalve genus, *Lasaea* (Mollusca). A subspecies of *Lasaea subviridis* (548) from the west coast of Canada is a simultaneous hermaphrodite. The gonad is an ovotestis that produces mostly ova but also a small number of sperm cells. The diploid eggs of these bivalves require sperm activation, but unlike the gynogenesis in the Amazon molly, the activating sperm for

the eggs of *L. subviridis* come from the animal's own ovotestis! This is not a case of self-fertilization, but rather a case of "self-activation" of diploid eggs. After sperm-egg fusion, sperm nuclei degenerate within the egg cytoplasm.

At present, all known cases of hybridogenesis are restricted to a few species of fish and amphibians, including a hybrid form of European water frog. The hybrid, *Rana esculenta* (1382), results from cross-fertilizations between *Rana ridibunda* (1384) and *Rana lessonae* (1383). The hybrid offspring. carry chromosomes from both parents, but when females of *R. esculenta* produce eggs, they discard the chromosomes from *R. lessonae*. These eggs must then be fertilized by sperm from one of the two parental species to reconstitute the *R. esculenta* genotype. Apparently, *R. esculenta* is competitively superior to both parental species because it tolerates a wider range of physical conditions. Indeed, *R. esculenta* often outnumbers either of the parental species where the two co-occur. However, the dependence of *R. esculenta* on sperm from one of the parental species means that the hybrid form can never totally displace either of the parental species. Incidentally, hybrid males of *R. esculenta* are actively avoided as mates by both *R. esculenta* and *R. ridibunda*, and, when matings do occur with males of *R. esculenta*, the fertilized eggs have low viability.

14.2d Why Bother With Sex?

Why don't asexual and unisexual species overwhelm and displace sexual species? A simple mathematical calculation reveals the relevance of this question. A female of a bisexual species will produce a given number of offspring of which, on average, half will be male and half female. Only her female offspring will go on to produce subsequent broods of female offspring, of which only half will be female. Compare this to unisexual reproduction. Here all of a female's energy allocation for reproduction is channelled into female offspring because males are not needed. She should be able to generate twice as many egg-producing offspring relative to the bisexual species. All things being equal, unisexuals should double their output of progeny-producing offspring, relative to a bisexual species, with each generation. Furthermore, an individual's genes are not diluted by the genes of the sexual partner at each generation.

However, despite the fact that unisexual species have twice the reproductive potential of bisexual species, the latter is overwhelmingly more common among animals. This situation is usually justified by two arguments. First, offspring resulting from bisexual reproduction represent a greater diversity of genetic material presented to an unpredictable environment (such as a freak storm or virulent disease). One hypothesis even suggests that bisexuality originated in response to increasing threat from pathogens,

Figure 14.13 Amazon molly (*Poecilia formosa* (1207)), a fish that reproduces by apomictic gynogenesis.

because genetic recombination during meiosis and mingling of genes from two parents allows host defenses to evolve faster than mechanisms to overcome those defenses by asexually reproducing pathogens. Second, bisexual reproduction provides an avenue for ridding the genome of "mutation load"—accumulated genetic defects caused by environmental mutagens and errors in gene duplication. Accumulation of these mutations among unisexual species should eventually lead to a decline in fitness. The fact that most unisexual lineages are very young (recently evolved) is consistent with the expectation of higher extinction and lower speciation rates for unisexuals relative to bisexual species; their relatively low genetic heterozygosity compromises their ability to respond to environmental change.

Despite these arguments for the long-term superiority of bisexual reproduction, asexual and unisexual reproduction has been remarkably successful under certain circumstances. For example, by eliminating sex, small animals such as water fleas and aphids with short generation times and seasonally short-lived habitats can more rapidly convert available resources into many offspring while conditions are good. Nevertheless, most water fleas, aphids, and rotifers have a phase of bisexual reproduction in their annual life histories. Justifying unisexual reproduction in one particular group of rotifers, the bdelloid rotifers, has been much more challenging because males are completely unknown for all species within this group and bisexual reproduction never occurs. Despite this, bdelloid rotifers have been identified in amber dating from 30–40 million years ago. During that time, they have diversified into more than 300 species that collectively inhabit every possible type of freshwater habitat. Evolutionary theory and population genetics have yet to provide a satisfactory explanation for the long-term persistence and speciation of bdelloid rotifers, despite the lack of opportunity for genetic recombination.

STUDY BREAK

1. Give two examples of asexual reproduction. For each of your examples, provide an explanation for a possible functional value of this reproductive strategy.
2. Describe three mechanisms for producing diploid eggs in species that show unisexual reproduction.
3. Justify the long-term evolutionary value of bisexual reproduction.

14.3 Determining Gender

Most bisexual animals are **gonochoristic** (also known as **dioecious**), meaning that individuals are either male or female. However, a considerable number of animal

Figure 14.14 Two colour morphs of the sea slug (Mollusca, Gastropoda, Opisthobranchia). The gonad of these animals is an ovotestis and they are hermaphroditic.

species (many invertebrates and a few vertebrates) are **hermaphroditic** (also known as **monoecious**), meaning that every individual is capable of producing both eggs and sperm within a gonad known as an **ovotestis**. In some cases, both the sperm and eggs of hermaphrodites are produced simultaneously. The often beautiful sea slugs (Mollusca, Gastropoda, Heterobranchia, Fig. 14.14), as well as acorn barnacles (Arthropoda, Crustacea, Cirripedia), and both earthworms and leeches (Annelida, Clitellata), are examples of simultaneous hermaphrodites. During mating between simultaneous hermaphrodites, gametes are exchanged simultaneously or one partner performs the male role while the other is the functional female and then the two reverse these roles. In other cases, hermaphrodites spend part of their lives as one gender before switching to the other. Another gastropod mollusc, the slipper limpet *Crepidula fornicata*, is a functional male at small body size, but gradually switches to egg production as it grows larger. The remainder of this section on determining gender deals with dioecious animals.

Sexual phenotype refers to all the various characteristics of bisexually reproducing animals that are expressed differentially in males and females. These are traits associated with gender. The **primary sex characteristic** is the type of gonad: testes producing sperm or ovaries producing ova. **Secondary sex characteristics** are the extra-gonadal traits that distinguish males from females. They may be any aspect of phenotype—cytological, morphological, physiological, or behavioural—that differs between the two genders. Gender-specific secondary sex characteristics are minimal or absent altogether in some species, such as sea urchins and sea stars (Echinodermata) that broadcast spawn their gametes into surrounding seawater where fertilization occurs (Fig. 14.15A). However, species in which eggs are internally fertilized generally show some if not many secondary sex characteristics. At the very least, there are differences in male and female reproductive tracts, the organ systems directly involved with transport and delivery of gametes and with egg fertilization. Many animals with internal fertilization also show

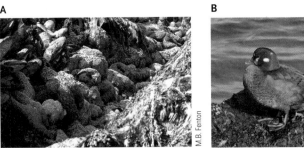

Figure 14.15 **A.** Sea stars. **B.** Male and female Harlequin Ducks (*Histrionicus histrionicus*) have different plumage colours (male on right).

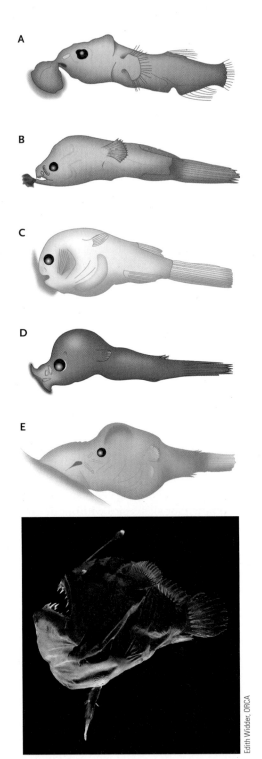

pronounced differences in many other body traits of males and females. Species that exhibit externally visible differences in male and female morphology or behaviour are termed **sexually dimorphic** (Fig. 14.15B). The differences may involve such things as body size, ornaments, weaponry, vocalizations, or chemical signals. These traits enhance an animal's reproductive fitness by, for example, advertising sexual availability and quality, facilitating courtship, or competing for mates. Sexually dimorphic traits are the evolutionary result of **sexual selection**, which will be discussed in Section 14.7.

Deep sea anglerfish (Lophiiformes, Ceratioidei) (Fig. 14.16) illustrate an extreme case of sexual dimorphism. The differences between anglerfish males and females are ultimately valuable for promoting reproductive success, but not by increasing an individual's attractiveness to a potential mate or improving competitiveness among sexual rivals. Instead, the sexual dimorphism of anglerfish probably relates to the difficulty of finding mates and the low food abundance in their deep ocean environment. Male anglerfish are typically much smaller than females (**dwarf males**), and in most species they are permanently attached to the female's body. Head tissues of the males often grow into and fuse with the body wall of the female (Fig. 14.16), and the female circulatory system may even become confluent with that of the male. The male essentially becomes a sperm-producing appendage on the female's body. With body lengths of 6.7–7.4 mm, male anglerfish are among the smallest of vertebrates and live on females that may be eight times larger. Disparity in size between males and females may reflect limited energy resources in the habitat. By linking a male to the body of a female, both physically and metabolically, resources acquired by a mating pair can be preferentially channelled into producing energy-expensive eggs, while relatively few resources need to be allocated to the production of inexpensive sperm. By sacrificing his freedom upon finding a female, the male anglerfish loses the opportunity to find and mate with other females; but because mates are so difficult to find in this habitat, the better strategy may be to

Figure 14.16 Anglerfish: *Melanocetus johnsonii*, (1226), a 75-mm-long female with a 23.5-mm-long attached male. Also shown are parasitic males of other species attached to females; **A.** *Caulophryne* (1221), **B.** *Borophryne apogon* (1220), **C.** *Haplophryne mollis* (1224), **D.** *Linophryne agyresca* (1225), and **E.** *Photocorynus spiniceps* (1228).

SOURCE: Based on Pietsch, T.W. 2005. "Dimorphism, parasitism, and sex revisited: modes of reproduction among deep-sea ceratioid anglerfishes (Teleostei: Lophiiformes)," *Ichthyological Res.* 52: 207–236.

remain with a single female once she is located. Although anglerfish are well known for their use of bioluminescent lures to attract their prey, they should perhaps be better known for their reproductive strategies.

14.3a Chromosomal Sex Determination

The mechanism for specifying gender is simply unknown for most animals. However, what is known indicates that mechanisms for gender specification can differ considerably among different animal groups, so sweeping generalizations about gender specification are dangerous. Nevertheless, for many animals gender is specified by genes located on special **sex chromosomes**. Two systems for chromosomal sex determination will be described: the XY system used by most mammals and the ZW system used by many reptiles, including birds. A third system, the haplodiploid system of social insects, is one in which females are diploid individuals and males are haploid.

Mammalian gender is determined genetically at the time of egg fertilization, although the actual differentiation of the male or female phenotype depends on gene-directed hormonal signalling during development. Mammals inherit two **heteromorphic chromosomes** (sex chromosomes) from their parents. The two possible types, X and Y, are morphologically distinct. Females have two X chromosomes, and males have an X and a Y. Gender specification in mammals is entirely dependent on a group of genes that reside on the short arm of the Y chromosome. Although the details of how these genes and their products interact to produce maleness have not been completely resolved, at least one of the genes is the *SRY* gene, an acronym for "sex determining region on the Y chromosome." If the short arm of the Y chromosome is present in an embryo, then testes will differentiate. If only X chromosomes are present or if neither of the sex chromosomes is present, the gonad will become an ovary. Femaleness is the default sexual phenotype among mammals.

When primordial germ cells first differentiate into the future gonads of a developing mammalian embryo (see Section 14.4a), they are not yet committed to a fate of sperm or ova production. Similarly, the rudimentary urogenital system has not committed to becoming either a male or female system (Fig. 14.17). However, if the embryonic gonad receives a signal to become a testis (a signal that depends on the expression of critical genes on the short arm of the Y chromosome), mesenchyme cells within the gonad begin to secrete two types of hormones that will specify primary and secondary characteristics of the male phenotype. Anti-Müllerian duct hormone (AMH), which is secreted by Sertoli cells of the testes, causes degeneration of the Müllerian ducts and differentiation of the Wolffian ducts into vas deferens (Fig. 14.17). AMH also inhibits development of mammary glands. Testosterone secreted by Leydig cells of the testes directs development of sperm and many secondary sex characteristics of the male phenotype, including external genitalia. A developing mammalian gonad that does not receive a specific signal to become a testis and secrete androgens will, by default, become an ovary. Follicle cells

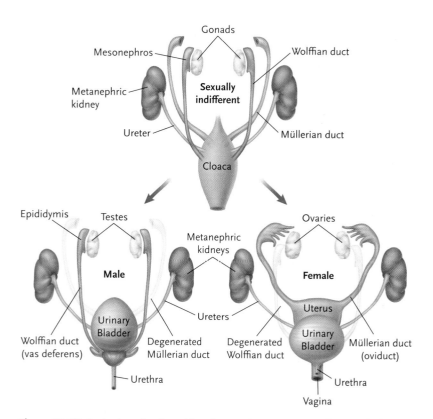

Figure 14.17 Derivation of male and female reproductive systems of mammals from the "indifferent" stage of the developing urogenital system.

within the differentiating ovary surround individual germ cells and secrete estrogens that promote differentiation of the Müllerian ducts into oviduct, uterus, cervix, and the upper vagina (Fig. 14.17). The Wolffian ducts degenerate. Phenotypic traits associated with sexual maturity of each gender, such as growth of mammary glands in females and antler growth in male deer and positive allometry of body size in males of many pinnipeds, are initiated by a pulse of sex hormones at the onset of puberty.

Marsupials are an exception to this basic mammalian pattern because not all aspects of male phenotype are specified by genes on the Y chromosome. Aberrant marsupials with an XXY set of sex chromosomes have testes but no scrotum, yet they have a pouch with mammary glands. Alternatively, those with only a single X chromosome lack a pouch and mammary glands but have a scrotum. Presumably, one or more genes on the marsupial X chromosome control some aspects of sexual phenotype.

In birds and many reptiles, sexual identity is determined by a ZW system, where males are ZZ and females ZW (homozygotic males, heterozygotic females). However, a recent study has revealed an even more profound difference between mammals and birds regarding sexual determination. In mammals, although genes determine sexual identity, embryos are sexually indifferent until the sex-determining genes initiate hormonal secretion by the gonad, which in turn determines gamete type and secondary sex characteristics. In birds, however, it appears that both gonadal and somatic cells have an inherent sexual identity that is independent of hormonal action. This conclusion was suggested by the occurrence of a rare mutation among chickens that results in a "gynandromorph," where one half of the body is male and the other female (Fig. 14.18). The notion of cell-autonomous sexual phenotype in chickens was supported by experiments in which a small piece of an embryo's gonad was transplanted into the gonad of recipient embryos of the opposite sexual genotype. The donated tissue retained the sexual identity of the donor's sexual genotype, regardless of the sexual genotype of the recipient. When these sorts of transplants are done with mammals during the sexually indifferent stage (before hormonal secretion by gonads), the donated tissue acquires the sexual identity of the recipient.

14.3b Environmental Sex Determination

In many animals, gender is specified by an environmental (not chromosomal) cue. In these groups, a variable component of the environment determines what type of gametes and what secondary sex characteristics will develop. The type of environmental signal and the response depends on the species. Instances of environmental sex induction are most likely to occur among animals that live in patchy environments, where environmental factors that change in time or space have different consequences for the fitness of the male and female gender.

The small amphipod crustacean, *Gammarus duebeni* (793), provides an example of environmental sex determination. In the north of England, mating and egg laying by *G. duebeni* occur repeatedly throughout the spring and most of the summer by adults that were born and grew in body size the previous year. The juveniles overwinter in a diapause state, when metabolism is maintained at a very low rate. Gender of the developing offspring is determined by the photoperiod prevailing when the eggs are laid. Long days of early spring specify males, whereas the shorter days later in the summer specify females. This timing means that males have longer to feed and grow before the onset of winter than do females. As a result, the adult popula-

tion that emerges the following summer consists of larger males and smaller females. Large size increases fitness for both genders, because large females produce more eggs and large males can mate with larger females. However, a male can only mate with a female that is smaller than him, so a large female does not have a fitness advantage if all available males are too small to mate with her. Conversely, small males cannot mate because all females are the same size or larger.

A second example illustrating an obvious adaptive value of environmental sex determination is provided by the green spoonworm, *Bonellia viridis* (371), an echiuran annelid. The gender of a green spoonworm is not determined genetically at fertilization, but much later when the free-swimming larva settles on the ocean floor. If a larva settles a distance from other members of its species, it becomes a large female that lives with its bulbous trunk buried in sand and its long "proboscis" extended over the surface of the sand to pick up the organically filmed sand particles that it eats. However, if a settling larva encounters an adult female spoonworm, it develops into a male. Like the anglerfish, males of spoonworms are tiny, dwarf parasitic males. They live within the body of the female doing little more than manufacturing sperm. A chemical component of adult female spoonworm, currently unidentified, is the cue that induces "maleness" in settling larvae of this species.

Recent research has documented an increasing number of reptile species in which gender is specified by the temperature experienced during incubation of the eggs. Jacky dragon lizards (*Amphibolurus muricatus* (1407)) (Fig. 14.19) have been used to test the expectation that temperature of egg incubation differentially influences the fitness of males versus females of this species. Normally, low and high temperatures during egg incubation produce female jacky dragons, whereas intermediate temperature produces males. If this is truly adaptive, then the fitness of males when eggs are incubated at intermediate temperatures must be

Figure 14.19 Jacky dragon lizard, *Amphibolorus muricatus* (1407), a species in which gender is determined by temperature during egg incubation.

Figure 14.18 This gynandromorph chicken has female characteristics on its right side and male on its left (note differences in colour, size of breast musculature and wattle, and presence of spur).

SOURCE: Photo courtesy of The Roslin Institute, The University of Edinburgh

greater than when males are incubated at low or high temperatures. Females should enjoy greater fitness when they hatch from eggs incubated at either low or high temperatures. How could such a prediction be tested? An opportunity arose with the discovery that drugs that interfere with reproductive hormones can override temperature-induction of gender. Experimenters were able to produce both males and females at all three egg incubation temperatures. They subsequently maintained these animals over three breeding seasons (the full life span of these lizards), and their lifetime fitness was quantified by genotyping all produced offspring to determine parentage. The results dramatically corresponded with predictions. Males that hatched from eggs incubated at intermediate temperatures, the normal situation, produced more offspring than males that were artificially forced at low and high incubation temperatures. Similarly, females that hatched at high temperatures, also the normal situation, produced more offspring than females incubated at intermediate temperature. This result dramatically demonstrates that environmental sex induction in this system is indeed adaptive, but the reason for the temperature effect on the fitness of each gender remains unclear.

Some coral reef fish exhibit an extraordinary ability to switch genders. The gonad of these fish is an ovotestis, but the type of gametes produced depends on hormonal signals that can be changed by social cues. For example, large male bluestreak cleaner wrasses, *Labroides dimidiatus* (1270) (Fig. 14.20A), normally oversee a harem of six or more females. If the male dies or disappears, the largest female in the harem switches gender and becomes the male for the harem. Conversely, family groups of clown fish in the genus *Amphiprion* (1252) (Fig. 14.20B) consist of a large female, a smaller reproductive male, and a collection of even smaller non-reproductive males. If the large female disappears, the reproductive male assumes the female role and one of the non-reproductive males becomes reproductive.

STUDY BREAK

1. Define "secondary sex characteristic" and provide examples of the functional value of such characteristics.
2. How is sex specified in mammals? What roles are played by the androgenic hormones AMH and testosterone in the differentiation of secondary sex characteristics?
3. Under what circumstances is environmental sex determination appropriate?
4. Describe the design and results of the experiment on jacky dragon lizards that demonstrated an adaptive value for environmental sex determination.

 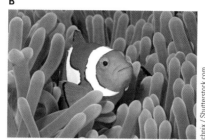

Figure 14.20 A. Bluestreak cleaner wrasse (*Labroides dimidiatus* (1271)). **B.** Clown fish (*Amphiprion* (1252)).

14.4 Gametes

Functional gametes must be much more than cells with a haploid set of chromosomes. Sperm and ova are exquisitely specialized cells designed to play essential roles in the start of a new organism, a process that requires each gamete to cooperate with a genetically foreign cell (the other gamete). The respective tasks of sperm and ova in this cooperative enterprise are quite different, which may explain why the process of meiosis is virtually the only cytological process that is shared by these gametes during their differentiation. This section will focus on the embryological origin of cells destined to produce gametes; the cytological, morphological, and functional differences between sperm and ova; and how these differences arise during their respective cytodifferentiation processes.

14.4a Primordial Germ Cells

Primordial germ cells (PGCs) are the gamete stem cells of a bisexual animal. These cells acquire their gamete-producing fate during early development, even though the actual production process will not begin until the onset of sexual maturity. Primordial germ cells have been identified in many vertebrates and in some invertebrates such as the fruit fly, *Drosophila elegans* (960), and the nematode, *Caenorhabditis elegans* (585).

Ironically, primordial germ cells do not actually originate within the developing gonads of embryos. In fish, frogs, fruit flies, and nematodes, the special cytoplasm that will become incorporated into primordial germ cells is recognizable within fertilized eggs even before they begin to divide at the onset of embryo development. In these cases, directions from the mother's genome places molecular components in her developing eggs and provide this particular cytoplasmic region, known as **germ plasm**, with a "special potency." When cytoplasm of the fertilized egg becomes progressively subdivided during embryonic cleavages (mitoses) at the onset of embryogenesis, the germ plasm becomes partitioned into a discrete set of nucleated cells—the primordial germ cells. The primordial germ cells then undertake a complex migration through the early embryo to reach the future gonads.

In mammals, no germ plasm is identifiable within fertilized eggs prior to onset of embryonic cleavage. Rather, the special potency that characterizes primordial germ cells is induced during later development by signals received from other embryonic cells. In mice, cells situated close to the boundary between the embryo proper (the **epiblast**) and cells of the extraembryonic membranes (yolk sac and allantois) are induced to become primordial germ cells (Fig. 14.21) when they receive signals from neighbouring cells. At least one of these signals has been identified as a member of the bone morphogenetic protein (BMP) family. The newly induced primordial germ cells migrate through the developing gut and dorsal mesentery of the embryo to eventually colonize the paired genital ridges (future gonads). Once within the gonad, these germ cells, which have already undergone a number of mitoses during their migration, differentiate into spermatogonia or oogonia—the precursor cells for gametes. Which fate is followed depends on the type of hormonal signals present within the gonad.

What do we mean by a "special potency" embodied within primordial germ cells? To answer this, you must consider what these cells need to accomplish compared with other cells of the developing embryo. From the onset of development, other cells within the embryo become progressively channelled into a particular cell fate by responding to various inductive signals within the embryonic environment. This process is orchestrated by a highly regulated sequence of switches that turn on specific genes while shutting down others.

Primordial germ cells must protect themselves from this environment of inductive signals that would channel their fate along a restricted pathway. Their entire genome needs to retain the ability to direct the development of a whole organism and all its parts, a property termed **totipotency**. On the other hand, primordial germ cells must be capable of selectively activating a few specific genetic programs, such as those that allow pathfinding and cellular locomotion during their migration to the gonad. Early segregation of primordial germ cells to remote sites (e.g., the extraembryonic membranes of mammals) may isolate these cells from any cues that induce differentiation within the embryo proper. The molecular mechanisms that maintain the delicate balance between highly regulated genome transcription and genetic totipotency are currently subjects of active investigation. The strong impetus for this research comes from the potential use of **embryonic stem cells** that retain unlimited differentiation potential for therapies that might hold cures for various degenerative diseases.

14.4b Sperm and Spermatogenesis

The functions of a sperm cell are 1. to activate an egg, signalling it to complete meiosis if it hasn't already done so and to commence embryonic cleavages, and 2. to contribute genes to the genome of the zygote. For this job, a sperm needs a haploid set of chromosomes as well as mechanisms to penetrate membranes surrounding the egg and then to fuse with the egg membrane. The sperm of many animals are motile and travel to the egg usually by means of an undulating flagellum powered by mitochondria. However, not all sperm move by a beating flagellum. Sperm cells of nematodes move by pseudopod-like cytoplasmic extensions. Nevertheless, the following description of sperm and spermatogenesis is based largely on what has been learned from studies on the sperm of sea urchins, clams, and rodents, the model systems for a great deal of the research done on the male gamete of animals.

Sperm of most animals have three main parts, a head, midpiece, and tail, all enclosed within a cell membrane (Fig. 14.22). The sperm head contains the all-important haploid nucleus and an **acrosome** (acrosomal vesicle), which contains materials that, depending on the species, may function to penetrate the egg membranes and to recognize and fuse with the ovum (see Section 14.5). The midpiece contains proximal and distal centrioles and mitochondria that will provide energy for the swimming tail. The tail is a motile flagellum containing microtubules generated by the distal centriole. Sperm typically contain very little cytoplasm. They are stripped down fertilization machines that are cheap to produce—so many are made. With very few exceptions, the number of sperm produced by the males is much greater than the number of ova produced by the females of a species.

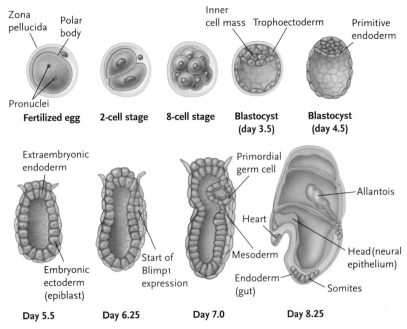

Figure 14.21 Origin of the primordial germ cells in mice. A patch of cells embedded in tissue of the extraembryonic membranes is induced to become primordial germ cells. During later development, these cells migrate through the embryo to enter the tissue of the future gonads, where they are induced to become either spermatogonia (sperm stem cells) or oogonia (ova stem cells).

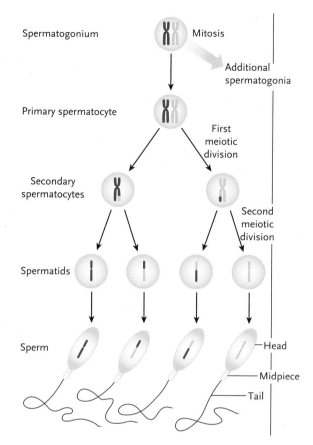

Figure 14.22 Stages of spermatogenesis. Recombination (chromosomal cross-over) occurs during the primary spermatocyte stage.

The process of sperm formation is called **spermatogenesis** (Fig. 14.22). The starting point is the population of cells known as **spermatogonia**, the direct descendants of the primordial germ cells that populated the testes during development. Spermatogonia are generally capable of proliferating by mitosis throughout the sexually mature life of the organism, even if this occurs in bursts associated with restricted reproductive seasons. At some point, however, subgroups of spermatogonia undergo meiosis, a phase when they are called **spermatocytes**. When meiosis has finished, each spermatogonium has produced four haploid cells called **spermatids**. Spermatids are not yet ready to fertilize eggs; they must first undergo a differentiation process known as **spermiogenesis**.

An important event of spermiogenesis is formation of the **acrosome**, which is a membrane-bound packet of proteins that will facilitate fertilization. Acrosomal proteins manufactured in the rough endoplasmic reticulum are packaged into pro-acrosomal vesicles by the Golgi apparatus (Fig. 14.23A). The vesicles fuse together to form a single large acrosomal vesicle. In some species, such as mammals, the acrosome forms at the end of the nucleus opposite to the position of the centrioles. Alternatively, in echinoderms the acrosome is manufactured close to the centrioles and must migrate to the opposite end of the nucleus (Fig. 14.23B). In the sperm of echinoderms, a concentration of monomeric actin (peri-acrosomal material) accumulates between the nucleus and the acrosomal vesicle; this plays an important role in fertilization. The distal centriole of the pair of centrioles

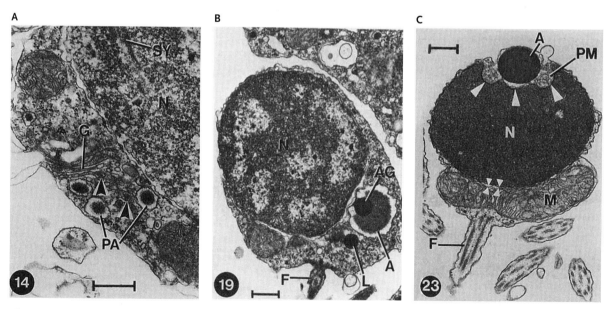

Figure 14.23 Transmission electron micrographs showing sperm maturation stages of an echinoderm (the crinoid *Florometra serratissima* (1066)). **A.** Pro-acrosomal vesicles arising from the Golgi body. **B.** Maturing acrosomal vesicle during its migration to the anterior end of the nucleus. The section also passes through part of the sperm flagellum, and the sperm chromosomes are not yet fully packaged by protamines. **C.** Mature sperm. Labels: F = flagellum; G = Golgi body; M = mitochondrion; N = nucleus; PA = pro-acrosomal vesicles; PM = periacrosomal material. Scale bars = 0.5 μm.

SOURCE: Bickell, L.R., Chia, F.S., and Crawford, B.J., "A fine structural study of the testicular wall and spermatogenesis in the crinoid, Florometra serratissima (Echinodermata)," *Journal of Morphology* Volume 166, Issue 1, pages 109–126. John Wiley and Sons. Copyright © 1980 Wiley-Liss, Inc.

elaborates the microtubules of the flagellar axoneme, whereas the proximal centriole will be delivered into the egg cytoplasm during fertilization and will help form the microtubule organizing centres of the first cleavage spindle of embryogenesis. Also during spermiogenesis, mitochondria fuse together to form one or only a few large mitochondria that encircle the centrioles at the base of the flagellum. Mitochondria provide energy to fuel the sperm's brief but frantic life after motility is initiated (they are not motile during storage in the testis). The size of the spermatid's nucleus becomes progressively more compact during spermiogenesis by tight packaging ("condensation") of the chromosomes. Toward this end, histones, which are the conventional DNA packaging proteins of animal somatic cells, are replaced by protamines, which are able to bind up DNA even more tightly. Finally, all but a thin layer of cytoplasm is expelled from the spermatid. The result of these cytological events is a mature sperm, ready to fertilize an egg.

14.4c Ova and Oogenesis

Mature female gametes are called ova or eggs. Progenitor cells for ova are **oogonia**, and the process that manufactures fertilizable ova from oogonia is called **oogenesis**. Meiosis is almost the only process that oogenesis and spermatogenesis have in common, which is not surprising, given the different roles played by the two types of gametes at fertilization and beyond. Although the cytoplasmic differentiation of sperm (spermiogenesis) occurs *after* meiosis by the male gametes, the cytoplasmic differentiation of ova during oogenesis occurs *before* meiosis is completed. Furthermore, an individual ovum is typically larger, sometimes much larger, and more expensive to manufacture than an individual sperm of a given species (Fig. 14.24), so fewer ova can be produced. As a result, oogonia undergo fewer cycles of mitosis than spermatogonia.

The lifetime capacity for oogonial mitoses is species-specific. Oogonia within female sea urchins and clams, which produce millions of ova during repeated reproductive seasons, retain their ability to undergo mitoses throughout the life of the organism. In humans, however, oogonia have stopped dividing by the 7th month of gestation and all have entered prophase of meiosis I by the time the baby girl is born. Meiosis and further differentiation of the primary oocytes is arrested during childhood, but beginning at puberty, groups of these **primary oocytes** proceed to meiosis II at approximately monthly intervals (Fig. 14.25). The last batch will not make this transition until just before menopause, a time interval that can extend to 50 years—a very long time for a cell to retain totipotency and maintain itself in good repair!

Although the nucleus of egg and sperm make an equal contribution to the genome of the offspring, the cytoplasm of the egg plays a much more momentous role in the early development of the embryo. The egg cytoplasm must acquire a host of provisions to support the process of early embryonic development, including nutrient reserves for the embryo, cytoskeletal proteins required for cytological events of fertilization and for embryological cleavages, organelles, nuclear and cytoplasmic enzymes, messenger RNA transcripts encoding instructions for early developmental events,

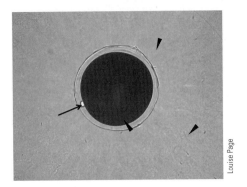

Figure 14.24 Recently fertilized egg of the blue topsnail *Calliostoma ligatum* (410) (Mollusca, Gastropoda). Note the large size of the egg compared to the tiny sperm (arrowheads); also note an extruded polar body (arrowhead) within the space between the vitelline envelope and the egg plasma membrane.

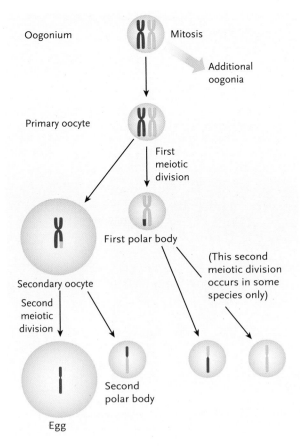

Figure 14.25 Stages of oogenesis. Crossing-over occurs during the primary oocyte stage, which is often prolonged as the oocytes accumulate provisions to support early embryonic development of the fertilized egg.

ribosomes, and transfer RNAs. Some of these provisions are acquired under direction from the oocyte nucleus, but most are delivered to developing primary oocytes from elsewhere. The accumulation of egg provisions mostly occurs during prophase of meiosis I, when developing female gametes are in the primary oocyte stage.

Somatic cells of the ovary that serve to provide support, possibly nutrients, and hormonal signals to developing oocytes have been given a variety of names, depending on the animal group, but **follicle cells** is a suitable general term. Even cnidarians have cells that have been interpreted as follicle cells for developing oocytes. The gametes of these structurally simple, basal metazoans differentiate from gastrodermal cells (the cells lining the gastrovascular cavity), and they bulge into the mesoglea as they mature. A study on oogenesis by the starlet sea anemone, *Nematostella vectensis* (173), showed that a cluster of morphologically specialized gastrodermal cells is invariably associated with each developing oocyte (Fig. 14.26). This cluster may function to convey nutrients from the gastrovascular cavity to the maturing ovum. In many other animals, such as mammals, developing oocytes within the ovary are completely surrounded by follicle cells.

Unlike the process of meiosis during spermatogenesis, meiosis during oogenesis often takes a long pause during the primary oocyte stage. The duration of this pause varies greatly between species and may have different reasons. We have already described the situation in humans, where the pause at prophase I of oocyte maturation extends from 12 to 50 years. In humans, the delay is related to the long time period to achieve sexual maturity and to remain sexually competent during a large part of a long life span. In frogs such as the African clawed frog (*Xenopus laevis* (1393)), the pause at primary oocyte stage is three years, during which the egg accumulates provisions for embryonic development. In some invertebrates, the duration of the primary oocyte period may be only months or weeks. However, although meiosis may be arrested during this period, a great deal of other metabolic activity may be occurring.

Prophase I of oogenesis is often subdivided into two phases: **previtellogenesis** and **vitellogenesis**. A major event during previtellogenesis is the secretion by the primary oocyte of an extracellular coat, typically consisting of glycoproteins, called the **vitelline envelope** in most invertebrates. In mammals this egg coat is called the **zona pellucida**, whereas in fish the traditional term is **zona radiata**. As described in Section 14.5, this investment plays a role during interaction between the sperm and egg at fertilization. Other egg coverings, such as jelly coats or sometimes very elaborate hulls, are species-specific characteristics, and their manufacture may involve oocytes, follicle cells, or both (Fig. 14.27).

During the vitellogenic phase of prophase I, the oocyte accumulates yolk and many other cytoplasmic constituents that will nourish and direct early develop-

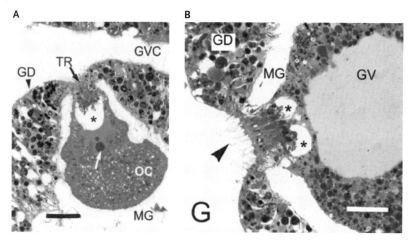

Figure 14.26 Follicle-like cells (labelled TR) possibly involved in transfer of provisions to a developing oocyte of the sea anemone, *Nematostella vectensis* (173) (Cnidaria). **A.** Oocyte (OC) bulging into the mesogleal layer (MG) from the gastrodermal epithelium (GD) lining the gastrovascular cavity (GVC). **B.** Enlarged view of the follicle-like cells; GV is germinal vesicle (nucleus) of the oocyte.

SOURCE: Eckelbarger, K.J. et al., "Ultrastructural features of the trophonema and oogenesis In the starlet sea anemone Nematostella vectensis (Edwardsiidae)," *Invertebrate Biology* Volume 127, Issue 4, pages 381–395, Fall 2008, John Wiley and Sons. © 2008, The Authors. Journal compilation © 2008, The American Microscopical Society, Inc.

ment of the embryo. Species-specific differences in the amount of yolk that accumulates within primary oocytes are largely responsible for the marked differences in animal egg sizes, which may measure less than 100 µm in diameter (barely detectable with the unaided eye) for many invertebrates, to 5 mm for some frog and elasmobranch eggs, and up to 20 cm for eggs of the North African Ostrich (*Struthio camelus*) (Fig. 14.28).

Egg yolk is a combination of nutrients, particularly lipoprotein and phosphoproteins. The eggs of many animals receive much, although not all, yolk material

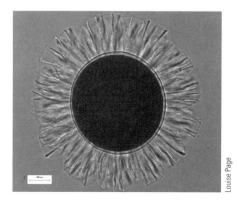

Figure 14.27 Egg of a black katy chiton, *Katharina tunicata* (396) (Mollusca, Gastropoda) showing the elaborate hull surrounding the egg.

Figure 14.28 Comparison of animal egg sizes from **A.** a sea urchin (phylum Echinodermata) and **B.** three birds: ostrich, elephant bird, and hummingbird. The Canadian $2 coin is 28 mm in diameter, the ostrich egg, 200 mm.

as **vitellogenin.** In vertebrates, vitellogenin is manufactured in the female's liver and transported to the ovary, where it is delivered to oocytes. Vitellogenin is a high-molecular-weight lipoglycophosphoprotein. This large, complex molecule begins as a protein encoded by the vitellogenin gene (highly conserved during the evolution of animals) that includes multiple binding sites for lipids, carbohydrates, and phosphate groups (see Box 14.1).

After the primary oocyte has completed vitellogenesis, it is activated to continue meiosis, which eventually results in the sequential release of two **polar bodies.** The 1ˢᵗ polar body may subsequently divide so that a total of three polar bodies are produced (Fig. 14.25). Formation of polar bodies reveals yet another difference between oogenesis and spermatogenesis. During spermatogenesis, each of the four cell products of meiosis receives an equal amount of cytoplasm, so four equal-sized sperm are formed from each spermatogonium. This is not the case with the four cell products of oogenic meiosis. The materials that are accumulated within the cytoplasm during prophase I are enough to provision only one functional gamete. Therefore, when meiosis is completed, one of the haploid products has most of the cytoplasm (the ovum); the other three (the polar bodies) have only a thin rind of cytoplasm. The polar bodies eventually degenerate.

Ova are released from the ovary and fertilized either externally or within a specialized compartment within the female's body. Another feature that is species-dependent is the stage of meiosis at which ova are capable of being fertilized. For example, in broadcast spawning abalone (Mollusca, Gastropoda) and razor clams (Mollusca, Bivalvia), eggs are still primary oocytes when spawned from the female. These eggs require fertilization to activate the remainder of meiosis so three polar bodies appear only after the sperm has fused with an egg. Alternatively, the internally fertilized eggs of humans and most other mammals are ovulated at the secondary oocyte stage, so fertilization activates the release of the second polar body. Finally, the eggs of sea urchins (Echinodermata) and most hydrozoan jellyfish (Cnidaria) have fully completed meiosis at the time of spawning, so no polar bodies are extruded from the eggs at fertilization. There appears to be no obvious phylogenetic pattern for the time of fertilization relative to the stage of ova meiosis.

STUDY BREAK

1. What is special about primordial germ cells relative to somatic cells?
2. What are some major differences between the processes of spermatogenesis and oogenesis?
3. What is vitellogenin?

14.5 Fertilization

Fertilization involves a precisely orchestrated cascade of signals and responses between an egg and sperm. Although we may be most interested in how the process works in mammals such as ourselves, fertilization is very difficult to study in mammals because the event occurs within the body of the female and involves relatively few eggs at any one time within any one female. For practical reasons, the elegant details of how fertilization is orchestrated were first worked out on sea urchins. An individual sea urchin spawns many thousands or millions of gametes at once and spawning can be induced on demand by simply injecting a weak potassium chloride solution into the animal's perivisceral coelom. Therefore, fertilization can be observed in a petri dish containing seawater, and many gametes can be obtained for molecular analysis. We will therefore draw on the large body of research on sea urchins to describe the essentials of the fertilization process, but occasional information on other species will be provided for comparison.

14.5a Sperm Attraction and Activation

Fertilization begins with the ovum "talking" to the sperm. The jelly coat that invests sea urchin eggs contains a small, diffusible peptide that attracts sperm (Fig. 14.29). Egg attractants that act on conspecific sperm have also been identified among many other animals, particularly those that broadcast spawn gametes. The jelly coat of sea urchin eggs also initiates the so-called **acrosome reaction** by sperm, which sets in motion a train of events allowing sperm to talk back to the egg. As shown in Fig. 14.30, the acrosome reaction begins with fusion between the sperm plasma membrane and the membrane surrounding the acrosomal vesicle (= acrosome). This event exposes the contents of the acrosomal vesicle to the surrounding environment. Concurrently, the monomeric (non-polymerized) actin units that constitute the peri-acrosomal material polymerize to form a rod-like bundle of microfilaments that pushes out the basal membrane of the acrosomal vesicle. This protruded membrane with its core of microfilaments is the acrosomal process. In sea urchins, a protein known as **bindin**, which is linked to the inner membrane of the acrosomal vesicle, is exposed on the surface of the acrosomal process. Receptors for bindin are located on the vitelline envelope of conspecific sea urchin eggs and allow sperm to adhere to this egg coat. After penetrating the vitelline envelope, microvilli from the egg membrane reach up to the sperm and the membranes of sperm and egg fuse. In abalone, material within the acrosomal vesicle actually punches a hole in the vitelline envelope of conspecific eggs to allow the sperm to swim through.

BOX 14.1 Molecule Behind Animal Biology

Vitellogenin

During oogenesis, egg nutrients manufactured outside the oocyte are transported through the circulatory system of the animal and delivered to the oocytes. Vitellogenins (Fig. 1), a highly conserved class of molecule, serve this nutrient transport function in both invertebrates and vertebrates. Vitellogenins are proteins encoded by large mRNA transcripts of 6–7 kb and modified after synthesis by the addition of carbohydrates, lipids, and phosphate groups. These materials are the cargo that vitellogenin delivers to oocytes. However, vitellogenin is the ultimate in efficient cargo delivery, because the major nutrient it delivers is the cargo truck itself—the protein backbone of vitellogenin. Within the ovary, vitellogenin squeezes between the follicle cells and is imported into oocytes by receptor-mediated endocytosis. Vitellogenin is then processed to release the various subcomponents of this large molecule for separate storage within the oocyte cytoplasm.

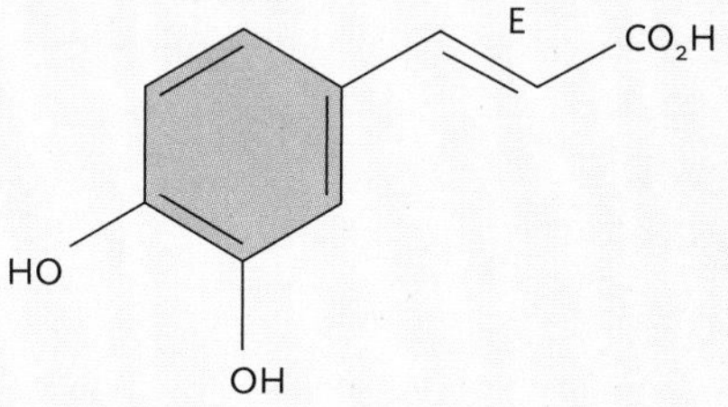

In some interesting cases vitellogenin has been co-opted to serve other functions in animals. One case comes from marine teleost fish, by far the most diverse and species-rich group of vertebrates. The spectacular diversification of teleosts within the ocean first required a host of adaptations to deal with a hyperosmotic environment, because freshwater was the ancestral habitat for this group of fish. In particular, eggs in seawater must deal with an osmotic flux that is opposite to the osmotic flux presented by a freshwater environment. Without corrective measures, teleost eggs in saltwater would rapidly lose water. The vitellogenin gene in teleost fish has duplicated. One paralogue generates vitellogenin for nutrient storage in eggs while the other generates vitellogenin that is hydrolyzed in late oocytes to provide a pool of free amino acids that function as organic osmolytes. Thus oocytes of marine teleosts become inflated with osmotically imbibed water just before they are laid. Furthermore this freshwater store also confers buoyancy to these eggs.

Honey bees living in large, complex social groups provide another example of a derived role for vitellogenin. Again the functional theme of nutrient transfer is retained. In a honey bee society the queen is the egg-laying female, while other females are workers. Younger workers are nurses that feed others in the colony with secretions from their hypobranchial glands. Older workers leave the hive to forage for nectar and pollen. Drones are haploid males. The nurses have been called the "social stomach" for the colony because they eat pollen collected by the foraging workers and digest this protein-rich meal. Vitellogenin then converts the amino acids within the hypobranchial gland of the nurses into the food fed to others in the colony. Expression of the vitellogenin gene within the hypobranchial gland is shut down when a pulse of juvenile hormone from the brain signals the transformation of nurses into foraging workers. It has been suggested that this "off" switch for vitellogenin synthesis in foragers prevents any of the colony's precious communal cache of protein from accumulating within foraging bees, because foragers have a high risk of succumbing to the many mortality factors in the world outside the hive.

FIGURE 1
Vitellogenin from carp.

When the vitelline envelope has been penetrated by a sperm, the sperm cell membrane fuses with that of the ovum.

14.5b Gamete Fusions and Prevention of Polyspermy

After a sperm has fused with an ovum, it is critically important that no additional sperm fuse with that egg. This is because a sperm cell contributes both its haploid nucleus and its proximal centriole to the egg at fertilization. The sperm's proximal centriole initiates the formation of a pair of microtubule organizing centres that generate the mitotic spindle for the first embryonic cleavage. If an egg receives more than one haploid nucleus and centriole, the additional centriole will seriously disrupt embryonic cleavages.

To prevent **polyspermy**, the fatal fusion of more than one sperm with an ovum, the fertilizing sperm signals the ovum to create first a fast and then a slower block to additional sperm fusions. The fast block, accomplished within 1 to 3 seconds, is a depolarization of the egg cell membrane created by an opening of sodium channels, much as a nerve cell depolarizes during transmission of an action potential. This depolarization prevents additional sperm fusions with the egg membrane. However, because the depolarization is transitory, it is backed up by a slower block to polyspermy, which involves exocytosis of cortical granules residing beneath the plasma membrane of the egg.

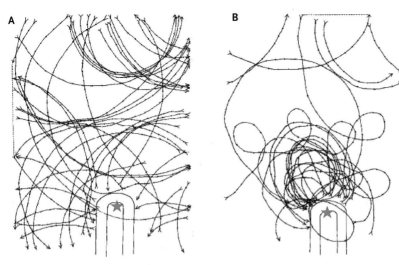

Figure 14.29 Demonstration of sperm attractant from eggs of the hydrozoan medusa *Phialidium (Clytia) gregarium* (133). An open pipet (asterisk) filled with either **A.** seawater or **B.** an extract from eggs was placed in a petri dish with conspecific sperm. Swimming trajectories of individual sperm between 7.5 and 12.5 sec after introducing the pipet were traced from successive film frames.

SOURCE: There are instances where we have been unable to trace or contact the copyright holders. If notified the publisher will be pleased to rectify any errors or omissions at the earliest opportunity. With kind permission from Springer Science+Business Media: *Mar. Biol.*, "Sperm chemotaxis by hydromedusae. I. species specificity and sperm behavior," Vol. 53, 1979, pp. 99–114, Miller, R. L.

Secretory material within these cortical granules lifts the vitelline envelope away from the egg plasma membrane and hardens this envelope by cross-links between proteins to prevent additional sperm from penetrating.

14.5c Egg Activation

The exocytosis of cortical granules, which constitutes the slow block to polyspermy, is facilitated by a sudden increase in calcium ions within the cytoplasm of an ovum, an event initiated by sperm-egg fusion. These calcium ions may enter the ovum across its plasma membrane, or they may be released from the endoplasmic reticulum. Concurrently, intracellular hydrogen ions decrease within the ovum's cytoplasm. This combination of high intracellular calcium ion and low intracellular hydrogen ion concentrations initiates a host of processes within egg cytoplasm. One of these is the completion of meiosis if the ovum has not already done so (a species-dependent trait). A second important process is the combining of sperm and egg nuclei, now called pronuclei, to form a single zygotic nucleus.

STUDY BREAK

1. What signals are communicated by the egg to the sperm prior to sperm fusion with the egg?
2. What tasks are achieved by the acrosome reaction? Describe the events of the acrosome reaction.
3. Why does fusion of more than one sperm with an egg disrupt development?
4. What cytological events are responsible for the fast and slow block to polyspermy?

14.6 Timing of Reproduction

Synchronization of reproductive activities among individuals in a population is a vital component of bisexual and some forms of unisexual reproduction. Timing of gametogenesis for males and females needs to be coordinated so sperm and eggs reach maturity within the same time window, even if one type of gamete requires more time for gametogenesis than the other. Reproductive behaviours must also be initiated at a time that is most conducive to survival of offspring. This is particularly important when there are seasonal differences in availability of food and risk from predators, particularly predators of offspring.

Timing of the various physiological and behavioural traits related to reproduction is often controlled by a hierarchy of cues that initiate different phases of the whole process. Cues often take the form of physical attributes of the environment that vary seasonally, but they may also come from social signals communicated as pheromones or behavioural displays by conspecifics. One type of environmental cue may coordinate gametogenesis, whereas other cues may initiate activities related to mating or spawning. Potential environmental cues for enhancing reproductive success are many and varied; they depend on what is available within an animal's habitat and what can reliably signal temporal events of importance to an animal's reproductive success. The following section gives examples.

14.6a Synchronization of Spawning or Mating

Synchronized gamete release by males and females is particularly critical for marine invertebrates that broadcast spawn their gametes into surrounding ocean water. Individuals that release their gametes a little too far from other spawning individuals or a little before or after other spawners will waste their entire investment in gamete production.

Research on stony corals of the Great Barrier Reef has documented a hierarchical cascade of environmental cues for controlling the timing of gametogenesis and spawning and the functional value of that timing. Most species of reef-forming stony corals broadcast spawn their gametes. Each polyp of a colony releases a bundle of both sperm and eggs (these cnidarians are hermaphroditic), but gametes released from a single polyp or colony cannot self-fertilize. Remarkably, the many coral species that populate the Great Barrier Reef all release their gamete bundles in a mass spawning event that occurs 5-8 days after the full moon during the Austral Spring (October or November) (Fig. 14.31). Spawning lasts for only one to several nights and occurs only after sunset. The mass of buoyant gametes forms an oily pink slick that stretches for many kilometers. The precise timing of the mass spawning suggests that all

BOX 14.2 Person Behind Animal Biology

John Rocks, M.D.

John Rocks (1890–1984) received his M.D. from Harvard Medical School in 1918. He taught obstetrics at Harvard Medical School for more than 30 years and was a pioneer in the freezing of sperm cells and in *vitro* fertilization.

John Rocks is best known as one of the inventors of "The Pill," an oral contraceptive approved for use by the Federal Drug Administration in 1960. The details of how the pill worked were determined by John Rocks' colleagues Gregory Pincus and Min-Cheuh Chang, and John took it through the many clinical trials that preceded its approval. Through a dose of progesterone, one of a class of progestin hormones, The Pill emulated pregnancy in women and prevented ovulation, making it effective in birth control. A 28-day-long course of The Pill included 21 days of tablets with progesterone, and 7 days of tablets without it, maintaining a cycle of menstruation, the shedding of the endometrium (the lining of the uterus).

Menstruation is not unique to humans; it occurs in other animals including some other primates and several species of bats. Like humans, some bats (some free-tailed bats (Molossidae) and Old World fruit bats (Pteropodidae)) that menstruate have a simplex uterus. Like humans, bats that menstruate show advanced endometrial growth and differentiation in advance of pregnancy. Menstruating species develop a thickened endometrial after ovulation in preparation for implantation of the blastocyst. Menstruation is a means of rapidly eliminated a well-developed endometrium in the absence of pregnancy. The metabolic cost of maintaining the endometrium means that menstruation results in an energetic advantage for species with low reproductive potential, like both bats and humans.

The Pill allowed women using it to be continuously receptive to copulation without risk of pregnancy. This form of contraception dramatically changed the behaviour of many people. In other animals, we have seen the impact of disrupting a "normal" reproductive cycle. For example, in a captive colony of rhesus monkeys (*Macaca mulatta* (1863)), ovariectomized females that received daily shots of estradiol were continuously receptive to males. Over time, males in the colony mated less and less with the females, reflected by fewer attempts to mount them and a reduction in the numbers of ejaculations. Replacing this group of females with others that were also ovariectomized and received daily injections of estradiol restored previous levels of mounting and ejaculation (Fig. 1).

One unexpected consequence of women taking oral contraceptives is rising levels of progesterone in watersheds. Passed in urine, the progesterone has a feminizing effect on developing fish, an example of disruption to the reproductive patterns of other, non-target animals. In some lakes, fathead minnows (*Pimephales promelas* (1199)) have been virtually extirpated because progesterone causes the development of intersex males and alters the patterns of oogenesis in females. Many other aquatic animals also have been affected. There also is evidence of human males being affected by progesterone acquired in drinking water.

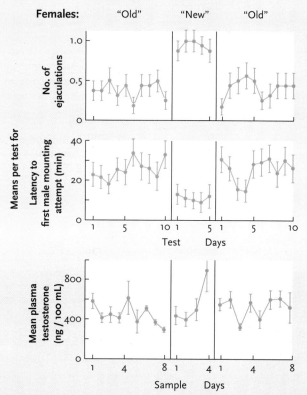

FIGURE 1

Responses of male rhesus monkeys to familiar ("old") and unfamiliar ("new") females over a period of eight weeks. Males more quickly mount and ejaculate into unfamiliar females, coinciding with an increase in levels of plasma testosterone.

SOURCE: From Michael, R.P. and D. Zumpe, "Potency in male rhesus monkeys: effects of continuously receptive females," *Science* 28 April 1978: Vol. 200 no. 4340 pp. 451–453. Reprinted with permission from AAAS.

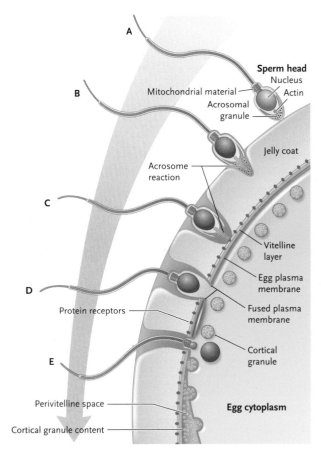

Figure 14.30 Acrosome reaction and fertilization in sea urchins (Echinodermata). **A.** Factors in the jelly coat of the egg initiate the acrosome reaction, which involves fusion between the membranes of the sperm and acrosome to expose the contents of the acrosomal vesicle to the exterior. **B.** Monomeric actin of the subacrosomal granule polymerizes to form a rod of filamentous actin, which extends the basal membrane of the acrosomal vesicle anteriorly as the acrosomal process. Bindin molecules on the acrosomal process bind to species-specific receptors on the vitelline envelope, and acrosomal enzymes lyse a hole in the vitelline envelope. **C & D.** The membrane of the acrosomal process fuses with the egg membrane, which initiates fast and slow blocks to polyspermy (note exocytosis of cortical granules within the egg). **E.** The sperm nucleus and centrioles move into the egg cytoplasm.

SOURCE: to come

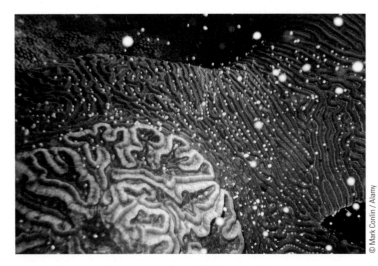

Figure 14.31 Broadcast spawning by corals.

corals of the Great Barrier Reef are responding to the same set of triggering cues. But what are these cues?

Increasing photoperiod and seawater temperature in the early Austral spring appears to initiate gametogenesis among Great Barrier Reef corals. However, once gametes are mature, the actual day(s) when spawning occurs is signalled by a lunar cue, and finally by time after sunset. Researchers have speculated that because 5 to 8 days after a full moon is a period when tidal currents are minimal, released gametes may be less likely to be swept away and diluted before fertilization is achieved.

Although corals forming the Great Barrier Reef spawn in spring during a very restricted time interval, corals in other geographic locations spawn during other seasons and often spawn over a more extended period. Recent research has found a correlation between timing and duration of coral spawning in various locations and local patterns of monthly wind speeds. Wind speeds greater than 6 m/sec greatly magnify the dispersion of particles suspended in surface water, which would rapidly dilute spawned gametes and reduce fertilization success. Along the Great Barrier Reef, waves less than 6 m/sec occur during only 17% of the year and corals spawn within this narrow window of opportunity. By contrast, the sea around the Galapagos Islands is calm for almost half the year and corals in this area have the longest spawning season of anywhere on the planet. Of particular interest is the fact that coral reefs along western Australia, which are populated by many of the same coral species that occur on the Great Barrier Reef of eastern Australia, spawn mainly during the Austral Fall rather than Spring. This difference in timing is consistent with seasonal differences in wind speeds in these two regions.

Studies have shown that many birds also use a hierarchy of cues to coordinate reproduction, and both environmental and social cues may be important. Sparrows in the genus *Zonotrichia* (1622) (Fig. 14.32) cue on seasonal change in daylength to initiate the first phase of gonadal development, but they will not progress to the second phase unless a threshold in local temperature is reached. Social cues also play a role; for example,

Figure 14.32 White-crowned Sparrow (*Zonotrichia leucophrys* (1622)). A succession of environmental cues coordinates physiological events associated with its reproduction.

the second phase is arrested if female sparrows are isolated from males. Photoperiod cues for initiating gametogenesis are even used by some species of birds in tropical latitudes, where change in photoperiod throughout the year is minimal. For Spotted Antbirds (*Hylophylax naevioides* (1604)) in Panama, a photoperiod change of only 17 minutes is sufficient to trigger gonadal development, although the rate of gonad development is fine-tuned by rain and food availability.

Most aquatic and terrestrial animal habitats provide plentiful environmental cues that might be suitable for triggering synchronous reproductive events at times appropriate for successful reproduction. By comparison, however, habitats such as the deepest, darkest depths of the ocean are characterized by a monotonous sameness. Yet some deep ocean inhabitants are invertebrates that broadcast spawn, for example, the giant deep ocean clams, *Calyptogena soyoae* (546), which form dense populations off the coast of Japan at depths greater than 1000 m. The answer to the question of how these gonochoristic clams achieve synchronous gamete release was answered when video data from a camera submerged adjacent to a clam bed was compared with concurrently collected records from an *in situ* temperature recorder and other sensors for physical parameters. Erratic temperature spikes of only 0.2 °C, generated by geothermal activity, corresponded with release of sperm by male clams. Frequently, female clams spawned approximately 10 min. later, a response possibly initiated by the sperm directly or by a pheromone released with the sperm. These temperature spikes probably had no direct effect on gamete or offspring survival; they were simply an occasional environmental anomaly that male clams responded to and then females responded to the males, hence achieving synchronous gamete release.

14.6b Promoting Offspring Survival

Timing of reproduction to ensure a ready supply of seasonally available food for offspring is vital for many species. The barnacle-eating nudibranch (*Onchidoris bilamellata* (457)) copulates and lays eggs in winter, rather than in spring and summer like many other temperate latitude nudibranchs. The timing of reproductive activity in this species means that larvae of *O. bilamellata* require several months of feeding on phytoplankton to develop to the point of metamorphosis. During metamorphosis barnacle larvae settle onto rocks in the intertidal zone. Young juvenile (= not sexually mature) *O. bilamellata* eat just-settled barnacle larvae (= spat) as food.

Precise timing of reproductive activities by environmental cues to ensure a burst of seasonal food for young can be the nemesis of a species if the functional reason for using the environmental cue changes independently of the cue. In the Netherlands, the number of Pied Flycatchers (*Ficedula hypoleuca* (1601)) and other insect-eating birds that overwinter in northern Africa and migrate north to breed have decreased substantially since 1984. Normally these birds return to the Netherlands from their winter range, build nests and lay eggs that hatch just as local caterpillars reach their peak abundance in the Spring. But that was before the climate warmed in northern Europe. Now the leaves of trees appear earlier and caterpillar abundance peaks well in advance of the hatching of young flycatchers, many of whom starve to death.

Avoiding predators that target small or relatively defenseless offspring is another factor that can influence reproductive timing. On the prairies of Canada and the USA, groups of female pronghorn antelope (*Antilocapra americana* (1741)) give birth to their young within a period of only a few days. This has been interpreted as "predator swamping." Newborn antelope are easy prey for coyotes, but coyotes can eat only a small proportion of the many fawns available during the brief interval after mass birthing, when newborn fawns are particularly vulnerable to predators.

14.6c Timing Conflicts for Reproductive Activities

For some species, the best time for mating activity may not result in the best time for laying eggs or giving birth to offspring. These kinds of timing mismatches are resolved in a number of ways, typically involving storage of sperm by females or by various methods to delay development of fertilized eggs. As illustrated by the following examples, mechanisms to resolve these mismatches in reproductive events have arisen many times independently among both vertebrates and invertebrates.

Plain-nosed bats (Vespertilionidae) of the north and south temperate parts of the world (North America and Eurasia) mate in late summer and early autumn before entering winter hibernation. Females store sperm until they emerge from hibernation, when ovulation and fertilization occur. Young are born later in the spring. This pattern means that mating occurs when adults are in their prime at the end of the summer and best able to compete for mates, but the young are not born until the following late spring, a time when insect prey for both offspring and lactating females is bountiful.

Sperm storage by females also occurs in many reptiles, birds, and internally fertilizing invertebrates. For example, marine turtles (Fig. 14.33) that forage over vast areas of the ocean during most of the year may encounter mates infrequently. Sperm transferred during these matings must be stored by females, sometimes for long periods, before the females deposit eggs at beach nesting sites. A study using microsatellite analysis found that one third of loggerhead egg clutches contained eggs fertilized by more than one male—the result of promiscuous mating and sperm storage while the females were out at sea.

Figure 14.33 Loggerhead turtle (*Caretta caretta* (1420)). An example of a marine turtle that mates with multiple partners at sea and crawls onto a beach to bury a clutch of eggs.

Mihai Dancaescu / Shutterstock.com

In mammals, delayed implantation of young embryos is another mechanism by which time of copulation can be uncoupled from the time when offspring are produced. The phenomenon is common among pinnipeds, for example, the northern fur seal, *Callorhinus ursinus* (1707). Northern fur seals migrate north in the spring to their breeding grounds, called rookeries, which are primarily the Pribilof Islands in the eastern Bering Sea (Fig. 14.34). Males arrive first in May and fight for territories, which they maintain throughout the ensuing four summer months without foraging out to sea. Females begin arriving in June, become incorporated into a male's harem, and within a day or two give birth to a pup conceived the previous year. Approximately 10 days after birth, a female enters oestrus and is inseminated by the male leader of the harem. A dominant male may mate with up to 50 females, whereas males unable to defend a territory will mate with none. The female then spends the summer foraging out to sea for food and returning to the island at regular intervals to nurse her pup. All animals leave the rookery in late fall and migrate south. During the four months that a female spends nursing her pup, the embryo she conceived shortly after the current pup's birth floats in her uterus without implanting. Once implantation has occurred (an event triggered by a threshold photoperiod), a further 7–8 months of gestation results in birth coinciding with her arrival at the rookery the following spring. As a result of this arrangement, females mate after the males have established their superiority through dominance contests, but females can still take advantage of rich northern feeding grounds to provide milk to offspring, albeit those conceived in the previous year.

Delayed development is often exhibited by invertebrates as part of an overwintering strategy. For example, many species of copepods release encapsulated, fertilized eggs that, like sponge gemmules, arrest development until conditions are suitable for survival of the larval stage that eventually hatches out of the capsule.

STUDY BREAK

1. Give examples of factors that require precise timing of events associated with reproduction.
2. Give an example of where climate change can have an indirect effect on the success of an animal's reproductive activities.
3. Name two mechanisms that can uncouple time of mating from time of hatching or birth of offspring.

14.7 Sexual Selection

Surviving the many challenges that an animal faces within its environment is certainly a critical criterion for determining which animals will survive to sexual maturity and be sufficiently healthy to participate in reproductive activities. This contest with the environment is called natural selection. However, as Darwin himself recognized, a second type of selection influences a bisexually reproducing organism's success at passing its genes to a subsequent generation. This second type is known as sexual selection. Sexual selection acts on traits not directly related to survival, but rather to maximizing genetic contribution to offspring during reproduction. Indeed, natural selection and sexual selection can lead to conflicts. For example, flamboyant colouration of male plumage among birds may be promoted when females assess male quality on the basis of male decoration (sexual selection), but flamboyant plumage makes these males more conspicuous to visual predators (natural selection).

There are two basic types of sexual selection: **intersexual selection** and **intrasexual selection**. Intersexual selection occurs when one sex discriminates among members of the other sex for potential mating partners. Intrasexual selection arises when members of a single gender compete among themselves (either the whole animals or just their gametes) for access to more of the gametes produced by the other gender (Fig.14.35). As described below, the single most important factor for determining the strength of sexual selection and whether intrasexual or intrasexual selection plays a dominant role is the extent of the imbalance between the two genders in the energy they invest in reproduction. Males may only produce sperm, while females more often care for and feed the young. In this situation, investment by females exceeds that of males.

Figure 14.34 A rookery of the northern fur seal (*Callorhinus ursinus* (1707)). Competition for territories within the rookery is intense.

DPS / Shutterstock.com

Figure 14.35 Two male Egyptian Geese (*Alopochen aegyptiacus* (1651)) in a territorial dispute.

In general, an egg is more energetically costly than a sperm cell, because eggs must be provisioned with resources to fuel and direct early development. A single egg is a more precious commodity than a single sperm. Furthermore, females of internally fertilizing organisms often encapsulate eggs with expensive protective coatings or nourish and protect offspring for a prolonged period. On the other hand, inexpensive male gametes can be produced in large quantities. This fundamental asymmetry between the sexes sets up different sexual selection pressures for the two genders. The gender that invests more in reproduction (usually the female) is more limited by resources than by opportunities to have gametes fertilized, and is therefore more discriminating about the quality of prospective mates (intersexual selection). Conversely, the gender that invests less in reproduction (usually the male) is primarily limited by success at obtaining fertilizations, shows greater variance in success, and is more subject to intrasexual selection. The foregoing statements are generalizations; in some species, investment in reproduction by the two genders is fairly equal or males may actually invest more.

14.7a Mating Systems

Heavy investment of resources into a few eggs is generally not an option for broadcast spawners, because the risk of excessive gamete dilution and failed fertilization is too high. Strategies to reduce this risk include building protected nests where both sexes deposit gametes, consolidating sperm into discrete sperm packets (spermatophores) that are delivered to or picked up by the female's gonopore, or best of all, copulation—delivery of sperm directly into the female's reproductive tract by means of a male intromittent organ (penis). Strategies that allow close to 100% fertilization of eggs make it possible for females to invest heavily in fewer eggs, which in turn intensifies sexual selection. Furthermore, whereas females generally know that their offspring carry their genes, males

cannot be totally sure that their investment has not been usurped by another male or somehow blocked by the female. Sexual selection acting on different animal body plans within different environmental settings has generated a number of standard mating systems to help resolve conflicting optimal reproductive goals of the two sexes. Mating systems help provide protection of investment in mate choice.

Monogamous mating systems involve a pair bond between one male and one female, which may last for one season of reproduction or throughout the reproductive life of the individuals. **Polygamous** systems involve mating bonds between multiple individuals and are called **polygyny** when the bond is between one male and multiple females and **polyandry** when the bond is between one female and multiple males. **Promiscuous** systems do not involve pair bonds. A **lek** is an example of a promiscuous system in which males aggregate in a display area which is visited by females who mate selectively with one or more of the males.

The situations that lead to monogamous, polygamous, or promiscuous mating systems can differ widely. Polygamous systems can arise when a dominant male can monopolize access to many females by overpowering and excluding weaker males. In other cases, males or females may control access to resources, indirectly controlling access by the other gender. Monogamous systems may be favoured when males have a high probability of being a parent of the female's offspring and offspring survival is enhanced by joint parental care.

African lions (*Panthera leo* (1726)) have a polygynous mating system. They live in social groups, called prides, consisting of a dominant male along with a group of females and their offspring. When a male lion takes over a pride by successfully battling a pre-existing pride leader, he routinely kills unweaned young. As a result of this infanticide, the females stop lactating, come into heat (oestrus) and are ready to mate. The pride male takes advantage of this opportunity, and the lionesses bear his young. The genetic advantages to the male (his genes into the next generation) are obvious, but why do the females tolerate the male's behaviour? The reason probably relates to the fact that lionesses in a pride are usually sisters, and a lioness and her cub are more likely to survive in a group than in isolation. Lionesses hunting in a group are more efficient than lone hunters, and groups of lions are better able to resist competition from spotted hyenas (*Crocuta crocuta* (1710)). Further, lionesses living in the same pride usually come into heat at the same time, so their young are the same age, and females share motherly duties, including nursing the young of other females. This is genetically advantageous if offspring of other females in the pride are nieces and nephews. Ironically, the aggressive behaviour of male lions may be ultimately due to sexual selection by females. Females get the strongest, most fit males to father their offspring. Infanticide is a tolerated expense among lionesses because it ultimately promotes more of their genes into the next generation in combination with those of a male of demonstrated physical prowess.

Animals use courtship behaviour (see page 479) to set the stage for mating, be it monogamy, either form of polygamy (polygyny or polyandry), or promiscuity. Courtship displays are designed to serve at least three roles:

1. confirm species' identity and appropriate gender
2. assess genetic quality or health of a potential mate
3. align bodies and motivational states to ensure transfer of sperm.

Many secondary sex characteristics, including sexual dimorphisms in weaponry, ornaments, body size, and behavioural traits, are integral to the courtship process and to the different types of mating systems.

14.7b Paternity Guards

Males generally aim to increase the number of eggs they fertilize, but males cannot be totally sure that the female's eggs were fertilized by his sperm. Several factors can complicate the situation. A female may accept sperm from multiple males, perhaps reducing the chance of any one male siring young. A female may reject a male's sperm by mechanisms acting within her reproductive tract (see Section 14.7e), or another male may remove sperm previously transferred to a female by a different male. Males may kill offspring sired by another male. When mating, male damselflies (*Calopteryx maculata* (852)) use their copulatory organ to first remove any spermatophores (sperm packets) in the female's reproductive tract before depositing their own spermatophore. These situations have led to various paternity guards that help protect a male's investment in a mated female. These include avoidance of already mated females, mate guarding, territoriality, frequent copulations, and copulation plugs.

Mate guarding occurs when a male remains close to a female when she approaches sexual receptivity to physically prevent access by other males. The small copepod, *Tigriopus japonicus* (737), which inhabits tidepools in the high intertidal zone, shows particularly zealous mate-guarding behaviour. Juveniles of *T. japonicus* pass through six copepodite larval stages before they moult to the sexually mature adult stage. Mature males stroke female copepodites with their antennules (anterior pair of sensory appendages) before grabbing the young female's carapace and carrying her around on the ventral side of his body (Fig. 14.36). He will remain with her until she moults to the adult stage, at which time he transfers a spermatophore to her gonopore. Copepodites increase in attractiveness to males as they proceed through the moults leading to the adult stage, but even 1st-stage copepodites will be grabbed if later stages are not

Figure 14.36 Mate guarding behaviour. A large male of the copepod *Tigriopus* sp. (737) holds a still immature female as he swims around in high tide pools.

available. The attractive factor has been identified as a compound exposed on the female's exoskeleton. After mating, females lose their attractiveness to males. Similarly, when males of many species of decapod crabs detect that a conspecific female is about to moult, he will clasp her to the ventral side of his body, assist her with escaping from her exoskeleton, then immediately transfer sperm by spermatophore (anomuran crabs) or intromittent organ (brachyuran crabs). Female crabs can be inseminated during only a brief window of opportunity immediately following moulting.

Copulatory plugs (de facto chastity belts) are devices that prevent sperm of other males from being placed within a female's reproductive tract. An example comes from the Australian redback spider (*Latrodectus hasselti* (671)) (Fig. 14.37). Males of all spiders transfer sperm to the female's sperm storage organ using specialized structures at the distal end of one or both of their **pedipalps**, a pair of appendages located immediately behind a spider's fangs. During mating, the male inserts the sperm-laden, distal end of one of his pedipalps into one of the female's two sperm storage organs and simultaneously does a "copulatory somersault" to place his abdomen immediately above the female's fangs. The female then begins feeding on the male as he delivers sperm into one of her sperm storage organs. Surprisingly, most males survive the first copulation and go back for more of this grisly mating routine, during

Figure 14.37 Australian red-backed spider female and much smaller male.

which sperm from the male's second pedipalp is delivered into the second sperm storage organ of the female. Most males do not survive the second sperm transfer. Sexual cannibalism in redback spiders serves to enhance the male's success at fertilizing all the female's eggs because, by offering his body as a snack for the female, he can prolong the time for transferring a maximum amount of sperm. The self-sacrificial mating behaviour of male redback spiders must impose strong selection to ensure that females don't subsequently mate with other males, which would dilute the first male's paternity of the eggs. This may explain why the distal tip of the male's pedipalps break off at the entrance to the sperm storage organs of the female during mating. This broken fragment of male copulatory organ wedged into each sperm storage organ of the female prevents the sperm of other males from entering.

A species of acanthocephalan worm, *Moniliformis dubius* (311), has found an additional use for copulatory plugs in the battle among males to monopolize access to the eggs of females. All species within the phylum Acanthocephala are endoparasites that reach maturity within the alimentary tract of a vertebrate host. Anchorage within the host's intestine is accomplished by a proboscis covered with recurved hooks that is driven into the intestinal wall. *M. dubius* lives within the small intestine of rats, which may acquire heavy infections by eating cockroaches infected with the larval stage of this species of acanthocephalan. A single male of *M. dubuis* can mate with many mature females, and the sperm are long-lived within the female's reproductive tract. Therefore, the potential for female inseminations by multiple males is high; this would substantially dilute the paternity of any male among the fertilized eggs. To prevent this, each male has a cement gland at the base of the copulatory organ that secretes a seal over the female **gonopore** (reproductive opening) after sperm have been delivered, which prevents the female from receiving sperm from other males. The seal eventually dissolves so that fertilized eggs can leave the female's body. Males of *M. dubuis* will also "pseudo-copulate" with other males, not to deliver sperm but rather to seal shut the gonopores of other males with secretions from the cement gland. This is an inordinately aggressive strategy to neutralize rivals.

14.7c Sperm Competition

Competition among males for access to female gametes can occur even at the level of individual sperm cells. In species that mate promiscuously, ejaculates of different males regularly occur in the reproductive tract of females. Fertilization success often reflects the numbers of sperm available, the male with the most sperm achieving the most fertilizations. But research has shown that sperm can have features that enhance their swimming speed along the female reproductive tract. The sperm head of rodents has a prominent crest and

hook (Fig. 14.38A) that facilitates the formation of motile aggregations of sperm that swim more rapidly than individual sperm (Fig. 14.38B). A recent study used two colours of fluorescent mitochondrial labels to ask if sperm from the same individual preferentially formed speedy aggregates in the presence of sperm from another individual. Two species of deer mice were used in these experiments: *Peromyscus polionotus* (1899) is a monogamous species, and *P. maniculatus* (1898) is a promiscuous breeder. A first experiment found that when sperm from the two different species were mixed together, sperm preferentially formed aggregates with conspecific sperm (Fig. 14.38B). More importantly, further experiments showed that sperm from the polygamous species were able to recognize and selectively form aggregates with other sperm from the same individual or from a close relative, whereas sperm from the monogamous species was less able to discriminate self-sperm or familial relatedness among other sperm. This suggests the promiscuous species has experienced stronger sexual selection; greater competition among self and non-self sperm has fostered the ability to preferentially form high-speed aggregates with self sperm.

14.7d Cheaters within Mating Systems

Mating activity outside of a male-female pair bond established through contest and courtship between individuals ("extra-pair copulation" (EPC)) is quite common. The behaviour is described as "cheating" because it may deceive a mating partner into caring for young that are not theirs or may usurp mating opportunities that have not been won through honest competition. However, EPCs may confer significant

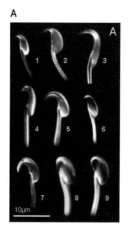 

Figure 14.38 Rodent sperm and sperm competition. **A.** The sperm head of rodents has a prominent crest and hook that promote the formation of fast-swimming aggregations. **B.** Sperm aggregation where the hooked heads have linked together. The sperm from each donor can be distinguished because the mitochondria within the midpiece were labelled with different coloured fluorescent labels.

SOURCE: A. S. Immler, H.D.M. Moore, W.G. Breed, T.R. Birkhead. 2007. "By Hook or by Crook? Morphometry, Competition and Cooperation in Rodent Sperm," *PLoS ONE* 2(1): e170; B. Heidi S. Fisher/Harvard University

advantages to one or both sexes. In particular, a male may use EPCs to increase the number of young he sires. A female may use EPCs to increase the genetic diversity or quality of her offspring. Both sexes can ensure against possible infertility of their chosen mate by occasionally mating outside the pair bond. Nevertheless, the behaviour can also carry risks. Both sexes may expose themselves to disease or injury from partner retaliation and males pursuing females elsewhere may be cuckolded at home during their dallying absences.

Measuring the extent of extra-pair fertilizations and their significance for sexual selection has emerged from the advent of molecular techniques, such as DNA fingerprinting. It is now possible to unambiguously determine parentage of offspring. Before DNA fingerprinting, most studies of the reproductive behaviour of birds suggested that these animals were monogamous—a male and female build a nest together, mate, lay eggs, and raise young. DNA fingerprinting has revealed a surprisingly high incidence of extra-pair fertilizations among "monogamous" bird species. A female's mate is not always the father of all her offspring. Since males of monogamous bird species typically contribute substantially to parental care of offspring, instances of extra-pair copulations can be viewed as sexual parasitism.

The Purple Martin (*Progne subis* (1612)) provides an extreme example of sexual parasitism within a seemingly monogamous bird. Early in the breeding season, aggressive older males secure multiple nesting boxes by excluding other older males. The male then attracts and mates with a female, who commences nesting in one of the boxes. After the female begins egg-laying, the male uses vocalizations to attract younger males and females to the other nesting boxes. After the young pairs have begun nesting, the older male parasitizes the parental care by the younger males by swooping in and mating with the younger males' mates. DNA fingerprinting shows a substantial fitness advantage for the aggressive behaviour of the older male. Older males fathered, on average, 8.1 offspring: 4.1 with their mates and 3.6 from the nests of younger parasitized males. The young parasitized males fathered only 29% of the eggs in their nests.

Another tactic used by males of some species to achieve copulations without honest competition for mates is the "sneaker male" strategy. Bluegill sunfish (*Lepomis macrochirus* (1271)) (Fig. 14.39) show two forms

of this strategy. Conventional males of this species, so-called parental males, achieve maturity at seven years of age. They make nests on the lake bottom and attract females. When a female enters a parental male's territory, he guides her to his nest and fertilizes her eggs as she releases them into the nest. Parental males may breed with several females a day and provide sole parental care for the developing eggs, which involves fanning eggs to keep them aerated and protecting eggs from predators. A parental male does not feed while caring for eggs in his nest. However, two kinds of other male phenotypes may cheat the parental male. Small-sized sneaker males, which mature precociously at three years of age, release sperm during rapid swims through nests where females and parental males are spawning. Another class of males, so-called satellite males, impersonate females in size, colouration, and behaviour to the extent that parental males will tolerate satellite males in their nest. These satellites can release sperm over the eggs of an ovipositing female. Nevertheless, bluegill sunfish provide evidence of an evolutionary arms race within the context of sexual selection. The cheating tactics of the sneaker and satellite males are partially foiled because the parental males are able to recognize their own young through a pheromonal cue and preferentially care for them.

Sneaker male tactics have been documented among a few invertebrates. For example, the giant Australian cuttlefish (*Sepia apama* (516), Mollusca) has a mating behaviour not unlike the bluegill sunfish, in that males construct nests to attract females. Competition for mating opportunities with females is intense: males guard females and aggressively chase off other males. Sneaker males in this species are sexual mimics of females; they hide their two male-type arms, adjust their chromatophore pattern to imitate female mottling, and assume a posture that mimics that of egg-laying females. "Honest" males tolerate the sexual mimics, but when they briefly leave the nest to chase off more obvious male competitors, the sneakers can transfer a spermatophore into the female's mantle cavity.

14.7e Female Choice

As previously discussed, female success often depends on her choice of high-quality mating partners. Females choose mates according to diverse criteria, for example, resources offered by males (territory for foraging, nuptial gifts), good parent assessment, and/or genotype quality (assessed by symmetry, male ornaments, or competitive success in battle with other males). Indeed, virtually all secondary sexual characteristics have arisen due to sexual selection, and the most elaborate of these are generally features of males that have been driven by female assessment of male quality. The flamboyant plumage of male peacocks is a classic example of the

Figure 14.39 Bluegill sunfish (*Lepomis macrochirus* (1271)).

Bruce MacQueen / Shutterstock.com

remarkable degree of male decoration driven by female preference. Similarly, the large size and combat ability of northern fur seals, which directly translates into larger harems of females, can be attributed to female preference for physically superior males.

Lemon tetras (*Hyphessobrycon pulchripinnis* (1183)) provide another example of female choice among potential mating partners (Fig. 14.40). Males of this small tropical fish species defend territories and attempt to entice a female to enter the territory for spawning. After a female enters a male's territory, she will spawn approximately 23 times, requiring multiple spawn ejaculates from the male. In this promiscuous mating system, males continuously attempt to entice females to their territory. Research has shown that, contrary to the popular notion that males have an inexhaustible supply of sperm, male lemon tetras become progressively more infertile with successive spawn ejaculates. After 20 spawning events during a single day, they are able to fertilize only about 50% of additional eggs spawned by females. Perhaps not surprisingly, female lemon tetras prefer to mate with males that have not mated previously. How the females recognize males that have recently spawned remains open to conjecture.

STUDY BREAK

1. What is the difference between natural selection and sexual selection?
2. Explain why intrasexual selection is most often the dominating form of sexual selection for males of a species, but not for females.
3. What are the three main types of mating systems?
4. What functions are served by courtship displays?
5. Give an example of an animal trait (any aspect of phenotype) that has likely resulted from intrasexual competition. Give an example of an animal trait that has likely resulted from intersexual competition.

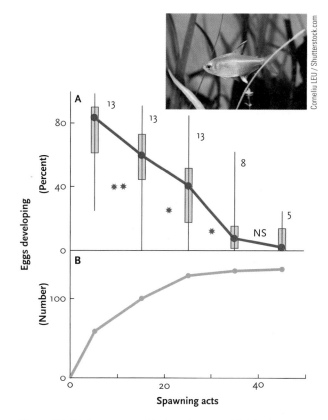

Figure 14.40 Lemon tetra (*Hyphessobrycon pulchripinnis* (1183), inset). **A.** The relationships between daily spawning acts and fertility and **B.** between daily spawning acts and predicted number of offspring. Sample sizes are shown along with probability values for between-pair tests (NS = not significantly different; $*$ = $p < 0.05$; $**$ = $p < 0.01$).

SOURCE: From Nakatsuru, K. and D.L. Kramer 1982. "Is Sperm Cheap? Limited Male Fertility and Female Choice in the Lemon Tetra (Pisces, Characidae)," *Science* 14 May 1982: Vol. 216 no. 4547 pp. 753–755. Reprinted with permission from AAAS.

REPRODUCTION **IN PERSPECTIVE**

This is a good time to consider the links between reproduction and other chapters in this book, particularly evolutionary processes, larvae and life cycles, symbiosis and parasitism, behaviour, and interactions between animals and humans. Consider the impact that the drive to reproduce has on the behaviour of your pet, or on your own behaviour.

Questions

Self-Test Questions

1. What type of reproduction is demonstrated by organisms produced by fertilization of haploid ova by male sperm to produce organisms that use genes from both genomes for somatic development, but only pass along the female genes to the following generation?
   a. bisexual reproduction
   b. unisexual reproduction
   c. sexual reproduction
   d. asexual reproduction

2. Which of the terms describing gamete production and gonad type are correctly paired with pattern of reproduction?
   a. unisexual reproduction: monoecious animals, ovotestis
   b. unisexual reproduction: gonochoristic animals, ovotestis
   c. bisexual reproduction: dioecious animals, ovotestis
   d. bisexual reproduction: monoecious animals, ovotestis

3. Which of these egg–sperm interactions is critical for the prevention of polyspermy?
   a. The sperm contributes both its haploid nucleus and its proximal centriole to the egg.
   b. Transitory depolarization of the egg membrane elicits the exocytosis of cortical granules.
   c. The egg jelly coat releases species-specific sperm attractants and triggers the acrosomal reaction of the sperm.
   d. Polymerization of actin granules form filamentous actin in the extension of the acrosomal process to bind to species-specific bindin on the vitelline envelope.

4. What is characteristic about the environmental cues important in synchronizing gametogenesis, spawning and mating?
   a. Cues are linked to seasonal changes such as temperature and photoperiod.
   b. Spawning in marine species is linked to high tidal current to maximize dispersal of eggs and sperm.
   c. A hierarchy of cues involving environmental cues, local conditions and social cues serve to fine-tune the timing of reproductive events.
   d. Synchronization is important primarily for species that rely on external fertilization.

5. Which of the following examples best demonstrates intra-sexual selection among males?
   a. Male damselflies use their copulatory organ to first remove any previously deposited spermatophores from the female's reproductive tract before depositing their own.
   b. The flamboyant plumage of male birds is attractive to females, but puts males at risk for predation.
   c. Self-sacrifice of male redback spiders occurs during sperm transfer.
   d. Extra-pair copulation may occur outside an established pair bond when females mate with other males.

Questions for Discussion

1. How would you measure the fitness of an individual? How would knowledge of parentage influence your measure? How would your calculations differ among sexually and asexually reproducing species?

2. How does seasonality influence reproduction? What cues do animals use to trigger reproduction? Are humans seasonal in their reproduction?

3. What effect can climate change have on the incidence of males and females in a population? Why is temperature-dependent sex determination not known in birds and mammals?

4. Use the electronic library to find a recent paper that explores some aspect of animal reproduction. Does the new information revolutionize any of the information in this chapter?

Figure 15.1 A spitting cobra (*Naja mossambica* (1510)) lets fly with a spray of venom. These snakes can spit accurately at moving targets more than 1.5 m away.

SOURCE: Stu Porter / Shutterstock.com

STUDY PLAN

15 Neural Integration

WHY IT MATTERS

The co-ordination of action is a signal feature of animal behaviour. The several species of spitting cobras (Fig. 15.1) from Africa and Asia are notorious examples. These snakes are adept at spitting venom into the eyes of vertebrate attackers. To be effective, their aim must be good enough for the venom to contact the cornea. The challenge to the cobra is fourfold:

1. the target is usually moving
2. targets are often >1.5 m away
3. spitting takes ~50 ms
4. the hole in the fangs through which the venom is expelled is fixed

This means that spitting cobras aim their heads (mouths) by rapid lateral and vertical movements. When studied in the laboratory, three species of spitting cobras (*Naja pallida* (1512)), *N. nigricollis* (1511)), and *N. siamensis* (1513)) showed an extraordinary ability to track and anticipate a target's movements. This level of neural processing had not previously been documented in reptiles.

Archerfish (*Toxotes chatareus* (1294)) look from water into air to locate potential prey, usually insects (Fig. 15.2). They then spit droplets

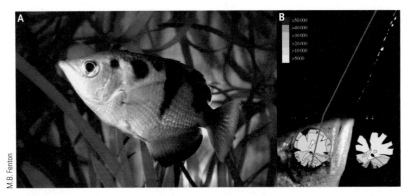

Figure 15.2 A. An archerfish (*Toxotes chatareus* (1294)). **B.** Specializations of its eyes. Shown are the line of sight (red line) and the direction of a spit. The distribution of cones (inset on eye) and ganglion cells (right inset) are overlays on the image of the archerfish eye. Densities of cells are shown by colour code (cells/mm²). Note the bending of the red line as it passes through the water's surface.

SOURCE: B. Temple et al. 2010. "A spitting image: specializations in archerfish eyes for vision at the interface between air and water." *Proc. R. Soc. B* 7 September 2010 vol. 277 no. 1694 2607–2615, by permission of the Royal Society.

animals appear to lack image-forming eyes, the spit aiming mechanism remains unknown.

SETTING **THE SCENE**

As noted above in the case of fly balls or spitting cobras, animals must co-ordinate their actions in order to succeed. The nervous system and a multitude of nerves, sensors, effectors, and connections are fundamental to co-ordinated activity that underlies virtually all activities. We progress from the components of the nervous system including activity of nerves, signalling, networks, transduction, and amplification. We also present information about a variety of receptors that provide input about light and infrared, as well as vibrations including both seismic and sound. Chemoreception, electroreception, and intrinsic receptors are also treated. We use some examples of animals in action, and finish with questions about how (if?) animals detect earthquakes or navigate.

of water at the prey, knocking it into the water where they can eat it. Archerfish eyes are specialized for operating at the interface between water and air, with rods and cones showing quantitative differences correlated with aquatic versus aerial views. These differentially tuned rods and cones are distributed across the retina. Many other predators look in the other direction, from air into water (Fig. 15.3), so also require visual correction for the bending of light at the air-water interface.

Catching a fly ball in baseball requires considerable hand-eye co-ordination (Fig. 15.4A), even when the would-be fielder is in the right place at the right time. To resolve the problems of space and time, outfielders convert the position in time to a position in space. They appear to do this by running a path (Fig. 15.4B) that maintains the same linear optical trajectory. This approach gives the fielder some control over the relative direction of the ball, rather like the tracking behaviour of some predators.

Integration of sensory input is crucial to co-ordination, but we do not always know the details. For example, many spiders "spit silk" (see page 501) and presumably rely on vision to aim the silk. Onychophorans are well known for their ability to spit a fast-drying adhesive for both prey capture and defense (Fig. 17.42). Since these

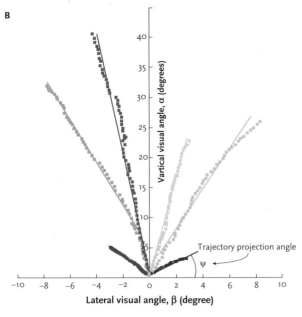

Figure 15.4 A. The fielder about to catch a fly ball maximizes the chance of being in the right place at the right time by running a linear optical trajectory. **B.** Six linear optical trajectories are shown: two line drives (projection angle <5°) and four high fly balls. The origin is home plate.

SOURCE: From McBeath, M.K., Shaffer, D. M., & Kaiser, M. K., "How baseball outfielders determine where to run to catch fly balls," *Science* 28 April 1995: Vol. 268 no. 5210 pp. 569–573. Reprinted with permission from AAAS.

Figure 15.3 Pied Kingfishers (*Ceryle rudis* (1575)) find prey by looking from air into water.

15.1 Fundamentals of Nerve Activity

Like most other cells in the body, nerve cells or **neurons** have a nucleus and a variety of organelles that perform different functions (Fig. 15.5). Neurons also share the property of the **membrane potential**, an electrical potential difference across the cell membrane, which is a fundamental property of all cells. All living cells must have this charge difference across their membranes because it is essential for the cellular functions carried out by the different organelles. Brief changes in membrane potential are the basis for electrical signalling, characteristic of nerve cells, and integral to the function of muscle and glandular cells.

The intracellular fluid (cytoplasm) of all cells consists primarily of water, protein, and inorganic salts (Fig. 15.6). The proteins range in size from large structural macromolecules and enzymes down to smaller polypeptides and amino acids. Many of these molecules have terminal groups that are dissociated (ionized) in the aqueous medium of the cytoplasm, and their net electrical charge makes them ions. In the squid giant axon the specific nature of these organic ions can be determined by simply squeezing out the axoplasm and assaying it. The main organic ion is **isethionate**, which has a net negative charge making it an organic **anion** (A^- in Fig. 15.6). Other nerve cells include organic ions such as glutamate, aspartate, and organic phosphates. Whatever their identity, the net charge on these molecules is negative with respect to the outside of the cell, so they, too, are anions.

Inorganic ions in the cytoplasm are required for the operation of many enzymes and the maintenance of electrical, chemical, and osmotic equilibrium within the cell and between the cytoplasm and the extracellular fluid. The primary intracellular cation is potassium (K^+ in Fig. 15.6), along with small amounts of calcium (Ca^{++}) and magnesium (Mg^{++}). The inorganic anions include chloride (Cl^-), phosphate (PO_4^-), and sulphate (SO_4^-). The electrolyte composition differs between inside and outside the cell: the main extracellular cation is sodium (Na^+), and the main anion is chloride. This is summarized more quantitatively in Figure 15.6 for an invertebrate cell (the squid giant axon) and a vertebrate cell (frog muscle fibres). Both cell types have been well studied and are thought to be characteristic of invertebrate and vertebrate neurons. Although the specific ion concentrations are different, you will notice that the **ion gradients**, the differences across the membrane, are similar in both relative magnitude and direction in both types of cells. Differences in ionic concentrations between vertebrates and invertebrates suggest different ways to achieve the same end.

The electrical activity of neurons is derived from two factors:

1. the ionic gradients across the membrane, determined by concentration and electrical differences (the **electrochemical gradient** for each ion)
2. changes in the permeability of membranes to these ions, based upon ion channel activity. The

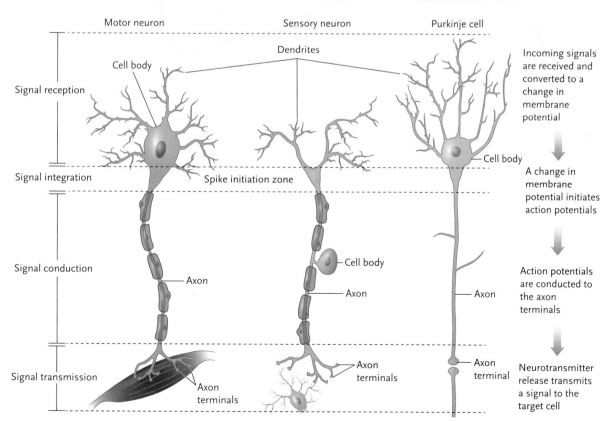

Motor neuron   Sensory neuron   Purkinje cell

Dendrites

Cell body

Signal reception

Signal integration   Spike initiation zone

Cell body

Signal conduction   Axon   Axon   Axon

Cell body

Signal transmission   Axon terminals   Axon terminals   Axon terminal

Incoming signals are received and converted to a change in membrane potential

A change in membrane potential initiates action potentials

Action potentials are conducted to the axon terminals

Neurotransmitter release transmits a signal to the target cell

Figure 15.5 Neurons have many shapes and sizes, but share common cellular features, including the cell body or soma (s), slender cellular extensions called dendrites (d), and a single axon (a) through which signals are transmitted to adjacent cells.

SOURCE: From RUSSELL/WOLFE/HERTZ/STARR. *Biology*, 2E. © 2013 Nelson Education Ltd. Reproduced by permission. www.cengage.com/permissions

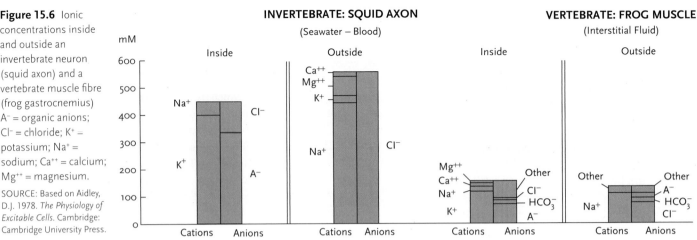

Figure 15.6 Ionic concentrations inside and outside an invertebrate neuron (squid axon) and a vertebrate muscle fibre (frog gastrocnemius) A^- = organic anions; Cl^- = chloride; K^+ = potassium; Na^+ = sodium; Ca^{++} = calcium; Mg^{++} = magnesium.

SOURCE: Based on Aidley, D.J. 1978. *The Physiology of Excitable Cells.* Cambridge: Cambridge University Press.

changes in permeability allow or inhibit ions to move along their gradients across the membrane, which creates electrical currents.

Since the concentration gradients for the major ions (sodium, potassium, and chloride) involved in neural signals are known, we can measure the electrical forces (E_{mem}) and calculate the membrane potentials at which these forces are balanced such that no net movement of that ion will occur. The value of E_{mem} for an ion is its **equilibrium potential** (E), calculated using the **Nernst equation** (named for Walther Nernst, a German chemist, who derived it in 1888). The Nernst equation allows us to express the concentration gradient of an ion in mV, so that it can be compared to the membrane potential, also expressed in mV.

The Nernst equation looks like this:

$$E = \frac{RT}{zF} \ln \frac{[\text{ion outside cell}]}{[\text{ion inside cell}]} = 2.303 \frac{RT}{zF} \log_{10} \frac{[\text{ion outside cell}]}{[\text{ion inside cell}]}$$

where:

E = equilibrium potential

R = gas constant (a measure of the energy of the substance)

T = absolute temperature (a substance is more active as the temperature is raised)

z = valence of the ion (+1 for potassium)

F = Faraday's constant (magnitude of electric charge per mole of electrons)

RT/zF is the potential needed to balance an e-fold concentration gradient across the membrane, and at room temperature RT/zF = 25 mV. The equilibrium potential for K^+ is E_K = 25 ln (K_o/K_i) mV. If we convert from natural logarithms to base 10 logarithms, E_K = 58 log (K_o/K_i) mV. For a typical cell, the concentration ratio for K is 1:30, and E_K = −85 mV. This is the membrane potential that balances the outward concentration gradient of K so that there is no net movement of K^+. A typical cell with a resting membrane potential of around −70 mV has a small outward force on the K^+, but most nerve cells have a high resting K^+ permeability (known as a "leakage" current), so K^+ ions are the major determinant of the resting membrane potential.

What are the resting conditions on other ions? If we apply the Nernst equation to sodium, E_{Na} = 58 log (Na_o/Na_i) mV. For a typical cell in which the extracellular concentration ratio of Na^+ is about 10 times higher than intracellular Na^+, E_{Na} is typically 60 mV. So, for a typical resting cell, there is a strong inward force on Na^+, which only balances out when sufficient Na^+ enters the cell to bring the membrane potential up toward 60 mV, at which point Na^+ reaches its equilibrium potential. Similarly, all other ions have equilibrium potentials produced by the same two forces. Of particular interest is Ca^{++}, which has an even higher concentration gradient than Na^+, so E_{Ca} is typically 155 mV. Also of interest is Cl^-, which is distributed approximately equally in concentration across the membrane, so E_{Cl} is typically near the value of the resting membrane potential, around −65 mV.

The Nernst equation allows us to predict the movement of ions across the cell membrane. Because we can express the concentration gradient in mV, we can compare the force on the ions to the electrical force provided by the membrane potential. In effect, the concentration gradients for ions in living cells do NOT change, so ion movement across the membrane is largely a function of the membrane potential.

STUDY BREAK

1. What is responsible for electrical activity in cells? In nerve cells?
2. How does the Nernst equation inform us about neural activity?

15.2 Signalling

15.2a Membrane Potentials

The signals used by neurons to integrate and transmit information are electrical in nature, consisting of potential changes across the cell membrane that are produced by the electrical currents of ion movements. The currents are carried primarily by Na⁺ and K⁺ ions. All cell membranes, including those of neurons, have ion channels that, at rest, exhibit a random pattern of opening and closing. This channel activity and the resultant ionic currents have been observed directly, through the technique of **patch clamping** (see Box 15.1). Recordings of ion movement through the channels provide information about channel opening and closing and about the contributions of different ions to changes in membrane potential.

Although stimuli applied to the membrane of any cell create transient changes in channel activity and ion movements, these signals are severely reduced or "attenuated" over a relatively short distance across a membrane. These signals are therefore referred to as **graded** or **localized potentials**, and fall into a category of signals that neurons share with all other types of cells, including plant cells! While they encode stimulus strength in terms of the change in membrane potential, these signals degrade quickly both in time and space. The main characteristic of local potentials is that they can be graded continuously in size. In sensory endings, such potentials are known as generator potentials or **receptor potentials**. For example, the receptor potential of a sensory nerve ending that is sensitive to pressure on the skin is proportional in size to the magnitude of the applied pressure. More pressure opens more ion channels and generates a larger local current. The nervous system includes many types of such endings or receptors, each responsive to one type of physical stimulus such as distortion of a membrane, bending of a hair, changes in temperature, changes in the angle of a joint, or (in photoreceptors) light. Neural receptors and their ability to transduce physical cues

into electrical signals (see Section 15.5) form the basis for the sensory systems described in this chapter.

While patch clamp technology was originally designed to explore the action of Na⁺ ion channels, K⁺ channels are also important in the generation of graded potentials. While the electrochemical gradient for Na⁺ ions is strongly inward in a resting cell, the resting forces on K⁺ ions are directed outwards (Fig. 15.7). If a stimulus opens sodium channels in a membrane, the sodium ions move along their electrochemical gradient into the cells until they reach their equilibrium potential (E) as calculated by the Nernst equation. This **sodium influx** makes the membrane less negative at the point of entry and is called a **depolarization** (Fig. 15.8). However, if potassium channel activity is triggered by the stimulus at the membrane, the opening of potassium channels will allow a **potassium efflux** as

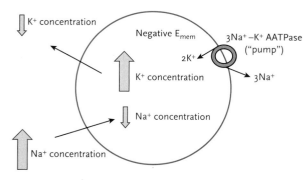

Figure 15.7 Electrochemical gradients for sodium and potassium. The resting membrane potential (E_{mem}) is the charge across the membrane, which ranges in magnitude from −55 mV to −110 mV, depending upon the cell type.

BOX 15.1
Patch Clamps and Ion Currents

The technique of patch clamping was developed by E. Neher and B. Sakmann and their colleagues to record from small cells, and earned them the Nobel Prize for Physiology or Medicine in 1991. Basically, the patch clamp is an electrode that forms a seal in contact with the cell membrane (Fig. 1), such that only two or three membrane channels are included in the "patch" of membrane covered by the electrode. In this way, ion movement through the channels can be recorded, which provides information

about channel opening and closing. Voltage pulses applied to the patch serve as stimuli to open ion channels, with more channels opening for longer periods of time as the stimulus.

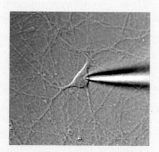

Figure 1
Photograph of a cultured hippocampal neuron under a somatic whole-cell patch clamp

SOURCE: Courtesy of Arne Battefeld and Ulf Strauss, Charité-Universitaetsmedizin Berlin

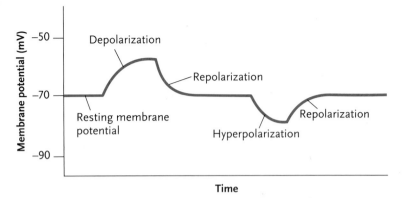

Figure 15.8 Changes in membrane potential in a cell due to changes in ion permeability. The resting membrane potential (E_{mem}) in this cell is –70 mV. During depolarization, E_{mem} becomes less negative. During hyperpolarization, E_{mem} becomes more negative. During repolarization, E_{mem} returns to its resting level.

SOURCE: From RUSSELL/WOLFE/HERTZ/STARR. *Biology*, 2E. © 2013 Nelson Education Ltd. Reproduced by permission. www.cengage.com/permissions

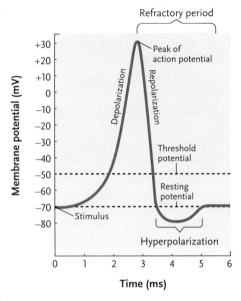

Figure 15.9 Changes in membrane potential during an action potential.

SOURCE: From RUSSELL/WOLFE/HERTZ/STARR. *Biology*, 2E. © 2013 Nelson Education Ltd. Reproduced by permission. www.cengage.com/permissions

the ions move toward their more negative equilibrium potential, and this movement of potassium out of the cell increases the negativity of the membrane potentials, called a **hyperpolarization**.

15.2b Membrane Potentials in Neurons

Unlike other cells, neurons can generate a second type of impulse that allows them to transmit a signal, in the form of a rapid wave of depolarization (a nerve impulse), along the neuron membrane to target cells and tissues (effectors). These signals are made possible by **voltage-sensitive ion channels** located in the axonal membranes of neurons. Some other excitable cells, such as muscles and glands, have similar channels, but only neurons have the cell structure and axons that allow transmission of signals from one cell to the next. The voltage-sensitive channels respond to the membrane potential (E_{mem}). If E_{mem} is increased above the resting potential by a difference referred to as the **threshold potential**, then voltage-sensitive sodium channels open to generate a signal, an **action potential** (Fig. 15.9). The membrane depolarizes so rapidly that the inside of the neuron becomes transiently positive, so that when Na⁺ channels close and voltage-sensitive potassium channels open, the rapid efflux of potassium repolarizes the cell just as rapidly. This entire event typically lasts 2–3 ms. In most neurons, the action potential is terminated by a brief hyperpolarization, during which the potassium channels close.

In contrast to local graded potentials, an action potential is a brief event that travels unattenuated along an axon. The action potential is often called "all-or-none," meaning that its amplitude remains relatively constant along the axon. The entire action potential sequence must occur before another action potential can be initiated at that point in the membrane.

Evidence for channel activity and ion movements during an action potential, particularly the role of Na⁺ ions, has resulted from the development of the **voltage clamp**, devised by Cole in the 1950s and further developed by A.L. Hodgkin, A.E. Huxley, and B. Katz. Using electronic technology, voltage clamps hold the membrane potential at a constant value and measure ion channel activity (conductance) by recording ion movements (currents). Cole and his colleagues worked with squid giant axons bathed in seawater. They inserted fine silver wires longitudinally into the axon and passed a set current across the membrane of the axon to "clamp" the membrane potential at a set voltage. With this clamped voltage they could determine which currents are carried by individual ions while monitoring rapid changes in membrane potential and ion movements associated with an action potential. They then estimated the magnitude and time course changes in membrane permeability created by changes in ion channel activity (Fig. 15.10).

A **refractory period**, during which a second impulse cannot be initiated, is the time it takes to reset the voltage-sensitive Na⁺ channel. When depolarization of a nerve above the threshold outlasts the refractory period, a second action potential may be initiated. In many neurons, prolonged depolarization may produce a train of action potentials that last as long as the stimulating depolarization. The frequency of the repeated action potentials is limited by the refractory period and ion channel activity.

Propagation of the signal along an axon occurs because of voltage-sensitive ion channels. Sodium influx and the resultant depolarization of an action potential at one point along the axonal membrane stimulate the sodium channels at adjacent regions of

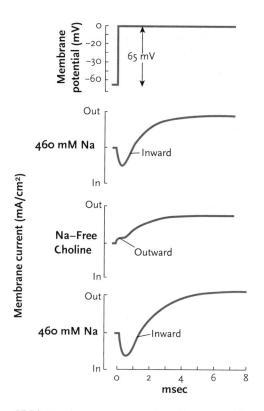

Figure 15.10 Membrane current contributed by an inward flow of sodium ions when the membrane of a squid axon is depolarized by 65 mV. In sodium-free solution, the inward current disappears, replaced by a small transient outward current as sodium moves out of the axon. The choline in this trace does not permeate the membrane, but replaces sodium to maintain osmotic balance. The inward current is restored when sodium is reintroduced.

SOURCE: Based on Hodgkin, A. L. and A. F. Huxley. 1952. "A quantitative description of membrane current and its application to conduction and excitation in nerve," *J. Physiol.* (London) 117:500–544.

the axon and move the depolarization down the axon (Fig. 15.11).

The action potential is propagated along the axon and does not decline with distance. As well, action potential amplitudes do not vary much from one kind of nerve fibre to the next. The speed of propagation of the action potential, however, does vary, being greater in larger-diameter axons than those of smaller diameters. In many situations, there is selection pressure for faster nerve conduction; sensory inputs about a threat or potential prey must be integrated, and an appropriate motor response triggered. The faster this happens, the greater the advantage to the animal. Action potential conduction speed depends on two physical properties of the axon: its internal electrical resistance and the capacitance of its membrane. The greater the electrical resistance or the membrane capacitance, the longer it takes for an action potential to charge the next region of the axon and the slower the impulse travels. Reducing the internal resistance or the membrane capacitance (or both) increases the speed of transmission. The simple way of effecting such changes is to increase the diameter of the axon; larger

axons have lower resistance and membrane capacitance and conduct action potentials faster. Increases in neuron diameter increase action potential conduction in proportion to the square root of the axon diameter. Clearly if all the axons in a bundle were enlarged to provide faster conduction, then nerve cords would become too big to fit into available space. Giant axons have evolved as elements in the nervous system of many invertebrates where rapid coordination of a response is important. Crayfish giant axons coordinate an escape response, and these axons conduct at a rate of about $2 \text{ m} \cdot \text{s}^{-1}$. Squid giant axons may be as much as 2 mm in diameter and coordinate mantle contraction in rapid jet propulsion (see page 199); squid axons can conduct an impulse at up to $25 \text{ m} \cdot \text{s}^{-1}$. Giant axons are not limited to large invertebrates coordinating dramatic responses. Giant axons are found in the antennal nerves (sensory) of copepods and in the motor connections involved in flight initiation in *Drosophila* (960)); even in these small organisms speed of response is a part of survival.

In vertebrates (except lampreys and hagfish), impulses do not move steadily along an axon. Instead they jump from point to point, a process known as **saltatory** conduction. In Figure 15.5 both sensory and motor axons are diagrammed with a series of purple-coloured wraps around the axon. This sheathing is a spiral-bound covering of myelin, and each gap between regions of the myelin sheath is a **node of Ranvier** (Fig. 15.12A). The physiology of saltatory conduction is explained in Figure 15.12B.

Saltatory conduction allows high-speed conduction along small axons. Small internal sensory neurons may propagate a signal at a rate of 5–$10 \text{ m} \cdot \text{s}^{-1}$, whereas signals in large mammalian motor neurons can travel as quickly as $120 \text{ m} \cdot \text{s}^{-1}$.

Although myelination is frequently viewed as being restricted to vertebrates, myelin-wrapped nerves have long been known in some groups of invertebrates. Although the morphology of myelination in invertebrates differs from that of vertebrates, the physiology of saltatory conduction is the same. Invertebrate myelination does not follow any phylogenetic pattern. For example, many shrimps have myelinated nerves while crabs and lobsters (also decapods) do not. Some calanoid copepods have myelinated fibres while others lack myelin. Myelin is found in some annelids (e.g., present in earthworms but absent in leeches) but has not been reported in insects or molluscs. Large myelinated fibres in a shrimp achieve the highest conduction speeds recorded for any nerve: $200 \text{ m} \cdot \text{s}^{-1}$.

15.2c Synapses

Each nerve cell has at least one **synapse**, a region at its axon terminal specialized for communication between neurons or between a neuron and its target. Synaptic transmission can be electrical or chemical (Fig. 15.13).

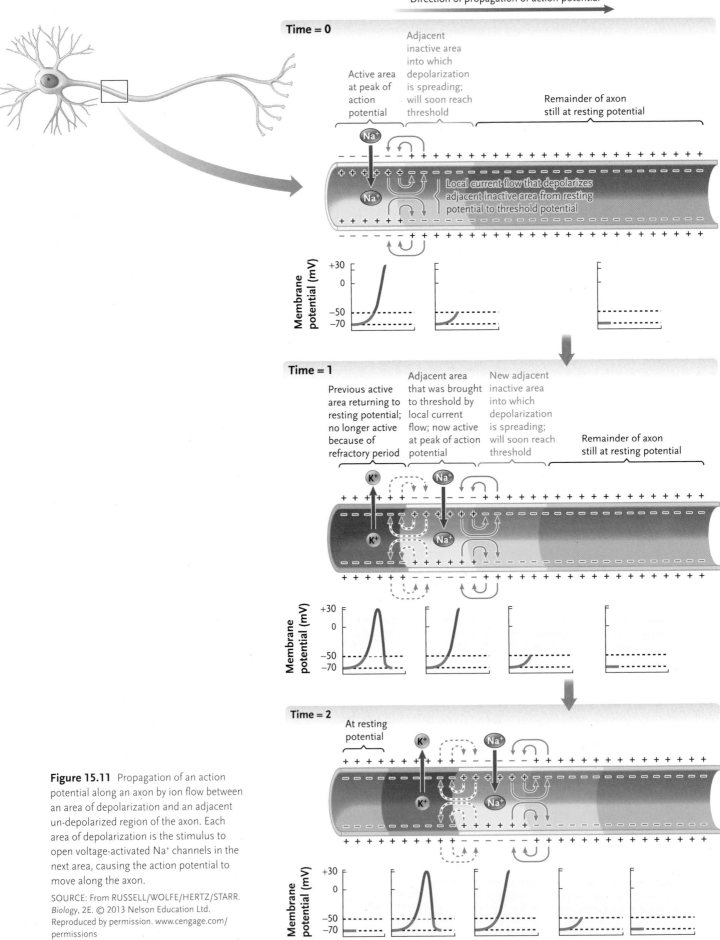

Figure 15.11 Propagation of an action potential along an axon by ion flow between an area of depolarization and an adjacent un-depolarized region of the axon. Each area of depolarization is the stimulus to open voltage-activated Na⁺ channels in the next area, causing the action potential to move along the axon.

SOURCE: From RUSSELL/WOLFE/HERTZ/STARR. *Biology*, 2E. © 2013 Nelson Education Ltd. Reproduced by permission. www.cengage.com/permissions

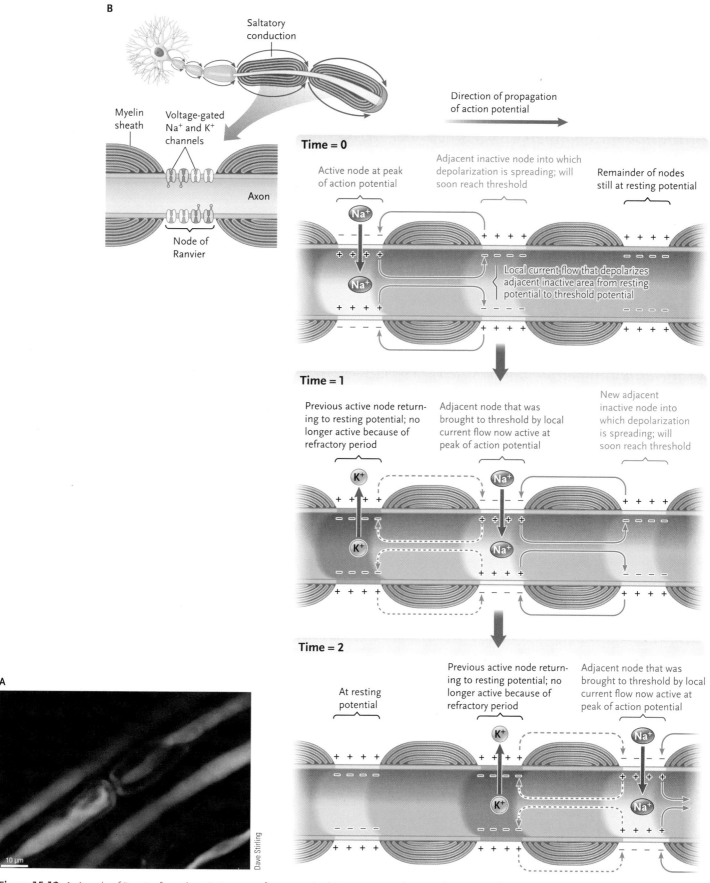

Figure 15.12 A. A node of Ranvier from the sciatic nerve of a mouse. In this preparation, the axon shows green fluorescence, and the myelin sheath appears red. **B.** The physiology of saltatory conduction.

SOURCE: From RUSSELL/WOLFE/HERTZ/STARR. *Biology*, 2E. © 2013 Nelson Education Ltd. Reproduced by permission. www.cengage.com/permissions

Gap junctions (Fig. 15.13A) are synapses specialized for **electrical transmission** of the action potential via membrane proteins that span the gap between the **presynaptic** and **postsynaptic cells** (Fig. 15.13A). Thus current can flow directly from one cell to the next through the gap junctions. This type of direct electrical transmission is very fast, and is characteristic of many circuits that require simultaneous rapid and coordinated activity among many cells. Electrical synapses can conduct action potentials in either direction so a signal can arise from any of the electrically-connected cells and travel throughout the circuit. This can be extremely beneficial in coordinated rapid responses to external stimuli, for example, in the defense systems of many insects. The disadvantage of gap junctions is illustrated by the uncoordinated electrical events that can arise in cardiac muscle during fibrillation (abnormal rapid and irregular contraction or "twitching").

Chemical transmission involves the release of a chemical **neurotransmitter** from the presynaptic cells (Fig. 15.13B). Chemical transmission is slower than electrical transmission, but it has the advantage of directionality (communication only passes from presynaptic to postsynaptic cells). As well, the communication is not fixed but is **plastic** because it can strengthen or weaken over time.

Communication across a chemical synapse starts with depolarization of the axon terminal (Fig. 15.14). Depolarization opens voltage-activated Ca^{++} channels

in the presynaptic membrane. The driving force on Ca^{++} ions is very high because of a large concentration gradient directed inward and the negative electrical potential of the neuron. Therefore the influx of Ca^{++} serves as a powerful signal in the axon terminal. Neurotransmitters are synthesized in the cell body of the neuron, transported down the axon, and stored in membrane-wrapped **vesicles** in the axon terminal. The entry of Ca^{++} into the presynaptic cell initiates a chain of events that results in vesicles docking in specialized release sites in the presynaptic membrane, and release of neurotransmitter contents by exocytosis. At the postsynaptic cell, the receptor to which the neurotransmitter binds determines the response, which may be excitatory (EPSP) or inhibitory (IPSP) (Fig. 15.15). Both of these synaptic potentials are transient, lasting only 5 to 10 ms.

Whether an EPSP or an IPSP is generated in response to neurotransmitter binding is a property of the **receptor neuron**, not the transmitter. The same chemical transmitter can produce an EPSP when binding to receptors at one location on a post-synaptic cell, and an IPSP when binding to receptors at other postsynaptic cells. Neurotransmitter effects show this variation both among organisms and within a single organism!

Initiation of the nerve impulse and its mechanisms of propagation are essentially the same in neurons from many animals—both invertebrate and

A Electrical synapse

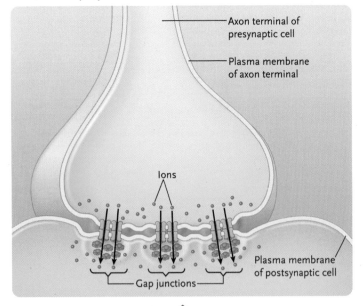

In an electrical synapse, the plasma membranes of the presynaptic and postsynaptic cells make direct contact. Ions flow through gap junctions that connect the two membranes, allowing impulses to pass directly to the postsynaptic cell.

B Chemical synapse

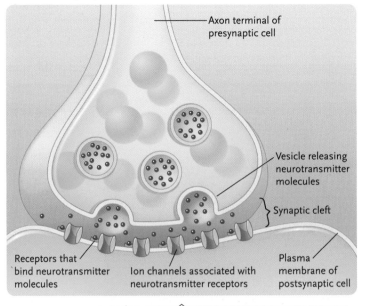

In a chemical synapse, the plasma membranes of the presynaptic and postsynaptic cells are separated by a narrow synaptic cleft. Neurotransmitter molecules diffuse across the cleft and bind to receptors in the plasma membrane of the postsynaptic cell. The binding opens channels to ion flow that may generate an impulse in the postsynaptic cell.

Figure 15.13 A. An electrical synapse. **B.** A chemical synapse.

SOURCE: From RUSSELL/WOLFE/HERTZ/STARR. *Biology*, 2E. © 2013 Nelson Education Ltd. Reproduced by permission. www.cengage.com/permissions

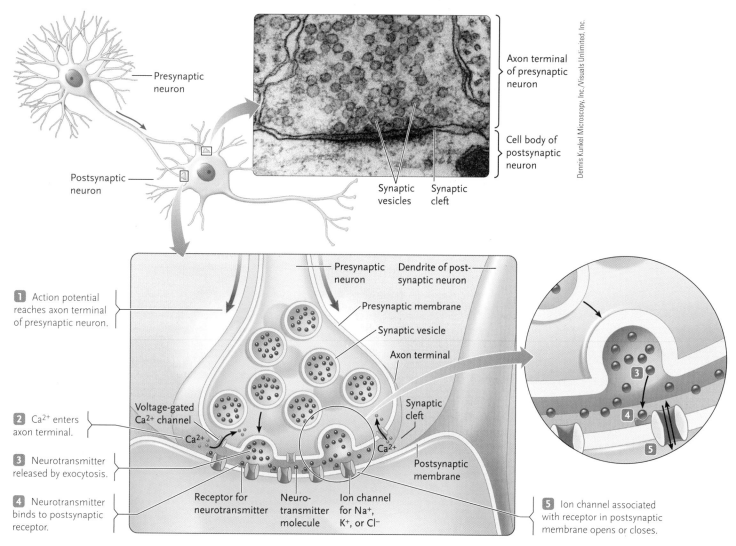

Figure 15.14 Structure and function of a chemical stimulus.

SOURCE: From RUSSELL/WOLFE/HERTZ/STARR. *Biology*, 2E. © 2013 Nelson Education Ltd. Reproduced by permission. www.cengage.com/permissions

vertebrate—as are the mechanisms underlying receptor potentials and postsynaptic potentials. Thus, the combination of local potentials and propagated action potentials constitutes the universal language of all known nervous systems.

STUDY BREAK

1. What is patch clamping, and how does it work?
2. What is an action potential? What causes it?
3. What is a synapse? How does it work?

15.3 Neural Networks and Nervous Systems

The fundamentals of neurons, membrane potentials, and synapses come to life when we turn our attention to neural circuits. The nervous systems of different groups of animals are organized in different ways, reflecting differences in lifestyle and habitat. All nervous systems are similar, however, in that they sense stimuli, integrate this information, and generate appropriate responses. All responses to stimuli, whether intrinsic or extrinsic, are generated by the actions of interconnected sets of neurons. These neurons fall into three broad classes: **sensory (afferent) neurons, interneurons**, and **motor (efferent) neurons**. Sensory information is detected by specialized cells called receptors. These cells act as **transducers** that convert stimulus energy into neural responses. Sensory neurons transmit these neural signals to processing centres such as a brain or another collection of neurons in clusters called **ganglia** (singular = ganglion). Motor neurons convey the output signals from the processing centres to "effectors" such as muscles and glands. They are responsible for generating a response to neural signals.

Interneurons serve as the simplest level at which sensory and motor information is **integrated**. All neurons receive input from a variety of sources, and it is the combination or summation of those inputs that determines the cellular response (Fig. 15.15). All neurons make multiple connections with others, and that

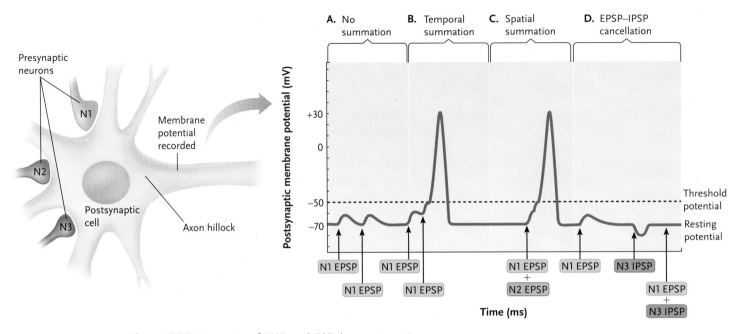

Figure 15.15 Summation of EPSPs and IPSPs by a postsynaptic neuron.

SOURCE: From RUSSELL/WOLFE/HERTZ/STARR. *Biology*, 2E. © 2013 Nelson Education Ltd. Reproduced by permission. www.cengage.com/permissions

number can be especially large for interneurons. Some single interneurons in the human brain, for example, form as many as 100 000 synapses with other neurons. The combined activities of all the neurons in an animal provide the flow of information on which the integrated functioning of increasingly complex organisms depends.

The arrangement of neural networks varies among animal groups, but follows some patterns in both invertebrates and vertebrates. Cnidarians, including jellyfish and sea anemones, are radially or biradially symmetrical. The primitive nervous system of a cnidarian is thought to have been a **nerve net** (Fig. 15.16A, left) which shows a complex system of interactions. All neuronal processes can conduct action potentials, and there are synapses where processes of neurons cross other neurons as well as at axon terminals. All processes involved in a synapse may produce transmitters and have receptors for transmitters. When part of the animal is stimulated, impulses are conducted throughout the nerve net in all directions from the point of stimulation. Some pathways through a nerve net, for example, around the oral opening of *Hydra* (135), conduct signals faster than other parts of the net (Fig. 15.16B). In medusae, the nerve net may innervate the muscle fibres (Fig. 15.16A, right) while sensory inputs go to a nerve ring that surrounds the bell (Fig. 15.16C). This ring is viewed as analogous to the coordinating ganglia of other invertebrate groups and may be likened to the circumoral nerve ring of the pentaradial echinoderms. That nerve ring coordinates the activities of the arms and tube feet. It is interesting that the brittlestars (Ophiuroidea), while having pentaradial symmetry like other echinoderms, walk as though they are bilaterally symmetrical with one arm leading and a pair of arms on each side working as though the animal is four-legged.

In the evolution of Bilateria, cephalization and the development of mobility are linked to the concentration of sensory structures at the anterior end. There the cephalic ganglia (concentrations of neurons) receive sensory inputs and coordinate body activities. In groups with segmentation, longitudinal nerve cords are augmented by segmental ganglia that coordinate the activities of the segment in relation to its anterior and posterior neighbours.

Separation of protostome and ecdysozoan lineages from the line leading to Deuterostomata coincides with a major difference in the arrangement of nervous systems. In the protostome-ecdysozoan line, the cephalic ganglia are dorsal and connected to a mid-ventral nerve cord by a nerve ring around the gut. In deuterostomes both the ganglia and brain are dorsal.

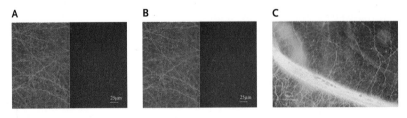

Figure 15.16 Nerve nets in Cnidaria **A.** Two views of the same area of nerve net in *Aurelia*. Left: frame staining for FMRF-amide shows all fibres in the net. Right: the filter was changed to show only the tubulin stained elements in the motor part of the net. **B.** A concentration in the nerve net following a radial canal. **C.** The nerve ring around the edge of the bell of a cubozoan medusa.

SOURCE: Reproduced with permission from R. A. Satterlie. 2011. "Do jellyfish have central nervous systems?" *Journal of Experimental Biology* 214, 1215–1223. The Company of Biologists.

15.4 Central Nervous System

Central nervous systems co-ordinate the activities of organisms, linking afferent input and efferent activities. Invertebrate central nervous systems are not simple. The brain of of *Drosophila* (959)) (Fig. 15.17), like the brains of vertebrates (Fig. 15.18), is divided into discrete regions. Behind the compound eye, the lamina, medulla, and lobula are responsible for the integration of visual information from the compound eye. The paired mushroom bodies (corpora pedunculata), prominent features in the brains of some annelids and arthropods (Fig. 15.19), are closely linked to the antennal lobes and antennae. They appear to be involved with olfaction and behaviours associated with olfactory learning (courtship, for example). Mushroom bodies also have a role in visual learning and are involved with sleep and memory. Among annelids, mushroom bodies occur more often in free-living and predatory polychaetes than in species with other life styles. There are suggestions that the mushroom bodies are involved in higher cognitive interactions.

Mushroom bodies are composed of Kenyon cells, small neurons atop the calyx, a tangled zone of synapses that receives sensory inputs. A stalk consisting of axons from Kenyon cells projects away from the calyx toward the efferent lobes.

Insect motor programming controlling walking lies in the thoracic ganglia. The central complex of thoracic ganglia includes the fan-shaped and ellipsoid bodies, and integrates sensory and mechanical information needed to regulate walking behaviour. Cephalopod brains regulate the squids' and octopuses' complex behaviours and are comparable in size to vertebrate brains (Fig. 15.20). The large cells in some molluscan central nervous systems have allowed distinction between individual brain cells involved in learning and in memory (see Box 17.2).

The vertebrate CSN consists of a dorsal nerve tube (spinal cord) that expands in the head to form a highly complex **vesicular** brain. Although all vertebrates show

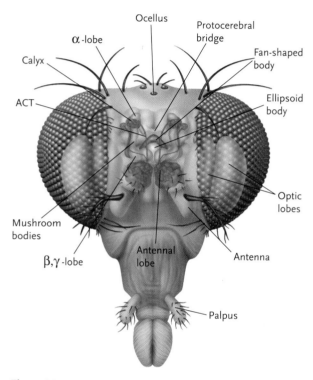

Figure 15.17 Brain regions in *Drosophila* (960).

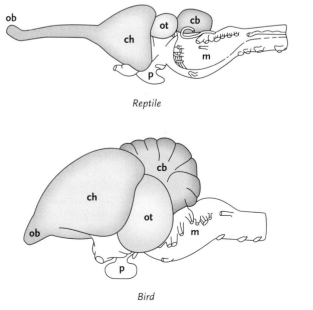

Reptile

Bird

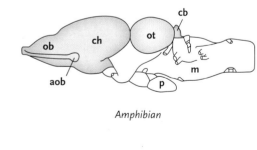

Amphibian

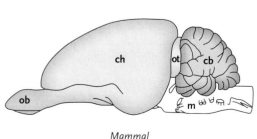

Mammal

Figure 15.18 Overall, total brain size and the sizes of most brain divisions vary among vertebrate taxa. A typical vertebrate brain is composed of three main sections: hindbrain, midbrain, and forebrain. The **hindbrain** contains the medulla oblongata and cerebellum; the **midbrain** contains the optic tectum; and the **forebrain** includes the cerebral hemispheres, olfactory bulb, accessory olfactory bulb, and pituitary gland.

SOURCE: Based on Orthcutt, R.G. 2002. "Understanding vertebrate brain evolution," *Integrative and Comparative Biology* 42(4): 743–756.

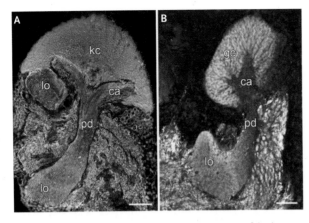

Figure 15.19 Mushroom bodies. Sagittal sections of the brains of **A.** a cockroach (*Leucophaea maderae* (859)) and **B.** a polychaete (*Nereis diversicolor* (322)) show neuropil organization through reaction to a stain (horseradish peroxidise). Legend: kc = Kenyon cells, ca = calyx, pd = peduncle, lo = vertical lobes. Scale bars = 80 μm.

SOURCE: Heuer, C.M., C.H.G. Muller, C. Todt and R. Loesel. 2010. Comparative neuroanatomy suggests repeated reduction of neuroarchitectural complexity in Annelida. *Frontiers in Zoology*, 7:13.

a similar neural organization, there is considerable variation in brain size and divisions (Fig. 15.18). Variation in the sizes of different parts of the vertebrate brain is usually associated with different lifestyles and levels of overall neural development. As we will see below, among many other things, these differences can reflect sensory input or social networks.

Neural anatomy can reflect sensory input. For instance, the star-nosed mole (*Condylura cristata* (1918)) uses the 22 rays of its fleshy nose star to detect mechanosensory stimuli. Each ray is covered with specialized, highly innervated **Eimer's organs**, making the star one of the most sensitive organs found in any mammal. The moles use these rays to search for food as they move around in darkness underground and under water. Stimulation of the rays results in activity in the somatosensory cortex, the area of the brain detecting mechanical stimulation. Compared to the actual size of the star, a disproportionate area of the somatosensory cortex is devoted to receiving information from the rays (Fig. 15.21).

Motor responses to stimuli may be voluntary or involuntary. Involuntary responses (**reflexes**) occur through **reflex arcs**. Each reflex arc consists of a receptor and afferent neurons connecting to a reflex centre that mediates a response and stimulates efferent neurons, resulting in a muscular or glandular response. In **spinal reflexes**, the spinal cord is the reflex centre. In other reflex arcs, the brain serves this function. Reflex responses may be simple, involving one action, or may be complex, involving multiple muscle groups.

Sometimes when a reflex arc is repetitively stimulated, it becomes less sensitive and requires increasingly greater stimuli to trigger a reflex response. This process is known as **habituation**, and one of the best studied examples is the escape response. For example, crayfish have two sets of giant neurons that run the length of the ventral nerve cord, **medial giants** and **lateral giants**. The giant fibres receive afferent information from both the front (medial giants) and the back (lateral giants) of the crayfish. Excitation of any of the giant fibres results in a reflex tail flip (lateral giants) or backwards darting movement (medial giants), propelling the crayfish swiftly away from the source of the stimulus. Persistent triggering of the escape reflex results in an inhibition response, where reflex responses do not occur as readily. Several complicated mechanisms drive flexibility in the crayfish escape response, including intrinsic factors such as changes in transmitter release at the synapse and the presence of neuromodulating chemicals. Extrinsic factors such

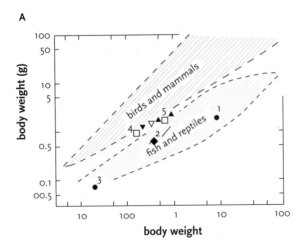

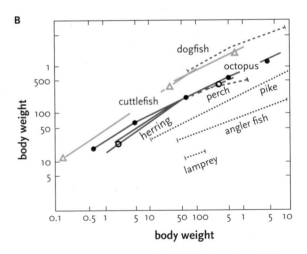

Figure 15.20 A. Brain:body weight ratios for selected cephalopods compared to vertebrate groups (A = *Sepia* (517); V = *Loligo* (503); 1 = *Octopus vulgaris* (514); 2 = *O. salutii* (513); 3 = *O. defilippi* (511); 4 and 5 = squids *Illex* (426) and *Todarodes* (521)). **B.** Brain:body weight ratios of cuttlefish, octopus, and selected fish.

SOURCE: Packard, A., "Cephalopods and fish: the limits of convergence," *Biological Reviews* Volume 47, Issue 2, pages 241–307, May 1972, John Wiley and Sons. Copyright © 2008, John Wiley and Sons.

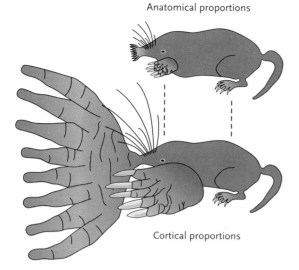

Anatomical proportions

Cortical proportions

Figure 15.21 The upper drawing shows the actual proportions of a star-nosed mole (*Condylura cristata* (1918)). The lower drawing shows the anatomical proportions as represented on the somatosensory cortex.

SOURCE:Kenneth C. Catania Jon H. Kaas, "The Unusual Nose and Brain of the Star-Nosed Mole," *BioScience* Vol. 46, No. 8 (Sep., 1996), pp. 578–586, University of California Press. Copyright © 1996, American Institute of Biological Sciences.

In vertebrates, neural mechanisms that underlie social networks and associated decision-making are shared among the ancestors of both actinopterygian and sarcopterygian fishes, as well as tetrapods. These provide another example of a co-ordinating role played by the CNS. A comparison of neurochemical genes across 12 brain regions in 88 species shows impressive conservation across fish and tetrapods. In the context of social decision-making (SDM) networks, the most conserved areas are the pre-optic area of the optic tectum in the cerebral area, the hypothalamus, and basolateral amygdala (Fig. 15.18). The subcortical striatum is the least conserved area in the context of SDM networks. The presence of neural networks that respond to sex steroids and neuropeptide hormones associated with reproduction, aggression, and parental care in such a range of vertebrates raises interesting questions about the presence of analogous systems in social invertebrates.

STUDY BREAK

1. What is the central nervous system?
2. What are reflexes?

as whether the individual crayfish is alone or in the presence of another crayfish also affect the response.

Animals may learn to associate neutral stimuli with a reflex reaction if the neutral stimulus frequently accompanies a reflex stimulus. Conditioned responses occur when a reflex reaction happens in response to the neutral stimulus, even in the absence of the original reflex stimulus. Pavlov's dog is a famous example. Pavlov, a Russian physiologist, observed that a dog salivated (reflex response) in the presence of food (reflex stimulus). He then rang a bell as he presented the food. Eventually, just the ringing bell caused the dog to salivate, a conditioned response.

The escape response in teleost fish is a well-studied, stereotypical behaviour called a **C-start.** In response to auditory, visual, or lateral line stimuli, fish startle by first bending their body laterally to form a "c" shape and then propel themselves forward with the caudal (tail) fin. C-starts occur when one of two giant sensory neurons, called **Mauthner neurons,** are stimulated. Two Mauthner neurons are located symmetrically on either side of the fish's hindbrain. They have axons that cross each other and then move down the spinal cord. If a Mauthner neuron receives afferent stimuli, the resulting action potential results in a series of physical responses including contraction of muscles on one side of the fish and simultaneous inhibition of muscles on the other side to allow for the formation of the "c" shape. C-starts are normally an efficient defensive behaviour. In Southeast Asia, tentacled snakes (*Erpeton tentaculatus* (1495)) prey on fish. When hunting, tentacle snakes exploit the C-start (Fig. 15.22).

15.5 Transduction

Sensory receptors are transducers that convert stimuli about intrinsic and extrinsic events to electrical information. The impact of the stimulus on the transducer results in a change in neural activity, converting stimulus energy into an electrical signal. This is the process of **sensory transduction.** Stimuli leading to transduction can involve mechanical, thermal, chemical, light, and other forms of energy, depending upon the receptor involved. In all cases, transduction is effected by a change in membrane potential triggered by the stimulus, which is then interpreted by the CNS.

Some transducers provide digital (on/off) information, while others generate analog (graded) data. The manner of sensory transduction varies with receptor type. Some reactions are relatively simple. For example, the surface membrane of a **chemoreceptor** may respond when a molecule of a chemical stimulant reacts with one of its receptor molecules. Binding of the stimulus molecule to the receptor opens ion channels, inducing an influx of ions, with a resulting change in neural activity. Other reactions are more complex. Many **nociceptors** (sensitive to noxious stimuli) contain transient receptor potential vanilloid 1 (TRPV1) receptors on their surface membranes. TRPV1 receptors can be activated by capsaicin (the active ingredient of chili peppers), noxious heat, or lowered pH. In all cases, the open ion channel allows an influx of Ca^{++} that triggers movements of Na^+ and K^+ across the nociceptor membrane, depolarizing the receptor membrane and activating a number

Figure 15.22 The tentacled snake specializes in eating fish and has developed a unique response to the C-start escape responses used by its prey. These snakes form a "j" shape while hunting, with their head as the curve of the "j". When fish swim into the concave portion of the "j", the snake startles the fish by moving a portion of its lower body. In these photos, the fish C-starts away from this movement, right toward the snake's mouth! In situations where the initial position of the fish is already perpendicular to the snake's jaw, a C-start would not move the fish toward the snake. In these cases, the snake is able to predict the fish's movement and adjust its strike accordingly. Once the fish begins a C-start, it cannot stop.

SOURCE: Catania, K.C. 2011. "The brain and behaviour of the tentacled snake," *Annals of the New York Academy of Sciences* 1225: 83–89, John Wiley and Sons. © 2011 New York Academy of Sciences.

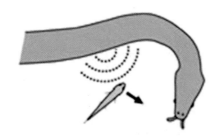

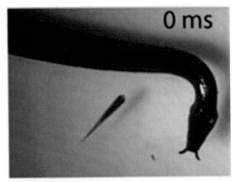

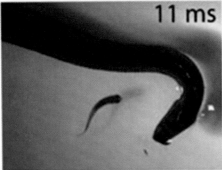

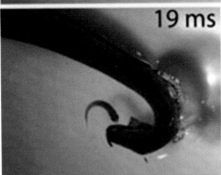

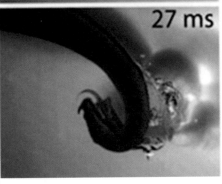

ion channels are open in the dark and close when photons are absorbed by pigment on the disc membrane. The signal for the closure of photoreceptor ion channels is transmitted by second-messenger molecules that are produced in response to light-induced changes in the pigment molecules.

The details of sensory transduction vary with receptor type, but transduction normally depends on changes in current through ion channels in response to activation of receptor molecules by the stimulus. Receptor molecules and ion channels may be directly coupled, or may be coupled indirectly through second-messenger cascades.

Sensory transduction generally induces a receptor potential in the peripheral terminal of the associated **primary afferent sensory neuron**. A receptor potential is usually a graded depolarizing event resulting from inward current flow, and it may bring the membrane potential of the sensory receptor toward or past the threshold needed to trigger a nerve impulse. The size or intensity of a stimulus can thus be encoded by the size of the graded receptor potential and the resultant number of action potentials generated, with larger stimuli resulting in the generation of a burst of action potentials. This is how the same receptor can send information about stimuli of different sizes. **Thermoreceptors**, for example, respond to differences in temperature over a fairly broad range, but smaller changes result in the production of fewer action potentials than larger changes. Thus, the CNS can differentiate between stimuli that are warm as opposed to hot, or cool as compared to cold.

In some sensory receptor organs, the peripheral terminal of a primary afferent fibre (a sensory neuron) contacts a separate sensory cell, complicating the relationship between stimulus and response. For example, a sound of a given frequency depolarizes the membrane of a hair cell in the cochlea, causing release of an excitatory neurotransmitter into the primary afferent terminal. The resulting inward current depolarizes the primary afferent neuron terminal, producing a **generator potential**. This depolarization brings the membrane potential of the primary afferent sensory neuron toward or beyond threshold for firing nerve impulses.

Sensory receptors can habituate to maintained stimulation. A long-lasting stimulus may produce a prolonged repetitive discharge of action potentials, or it may result in a brief response (one or a few discharges), depending upon whether the sensory receptor is slowly or rapidly adapting. The adaptation rates differ because a prolonged stimulus may produce either a maintained or a transient receptor potential in the sensory receptor. The functional implication of the adaptation rate is that different temporal features of a stimulus can be analyzed by receptors with different adaptation rates. In this way, sensory receptors encode different features of stimuli. For example, during indentation of the skin, some slowly adapting mechanoreceptors simply respond repetitively at a rate

of signal-transduction pathways. **Mechanoreceptors** have a more straightforward effect. Here, mechanically sensitive ion channels on the surface membrane open in response to the application of a mechanical force along the membrane. In **photoreceptors**, however,

proportional to the amount of indentation, encoding the location of the stimulus. Meanwhile, rapidly adapting receptors in the same region respond best to transient mechanical stimuli, transducing information about the onset and cessation of the stimulus, rather than the specific location. Working together, the pattern of neural activity received by the CNS from all the different receptors allows interpretation of many specific stimulus features.

STUDY BREAK

1. What is a transducer?
2. What is a nociceptor?

15.6 Amplification

If you think of the difference between the ear-splitting din of a siren and a soft whisper in your ear, you can appreciate one of the challenges facing sensory systems. Perhaps it's easier to remember how a bright flash of light at night disrupts your night vision, or that commercials on television seem to boom at you when you can hardly hear the dialogue in the program.

Because animals contend with a wide diversity of environments, there are many different transducers and other mechanisms for ensuring appropriate sensitivity. One indication of the degree of amplification of a stimulus is the number of heterotrimeric GTP-binding proteins (G proteins or transducins) activated by the stimulation of a single membrane receptor; in other words, the more G proteins activated, the stronger the signal. In the olfactory system of frogs, receptors are ligand-bound because they rely on intermolecular forces. An odorant receptor has a low probability of activating even one G protein molecule. This is strikingly different from the visual system where a long-lived photoisomerized rhodopsin molecule in a rod photoreceptor in the retina activates many G protein molecules. This represents a large response to stimulation by a single photon and indicates a fundamental difference between olfactory and visual systems, at least in vertebrates.

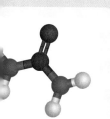

BOX 15.2 Molecule Behind Animal Biology

Heterotrimeric GTP Binding Molecule (G protein)

G proteins, composed of α, β, and γ subunits (Fig. 1), transduce signals from receptors on membranes to many different types of intracellular effectors. The steps involved in the activation of G proteins and their return to the inactive state are shown in Figure 2.

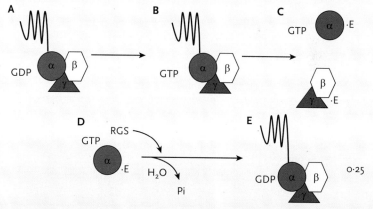

FIGURE 2

*Activation of a G protein from the inactivated state. **A.** A seven-transmembrane-spanning receptor protein is associated with the Gαβγ heteromere. Here GDP is associated with the Gα. **B.** A ligand (*), activated at the receptor, acts as a guanine nucleotide exchange factor (GEF) for the α subunit after a conformational change. This stimulates a change from GDP for GTP. **C.** GTP binds with Gα, which dissociates from Gβγ. Different effector proteins (E) interact with the dissociated Gα and Gβγ subunits. **D.** Gα subunits hydrolyze GTP, resulting in the attenuation of the signal, activity accelerated by proteins that activate GTPase and **E.** return the G protein to its resting state.*

SOURCE: Bastiani, C. and Mendel, J. Heterotrimeric G proteins in *C. elegans* (October 13, 2006), *WormBook*, ed. The *C. elegans* Research Community, WormBook, doi/10.1895/wormbook.1.75.1, http://www.wormbook.org.

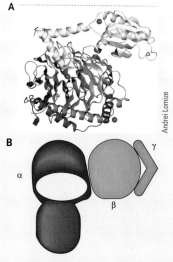

Andrei Lomize

FIGURE 1
A. *Detailed and* **B.** *general view of a heterotrimeric GTP binding molecule, or G protein.*

Differences in amplification indicate the importance of gain control (analogous to the volume control on a sound system) in the system. Along the mammalian auditory pathway there appear to be multiple stages of gain control, ensuring that we can hear both the wail of a siren and a companion's whispered comment (perhaps not at exactly the same time). In mammals, the external ears (pinnae) play an initial role, as do the auditory ossicles (malleus, incus, and stapes), the basilar membrane of the cochlea, the auditory midbrain, and the auditory cortex.

STUDY BREAK

1. What are G proteins?
2. How are they involved with amplification?

15.7 Why Are Receptors in Pairs?

Animals typically have pairs of sensory organs, allowing them to determine the sources of stimuli more precisely (Fig. 15.23). Pioneering work by Spallanzani on the orientation behaviour of bats hinged on the effect of blocking one of a bat's ears. Laboratory rats smell in stereo because the olfactory stimuli arrive at the left and right sensors at different times. Forked tongues, which are common (but not ubiquitous) in lizards and snakes, allow these animals to follow a scent trail by **tropotaxis**: simultaneously comparing odour intensities at two locations. The discovery that some preying mantises have a single ear (used as a bat detector) provides an interesting exception. Typically, males of some species of mantises have ears because they fly in search of females and are more vulnerable to hunting bats.

Can we infer function from structure? By now you should expect that biologists have found different explanations even for features that are superficially similar. Eyes on stalks provide a good example.

Although many species from phyla as a diverse as Mollusca, Arthropoda, and Chordata have eyes on stalks, those of larval black dragonfish are very striking. A range of eye conditions is seen in fishes: eyes that do not protrude from the skull, those that do, and eyes located at the ends of stalks of various lengths. Extending the eyes away from the head should enlarge the volume of water the fish can scan, allowing detection and tracking of prey at greater distances. These fish illustrate variation in eye structure that correlates with diet and lighting and show that these traits can change over the life of an individual.

Flies in the family Diopsidae often have their eyes on stalks (Fig. 15.24). Where the behaviour of these flies has been studied, the dimorphism arises from sexual selection. Female stalk-eyed flies prefer males with longer eye stalks. During mating season, the flies (e.g., *Cyrtodiopsis whitei* (972)) aggregate at display sites where males compete for female attention. Males with longer eye stalks attract more females than those with shorter eye stalks. Sexual dimorphism in eye stalks has evolved independently at least four times within this one family.

The examples of fish larvae and flies demonstrate that eyes on stalks can serve different functions. Larval fish are, by definition, not reproductive, so sexual selection is not a credible explanation; increased feeding ability, with the idea that their stalked eyes provide improved vision, is more likely. This leaves questions about the function of stalked eyes in other animals. For example, stalk-eyes appeared independently in several clades of trilobites (Fig. 15.25A), including three orders and at least three families. In these animals, being able to extend or retract stalked eyes gives them a greater potential field of view.

Figure 15.23 Pairs of sense organs (eyes, ears, nostrils) are obvious on **A.** the faces of some animals or on **B.** the tongues of others. **C.** However, while preying mantises have a pair of compound eyes and a pair of antennae, they have only a single ear.

Figure 15.24 A male stalked-eyed fly has much longer eye stalks than this female (see Fig. 4.19).

BOX 15.3 People Behind Animal Biology

Kenneth D. Roeder

Kenneth Roeder may be best known for his work on the ability of moths to hear the echolocation calls of bats, but his contributions extended to other aspects of insects' nervous systems (as presented in his book *Nerve Cells and Insect Behaviour*). Ken established that by attaching an electrode to the auditory nerve of a moth, he could monitor what the moth heard. He demonstrated that in response to a faint bat call (distant bat), moths with ears turned and flew away from the bat (negative phonotaxis). Closer bats— louder echolocation calls—caused moths to dive toward the ground or fly erratically. Although people had known about moth ears for a long time, it was Roeder's work that demonstrated what they were listening to. Other researchers have shown that some moths use sounds in courtship and others use them to warn bats that they taste bad.

For most of his career, Ken was at Tufts University near Boston. He lived in Concord where he used a garden shed as a summer lab. In the early 1960s, after he suffered a stroke, his graduate students built a lab on the end of his garage. The expanded facility made it easy for Ken to remain active as he recuperated. Furthermore, a colony of big brown bats (*Eptesicus*

fuscus (1791)) lived in a barn on a neighbouring property. The bats left the barn and flew across his yard every evening in the summer, giving him ample opportunity to test moth hearing with wild bats.

Don Griffin worked in Boston. Don was always trying to come up with a better bat detector, and he and Ken would meet at Ken's place to test Don's latest detector against a moth ear. The moths were always better at detecting the bat calls than the bat detector was.

Once at the University of Texas in Austin, Ken had given a seminar about his work on bats and moths. In the evening after the seminar, the group of biologists sat on an outdoor patio. As they spoke, hawk moths (Sphingidae) visited the flowers around the patio to get nectar. One of the students asked Ken if those moths could hear bats. He answered, "No" because he had looked carefully at those moths and found no sign of an ear. So he said, "Watch," and shook his keys (a good way to produce high frequency sounds). The moths took off in response to the jangling keys suggesting that they could hear high-frequency sounds!

Ken later said that this experience taught him not to make generalized

statements about what moths could and could not hear. It also led to further research which revealed that the hawk moths in question used a mouthpart as the ear, rather than the more usual eardrum arrangement.

The following poem, published in 1968, arose from a limerick contest between Ken and English zoologist David Pye. The poem immortalized the bat-moth interaction.

In days of old and insects bold
Before bats were invented
No sonar cries disturbed the skies
Moths flew uninstrumented.
The Eocene brought mammals mean
And bats began to sing
Their food they found by ultrasound
And chased it on the wing.
Now deafness was unsafe because
A loud, high-pitched vibration
Came in advance and gave a chance
To beat echolocation.
Some found a place on wings of lace
To make an ear in haste
Some thought it best upon the chest
And some below the waist.
Then Roeder's keys upon the breeze
Made sphingids show their paces
He found the ears by which they hear
In palps upon their faces,
Of all unlikely places.

SOURCE: David J. Pye 1968, *Nature* 218:797

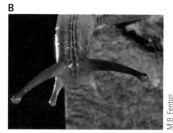

Figure 15.25

A. A reconstruction of a trilobite shows distinctive stalked eyes. The slug *Limax maximus* (478) can extend (**B**) or retract (**C**) its stalked eyes.

3drenderings / Shutterstock.com

M.B. Fenton

M.B. Fenton

STUDY BREAK

1. What is tropotaxis? Who uses it?
2. What advantage(s) is (are) conferred by eyes on stalks?

15.8 Photoreception: Light

Organisms from Cyanobacteria to whales are able to perceive light. This ability may occur in an organelle within a cell or in photoreceptive organs that may be called **eyes** if they have a lens and form an image. The transduction of photons to signals within a photoreceptor involves visual pigments housed in microvilli or infoldings of a cell membrane along with melanin that shields the basal side of the cell.

In bilaterian animals, visual pigments are embedded in the membranes of retinula cells. Visual pigments consist of a chromophore, a coloured molecule (usually the carotenoid **retinal**), along with an opsin, a protein. Retinal$_1$ is derived from vitamin A$_1$, retinal$_2$ from vitamin A$_2$. Visual pigments are activated by photons of light that trigger a phototransduction cascade involving more than 10 proteins. Many opsins have been described (at least 50 in vertebrates alone), and they, in combination with visual pigments and screening or sensitizing pigments, determine the range of wavelengths of light to which the photoreceptor responds (see Box 4.3). Opsins are not found in sponges, but many sponge larvae have a directional sensitivity to light; a blue-light sensitive cryptochrome has been identified as the photoreceptor molecule in these animals.

Some animals have **extraocular** (nonocular) pathways for detecting light. Larval fruit flies have eyes on the head as well as photosensitive structures along the body that allow detection and avoidance of bright light levels. These nonocular photosensors are associated with dendritic arborization neurons throughout the body wall. The dendrites contact epithelial cells, but their cell bodies and axons are wrapped in glia. Other animals, including some reptiles, birds, amphibians, and fish, also have light sensors not associated with their eyes, called **extraocular sensors**. Some magnetic sensors in birds are also light-sensitive, extraocular sensors.

Light enters an image-forming eye through the pupil, passes through the lens, and arrives at the retina (Fig. 15.26). Here, photosensitive pigments in rod cells and cone cells transduce the energy of photons to electrical stimuli via bipolar cells and then ganglion cells whose density and dispersion in the retina define its visual channels. Each ganglion cell receives input from a number of photoreceptor cells, for example cones in diurnal birds of prey, rods in deep-sea fishes. Increased concentration of photoreceptor cells occurs at the pit in the retina—the **fovea**—in vertebrates from birds and mammals to fishes. The fovea is quite consistent in absolute size across a large variety of animals: about 14 mm in diameter in both larval fish (*Alepocephalus agassizii* (1238)) and Wedge-tailed Eagles (*Aquila audax* (1578)). Foveae are important for high visual acuity. Some deep-sea fishes achieve high visual acuity with layers of photosensitive cells (Fig. 15.26C).

The **tapetum lucidum**, a reflective layer behind the retina, is a common strategy for increasing the light reaching the retina (Fig. 15.27). Light is absorbed as it passes through the retina, and again as it reflects back from the tapetum lucidum. Many nocturnal land animals have tapeta lucida (plural), as do many deep-sea fish.

There are three basic types of image-forming eyes: camera, mirror, and compound (Fig. 15.28). Camera eyes are found in vertebrates and molluscs (gastropods and cephalopods), as well as in some annelids and copepod crustaceans. At least one mollusc, a cham-bered nautilus (*Nautilus pompilius* (485)) has a pinhole eye in which the small pupil projects an image onto the retina. Mirror eyes, with a lens and two retinal layers, are widespread in molluscs such as scallops.

The compound eye is the most common eye type, occurring in most arthropods, some molluscs (clams), and some polychaete worms. Each element (an ommatidium) of the apposition compound eye consists of a cornea and a crystalline cone that focus incident light on the underlying photoreceptors (rhabdomeres). Superposition compound eyes have a wide clear zone between the lenses and retinae, allowing a wider pupil diameter. Recent reports about *Anomalocaris* (Fig. 6.22) demonstrate that this Cambrian arthropod had well-developed compound eyes. This means that compound eyes appeared in arthropods before a hardened exoskeleton. The presence of a mobile, visually orienting predator (*Anomalocaris*) would have had repercussions in the community of Cambrian animals.

Some photoreceptors are multicellular organs that lack lenses and do not form images. Planarians such as *Polycelis auricularia* (233) have photoreceptors consisting of two cells, a photoreceptor and a pigment cell. Some leeches (Annelida) have pigment cup eyes, a pigment-coated cup of photoreceptors. The ocelli of insects (Fig. 15.29) are also photoreceptors that do not form images. They are often used to assess light levels and brightness.

The diversity of eyes, which he defined as organs consisting of at least two different cells, challenged Charles Darwin when he was writing *The Origin of Species*. Making sense of the many varieties of eyes that occur in animals was simplified by the discovery that the single master control, the *Pax6* transcription factor gene, is responsible for the development of eyes in planarians, arthropods, annelids, molluscs, and chordates (Fig. 15.30). *Pax6* genes are highly conserved because the pleiotropic gene specifies development of the eye(s) as well as the nose and parts of the brain. The blind nematode (*Caenorhabditis elegans* (585)) lacks rhodopsin but maintains *Pax6*.

Box jellyfish (cubozoans) are agile swimmers and use visual information in orientation and when hunting prey. Their visual system consists of four rhopalia, each with six eyes. Two of the eyes are slits, two are in pits, and there is one small upper lens eye and one larger lower lens eye (Fig. 15.31). Although a *Pax6* gene has not been found in Cnidaria, including those with eyes, a *PaxB* gene has been reported from a box jelly (*Tripedalia* sp. (157)). There is an 82% similarity between this *PaxB* gene and the mammalian *Pax2* gene, and 75% similarity with *Pax6* genes. In short, the genetic mosaic of *PaxB* genes in box jelly cubozoans strongly suggests extension of control of eye development to some cnidarians. It remains to be determined if other lineages of jellyfish have lost their eyes.

The details of expansion of the photoreceptor cell membranes to accommodate stored photopigments

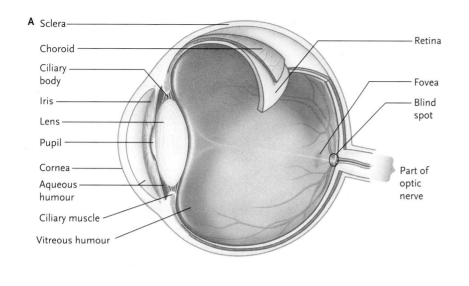

A
- Sclera
- Choroid
- Ciliary body
- Iris
- Lens
- Pupil
- Cornea
- Aqueous humour
- Ciliary muscle
- Vitreous humour
- Retina
- Fovea
- Blind spot
- Part of optic nerve

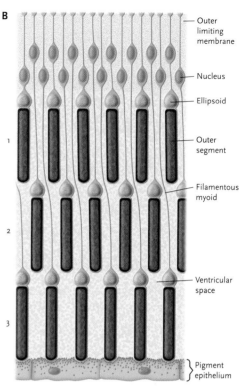

B
- Outer limiting membrane
- Nucleus
- Ellipsoid
- Outer segment
- Filamentous myoid
- Ventricular space
- Pigment epithelium

1
2
3

C

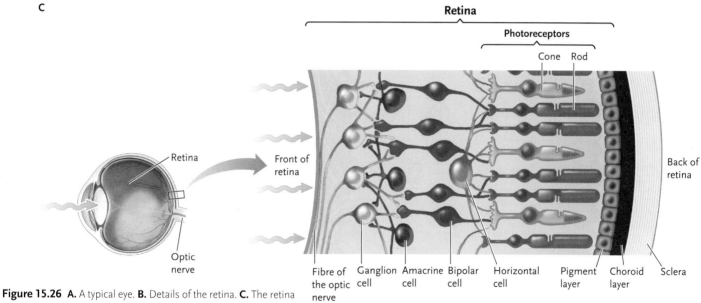

Retina

Photoreceptors

Cone   Rod

Retina

Optic nerve

Front of retina

Back of retina

Fibre of the optic nerve | Ganglion cell | Amacrine cell | Bipolar cell | Horizontal cell | Pigment layer | Choroid layer | Sclera

Direction of light

Direction of retinal visual processing

Figure 15.26 A. A typical eye. **B.** Details of the retina. **C.** The retina of a deep-sea fish. Legend: 1, 6, 8 = three layers of rods; 2 = ellipsoid; 3 = outer limiting membrane; 4 = outer segment; 5 = pigment epithelium; 7, 9 = filamentous myoids; 10 = ventricular space; 11 = radial fibre cytoplasm; 12 = rod nuclei; 13 = synaptic spherules.

differ between vertebrates and invertebrates. **Rhabdomeres** are the photoreceptor cells of invertebrates, characterized by microvilli in the apical cell surfaces. **Ciliary** photoreceptor cells are typical of the rods and cones of vertebrates. The folding of the ciliary membrane provides the necessary enlargement of the surface area. This also applies to photosensitive cells of the pineal eye, part of the diencephalon in vertebrates such as the tuatara (*Sphenodon* (1458)) that have

Figure 15.27 The eyeshine from the tapetum lucidum of a cat

Sean F. Werle

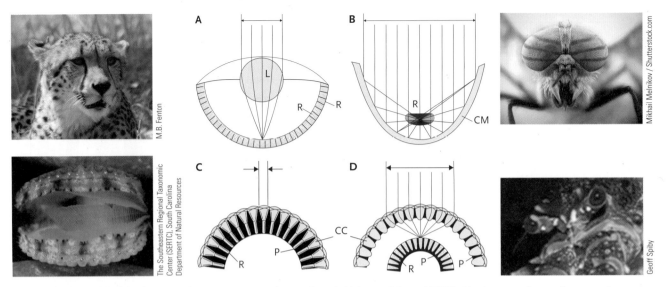

Figure 15.28 Three basic eye designs: **A.** camera, from a cheetah (*Acinonyx jubatus* (1705)); **B.** mirror or reflector, from a scallop (*Argopecten gibbus* (531)); and compound (**C.** apposition, from a tabanid fly; and **D.** superposition, from a Cape rock lobster (*Jasus lalandii* (821)). The diameter of pupils is indicated by the arrows. Legend: R = retina; L = lens; P = screening pigment; CM = a concave mirror; and CC = a crystalline cone.

SOURCE: Eric J. Warrant, N. Adam Locket, "Vision in the deep sea," *Biological Reviews* Volume 79, Issue 3, pages 671–712, August 2004, John Wiley and Sons. Copyright © 2007, John Wiley and Sons.

a "third" eye. The discreteness of the distinction between rhabdomeres and ciliary photoreceptors is challenged by the situation in a polychaete annelid, *Platynereis dumerilii* (324). In the larval and adult eyes, the photoreceptors are rhabdomeres, while in the brain the photoreceptor cells are ciliary.

But eyes can be more than photoreceptors. If you have ever looked on a winter landscape and thought that it looked cold, your perception may reflect the fact that, if you are cat-like, you have eyes that include cold receptors. Cat eyes have several kinds of nociceptors that respond to noxious and non-noxious stimuli (mechanical, chemical, and thermal). The cold sensors are located in the area of the iris and generate changes in local blood flow, which may ensure adequate blood supply to retinal cells, allowing continued function even under cold environmental conditions.

Animals such as sea urchins change their behaviour according to lighting conditions and have some ability to discriminate objects. However, in spite of concerted searches, the photoreceptors in these animals have not been discovered. The publication of the genome of the purple sea urchin (*Strongylocentrotus purpuratus* (1092), Fig. 15.32) made possible a different approach to the quest for photoreceptors in these animals. Data in the genome allowed researchers to design antibodies against *Sp-Opsin4* and to use *in situ* hybridization of *Sp-opsin4* and *Sp-pax6* which regulate phototaxis. This approach indicated that the tube feet of purple sea urchins contained two distinct groups of photoreceptor cells. The sea urchin photoreceptive cells are microvillar, r-opsin, previously known only from protostomes. The absence of photoreceptive cells in larval echinoderms suggests that the adult skeleton may operate as a screening device, making it a huge compound eye!

STUDY BREAK

1. What do *Pax6* genes have to do with photoreception?
2. What is extraocular perception?

Figure 15.29 Two ocelli (arrows) are obvious between the compound eyes of this honey bee (*Apis mellifera* (984)).

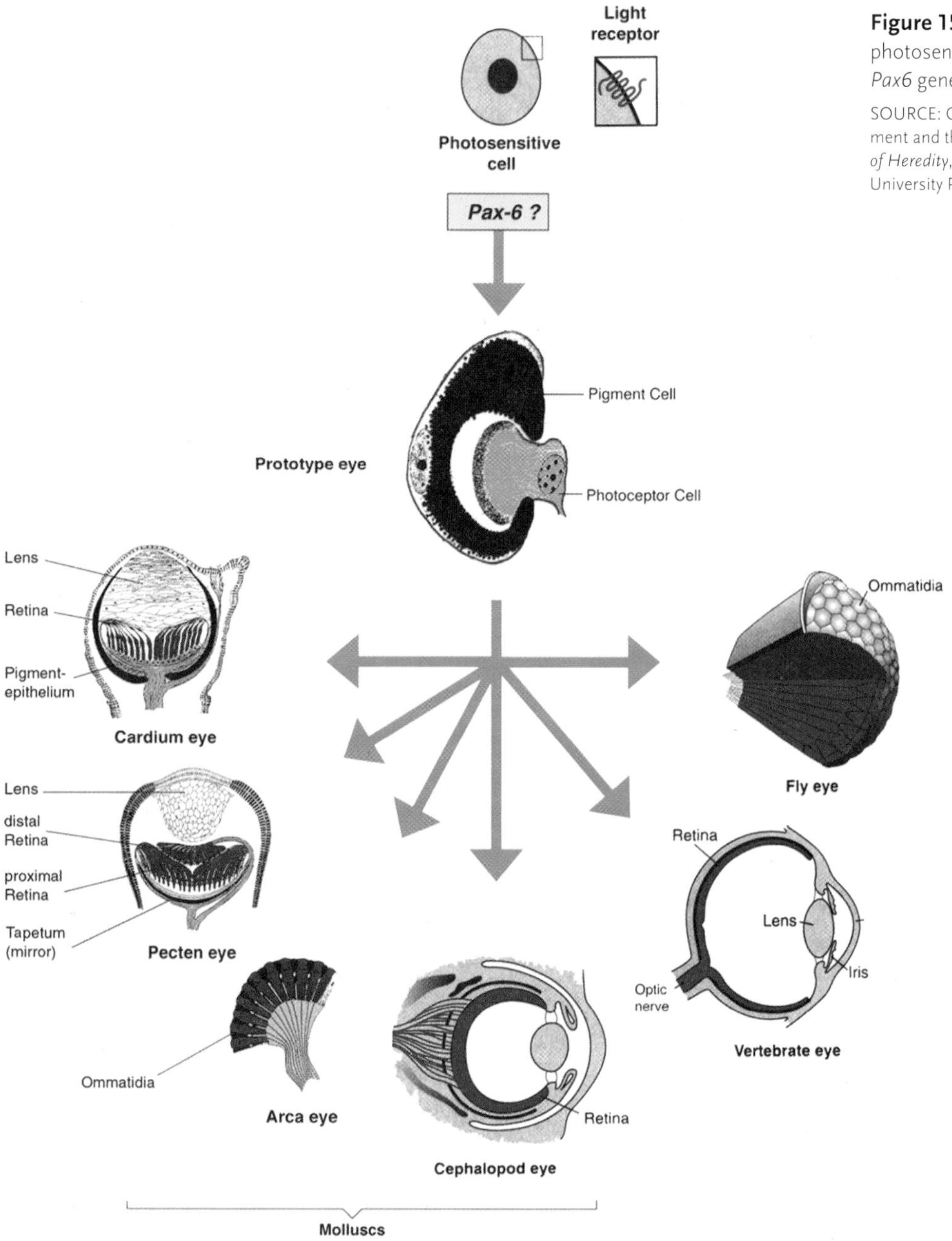

Figure 15.30 A possible scenario for evolution of photosensitive cells assembled under the control of *Pax6* gene.

SOURCE: Gehring, W. J., "New perspectives on eye development and the evolution of eyes and photoreceptors," *Journal of Heredity*, 2005, Vol. 96:171–184, by permission of Oxford University Press.

Light receptor

Photosensitive cell

Pax-6 ?

Pigment Cell

Prototype eye

Photoceptor Cell

Lens
Retina
Pigment-epithelium

Cardium eye

Ommatidia

Fly eye

Lens
distal Retina
proximal Retina
Tapetum (mirror)

Pecten eye

Retina
Lens
Iris
Optic nerve

Vertebrate eye

Ommatidia

Arca eye

Retina

Cephalopod eye

Molluscs

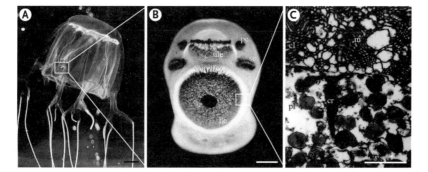

Figure 15.31 Three progressively closer views of the visual system of the box jelly *Chiropsella bronzie* (156). **A.** The whole medusa, **B.** a close-up of a rhopalium, and **C.** a close-up view of the lower lens eye retina. A ciliary rootlet and a cilium with microvilli are typical of ciliary photoreceptors. Legend: cl = ciliary layer; cr = ciliary rootlet; lle = lower lens eye; m = microvilli; pe = pit eyes; pl = pigment layer; se = slit eye; ule = upper lens eye. Scale bars: **A** = 1 cm, **B** = 100 μm, and **C** = 1 μm.

SOURCE: O'Connor, M., A. Garm, J.N. Marshall, N. S. Hart, P. Ekstrom, C. Skogh and D-E. Nilsson, "Visual pigment in the lens eyes of the box jellyfish Chiroptsella bronzie," *Proceedings of the Royal Society B*, 2010, Vol. 277:1843–1848, by permission of the Royal Society.

Figure 15.32 A purple sea urchin, *Strongylocentrotus purpuratus* (1092).

NatalieJean / Shutterstock.com

BOX 15.4
The Deep Blue Sea

In the open ocean, the epipelagic zone (sea level to 150 m) is, for the most part, well lit and oxygen rich. As you go deeper into the mesopelagic zone (150–1000 m), less and less light filters down from above that would create an extended source of light, and point sources of light from bioluminescence appear. Below 1000 m, the bathypelagic zone is a dark, cold (1–5°C) place where animals (predators, prey, mates) can be few and far between. The only light is from point sources. Hydrothermal vents are islands of diversity in the benthic zone, where hot water and chemosynthetic bacteria combine to provide a rich setting supported by oxidization of sulphides coming from the vents.

The longer wavelengths of light do not penetrate far through water, consequently with increasing depth, the light is increasingly blue. The open ocean is a featureless, infinite, monochromatic blue scape, bright above, dark below and offering residents no places to hide. This situation generates interesting challenges around seeing, being seen, and avoiding being seen. These challenges translate into differences in visual systems as well as in body form and appearance.

There is a good match between eye design and the visual scene at different depths in the ocean. In the mesopelagic zone, some species use spatial and temporal summation to take advantage of extended sources of light, while others specialize in detecting bioluminescence. In some species, different areas of the retina are specialized for each situation. Eyes of animals in the bathypelagic zone tend to be smaller with excellent spatial resolution. The eyes of animals in the benthic zone tend to be larger than those of pelagic species, and many of the fish there have retinal ganglion cells arrayed in a horizontal visual streak, allowing them a good view of the wide, flat space.

Three fish exemplify different visual specializations for conditions from epipelagic to bathypelagic zones. In its distribution, *Howella sherborni* (1268) overlaps the epipelagic and mesopelagic light zones. This fish has an extensive frontal field of view, with an average of 14 000 ganglion cells·mm^{-2}, 24 500 cells·mm^{-2} in the fovea where there are 22 layers of rods. These fish use surrounding light to illuminate prey. *Searsia koefoedi* (1242) is a bathypelagic species with eyes generally similar to those of *H. sherborni* except that the fovea is smaller and steep-sided, allowing precise detection and localization of objects in its visual field. *Scopelarchus michaelsarsi* (1167) is another bathypelagic form with dorsally pointing tubular eyes that have two retinal areas, giving the fish good sensitivity to both extended and point sources. Lanternfish such as *Lampanyctus macdonaldi* (1236) are bathypelagic species with exceptional visual capability. Their eyes absorb all of the incident light, and their visual sensitivity is 100 times greater than that of a nocturnal toad for extended sources and 120 times more sensitive than human eyes.

Deep-water mesopelagic fishes tend to have large eyes, some rounded, others tubular (Fig. 1). The more tubular eyes allow greater binocular overlap and permit more effective resolution of distances. An accessory retina on the side of the tubular part of the eye does not receive a focused image, instead it receives light from close to the fish's side, resulting in an extended field of view (by up to 70°). Other fish have lens pads that provide a separate field of view. At least one species, *Bathylychnops exilis* (1239), is more specialized, having an eye with two lenses, the main one that points up and a small secondary one that points down (Fig. 2).

Rods in the eyes of many species of deep sea fishes are 100 μm in length, about four times longer than rods in a human eye (26 μm), and they absorb twice as much light. While the rods of most deep sea fishes have one pigment, some, such as dragonfish (Section 15.7), have three or four. The details of multilayered retinas (Fig. 15.26C) remain unclear but could

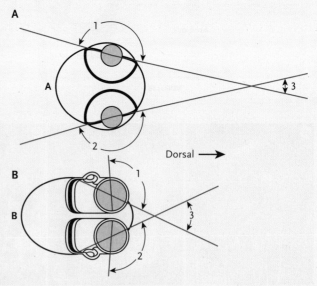

Figure 1
A. Round eyes versus B. tubular eyes of different species of mesopelagic fishes. Shown are the visual fields of left (1) and right (2) eyes, as well as the binocular visual field (3).

SOURCE: Based on Eric J. Warrant, N. Adam Locket, "Vision in the deep sea," *Biological Reviews* Volume 79, Issue 3, pages 671–712, August 2004, John Wiley and Sons.

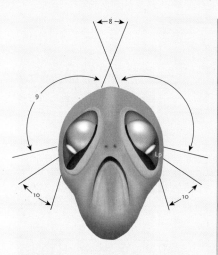

Figure 2

The arrangement of eyes in a scopelarchid fish. 1 = the view of each main right eye, 2 = zone of binocular overlap, and 3 = the field of view of the accessory retina, through light arriving through the lens pad (lp).

SOURCE: Based on Eric J. Warrant, N. Adam Locket, "Vision in the deep sea," *Biological Reviews* Volume 79, Issue 3, pages 671–712, August 2004, John Wiley and Sons.

occur through regeneration and growth and also may provide some filtering of light.

The camera eyes of deep-sea cephalopod molluscs are often very large and show similar adaptations to those of the deep-sea fish, often with one visual pigment. Other species have two or three visual pigments. The mesopelagic *Watasenia scintillans* ((522), firefly squid) has a multilay-ered retina and three visual pigments, each located in different chromato-phores. One pigment occurs through-out the retina, but others only in the ventral region. Firefly squid have the potential for detailed colour vision, perhaps important during courtship displays.

Compound eyes give other marine animals a great advantage when operating under low-light conditions because the size of each pupil can be very large relative to the focal length of the lens. Slower photoreceptor responses allow animals with compound eyes to use longer exposures to collect more light. In some cases the visual integration times are 160 ms, 20 times longer than those of house flies (*Musca domestica* (966)).

In water where blue is the prevailing colour, being blue can be effective camouflage. Many crusta-ceans and cnidarians have chromato-phores with blue carotenoprotein pigments. In the upper mesopelagic zone, red is a common colour because it absorbs incident blue light. In the bathypelagic zone, most animals are dark or black because these colours absorb bioluminescent flashes. Viewing angle and depth strongly influence the success of any camouflage strategy.

Flattened bodies with silvery sides are virtually invisible except from directly above or below them. Mesopelagic hatchet fishes use multiple stacks of guanine crystals to reflect almost 100% of incident light. At night, however, with reduced incident light, a bright bioluminescent flash will reveal them. At night, these fishes often disperse dark chromato-phores over their reflective bodies.

Transparent bodies have the same reflective index (100%) as the water, and the critical issue is the distance at which they can be detected. The body plans of cnidarians make it easier for them to be transparent (91%) partly by incorporating sea water into their bodies. Ironically, retinas must retain pigments to operate, and their pigments decrease their refractive indices. The copepods *Cystisoma* (790) eliminate screening pigments from their eyes, and amphipods such as *Phronima sedentaria* (795) have eyes with tiny retinas, both strategies increasing the animals' refractive index. Still other animals, for example larval lobsters, have very thin bodies, dispersing the reflection of internal organs over as large an area as possible.

Counter-illumination is achieved by bioluminescence, and many animals (fishes, cephalopods, and crustaceans) have a battery of bioluminescent organs along their ventral surface. Viewed from below, the bioluminescence of these organs breaks up the outline of the animal, making it virtually invisible. Ventral bioluminescence is typically turned on by day, off at night.

15.9 Infrared Detection: Heat

Finding the "correct" temperature is vital to the survival of animals so we expect them to be able to recognize heat and cold and move in the right direction to remain in a favourable situation. Fire-loving (pyrophilous) insects such as some jewel beetles (*Melanophila* (941) and *Merimna* (942), Coleoptera) and flat bugs (*Aradus* (867), Hemiptera) are good examples. These insects lay eggs in recently burned wooded areas and use a combination of smoke and infrared (IR) detectors to recognize those areas. Pyrophilous beetles in the genus *Melanophila* are widespread in the world except Australia, where *Merimna* occurs. The IR detectors differ between *Melanophila* and *Aradus* (Fig. 15.33), but their mode of operation may be similar. The sensitivity of the IR receptors allows some beetles to navigate to a moderate-size fire from distances of over 10 km.

Leaf-cutter ants (see "Why It Matters", Chapter 1) have a thermosensitive neuron in the sensilla coelo-conica on each antenna (Fig. 15.34). This sensor responds to temperature increases of 0.005°C and appears to allow the ants to detect very small changes in their thermal environment. This information may be used in general orientation. In triatomine bugs (e.g., the kissing bug (Hemiptera, *Rhodnius prolixus* (876),

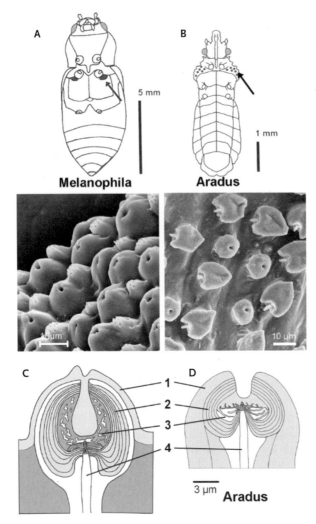

Figure 15.33 **A.** Jewel beetles (*Melanophila* (941)) and **B.** flat bugs (*Aradus* (867)) both have infrared receptors (red structures at tips of arrows) located on the thorax and shown in detail in scanning electron micrographs below. In *Melanophila* the sensors are located in pits and associated with wax glands. *Aradus* has about 12 IR sensors interspersed among mechanosensory bristles. Schematic drawings of a single IR receptor for **C.** *Melanophila* and **D.** *Aradus*. Legend: 1 = outer exocuticle; 2 = the exocuticular shell of the inner sphere; 3 = the microfluidic core; 4 = the tip of a mechanosensory dendrite.

SOURCE: Klocke, D., A. Schmitz, H. Soltner, H. Bousack and H. Schmitz. 2011. "Infrared receptors in pyrophilous ("fire loving") insects as a model for new un-cooled infrared sensors," *Beilstein Journal of Nanotechnology*, 2:186–197.

page 355), a cave-like sense organ in each antenna is a heat-detector, allowing these insects to find their prey, birds and mammals. Many other invertebrates have heat-sensing organs which may be used to find prey (e.g., bed bugs, lice, ticks), or appropriate places to live.

In vertebrates, transient thermal potential (TRP) channels operate by controlling the movement of Ca⁺⁺ ions. Humans, for example, have 28 known TRP channels grouped in six families. Three of these channels are involved in thermoreception: TRPV (TRP vanilloid), TRPM (TRP melatonin), and TRPA (TRP ankyrin). Humans are not capable of infrared

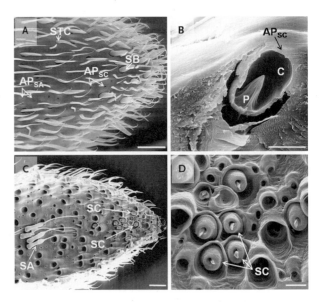

Figure 15.34 **A.** Leaf-cutter ants have sensilla coeloconica (SC) near the tip of each antenna. **B.** A longitudinal section of the chamber (c) of the sensilla coeloconica shows the sensory peg (p). **C.** An internal view of the last flagellar segment shows the location of sensillae coeloconicae. **D.** Sensillae coeloconicae embedded in the cuticle of the antenna. Legend: SA = sensilla ampullaceum, SB = sensilla basiconicum, and STC = sensilla trichodeum curvatum. Scale bars: A, C = 20 μm, B, D = 5 μm.

SOURCE: Reprinted from *Arthropoda Structure and Development*, 38, Ruchty, M., R. Romani, L.S. Kuebler, S. Ruschioni, F. Roces, N. Isidoro and C.J. Kleineidam, "The thermo-sensitive sensilla coeloconica of leaf-cutting ants (Atta vollenweirderi)," Pages 195–205, Copyright 2008, with permission from Elsevier.

(IR) detection, but it has been reported from some snakes (pit vipers, some boas, and some pythons) and from vampire bats (*Desmodus rotundus* (1788), Fig. 15.35).

In all vertebrates, the trigeminal nerves provide sensory input from the head and face, but in some snakes and vampire bats they also serve specialized heat detectors. In the snakes, TRPA1 is the cell-surface ion channel used in IR detection; in vampire bats it is TRPV1. In other mammals, TRPV1 is a nociceptor that detects noxious heat (>43°C), while in vampire bats TRPV1 is activated at about 30°C.

STUDY BREAK

1. What animals have the ability to detect infrared reception?
2. How do they use it?

15.10 Vibrations

15.10a Seismic

In some parts of the southern United States, people traditionally use "worm grunting" to attract earthworms for fishing bait. Grunting involves producing broadband low frequency vibrations through the

M.B. Fenton

Figure 15.35 A. Infrared (IR) sensors in pits (arrow) on the face of a pit viper (fer de lance, *Bothrops asper* (1474)) are used to locate warm prey. **B.** IR sensors on the central part of the noseleaf or the upper lip (yellow arrows) of common vampire bats are used to locate places where blood flows close to the skin.

ground, sometimes by rhythmically scraping a wooden stake that has been driven into the ground. Within minutes of starting the grunting vibrations, earthworms (*Diplocardia mississippiensis* (356) or *D. floridana* (355)) surface and emerge from the ground. Two hypotheses have been suggested to explain the worms' behaviour: first, the worms detect vibrations associated with rain and emerge from the ground to avoid drowning. Second, vibrations associated with digging by a predator could cause the worms to surface. Although the two hypotheses are not necessarily mutually exclusive, predator evasion may be more likely. In any event, the effect of human grunting behaviour on earthworms demonstrates that vibrations can provide important information to animals.

Drumming or stamping one's feet on the ground is another way to produce seismic vibrations. Banner-tailed kangaroo rats (*Dipodomys spectabilis* (1883)) use foot drumming to indicate ownership of territory. This signal includes both seismic and acoustic elements and is often produced when the animal is in its burrow. Within a burrow, an individual will drum in response to the drumming of another, and playback experiments demonstrate that the seismic vibrations carry farther than acoustic vibrations, the actual distances depending upon the force of drumming and the type of substratum. Many other animals, from burrowing crickets to frogs, use a combination of seismic and acoustic vibrations to advertise presence and territory ownership.

Red-eyed frogs, *Agalychnis callidryas* (1363), lay their eggs on a variety of substrata from rigid to flimsy, from tree trunks to the undersides of leaves. Vibrations associated with attacks by predatory snakes and wasps on clutches laid on leaves stimulate early hatching of the eggs, increasing the chances of survival of the young. Comparing the responses of embryos to the frequencies of the vibrations associated with attacks by snakes and wasps with vibrations from benign events (wind, rain, movements of embryos) demonstrate that larval frogs could distinguish among different stimuli (Fig. 5.38). Embryos exposed to vibrations associated with attacks by predators hatched about three days

earlier than control clutches. Tent-making bats and bats roosting in unfurled leaves also use vibrational cues to detect approaching predators; these cues can be amplified by the length of the leaf's petiole.

Giant dune scorpions (*Paruroctonus mesaensis* (694)) in sandy deserts detect compressional and surface waves generated by digging crickets (*Arenivaga investigata* (857)) at 50 cm away. Each of the scorpion's legs has a combination of compound slit sensilla on the basitarsal leg segments and tarsal hairs (Fig. 15.36B). These allow the hunter scorpion to detect both surface

© Bruce Farnsworth / Alamy

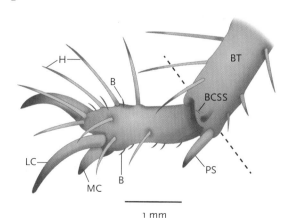

Figure 15.36 A. The giant dune scorpion (*Paruroctonus mesaensis* (694)) of the Mojave Desert uses sensors on its legs to detect seismic waves in sand, generated by the burrowing mole crickets on which it feeds. **B.** The fourth tarsus of the scorpion. Legend: H = tarsal hairs; BCSS = basitarsal compound slit sensillum; BT = basitarsus; T = tarsus; B = bristle hairs; LC and MC = lateral and medial claws; PS = pedal spur.

SOURCE: Based on Brownell, P.H. 1977. "Compressional and surface waves in sand: used by scorpions to locate prey," *Science*, 197:479–482.

1 mm

waves (slit sensilla) and higher frequency compressional waves (tarsal hairs) to locate and direct their attack on the digging cricket.

15.10b Sound

An obvious extension of seismic vibrations, sounds are vibrations travelling in air or in water. The ability of some species of golden moles (Chrysochloridae) to sense seismic vibrations with their ears illustrates the link. In the ears of these moles, the head of the **malleus**, one of the auditory ossicles (malleus, **incus**, **stapes**) is extremely enlarged (hypertrophied) (Fig. 15.37). The head of the malleus is attached to the tympanic membrane, and so this enlargement makes it sensitive to the very low frequencies that characterize seismic vibrations, and may allow golden moles to detect and avoid predators or to find prey. Variation in the development of the head of the malleus among golden mole species may reflect differences in the soils occupied by these tunnelling (fossorial) mammals. Loose, sandy soils may transmit seismic waves better than dense soils. Having a specialized malleus simply extends the range of frequencies that golden moles can hear, from the range of audible frequencies downward into the range of seismic frequencies.

15.10b.1 Sound Detection.

Whether or not detecting seismic vibrations is a form of hearing, acoustic detection is very common among animals. Sounds can be used to detect predators or prey and often serve a major role in courtship displays. Auditory sensory systems also play a central role in orientation of the animal with respect to gravity. In fish, this ability extends to detection of the hydrodynamic forces surrounding the animal.

Saccules, small chambers within the inner ear containing sensory hair cells, are presumed to be the primary auditory receptors in fish. An axis from front (rostral) to back (caudal) in the saccules of goldfish (*Carassius auratus* (1194)) is organized by tones (= **tonotopic**), with lower frequencies (100 Hz) in the posterior region and higher frequencies (4000 Hz) in the rostral area.

Some bony fishes have **Weberian ossicles** that connect the anterior part of the swim bladder (a buoyancy organ) to the labyrinth and facilitate hearing. In catfish (Siluriformes), species with larger swim bladders and more Weberian ossicles have better hearing ability, especially at higher frequencies. The 3100 species in the Siluriformes show considerable variation in swim bladders and Weberian ossicles. Some species have a large, single, free swim bladder, usually with four Weberian ossicles, while in others the swim bladders are small, paired, and encapsulated by bony capsules. Encapsulated swim bladders typically are associated with fewer Weberian ossicles (Fig. 15.38).

To what do fish listen? Acoustic displays accompany many facets of the behaviour of some fish, from interactions around feeding and aggression to courtship. Some herring use the echolocation clicks of foraging porpoises to detect and avoid these predators. Pacific herring (*Clupea pallasii* (1190)) swim in large schools and communicate with one another at night, when visual cues would be negligible, using fast repetitive tick sounds. The importance of the signals remains unclear, but they are produced by expelling bubbles of air from the anal tract; in other words they are fart-like sounds not associated with digestive gas or gulped air.

Fish such as mudskippers (periophthalmines of the family Gobiidae) are amphibious and well-adapted to intertidal zones, spending some time in water and some on land. During courtship and territorial interactions, individuals communicate with sounds that are also transmitted as vibrations in the mud on which they live. The source of the sounds appears to be the opercular chambers. Mudskippers may detect acoustic vibrations via their pectoral fins (in contact with ground or water). Amphibious fishes may provide useful subjects for exploring sensory implications for extinct animals such as *Ichthyostega* (see page 156).

The movement of animals from water to land had to be accompanied by important sensory changes related to the difference in density (and thus sound conductance) between air and water. If, when swimming, you duck your head underwater, you can "hear" or even "feel" motor-boat engines at a greater distance than when your head is out of water. Sound travels about three times faster in water than in air.

Among terrestrial animals we see a variety of ears, structures that receive (transduce) sounds. Insect ears range in complexity from simple hairs to tympanal organs that are sensitive to a wide range of frequencies, and they can be located in several different places, such as head, mouthparts, wings, abdomen, and legs. Regardless of their complexity and neural connections, insect and crustacean ears are mechanoreceptors known as **chordotonal organs**. Chordotonal organs consist of one or more **scolopidia** (singular = scolopidium), simple

Figure 15.37

A comparison of the left malleus and incus of two species of golden moles, *Chrysochloris asiatica* (1699) (left) and *Eremitalpa granti* (1702) (right). The exceptional enlargement of the head of the malleus is obvious in *E. granti*.

SOURCE: Photograph courtesy of Dr. Matthew J. Mason, University of Cambridge

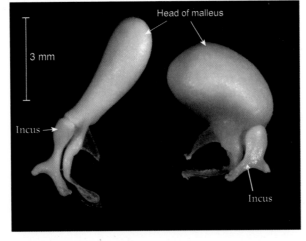

Head of malleus

3 mm

Incus

Incus

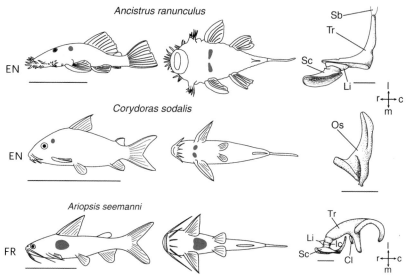

Ancistrus ranunculus

EN

Corydoras sodalis

EN

Ariopsis seemanni

FR

Sb
Tr
Sc
Li

Os

Tr
Li
Sc Ic
Cl

Figure 15.38 Lateral and ventral views of three species of catfish, two with encapsulated (EN) swim bladders and one with a free (FR) swim bladder. Swim bladders are shown in blue. The Weberian ossicles of the three species are also shown. Legend: Os = ossicle; Sb = swim bladder; Sc = scaphium; Tr = tripus. Ossicle sizes are standardized, scale bar = 5 cm for fish, 5 mm for ossicles.

SOURCE: Lechner, W., and F. Ladich. 2008. Size matters: diversity in swimbladders and Weberian ossicles affects hearing in catfishes. *Journal of Experimental Biology*, 211:1681–1689.

mechanoreceptors that detect rapidly alternating pressures (Fig. 15.39). There is considerable variation among scolopidia and chordotonal organs, which is accompanied by differences in the numbers of auditory nerves underlying the exoskeleton that conduct acoustic information from transducer to the central nervous system.

The variety of ears among arthropods indicates that these animals may respond to different sounds. Many arthropods use sounds to attract prey or detect and avoid predators. Others, such as crustaceans, appear to use sounds to identify appropriate habitats. Playback experiments involving over 600 000 individuals demonstrated that a range of species use reef sounds to find appropriate habitat, either avoiding reefs or moving toward them.

Cephalopods such as longfin squid (*Loligo (Doryteuthis) pealei* (505)) appear to hear as well as most fish, but how they do so remains unclear. Detection of evoked action potentials in neurons generated by sounds (auditory evoked potentials) supports the suggestion that squid hear with their statocysts, structures usually used to monitor the Earth's field of gravity. Statocysts of squid are complex. Such a system provides information about orientation in the gravitational field (the statocyst itself), and a second system (crista-cupula) performs like an angular accelerometer. Squid appear to detect sounds through the particle motion component of the sound field rather than the pressure component.

15.10b.2 The Tympanic Middle Ear.
Most terrestrial vertebrates have a **tympanic middle ear** well suited to receiving airborne sounds and converting sound energy (mechanical vibrations) in air to electrical stimuli (Fig. 15.40). The eardrum (**tympanum**), located at the opening to the middle ear (auditory meatus), receives sound vibrations and transfers them to an ear bone (amphibians, reptiles, and birds) or ear bones (mammals). The bones transfer the vibrations to inner

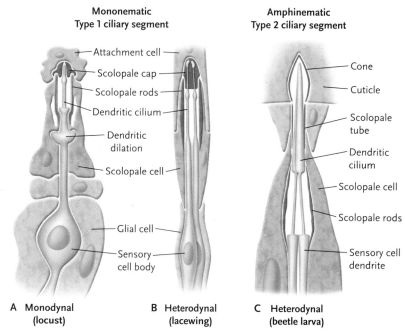

Mononematic Type 1 ciliary segment

Attachment cell
Scolopale cap
Scolopale rods
Dendritic cilium
Dendritic dilation
Scolopale cell
Glial cell
Sensory cell body

Amphinematic Type 2 ciliary segment

Cone
Cuticle
Scolopale tube
Dendritic cilium
Scolopale cell
Scolopale rods
Sensory cell dendrite

A Monodynal (locust) **B** Heterodynal (lacewing) **C** Heterodynal (beetle larva)

Figure 15.39 The arrangement of scolopidia in different types of chordotonal organs in the ears of animals: **A.** a locust, **B.** a lacewing, and **C.** a beetle larva. The nature of the dendritic cilium distinguishes Type 1 from Type 2 scolopidia, and mononematic versus amphinematic based on extracellular structure of the scolopale cell. Monodynal are distinguished from heterodynal by the number of sensory neurons per scolopidium.

SOURCE: Based on Yack, J.E. 2004. "The structure and function of auditory chordotonal organs in insects," *Microscopy Research and Techniques*, 63:315–337.

ear fluids where they are converted, via hair cells, to electrical impulses.

The evolution of the tympanic middle ear made animals more sensitive to higher frequency sounds because refinement of micromechanical tuning extended the frequency range. This trend is best displayed in birds and mammals where the articulation of the ear ossicles provides a mechanism to increase sensitivity to low-level sounds as well as amplification of sounds at the cochlear level. The role of the ear

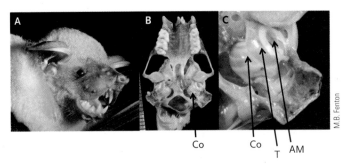

Figure 15.40 A greater bulldog bat (*Noctilio leporinus* (**1803**)). **A.** The bat's face, **B.** a ventral view of the skull, and **C.** a close-up of the cochlea (Co) and opening of the ear canal (auditory meatus, AM) into the middle ear, which is enclosed by the tympanic bone (T). The spirals of the cochlea are visible in C.

ossicles within the air-filled middle ear is crucial for effective transfer of acoustic energy from air to the fluid of the inner ear. The two media differ in both density and **impedance**, so the movement of sound waves from air to fluid is difficult. Air has much lower impedance than the fluid of the inner ear, so if airborne sounds impinged directly on the oval window (connection between middle ear and inner ear), <5% of the energy would be transferred. Instead, sound waves are converted to physical vibrations at the tympanum, which is surrounded by air on both sides. The impedance of the tympanum is similar to that of air, permitting effective transfer of energy into the auditory system rather than reflecting it back out. The physical vibrations are transferred through the ear ossicle(s) within the air-filled middle ear directly to the oval window.

Knowing this should mean that you will not be surprised to learn that cetaceans have lost the ear canal (**auditory meatus**). They depend upon the lower jaw to transmit vibrations in the water to the auditory system. Modifications in the inner ear include elongation of papillar structures, which increases the number of hair cells, allowing specializations among these cells for the detection of more specific sound frequencies. Birds have more stereovilli (projections or microvillii on surfaces of epithelial cells) per hair cell than mammals do, as well as a thin basilar membrane and papillae with many specialized cells. These structural modifications effectively produce high hearing sensitivity as well as frequency selective hearing.

15.10b.3 Vertebrate Hearing. External ears or **pinnae** are obvious sound collectors (Fig. 15.41). The role of pinnae in hearing is similar to the use of ear trumpets (Fig. 15.42), or cupping your hand behind your ear to catch a whispered comment.

Vibrations at the eardrum are transferred to the bones of the middle ear. Frogs, reptiles, and birds have only one auditory ossicle, the **stapes** (sometimes called the **columnella**). In birds, an extra stapes attaches to the tympanum. The stapes in turn leads to the footplate that abuts the oval window, which is the connection to the cochlea. Mammals have three auditory ossicles, the **malleus**, **incus**, and **stapes**. The manubrium (Fig. 15.39B) or head of the malleus abuts the tympanum and starts the chain of amplification leading to the stapes and the oval window (Fig. 15.39). Like Weberian ossicles in fish, the stapes and auditory ossicles of mammals amplify vibrations of the tympanum.

Most species of frogs (Anura) have a tympanic middle ear; but such an ear is absent in salamanders (Urodele) and caecilians (Apoda). Receiving seismic vibrations or vibrations in water does not require a tympanic middle ear, and many aquatic and fossorial species of amphibians and reptiles lack both the tympanum and the middle ear. Some amphibians, including some species of frogs, do not rely on their ears to detect airborne sounds. Tadpoles also lack tympanic middle ears and may directly perceive vibrations in water via the oval window.

In most frogs, the auditory system consists of three components that respond to different frequencies, reflecting the tuning of hair cells. The **sacculus** responds to low frequencies, the **amphibian papilla**

Figure 15.41 External ears in vertebrates. External pinnae are obvious in some mammals (**A.** *Marmosa mexicana* (**1685**) and **B.** *Otonycteris hemprichii* (**1805**)), while in some birds (**C.** *Struthio camelus* (**1674**)) and reptiles (**D.** *Varanus komodoensis* (**1528**)) the opening to the auditory canal is relatively simple. In some aquatic mammals (**E.** *Delphinapterus leucas* (**1752**)) there is no external opening (but there is a small scar behind the eye). In some reptiles and amphibians (**F.** *Dermatemys mawii* (**1425**) and **G.** *Bufo marinus* (**1367**), respectively), there is an obvious tympanum or ear drum.

SOURCE: M.B. Fenton

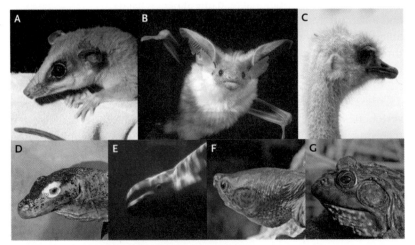

Figure 15.42 A brass ear trumpet, precursor of today's hearing aids.

SOURCE: pixelfabrik / Shutterstock.com

to mid-frequency, and the **basilar papilla** to higher-frequency sounds (Fig. 15.43). The sacculus, amphibian papilla, and basilar papilla are housed in the otic capsule of the braincase. In addition to low-frequency sounds, the sacculus provides the frog's gravitational sense. While most frogs use calls well within the range of human hearing, at least one species, *Huia cavitympanum* (1501) of Borneo, emits exclusively high-frequency calls (>20 kHz), which play an important role in courtship. Male frogs of the species *Odorrana tormota* (1374) from China produce courtship calls with high-frequency components, and the females also produce these signals just before they ovulate. In these two frog species, the eardrum is below the surface, so they have a distinct ear canal (auditory meatus) leading from the outside to the eardrum. It is possible that the tympanic middle ear of anurans evolved separately from that of amniotes.

Among reptiles, the ears of turtles and the tuatara lizard (*Sphenodon* (1458)) are the least specialized. Their hair cells are not specialized, but do show some variation in hair lengths. Ion channels on hair cells generate electrical signals when hairs are deflected by movement of the **basilar membrane**, but membrane movement shows little frequency selectivity. On the other hand, snake and lizard ears show considerable diversity and frequently have tonotopic organization, like that seen in birds and mammals. They show a greater diversity of hair cells than do mammals (Fig. 15.44). Sometimes hair cells are covered by a **tectorial membrane** (Fig. 15.44), either continuous or divided into different sections, coinciding with a wider range of hearing sensitivity to different frequencies.

Birds (and crocodiles) and mammals have independently evolved populations of specialized hair cells distributed across papillae within the cochlea, along the basilar membrane. These provide more or less continuous tonotopic organization and sensitivity to high frequencies: in birds and crocodiles to about 10 kHz, in mammals up to and beyond 100 kHz. The papillae in these two groups can be 11 mm long (owls), and >100 mm (whales), compared to 2 mm in lizards and snakes. The coiled cochlea of marsupial and placental mammals was instrumental in allowing elongation of papillae. There is an obvious trend to elongation of the basilar membrane among amniotes, (Fig. 15.45).

Hearing high-frequency sounds, however, is not limited to birds and mammals. Pygopod geckos from Australia are snake-like in body form and capable of astonishing high-frequency hearing extending to

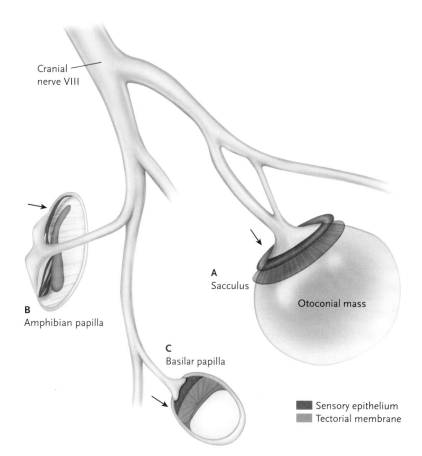

Figure 15.43 Three components of the frog's auditory system sensitive to low (sacculus), medium (amphibian papilla), and high (basilar papilla) frequency sounds. Arrows indicate how the hair bundles protrude in situ. N = auditory nerve.

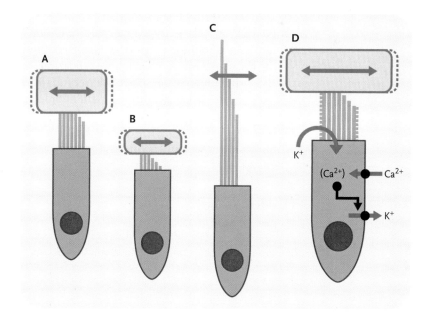

Figure 15.44 A.–D. Hair cells contributing to micromechanical tuning by frequency in reptiles. Hair cells show variation in a covering tectorial structure (present, A, B, D; absent C). Taller stereovillar bundles (A versus B) are more sensitive to lower frequencies. **C.** Absence of a tectorial covering coincides with taller stereovillar bundles. **D.** Flow of mainly K⁺ ions depolarizes cell, activating Ca⁺⁺ channels, raising Ca⁺⁺ concentration in the cell, and initiating firing of nerve fibres in the auditory nerve.

SOURCE: Manley, G.A. 2000. "Cochlear mechanisms from a phylogenetic viewpoint," *PNAS* 97:11736–11743. Copyright 2000 National Academy of Sciences, U.S.A.

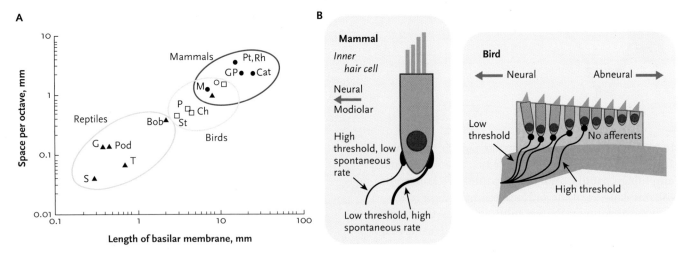

Figure 15.45 **A.** Lengths of basilar membranes in reptiles (solid triangles), birds (□) and mammals (●). **B.** Patterns of innervation in mammals and birds. Legend: Bob = Australian bob-tailed lizard; Cat = cat; Ch = chicken; G = alligator lizard; GP = guinea pig; M = mouse; O = barn owl; P = Pigeon; Pod = European lizard; Pt = bat *Pteronotus* (1810); Rh = bat *Rhinolophus* (1813); S = granite spiny lizard; St = starling; T = turtle.

SOURCE: Manley G A (2000) in *Auditory Worlds: Sensory Analysis and Perception in Animals and Man*, eds Manley G.A., Fast l.H., Köss l.M., Oeckinghaus H., Klump G.M. (Wiley, Weinheim), pp 7–17.

14 kHz and in some cases to >20 kHz. Specializations in the basilar papillae account for their hearing performance. Groups of hair cells within the papilla minimize the chances of their being deafened by their own calls. While their hearing is generally similar to that of other geckos, it is still not clear what role high-frequency hearing plays in their lives.

Each of the two middle ear cavities in tetrapods is connected to the pharynx (throat) by a **Eustachean tube**. The middle ear cavities of many birds, lizards, frogs, turtles, and crocodiles are broadly connected by wide Eustachean tubes, sometimes with additional interaural tubes. In these animals, acoustic coupling of the two ears allows directionality of hearing. In mammals, the Eustachean tubes are very narrow, resulting in relatively isolated left and right middle ear cavities. Mammals determine the direction from which a sound arrives by analyzing the relative differences in the timing and intensity of a sound as it arrives in each ear (binaural cues).

STUDY BREAK

1. What is a seismic detector? How is it used?
2. Compare the ears of bats with the bat-detecting ears of some insects.

15.11 Proprioception

Sensing the relative position of body parts (proprioception) is an important aspect of movement, whether in water or on land. This ability stems from receptors associated with joints and muscles that provide information about the relative forces exerted in movement, as well as more specialized receptors that provide information about equilibrium (balance). Proprioception involves some integration of exteroception (information from the near environment) with more internal cues.

Fish sense near-field hydrodynamic movement across their bodies using the lateral line, which consists of neuromast cells (Fig. 15.46). Pit organs, also known as free neuromasts, occur across the skin and are directly exposed to water. Pit organs encode water velocity and are associated with grooves in the skin in rays, while in sharks they occur between modified denticles. In teleosts, the distribution of pit organs varies with the life cycle, while in elasmobranchs they are consistent.

Canal neuromasts form a network below the epidermis and occur as pored and non-pored canals. Pored canals connect to the surrounding water. In elasmobranchs, pored canals detect acceleration of water flow near the skin's surface, while non-pored canals are used to measure hydrodynamic velocity. In benthic rays, non-pored canals also respond to direct displacement of the skin and may be used as tactile sensors on the ventral surface.

Just as in the longfin squid, the sense of hearing in vertebrates is associated with the ability to sense position relative to the Earth's field of gravity. In vertebrates, the **labyrinth** (Fig. 15.47) includes **semicircular canals** arranged at right angles to each other, enlargements called **ampullae** associated with each canal, and larger chambers containing sensory hairs known as **lagenae**, **saccules**, and **utricles** that vary among groups and habitat. Each of these structures is fluid-filled and continuous with the inner ear structures involved with hearing. In fish, the labyrinth organ also connects to the lateral line system (Fig. 15.47), which senses pressure and receives sounds.

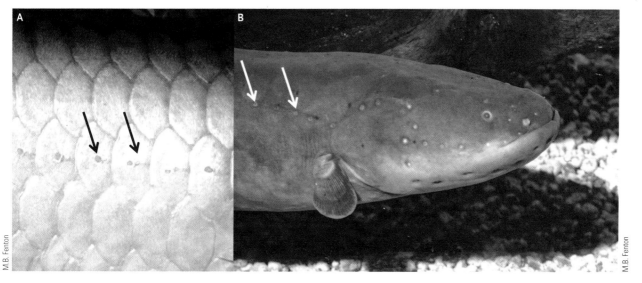

Figure 15.46 The lateral line system of a fish is visible as a line of pores (yellow arrows) on **A.** an arowana (*Osteoglossum* sp. (1247)) and **B.** an electric eel (*Electrophorus electricus* (1223)). The pores contain neuromasts and connect to the labyrinth organ (Fig. 15.47)

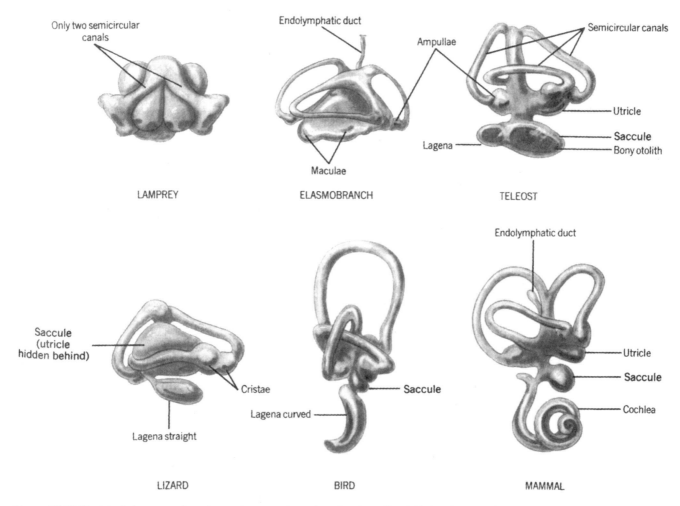

Figure 15.47 The labyrinth organs of vertebrates: lamprey, elasmobranch, teleost, lizard, bird, and mammal.

SOURCE: Hildebrand and Goslow fig. 19.8 *Analysis of vertebrate structure* 5th edition

1. What are the components of the lateral line system in fish?
2. How does a statocyst work? What information does it provide?

15.12 Mechanoreception

The ability of animals to detect vibrations (see above) is one aspect of mechanoreception. Sensory "whiskers" are common among mammals (Fig. 15.48), and their equivalents occur widely among animals. The motor cortex of mice controls protraction of whiskers, an exploratory movement effected through direct connections between the whiskers and the primary somatosensory cortex. The cortex also regulates retraction of whiskers through a rapid negative feedback system.

The pectines of scorpions (Fig. 15.49) serve to detect vibrations and also are sexually dimorphic, allowing us to distinguish males from females. Coating the pectines with paraffin wax reduces the animals' ability to locate prey and presumably interferes with interactions between males and females. Note the variation in detection of vibrations among scorpions (see *Paruroctonus mesaensis* (694), Fig. 15.36).

Tentacled snakes (Fig. 15.22) live in murky, slow-moving waters in Southeast Asia and use mechanoreceptors on their tentacles to detect water movements made by small fish. Sensory input from the tentacles goes to the optic tectum where large somatosensory receptive fields occur in deeper layers where they approximately overlay the visual fields. The tectum appears to be the site of integration of visual and mechanosensory cues. The result of this integration is the snakes' ability to accurately strike at swimming fish.

The wings of bats have considerable sensory capacity. The tactile sensitivity of the wing surface of big brown bats (*Eptesicus fuscus* (1791)) is 0.2–1.2 mN, close to or more sensitive than that of human fingertips. Furthermore, mechanoreceptive domed sensory hairs on the wing surface (Fig. 15.50) provide feedback about airflow and flight speed. Experimental removal of these hairs with a depilatory cream reduced the bats' capacity for manoeuvrable flight. These findings demonstrate that the flight of bats involves more multiple sensory input than previously thought.

1. What do the pectines of scorpions have in common with the tentacles of tentacle snakes?

15.13 Chemoreception: Olfaction and Taste

Animal chemoreception involves olfaction (smell) and gustation (taste). In spite of its fundamental importance in many aspects of the lives of animals, only a

Figure 15.48 Sensory whiskers of a naked mole rat (*Heterocephalus glaber* (1886)) even though the animal lacks fur.

Figure 15.49 Ventral view of a scorpion showing the pectines that are mechanoreceptors specialized for detecting vibration. Sexual dimorphism suggests that they also play a role in mate selection.

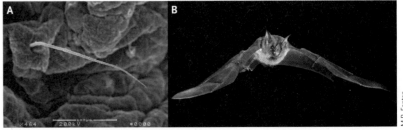

Figure 15.50 A. Scanning electron micrograph of a sensory hair on the wing membrane of a short-tailed fruit bat (*Carollia perspicillata* (1787)). **B.** These hairs contribute to the exceptional maneuverability of a flying bat (short-tailed fruit bat) by providing input about airflow and flight speed.

SOURCE: A. Courtesy of Susanne Sterbing-D'Angelo in the Batlab at the University of Maryland

few studies provide even limited information about the organs or transducers involved. Chemoreception plays a basic role, for example, in many aspects of behaviour (feeding, defense), reproduction, and ecology, and may involve interspecific (even interphylum) chemical cues. Although organs with chemoreceptive transducers are often concentrated around the mouth, they may also occur on different parts of the body, such as the feet. We have more information about chemoreception in metazoans than in many other animals.

Unlike its blood-feeding relatives, the horse leech (*Haemopis marmorata* (366), Annelida), is carnivorous and forages at night, usually just above the shoreline where they eat earthworms, gastropods, and larval chironomids (flies). Behavioural experiments demonstrate that horse leeches use olfactory cues, probably a metabolite produced by earthworms, to identify them. But the olfactory receptors used by horse leeches remain unknown, although the receptors appear to be located on the leech's dorsal lip, above the mouth. This situation reinforces the importance of olfactory cues, and the reality that demonstrating behaviour that involves olfaction does not necessarily identify the receptors involved.

In arthropods, peripheral olfactory receptor neurons are compartmentalized and arranged in **sensilla**, specialized cuticular structures often located on antennae or mouthparts such as maxillary palps. *Drosophila* sp. (960) have three distinct types of olfactory hairs: basiconic sensilla that resemble a club, trichoids with sharp tips, and coeloconics that are dome-shaped with a grooved surface. These hairs are differentially distributed on the arista of the antennae, and appear to respond to different concentrations of odorants. Basiconic and grooved-peg sensilla (Fig. 15.51) on the antennae of fifth-instar nymphs of the blood-feeding barber bug (*Triatoma infestans* (879)) respond to the

odours of their hosts. Basiconic sensillae are stimulated by nonanal, a constituent of sheep wool and chicken feathers, while isobutyric acid in rabbit odour stimulates one receptor type in the grooved-peg sensilla. In some cases, CO_2 detectors also guide blood-feeding insects to food, for example, biting midges, mosquitoes, and ticks. These insects respond to these odours by moving toward the sources. In crustaceans, the cuticular sensilla are called aesthetascs. In the spiny lobster (*Palinurus argus* (830)), the cuticle of the aesthetascs has the capacity to sieve some molecules, determining which ones generate an olfactory response.

Olfaction plays a central role in mollusc feeding. The terrestrial snail *Helix aspersa* (477) uses olfactory cues when choosing its food plants. Furthermore, as in mammals, learning and olfactory mechanisms in molluscs are strongly linked. In molluscs such as *Limax maximus* (478), the procerebral lobe of the brain is the site of activation through learning associated with odours. These slugs have two pairs of antennae (Fig. 15.25B) and two procerebra, but one pair of antenna and one procerebrum is sufficient for learning. This suggests a level of structural redundancy in both peripheral and central organs. Although the procerebrum is quite different from the mushroom body of annelids and arthropods, it may serve the same function.

Although our experience tells us that odours have remarkable power to invoke memories, we still know relatively little about the role that odours can play in our lives. In 2006, the discovery of trace amine-associated receptors (TAARs), encoded by genes that occur in humans, mice, and fish, was astonishing—animals, including humans appeared to have the capacity to detect previously unknown odours. In mice, at least three TAARs allow recognition of volatile amines in urine. One detects a compound linked to stress, the other two respond to compounds enriched in male urine versus female. This discovery emphasizes the role of chemosensory cues in social interactions, even in humans.

Can terrestrial vertebrates also use olfaction underwater? Star-nosed moles (*Condylura cristata* (1918) Fig. 15.21) exhale bubbles while diving and then use the papillae around their nostrils to capture bubbles, direct them into their noses, and allow the nose to smell airborne odours under water.

In any sensory situation, questions about how the brain and sensory systems classify input are very intriguing. To classify odours, zebrafish (*Danio rerio* (1197)) depend upon overlapping patterns of sensory activation by structurally similar odours. Classification of odours occurs in the glomeruli that provide input to the olfactory bulbs. Neuronal circuits in the glomerulus amplify small differences between signals generated by similar odours but change their activity patterns with odour concentrations. Classification occurs even when odours are identified as the same stimulus over

Jany Sauvanet / Photo Researchers, Inc.

Figure 15.51 A. Basiconic sensillum (upper left arrow) and grooved-peg sensillum (lower right arrow) on the antenna of a fifth instar barber bug (**B.** *Triatoma infestans* (879)). Scale bar = 20 μm.

SOURCE: A. Reproduced with permission from Guerenstein, P.G. and P.M. Guerin. 2001. "Olfactory and behavioural responses of the blood-sucking bug *Triatoma infestans* to odours of vertebrate hosts," *Journal of Experimental Biology* 204:585–597. The Company of Biologists.

a range of concentrations. These data suggest mechanisms of pattern classification that involve changes in small ensembles of neurons rather than shifts at the network level.

More recent work demonstrates that mice can encode odours by time across a series of sniffs. This allows them to read temporal patterns of olfactory stimuli.

STUDY BREAK

1. Why does a head cold affect your sense of taste?
2. What are TAARs? What do they do?

15.14 Electrodetection

Electrodetection allows some animals to detect potential predators or prey through the electrical fields generated by the operation of their nervous systems. For example, ampullae of Lorenzini are passive electroreceptors found in elasmobranchs. These receptors are located under the skin and, via a somatic pore and canal, link to the external environment. Ampullae of Lorenzini are often clustered and encapsulated by connective tissue to eliminate interference from the electric field of the animal itself. There are at least three kinds of ampullae: macro-ampullae with elongated canals that end in ampullary bulbs, micro-ampullae that are very small, and microscopic mini-ampullae. Macro-ampullae are widespread in marine elasmobranchs and micro-ampullae in hexanchid sharks and holocephalians, while mini-ampullae occur in freshwater rays (*Potamotrygon* spp. (1142)). The variety of ampullae of Lorenzini reflects skin resistance, electrical loading, and conductance of the medium (freshwater versus saltwater).

Some bony fish have electroreceptors and generate electric signals that they use in both communication and orientation. To detect electric signals, adult elephant fish (mormyrids) use ampullary organs in addition to two tuberous organs, mormyromasts and knollenorgans. Larval mormyrids use promormyromasts in the lateral line system as electroreceptors. These larvae have two other electroreceptors, and the three receptors appear to be a transition to the adult electrosensory system.

Other animals known to detect and respond to electric fields include amphibians and monotreme mammals. There is also evidence that some crustaceans detect and respond to electrical fields. The topic, like many others in sensory biology, remains one of active research.

STUDY BREAK

1. What is electroreception? How does it work?
2. In what habitats do you find species with electroreceptors? Why?

15.15 Intrinsic Sensing

In addition to sensors that provide animals with information about their surroundings (extrinsic input), other sensors monitor their internal situations (intrinsic input). The array of intrinsic sensors is as impressive as the extrinsic ones, but not as obvious at first glance. Stretch receptors provide information about the extension of muscles. Here, muscle spindles are the sensors, and they provide continuous feedback. Think of the complex muscle activity patterns that must be monitored when you play an instrument or run. Another sensor in mammals is located in the carotid bodies, which lie at the branching between the internal and external carotid arteries. These sensors monitor the concentrations of oxygen and carbon dioxide. Input from these sensors is received by the medulla and pons in the brain and affects the action of the circulatory system and the pattern of ventilation of the sites where gaseous exchange takes place (see page 314). Other sensors provide information about intrinsic temperature conditions, while still others information about the amount of food in the stomach.

STUDY BREAK

1. What range of intrinsic receptors do you expect to find in *Hydra* (135)? In *Dugesia* (230)? In *Helix* (477)?

15.16 Animals in Action

Most animal behaviour is influenced by sensory input. Co-ordinating the sensory input with the response is critical in these situations, emphasizing the importance of the central nervous system.

Box jellyfish, *Tripedalia cystophora* (157) (Cubozoa, Section 15.8), have a complex set of eyes that are somewhat similar in structure to those of vertebrates and cephalopods. In mangrove swamps where they occur, these jelly animals detect the canopy of mangroves seen through the water's surface and use this input to navigate their way through the swamps. Remaining in the swamps and away from open lagoon waters ensures better access to food. To accomplish this, *T. cystophora* look up using four specialized eyes (on the rhopalia).

Sensing the presence of a predator can influence and even interrupt the mating behaviour of a prey species. For example, males of many species of katydids (Orthoptera: Tettigoniidae) use songs to attract females. But these calls also attract predatory bats, leaving the males in a difficult situation: calling can attract females but may also reduce life span. In the Neotropics, the males of many species attract females by calling from plants, producing very short songs and using strong vibrations (tremulations) that shake the plants (Fig. 15.52). The tremulations do not attract foraging bats. Many other animals that use songs

Figure 15.52 Katydids. **A.** Male *Neoconocephalus ensiger* (901) call to females. **B.** Male *Docidocercus gigliotosi* (897) both sing and tremulate.

SOURCE: Hannah ter Hofstede

to attract mates suffer the same risk of exposure to predators as well as to parasites.

Beetles are not able to take flight as readily as most other insects, making them more vulnerable to attackers. Some bombardier beetles (e.g., *Brachinus* spp. (924)) use a hot spray emitted from the tip of the abdomen to deter attackers. These beetles rely on mechanosensory input to direct their defensive secretions. When attacked by ants, for example, an African bombardier beetle (*Stenaptinus insignis* (947)) aims its spray directly at the ant based on sensing where the ant is biting. The spray, a mixture of hydrogen peroxide and two hydroquinones, is ejected at about 100°C. The combination of heat and the irritant effect of the hydroquinones makes the defence effective.

Detecting movement is important for animals visually tracking a target, although visual clutter can be a problem. In male hoverflies (e.g., *Eristalis tenax* (961)), small moving targets are tracked by target neurons, some of which are inhibited by motion in the background. Others focus on movement of the target within the visual field. In these flies, rejection of background motion is probably achieved by selectivity of target neurons for small targets. These flies, and others, are very sensitive to moving targets (Fig. 15.53).

Figure 15.53 Had the crab spider (*Misumenoides formosipes* (675)) moved, the hoverfly (*Toxomerus marginatus* (973)) would likely have detected it. Motion detection is the key to prey species avoiding a predator, as well as the other way around. As it turned out, the spider dined on the fly.

STUDY BREAK

1. What do the animal examples in this section have in common?
2. What would be another good example to have included? Why?

15.17 How on Earth?

15.17a Predicting Earthquakes

Some animals are said to change their behaviour immediately before an earthquake. In Italy, common toads (*Bufo bufo* (1366)) at a mating site showed a reduction in activity before an earthquake. What cues were associated with the behaviour and how they were perceived remain a matter of conjecture.

15.18b Navigation and Way-Finding

Finding your way demonstrates how navigation behaviour, widespread among animals, requires integrated information from several sources. As soon as animals could move around, they needed to be able to orient themselves to find important places (such as nests or concentrations of food) and avoid dangerous ones. Orientation and navigation are tied to spatial memory, and some animals use many cues and different sensory modalities to navigate.

Some animals demonstrate behaviour related to the ability to detect magnetic fields. These data are usually derived from experiments about migration. Detecting a magnetic field can provide directional information, permitting an animal to follow a consistent course relative to magnetic north. Some animals even appear to have a magnetic map and can determine their approximate position, or their position relative to a roost or nest. What transducers are involved in magnetoreception? One suggestion is deposits of magnetite within animals' bodies. These crystals could affect secondary receptors, perhaps mechanoreceptors, as particles of magnetite orient with the magnetic field. Alternatively, magnetite particles adjacent to nerve cells could open Ca^{++} channels. Another possibility is that some biochemical reactions are influenced by magnetic fields.

Recent experiments with mouse-eared bats (*Myotis myotis* (1801)) revealed how these bats use a sun compass to reset their magnetic compasses. The field remains one of active investigation with many more questions than answers.

Innovative use of tags with GPS (global positioning systems) has allowed researchers to track the flights of Egyptian fruit bats (*Rousettus aegyptiacus* (1815)) from cave roosts to fruit trees and back. The behaviour of the bats demonstrated that they had a large-scale cognitive map of the areas in which they operated, and they used this in orientation. But other cues could also be involved.

Route-following is an obvious means of proceeding from A to B. However, think of giving (or receiving) directions from a passer-by, and the simplicity of route following becomes open to question. Some limpets (Mollusca, Gastropoda) have a fixed home point. They leave a trail of mucus as they move from their home location to feed on algae and then follow their trails home. Snakes and lizards with forked tongues provide an example of how a trail might be recognized and followed, by simultaneously comparing stimulus intensities at two points. When many species of ants follow a route, they use memorized visual snapshots of the landmarks and panoramas that they have learned to make a trajectory that takes them to and from their destinations.

How do animals measure distance and direction? In some cases, sensors such as those in the inner ears of vertebrates integrate direction and rotational accelerations. Spiders monitor the movements of their legs, apparently with elytriform slit sense organs at the leg joints. Celestial and magnetic compasses provide further cues, although the transduction of magnetic cues remains relatively unknown.

Everyone who has looked at a map to a buried treasure or other prize knows that measurements of distance over time (odometry) can be critical in orientation. The movement of images of surroundings across the retina (optic flow) can indicate self-motion as well as direction. Honey bees use optic flow in orientation, but optic flow depends upon the distance from the eye to the visual background. Honey bees solve this problem by flying at a consistent height above the ground (raising the question of how this is achieved!).

Stride length is another source of information about distance covered, although this can be affected by terrain and mode of locomotion. Your stride length can also differ according to your behaviour—walking, jogging, or running. Experiments involving surgical lengthening or shortening of an ant's stride legs demonstrated how they measure distance by stride length.

The challenges of orientation and navigation epitomize the importance of synchronization and integration of efferent and afferent neural input from sensors throughout the body.

STUDY BREAK

1. Survey your friends. Do those who come from the "city" use different cues in navigation and wayfinding than those from the "country"? Do they give the same directions to get from one place to another?

NEURAL INTEGRATION **IN PERSPECTIVE**

From spitting cobras to foraging kissing bugs, animals collect and process information from outside and inside their bodies. The diversity of animals, from structure and function to lifestyle and habitat, translates into a wealth of specializations. From this chapter, think of connections to others—endocrine systems, behaviour, or reproduction to name just three. Think of how our own sensory limitations have limited our ability to appreciate the behaviour of other animals—from ultraviolet nectar guides to ultrasonic signals.

Questions

Self-Test Questions

1. Which of these characteristics distinguishes neurons from other types of animal cells?
   a. ion gradients across the membrane
   b. negative intracellular electrical charge relative to extracellular environment
   c. ion movements across the membrane to create attenuated "local" potentials
   d. ion channels that are sensitive to changes in membrane potential

2. What advantage do giant axons provide to animals?
   a. They allow animals to be selected for scientific research exploring the properties of neurons.
   b. Larger diameter axons propagate signals more quickly than smaller axons, and have been selected for to enhance rapid coordination in invertebrates.
   c. Giant axons have developed as species-specific characteristics of squid to direct the mantle activity necessary for rapid jet propulsion.
   d. Saltatory conduction is enabled by giant axons to increase the speed of conduction.

3. Which characteristic differentiates chemical and electrical synapses?
   a. Chemical synapses allow neural transmission in vertebrates.
   b. Chemical synapses allow for more rapid conduction of signals between cells than do electrical synapses.
   c. Chemical synapses are unidirectional, whereas electrical synapses are not.
   d. Neurotransmitters are employed for rapid synchronization of events at electrical synapses.

4. Which neural structure is correctly paired with its function?
   a. Mauthner cells: invertebrate olfactory integration areas
   b. mushroom bodies: receptors sensitive to vibratory stimuli
   c. nociceptors: invertebrate motor centres
   d. Eimer's organs: highly sensitive mechanoreceptors in the star-nosed mole nose

5. Which of the following distinguishes between the sensations of vibration and proprioception?
    a. Detection of seismic or airborne vibrations involves similar vibration of some aspect of the animals' anatomy, which serves as a receptor, whereas proprioception involves sensing the relative position of body parts.
    b. Both types of sensory perception involve some integration of information from the near environment with more internal cues.
    c. Semicircular canals in the inner ear of vertebrates are sensitive to acoustic vibration.
    d. Tympanal organs and chordotonal organs are important positional receptors in insects and crustaceans that sense vibration.

Questions for Discussion

1. Cubuzoan jellyfishes have different kinds of eyes. What differences would you expect in their structures? Why?

2. What role might bioluminescence play in explaining the different eye structures of animals (e.g., fishes, cephalopods, crustaceans) from the oceans' depths?

3. Do animals have primary senses? If so, is the primary sense of humans the same as that of earthworms? Of scorpions? Of bats? How would you "prove" the existence of a primary sense?

4. Using the electronic library, find a recent paper about animal sensory systems. How does the paper advance our knowledge?

Figure 16.1
Burmese python (*Python molurus bivittatus* (1520)).
SOURCE: John Mitchell / Science Source

16 Endocrine Integration

WHY IT MATTERS

"Hey Mom! I let Fluffy go outside a few minutes ago, and now he's gone." Lost pets are a common occurrence, but if Fluffy lives in the Florida Everglades, his problems may be greater than poor navigational skills. Burmese pythons (*Python molurus bivittatus* (1520)), are among several snake species from around the world that have been turning up in and around Everglades National Park. Often, exotic snakes live in the park because pet owners released them after discovering that the small (0.1–1.0 kg) snake they bought at the pet store is actually one of the largest snake species on earth. Burmese pythons can grow to 6.5 m long and weigh 100 kg. These carnivorous snakes thrive in habitats such as the Everglades, and this invasive species is of great concern because of its rapid spread. Since 2002, over 1000 pythons have been removed from the park and surrounding areas, and this is likely just a fraction of their total population there. Burmese pythons eat a variety of birds and mammals in the Everglades. They are a threat to native wildlife and competition for native predators, and they have a serious impact on the ecological community.

In the Everglades, large Burmese pythons may kill and ingest large prey, even alligators. These snakes are adapted to consuming large meals at long, erratic intervals. Field studies of snake species that hunt from ambush have shown that the mass of a prey animal, swallowed whole, averages 25% of the snake's body mass, but can be up to 1.6 times the snake's mass. (This is analogous to a person weighing 62 kg swallowing a 100-kg meal in one gulp!) Such consumption requires time for digestive processing. Typical feeding intervals for such snakes in the wild are one or two months, but can exceed one year. As a result, pythons experience remarkable physiological and anatomical changes between states of fasting and feeding.

Pythons are a model of extreme metabolic regulation in which many organs, including the heart, increase in mass after a large meal. Many animals—both vertebrates and invertebrates—show changes in heart size (hypertrophy) with increased cardiac function in response to increased metabolic demands. Whereas most mammalian examples of physiological cardiac hypertrophy typically demonstrate modest growth (10–20% increase in mass) after weeks of stimulation, the python heart grows in mass by up to 40% within 2–3 days of consuming the large meal (Fig. 16.2)! The cardiac hypertrophy allows for increased cardiac output (volume of blood pumped per minute) and appears to be adaptive because it supports the large increase in postprandial metabolic rate. The change in metabolic rate is accompanied by increased systemic nutrient transport and widespread organ growth (especially of intestines, liver, and kidneys) required to process such a large meal.

What coordinates these extreme changes in the python's metabolism and anatomy? We might expect the nervous system to play a role in providing sensory information so that the python "knows" that it has food in the digestive tract and must ramp up its metabolic rate. However, long-term changes in the body more typically involve chemical signals. Molecules are released into the circulatory system and travel throughout the body, simultaneously stimulating changes in many tissues. These molecules are **hormones**, the chemical messengers produced by **endocrine organs**, and are the subject of this chapter.

With the story of the kissing bug, *Rhodnius prolixus* (876), on page 355, we saw an example of how an invertebrate animal deals with large meals consumed at infrequent intervals. Our focus there was on the incredible excretory processes required for an animal to consume 10 times its body weight in blood during a single feeding. The extraction and excretion of water from the blood meal is under the control of an amine—5-hydroxytryptamine (5-HT, serotonin)—as well as a peptide diuretic hormone (DH) that is similar in structure to the corticotropin-releasing factor (CRF) family of peptides. Both hormones are released as chemical messengers from the cells of the mesothoracic ganglion into the hemolymph and act on the cells of the Malpighian tubules to stimulate fluid excretion. This stimulation is mediated by **second messengers** in the cells of the tubules, which are characteristic of the action of peptides and amines, as shown later in this chapter.

Postprandial changes in heart muscle in the Burmese python do not reflect an increase in the *number* of cardiac cells. Rather, expression of genes within the cells and the synthesis of cardiac muscle proteins result in an increase in cell size (Fig. 16.3). These types of changes are typical of a cellular response to a chemical messenger rather than to a neural signal. Comparison of gene transcription in fasting and fed python hearts identified 464 genes differentially expressed after feeding. Of these, genes expressed one day after feeding were mostly involved in metabolizing the products of digestion (glucose, proteins, lipid, and nucleic acids) and activating mitochondrial processes that transfer energy to ATP. By three days post-feeding, gene activity in the python was related more to processes involved in the generation, organization, and localization of functional cellular components, and the assembly of structural proteins, including processes responsible for cardiac function such as myofibril assembly and contraction.

Treating rat or mouse cardiac muscle with fed-python plasma generates a similar pattern of cardiac hypertrophy, providing further evidence for the presence of a circulating chemical signal. Chemical analysis of the components in python plasma identified three fatty acids that changed in plasma concentration during the same time sequence of genetic and morphological changes. Myristic acid, palmitic acid, and palmitoleic acid triggered cardiac growth *in vivo* when injected into unfed pythons (Fig. 16.4). This mixture of fatty acids was as effective at stimulating cardiac growth as either feeding the snake directly or infusing plasma from a fed snake. Similarly, infusion of the same fatty acid mixture in mice over a 7-day period generated a significant increase in the left ventricular mass, a pattern typical with mammalian cardiac hypertrophy.

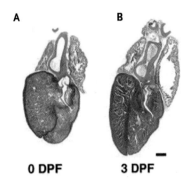

A **B**

0 DPF **3 DPF**

Figure 16.2 Cardiac growth in the python. **A.** Longitudinal section of the heart before feeding (0 DPF) and **B.** three days post-feeding (3 DPF), when the maximum size was observed.

SOURCE: From Riquelme, C. A. J. A. Magida, B. C. Harrison, C. E. Wall, T. G. Marr, S. M. Secor and L. A. Leinwand, "Fatty acids identified in the Burmese python promote beneficial cardiac growth," *Science* 334: 528–531. With permission from AAAS.

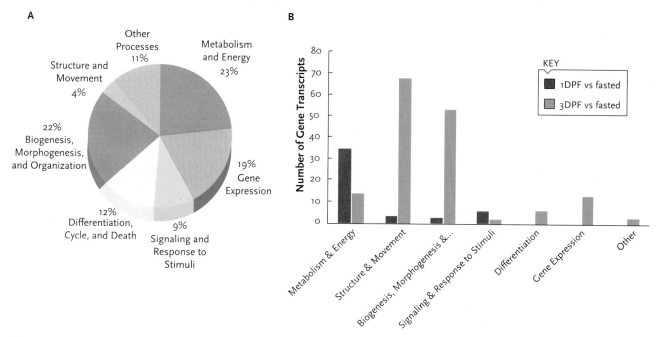

Figure 16.3 Gene ontology analysis was used to sort gene transcripts into general descriptive categories of biological processes. **A.** This chart shows the percent of 464 gene transcripts that were identified, sorted into each of the broad descriptive categories analysed. **B.** Of those, gene expression at one day post feeding (1 DPF) compared to fasting showed increased activity of genes involved in metabolism and energetics (red bars). By three days post feeding (3 DPF), these genes were down-regulated, at which point gene expression reflected processes involved in biogenesis and structural remodelling (green bars).

SOURCE: A. Wall, C. R., , S. Cozza, C. A. Riquelme, W. R. McCombie, J. K. Heimiller, T. G. Marr and L. A. Leinwand. 2011. Whole transcriptome analysis of the fasting and fed physiological cardiac adaptation Burmese python heart: insights into extreme physiological cardiac adaptation. *Physiol Genomics* 43: 69–76. Copyright © 2011, The American Physiological Society; B. Data from Wall, C. R., S. Cozza, C. A. Riquelme, W. R. McCombie, J. K. Heimiller, T. G. Marr and L. A. Leinwand. 2011. Whole transcriptome analysis of the fasting and fed physiological cardiac adaptation Burmese python heart: insights into extreme physiological cardiac adaptation. *Physiol Genomics* 43: 69–76.

SETTING **THE SCENE**

Unravelling the mechanisms of communication that integrate digestion and metabolism in the python illustrates some of the typical questions and methodologies involved in studying chemical

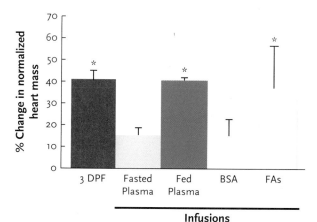

Figure 16.4 Python plasma fatty acids induce cardiac growth *in vivo*, comparable to that observed after ingestion of a meal (3 DPF). Infusing fasting pythons with fatty acids (FAs) or plasma from fed animals resulted in increased heart mass (heart weight/body weight). Bovine serum albumin (BSA) was used to dissolve the FAs.

SOURCE: From Riquelme, C. A. J. A. Magida, B. C. Harrison, C. E. Wall, T. G. Marr, S. M. Secor and L. A. Leinwand, "Fatty acids identified in the Burmese python promote beneficial cardiac growth," *Science* 28 October 2011: Vol. 334 no. 6055 pp. 528–531 . Reprinted with permission from AAAS.

communication in animals. Chemical signals are integral to all aspects of animal biology, from development and growth to metabolism, reproduction, and behaviour. In this chapter, we will use diverse examples to explore the distinctions among different types of chemical signals, the patterns of internal communication, and the regulation of these signals. This is not a comprehensive review of all aspects of endocrinology in all animals, as that would be a book all on its own. Instead, examples illustrating the principles of endocrine regulation have been chosen from the wealth of research in this field.

16.1 Principles of Chemical Signalling

Our story of python metabolism has introduced some aspects of the complexity of chemical signalling and integration in animals. We have already considered the signals released by neurons to trigger changes in their effector organs (see Chapter 15), as well as the signals that provide important means of communication between animals, particularly for the interaction between parasites and their host (see Chapter 10) and the recognition of individuals and the physiological changes necessary for reproduction (see Chapter 14). In this chapter, we will consider chemical coordinators that are released by animal cells to have an effect on **target cells** within that same animal.

Generally, four types of communication are mediated by chemical messengers (Fig. 16.5). **Endocrine cells** (Fig. 16.5A) secrete chemicals called **hormones** into the circulatory system (bloodstream or hemolymph), in which they may travel some distance to affect target cells or tissues. Typically arranged in **endocrine glands**, these endocrine cells receive input that regulates the synthesis and secretion of hormones through either neural mechanisms and/or other chemical sensors. In contrast to exocrine cells, endocrine cells are ductless, secreting their products directly into the bloodstream. While this type of chemical coordination has been well studied in vertebrates, we know much less about it in invertebrates. Our examples will include work on chemical integration in insects and crustaceans, although the roles of hormones in growth and sexual maturation have also been elucidated in arthropods and annelids. We will also consider evidence for hormonal activity in other invertebrate taxa.

Neurosecretory nerve cells release their chemical messengers into the circulatory system, rather than into a synapse (Fig. 16.5B), and their products are known as **neurohormones**. In vertebrates, neurosecretory cells are associated with the brain and typically arise from the hypothalamus (see Section 16.4c) or pineal gland. In invertebrates, neurosecretory cells are the primary source of chemical signals, as few discrete endocrine organs or tissues have been identified. Secretory neurons are typically grouped in a cluster called a **ganglion**, identified by its location (e.g., cerebral, subesophageal, and thoracic ganglia). They synthesize and secrete neuropeptides associated with a wide range of functions.

Not all cells secrete their chemical signals into the bloodstream. **Paracrine** cells and **autocrine** cells release their messengers into the intracellular fluid, where they affect adjacent cells (Fig. 16.5C) or the same cell (Fig. 16.5D) respectively. While these also constitute chemical signalling, they are not emphasized in this chapter.

A **hormone** is a chemical messenger released into the bloodstream by a structure that is part of the endocrine system. The chemical messenger circulates throughout the body and is received by cells of a **target organ**. Many molecules that are released in neuron–neuron communication as neurotransmitters are also active as hormones, being released by neuroendocrine cells into the bloodstream. Strictly speaking, the fatty acids in our python story are the products of digestion, not produced by the python's cells. But they are released from the cells of the digestive tract to

Figure 16.5 General types of cell-to-cell communication in animals.

SOURCE: From RUSSELL/WOLFE/HERTZ/STARR. *Biology*, 1E. © 2010 Nelson Education Ltd. Reproduced by permission. www.cengage.com/permissions

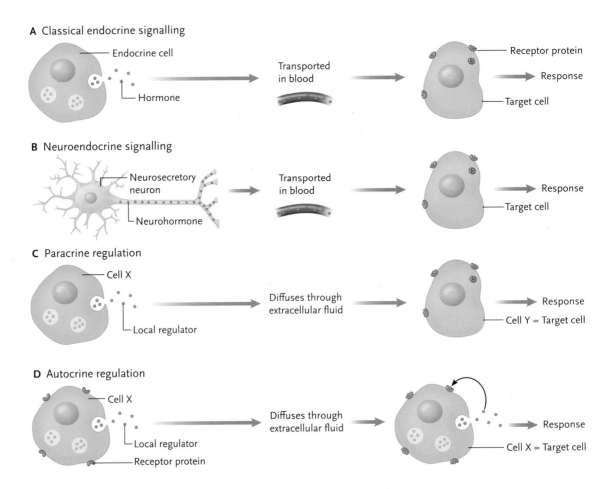

A Classical endocrine signalling

Endocrine cell — Hormone — Transported in blood — Receptor protein — Response — Target cell

B Neuroendocrine signalling

Neurosecretory neuron — Neurohormone — Transported in blood — Response — Target cell

C Paracrine regulation

Cell X — Local regulator — Diffuses through extracellular fluid — Response — Cell Y = Target cell

D Autocrine regulation

Cell X — Local regulator — Receptor protein — Diffuses through extracellular fluid — Response — Cell X = Target cell

travel through the bloodstream to act on the heart and other organs. Therefore they follow the typical path of hormonal action.

The study of chemical messengers is further complicated by changes in endocrine glands associated with developmental or seasonal influences. This complicates identification of glands and their products as well as recognition of their effects on target cells! As so often happens in scientific research, these challenges have fueled many advances in the technology used to study chemical messengers (see Box 16.1).

STUDY BREAK

1. What is a hormone? How are hormones different from other chemical messengers?
2. What are endocrine organs?

16.2 Hormone Structure and Function

Hormones—organic chemical molecules—are typically grouped into classes based upon their structure (protein- or lipid-based) and mode of action. A single endocrine gland produces just one class of hormone, with some notable exceptions. It is important and useful to discriminate among the basic types of hormones because they differ in the mechanisms of their synthesis and release and in their mode of travel through the circulatory system. Target cells must have receptors for the hormones to bind to. The physiology of hormone action also is influenced by its structure.

16.2a Protein, Peptide, and Monoamine Hormones

Most hormones are peptides, composed of amino acids. Like other proteins, a **preprohormone** is synthesized on ribosomes. This preprohormone is longer, containing the specific hormone amino acid sequence. It also includes a signal peptide sequence that permits movement of the protein across the ER membrane into the Golgi apparatus. There the preprohormone undergoes further enzymatic processing before being packaged in secretory granules. Vesicles containing the final protein product are pinched off the terminal cisternae of the Golgi apparatus, and stored within the cytoplasm prior to release.

BOX 16.1
Techniques for Studying Endocrine Systems

How do endocrinologists gather evidence to support the hypothesis that a particular physiological response is triggered by hormone action? Classical experiments involved removing the source of the hormone: techniques that included castration (testes), eyestalk ablation (crustacean Y-organ), or decapitation (insect brain ganglia), and noting the correlation between absence of the hormone and the response to which the hormone was linked. As more information became available about the chemical nature of the hormones in question, techniques became somewhat more refined, including pharmacological blocking or stimulation of hormone receptors. **Bioassays** involve a test of the effects of a hormone on a living animal. This technique is often important in the initial stages of identifying a chemical signal, as the living animal can serve as a reliable, quantifiable response system on which to test extracts and chemical fractions for biological activity. The level of hormone or tissue extract tested may even be from a different source from the test animal. For example, the levels of prolactin from pituitary extracts of rats (or other animals) were traditionally bioassayed in pigeons by noting changes in crop cell epithelium (Figure 1).

Probably the most famous endocrine bioassay was the Friedman test, better known as the "rabbit test," commonly used in North America until about 1950 to test for the

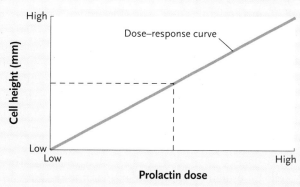

Figure 1
A bioassay for prolactin. Increase in crop sac epithelial cell height can be correlated with known amounts of prolactin to create a dose-response curve.

SOURCE: Nelson. *An Introduction to Behavioral Endocrinology*, 4th edition. Sinauer Associates: 2011. Reprinted by permission.

presence of human chorionic gonadotropin (hGC), a hormone released from the implantation site of a blastocyst. This test for the early signs of pregnancy involved injection of a urine sample from a potentially pregnant woman into a rabbit. If hCG was present, the rabbit's ovaries would show the results within 48 hours. This was considered a remarkable improvement over the earlier mouse test (developed in 1928) which required several mice per test and at least 96 hours to complete! Today, home pregnancy tests make use of an **immunoassay (IA)**, based upon the principle of competitive binding of an antibody to an antigen. Specific antibodies can be prepared for the hormone of interest and labelled through a variety of techniques, including radioactive tags (RIA) or an enzyme (EIA) that changes the optical density (colour) of the molecule when bound. Hormone levels can then be determined by measuring the amount of bound hormone to labelled antibodies.

The ability to chemically identify hormones and generate specific antibodies against these molecules has also led to the development of techniques involving **immunocytochemistry** to determine the location of a hormone in the body. By linking antibodies to marker molecules, such as fluorescent dyes, or by following the uptake of labelled hormones by target cells, researchers can trace the source of potential hormones in many animal tissues.

Techniques common to molecular biology are also employed in endocrinology, particularly to determine the presence of a specific nucleic acid or protein in a specific tissue that is responsive to a hormone. These techniques involve separation of nucleic acids or proteins by electrophoresis and identifying genes or proteins of interest through the use of labelled nucleic acid markers of antibodies. New advances in molecular biology have also made genetic manipulations of organisms possible, so that genes encoding hormones or receptors can be inserted into (**transgenic**) or removed from (**knockout**) laboratory animals. We have seen the results of such studies in our consideration of the role of leptin (see Box 7.2).

Biogenic amines (e.g., octopamine, histamine), including the catecholamines (dopamine, norepinephrine, and epinephrine) and other monoamines (e.g., serotonin, melatonin) are synthesized within the axonal endings of neurons. The process is similar to the synthesis of neurotransmitters. These molecules are also derived from amino acids, and are highly conserved in animals, with demonstrated roles in modulating complex behaviours centrally through synaptic actions or in conjunction with peripheral actions mediated by neurohormones. Release of the hormone from the cell occurs through **exocytosis**, a process involving fusion of the secretory vesicle with the plasma membrane and release of its contents. This process is remarkably similar to the synthesis and release of neurotransmitters (see Figure 15.13), following the same pattern of depolarization and release.

The **half-life** of a hormone is defined as the amount of time required for half the molecules to become inactivated or cleared from circulation. Shorter peptides usually have short half-lives (2–30 min), whereas larger proteins may persist for as long as 60 minutes. Peptidases, enzymes responsible for metabolizing peptides and removing them from circulation, play an essential role in biological regulation because of the close involvement of peptide messengers in the maintenance of homeostasis.

Target cells for protein, peptide, and monoamine hormones are distinguished by hormone receptors in the cell membranes. Membrane receptor molecules include G-protein-coupled receptors, cytokine receptors, and receptor tyrosine kinases. The specific structure and function of these receptors and the coupling of receptor activation to signal transduction in cells is beyond the scope of this text. The common element in these processes is that the hydrophilic peptide hormones, which represent the "first messengers," cannot cross the plasma membrane. Binding of hormone to receptor therefore activates enzymatic processes that result in the synthesis of a **second messenger** that activates specific enzymatic events within the target cell (Fig. 16.6). Stimulation of many membrane receptors leads to the activation of cyclizing enzymes that initiate the synthesis of a cyclic nucleotide, either cAMP or cGMP. Other membrane receptors are coupled to enzymes that hydrolyze membrane lipids such as phosphoinositol, which results in the production of diacylglycerol and inositol triphosphate (IP_3). These common second messengers generally activate one or more specific protein phosphorylation events within the target cell. The addition of a phosphate group to an enzyme, for example, by a phosphoprotein kinase, may result in its activation. Phosphorylation of another protein might enhance its contractile activity. Protein substrate phosphorylation, in contrast, may also inactivate enzymes. Phosphoprotein phosphatases, on the other hand, remove phosphate groups from proteins, which can result in either activation or inactivation of the particular protein.

16.2b Steroids and Lipid-Based Hormones

Unlike peptide hormones, steroid hormones are not synthesized through the expression of specific genes. Steroid hormones are synthesized from enzymatic processing of **cholesterol**, their common precursor molecule. Cholesterol is primarily available from ingested fats. Invertebrates cannot synthesize cholesterol, depending entirely on dietary intake, whereas vertebrate liver cells can synthesize cholesterol from acetate. Cholesterol is conveyed to cells via the circulatory system, transported into the cytoplasm, and delivered to the mitochondria. **Steroidogenic** enzymes in the mitochondria and cytoplasm of endocrine cells make sequential modifications to the cholesterol, leading to the formation of specific steroid hormone molecules such as progesterone, cortisol, or ecdysteroids. Steroid hormones are fat-soluble and move easily through cell membranes. Consequently, steroid hormones are never stored, but almost immediately leave the cells in which they are produced. A signal to synthesize steroid hormones is also a signal to release them, although the delay between the stimulus and the response of biologically significant steroid production may be slow.

Because steroid hormones are not very soluble in water (hydrophobic), they usually bind to water-soluble carrier proteins in the circulatory system. These carrier proteins increase their solubility, facilitate transport through the blood to their target tissues, protect them from being degraded prematurely, and increase their half-life by about several hours in the body. Target tissues have **cytoplasmic** (intracellular) receptors for steroid hormones and accumulate steroids against a concentration gradient.

Upon arrival at target tissues, steroid hormones typically dissociate from their carrier proteins, although some carrier-bound steroids can selectively enter some target tissues by receptor-mediated endocytosis. The mechanism for this is similar to the clearance of cholesterol via lipoproteins. In either case, once inside the cell, the steroid interacts with receptors in either the cytoplasm or nucleus. The amino acid sequence of steroid hormone receptors is highly conserved among vertebrates, and also shows remarkable similarity to invertebrate receptors. Once formed, the steroid-receptor complex binds to DNA sequences called **hormone response elements (HRE)** and stimulates or inhibits the transcription of specific mRNAs. These messages result in the synthesis of specific structural proteins or enzymes that produce the physiological response (Fig. 16.7). Changes in the types of proteins a cell makes (the gene products) can often be observed within 30 minutes of arrival of a steroid within its target cell.

As yet we do not know the precise mechanism by which the binding of a steroid-receptor complex to a specific HRE evokes activation or suppression of gene transcription. It appears that coactivator proteins are often necessary to initiate this process. The effects of environmental, social, or other extrinsic or intrinsic factors on the regulation of specific coactivators have not been well studied and represent a potential process

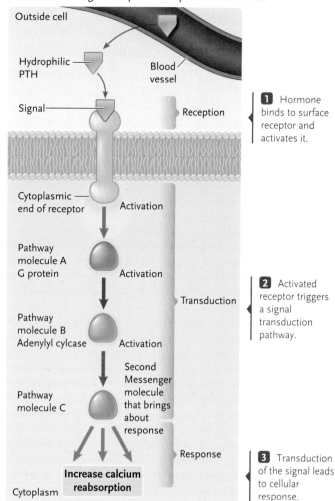

Hormone binding to receptor in the plasma membrane

1 Hormone binds to surface receptor and activates it.

2 Activated receptor triggers a signal transduction pathway.

3 Transduction of the signal leads to cellular response.

Figure 16.6 The mechanism of action of hydrophilic hormones that bind to receptor proteins in the plasma membrane, as shown by the action of parathyroid hormone (PTH) on a kidney tubule cell, and act through second messengers

SOURCE: From RUSSELL/WOLFE/HERTZ/STARR. *Biology*, 1E. © 2010 Nelson Education Ltd. Reproduced by permission. www.cengage.com/permissions

by which individual variation in hormone-behaviour interactions may be mediated.

STUDY BREAK

1. Compare the synthesis, transport, and mode of cellular action of protein-based (hydrophilic) and lipid-based (hydrophobic) hormones.
2. What is the importance of second messengers in hormone action? Why are these molecules not required to mediate steroid hormone action?

16.3 Feedback Mechanisms in Chemical Regulation

The levels of hormone(s) circulating in the body are typically under one of two basic patterns of control. Secretion of hormones can be regulated by the

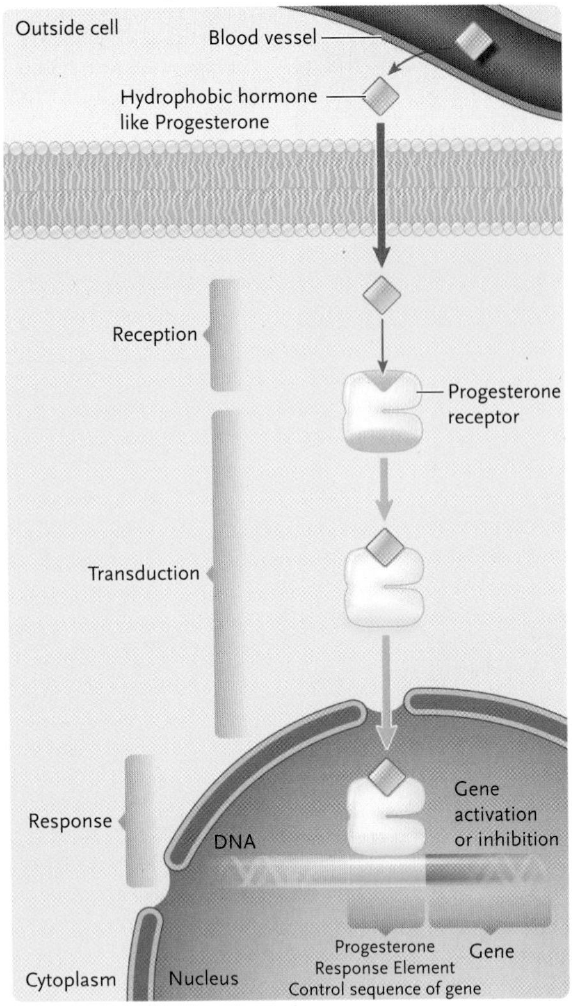

Hormone binding to receptor inside the cell

Outside cell

Blood vessel

Hydrophobic hormone like Progesterone

1 Hydrophobic hormone passes freely through plasma membrane.

Reception

Progesterone receptor

2 Hormone binds to receptor, activating it.

Transduction

3 Activated receptor binds to control sequence of a gene, leading to gene activation or inhibition.

Response

DNA

Gene activation or inhibition

Cytoplasm   Nucleus

Progesterone Response Element   Gene
Control sequence of gene

Figure 16.7 The mechanisms of action of hydrophobic hormones that bind to intracellular receptors, as shown by the action of progesterone.

SOURCE: From RUSSELL/WOLFE/HERTZ/STARR. *Biology*, 1E. © 2010 Nelson Education Ltd. Reproduced by permission. www.cengage.com/permissions

Figure 16.8 Crustacean hyperglycemic hormone (CHH) acts in carbohydrate metabolism in species like the Christmas Island red crab (*Gecarcoidea natalis* (817)).

The primary source of CHH is the neurosecretory neurons of the eyestalk X-organ and its associated neurohemal organ, the **sinus gland**. Although other sources of CHH-like peptides have been identified in other tissues, CHH released from the sinus gland triggers the rapid mobilization of glucose from reserve carbohydrates such that circulating glucose levels rise within 5 to 10 minutes of an increase in circulating CHH. Cells of the X-organ that synthesize CHH are inhibited by elevated glucose in the hemolymph, which stops CHH release. Thus, glucose levels are maintained in an optimal range (Fig. 16.9A). This simple **negative feedback** system is characteristic of the physiological regulation of hormone levels found in many animal systems, and it works much the way a thermostat controls the temperature in your home.

Simple negative feedback systems can be temporarily over-ridden in times of need, similar to how an animal's heart rate is typically regulated within a predictable range but can escalate during periods of exercise. Such is the case with CHH, which appears to be involved not only in the maintenance of resting glucose levels, but also in temporary increases related to physiological needs. Stress hyperglycemia, a common response of decapods to environmental stressors, is mediated by CHH, and acclimation to higher temperatures is correlated with hyperglycemia in intact but not in animals with their eyestalks removed. CHH may also play a role in metabolic adaptation to changing environmental conditions by making easily metabolizable energy available.

Gonadotropic releasing hormone (GnRH) is an example of a hormone that is regulated by a more complex system of negative feedback controls (Fig. 16.9B). In response to external or endogenous stimuli, GnRH is released by the vertebrate hypothalamus and stimulates the release of **gonadotropins** (see Section 16.5) from the anterior pituitary. These hormones stimulate steroid and gamete production in the gonads. The synthesis and secretion of steroid hormones from the gonads (shown here as testosterone secretion from the

physiological by-products generated in response to their actions, and hormone levels are often regulated by the stimulatory or inhibitory effects of other hormones. Within this second type of control system, there may be one, two or three hormones in a regulatory chain, or there may be **autoregulation** ("self" regulation) provided by the hormone itself. Both of these mechanisms involve feedback to the hormone-releasing tissue, as illustrated in the following examples.

Crustacean hyperglycemic hormones (CHHs) are neuropeptides involved in the regulation of glucose levels in the hemolymph. Although the first role of CHHs was described in the regulation of carbohydrate metabolism in species such as the Christmas Island red crab (*Gecarcoidea natalis* (817), Fig. 16.8), it is now widely known that these hormones are **pleiotropic**—they have a diverse array of actions in a number of systems such as moulting and regulation of water and pH. We will focus on the regulation of glucose levels.

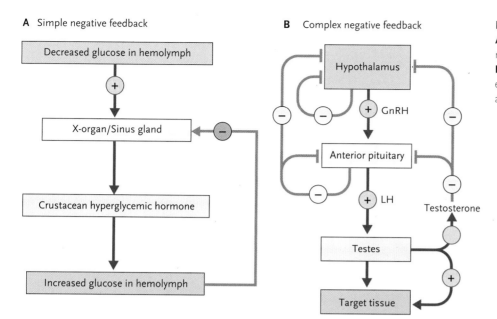

A Simple negative feedback

B Complex negative feedback

Figure 16.9 Negative feedback in hormonal control. **A.** Release of crustacean hyperglycemic hormone is regulated by a simple negative feedback mechanism. **B.** A more complex negative feedback relationship exists among the hypothalamus, anterior pituitary, and testes in the regulation of testosterone levels.

testes) feeds back to turn off GnRH production in the hypothalamus. This, in turn, shuts down gonadotropin secretion from the anterior pituitary. The gonadotropins also feed back to shut down GnRH production, and GnRH feeds back on the hypothalamus to regulate its own secretion (autoregulation). Peptide hormones from the gonads, such as activins and inhibins, may also be involved in the regulation of gonadotropins (not shown in Fig. 16.9). Each level of negative feedback provides a layer of tight control on the level of circulating hormones.

STUDY BREAK

1. What is negative feedback? Use an example to illustrate the role of negative feedback in endocrine regulation of animal functions.

16.4 Neural and Chemical Integration

Physiological processes in animals are affected by a variety of stimuli, including signals that are **extrinsic** (external or environmental) and **intrinsic** (within the body) in origin. The recognition of these factors is accomplished by sensory receptors (Section 15.5) that may be cellular components of glands, such as the pancreas, or may be specific neural elements. The sensory cells may respond to these stimuli by the direct release of a hormone or by transmitting nerve impulses to other neurons or cellular elements that then release one or more chemical messengers. There is no one simple pathway! The important part is that the nervous system, working through the endocrine system, plays an important role in homeostasis of a number of physiological processes such as water balance, temperature regulation, feeding behaviour, and even reproductive

behaviour. The following examples of neuroendocrine integration describe the underlying mechanisms for some topics that have been encountered in previous chapters.

16.4a Moulting and Metamorphosis in Insects

The physiological control of moulting and metamorphosis in insects has been studied for almost a century, a reflection of the complexity of the neural and endocrine integration of this event (see also page 120). The primary control is mediated by the brain. Specifically, the peptide **prothoracicotropic hormone** (PTTH) acts on the prothoracic gland (PG) through the second messengers calcium and cAMP, and triggers synthesis and release of the steroid hormone **ecdysone** (Fig. 16.10). Acting together, PTTH and ecdysone trigger every moult: both larva to larva and larva to adult. The transition from larva to adult is linked to changes in secretion of a second peptide signal, **juvenile hormone** (JH), from a pair of cerebral ganglia called the **corpora allata**. As long as the concentration of JH remains high, moulting in insects proceeds from one larval stage to the next. The moult from larva to pupal or adult stages is accompanied by decreased levels of JH (Fig. 5.34). This interaction between JH and ecdysone also affects both the appearance and behaviour of insect larvae (see Figure 17.4).

Recent research has indicated that, in addition to control exerted by PTTH from the brain, the prothoracic gland may be controlled by several other interacting stimulatory and inhibitory factors. Small mono-amine-like peptides known as **bombyxin**, released from the first thoracic ganglion (T1G), also act on the PG to stimulate steroid release. Bombyxin was the first insulin-like peptide isolated from insects, and

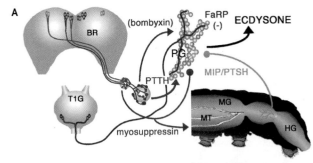
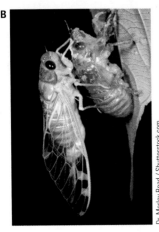

A

B

Figure 16.10 Control of moulting in insects.
A. Ecdysone secretion by the prothoracic gland (PG) is primarily under direction of prothoracicotropic hormone (PTTH) from the brain (BR), but the entire process is modulated by a host of other chemical signals (see text).
B. Moulting in a cicada, *Magicicada* spp. (874).
FaRP = amide-related peptides; HG = hindgut; MG = midgut; MIP/PTSH =myoinhibitory peptide/prothoracicostatic peptide; T1G = first thoracic ganglion; MT = Malphigian tubules.

SOURCE: Truman, J. W. 2006. Steroid hormone secretion in insects comes of age. *PNAS* 103 (24): 8909–8910. Copyright 2006 National Academy of Sciences, U.S.A.

like its counterpart in vertebrates it mediates nutrient-dependent growth in target cells and tissues. This evidence suggests that insulin-dependent growth of the PG is an important endocrine component to trigger metamorphosis. In addition, several inhibitory factors, including myoinhibitory peptide/prothoracicostatic peptide (MIP/PTSH and myosuppressin, Fig. 16.10), appear to suppress ecdysone secretion, even in the presence of PTTH stimulation. While the physiological role of these inhibitors is not clear, their pattern of release suggests a role in balancing the growth of larval stages and timing of moulting. MIP/PTSH also acts on neurosecretory cells of the hindgut, and the timing of its release suggests that it may act to lower steroid levels between moults. Myosuppressin levels are higher during intermoult stages, and the actions of this chemical signal on the midgut and Malpighian tubules suggest that it may prolong larval growth periods, until PTTH release triggers a new moult. As if this story were not already complex enough, it also appears that some of the signal molecules, such as bombyxin, are delivered to the PG through direct innervations, rather than through the circulation! Clearly we still have a great deal to learn about the collaboration between neural and endocrine systems in insect metamorphosis and moulting.

16.4b Photoperiod and Hormonal Influences on Smoltification in Salmonids

In many salmonids, parr-to-smolt transformation or **smoltification** occurs prior to seaward migration (see page 105). This process is a metamorphosis from a sedentary, cryptically marked fish (the **parr**) with a freshwater physiology suited to its natal stream, to an active, silvery **smolt** that is equipped for osmoregulation in salt water (Fig. 16.11). This transformation includes sequences of endocrine, physiological, and behavioural changes that are characterized by increased cortisol secretion, along with surges in thyroid hormone (T_4), insulin, growth hormone (GH), and prolactin.

Photoperiod appears to be the primary environmental cue that triggers and regulates the process of smoltification, coordinating neural detection of changes in day length with hormonal changes in growth and development. Photoperiod influences hormone levels in many organisms through a pathway described as the light–brain–pituitary axis, and this is true for salmonids as well (Fig. 16.12). The synthesis of melatonin, a hormone produced by the pineal gland, is regulated by photoperiod. In salmonids, melatonin is at elevated levels throughout the dark phase of the light cycle. Direct photic influence to various salmonid brain regions may also be mediated via neural projection from both the retina and the pineal gland, which include direct inputs to neurons that control pituitary function. Immunocytochemical studies have provided evidence for structural reorganization in the light–brain–pituitary axis during smoltification, as projections from the retina in smolts expand into new territories of brain nuclei that have input to the pituitary. These structural changes are followed by sequential surges of brain neurotransmitters and receptors, and occur prior to the major surges of circulating thyroid hormone and growth hormone levels that are central to the smolt-parr transformation.

The correlation between hormonal levels and the modulation of osmoregulatory machinery that is critical for the transition from freshwater to seawater is clear, but doesn't provide information about the mechanisms involved in adaptation. Recent research has demonstrated differential expression of genes for **claudins**, the

Figure 16.11 Hormones that increase during smoltification (e.g., growth hormone (GH), thyroid hormone (T_4), prolactin, and cortisol) integrate the physiological and behavioural changes in salmonids from **A.** the stream-dwelling, cryptically coloured territorial parr to **B.** the seaward-migrating silver smolt.

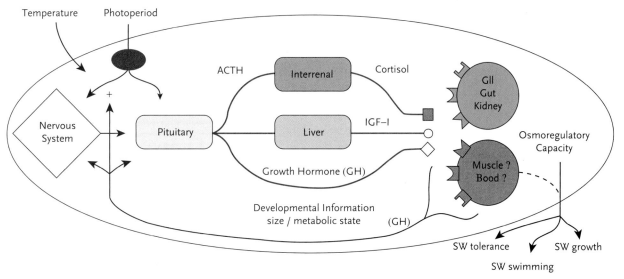

Figure 16.12 Integration of photoperiod and other environmental factors in endocrine control of growth and development in juvenile salmon in the transition to seawater. Body size is an important factor in the responsiveness of the growth hormone (GH) / insulin-line growth factor (IGF-1) IGF-1 and cortisol pathways to photoperiod, and the adaptation of osmoregulatory (and perhaps other) organs for the transition to seawater.

SOURCE:McCormick, S. D. 2009. Evolution of the hormonal control of animal performance: Insights from the seaward migration of salmon. *Integrative and Comparative Biology*, 49 (4): 408–422, by permission of Oxford University Press.

membrane proteins directly involved in regulating ion transport across epithelial membranes as components of tight junctions between epithelial cells. There are numerous isoforms of claudin proteins throughout the body, but changes in salinity and photoperiod have been shown to be important specifically in the regulation of claudin gene expression in the gill epithelium of Atlantic salmon. These genetic changes correlate with the osmoregulatory adaptations of gill epithelia when faced with changes in salinity. While there is still much to be learned about the dynamics of claudin functions, these findings provide another illustration of the specific effects of hormones in mediating physiological changes important in the transition from parr to smolt.

16.4c Adrenal Hormones and Stress

The paired adrenal glands of most vertebrates are composed of two tissues of different embryological origin, each of which produces a class of hormones that are unrelated to each other in structure. Nevertheless, the hormones of both the adrenal **steroidogenic** tissue and the adrenal **chromaffin** tissue play important roles in response to stress. The chromaffin cells are derived from neural crest, whereas steroidogenic tissue arises from the coelomic mesoderm in the genital ridge of the embryo. In many mammals, the steroidogenic tissue forms a cortical mass surrounding an inner medullary component of chromaffin tissue. In humans and many other mammals, these tissue components are referred to as the **adrenal cortex** and **adrenal medulla** respectively (Fig. 16.13). In many vertebrates, including some mammals, these two tissues are intermingled, whereas in some poikilotherms, the two endocrine components are totally separate masses of tissue. For these reasons, the comparative endocrinologist typically prefers the functional designations **adrenal steroidal** and **adrenal chromaffin** tissue.

Epinephrine and norepinephrine are catecholamine hormones of the adrenal chromaffin tissue, and

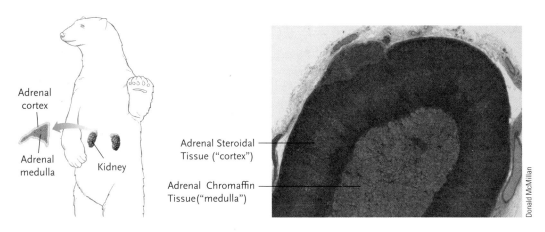

Adrenal cortex

Adrenal medulla

Kidney

Adrenal Steroidal Tissue ("cortex")

Adrenal Chromaffin Tissue ("medulla")

Figure 16.13
Cross-section of monkey adrenal gland showing steroidogenic tissue of the cortex and chromaffin tissue of the medulla.

SOURCE: From RUSSELL/ WOLFE/HERTZ/STARR. *Biology*, 2E. © 2013 Nelson Education Ltd. Reproduced by permission. www.cengage.com/ permissions

Donald McMillan

norepinephrine also serves as a transmitter of autonomic neurons. Both play a role in maintaining the constancy of the internal environment of the body. The integration of neural and endocrine activity is therefore an inherent feature of adrenal chromaffin secretions. The synthesis and secretion of adrenal steroidal hormones is regulated by the hypothalamic–pituitary pathway, so neural input is mediated through input to the hypothalamus, which may arise from a variety of sources. It is this hypothalamic–pituitary controlled aspect of adrenal function that we will explore through the following example.

As demonstrated throughout this text, all animals are a complex package of systems in dynamic equilibrium or homeostasis. Any perturbation to homeostasis requires the animal to expend energy to restore the original steady state; these perturbations may be described as **stressors**. There are many sources of stressors, including physiological factors such as insufficient food quality or the physical demands of staying alive, environmental factors such as temperature extremes or noise, and psychosocial factors such as fighting, social subordinance, or even novel situations. A **stress response** is a suite of physiological and behavioural responses that help to re-establish homeostasis. The stress response is relatively nonspecific, in that many different stressors elicit a similar response.

The major components of the stress response in most vertebrates involve the adrenal glands. Within seconds of perceiving a stressor, catecholamine levels increase as the sympathetic nervous system begins to secrete norepinephrine and the adrenal chromaffin tissues begin to secrete epinephrine. This response brings about physiological changes in cardiovascular tone, respiration rate, and blood flow to the muscles that are characteristic of what has been called the **fight-or-flight** response. These catecholamines also increase blood glucose levels, which fuel the physical responses, increase alertness, and enhance learning and memory.

Stressors also initiate another suite of endocrine responses within minutes of perceiving a stressor, which results in the secretion of **glucocorticoids** (corticosterone in most rodents, birds, reptiles, and fish; cortisol in most primates, large mammals, and carnivores). This response involves the **hypothalamic–pituitary–adrenal (HPA) axis**, which is activated so that corticotropin releasing hormone (CRH), adrenocorticotropic hormone (ACTH), and glucocorticoids are released in response to stressors (Fig. 16.14). While CRH released from the hypothalamus into circulation acts as a hormonal trigger for ACTH release from the pituitary, it is interesting to note that CRH also acts as a neural signal from the hypothalamus to the amygdala, where it has a role in mediating an animal's anxiety response. Identification of specific stressors is a critical factor in exploring the impact of human activities on animal communities (Fig. 16.15).

In addition to hormones of the HPA axis, the anterior pituitary may also release two important peptides:

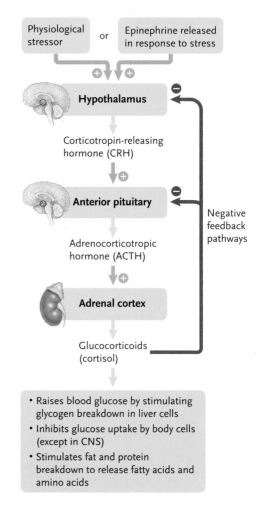

Figure 16.14 When the hypothalamic–pituitary–adrenal (HPA) axis is activated by stressors, the hypothalamus releases CRH and other releasing hormones, which stimulate ACTH and β-endorphin release from the anterior pituitary gland. ACTH stimulates corticosterone secretion from the adrenal cortex. Prolactin is often released from the anterior pituitary during stress as well. Vasopressin is released from the posterior pituitary.

SOURCE: From RUSSELL/WOLFE/HERTZ/STARR. *Biology*, 1E. © 2010 Nelson Education Ltd. Reproduced by permission. www.cengage.com/permissions

Figure 16.15 Human activities serve as stressors for many animal species. Michael P. Muehlenbein and his students of Indiana University in Bloomington, Indiana, study the effects of ecotourism on species such as the orangutans of Borneo (*Pongo pygmaeus morio* (1867)).

prolactin and β-endorphin. Prolactin has been shown to have multiple actions, including a role in adaptation to stressors, and β-endorphin binds to opioid receptors that have documented analgesic effects (Box 16.2). Stressors also frequently result in the release of **vasopressin**, a peptide produced in the hypothalamus and released through neurosecretory pathways of the posterior pituitary. The osmoregulatory action of vasopressin may be important in maintaining elevated blood pressure, important to sustain cardiovascular function during stress responses.

How does one study the role of hormones in the natural stress responses of animals without adding to the stress of the animals being studied, either through handling or restraint or simply the stress of being removed from their normal routine? Adrenal glucocorticoid levels in biological sources such as saliva, cerebral spinal fluid, blood, urine, and feces have been used to assess stress levels. Of these sources, feces is the most accessible in free-ranging animals, and has been particularly important in assessing the effects of stress on endangered species. Both white rhinoceros (*Ceratotherium simum* (1838)) and black rhinoceros (*Diceros bicornis* (1839)) are endangered species that do not reproduce well in captivity. Monitoring glucocorticoid levels in the feces of these animals under various conditions of captivity has been useful in monitoring physiological stress both in captivity and during and after translocation. Similarly, studies with free-ranging African elephants (*Loxodonta africana* (1871), Fig. 16.16) have shown a correlation of fecal glucocorticoid levels with the physiological stress of pregnancy, as well as the ecological stress induced by no rainfall (Fig. 16.17). Glucocorticoid levels were highest during the harshest part of the dry season, when food, water availability, and body condition were at their lowest. Monitoring hormone levels through such non-invasive techniques ensures that the study itself is not contributing to the stress levels of the animal.

In order to affect behaviour, endocrine mediation of the stress response must affect the brain. Epinephrine does not cross the blood-brain barrier, so glucocorticoids are the best candidates for mediating the behavioural effects of stress.

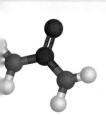

BOX 16.2 Molecule Behind Animal Biology
Opioids and Their Receptors

Opioids are endogenous (from within) painkillers. They interact with opioid receptors throughout the nervous system to ameliorate pain sensations. Three natural groups of opioids are traditionally recognized: **endorphins**, **enkephalins**, and **dynorphins**. Recently two new opioid neuropeptides have been identified: **nociceptin** and **endomorphin**. Each opioid has its own specific membrane receptor, although all are structurally similar and have the same basic mode of action. Exogenous opiates such as heroin and morphine interact with these same receptors.

During the stress adaptation response (Fig. 16.14), ACTH and β-endorphin are secreted simultaneously from the anterior pituitary in response to CRH from the hypothalamus. As we have seen, ACTH stimulates glucocorticoid secretion, which fuels the animal in stressful conditions. These same conditions may also trigger the release of β-endorphin to reduce pain, through the same mechanisms by which both endogenous and exogenous opioids inhibit the excitability of neurons.

Natural opioids are involved with **affiliative behaviours**, social behaviours that bring animals together, such as reproductive and mating behaviours, pair-bond maintenance, and parental behaviours. Specifically, opioids are important in the development of social bonds in rhesus monkeys and other primates, and mediate social grooming in primates and rodents. The role of these peptides in mediating social contact was first suspected upon observations of similarities between narcotic addiction and the comfort of social contact. Both behaviours involve strong emotional attachments and share physiological symptoms of withdrawal. Endogenous opioids mediate isolation-induced distress vocalizations, and social contact increases endogenous opioid concentrations. These similarities suggested a common physiological mediation, which has been confirmed through research that involves blocking opioid receptors with antagonists such as naloxone or naltrexone, or stimulating receptors with exogenous opioids.

How do we get from drug addiction to social behaviours (or at least those that are socially acceptable)? Morphine itself is produced naturally by some neurons, although in much lower concentrations than in the opium poppy *Papaver somniferum*. The biosynthetic pathway is similar between poppy and neuron, however, and although the physiological role of endogenous morphine is still obscure, it undergoes Ca^{2+}-dependent release which suggests that it has a role as a neurotransmitter or neuromodulator. Neurons responsive to opioids are typically involved with brain centres that form what is known as the primary reward pathway. This relationship is thought to be critical in the development of addictive behaviours in the use of exogenous opioids.

Glucocorticoid receptors have been identified in numerous regions of the vertebrate brain, including the hippocampus (which has important roles in memory and spatial navigation) as well as on neurons of the hypothalamus. In the hypothalamus, the presence of glucocorticoid receptors is likely related to the role of negative feedback in regulating the stress response. Recent studies have shown that these receptors become less sensitive to glucocorticoids in chronically stressed animals, which may help improve our understanding of the underlying mechanisms that result in the dysregulation of the HPA axis in animals exposed to chronic stress. The multiple roles of hypothalamic and other endocrine secretions in regulating both endocrine and neural activity are characteristic of the complicated interaction of neural and endocrine mechanisms.

STUDY BREAK

1. Is there a "stress hormone"? If so, what is it, and where is it produced?
2. What is the HPA axis and what does it regulate?

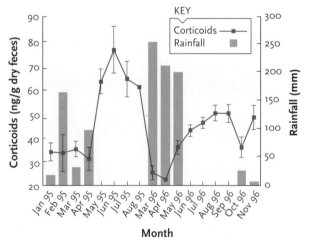

Figure 16.17 Mean corticoid concentrations for 16 wild female African elephants matched to rainfall data. The peak corticoid levels corresponded with periods of environmental stress brought on by low rainfall, poor food and water availability, and poor body condition.

SOURCE: Foley, C. A. H., S. Papageorge and S. K. Wasser. 2001. "Noninvasive stress and reproductive measures of social and ecological pressures in free-ranging African elephants," *Conservation Biology* 15(4): 1134–1142, John Wiley and Sons. Copyright © 2002, John Wiley and Sons.

16.5 Chemical Integration of Reproduction: Gonadotropins

Reproduction includes the process of sex determination and sexual differentiation (conversion of the indifferent gonads into testes or ovaries), embryonic development and birth, sexual maturation, development of gametes, physiological and behavioural aspects of courtship and mating, fusion of gametes, and development of the resulting zygote (see Chapter 14). In addition, a period of complex parental care may be intercalated between birth and sexual maturation. Every step in this complicated reproductive process is controlled directly or modified by chemical regulators secreted within the body, often in combination with pheromones from other members of the species. An understanding of animal reproductive patterns and their hormonal control is central to both the perpetuation of a species through evolutionary time as well as our immediate concerns about environmental quality and the future of ecosystems that are affected adversely by human activities. Our attention in this section will be limited to common, comparative aspects of the control of gamete maturation and release, but clearly many other connections can be formed!

The **hypothalamus–pituitary–gonad (HPG)** axis regulates the reproductive success of all vertebrates. It is influenced by a variety of internal cues, such as hormones from the thyroid and adrenal glands and other regulatory molecules, and external cues, such as temperature, photoperiod, and pheromones. These cues trigger the synthesis and release of **gonadotropins**, which play important regulatory roles. In vertebrates and protochordates, the release of gonadotropins is under the control of the hypothalamus through the action of gonadotropin-releasing hormone (GnRH) (Fig. 16.9b). The vertebrate gonadotropins—**follicle-stimulating hormone (FSH) and luteinizing hormone (LH)**—are glycoproteins that are structurally and functionally conserved across a wide range of vertebrate species, although no such specific group of molecules has been identified in invertebrates. In vertebrates, FSH initiates gamete production through the stimulation of follicle growth in females, facilitates the conversion of androgen precursors to estrogens in females, and stimulates spermatogenesis in males. LH stimulates the synthesis and secretion of androgens in both sexes, as well as ovulation in females with the subsequent secretion of progesterone from the corpus luteum.

Control of gonad development in insects requires juvenile hormone, ecdysteroids (see Section 16.4a), and a peptidic brain gonadotropin, but the molecular structure of the insect gonadotropins is far less uniform across the class. None are glycoproteins, nor can they be grouped into a single peptide family. Some of the identified gonadotropins speed up **vitellogenesis**, the synthesis and uptake of yolk proteins, or stimulate ecdysteroid production by the ovaries or testis, and so, in essence, they are the physiological counterparts of LH and FSH in vertebrates (see Box 16.3). In other invertebrates, some gonadotropic hormones have been identified, such as the egg development neurosecretory hormone of crayfish, brain hormones of nereid polychaetes, the egg-laying hormone in *Aplysia* (440), and the androgenic gland hormone in isopods. Despite their chemical diversity, all of these chemicals regulate different aspects of gametogenesis.

In 1964, the gonad-stimulating substance (GSS) of an echinoderm, the starfish *Asterias amurensis* (1071) (Fig. 16.18) was the very first gonadotropin to be identified in invertebrates. GSS mediates oocyte maturation

in starfish by acting on the ovary to produce the maturation-inducing hormone (MIH), which in turn induces the maturation of oocytes. In this sense, GSS is functionally similar to vertebrate LH, especially when compared to piscine and amphibian LHs, which also act on the ovarian follicle to produce MIH and induce the final maturation of the oocyte.

Comparatively little is known about chemical bioregulation in cnidarians. In contrast to vertebrates, cnidarian cells are differentiated into tissues but not generally organized into organs or systems (see Chapter 2). The cnidarian homologue of the nervous system is a nerve net that contains chemical synapses and releases neurotransmitters, including biogenic amines and peptides. Bioregulation in cnidarians has

BOX 16.3 People Behind Animal Biology

Kenneth George Davey

Kenneth G. Davey, a Distinguished Research Professor of Biology at York University, Ontario, is one of Canada's most prominent scientists, teachers, and science administrators. Ken Davey graduated from the University of Western Ontario with a B.Sc. (1954) and M.Sc. (1955) in Zoology and earned a Ph.D. at Cambridge University in 1958. He served as Director of the Institute of Parasitology at McGill University and was Chair of the Biology department and Vice President (Academic Affairs) at York University. Ken Davey's contributions to Biology have had both national and international impact, and he is the author of a book on the reproduction of insects and of nearly 200 scientific papers.

Ken Davey's research repeatedly reminds us about the importance of invertebrates as subjects for on the study of endocrine control. In a record of publication that began in 1955, Ken has worked on everything from describing a peptide hormone in an insect to elucidating the impact of juvenile hormone on membranes. His 1961 paper in *Nature* presented evidence of a peptide hormone in insects, the first non-steroid insect hormone to be reported.

Along the way, Ken and his colleagues studied hormone control in

the blood-feeding bug *Rhodnius prolixus* (876) (Insecta, Hemiptera; see page 355). Juvenile hormone is the main hormone controlling egg production. This hormone mediates the control that feeding and nutrition exert on vitellogensis. A myo-active hormone controls ovulation and oviposition, and its release is, in turn, affected by hormonal input from both mating and the ovaries. One clear message emerging from this work is the level of redundancy involved in both inhibitory and excitatory aspects of the interactions among feeding, nutrition, mating, and egg production. Specifically, through a membrane receptor, juvenile hormone acts on the follicular epithelium of the ovary. This connection operates by controlling access of yolk proteins to the surface of the egg.

Although it was commonly assumed that juvenile hormone acted at the genomic level, Davey demonstrated that it also worked directly on cell membranes. His discovery that juvenile hormone can bind to the thyroid membrane receptor in mammals was unexpected. But thyroid hormones in mammals mimic the mode of operation of juvenile hormone through the same receptor. The discovery that both juvenile

hormone and a thyroid hormone (T3) affect the membranes of sheep erythrocytes through the same membrane receptor confirmed this.

Work on how juvenile hormone exerted its effects led Ken to investigate endocrine cascades in nematodes. The convergence of ideas arising from results with insects and with nematodes in turn stimulated ideas about the origins of the receptors. Specifically, the work generated the theory that hormones are biotic signals that originally operated through receptors for CO_2. It was this factor that underlay the coincidence of action between juvenile hormones and some mammalian thyroid hormones.

Perhaps the most important "product" of Ken's research was his contributions with students and their research. In seven countries, Davey's academic children (graduate students over six decades) are now busy furthering our knowledge of biology in general and endocrine systems in particular. His contributions to biology and science extend from research through teaching to administration and science policy.

Ken Davey's energy and contributions set a high standard for academic excellence.

Figure 16.18 Flatbottom sea star, *Asterias amurensis* (1071), from which gonad-stimulating substance (GSS) was first isolated.

Jan Haaga

been thought to be primarily neurochemical, making used of distinct signal molecules, although much of the immunochemical identification of vertebrate-like receptors has failed to show a link between immuno-receptivity and the actual production of similar molecules by cnidarian cells. So, while putative GnRH and sex steroids have been identified in cnidarian tissues, it is unknown whether these compounds are components of a larger signal cascade comparable to the vertebrate HPG axis.

Spawning events in cnidarians may be coordinated by chemical signals triggered by environmental cues such as temperature and sensitivity to phases of the lunar cycle (Figs. 14.31 and 16.19). Similar studies on sea urchin spawning cues suggest that the lunar cycle, temperature change, phytoplankton blooms, and sperm maturation may all play roles in synchronizing spawning events. The potential roles of chemical cues remain to be elucidated.

16.6 Chemical Integration of Osmoregulation and Metabolism

Survival of any animal requires ingestion of a nutrient source, enzymatic digestion of its macromolecules, and absorption of the end products of digestion. Once absorbed, nutrients may be used as energy sources, as raw materials for the synthesis of various molecules, or stored for later use. In vertebrates, the processes responsible for these events are closely regulated by the nervous system and by hormones that are produced by the gastrointestinal tract and associated organs (such as the endocrine pancreas) and the hypothalamus–pituitary system. Similar chemical signals are involved in invertebrate metabolism, although the specific mechanisms have not been well studied, and the role of endocrine control is not clear. For example, energy metabolism in malacostracan crustaceans is known to be controlled by crustacean hyperglycemic hormones (CHHs) (Fig. 16.9A), but similar mechanisms have not been elucidated in all invertebrate groups.

Similarly, while many animals have great capacity for osmoregulation (see Section 16.4b and page 356 in Chapter 13), the role of chemical signals in controlling osmoregulatory mechanisms is not clear in many invertebrates. The well-documented role of hypothalamic peptide neurohormones in vertebrate osmoregulation is well-conserved across vertebrate groups, involving nonapeptide hormones, which are synthesized in hypothalamic nuclei as part of larger propeptides (see Section 16.2a). These molecules, collectively known as **vasopressins**, are widely distributed in both vertebrates and invertebrates. The main physiological role for vasopressins appears to be an antidiuretic action, with a secondary role in elevation of blood pressure through effects on vascular smooth muscle. Vasopressins in vertebrates may also act as neurotransmitters and neuromodulators in the central nervous system, and some have been shown to enhance the release of ACTH from the pituitary. While the antidiuretic action of vasopressins in vertebrates is achieved through specific actions at the kidney tubules, similar changes in osmoregulation have been documented in platyhelminths, annelids, arthropods, molluscs, and tunicates through actions on a wide range of epithelial tissues.

The diuretic action of a neuropeptide diuretic hormone (DH) was discussed for the blood-feeding bug *Rhodnius prolixus* (876) in the opening section of this chapter. In addition to this neuropeptide, the Malpighian tubules of *Rhodnius* have also been shown to respond to 5-hydroxytryptamine (5-HT, serotonin), a naturally occurring biogenic amine, also of neural origin, that travels through the hemolymph to act with the peptide DH to trigger diuresis.

Prolactin (PRL), a peptide hormone better known for its reproductive roles in mammals, has been shown to have more than 300 different actions in vertebrates! Additionally, expression of PRL or prolactin-like hormones has been documented in invertebrates. Actions of PRL related to water and electrolyte balance include changes in the permeability of gill, gut, renal, and integumentary epithelia. These actions have been applied in systems ranging from osmotic balance in parasitic nematodes to the production of a suitable osmotic environment for the developing young in the brood pouch of male seahorses.

While the basic principles of action for chemical signals are fairly well understood, there is still much

Figure 16.19 Spawning in cnidarians may be coordinated by chemical signals.

© Mark Conlin / Alamy

to be learned about the specific actions of specific hormone molecules in animals.

ENDOCRINE INTEGRATION
IN PERSPECTIVE

The aspects of chemical communication explored in this chapter range from morphogenesis and reproduction to stress responses and osmoregulation. And yet this selection represents just a few examples of the range of functions that chemical signals play in integrating animal systems. While the basic principles of action for chemical signals are fairly well understood, we have much to learn about the specific actions of specific hormone molecules in animals.

Questions

Self-Test Questions

1. Which of these characteristics distinguishes peptide hormones from other types of chemical messengers?
   a. Peptide hormones are released from endocrine glands into the circulatory system.
   b. A target tissue for a peptide hormone has receptors specific for that hormone.
   c. Peptide hormone receptors in target cell membranes are linked to enzymes that synthesize second messengers that bring about the response of the target cell.
   d. Intracellular hormone receptors bind peptide hormones and allow target tissues to accumulate the hormone against a concentration gradient.

2. Which of the following best illustrates feedback control of hormone action?
   a. Crustacean hyperglycemic hormones (CHH) are pleiotropic, with a diverse array of actions in many body systems that stimulate other physiological responses.
   b. CHH released from the sinus gland triggers the rapid mobilization of glucose from reserve carbohydrates such that circulating glucose levels rise.
   c. CHH is synthesized by neurosecretory neurons of the eyestalk X-organ and released through the sinus gland in response to low circulating glucose levels.
   d. X-organ cells are inhibited by elevated circulating glucose levels.

3. Why is neuroendocrine control important in the physiological processes of animals?
   a. The nervous system, working through the endocrine system, can integrate external and internal stimuli with widespread chemical responses to maintain homeostasis.
   b. Whole-body changes, such as those involved with moulting and metamorphosis, are directed by chemical signals.
   c. Sensory receptors can be tuned to a wide variety of stimuli, including extrinsic and intrinsic changes.
   d. Behavioural changes, such as those exhibited during different life stages of Atlantic salmon, can be elicited by hormones.

4. How is the hypothalamus–pituitary–gonad (HPG) axis similar in vertebrates and invertebrates?
   a. The HPG axis mediates the response of animals to physiological stressors.
   b. The HPG axis coordinates internal and external stimuli with gametogenesis.
   c. Gonadotropins, such as follicle stimulating hormone (FSH) and luteinizing hormone (LH), are glycoproteins that are structurally and functionally conserved across all species.
   d. Gonadotropins act on the ovaries to induce the final maturation of oocytes through the production of maturation-inducing hormones.

5. What is the primary role of vasopressins in animal systems?
   a. They act as neuromodulators.
   b. They increase blood pressure.
   c. They trigger the reabsorption and retention of water.
   d. They act as diuretics.

Questions for Discussion

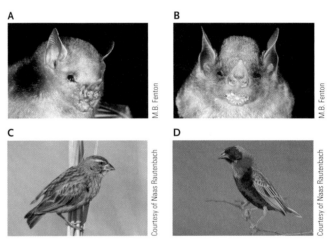

Figure 16.20 Two male yellow-shouldered bats (*Sturnira lilium* (1816)) caught at the same time on the same night in Belize. **A.** Adult male in reproductive condition differs in colour and development of shoulder patches from **B.** a non-reproductive male. Similarly, the Red Bishop (*Euplectes orix* (1599)) shows remarkable changes between **C.** non-breeding and **D.** breeding males.

1. Examine the images in Figure 16.20. What hormonal influences would you expect to be important in these morphological changes? What signals would trigger these changes? What techniques would be useful in determining the roles of hormones in these examples?

2. How could changes in a hormone level in a host organism influence a blood-feeding ectoparasite? Could such changes also affect endoparasites?

3. Why are the postprandial changes in the Burmese python not hormonal? What other chemical signals have we encountered throughout this text that are not hormonal?

4. Why are chemical mechanisms so prevalent in integrating physiological processes in animals? Illustrate your answer with specific examples.

5. Using the electronic library, find a recently published example of interactions between neural and endocrine systems.

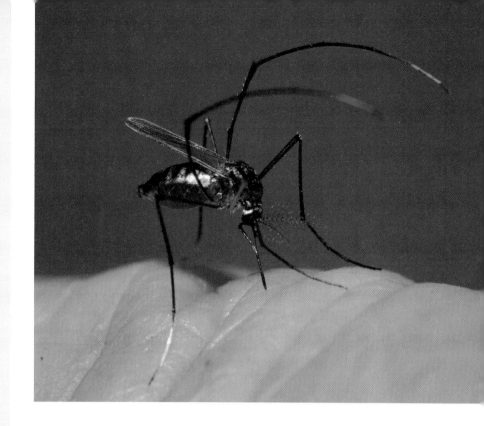

Figure 17.1
Mosquito biting human.
SOURCE: M.B. Fenton

17 Behaviour

WHY IT MATTERS

While the behaviour associated with predation is fascinating, it may also have practical importance. For example, a predator that specializes on prey that are considered to be pests or a danger to people may have considerable economic value because it could be useful in biocontrol operations. When the prey are mosquitoes, some of which are vectors for diseases such as malaria that kills millions of people every year, biological control by predation is a tantalizing concept.

Many people think that bats are important predators of mosquitoes. In the 1920s, Dr. Charles A.R. Campbell erected bat towers in several southern American states with the rationale that housing for bats might increase their populations and lead to a reduction in the mosquito population. In the early 1950s, Edwin Gould shot several bats and examined their stomach contents. Two tricoloured bats (*Perimyotis subflavus* (1806)), one weighing 5.8 g and the other 6.7 g and shot within 30 minutes of starting to hunt, had caught 1.4 g and 1.7 g of insects, respectively, but the insects were unidentifiable. In the late 1950s, Donald Griffin and colleagues demonstrated in the laboratory that little brown bats (*Myotis lucifugus* (1800)) used echolocation

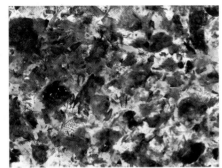

Figure 17.2 Insect remains in bat feces.

SOURCE: ELIZABETH L. CLARE, ERIN E. FRASER, HEATHER E. BRAID, M. BROCK FENTON, PAUL D. N. HEBERT. "Species on the menu of a generalist predator, the eastern red bat (Lasiurus borealis): using a molecular approach to detect arthropod prey." Molecular Ecology, Volume 18, Issue 11, pages 2532–2542, June 2009.

(biosonar) to detect insects the size of mosquitoes and fruit flies. In this laboratory setting, these bats could catch 14 fruit flies (*Drosophila* (960)) a minute. These two otherwise unconnected sets of data led to the widely held view that bats can catch more than 500 mosquitoes (or fruit flies) per hour.

It is clear that some bats eat large quantities of insects, but just what insects do they eat? Bats chew their food thoroughly and it has been difficult to identify insect pieces beyond order. The advent of DNA barcoding has allowed us to identify the species of insects eaten by bats just from a sampling of fragments in bat feces (Fig. 17.2). To the relief of those who believe that the bats in their roof protect them from mosquitoes, the DNA barcoding data reveal that little brown bats occasionally eat mosquitoes (Fig. 17.3). But they more commonly eat mayflies, caddisflies, and midges. It is not clear that little brown bats would qualify as effective agents of mosquito control.

Do any predators specialize on mosquitoes? As noted on page 223, the jumping spider *Evarcha culicivora* (668) (Fig. 8.40) from the Lake Victoria region of East Africa qualifies as a mosquito specialist, eating mosquitoes even when faced with many

mosquito-sized prey such as midges. Further, *E. culicivora* eats mainly mosquitoes that have taken a blood meal. Their selection for female mosquitoes engorged with blood could reflect their exceptional visual acuity and/or the fact that mosquitoes have a different odour after they have eaten blood. After a blood meal, *Anopheles* (954) mosquitoes have distended (and red) abdomens and rest with their abdomens tilted up.

Mosquito fish (several species in the genus *Gambusia* (1203)) eat mosquito larvae in addition to other prey. Because *Gambusia affinis* is said to be a specialist on mosquito larvae, it has been introduced to many parts of the world. There are two drawbacks to these fish as agents of biological control. First, they readily turn to other prey when mosquito larvae are not abundant and thus become threats to some other native species. Second, they are vulnerable to contaminants such as hormones that enter the water from human or veterinary sources, effectively sterilizing males and threatening the survival of the fish (see pages 517–518).

Does eating mosquitoes automatically make a predator a potential biocontrol agent for mosquitoes? Does predation by bats, jumping spiders, or mosquito fish affect the population of mosquitoes? To address these questions, we need to know the sizes of the populations of mosquitoes and predators, the numbers of mosquitoes taken by each predator, and the rates at which the two species reproduce. Smaller specialist predators such as the spiders may be more focused on mosquitoes, while larger predators, especially warm-blooded ones such as bats, have a more urgent need for energy and readily switch to the largest and/or most abundant prey they can handle.

Information about the behaviour of predators and prey may inform conservation and public-health strategies, but without the full picture, salvation from mosquitoes by predators may be elusive. Do we want a world without mosquitoes? Mosquitoes are vital pollinators for some plants and, ironically, are prey for many species of birds and other predators. If we focus only on malaria, worldwide about US$1990 million is required annually for spraying to control mosquitoes and an additional US$2090 million for insecticidal nets. In July 2010, RNA interference, male sterilization, improved insecticides that kill mosquitoes, and mosquito traps were listed as the most promising approaches for mosquito control. Biocontrol was not on the list.

There are ~3500 named species of mosquitoes, and the first mosquito fossils are more than 100 million years old. Mosquitoes are an old and successful group. Controlling species that pose a threat to us may be most effectively done by changing behaviour of people to minimize their exposure and of mosquitoes to interfere with their life cycles or ability to detect prey.

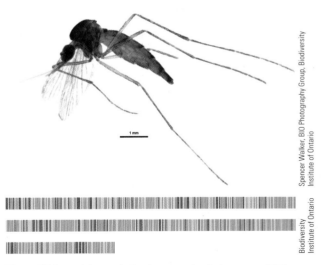

Spencer Walker, BIO Photography Group, Biodiversity Institute of Ontario

Biodiversity Institute of Ontario

Figure 17.3 DNA bar code for the mosquito *Aedes vexans* (953).

Through their behaviour, animals can operate well outside the apparent limitations of morphology and physiology. Behavioural adaptations are fundamental to the success of animals in their environments and behaviours are intrinsic to many aspects of the biology of any species. After reading this chapter you might look back over other chapters in the book and realize that many examples of animal behaviours, for example behaviours associated with feeding or with parasitic life cycles, have already been discussed. In this chapter we have grouped behaviours under subheads that describe various types of behaviour (e.g., defensive, migratory). Nature does not respect this type of organization and many examples might have appeared under different subheads or in other chapters of the book. Consider the broader implications of each behaviour we have selected as an example.

17.1 Introduction

In this chapter we use behaviour to illustrate the diversity of animals. We will begin by considering the impact of hormones on behaviour and progress to behaviour associated with signalling and communication. We will review material relating to defensive behaviour that we will consider from the standpoint of attacker and defender. In the section about migration we will examine details of movements as well as the challenges of navigation. Behaviour relating to sex and reproduction is a good way to explore the topic of social behaviour. We will consider some of the many examples of learning by animals and wrap up by examining different patterns of spider behaviour.

17.2 Hormones and Behaviour

On pages 393 and 395, we saw how levels of hormones affected the external genitalia, and how hormones associated with pharmaceuticals could disrupt behaviour and sexuality in a variety of animals. It is common to think of hormonal changes in association with slow, long-term changes, but, as we shall see, this is not always the case.

Hormone levels may change rapidly and affect behaviour as animals experience and respond to different situations. In a study on male chimpanzees (*Pan troglodytes* (1866)) and bonobos (*Pan paniscus* (1865)), it was found that they differ in their hormonal responses to situations associated with food availability. The hormones in question are steroids (glucocorticoids (cortisol) and testosterone), thought to mediate energy allocation. Increased cortisol is characteristic of a stress response while testosterone plays a role in competition and dominance. In one situation, two animals (two chimps or two bonobos) were presented with food that could be shared. In another situation, the two animals, one of them dominant, were presented with food that could not be shared.

Hormone levels (testosterone and cortisol) were measured in the animals' saliva. Both chimps and bonobos showed changes in hormone levels that differed according to food accessibility (shared or not). In bonobo males, cortisol levels changed, whereas testosterone levels changed in chimps. The data suggest that while bonobos responded to the competition situation as if it were a stressor, chimpanzees responded as expected for a dominance contest.

In the laboratory, the behaviour and cortisol levels of Red Crossbills (*Loxia curvirostra* (1606)) are influenced by the feeding behaviour of conspecifics they can see but not interact with. Cortisol levels rise in Red Crossbills that observe conspecifics behaving as if there is a shortage of food. In the wild, Red Crossbills often respond to extreme fluctuations in food abundance by migrating long distances (irruptive behaviour). In a sense, the behaviour of Red Crossbills searching for food is "public information" readily available to conspecifics. When Red Crossbills think that there is a shortage of food, they show immediate responses in behaviour and levels of cortisol.

Hormones work because cell-membrane receptors for those hormones initiate cellular responses. Territorial male California mice (*Peromyscus californicus* (1896)) that win aggressive encounters show different levels of expression of androgen receptors (AR) in their brains depending on the location of the encounter. AR expression, in key areas of the brain associated with social aggression (medial anterior bed nucleus of the stria terminalis), is stimulated by winning. Winning on the animal's home range stimulates AR expression in the areas of the brain associated with motivation and reward (nucleus accumbens and ventral tegmental area); winning in an unfamiliar area does not have this effect. In short, winning fights changes the brain in ways that probably promote future victory and primes individuals to fight.

In insects, changes in levels of juvenile hormone (JH), ecdysteroids, and neurotransmitters such as octopamine could be responsible for changes in host behaviour induced by parasitoids (other insects that lay their eggs on or in animal hosts). Caterpillars of a geometrid moth (*Thyrinteina leucocerae* (1049)) hosting larvae of a parasitoid wasp (*Glyptapanteles* spp. (997)) change their behaviour after the wasp larvae leave the host and pupate (see page 291, Fig. 10.57). Juvenile hormone also can regulate the appearance of insects. Over its development, larval (caterpillar) swallowtail butterflies (*Papilio xuthus* (1043)) change from mimicking bird droppings to being cryptically coloured

Oxytocin and testosterone are two hormones that fundamentally affect the lives of animals (Fig. 1).

Oxytocin is a neuropeptide produced in the hypothalamus. CD38 is a transmembrane glycoprotein and its ADP-ribosyl cyclase activity catalyses the release of Ca^{++} molecules involved in signalling. Adult mice lacking CD38 show

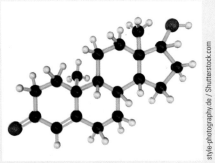

FIGURE 1
Molecules of oxytocin and testosterone.

marked defects in maternal nurturing and social behaviour. In these mice, oxytocin but not vasopressin is strongly decreased. However, administration of oxytocin restored some social memory and maternal behaviour, suggesting that CD38 has a key role in release of neuropeptides, including oxytocin. In humans, oxytocin has been linked to regulation of intergroup conflict, driving a "tend and defend" response that promotes in-group trust and cooperation as well as defensive aggression toward competing groups rather than offensive aggression. Administered to people as a nasal spray, oxytocin promotes general trust and co-operation, reducing the likelihood that the recipient(s) will exploit others` cooperative behaviour. In short, oxytocin can promote parochial altruism, self-sacrifice that contributes to in-group welfare.

In mammals, testosterone is a steroid hormone produced mainly in the testes and, to a lesser extent, in the ovaries. Testosterone affects brain development and sexual behaviour and is present in the central nervous system throughout life. Testosterone is alleged to cause antisocial egoistic and aggressive behaviour in humans, and

in some legal defenses a "steroid-induced rage" has been considered a legitimate defense. Other points of view hold that testosterone in humans is more involved with status-related behaviour and challenging behaviour. Sublingual administration of a single dose of testosterone to women elicits higher levels of fair-bargaining behaviour and fewer incidents of conflicts over bargaining. But subjects who believed that they had received a dose of testosterone behaved more unfairly than those who believed that they had received a placebo—demonstrating how perception can influence behaviour.

Earlier we have seen how testosterone levels can masculinise the genitalia of some female mammals, and that changes in hormone levels and receptors associated with testosterone and cortisol also affect behaviour. The next several years should see further data about the impact of hormones on behaviour. Increasingly, researchers are documenting hormonal changes in the animals they study. Lessons from the impact of hormones on our own species will help us to appreciate details of the behaviour of other species.

(Fig. 17.4). Changes in appearance and behaviour coincide with different levels of JH, which decreases in concentration as the larva ages. At high levels of JH, specific cuticle proteins generate tubercle structures, while at low levels bilin-binding proteins are responsible for the green colouration that contributes to camouflage.

STUDY BREAK

1. Name three ways in which hormones influence the behaviour of animals.
2. How do hormone receptors mediate behaviour?
3. Which hormones work in antagonistic pairs?

17.3 Signals and Communication

The behaviour of communicating can be focused on the signals animals use when interacting with others. Individuals of many different species are distributed spatially in suitable habitat and their density often reflects the availability of resources such as food and shelter. Anyone who has walked a dog will recognize some of the signals that can indicate the use of space by individuals of the same species. The dog that sees or hears another dog reacts in one way, but responds quite differently when visiting a scent post that has been used by another dog. One difference between the responses is the distances between dogs. **Territories** are a spatial subset of the **home range**, the area regularly traversed by a resident animal.

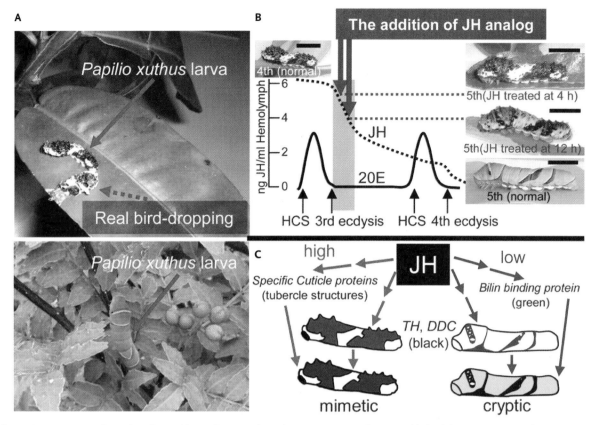

Figure 17.4 A. Caterpillars of swallowtail butterflies (*Papilio xuthus* (1043)) can either resemble bird droppings (upper photo) or be cryptic (lower). These changes are mediated by juvenile hormone (JH). **B.** Treatment of fifth-instar larvae with JH analogs at a time when these levels normally decline maintains the production of cuticle proteins at a time when they would normally produce bilin-binding protein and display a more cryptic green colour (**C**). HCS = slippage of the head capsule; TH = tyrosine hydrolase. 20E = hydroxyecdysone; DDC = dopa decarboxylase

SOURCE: From Fatahashi, R. And H. Fujiwara, "Juvenile hormone regulates butterfly larval pattern switches," *Science* 22 February 2008: Vol. 319 no. 5866 p. 1061. Reprinted with permission from AAAS.

Animals typically use signals to mark their presence in an area and perhaps their ownership of a territory. Bird songs are familiar advertising signals, and animals ranging from mammals to insects, frogs, and fish announce their presence with acoustic signals. Some caterpillars use vibrational signals to assert their presence in and ownership of a territory. For example, masked birch caterpillars (*Drepana arcuata* (1032)) construct silken shelters with leaves and make repetitive vibratory displays when approached by an intruding conspecific. The display has three main elements: anal scraping, drumming with mandibles, and mandible scraping (Fig. 17.5). Confrontations between caterpillars are highly ritualized and usually the intruder abandons its approach to an occupied shelter. The caterpillars' vibrational signals are derived from sounds associated with locomotion and the presence of a specialized morphological feature (an anal oar).

Visual displays involving rapid changes in colour are another way to dissuade conspecifics. Although it is easy to believe that chameleons use colour change for camouflage, male dwarf African chameleons (*Bradypodion* spp. (1475)) use dramatic colour changes during competitive encounters with other males.

Chameleons (and birds, potential predators) can see colour changes that produce high contrast against the environmental background and adjacent regions of the body (Fig. 17.6). Thus, a chameleon's camouflage works in both inter- and intraspecific situations (Fig. 17.7).

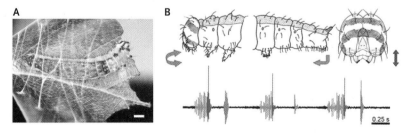

Figure 17.5 A. The silken leaf shelter of a masked birch caterpillar (*Drepana arcuata* (1032)). **B.** Diagram showing the three types of vibratory signals used in response to an intruder. The time/amplitude trace shows the vibrations associated with anal scraping (green), mandible scraping (orange), and mandible drumming (blue). Scale in A = 2.5 mm; time frame on graph = 0.25 s.

SOURCE: Reprinted by permission from Macmillan Publishers Ltd: *Nature Communications 1*, Jaclyn L. Scott, Akito Y. Kawahara, Jeffrey H. Skevington, Shen-Horn Yen, Abeer Sami, Myron L. Smith & Jayne E. Yack, "The evolutionary origins of ritualized acoustic signals in caterpillars," copyright 2010.

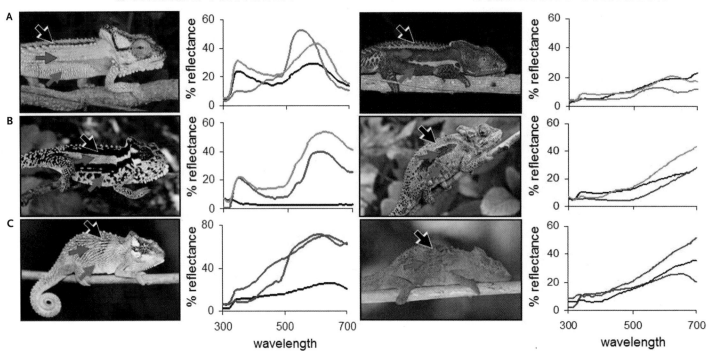

Dominant coloration

Submissive coloration

Figure 17.6 Dominant and submissive colouration of the dwarf chameleons from across the geographic range of the genus (*Bradypodion*). **A.** *B. damaranum* (1478) from the south of the range, **B.** *B. transvaalense* (1481) from the north, and **C.** *B. caffrum* (1477) from the central east. The percent reflectance compares the three species' dominant and submissive displays. Arrows indicate sites of reflectance measurements (black = top flank; blue = mid-flank; and red = bottom flank).

SOURCE: Devi Stuart-Fox, and Adnan Moussalli, "Selection for Social Signalling Drives the Evolution of Chameleon Colour Change," PLoS Biol. 2008 January; 6(1): e25. Courtesy of Devi Stuart-Fox.

Cephalopod molluscs (octopus, squid, and cuttlefish) can be exceptionally colourful (Fig. 17.8). They are also unique in that rapid colour changes can serve both in communication and camouflage. Colour can be produced by pigments or by structures. Pigmented material, localized in chromatophores, produces colour by selective absorbance that establishes the colours of incident light are reflected. Structural materials are colourless and create colour by coherent scattering of light. Cephalopods' ability to instantaneously change colour derives from the dual action of thousands of chromatophores and structural reflectors known as iridophores and leucophores. Proteins are responsible for some optical effects of iridophores and leucophores which operate in conjunction with overlying chromatophores. Cephalopods, many of which appear to be colour blind, are particularly sensitive to polarized light.

As demonstrated by the chameleons and cephalopods, animals can use the same or similar signals in different contexts. Elephant fish from Africa (family Mormyridae) use **electrolocation**: they produce weak electric organ discharges to collect information about their surroundings. In this way, they locate prey and communicate both intraspecifically and interspecifically. Electrolocation signals allow individuals to space themselves in available habitat. They also play a pivotal role in mate recognition and assortive mating among two closely related species, *Campylomormyrus compressirostris* (1231) and *Campylomormyrus rhynchophorus* (1232) that are sympatric in the rapids of the lower Congo River. The electric organ discharges of a third, more distantly related species (*C. tamandua* (1233)) are distinctly different from the other two (Fig. 17.9). Electrolocation is intriguingly analogous to echolocation in terms of how the animals both collect information and put it to a variety of uses.

Language, a specialized form of communication, involves very specialized signals, many of which are used

Figure 17.7 Camouflage and antipredator response. **A.** *Bradypodion taeniabronchum* (1480), **B.** *B. gutturale* (1479), and **C.** *B. atromontanum* (1476) in the presence of a stuffed Fiscal Shrike (*Lanius collaris* (1605)).

SOURCE: Devi Stuart-Fox

Figure 17.8 Colourful cephalopods. **A.** *Hapalochaena maculosa* (500) and **B.** *Octopus mototi* (512).

symbolically. The language of humans, an extreme example of specialized communication, was earlier thought to be restricted to our species. The dance language of bees was one of the first examples of "language" in animals other than humans and its discovery was instrumental in Karl von Frisch receiving a Nobel Prize.

Worker honey bees (*Apis mellifera* (984)) use a "waggle dance" to recruit other workers to new food sources or new nest sites that are away from the hive.

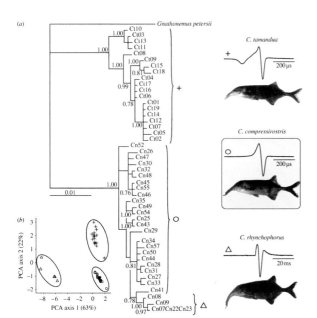

Figure 17.9 A. A phylogeny of mormyrid fishes in the genus *Campylomormyrus* based on 1999 base pairs of mitochondrial cytochrome b and the nuclear ribosomal S7 gene. Diagrams show each species, its arrangement in the phylogeny and its electric organ discharge pattern. **B.** Ten features of the electric organ discharges were used in a principal component analysis (PCA).

SOURCE: Philine Feulner, "Electrifying love: electric fish use species-specific discharge for mate recognition," *Biol. Lett.* 23 April 2009 vol. 5 no. 2, pp. 225–228, by permission of the Royal Society.

Inside the hive, the dance alerts other worker bees to the resource and gives them directions to its location. The liveliness and duration of the waggle dance reflect the attractiveness of the resource. The distance and duration of the waggle run correspond to flight distance between hive and resource. The angle of the waggle dance relative to the position of the sun provides directional information. By the dance, the worker bee communicates a flight vector that is the reverse of the vector she flew coming back to the hive from the resource—the one she will use to return to the resource.

Recent neurobiological investigations have identified four or five different sensory systems and their projections to the central nervous system of the bees (Fig. 17.11). Bees acquire the information by watching the dance using a mixture of sensory systems (see also Fig. 15.28). The dorsal part of the bee's medulla receives projections from sensory neurons on the dorsal rims of the compound eyes, which convey information from the sun-compass. The dorsal labial neuromere of the subesophageal ganglion receives projections from the neck hair plates and transposes information about gravity in the dark hive. The deutocerebral dorsal lobe, subesophageal ganglion, and posterior protocerebrum receive projections from antennal joint hair sensillae and Johnston's organ, which convey information about dance direction.

Signalling with language requires an effective combination of signals and the systems for encoding and interpreting them—whether the language is ours or the dance language of bees. Later in the chapter, we will see that some animals use specific alarm calls to "symbolize" different predators.

STUDY BREAK

1. Do the communication dances of honey bees count as "language"?
2. How do fish use electrical signals in orientation and communication?
3. What do colour changes tell us about an animal's behavioural state?

17.4 Defensive Behaviour

17.4a. Chemical Defense

At first glance, plants seem open and vulnerable to attacks from herbivores, but even a brief consideration of their various defenses challenges this view. Think of the outcome of tangling with a rose or with poison ivy. In addition to mechanical defenses such as spines and thorns, plants use chemical defenses, some that seriously threaten the well-being, if not the survival, of herbivores that contact or eat them. For our species, cooking is one way to denature dangerous chemicals to which we could be exposed when we eat plants such as milkweed or cassava. In response to the damage caused by herbivores, some plants emit green leaf volatiles (GLVs) which serve as signals that attract predators. When tobacco hornworms (caterpillars of *Manduca sexta* (1041)) feed on nicotine plants (*Nicotiana attenuata*), they stimulate the plants to release GLVs that attract a predatory bug, *Geocoris* sp. (869).

Large mammalian herbivores (Fig. 17.10) can consume huge amounts of vegetation, perhaps also ingesting smaller herbivores such as aphids along with the plants. Pea aphids (*Acyrthosiphon pisum* (865)) sense elevated heat and humidity in the breath of an approaching mammalian herbivore and synchronously drop from the plant before the herbivore begins feeding. Although individual pea aphids also drop from the plant in response to attacks by some insect predators, immediate mass dropping is triggered by the combination of increased temperature and humidity associated with the breath of a large herbivore. Mass dropping is not triggered by picking individual leaves or by casting a shadow over the host plant, it is an adaptation that minimizes the chances of aphids being ingested by the herbivores. Other invertebrates also detect and respond to mammalian breath. Mosquitoes, ticks, and tsetse flies locate hosts/prey by detecting the combination of carbon dioxide, exhaled organic volatiles, heat, and humidity. Still other arthropods respond to the odour of a predator's breath by secreting and releasing noxious chemicals, sometimes in combination with other defensive responses. The tenebrionid beetle, *Bolitotherus cornutus* (923) (Fig. 17.11), everts a pair of quinone-producing glands in response to the breath of mammals. Emissions from the glands provide some protection against attacks by mice (*Peromyscus* spp. (1895)).

17.4b Sound and Hearing

While some animals respond to the smell of a predator's breath, others take their defensive cue from a predator's sounds. Dogbane tiger moths (*Cycnia tenera* (1030)) have bat-detecting ears. When a moth hears the echolocation calls of foraging bats (Box 15.3), it responds with clicks that it produces from microtymbals on its thorax. Bats using echolocation adjust the details of their calls (frequencies in the calls, durations of calls, and time between calls) during their hunt, from searching for, detecting, assessing, and then attacking a flying insect. Dogbane tiger moths' patterns of click production demonstrate that they learn more quickly to ignore bat calls associated with searching for a target than bat calls produced during attacks (Fig. 17.12). The moths consider a bat searching for a target to be a lower risk than one actually making an attack. The moths also have a backup chemical defensive system against bats: the cardiac glycosides their caterpillars harvested from the dogbane (*Apocynum androsaemifolium*) on which they fed. The moths' clicks may warn the bat of bad taste, startle the bat, or interfere with (jam) its echolocation system.

In general, moths with bat-detecting ears are less vulnerable to attacks by bats than moths and other insects without these ears. Bats achieve 40% success rates in attacks on moths with bat-detecting ears, compared to >90% on non-hearing insects. But eastern red bats (*Lasiurus borealis* (1795)), whose calls should be entirely obvious to eared moths, prey heavily and consistently on moths able to detect and evade bats. Identification by DNA barcoding indicates that eastern red bats regularly eat the moths with bat-detecting ears (e.g., *Haploa confusa* (1037), *Campaea perlata* (1029), and *Pero ancetaria* (1045)).

In Europe, barbastelle bats (*Barbastella barbastellus* (1785)) also eat moths with bat detecting ears (>85% of the diet), but compared to eastern red bats, barbastelles use quieter echolocation calls that allow them to closely approach (sneak up on) and not trigger the moths' defensive behaviour. In the same airspace in Europe, Leisler's bat (*Nyctalus leisleri* (1804)) produces much stronger calls and rarely catches moths with bat-detecting ears. It remains to be seen how eastern red bats succeed against moths that should have detected and avoided them. Still other insectivorous bats foil

Figure 17.10 An African elephant (*Loxodonta africana* (1871)) munching on vegetation.

Figure 17.11 A 10-mm-long female tenebrionid beetle (*Bolitotherus cornutus* (923)). In defense these beetles feign death, but in response to the breath of a mouse (*Peromyscus* spp.) the beetle everts secretion-laden glands from the tip of the abdomen and wets the tibiae with material from the gland. Mice avoid the odours associated with the eversion of the gland.

A

B

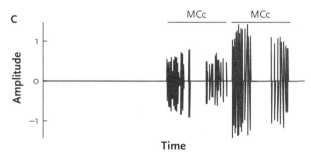

10 ms

C

MCc   MCc

Amplitude

Time

Figure 17.12 A. Dogbane tiger moth *Cycnia tenera* (1030).
B. Time-amplitude display of the echolocation call of an
attacking bat. **C.** The moth responds to the bat's call with trains
of clicks. Vertical axis is signal strength, horizontal axis is time.

SOURCE: John M. Ratcliffe, James H. Fullard, Benjamin J. Arthur, and Ronald
R. Hoy, "Adaptive auditory risk assessment in the dogbane tiger moth when
pursued by bats," *Proc. R. Soc. B.* © 2010 The Royal Society.

hearing-based defenses by using echolocation calls
that are outside the frequencies that insects hear best.
Some male moths (*Spodoptera litura* (1048)) have
sound-producing tymbals (Fig. 17.13) that produce
ultrasonic signals. Females of this species respond to
the males' signals as part of courtship behaviour and
do not mate with muted males. However, recorded
male moth signals or bat echolocation calls stimulated
female *S. litura* to mate with muted male moths. In
these moths, acoustic communication associated with
courtship has evolved from a defensive behaviour. As
usual, for every trick there is a counter-trick.

California ground squirrels (*Spermophilus beecheyi*
(1906)) have adopted an important way to thwart the
heat-sensors (see pages 437–438) of one of its predators, namely rattlesnakes (*Crotalus oreganus* (1488)).
When confronted by a rattlesnake, California ground

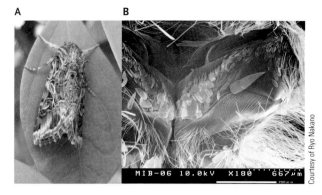

Figure 17.13 A. A male noctuid moth *Spodoptera litura* (1048).
B. Close-up SEM image of a metathoracic noisemaker (tymbals)
present in males but not females.

squirrels wave their tails at the snakes (tail-flagging)
and include an infrared component in the display (Fig.
17.14). The infrared component attracts more attention from rattlesnakes that strike at the "hot" tail. Rattlesnakes also appeared more cautious in the presence
of an infrared component to tail-flagging. The ground
squirrels also use tail-flagging when confronted by

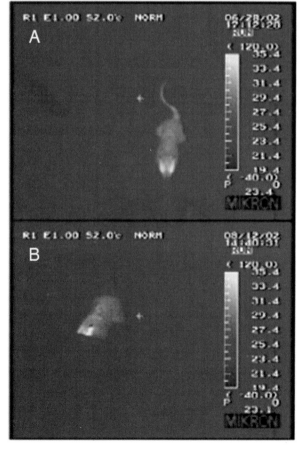

Figure 17.14 Two infrared images of a California ground
squirrel. **A.** The squirrel is interacting with a gopher snake that
lacks infrared sensors. **B.** The squirrel is interacting with a
rattlesnake that has infrared sensors.

SOURCE: Rundus et al., "Ground squirrels use an infrared signal to deter
rattlesnake predation," *PNAS* September 4, 2007 vol. 104 no. 36 14372–14376.
Copyright 2007 National Academy of Sciences, U.S.A.

Figure 17.15 The larva of a beetle, *Cassida* spp. (926), carries a fecal shield which it thrusts at the photographer or at other would-be predators.

Steve Marshall

gopher snakes (*Pituophis melanoleucus* (1516)) but do not add the infrared component to the display. Even if the infrared defence does not work, adult California ground squirrels also have defensive proteins that partially neutralize the venom of some rattlesnakes.

Signals ranging from breath to sounds to infrared signatures demonstrate how signals and signal modalities can be influenced by natural selection by providing selective advantages to predators and/or prey (page 85).

Larvae of some chrysomelid beetles (e.g., *Cassida rubiginosa* (926)) use a packet of trash (cast skins and feces, called a fecal shield) to fend off attackers (Fig. 17.15). The fecal shield is held in place by a two-pronged fork projecting forward from the tip of the abdomen. The larvae manoeuvre the fecal shield by flexing and rotating its abdomen, which permits them to thrust it into the faces of attackers such as ants. A marginal fringe of long, branched spines on either side of the larva's flattened body alerts the larvae to the location of the attacker. The effectiveness of the shield is obvious to anyone observing ants attacking these beetle larvae. Upon detecting a larva, an ant inspects it, but only when the ant moves to bite does the larva deploy

the fecal shield. Thus confronted, attacking ants usually broke off the engagement, some after biting into the shield. The addition of fresh wet feces at the base of the shield appears to make it more effective as an ant deterrent.

Larval tortoise beetles (*Hemisphaerota cyanea* (937)) conceal themselves under a thatch-like material that they make from long filamentous strands of their feces. The thatch is retained through moults of the larvae, increasing in size as the larvae grow. While some predators (e.g., *Cycloneda sanguinea* (930) and *Stiretrus anchorago* (878)) are effectively deterred by the thatch, another predator, a carabid beetle (*Calleida viridipennis* (925)), thwarts the defense by moving under the thatch or chewing through it to feed on the larva.

The use of feces in defense is a recurrent theme in the animal kingdom, and other animals camouflage themselves by looking like feces (see swallowtail butterflies, Fig. 17.4). This defense appears to work because predators fail to recognize feces as something edible.

17.4c Other Defensive Behaviour

Camouflage, a situation in which the predator fails to detect potential prey against the background, is another common defense (Figs. 17.7 & 17.16). Disruptive patterning, often colouration, is the underlying mechanism. The effectiveness of such defense mechanisms can be difficult to assess because of variation in the visual abilities of predators or potential predators. Computer simulations of variations in the appearance of moth-like prey (Fig. 17.17), presented to human observers, demonstrated that when patterned markings extended to the edges of the "moth's" wings, its body shapes were more difficult to detect. Disruptive colouration that did not match the background was the easiest to see.

Bad taste also can alert predators to dangerous toxins used in defense by insects. European Starlings (*Sturnus vulgaris* (1616)) learn to predict the amounts of toxin in individual prey by their bitter taste and to avoid consuming too many bitter (more toxic) food items.

A   B   C

Figure 17.16 A. In this setting, the horned adder (*Bitis caudalis* (1473)) is obvious, but **B.** against a different background it is well camouflaged. **C.** The walking stick (*Hydrometra* sp. (871)) is easily mistaken for an inedible piece of plant rather than an edible insect.
SOURCE: Courtesy of Naas Rautenbach

disruptive black   average colour

BCE   BCNE-high density   BCNE-low density

Figure 17.17 Five target types of moth disruptive patterning were designed for human observers. The top two were the easiest to detect, the bottom three more difficult. Disruptive black is a high-contrast disruptive pattern that does not match the background. Average colour also does not match the background. BCE is a bicolour with edges, a disruptive pattern with good matching to background. BCNE–high density is a bicolour with no edges and a high-density centre that is a nondisruptive pattern that matches the background. BCNE–low density is a nondisruptive pattern with better background matching.

SOURCE: Stewart Fraser, Alison Callahan, Dana Klassen and Thomas N. Sherratt, "Empirical tests of the role of disruptive coloration in reducing detectability," *Proc. R. Soc. B* 22 May 2007 vol. 274 no. 1615 1325–1331, by permission of the Royal Society.

A   B

Figure 17.18 Aposematic colouration. **A.** The monarch butterfly (*Danaus plexippus* (1031)) is a typical aposematic animal whose colour warns experienced predators of the bitter-tasting cardiac glycosides (obtained when the larvae fed on milkweed) within its body. **B.** The spotted maize beetle (*Astylus atromaculatus* (922)), protected by noxious chemicals, is also conspicuous.

Presented with beetle larvae as food, some of which have been treated with a bad-tasting substance such as quinine, European Starlings always eat the least bitter prey larvae available.

Animals that taste bad are often **aposematic** (Fig. 17.18), using warning displays such as bright colours, buzzing sounds, or aggressive behaviour to deter would-be predators. There are many examples of bright colours warning of danger: the bad taste of cardiac glycosides in adult monarch butterflies (*Danaus plexippus* (1031)) or the black and yellow or red markings of bees (accompanied by distinctive buzzing). Other species, however, take advantage of animals that are dangerous to prey by mimicking them. In **Batesian mimicry**, the dangerous animals are **models** to the **mimics**. A mimic is a species that resemble the model in appearance and behaviour but lack the model's defenses. One well-known example of Batesian mimicry involves monarch butterflies as models (protected by bad-tasting cardiac glycosides) and viceroy butterflies as its mimics (no cardiac glycosides, good tasting). The situation is complicated because not all monarchs have the same levels of cardiac glycosides and not all birds perceive the same taste.

Cardiac glycosides protect monarch butterflies because they are emetic, causing vomiting in birds such as Blue Jays (*Cyanocitta cristata* (1596)). Blue Jays typically reject monarchs as food once they have experienced the emetic reaction. Emetic responses of captive Blue Jays have been used to assess the levels of cardiac glycosides in monarch butterflies. During the fall migration, the cardiac glycoside content is widely variable among monarchs. Females tend to have higher concentrations than males, and butterflies from more southern locations have higher concentrations than those from the north. Blue Jays raised in captivity have not been exposed to "real" monarchs. Now, it is possible to rear "unprotected monarchs," varieties whose larvae eat and thrive on plants other then milkweed, plants that lack cardiac glycosides. This sets the stage for an interesting experiment. If an experienced Blue Jay watched a naive one eating monarchs and showing no ill effects, its prior experience was overridden. The mischievous experimenter ensured that the experienced bird got a "real" monarch and completely confused it!

Mullerian mimicry describes situations where several dangerous (model) species resemble one another. Stinging bees and wasps are good examples; they are usually black and yellow in colour and buzz as they fly. Predators such as birds learn to avoid these models. In summer, when young birds are fledging (leaving the nest and catching their own food), the only black and yellow insects they encounter are the stinging models. This ensures that young birds learn to leave bees and wasps alone. Flies that mimic bees and wasps are common in the spring and fall but not when young and impressionable fledglings are learning what to eat and what to avoid. With their black and yellow stripes and sometimes buzzing, these mimic flies appear to be protected by the birds' prior experience with the models.

Although we tend to think of mimicry in a defensive context, we have seen how some sexually mature male bluegill sunfish (*Lepomis macrochirus* (1271)) mimic females or immatures, enter nests and release sperm that may fertilize eggs (page 410). Mimics may also benefit from feeding opportunities not normally available to them. Two species of blennies provide an interesting example. Bicoloured fangblennies (*Plagiotremus laudandus* (1282)) are mimics, superficially resembling the model forktail blenny (*Meiacanthus atrodorsalis* (1273), Fig. 17.19). Bicoloured fangblennies and forktail blennies are nearly identical in colour: blue anteriorly, yellow posteriorly. Forktail blennies deliver a toxic bite with enlarged teeth connected to venom

Figure 17.19 A. Photographs of the model, the forktail blenny (*Meiacanthus atrodorsalis* (1273)) and the mimic, **B.** a bicoloured fangblenny (*Plagiotremus laudandus* (1282)).

SOURCE: Courtesy of Karen Cheney

glands. These fish eat small crustaceans and other invertebrates. Bicoloured fangblennies are not armed with fangs or venom and they feed by taking bites out of the fins and scales of other fish. Bicoloured fangblennies that associate with forktail blennies have more feeding opportunities than those operating in the absence of the models. In this complex, the mimic gains protection from predators and increased access to food. But although models and mimics are similar in colour, predators with three distinct spectral sensitivities (a trichromatic visual system) might be able to distinguish the mimics from the models.

Autotomy, the voluntary shedding of a body part, is generally considered to be a defensive (anti-predator) behaviour. One of the best known examples are lizards that shed their tails when attacked. Tail autotomy usually occurs along a breakage plane, and after shedding it, the lizard typically moves away from the shed tail that may continue to convulse and move. Tail autotomy is widespread among lizards and in some cases, for example some geckoes, the unshed tail is flaunted and may be the focus of attack (see tail-flagging, Fig. 17.14). Caudal autotomy remains a topic of ongoing investigation.

The description of **attack autotomy** was based on observations of two species of crabs in Panama. Initially, a tame Central American otter (*Lutra annectens* (1719)) attacked a freshwater crab (*Potamocarcinus richmondi* (835)). The crab grabbed and pinched the otter with a cheliped (pincer) that the crab then shed. The otter, with the crab claw still attached, terminated the attack and left the water. After a second attack, the otter again emerged with a crab claw attached to its right forefoot. In both cases the crabs escaped, but after the second encounter, the otter would not continue the experiment, it avoided crabs. Using a teddy bear as a substitute predator, two species of crabs (*P. richmondi* and *Gecarcinus quadratus* (816)) used a cheliped to counterattack, detached the cheliped and retreated. During bear attacks, 10 crabs

autotomized both chelipeds, 20 crabs one cheliped. The defensive value of the counterattack obviously outweighed the value of the cheliped. In species where chelipeds are used to attract mates, females do not appear to distinguish between males with original versus replacement claws. Autotomy also has been reported in other animals including earthworms and molluscs.

STUDY BREAK

1. What is autotomy?
2. How does camouflage differ from aposematism?
3. How do some insects detect hunting bats?

17.5 Migration

The traditional definition of animal migration involved long-distance, round-trip movements between areas used in winter and in summer. More recently, migration has been redefined as directed movements of an animal away from its home range to exploit food or other resources not locally available (in the home range). The distances covered are not germane to the definition, nor is movement to and from the area (bidirectional, unidirectional) required to constitute migration. The "new" definition recognizes that animals may move to remain in suitable habitat throughout the year. A principal challenge associated with migration is navigation and the cues that guide animals in this behaviour.

Under the traditional definition, sea turtles were the only reptiles recognized as migrants, but now the many species of reptiles that move between summer and winter habitats (e.g., snakes moving to and from hibernacula) are included as migrating species. Amphibians (frogs and salamanders), that move from terrestrial habitats where they spend the summer to water to breed and perhaps overwinter, are also recognized as migrants.

The movements of most monarch butterflies (*Danaus plexippus* (1031)) from their summer range in the United States and Canada to overwintering sites in Mexico is one of the best-studied insect migrations. From the original discovery of the wintering grounds in the 1970s, we can now identify some details of the migration. The citizen-science program Journey North engaged many people who looked for overnight roosts used by migrating monarchs during fall migration. The butterflies use two distinct flyways, the "central flyway" that begins in southern Ontario and crosses Kansas, Missouri, Oklahoma, and Arkansas, then passes through Texas to northern Mexico. The second and smaller flyway goes through eastern and coastal states, and the butterflies using it appear to be less likely to make it to Mexico. The distribution of overnight roosts (Fig. 17.20) illustrates the extent of the citizen science engagement in this project. Information about crucial resources for migrating butterflies can be used in conservation plans for these animals. The antennae of migrating monarchs are central to a time-compensated sun compass used by the butterflies to navigate during their migration. Unlike nonmigratory monarchs, migrating ones have a magnetic compass that also assists them in navigation.

The distances travelled (relative to the size of the animal) can affect the cues used for orientation and navigation. Short-distance movements may be aided by local navigational cues or landmarks, while longer distances involve magnetic and/or celestial cues (sun, stars). When the specific destination is important, for example an island in the ocean or a cave, the animal must be precise in its navigation. A 1° error would mean that at the end of its flight, the bat flying 100 km from summer to winter quarters would be 2 km away from its intended destination. At a flight speed of $5 \text{ m} \cdot \text{s}^{-1}$, flying the extra 2 km would take less than 7 min. If the migration were 1000 km, the bat would be 200 km off course, taking 11 hours of flying to correct. But a Marbled Godwit (*Limosa fedoa* (1554)) flying 11 000 km from Alaska to New Zealand would be 2200 km off course if its initial error was 1°. At $5 \text{ m} \cdot \text{s}^{-1}$ this would mean over 122 hours of extra flying.

Birds provide some of the most spectacular examples of animal migration. Based on band recoveries and at-sea surveys, Arctic Terns (*Sterna paradisaea* (1560), Fig. 17.21D) were thought to make extensive annual migrations from boreal and high Arctic breeding grounds to the southern oceans, an annual journey of >80 000 km. These ~125 g birds nest on the ground and eat mainly small fish and large zooplankton. Researchers tagged 11 Arctic terns with 1.4 g geolocators to document the details of their migrations more precisely (Fig. 17.21C). Going south, the birds used at least two migratory routes along the coast of

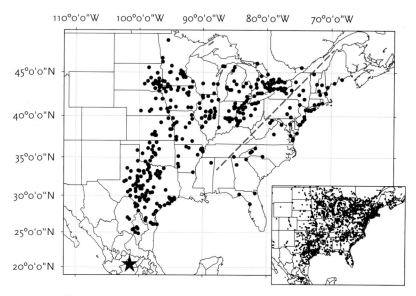

Figure 17.20 Geographic distribution of overnight roosts used by monarch butterflies between 2005 and 2009. The data were collected by the Journey North project. The dashed line identifies the divide between central and eastern flyways. Overwintering sites in Mexico identified by star.

SOURCE: With kind permission from Springer Science+Business Media: *Journal of Insect Conservation*, "The fall migration flyways of monarch butterflies in eastern North American revealed by citizen scientists," Volume 13, 2009, pp. 279–286, Howard, E., and A. K. Davis.

Africa or South America (Fig. 17.21) and used marine areas of high productivity for stopovers and wintering. Arctic Terns appeared to use prevailing global winds to reduce flight costs.

Geolocators have also revealed previously unavailable details of songbird migrations in eastern North America (Fig. 17.22). While Purple Martins (*Progne subis* (1612)) use similar route on their southwards and northwards flights, Wood Thrushes (*Hylocichia mustelina* (1603)) use quite different routes on their two annual migratory flights. As in the case of monarch butterflies, knowledge of the details of flight paths helps humans prepare conservation plans to protect critical resources along the migratory paths of animals.

Migratory bird species have smaller brains than sedentary species. Two hypotheses have been proposed to explain this. The behavioural flexibility/migratory precursor hypothesis suggests that sedentary bird species have larger brains to give greater flexibility in foraging and diet compared to migratory species. The energy trade-off hypothesis suggests that large brains are heavy and energetically expensive and therefore inappropriate for migratory species.

We know much less about the migration habits of bats than birds, but the available data from bats support the energy trade-off hypothesis. Specifically, the brains and neocortices of migratory bats are relatively smaller than those of sedentary bat species. This could reflect the differences in foraging flexibility between bats and birds, and suggest that the differences between migratory and sedentary bats are more subtle than those in birds.

Figure 17.21
Geolocators provided details of the movements of 11 Arctic Terns that were tagged at breeding colonies in Greenland (10) and Iceland (1). Two migration routes emerged: **A.** along the coast of Africa and **B.** along the coast of South America. **C.** The type of geolocator used. **D.** An Arctic Tern. Green lines = autumnal migration, yellow = spring migration, and red = winter range.

SOURCE: A. and B. Engevang, C., I.J. Stenhouse, R.A. Phillips, A. Petersen, J.W. Fox and J.R.D. Silk. 2010. "Tracking of Arctic terns Sterna paradisaea reveals longest animal migration," *PNAS*, 107: 2078–2081.

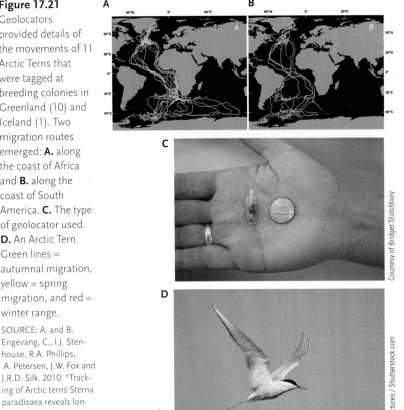

Courtesy of Bridget Stutchbury

nice_pictures / Shutterstock.com

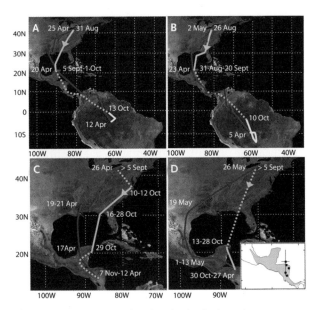

Figure 17.22 Geolocated tracks of individual Purple Martins (**A**, **B**) and Wood Thrushes (**C**, **D**) from their breeding grounds in northern Pennsylvania to their overwintering grounds. Blue lines = fall migration, red = spring migration, yellow = movements during winter. Winter territories of Wood Thrushes shown on inset map.

SOURCE: From Bridget J. M. Stutchbury, Scott A. Tarof, Tyler Done, Elizabeth Gow, Patrick M. Kramer, John Tautin, James W. Fox, Vsevolod Afanasyev, "Tracking Long-Distance Songbird Migration by Using Geolocators," *Science* 13 February 2009: Vol. 323 no. 5916 p. 896. Reprinted with permission from AAAS.

How much does migratory flight cost? Catching migrating birds one day, injecting them with doubly labelled water ($^2H^{18}O$) and then recapturing them the next morning allowed measurements of metabolic rates and the costs of migration in 12 Swainson's and Hermit Thrushes (*Catharus ustulatus* (1592), *C. guttatus* (1591), respectively). During migratory flight, the birds' flight cost was about 80 kJ above resting metabolism. The data generally support theoretical models for costs of migration, and demonstrated that cool weather increases the costs of sustained natural flight. Some migrating animals make stopovers during migration, pausing to rest and refuel. (We know most about this in birds.) However, not migrating on a cold night would involve thermoregulatory costs equivalent to those incurred in 2.5 h of flying.

While some migrating birds use stopovers during migration as an opportunity to refuel, some migrating bats do not (Fig. 17.23). Presumably the ability to enter daily torpor allows bats to save energy and means that they need not refuel during stopover.

Why do birds make long distance migrations? One reason would be gaining access to seasonally available food resources in areas that are only accessible at some times of the year. The behaviour of Arctic Terns demonstrates that it moves between rich patches of prey. One obvious way to increase fitness is to increase reproductive output. There is evidence of reduced predation on ground-nesting birds along a gradient from 50° to 85° N, which may help to explain the migration of many birds to the Northern Hemisphere during the breeding season.

STUDY BREAK

1. What are geolocators? What have they taught us about bird migration?
2. What cues do migrating animals use in navigation?
3. Compare three animals that migrate.

17.6 Sex and Reproduction

Earlier (page 408), we saw how monogamy, polygamy, and promiscuity (different mating systems) reflect the different environmental and ecological regimes under which animals operate. Male or female, the goal in reproduction is ensuring the representation of your genes in the next generation. In picking a mate, an animal may use some combination of behaviour, physiology, anatomy, and environmental setting, and there are examples of all of these options in animal reproduction. Animal species are often categorized by their approach to reproduction. Traditionally, birds have been considered monogamous because both males and females are often involved in caring for the young, but there is more and more evidence (below) that they are not as monogamous as they seem. Mammals have been considered polygynous because females invest more (gestation and lactation) in reproduction than males. Other species are obviously promiscuous in situations where neither males nor females can protect their investment in mate choice. Individuals of some species use more than one reproductive strategy.

Sexual ornaments are pervasive among animals, and may reflect a variety of selective factors such as mutual mate choice and same sex competition, often underlain by genetic variations. Usually it is difficult to determine if a female (for example) chooses a male because of his behaviour and sexual displays that are based on characteristics that reveal more about his overall condition. A 24-year-long study of over 8500 Collared Flycatchers (*Ficedula albicollis* (1600)) allowed a detailed examination of the features of males that attracted females. When all of the data were considered, including a variety of genetic correlations, female choice appeared to have a negligible effect on male features. The combined impact of environmental influences reduced the impact of sexual selection (see also 406). The results suggest that genes responsible for the behaviour associated with selection of an ornament evolved by their own pathways. The business of choosing a mate is complicated.

The African butterfly *Bicyclus anynana* (1027) shows variation in wing colouration triggered by temperature conditions prevailing during critical periods

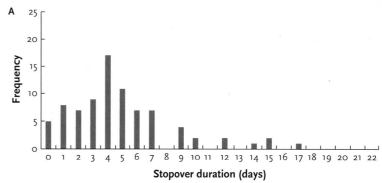

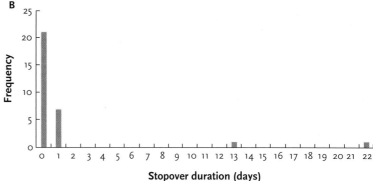

Figure 17.23 A. Migrating birds such as Black-throated Blue Warblers (*Dendroica caerulescens* (1597)) use stopovers to replenish their energy resources. **B.** Migrating bats such as silver-haired bats (*Lasionycteris noctivagans* (1794)) use torpor as a means of conserving energy and most apparently do not refuel during stopovers. These data are based on observation of 83 warblers and 30 bats.

SOURCE: Warbler data from Taylor PD, Mackenzie SA, Thurber BG, Calvert AM, Mills AM, et al. (2011) "Landscape Movements of Migratory Birds and Bats Reveal an Expanded Scale of Stopover," *PLoS ONE* 6(11): e27054, bat data from Liam P. McGuire, Christopher G. Guglielmo, Stuart A. Mackenzie and Philip D. Taylor, "Migratory stopover in the long-distance migrant silverhaired bat, Lasionycteris noctivagans," *Journal of Animal Ecology* 2012, 81, 377–385. Used with permission from Liam P. McGuire.

BOX 17.2
Migrations of Planktonic Crustaceans

The migration distances of turtles, birds, and monarch butterflies measure tens of millions of the animals' body lengths a year. The migration route of about 1.3 million wildebeest—1600 km around the Serengeti each year—involves movement of about 230 000 tonnes of ungulate. As impressive as they are, these numbers are barely significant compared to migration of zooplankton in the ocean's waters. There, a 2-mm copepod swimming 250 m up and down each day amounts to hundreds of millions of body lengths a year. If we take a round figure of 10 mg of dry weight of zooplankton per m³, and estimate that zooplankton might be concentrated in a 10-m-deep band, this amounts to over 67 billion tonnes

(dry weight) of planktonic animals migrating each year in the 335 million km² of the world's oceans!

There are many examples of diel migrations of zooplankton (Fig. 1) and at least six theories about why they expend the amount of energy required for the travel involved:

- Light avoidance, because strong light negatively affects zooplankton
- By leaving the surface, zooplankton allow phytoplankton to photosynthesize and rebuild a population that can be grazed at night
- Predation avoidance, by leaving the surface, zooplankton protect themselves from attack by visual predators
- Energy conservation, by diving into deeper, cooler waters during the day,

zooplankton reduce their metabolic rate and energy consumption
- Surface mixing, zooplankton can return to a patch of water that has been moved by the wind and contains refreshed phytoplankton resources
- Combinations of two or more of these possibilities.

While it is clear that light intensity is the immediate cue triggering vertical movement, determining its advantage or adaptive value is more challenging. Fresh water evidence contradicts the metabolic advantage theory. *Daphnia hyalina* (720) and *Daphnia galeata* are closely related species. *D. hyalina* performs diel migrations but *D. galeata* remains in the surface layer. If vertical migration had a metabolic advantage

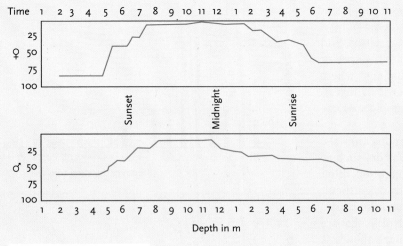

Time 1 2 3 4 5 6 7 8 9 10 11 12 1 2 3 4 5 6 7 8 9 10 11

♀ 25 50 75 100

Sunset   Midnight   Sunrise

♂ 25 50 75 100

1 2 3 4 5 6 7 8 9 10 11 12 1 2 3 4 5 6 7 8 9 10 11

Depth in m

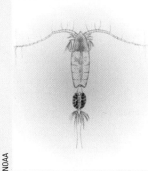

NOAA

Figure 1
The diurnal vertical migration of male and female Calanus finmarchicus *(728) and a female of the species bearing eggs (black areas).*

SOURCE: Based on McConnaughey and Zottoli, *Introduction to Marine Biology* 4th edition, 1983, Mosby.

© Mark Conlin / Alamy

Figure 2
A migrating line of spiny lobsters, Palinurus argus *(830), moving against the current.*

then *D. hyalina* should benefit, but the opposite is true. The migrating *D. hyalina* produce fewer eggs which develop more slowly than those of *D. galeata*.

Some evidence supports the predation avoidance hypotheses. Manipulating populations of copepods (*Acartia hudsonica* (725)) in the presence and absence of a predator (a stickleback, *Gasterosteus aculeatus* (1217)) revealed a clear diurnal migration in the presence of predator. No migration occurred when fish were excluded. When fish were allowed to enter the enclosures and predation was resumed, the non-migratory copepods became migratory within 12–24 hours.

This result parallels and extends data from a natural population of the copepod *Calanus newmani* (727). In the presence of visual predators (planktivorous fish), *Calanus* performed a normal migration, surfacing at night and diving during the day. At the times of the year when planktonic

predators fed on *Calanus*, the copepods reversed their migration, surfacing in the day and diving at night. In the absence of predators, *Calanus* did not migrate.

At some times of the year, "conga lines" of lobsters (Fig. 2) are a common sight in the Caribbean. Divers report that interrupting the movement of a conga line causes the lobsters to form a circle, tails in and claws outwards to defend themselves against the "attacker." When they move off again, a different lobster leads the line. This pattern of movement cannot be explained as either nomadic (random) wandering, or "homing," excursions followed by a return to the start point. Mass migration of the spiny lobster *Palinurus gilchristi* (831) occurs in South Africa and *Palinurus ornatus* (832) in New Guinea. In these populations, young lobsters move against prevailing currents and toward areas of breeding adults. This upstream movement reverses the

travel of planktonic phyllosoma larvae (the final larval stage in lobster development) that would have followed the reverse path, drifting downstream with the current.

There is debate about migration by American lobsters (*Homarus americanus* (819), Fig. 3). The lobster population in the Northumberland Strait divides into resident and migratory groups. The migratory group moved tens of kilometers, fitting a model involving rapid offshore movement in the fall, limited movement in the colder months of the winter, and then onshore movement in the spring. This movement is correlated with water temperature. The findings for the Northumberland Strait are confirmed in a lobster population in the Magdalen Islands. These data suggest that at least some populations of American lobsters migrate.

Strong/Buzeta

Figure 3
Many American lobster, Homarus americanus *(819), have a seasonal migration pattern that takes them out of shallow (colder) water in the fall to overwinter in deeper locations where the possibility of freezing temperatures is lower.*

of larval development. Butterflies that emerge during the wet season have large eyespots and a conspicuous band on the ventral surfaces of the wings, while those emerging during the dry season are cryptic and have reduced eyespots (Fig. 17.24). Dorsal eyespots on the forewings (Fig. 17.24D) are the male sex ornament in the wet season and are consistent among males. In the wet season, female *B. anynana* prefer males with dorsal eyespots with an ultraviolet centre. The dorsal forewing eyespots are the sex ornaments in the dry season as well, but now males choose females and, like wet-season females, base their choices on the eyespots. Like other insects, the act of mating involves the transfer of a spermatophore from male to female. Spermatophores often contain food resources for the female. Spermatophores of dry-season males appear to contain more resources for females than those of wet-season males because mating with dry-season males increases female longevity over mating with wet-season males.

Animals go to considerable lengths to maximize their chances of finding a mate. By controlling good resources such as access to food and water, some individuals control access to mates. During the rutting season, male topi antelopes (*Damaliscus lunatus* (1751)) establish territories into which they attract females. The more time a female spends in a male's territory, the more likely he is to mate with her. When danger approaches, such as a predator (lion or cheetah or human), male and females give alarm snorts, prick their ears, and stare at the predator, behaviour that reduces the chances of them being ambushed. In some situations, a male topi gives alarm snorts, accompanied by pricked ears and staring, to thwart a female's attempt to leave his territory (by making her think that she is moving toward a predator). This tactical use of an alarm signal appears to elicit the same behaviour as real alarm signals but increase the male's opportunities to mate. Future work may reveal that "crying wolf" may only work if it is not overused.

The data from topis suggest that males that mate more frequently sire more offspring than males that are less successful (at mating), but that is not always the case. Across two generations under natural field conditions, male field crickets (*Gryllus campestris* (898)) show more variation in reproductive success than females. Fundamentally, male and female crickets that mate with more partners have more offspring. Combined genetic and video studies of field crickets showed that the number of offspring produced by a male reflected greater success in postcopulatory sexual selection (the success of his sperm at fertilizing eggs) or differential survival of offspring. Lifetime reproductive success was measured across two field seasons by genetic analysis of offspring. The field crickets suggest that there is more to reproductive output than copulation.

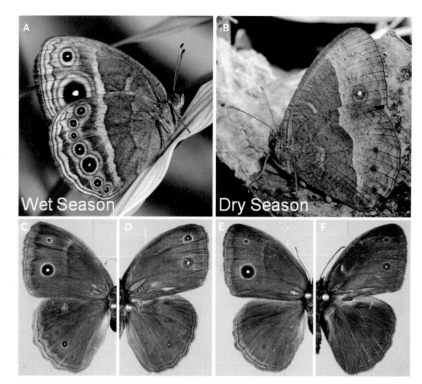

Figure 17.24 A and **B.** Ventral views of the wings of the African butterfly *Bicyclus anynana* from the wet (A) and dry (B) seasons. Note the eyespots and the vertical stripe in A compared with B. Dorsal views of the wings of male (D, F) and female (C, E) from wet (C, D) and dry (E, F) seasons. Note central spot (circled) with ultraviolet centre. SOURCE: Reprinted with permission from Prudic et al. 2011. Photo credit William Piel.

If reproduction is expensive, and what counts is genes in the next generation, then we should expect precision in mate selection and adaptive features that minimize wasting gametes. Arguably, internal fertilization should increase precision and decrease the waste of gametes. But the behaviour of males sometimes suggests that any mate will do. Evidence of this comes from male insects that will attempt to mate with anything that is the correct size and shape. The anecdote about a male ground squirrel (*Spermophilus richardsonii* (1906)) trying to mate with a female that had been shot and fallen forward in the presenting position suggests that mammals can also be indiscriminent. This and the behaviour of some male dogs suggest that wasting sperm might be common.

In juvenile female fossas (*Cryptoprocta ferox* (1711)) (fossas are the largest carnivores in Madagascar) the external genitalia are masculinized (Fig. 17.25). This is not linked to androgen levels in the females, but is transient and may protect juvenile female fossas from harassment by males or from aggression by females. In contrast, masculinization of female genitalia in spotted hyaenas reflects high levels of testosterone in pregnant females. Among spotted hyaenas, males compete for larger, dominant females and the behaviour of females is influenced by their levels of testosterone. Links between genitalia, behaviour, and reproduction make this topic endlessly intriguing.

Figure 17.25 The external genitalia of **A.** an adult male, **B.** an adult female, and **C.** a juvenile female fossa (*Cryptoprocta ferox* (1711)). The sheath of the male's penis has been pulled back to reveal the spines on the erect shaft. The labiae of the females have also been pulled back revealing the absence of a substantial clitoris in the adult, but a substantial spinescent clitoris in the juvenile female. The masculinized appearance of the genitalia of the subadult female may protect her from harassment from males.

SOURCE: Photos courtesy of Clare Hawkins

Mammals tend to be polygynous because internal gestation and lactation mean that females invest more heavily in young than males do. Male mammals are often larger than females because a male's size affects his competitive ability for access to females. In African lions, males are important to females as sources of sperm, and larger males are usually more effective at fending off attacks by other males attempting to usurp the females and gain a genetic advantage (see page 407). Comparing mammal species showed that larger males often experience higher rates of mortality than smaller males. There is a robust statistical association between male-biased parasitism and the degree of sexual dimorphism in size. Furthermore, male-biased mortality is linked to male-biased parasitism, For example, older male Soay sheep (*Ovis aries* (1763)) on St. Kilda Island in Scotland are more often infested with parasites that kill them than other Soay sheep.

House Wrens (*Troglodytes aedon* (1621)) are small songbirds originally thought to be monogamous. Detailed study of a population of House Wrens in Wisconsin revealed a more complicated situation. Male House Wrens could be monogamous and have a single brood, or sequentially monogamous (two broods, one after the other). Still other males were polygynous. Microsatellite markers provided data about the reproductive success of individual males, specifically the total number of young sired in within-pair and extra-pair fertilizations. Polygynous males were more likely to be cuckolded than monogamous males, but half of them sired a third brood. Despite paternities gained by extra-pair fertilizations by single-brooded males, males were more reproductively successful when they produced multiple broods in a season. These males used a combination of sequential monogamy and/or simultaneous polygyny. Longer term studies indicated that males that succeeded in producing more than one brood tended to arrive earlier in spring, providing them a better opportunity to produce more than one brood than later arrivals.

Females that mate with more than one male are likely to have the sperm of different males in their reproductive tract during a mating season. We have seen evidence (Fig. 14.36) that in some mice, sperm from genetically related males clump to increase mobility. In other cases there are mechanisms for sorting sperm, and these can be male-based (sperm competition) or female-based.

The size of sperm varies considerably among animals. Within any one group such as fruit flies (*Drosophila* (960)), there may be no obvious pattern in sperm size or other features. When more than one male mates with a female, there is the potential for competition among sperm. Possible factors determining which sperm fertilizes the egg(s) include sperm size, motility, and time since ejaculation. In house mice, *Mus domesticus* (1890), the length of the sperm midpiece is the only feature that accurately predicts the swimming velocity of sperm (Fig. 17.26).

Earlier we saw the colouring of an African butterfly changed depending upon when it emerged from the pupa. This mean different sexual ornaments between wet and dry seasons and how these changes appeared to reflect the size of spermatophores. Among 21 species of bushcrickets (Tettigoniidae, Orthoptera), the mass of the testes increased with the degree of polyandry (number of males mating with a female). But with increasing mass of ejaculate, the mass of testes decreased across these species. There was no significant relationship between the mass of the testes and either the number of sperm produced per ejaculate or the mass of the nuptial gift associated with spermatophores. The morphology of structures and details of behaviour associated directly (testes mass) and indirectly (sexual ornaments) with reproduction continues to be a fertile topic of research.

The presence of bystanders also can influence mating behaviour in mammals. Male rats (*Rattus norvegicus* (1903)) ejaculated significantly more sperm when copulating in the presence of another (rival) male

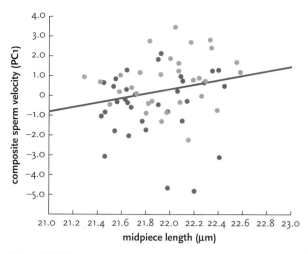

Figure 17.26 Relationship between composite sperm velocity and length of sperm midpiece in house mice. Blue circles = data from males from monogamous selection lines, green circles = males from polygamous selection lines.

SOURCE: Firman, R.C. and L.W. Simmons, "Sperm midpiece length predicts sperm swimming velocity in house mice," *Biology Letters* 2010, Vol. 5:513–515, by permission of the Royal Society.

compared to when copulating with no rival present. For any one of the 12 males tested, the magnitudes of adjustments to ejaculates were predictable and consistent.

A vital step in the process of reproduction is care provided by parents (see pages 132 and 408). The discovery that a deep-sea squid (*Gonatus onyx* (497), an abundant cephalopod of the Pacific and Atlantic Oceans) cradles and aerates its fertilized eggs was surprising. These squid, like others, were thought to deposit their eggs on the sea floor, and degeneration of musculature after reproduction was thought to limit locomotion and preclude effective parental care. Footage of squid were obtained from ROV Tiburon operated in Monterey Canyon off of the coast of California. At depths of 1500–2500 m, five *G. onyx* were observed, each holding an egg mass in its arms. At intervals of 30–40 s, the squid repeatedly extended their arms, apparently aerating the egg mass, but the adults also made escape manoeuvres and aggressive movements of their arms. Individuals with eggs were more effective at escaping than those carrying advanced embryos, suggesting degeneration of muscles. Squid with eggs and embryos are in the diving zones regularly covered by mesopelagic mammals such as whales and elephant seals, likely making them vulnerable to these predators.

Effective biparental care (both parents involved) often requires good co-ordination between the parents. In an African cichlid fish, *Pelvicachromis taeniatus* (1278) (Fig. 17.27), both males and females care for eggs and young. Both males and females prefer to mate with nonkin, but effective biparental care requires synchronous behaviour between parents. The costs of parental care generate conflict between parents, and related parents are more co-operative in delivering care

Figure 17.27
Pelvicachromis taeniatus (1278).

SOURCE: © Hippocampus Bildarchiv

to eggs and young than unrelated parents. To date there is no indication that this inbreeding is disadvantageous, so these fish seem to provide an example of a situation where inbreeding is advantageous because it increases production of young.

The examples we have considered so far are of animals that move about to find mates. The situation can be different in immobile species, and some barnacles provide interesting details about size of genitalia and mating opportunities. At five intertidal sites around Long Island (New York), the penis length (Fig. 17.28) in sedentary acorn barnacles (*Semibalanus balanoides* (751)) varies with the density of adults. Exposure to waves strongly influences barnacle density. The penises of individuals living at low density are larger in diameter and have greater ability to stretch compared to those of individuals living at higher densities. In these sessile, hermaphroditic animals, a longer penis gives an individual more access to potential mates with whom sperm can be exchanged.

STUDY BREAK

1. What cues do animals use in assessing potential mates?
2. What are sexual ornaments? How do they relate to sexual selection?
3. Do males and females use "honest" signals when advertising for mates?

17.7 Social Behaviour

There is considerable diversity in the social situations that occur in animals. Some species live in tightly-knit social groups, while others have a physiological continuum between individuals (e.g., corals). Many other species are aggressive and solitary. Furthermore, some species may be social for part of the year and solitary the rest. Prides of African lions (page 407) typically consist of a single male, a group of females, and their dependent young. Control of resources and the drive to reproduce often underlie this variety of social situations.

Defense is one reason to be part of a group. Being one of many individuals that generally look the same may make it difficult for a predator to single you out as

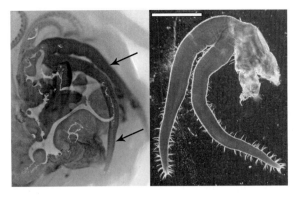

Figure 17.28 A barnacle, *Semibalanus balanoides* (751), with two penises.

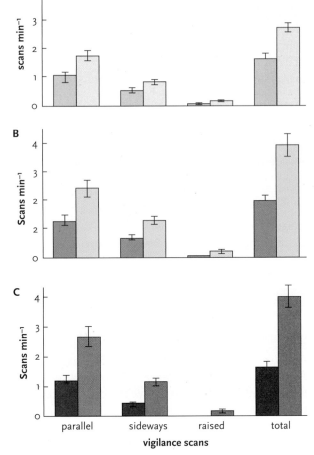

Figure 17.29 The frequency at which dwarf mongooses scanned for danger. **A.** Mongoose scanning behaviour in the presence (dark blue bars) or absence (light blue bars) of Fork-Tailed Drongos. **B.** Mongoose scanning behaviour when researchers played Drongo non-alarm calls (dark green) or mongoose contact calls (light green). **C.** Mongoose scanning behaviour during play-back of Drongo (dark red) and White-Billed Sunbird (light red) non-alarm calls.

SOURCE: Sharpe, L.L., A.S. Joustra and M.I. Cherry, "The presence of an avian co-forager reduces vigilance in a cooperative mammal," *Biology Letters* 2010, Vol. 6:475-477, by permission of the Royal Society.

a target to attack. Each member also benefits from the ability of others to detect approaching danger and alert others to the threat. We have seen how a male topi use alarm calls to gain access to more females. Other animals, such as Diana monkeys (*Cercopithecus diana* (1856)), respond differently to different predators. This is a relatively common phenomenon among birds and mammals because the same defensive behaviour may not work against all predators. Diana monkeys produce the "eagle alarm call" in response to Crowned Eagles (*Stephanoaetus coronatus* (1583)) and the "leopard alarm call" for leopards (*Panthera pardus* (1727)). In areas where they are sympatric, Yellow-Casqued Hornbills (*Ceratogymna elata* (1549)), which are vulnerable to Crowned Eagles but not to leopards, respond to the Diana monkeys' eagle alarm calls but not to the leopard alarm calls.

Groups of animals may form under a variety of circumstances, e.g., at localized nest sites or food resources or when parents and young remain together. In some cases group members include more than one species. Mixed-species foraging parties are common, and co-foragers may benefit from increased foraging efficiency or more effective defensive behaviour. Dwarf mongooses (*Helogale parvula* (1715)) are small (200–300 g), co-operatively breeding (share parental care beyond parents) mammalian carnivores that post sentinels who watch for predators and use alarm calls to signal their presence. Dwarf mongooses appear to prefer foraging in mixed-species groups, often with flocks of several species of birds. In mixed-species groups, dwarf mongooses respond to the alarm calls of their companion birds. When researchers played back tapes of non-alarm vocalizations of Fork-Tailed Drongos (*Dicrurus adsimilis* (1598)) to simulate the presence of these birds, the dwarf mongooses showed significantly less vigilant behaviour (Fig. 17.29).

What constitutes a social unit? Is a flock of birds a social unit? This question can be explored using small (up to 10 individuals) flocks of homing pigeons in which each bird is equipped with a high-resolution, lightweight GPS device. This approach allowed biologists to examine the relationships between individuals, data that were not previously available. The data reveal a well-defined hierarchy among flock members as indicated by each bird's position and directional choices relative to others. However, what applies to groups of 10 may not to groups of 100, 1000 or 1 000 000. Do the people leaving a high-rise apartment or office building represent a social unit?

Mixed-species associations may also involve sophisticated social behaviour such as **altruism** (behaviour that benefits a non-relative). Highly social animals may help unrelated individuals that never return the favour (reciprocate). The altruistic behaviour associated with helping might be explained if the helper gains in social image among bystanders (image-scoring) who witness the help. This presumes that the observers might, in turn, help the altruist at some time in the future. Such complex indirect reciprocity can appear only after two steps. First, bystanders must gain personal benefits from image-scoring. Second, altruistic behaviour in the presence of image-scoring bystanders is not altruistic if the "altruists" benefit from access to the bystanders. We usually expect such complicated behavioural interactions among members of our own species, so it is astonishing to find them in fish.

Cleaner fish (*Labroides dimidiatus* (1270)) establish a cleaning mutualism with a variety of other species of fish (the clients). The cleaners co-operate by removing ectoparasites from the clients but may cheat by taking mucus. In an experimental setup where one fish could watch others, bystanding clients spend more time with cleaners that co-operate (do not remove mucus) compared to unknown cleaners. Cleaner fish learned to be more co-operative feeders in an image-scoring versus a non-image-scoring situation.

When the social unit (group of individuals) is only one species, it is possible to define the units by interactions among individuals. In humans, individuals communicate frequently, allowing use of interactions within social communication networks (calls on mobile phones). Analysis of interactions among members of two groups revealed that larger groups persist longer when they can add or remove members from the group, suggesting that labile group composition can be an asset. In contrast, small groups are more stable if their composition is stable over time. The time commitment of members is a useful indicator of estimating the life span of the community. The same principles may well apply to interactions among members of groups of other species.

Observation of individual resident killer whales (*Orcinus orca* (1762), see page 5) in the waters off of British Columbia (Canada), revealed that 81 individuals formed 740 preferred companionships. The data were obtained by direct observation of orcas, and individual recognition by a photo identification catalogue. The were clear interactions within the 81 individuals best explained by matriline (identities of mothers) rather than by age or gender. Removal of members of the unit affected the social unit, with targeted removals having a greater effect than random ones (Fig. 17.30). Subadult females appeared to play a vital role in the cohesiveness of orca social units.

Studies of many species suggest that animals pay attention to the behaviour of others; we have seen several examples of such behaviour in this chapter. Societies of animals typically involve social hierarchies that reflect the status of each individual. In absolute dominance hierarchies, individual "a" always dominates individual "b" who, in turn always dominates individual "c". In relative dominance hierarchies, individual "a" may dominate individual "b" during access to foods, but "b" may dominate "a" in access to

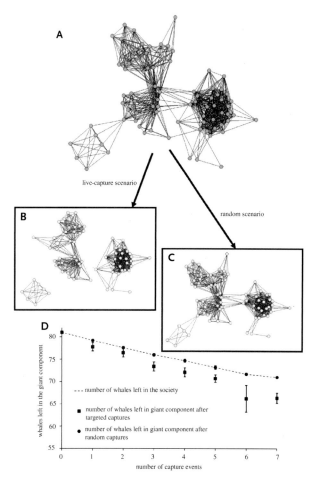

Figure 17.30 Social organization of 81 resident killer whales (*Orcinus orca*) based on individual associations. Targeted removals of 10 individuals were more disruptive to the social network than random ones.

SOURCE: Williams, R. and D. Lusseau, "A killer whale social network is vulnerable to targeted removals," *Biology Letters*, 2006, Vol. 2:497-500, by permission of the Royal Society.

mates. Hierarchies are usually more complicated in societies where social rank is relative rather than absolute. We have seen how the feeding behaviour of one Blue Jay can influence the eating behaviour of another (page 481). The impact of observation extends to other situations as well.

Observation alone is used by a variety of birds and mammals to make inferences about hierarchy. This behaviour, known as **transitive inference**, requires using known relationships to deduce unknown relationships. Experiments with captive African cichlid fish (*Astatotilapia burtoni* (1256)) reveal that males successfully make inferences about a hierarchy by observing pairwise fights between rival males. Bystander fish spend more time in closer proximity to conspecifics they view as weaker. When trained on socially relevant stimuli, and in familiar versus unfamiliar contexts, male *A. burtoni* use transitive inference. Transitive inference appears to be a widespread behaviour among many different animals and is likely that bystanders use many cues when making inferences.

17.7a Eusocial Animals

Eusocial animals live in aggregations that may number in the thousands of individuals, for example, insects such as many species of Hymenoptera (bees and ants). Eusocial species are those in which some individuals reduce, sometimes to zero, their own lifetime reproductive output in order to raise the young of others. This means that adult members of a colony include both reproductive and nonreproductive castes, the latter providing care for the young and the colony. Traditionally, the evolution of eusociality has been explained by kin selection: the nonreproductive castes actually promote their genes in the next generations because of a high degree of relatedness among them and with the reproductive female (the queen).

Modern eusocial animals have a colony or nest that they build and defend. Included on the list are arthropods (aculeate wasps, halictine and xylocopine bees, sponge-nesting shrimp, termophsid termites, colonial aphids and thrips, ambrosia beetles) and one mammal (naked mole rats, *Heterocephalus glaber* (1886)). Most eusocial systems begin with one inseminated queen (as in Hymenoptera), but sometimes two inseminated queens start the colony. Eusocial colonies grow by the addition of offspring that are nonreproductive, and the reproductive history of the queen(s) sets the genetic stage for eusociality.

Nest construction appears to be an important preadaptation for eusociality, and division of labour among colony members is a common feature. When some solitary bees are experimentally forced to live together, they show evidence of division of labour (foraging, guarding, and tunnelling). These observations support the "fixed-threshold model" for the development of eusociality, namely that variation in behaviour among colony members occurs through interactions, and individuals with the lowest threshold to assume a task are the ones that engage in it. Propensity for filling different roles may reflect some combination of morphological and genetic traits.

The development of eusociality appears to have five stages:

1. formation of group
2. occurrence of minimum and necessary traits that result in a cohesive group with a valuable and defensible nest
3. mutations that prescribe persistence of the group
4. emergent traits (shaped by natural selection) associated with the interactions between group members
5. multilevel drivers of selection that change colony life cycle and social structures.

Termite species (order Isoptera), with irreversibly wingless worker and soldier castes, are the most ecologically successful and, from a human perspective, do the most damage to wood. Genotype and environmental conditions together influence the development of these two castes in at least one species (*Reticulitermes speratus* (888)). In this species, an X-linked, one-locus/two-allele model explains the development of the two castes. The data were obtained by raising termites under uniform conditions and using four genetic crosses of soldier- and worker-derived secondary reproductive individuals. The discovery of a genetic influence supports the change towards eusociality postulated above.

In any event, close genetic relationship appears to be central to the evolution of eusociality. Analysis of female mating frequencies in 267 species of eusocial bees, wasps, and ants (Hymenoptera) revealed that a queen's mating with a single male maximizes relatedness of colony members. In eight independent eusocial lineages (Fig. 17.31), the queen's mating is with a single male, and mating with more than one male is always a derived character. Obligate or high levels of polyandry (queen mating with more than one male) are derived and occur in eusocial species whose workers have lost the ability to be primary breeders (Table 17.1). At the same time, honey bee (*Apis mellifera* (984)) queens may mate with many males which generates numerous genetically distinct patrilines within a colony. Swarms of bees from genetically diverse colonies (15 patrilines) found new colonies faster than swarms from uniform colonies (one patriline). Worker bees from genetically diverse colonies show higher rates of foraging and food storage, as well as more rapid population growth which leads to more production of drones and better winter survival. In some cases, therefore, eusocial species may be more successful because of diversity of patrilines, in turn reflecting lower genetic relatedness among colony members. This suggests that the high relatedness of colony members that may have been a driver for founding eusocial systems has been lost in some cases.

STUDY BREAK

1. How does eusocial differ from social? What animals are involved?
2. What pressures lead to social behaviour and the formation of groups?
3. What is monandry?

17.8 Learning and Tool Use

Not very many years ago, many people believed that using tools was behaviour unique to humans, an explicit divider between us and other species. As we shall see, evidence of other animals using tools is overwhelming, and animals that use tools include invertebrates as well as vertebrates. The diversity of tools and tool-users is astonishing.

We humans are familiar with the concept of traditions that can play a pervasive role in our behaviour.

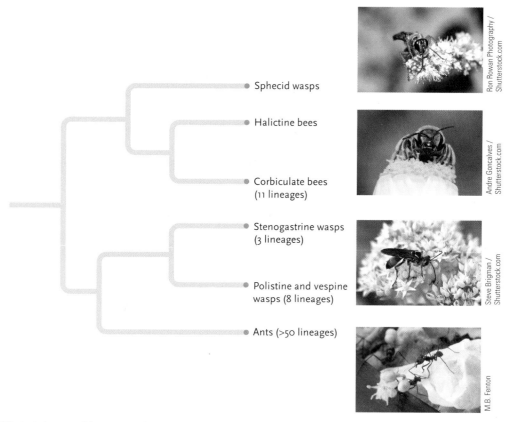

Figure 17.31 A phylogeny of the genera of eusocial Hymenoptera for which data on mating frequencies of females are available.

Table 17.1   **Summary of evidence for monandry in the independent origins of eusociality in the Hymenoptera (ants, bees, and wasps) and other eusocial lineages. _P_ values relate to ancestral state reconstructions by maximum likelihood. The first value indicates the probability of monandry not being the ancestral state, the second value the probability of polyandry being ancestral.**

Taxa	Eusocial Origins	Eusocial Species	Evidence
Sphecid wasps	1	1	This study; $P < 0.0001$; $P < 0.0001$
Halictid bees (*Augochlorella/Augochlora*)	1	Many of 140	This study; $P = 0.0014$; $P < 0.0001$
Halictid bees (*Halictus*)	1	Most of 217	This study; $P = 0.0014$; $P < 0.0001$
Halictid bees (*Lasiglossum*)	1	Most of 544	This study; $P = 0.003$; $P < 0.0001$
Allodapine bees	1		No data
Corbiculate bees	1	~1000	This study; $P = 0.015$; $P < 0.0007$
Stenogastrine wasps	1	~50	This study; $P = 0.0026$; $P < 0.0001$
Polistine and vespid wasps	1	~860	This study; $P = 0.034$; $P < 0.0001$
Ants	1	~12 000	This study: $P = 0.034$; $P = 0.0007$
Ambrosia beetle	1	1	Monoandry thought probable but no data
Aphids	17	~50	Eusocial colonies produced parthenogenetically by single female
Termites	1	~2800	Normally monoandrous, with only a few species exhibiting low polyandry
Thrips	1	7	Normally monoandrous
Snapping shrimps	3	6	Monoandrous
Mole rats	2	2	Facultative low polyandry

SOURCE: From Hughes, W.O.H., B.P. Oldroyd, M. Beekman and F.L.W. Ratnieks, "Ancestral monogamy shows kin selection is key to evolution of eusociality," *Science* 30 May 2008: Vol. 320 no. 5880 pp. 1213–1216. Reprinted with permission from AAAS.

BOX 17.3
Teaching and Learning

We have ample evidence from animal behaviour that many different animals adjust their behaviour based on experience and circumstances (i.e., learning). These changes confer direct advantages on individuals that show the change in behaviour, by providing more access to food, more chance of avoiding predators, or more opportunities to pass genetic material on to the next generation.

Are some of these examples of teaching as well as of learning? Caro and Hauser (1992, p. 153) defined teaching as "An individual actor **A** can be said to teach if it modifies its behaviour only in the presence of a naive observer, **B**, at some cost or at least without obtaining an immediate benefit for itself. **A**'s behaviour thereby encourages or punishes **B**'s behaviour, or provides **B** with experience, or sets

an example for **B**. As a result, **B** acquires knowledge or learns a skill earlier in life or more rapidly or more efficiently than it otherwise might do, or that it would not learn at all."

How many, if any, of these criteria are met by banded mongooses when pups learn from an escort about how to handle hard-shelled prey? If the escort is **A** and the pup **B**, **A** may modify its behaviour in the presence of **B**, and sets an example for **B** which appears to learn an important skill earlier and more rapidly than it would have otherwise.

Several studies have demonstrated that different species of bats learn to modify their foraging behaviour by observing experienced individuals. In captivity, little brown bats (*Myotis lucifugus* (1800)), big brown bats (*Eptesicus fuscus* (1791)),

pallid bats (*Antrozous pallidus* (1783)), and fringe-lipped bats (*Trachops cirrhosus* (1819)) adopt novel foraging behaviour by observing an experienced conspecific. In contrast, when inexperienced bats are placed alone in a chamber with a food source, they do not learn the novel foraging behaviour. By observing experienced conspecifics, fringe-lipped bats also learn to respond to novel acoustic stimuli that represent food. While learning novel foraging behaviour rapidly can be immediately beneficial, it is not clear that the bat or mongoose examples meet all of Caro and Hauser's criteria for teaching.

Setting aside remuneration for doing a job, how many of Caro's and Hauser's criteria for teaching are met by people in front of classrooms?

Whether the topic is food and eating or greeting and social interactions, we often have specific patterns of behaviour that we share with other members of our group. Demonstrating the presence of **behavioural traditions**, defined as "enduring behavioural practices that are shared by several individuals of a species and transmitted through social learning" (Muller & Cant 2010, page 1), in other animals can be challenging.

The problem-solving abilities of and responses to novelty by Black-capped Chickadees (*Poecile atricapillus* (1611), Fig. 17.32A) living in harsh, challenging environments exceed those of Chickadees living in milder climates. Black-capped Chickadees from the latitudinal extremes of the species' range show differences in memory and brain morphology. Behavioural performances of the birds in finding food they have cached demonstrated that morphological and memory differences may be inherited, and that environmental conditions can shape an individual's capacity for learning.

In Queen Elizabeth National Park in Uganda, wild banded mongooses (*Mungos mungo* (1723), Fig. 17.33 on page 497) pass preferences for foraging techniques to the next generation. Banded mongooses live in male-biased groups of 5–40 individuals. The small (<2 kg) dependent pups form one-on-one associations with conspecifics known as "escorts." Escorts may be genetically related to the pups they associate with, about the

same level of relatedness as pups in the same litter. Banded mongooses eat a range of prey, including hard-shelled items (birds' eggs and rhinoceros beetles) which they crack either by holding and biting or by throwing them against stones or other hard surfaces. Using artificial hard prey (Kinder® egg plastic containers filled with rice and fish), researchers explored preferences for different foraging behaviour and their transmission between escorts and pups (Fig. 17.33A). Escorts without pups were first presented with the novel food items and either opened them by biting or by hurling them against a hard surface. Preferences for mode of handling persisted over at least three months. The pups then learned their approach to hard prey items through their associations with escorts. The work demonstrates the importance of imitation in wild populations of animals and also shows that imitation in feeding behaviour does not result in behavioural homogeneity.

The reality that some animals learn by observing others makes an obvious link to animals using tools. We can use John Pierce's definition: "Tool use is the active external manipulation of a moveable or structurally modified inanimate environmental object, not internally manufactured for this use, which, when oriented effectively, alters more efficiently the form, position, or condition of another object, another organism, or the user itself." Some things do not count

BOX 17.4 People Behind Biology
Eric Kandel

Early in his career Eric Kandel switched from working on the mammalian hippocampus, part of a system with some million million nerve cells, to *Aplysia* (440), a marine snail (Fig. 1) in which ~20 000 central nerve cells are arranged in 10 ganglia, each containing around 2000 cells. Simple behaviours in *Aplysia* may be mediated by as few as 100 cells, offering an experimenter an advantage that is augmented by size—some Aplysia nerve cell bodies are 1 mm in diameter. Nerve cells in *Aplysia* are easily identified and can be isolated for biochemical studies. Furthermore it is easy to record from these nerve cells and to inject them with test compounds. When Kandel decided to take this reductionist approach, he could not have foreseen the advances he would achieve in our understanding of learning and memory. He received a Nobel Prize (2000) for Physiology or Medicine.

Kandel's first experiments showed that the gill withdrawal reflex in *Aplysia* could be sensitized and habituated, and could undergo classical conditioning. Furthermore, while the arrangement of intercellular neuronal pathways is fixed, the strength of the connections within those pathways can be modified, as originally suggested in 1894 by Santiago Ramóne y Cajal.

Kandel's experimental system is summarized in Figure 1. A light touch on the siphon triggers retraction of the gill under the mantle. This behaviour can be sensitized by applying a shock stimulus to the animal's tail. Repetition of the shock stimulus generates a short-term memory leading to a longer-lasting response. Serial repetition of the shock converts this short-term memory to long-term memory.

A simplified picture of the neural circuits involved is shown in Figure 2. The siphon is linked to 24 sensory neurons (SN(24)) that connect to the gill through six motor neurons (MN(6)). The sensory neurons also connect to a series of excitatory (Exc.) and inhibitory (Inh.) interneurons that, in turn, connect to the motor neurons and regulate their activity. A shock stimulus to the tail activates three types of interneurons (serotonergic (5-HT) neurons) that release a small cardioactive peptide (SCP) and L29 cells (which also release serotonin). Of these, the 5-HT neurons have the major effect on gill withdrawal and synapse with (i) the sensory neurons from the siphon, (ii) the synapses that these neurons make with the motor neurons, and (iii) the synapse between the sensory interneuron and Inh. and Exc. neurons. SCP and L29 cells also modulate the response of Inh. and Exc. cells.

Short-term memory results when a weak stimulus (thin arrows to the lower left) activates protein phosphorylation of ion channels, leading to an increase in transmitter release (Figure 3). A strong and longer-lasting stimulus (heavy arrows at centre left) stimulates increases in cAMP and this messenger activates protein kinases that affect the cell nucleus where new protein synthesis is initiated. The new proteins are involved in changes to the form (dotted outline) and/or function of the synapse with subsequent change to the amount of transmitter released.

FIGURE 1
A combination of a stimulus to the siphon and a shock to the tail of Aplysia *was basic to Kandel's experiments on the neural circuitry of learning in a simple system.*

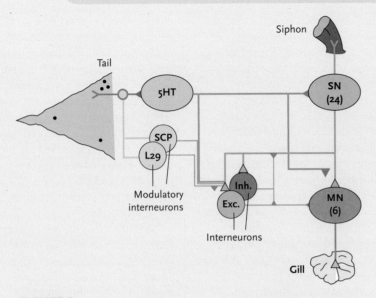

FIGURE 2
Neural pathways controlling the gill withdrawal reflex in Aplysia.

SOURCE: Based on Eric R. Kandel, "The Molecular Biology of Memory Storage: A Dialogue Between Genes and Synapses," *Science* 2 November 2001: Vol. 294 no. 5544 pp. 1030–1038.

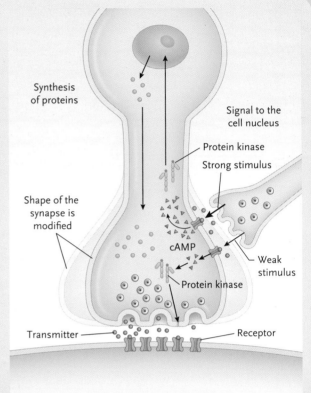

FIGURE 3 *A simplified view of short- and long-term memory formation at an* Aplysia *synapse.*

SOURCE: Based on Eric R. Kandel, "The Molecular Biology of Memory Storage: A Dialogue Between Genes and Synapses," *Science* 2 November 2001: Vol. 294 no. 5544 pp. 1030–1038.

Figure 17.32 Like many other species, **A.** Black-capped Chickadees, **B.** dwarf mongooses, and **C.** chimps (see below) show the capacity for learning.

as tool use, for example, hermit crabs' use of empty gastropod shells as shelters, because the shelter is effectively used all the time.

As we saw on page 235, some animals use tools to obtain food. However, animals such as gorillas (*Gorilla gorilla* (1857)) use walking sticks both to test the depth of the water and for stability when crossing a pool in a Congo swamp. Naked mole rats (*Heterocephalus glaber* (1886)) live underground in burrows (are fossorial) that they dig with their incisor teeth. When digging in a substratum that produces fine dust, some naked mole rats use a dust mask. The naked mole rat places a piece

A

Figure 17.34 A bottlenose dolphin (*Tursiops aduncus* (1771)) carrying a sponge that it uses to protect its skin from abrasion when it digs in the sand for fish.

Figure 17.33 **A.** The food, a modified Kinder® plastic container (left), that had been contained within a hollow chocolate (right), **B.** and banded mongooses.

of plant husk or a wood shaving behind its incisor teeth in front of its lips and molar teeth.

Meanwhile, in Shark Bay (Western Australia), some bottlenose dolphins (*Tursiops aduncus* (1771)), usually genetically related females, pick up sponges (Fig. 17.34) from the sea bed, wrap them around the snout, and then use the sponge to protect their skin from abrasion when digging fish out of sand. In this case, the tool (the sponge) provides a deferred benefit and is non-functional until used. Another example of deferred benefit is provided by veined octopus' (*Amphioctopus marginatus* (490)) use of shells and other structures as shelters. Divers observed individual veined octopus carrying empty shells or stacked half coconut shells under their bodies. When threatened, these octopuses stopped moving and occupied the shells (Fig. 17.35). Note that this behaviour involved assembling a functional shelter from two bivalve shells or two halves of a coconut shell.

Many birds (for example, gulls preying on crabs) carry food into the air and drop it to break it open. A reverse approach is found in Egyptian Vultures, Black Breasted Kites, and Bristle-thighed Curlews (*Numenius tahitiensis* (1557)). These species drop, or throw, rocks at their egg prey (ratite eggs for the kite and vulture, abandoned eggs of Black-footed Albatross for the curlew) to access the food in the egg.

There are many demonstrations of tool using by both African (*Loxodonta africana* (1871)) and Indian (*Elephas maximus* (1870)) elephants. African elephants often dig holes to reach drinking water. They then strip

bark from a tree, chew the bark into a ball, and use it to plug the opening to the water hole. The plug is covered by sand and the animal is able to return to its personal well for water at a later time.

Like lobster fishers, who use rotting fish as a bait to attract lobsters into their traps, Burrowing Owls (*Athene cunicularia* (1645), Fig. 17.36) use mammalian dung to attract dung beetles to the area of its burrow. The owls then eat the beetles that come to the dung. If the dung is experimentally removed, owls collect and replace their bait. This example was challenged on the basis that the smell of the dung might mask the odour of the owl's burrow and thus protect it from predators, but burrows with dung were attacked as often as those

Figure 17.35 **A.** *Amphioctopus marginatus* (490), the veined octopus. **B.** When disturbed, the octopus occupies a coconut shell shelter, **C.** which it transports by carrying it under its body.

Figure 17.36 A Burrowing Owl (*Athene cunicularia* (1645)).

without. Further, Burrowing Owls using dung as bait ate ten times the number of dung beetles as those with no dung around their burrow.

There are also many examples of tool use by insects. Myrmicine ants (*Aphaenogaster* spp. (982)) use leaf fragments to collect liquid food such as rotting fruit and body fluids of decomposing spiders and insect larvae. These ants dip pieces of leaf into the liquid food and carry them back to the nest. A dolichoderine ant (*Dorymyrmex bicolor* (995)) collects particles of soil and small stones and drops them into the nest openings of other species of ants. This behaviour appears to prevent other species of ants from foraging in the dolichoderine ants' territory. Sphecoid wasps (*Ammophila* (981) and *Sphex* (1009)) close their nest burrows with small stones and then use a stone, held in the mandibles, as a hammer to tightly pack the closure (Fig. 17.37). Some tree crickets (*Oecanthus* spp. (902)) build an amplifier to increase the volume of their stridulatory sound to attract mates. They do this by chewing a hole in a leaf and then positioning themselves above the hole so that the leaf acts as a sound baffle.

Consistent use of tools by some animals is a recurring theme. The list includes Galapagos Woodpecker

Figure 17.37 A sphecoid wasp (*Ammophila* sp. (981)) using a stone "hammer" to pound closing material into its burrow.

Finches (*Camarhynchus pallidus* (1590)), sea otters (*Enhydra lutris* (1712)), and Egyptian Vultures (*Neophron percnopterus* (1582)), species that specialize in the use of one kind of tool. New Caledonian Crows (*Corvus moneduloides* (1595)) make tools in the wild and in captivity and use them in extractive foraging (see also page 235). Capuchin monkeys use hammer and anvil tools to crack nuts and some stone tools to dig up roots. But these pale in comparison to tool use by chimps.

Chimpanzees' tool use provides one of the richest examples of learned behaviours (Fig. 17.32C). Chimps have been documented using about 20 types of tools in a variety of contexts from subsistence, sociality, and sex, to self-maintenance. Chimps from three populations in Uganda (Budongo, Kanyawara, and Ngog) use about half the kinds of tools reported from chimps from elsewhere. Chimps use tools for different purposes in different areas. At Goualougo in the Republic of Congo, chimps most commonly use tools to extract food, while at Ngog they are used for hygiene and courtship. Chimps everywhere in the wild use leaf sponges to obtain drinking water and show aimed throwing of missiles. Chimps also communicate by drumming on the buttresses of trees. Some chimps use sets of tools in sequence to achieve a single goal. In Gabon, chimps extract honey using a sequence of five tools: a pounder, a perforator, an enlarger, a collector, and an extractor. Successful use of tool sets requires that they be used in the correct order in the correct way. Examples of composite tool use include stones as hammers to crack nuts on anvils of stone or wood. Chimps also use compound tools involving two or more components combined into a single working unit. An example is leaf sponges made by crushing several fresh leaves into a ball that absorbs water and allows the chimps to extract water from otherwise inaccessible tree holes.

Orangutans also use tools, but their use is not as diverse as it is in chimps, coinciding with their more arboreal lifestyle. The other two living species of great apes, bonobos and gorillas, show no indication of tool use in the wild.

STUDY BREAK

1. Give three examples of tool use in animals other than humans.
2. Do you see any obvious phylogenetic distinction among species known to use tools?
3. Under what circumstances do animals use tools?

17.9 Spiders

Spiders provide a rich array of behaviour examples. The computer screensaver image of a spider web coated with water droplets glistening on the strands

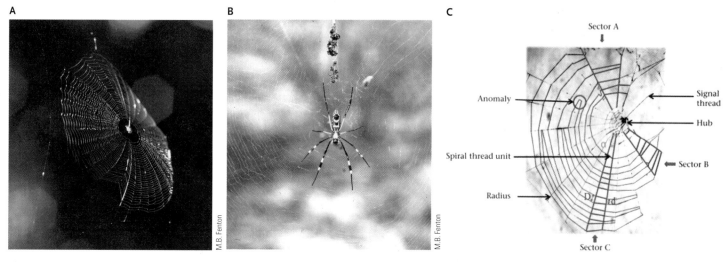

Figure 17.38 A. The orb web of a spider, **B.** a wood spider (*Nephila* (678)) at the hub of its web, and **C.** a labelled diagram showing the parts of an orb web.

SOURCE: Reprinted from *Animal Behaviour* Volume 84, Issue 5, M. Anotaux, J. Marchal, N. Châline, L. Desquilbet, R. Leborgne, C. Gilbert, A. Pasquet, "Ageing alters spider orb-web construction," Pages 1113–1121, Copyright 2012, with permission from Elsevier.

(e.g., Fig. 17.38A) somehow radiates a sense of peace. But some spider webs are lethal traps for insects. At first, a spider web appears to be a tool but because the silk was made by the spider, it does not fit the definition of a tool in Section 17.8. This distinction between tool and nontool becomes blurred when spiders build shelters by collecting leaves or other debris and incorporate them into the web (Fig. 17.39). Experiments in which predatory lizards (*Carlia* spp. (1482)) were placed in enclosures containing several species of spiders revealed that spiders with shelters were much less likely to be taken by the lizards than species without shelters.

Orb-web spiders such as *Argiope argentata* (660) use their webs to trap insects (Fig. 17.38C), but capture requires work by the spider. A hunting *A. argentata* sits at the hub of the web and, when an insect strikes the web, the spider localizes, immobilizes and then transports the prey. Localization is achieved by plucking web strands and using vibrations of the prey to localize the catch in the web. The spider immobilizes the prey by biting it, delivering venom that incapacitates the victim and starts the process of digestion. The spider then wraps the prey in silk before cutting it out of the web and transporting it to another location in the web. These spiders adjust their response according to the victim, biting moths or butterflies first, while wrapping other kinds of insects before biting them. These spiders usually feed at the hub of the web.

A large orb web of a spider from temperate regions may be ~50 cm across, much smaller than the web of Darwin's bark spider (*Caerostris darwini* (664)) from Madagascar. Darwin's bark spiders, less than 2 cm long, build webs across streams supported by 25-m-long frame members. These webs are the toughest biological materials known, with a strength of up to 520 MJ · m³, about 10 times stronger than Kevlar. The orb web itself

is up to nearly 3 m² and can catch as many as 32 mayflies at one time. The spider injects venom into each prey and wraps it in silk for a later meal.

The complexity of orb webs, with their sticky catching threads and strong and extensile frames and supports (Fig. 17.38) has attracted the attention of many investigators. In 1948, a scientist examined the effects of pharmacological agents on the diurnal clocks of web-spinning spiders. While his experiments did not change the spiders' diurnal clocks, treated spiders produced an elegant series of bizarre webs. Spiders treated with LSD made smaller, more regular webs than control spiders, while those treated with caffeine produced "webs" that were just random and disoriented threads (Fig. 17.40). A Skylab experiment with spiders demonstrated the influence of Earth's gravitational field on webs. In space, spiders built lighter webs with fewer turns.

Figure 17.39 A. Web shelters of *Cyrtophora hirta* (665) and **B.** *Phonognatha graffei* (680).

SOURCE: Manicom, C., Schwarzkopf, L., Alford, R. A., & Schoener, T. W. "Self-made shelters protect spiders from predation." *PNAS*, vol. 105 no. 39 14903–14907. Copyright 2008 National Academy of Sciences, U.S.A.

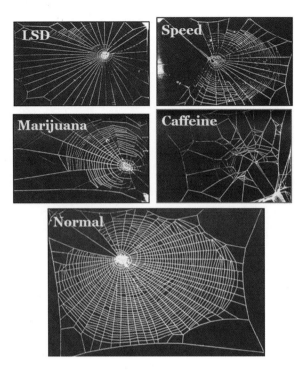

Figure 17.40 The effects of drugs on *Araneus diadematus* (659) web building.

SOURCE: WITT, Peter, N.; *SPIDER COMMUNICATION*. © 1982 Princeton University Press. Reprinted by permission of Princeton University Press.

While the many different forms of spider web probably evolved to catch prey, webs have come to serve other functions, often associated with mating. Sexual cannibalism is common in spiders, and male spiders have evolved ways to signal to females their identity and readiness to mate. Male spiders pluck a female's web. The resulting vibrations may serve at least four functions:

- species identification
- reducing the female's predatory behaviour
- preparing her for copulation
- indicating the reproductive fitness of the male

Some male wolf spiders express their virility through drumming behaviour. In *Stegodyphus lineatus* (683), web vibration stimulates a female to copulate. However, web vibration is not essential to mating, suggesting that species recognition was not involved. Furthermore, web vibrations do not provide information about male size. Males that find, and vibrate, webs of virgin females achieve mating success faster than non-vibrating males. Further, males vibrate webs recently abandoned by virgin females, suggesting that receptive females had produced a chemical signal (a pheromone).

Some spiders also use chemicals to defend their webs from attacks by other species. An orb-web spider, *Nephila antipodiana* (676), builds its web and then coats it with a pyrrolidine alkaloid to repel pharaoh ants (*Monomorium pharaonis* (1000)), a species that preys on spiders.

Some species of spiders use webs to disperse pheromones. The "bowl-and-doily" spider (*Frontinella pyramitela* (669)) builds an upwardly directed, bowl-shaped web that carries two pheromones to males contacting the web. One compound stimulates courtship behaviour, the other positive geotactic behaviour, drawing the male down into the bowl of the web.

Airborne pheromones evaporate from the webs of females. In the Sierra dome spider (*Linyphia litigiosa* (673)), a male attracted to the web then rolls up the web to eliminate further dispersal of the sex attractant.

Bolas spiders such as *Mastophora hutchinsoni* (674) combine a highly specialized "web" with a different type of pheromone. Bolas spiders secrete a single horizontal strand of silk and then hang from it. These spiders then secrete a second strand and, as it is formed, coat the web strand's end with sticky material so that the hanging thread forms a ball. The spider secretes a mimic of moth pheromone. When hunting, usually at night, the spider swings the ball while the pheromone it produces attracts male moths that are flying downwind of the spider. When the moth contacts the sticky ball, it is caught, and the spider then reels in the trapped moth. Manipulation of the system showed that the spider released the pheromone; the ball did not carry the odour. Bolas spiders produce a cocktail of pheromones that attract at least two species of noctuid moths. Bolas spiders are as effective in their approach to hunting with a web as orb web spiders are with theirs.

Spiders in the family Deinopidae, the "ogre-faced spiders" (Fig. 17.41), produce a small web that they hold between the first two pairs of legs. Ogre-faced spiders catch walking prey by a backward sweep of the net over the prosoma, trapping the prey against the ground. They catch flying prey by expanding the net forwards and

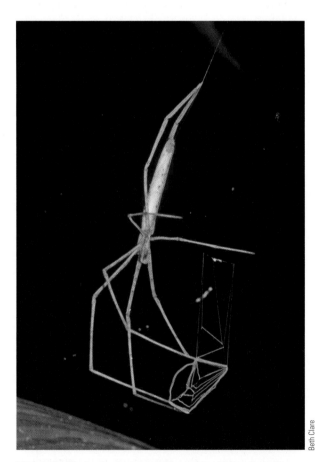

Beth Clare

Figure 17.41 A common net-casting spider (*Deinopis ravidus* (500)) with its net held between the front two pairs of legs.

downwards. They detect walking prey, often ants, by visual cues, flying prey by their vibrations. The net of a deinopid is unusual in that it is highly extensible, achieved by the folding and coiling of the fibres of sticky silk.

Spitting spiders (Family Scytodidae, Fig. 17.42) use a different approach to catch prey with silk. These spiders discharge a mixture of silk, glue, and venom from the opening of a venom duct near the base of each fang. The spiders' fangs oscillate laterally at high speed (up to 1700 Hz) and, in one 30-ms-long spit discharge, move in a dorso-ventral direction so that the spit is laid down in a zig-zag pattern one to two cm in front of the spider (= two to four spider body lengths). This behaviour ensures that spit is expelled at high speed (up to $28 \text{ m} \cdot \text{s}^{-1}$) from the venom duct. Prey are rapidly immobilized, probably through a combination of the action of a venom and contraction of the silk which dries rapidly and contracts to 40–60% of its expelled length.

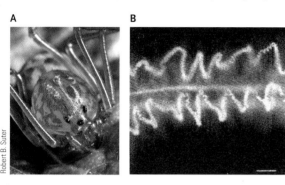

Figure 17.42 A. Spitting spider, *Scytodes thoracica* (681). The domed prosoma, typical of the spitting spiders, houses the large glands that supply the silk-glue-venom mixture of the spit. **B.** A silk strand from the net of a net-casting spider. Notice how the strand is folded, and at a finer scale, coiled. This gives the strand its extensible ability. Scale bar = 0.05mm.

SOURCE: B. Republished with permission of American Arachnological Society, from Coddington, J.A. and Sobrevila, C. 1987. Web Manipulation and Two Stereotyped Attack Behaviors in the Ogre-faced Spider Deinopis spinosus Marx (Araneae, Deinopidae). *Journal of Arachnology*, 15(2): 213–225; permission conveyed through Copyright Clearance Center, Inc.

STUDY BREAK

1. Name three types of webs that spiders build.
2. How do spiders differ in their use of webs?
3. What other animals produce and use silk? Do other animals use spider silk?

BEHAVIOUR **IN PERSPECTIVE**

Studying animal behaviour can provide a clear view of how animals respond to changes in their surroundings, whether the focus is how they learn to exploit food sources or find places to nest. Studying interactions between species increases our appreciation of the flexibility of animals, particularly when the focus is predator-prey interactions. Social lives of animals provide interesting glimpses into often complex societies, whether the subjects are eusocial insects or primates.

Questions

Self-Test Questions

1. Hormones have been shown to affect animal behaviours in many situations, but this works in reverse as well. Which of the following provides an example of how behaviour can alter an animal's response to hormones?
   a. Juvenile hormone alters the appearance of geometrid moth caterpillars.
   b. Winning aggressive encounters in their home range results in the expression of androgen receptors in specific brain cells of California mice.
   c. Cortisol levels rise in red crossbills that observe conspecifics searching for food.
   d. Testosterone levels rise in chimpanzees in response to food accessibility.

2. Which of the following provides an illustration of the adaptive nature of defensive behaviours?
   a. Blood-feeding insects, such as mosquitoes and tsetse flies, locate hosts/prey by detecting the combination of carbon dioxide and exhaled organic volatiles on the host's breath.
   b. Some insectivorous bats use echolocation calls outside the range of hearing of moths and so are undetectable by the insects.
   c. Mature male pumpkinseed sunfish mimic female colouration to gain access to nests where they can release sperm to fertilize eggs.
   d. Pea aphids sense elevated heat and humidity in the breath of an approaching mammalian herbivore and synchronously drop from the plant before the herbivore begins feeding.

3. Which of these statements is true about migratory behaviour?
   a. Migration always involves long-distance movements, such as those of monarch butterflies from their summer range in the United States and Canada to overwintering sites in Mexico.
   b. Bird migrations use similar routes on their southward and northward flights.
   c. Short-distance movements may be aided by local navigational cues or landmarks, while longer distances involve magnetic and/or celestial cues (sun, stars).
   d. Migratory bird species typically have larger brains than sedentary species, presumably to help coordinate navigational behaviour.

4. What advantage may be conferred to animals that exist as part of a mixed-species social group?
   a. Members of mixed-species foraging groups may benefit from increased defensive behaviour by responding to alarm calls of group members of other species.
   b. Groups composed of related individuals retain their cohesiveness over longer time periods.
   c. Social hierarchies develop within the group that reflect the status of each individual relative to other group members.
   d. Eusocial animals live in aggregations that may number in the thousands of individuals.

5. Which of the following is a common characteristic of tool use in animals?
   a. The tools are unmodified objects applied to a new task.
   b. Tool use is demonstrated primarily by vertebrate species.
   c. Tools are made by animals for immediate use.
   d. Animals can learn to use tools by observing others.

Questions for Discussion

1. Explore three examples that demonstrate intraspecific variation in the behaviour of animals. What explains or accounts for the variation?

2. Provide three examples of how genetics influences animal behaviour. What are the implications for the difference between the influence of nature as opposed to nurture?

3. How could we measure the intelligence of animals? Is tool-using an indication of intelligence? Why? Why not?

4. Use the electronic library to find newly published results about animal behaviour. How do these results influence your view of animal behaviour?

5. Which is the head end of this butterfly? Why would appearing to have two heads be an advantage? Is this an example of an animal "hiding" its head?

Sean F. Werle

6. This bulldog bat (*Noctilio leporinus* (1803)) from Belize appears to be a male. But, in fact, it is a pregnant female. What is the significance of the penis-like structure? What is it? What would be responsible for it? Where else do you find apparent masculinization of female genitalia?

M.B. Fenton

M.B. Fenton

Figure 18.1
Tom (male) turkey.
SOURCE: Jeff Banke / Shutterstock.com

18 Interactions with Humans

WHY IT MATTERS

Eating always brings us closer to other species. But from where in the world did our food come? Let's take turkeys, *Meleagris gallopavo* (1669) (Fig. 18.1), as an example. Although turkeys are a New World species, some were taken from southern Mexico (the subspecies *Meleagris gallopavo gallopavo*) to Europe in the early 1500s (from the same area that produced other well-known New World domesticates such as corn, beans, and squash). Generations of those turkeys were selectively bred and "domesticated," forming several stocks. In the 1700s, some domesticated turkeys were brought back to the American Atlantic coast where they became the basis for the commercial stocks we use today.

Were turkeys domesticated in Mexico or in Europe? Recently, DNA was extracted from 149 turkey bones and 29 turkey coprolites obtained at 38 archeological sites in the American southwest. These data were compared with genetic data from a wide variety of turkeys (wild and domesticated). The analysis revealed 12 mitochondrial haplotypes falling into three haplogroups (H1, H2, and H3). Most of the archeological material fell into H1 and some into H2, while H3 included all modern turkeys and some museum specimens from southern Mexico (Fig. 18.2). The H3 group supports the position that our domestic

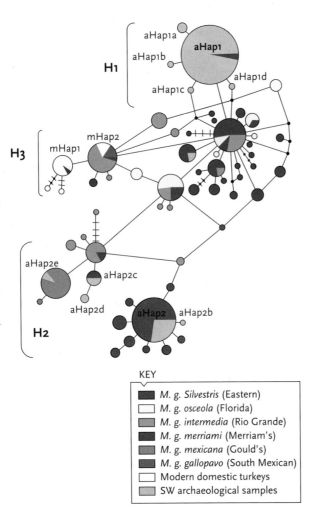

Figure 18.2 Genetic relationships among three haplotypes of turkeys. Solid colours represent modern wild turkeys, gray areas are haplotypes from archeological remains (coprolites and bones) from the American southwest, and yellow indicates haplotypes recovered from museum study skins.

SOURCE: Speller, C.F., B.M. Kemp, S.D. Wyatt, C. Monroe, W.D. Lipe, U.M. Arndt, and D.Y. Yang. 2010. Ancient mitochondrial DNA analysis reveals complexity of indigenous North American turkey domestication. *PNAS* 107:2807–2812.

KEY

- ◼ *M. g. Silvestris* (Eastern)
- ◻ *M. g. osceola* (Florida)
- ◻ *M. g. intermedia* (Rio Grande)
- ◼ *M. g. merriami* (Merriam's)
- ◼ *M. g. mexicana* (Gould's)
- ◼ *M. g. gallopavo* (South Mexican)
- ◻ Modern domestic turkeys
- ◻ SW archaeological samples

turkey stocks today were derived from *M. g. gallopavo*, supporting the domestication of turkeys in Europe.

However, the genetic uniformity of the H1 samples presents a genetic signature associated with selective breeding and indicates a second centre of domestication. This domestication event occurred in the American southwest as early as 2200 years ago and involved either *M. g. silvestris* or *M. g. intermedia*. Other archeological remains indicate that H1 domesticates were traded among peoples of the area who kept them in pens. At the archeological sites, remains associated with H2 genetic stock indicates that people also harvested local wild turkeys.

The evidence suggests that turkeys were domesticated at least twice, first in the American southwest and then in Europe from other stock. Turkeys provided protein (food) for people, their feathers and bones were used in ceremonies, and the feathers were used as insulation in covers. Today we think of turkeys as food, and they are big business.

In 2009, worldwide production of turkey meat was 5 062 000 metric tonnes, about 51% of it in the USA. That year in Canada, per capita consumption of turkey meat was 4.7 kg, compared with 8.0 kg in the USA. Although we may think of turkey meat at holidays such as Thanksgiving or Christmas, it is an important source of protein worldwide. Turkey meat is considered to be

a "healthy" alternative because it is lean. While turkey breast meat is high in salt, dark meat is not, and cholesterol is high in dark meat and low in white meat.

So a festive dinner that includes turkey, stuffing (bread), potatoes, turnips, and Brussels sprouts provides you with a mixture of domesticates from the Old World (wheat for bread, turnips, Brussels sprouts) and the New World (potatoes and turkey). The turkey on your table, however, is from stock domesticated in the Old World but originating in the New World. Molecular genetics has opened new windows on the history of our own species and our interactions with others.

The purpose of this chapter is to explore interactions between humans and other animals. This means considering positive and negative interactions, whether from the animals' or the humans' perspective.

SETTING **THE SCENE**

Humans depend extensively on animals, whether on species we have domesticated for our own ends or on ecosystems structured around interactions among species—from animals to bacteria. In this chapter we consider animals as resources, domestication, and other intereactions between humans and animals. Humans recognize that animals often play a central role in diseases to which we are vulnerable, perhaps an example of exploitation in the opposite direction. Although we exploit animals and other organisms, too often we are slow to appreciate how much our survival depends upon them. From food to employment and the very fabric of our structures, we cannot survive without other animals.

18.1 Animals as Resources

What do wool, silk, and leather have in common? All are animal products. In our day-to-day lives, we use many animal products such as feathers, eggs, cheese, and meat. While some of this use means killing animals (meat, leather), other animal products are harvested without killing and often without harming the donors. Ostriches, sheep, and llamas are examples of "multi-crop" animals, providing us with feathers, eggs, meat, and skin (ostriches), or wool, meat, and skin (sheep or llamas) (Figure 18.3).

As a species, we evolved as hunters and gatherers, depending on both plants and animals for food. Changes in diet may have been fundamental to the origin and adaptive radiation of species in the genus *Homo*. Our use of fire and tools, along with the consumption of aquatic animals rich in "brain food" (polyunsaturated fatty acids) and starches that are rich in energy (Fig. 18.4), could have heralded important changes in our ancestors. In 2012, microstratigraphic evidence from South Africa indicated controlled use of fire by *Homo erectus* (1858) by about 1 Ma.

Figure 18.3 Wool and fur, animal products that people use daily.

By 12 000 years ago, 100–200 people lived at Abu Hureyra, a collection of semi-subterranean dwellings in what is now Syria. Evidence from the site suggests that the surrounding area was a park woodland dominated by oak, and the people of Abu Hureyra ate the seeds of over 100 local species of plants as well as animals such as gazelles. This site represents a stage that is one step removed from a nomadic existence. Between 9400 and 7000 years ago when the climate was much drier, Abu Hureyra was home to 4000 to 6000 people living in multi-room dwellings made of mud and brick. At that time people depended on many fewer species of plants, many of which were cultivated on site, where cultivation meant systematic sowing of wild plant seeds and caring for (weeding and watering) the growing plants.

At some point in our history, our ancestors learned that by keeping some animals, they could collect their products repeatedly, such as eggs and milk, thus providing a reliable source of food. Evidence from isotopes indicates that by 5500 years ago, mares' milk was being processed in what is now Kazakhstan. By this time, evidence also indicates that horses were morphologically distinct from their wild relatives, indicating that they had been domesticated (see below).

Animal skins and bones have been used in applications ranging from clothing and housing to insulation and tools. The fur trade (Fig. 18.3) was one economic driver for the exploration and opening up of what is now Canada and the United States. In spite of vociferous opposition from some people and groups, the trade in furs and skins continues worldwide. Leathers may come from the skins of domesticated animals slaughtered for food, while the skins and fur of wild animals are also collected and used. In February 2010, for example, at Fur Harvesters Auction Inc. in North Bay, Ontario, Canada, 25 221 beaver (*Castor canadensis* (1877)) pelts were offered at an average price of US$21.33, ranging from US$14 to US$72. In the same period, 1940 otter pelts (*Lontra canadensis* (1718)) averaged US$42.20, and 1511 lynx (*Lynx canadensis* (1720)) from Canada averaged US$125. The top lot of lynx was purchased for a company in Greece; the top lots of raccoon (*Procyon lotor*

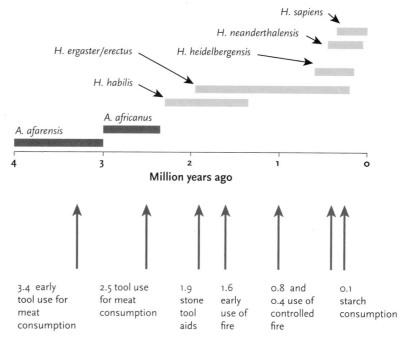

Figure 18.4 Changes in diet across 4 million years of hominid history (*A.* = *Australopithecus, H.* = *Homo*). Progression from using tools for meat consumption (3.4 and 2.5 million years ago) through using stone tools in the consumption of aquatic animals (1.9 million years ago), to the use of fire (1.6, 0.8 and 0.4 million years ago). Also shown is the consumption of catfish and other aquatic animals (0.19 million years ago) and then starches (0.5 million years ago). Brain size in hominids ranges from 385 cm³ in *Australopithecus* to 1350 cm³ in *Homo neanderthalensis* and *Homo sapiens*.

(1728)) and coyote (*Canis latrans* (1708)) for a company in Paris.

Today many species of animals are kept in captivity and raised for the products they provide, including:

- "ranched" mink (*Mustela vison* (1724)) raised for their fur;
- ostriches (*Struthio camelus* (1674)) raised for their feathers, leather, meat, and eggs;
- crocodiles (*Crocodylus niloticus* (1437) and other species) raised for their meat and skin
- pet-trade animals (e.g., ball pythons, *Python regius* (1521))

- animals such as leeches (*Hirudo medicinalis* (367)), used in medicine
- applications such as Geckle®, a hybrid adhesive combining the natural products that allow geckoes and mussels to adhere to substrates.

We have an astonishing variety of needs for animal products, and fortunately for us, the animal kingdom has an astonishing diversity.

STUDY BREAK

1. What evidence points to two domestication events for turkeys?
2. How are crocodiles and medicinal leeches used by people?

18.2 Ecosystems as Resources: Coral Reefs

Human interactions with coral reefs are complex and multi-level (Figs. 18.5 and 18.6). A scuba diver admiring individual coral polyps is at one end of the scale, and at the other is the global impact of anthropogenically induced climate change on the world's reefs. While reefs occupy only about 0.25% of the marine environment (about 600 000 km²), their biodiversity rivals or exceeds that of tropical rain forests.

Figure 18.5 Queensland (Australia) and the Great Barrier Reef seen from space (NASA).

SOURCE: Provided by the SeaWiFS Project, NASA/Goddard Space Flight Center, and ORBIMAGE

Figure 18.6 Diver and branched elkhorn (*Acropora palmata* (181)) coral head.

© imagebroker / Alamy

Between one and nine million species are estimated to occur in coral reefs, which annually generate some US$375 billion through their contributions to employment, fish supply, shoreline protection, recreation, and tourism, as well as food and pharmaceuticals. While fewer than 1% of the species on a reef are fish (about 4000 species), coral reefs are used by about 25% of the world's fish species. In general, fish eat the organisms that comprise the reef itself or other organisms associated with it. Through transport of ammonia, nitrogen, and phosphorus, fish contribute to the growth of the reef in nutrient-poor tropical waters. For example, schools of grunts (*Haemulon* spp. (1267)) shelter around coral heads during the day and at night move away from the reef to feed on sea grass beds. Exclusion experiments have shown that coral heads deprived of their normal supply of grunt feces grow significantly slower than those enjoying a normal balance with the fish schools.

As "homes" for fish, reefs are vitally important. A properly managed reef may produce some 15 tonnes of seafood per km² a year. Unfortunately almost none of the world's reefs are managed at a sustainable level. Indonesia loses about US $10 million each year from poison fishing (see below). Proper management of the same reefs could generate US $320 million and support 10 000 Indonesian fishermen. Destructive fishing practices, although frequently illegal, are common throughout Southeast Asia. Blast fishing, for example, uses the discharge of small explosive devices in the water to stun fish and make them easy for divers to collect. A typical blast breaks up about 5 m² of reef surface, and the rubble formed by the blast is not easily re-colonized by new coral growth.

Between 1960 and 2005, around a million kg of cyanide was used in fishing. In this process, a diver puts 2–3 cyanide tablets into a squeeze bottle and squirts the poison to stun fish on the reef. This practice began in the collection of tropical fish for aquaria and now is used to collect larger live fish (grouper and others) to ship to the luxury Hong Kong live-fish restaurant trade (estimated retail value about US$1 million a year). Cyanide fishing has three main negative effects. First, many fish are wasted because they are killed rather than stunned, making them useless for the aquarium and live-fish restaurant markets. Second, because affected fish frequently retreat into crevices in the reef, collectors attempt to retrieve them by breaking away coral using pry bars or hammers, thereby destroying large areas of the reef. Third, cyanide is toxic to corals, particularly to the symbiotic dinoflagellates that live within the cells of coral polyps. The photosynthetic activity of these symbionts (zooxanthellae) is essential for coral survival because they provide the organic carbon that the corals need (see page 269). In fact, most primary production that supports the reef ecosystem comes from the photosynthetic activity of coral zooxanthellae. Cyanide also affects other organisms in the area. Destruction of corals also can result from fish traps, by trawling near reefs, and by curtain lines that are used to scare fish into bag nets.

Over-fishing a reef environment can have rapid and long-term effects. One documented example, a coral pinnacle off Guam, was first fished in 1967. Within six months, fish populations were reduced below economically viable levels, and they had not recovered 34 years later. In a second example, a Hawaiian fisherman caught a school of large moi (*Polydactylus sexfilis* (1283)) at one location. Over the next 10 years, he did not see another moi there.

Marine protected areas (MPAs) are an important approach to reef preservation that can maintain a sustenance level of local fishery activity. The effectiveness of MPAs has been demonstrated in various ways. In St. Lucia, a small reserve increased the yield of fish caught in both large and small traps (Fig. 18.7). On the Florida coast, the size of record game fish increased within 200 km of a reserve compared with the rest of the Florida coast (Fig. 18.8). The effectiveness of an MPA in spreading fish throughout a significant area is shown by dispersal of the clownfish (*Amphiprion percula* (1254), Fig. 18.9) from an MPA on Kimbe Island (Papua New Guinea). Individually tagged larval clown fish marked in the reserve dispersed to other sites 30 km away.

Apart from fishery yields, the value of healthy reef systems cannot be overestimated. Tourism Florida estimates that its reefs bring in US$1.6 billion each year from tourism. Tourism to Caribbean reefs attracts about US$8.9 billion per year, and the direct economic worth of tourism in the Great Barrier Reef catchment and lagoon approaches US$4.3 billion per year.

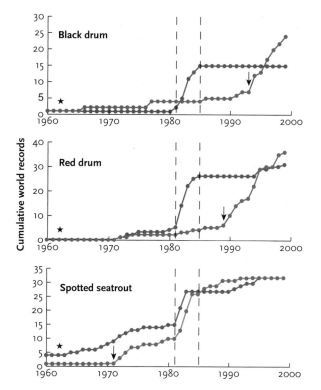

Figure 18.8 Cumulative world records for game fish in the 200-km coastal section adjoining the Merritt Island refuge (blue circles) compared with records from the rest of Florida (red circles). Asterisks mark the beginning of protection from fishing within the refuge. Vertical dashed lines show a period of rapid accumulation of new world records after addition of new fishing line strength classes by the International Game Fishing Association. Arrows mark the points at which there was a rapid increase in accumulation of new records for each species from areas around the Merritt Island refuge.

SOURCE: From Roberts, C. M., Bohnsack, J., A., Gell, F., Hawkins, J. P., & Goodridge, R., "Effects of Marine Reserves on Adjacent Fisheries," *Science* 30 November 2001: Vol. 294 no. 5548 pp. 1920–1923. Reprinted with permission from AAAS.

However, tourist activities in the water around reefs create new threats to reefs. Most tourists in reef areas protect their skin by applying commercial sunscreens. Lipophilic sunscreens can be taken up by coral tissues at levels as low as $33\mu l \cdot L^{-1}$ and can kill the corals. This effect is triggered by the activation of normally latent lytic viruses that destroy zooxanthellae. When the pigmented dinoflagellates are lost, the calcium carbonate exoskeleton platform of the coral colony is visible through the overlying transparent coral tissue, and the colony becomes white or bleached. Coral bleaching takes place in less than three days of exposure to sunscreen compounds. Estimates suggest that between 4000 and 6000 tonnes of sunscreen compounds are released over reefs each year, threatening 10% of the world's reefs.

Eutrophication is another source of damage to coral reefs. Nutrient enrichment may come from sewage run-off from human communities and/or farming operations, including aquaculture. Nutrient enrichment can promote the growth of phytoplankton,

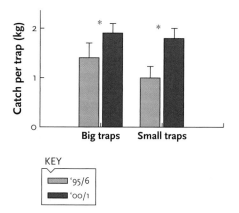

Figure 18.7 Comparison of fish catches between 1995–1996 and 2000–2001 for the two principal types of fishing gear used in the St. Lucian reef fishery. Differences were tested with the Mann-Whitney U test and were significant at P, 0.002 (*).

SOURCE: From Roberts, C. M., Bohnsack, J., A., Gell, F., Hawkins, J. P., & Goodridge, R., "Effects of Marine Reserves on Adjacent Fisheries," *Science* 30 November 2001: Vol. 294 no. 5548 pp. 1920–1923. Reprinted with permission from AAAS.

Figure 18.9 Orange anemone clownfish, *Amphiprion percula* (1254).

SOURCE: altug / Shutterstock.com

which reduces the clarity of the water and limits the light available for photosynthesis by dinoflagellates. Nutrient enrichment can also stimulate growth of macroalgae on the reef, which may outcompete coral colonies for living space by blocking light from reaching the corals. Heavy metal pollutants and other toxins generated by human activity also affect reefs. Changes in sediment loads arising from deforestation also harm reefs by reducing the light.

Climate change and acidification of the ocean may now be the most important threats to coral reef survival. Increased levels of atmospheric carbon dioxide, generated by combustion of fossil fuels, appear responsible for unusually high ocean temperatures, which also cause coral bleaching. While healthy corals can recover from partial bleaching, corals already under stress are frequently unable to recover. At the same time, increases in dissolved carbon dioxide have caused ocean acidification by swamping its carbonate buffering system. Acidification interferes with calcification, the basis of the growth of corals, and also promotes dissolution of previously secreted calcareous skeletons.

The world's coral reefs are in a perilous condition due to human activities. One estimate suggests that 30% of the world's coral reefs have been damaged beyond the point of no return, another 30% are seriously threatened, and only about 40% are stable and safe. Almost all of the reefs in the stable category are far away from centres of human population.

STUDY BREAK

1. How do fish enrich the environment of coral reefs?
2. Give three examples of destructive fishing practices.
3. Give three anthropogenic factors contributing to the degradation of coral reefs.

18.3 Use of Animals

Overexploitation of resources is a recurring theme in recent human history. Our expanding technological developments have increased our ability to easily harvest other species. When people recognize the need for sustainability in the local area, levels of fishing (or hunting) are often self-controlled. However, mobile fish harvesting units that operate outside territorial waters lack the local connection along with any motivation to achieve sustainable use. This can lead to a "tragedy of the commons," competitive depletion of a freely accessible resource that is owned by no individual consumer. The global fishery for sea urchins provides a good example (Fig. 18.10). In the absence of harvesting regulations, there is little prospect of sustainable use of the resource.

For those tempted to think of overexploitation of animal resources as a phenomenon limited to marine resources, consider the virtual extirpation of American bison (*Bison bison* (1743)) and Whooping Cranes (*Grus americana* (1586)) from western North America

A

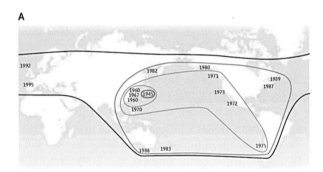

B

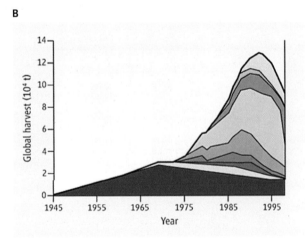

Figure 18.10 A. The spread of the sea urchin fishery in time and space, showing the dates of initiation of the fishery in chronological order from Japan and Korea to worldwide. **B.** The levels of harvest are shown in the same sequence as in A.

SOURCE: From Berkes, F., T.P. Hughes, R.S. Steneck, J.A. Wilson, D.R. Bellwood, B. Crona, C. Folke, L.H. Gunderson, H.M. Leslie, J. Norberg, M. Nystrom, P. Olsson, H. Osterblom, M. Scheffer and B. Worm, "Globalization, roving bandits, and marine resources," *Science* 17 March 2006: Vol. 311 no. 5767 pp. 1557–1558. Reprinted with permission from AAAS.

and the extermination of Passenger Pigeons (*Ectopistes migratorius* (1571)) in eastern North America. The same pattern of overuse to the edge of extinction occurred with the black rhino (*Diceros bicornis* (1839)) in Africa between 1960 and 1990, because of extensive poaching for horn to serve international rather than local demand.

Overexploitation of animals as resources has been pervasive in our history and not always associated with the technological ability for large-scale killing. Many species of birds were exterminated from South Pacific islands soon after humans arrived. These bird populations were small, and often people drove them to extinction by harvesting their eggs or damaging or destroying their habitat. Nevertheless, these examples of "low tech" methods are greatly exceeded by the many cases where increasingly efficient human technologies for animal harvesting have brought species to the brink or over the brink of extinction. The massacre of buffalo in western North America did not occur immediately after the arrival of humans in the New World (~20 000–30 000 years ago). First Nations people often used hunting techniques that killed many (tens) of buffalo at once, for example, at Head-Smashed-In Buffalo Jump in southern Alberta (Canada). The oldest buffalo remains suggest that the site had been used for killing buffalo for at least 5800 years. This mortality did not threaten the buffalo's survival as a species. Buffalo were brought to the brink of extinction in less than 50 years by the slaughter of tens of thousands after the arrival of horses and lever-action rifles, and the westward expansion of railways and farming.

We also use animals as labourers. For many people living in the "developed world" it is easy to forget just how much we depended upon animal labour only 100 years ago. Horses (Fig. 18.11), oxen, donkeys, mules, camels, llamas, dogs, and elephants come to mind as indispensable assistants. Today in the developed world, these animals are more often thought of as recreational or zoo animals. But some communities,

for example, some Mennonites and Amish, still rely on traditional horse power rather than horsepower generated by steam or internal combustion engines. In those communities and in many others in the developing world, animals still play a vital role in day-to-day operations, from transport and building to agriculture.

18.3a Animals in Warfare

The Carthaginian general Hannibal's use of elephants in his attack on the Roman Empire is a well-known example of humans exploiting animals in war. The feat of getting his elephants across the Alps on his way to invade Italy remains a stirring tale of human accomplishment. Soldiers have often used horses (*Equus caballus* (1840)) in warfare, and mounted units (cavalry) often turned the tide in battles. The same is true of camels (*Camelus dromedarius* (1747)) as mounts for soldiers. Less well known is the use we have made of arthropods.

Around 200 CE, the Roman Emperor Septimus Severus set out to control Mesopotamia. A key to this was capturing Hatra, a desert stronghold with a defensive perimeter 8 km long that consisted of two 12-m-high walls separated by a moat. As the Roman legions approached, King Barsamia and his citizens moved inside the walls of Hatra. According to the historian Herodian, these citizens prepared a defense that involved making bombs consisting of earthenware pots filled with stinging arthropods. It is not clear if the active ingredients in the bombs were scorpions or bees, but the impact of the bombs on Roman soldiers scaling the walls of Hatra was immediate and deadly. It took 20 days for the Romans to breach the defenses, at which point the Emperor called off the attack and took what was left of his legions away.

In 908 CE an army of Danes and Norwegians attacked Chester in England. The attackers dug tunnels under the walls in preparation for breaching them. Attackers in the confined tunnels were attacked by swarms of bees when the defenders hurled hives into the tunnels. Chester did not fall to the attackers. The same theme prevailed 700 years later in Sweden when General Reichwald led his army against the city of Kissingen. The Swedish defenders bombarded the attackers with hives of bees. In this case, armour protected the attackers from the outraged bees, but their horses were not so fortunate and the General abandoned his attack.

Many other records exist of insects and other arthropods being used in military endeavours, and the spectrum ranges from bombs (above) to instruments of torture. Anyone with a phobia about insects (or spiders) will easily imagine how vulnerable they would be to someone using such animals in torture. Between 1820 and 1850 in Central Asia, the Emir Nasrullah Bahadur-Khan of Bukhara used a "black hole"—a 6.5-m-deep pit—as a torture chamber. The Emir had the pit stocked with vipers and rats, but it was ticks and kissing

Figure 18.11 Horse and buggy.

bugs (Reduviidae, see page 355) that were more damaging to people placed in the pit. The combination of the painful bites of kissing bugs (especially species that normally did not bite humans) and the suppurating sores that developed at bite sites made a misery for victims.

During World War II, American forces developed Project X-Ray, designed to have Brazilian free-tailed bats (*Tadarida brasiliensis* (1817)) carry incendiary bombs into buildings in Japan. Cages filled with bomb-carrying bats would be released from a plane over a Japanese city. A pressure switch would open the cage at the right altitude, and the bats were expected to take flight and find refuge in the attics of Japanese buildings. Each 12-g bat would carry an 8-g bomb. Military tests succeeded in setting fire to a building, but bats with bombs were never released over Japan.

18.3b Pest Control

As we progress through this chapter, it should be obvious that there are very good reasons for the human reluctance to live too close to certain other animals. Many of the reasons relate to the health—and even lethal—consequences of sharing our homes with stinging, poisonous, or large carnivorous animals. But human activities often generate rich patches of food that attract other animals, so-called pests, as well as animals that come to eat the pests. The recent upsurge in infestations by bedbugs (*Cimex lectularius* (868)) is a current example of animals as unwelcome guests.

Our agricultural operations often attract pests because human food crops are tempting to other animals. This became clear to people experimenting in agricultural in the Negev Desert. Their crops were irrigated with water collected during run-off after rains, and fertilized with farm animal dung. The crops attracted many other consumers including porcupines (*Hystrix indica* (1887)), gazelles (*Gazella gazella* (1756), *Gazella dorcas* (1755)), Desert Partridges (*Alectoris graeca* (1663)), and many insects, notably black beetles. Stored foods also attract pests, often rodents. It is quite probable that cats (*Felis catus* (1713)) were domesticated by people who encouraged these predators to live among them for their rodent-control abilities.

Rats quickly come to mind when we prepare a list of pests that exploit human activities. But what do we mean by "rats"? More than 200 species of rats compete with humans for food. The Indonesian rice-field rat (*Rattus argentiventer* (1901)) causes the most damage as grains start to form, while *Bandicota bengalensis* (1876) does most of its damage just before harvest. Farmers must adjust their planting and harvesting strategies to mediate the rat damage. Farmers in the Philippines use a simple formula: plant 10 rows of rice, expect to lose two to rats and another to birds.

The "rat floods" experienced by farmers in the hill country of Bangladesh, northeastern India, and Myanmar occur about every 48 years in response to the fruiting of bamboo in local forests. The synchrony of the bamboo's (*Melocanna baccifera*) fruit production means a food bonanza for the rats and their populations explode because the animals breed two or three months earlier than usual. When the bamboo fruits have gone, the rats move on to the crops.

Changes in patterns of food availability can be reflected in the reproductive behaviour of rodents such as rats. Cyclone Nargis devastated the rice crop in the Ayeyarwady delta (Myanmar) in 2008. In response, farmers planted rice in more places and spread their planting over a longer period of time than usual. This resulted in a widespread and consistent food supply for rats, which led to an increase in the rat population and further depletion of food supplies for people.

As with rats, many species of insects are considered pests because their feeding activity causes economic loss and even widespread human starvation. Desert locusts (*Schistocerca gregaria* (904)) have long been recognized as agricultural pests and are identified as such in both the Bible and the Koran. Normally these insects are solitary, but under the right weather conditions, they form swarms (Fig. 18.12) that can comprise 80 million individuals×km^{-2}. A swarm may cover anywhere from a few hundred m^2 to >1000 km^2. Each locust eats about 2 g per day of almost anything that is green: leaves, flowers, fruit, seeds. The impact of a swarm of locusts can be huge, and outbreaks can be widespread (Fig. 18.13). In 1986 about 1 363 000 ha of arable land were under cultivation in Tunisia with the expected wheat and barley crop worth about US$72 million. At least 40% of this crop was at risk due to locusts. With this kind of loss, a country may not be able to feed its people.

The production of genetically modified (GM) organisms is one way that humans have tried to thwart pests. Cotton containing the *Cry/Ac* gene from the bacterium *Bacillus thuringiensis* (Bt) (Bollgard® cotton, a trademark of Monsanto) was toxic to insect pests such as pink bollworms (*Pectinophora gossypiella* (1044)). An upgraded Bollgard® cotton (Bollgard II®) with two Bt proteins was expected to protect the cotton crops of the 65% of farmers in Gujarat state in India who planted Bollgard II® cotton in 2009. In 2010 it

Figure 18.12
A swarm of locusts.

© Photoshot Holdings Ltd / Alamy

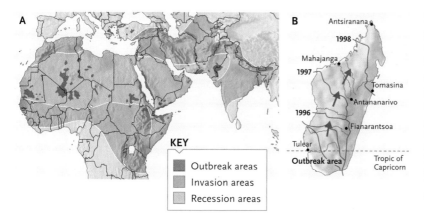

A

B
Antsiranana
1998
Mahajanga
1997
Tomasina
Antananarivo
1996
Fianarantsoa
Tulear
Outbreak area   Tropic of Capricorn

KEY
■ Outbreak areas
■ Invasion areas
□ Recession areas

Figure 18.13 **A.** Sequence of desert locust plagues across North Africa and east to India. Outbreak areas are shown in brown, invasion areas in green. Yellow identifies areas where locust plagues have receded. **B.** The spread of an outbreak of Malagasy migratory locusts in Madagascar.

SOURCE: Republished with permission of ORTHOPTERISTS SOCIETY - BIOONE, from Lecoq, M. 2001. "Recent progress in desert and migratory locust management in Africa. Are preventative actions possible?" *Journal of Orthopteran Research*, 10:277–291; permission conveyed through Copyright Clearance Center, Inc.

appeared that some pink bollworms had developed resistance to Bollgard II®, so the long-term viability of this particular GMO remains in doubt. In the US, GMO corn, soy, and cotton have been successful both in their impact on farming and the environment. The situation around the use of GMOs is complex and variable with evidence of successes and of failures. All too often it is clear that increasing agricultural productivity also increases the levels of pest activity, and can expose us to additional contaminants in our food—from pesticides to growth hormones and beyond.

18.3c Experimental Animals

Although rats and mice are arguably the most prevalent animal models in biomedical research, other model animals, such as fruit flies (*Drosophila melanogaster* (960)) and nematodes (*Caenorhabditis elegans* (585)), are used heavily. One advantage of "model" animals is that we can investigate phenomena such as the impact of diseases without exposing humans to their harsh realities. This approach to research is welcomed by many, but opposed by others who do not think that animals should be exploited in this way.

Molluscs have proved particularly important experimental models for studying nerves. Like many other invertebrates, molluscs have unmyelinated nerve fibres and the speed of conduction along their nerves depends on the size of the nerve process: the bigger the nerve, the faster the conduction. Squid are jet-propelled swimmers (see page 199) and the rapid movement of their escape response depends on all the mantle muscles contracting simultaneously. Giant axons provide the rapid transmission and make the escape response possible. In the common squid *Loligo* (503), giant axons are roughly 1 mm in diameter and conduct impulses faster than 25 m·s⁻¹, large enough to allow researchers to insert electrodes into the **axoplasm**. This led to the first voltage clamp experiments (later producing a Nobel Prize for Alan Hodgkin and Andrew Huxley, page 418). These revealed the imbalance of sodium and potassium maintained across the neural membrane and the sodium pumping activity of the membrane proteins.

Squid are not easy animals to keep in captivity, so other animals with giant axons became the experimental subjects of choice. Crayfish and lobsters have giant axons that coordinate their tail flip escape response, and giant axons trigger the withdrawal response in many tube-dwelling worms. The polychaete *Myxicola infundibulum* (336) (Fig. 18.14) and the sea slug (Fig. 18.15) have been used in a number of laboratories. Even small animals such as *Drosophila* (960) and

Figure 18.14
The polychaete *Myxicola infundibulum* (336).

Joao Pedro Silva / Shutterstock.com

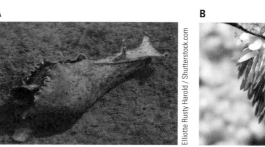

A   B

Elliotte Rusty Harold / Shutterstock.com   Natalie.Jean / Shutterstock.com

Figure 18.15 Sea slugs. **A.** *Aplysia californica* (441) and **B.** *Hermissenda crassicornis* (455) are important experimental animals (see Box 17.2).

BOX 18.1 Molecule Behind Animal Biology
Silk

A gala performance at an international opera house, the audience formally dressed, men in evening suits with silk faced lapels, many of the women in multi-coloured silk dresses. How many in the audience pause to remember that they are wearing the product of an insect? Their silk garments are versatile, cool, comfortable, and valuable.

The silkworm, *Bombyx mori* (1028), is a major source of silk. Adult silkworms are moths (Lepidoptera) whose caterpillars spin cocoons (Fig. 1) from silk (hence "silkworm"). Wild *B. mori* occur throughout China and eastern Russia and are now genetically different from domesticated strains. Silkworms apparently were domesticated in China about 5000 years ago. Genetic variability within the domesticated strain suggests a single domestication event that used a large number of different wild worms as original stock. The Silk Road (Fig. 2), a network of caravan routes, grew as the Roman Empire expanded. It eventually stretched over 11 000 km, from the Yellow River Valley in central China to Constantinople (now Istanbul). Silk and other goods were regularly transported, usually by camel, along the Silk Road until about 1400 CE

For 2000 years silkworms were closely guarded. Anyone caught trying to smuggle silkworms or their eggs out of China was executed. In the 6th century CE, two monks smuggled some silkworm eggs to Byzantium, which

became a centre of silkworm culture, an industry that quickly spread widely.

Silk production (sericulture) depends on raising silkworms on mulberry leaves, and then collecting them and allowing them to pupate. The cocoon is placed in hot water to soften the outer (sericin) layer, and then is unravelled. A single cocoon yields about 800 m of double-stranded silk fibre, each with a diameter of about 10 μm. The fibre is two-layered, having a protein core and a gum (sericin) cover. The fabric is woven from the silk fibres.

Domesticated silkworms have enriched expression of genes controlling the silk gland, a labial gland in *Bombyx* and other lepidopteran silk producers, such as the Mopani worm (*Gonometa rufobrunnea* (1036)) of southern Africa, which is now cultured to make coast silk cloth.

Secretion of silk may have evolved 23 times within the Insecta where it

may be produced in labial glands (as in *Bombyx*), colleterial glands (accessory female reproductive glands), or modified Malpighian tubules. Colleterial glands are dermal in origin but can occur in other locations. Colleterial glands at the base of the first pair of legs of male hilarine flies (dance flies; Diptera, Embioptera) secrete silk they use to wrap nuptial gifts that they present to a female during courtship. Another embiopteran, *Aposthonia gurneyi* (955), lines its burrows with the finest silk strand (65 nm) found to date, 100× thinner than silkworm silk. Spiders also make silk (page 499).

The protein core of a silk strand has a semicrystalline structure, an ordered molecular structure (a crystallite) in an amorphous matrix. The crystallite is a series of regularly arranged amino acids, arranged so that hydrogen bonds form networks within and

FIGURE 2
Trans Asia trade routes for moving silk. Blue are sea routes, red are land routes—the Silk Road.

SOURCE: Splette and NASA/Goddard Space Flight Center

FIGURE 1
Silkworm cocoons, with a silkworm and a butterfly of a silk worm.

SOURCE: holbox / Shutterstock.com

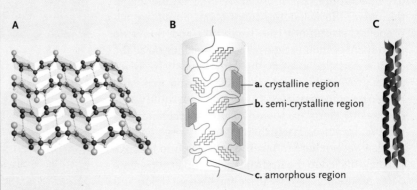

a. crystalline region
b. semi-crystalline region
c. amorphous region

FIGURE 3
Arthropod silk. **A.** *Pleated beta-sheet;* **B.** *arrangement in a spider silk thread; and* **C.** *coiled arrangement of a honey bee silk thread.*

between protein molecules. Silks may be formed from protein coils, strands, or sheets. *B. mori* silk is formed from beta-sheets (Fig. 3).

Honey bee silk is different because it has coiled (Fig. 3C) silk fibres, making it tougher and more extensible than the silk of *B. mori*. The silk of honey bees retains its properties when wet. Insect silks typically have high levels of glycine, alanine, and serine in their structure. These non-essential amino acids are somewhat hydrophobic. The use of non-essential amino acids may avoid the limitations on silk production resulting when essential amino acids are required as food. Silk proteins are hydrophobic because they need to be soluble in the gland but insoluble in their ordered form after secretion.

The larvae of caddis flies (Trichoptera) secrete and spin silk under water. Caddis fly silk has less alanine and more arginine than *B. mori* silk, as well as a higher level of phosphorylated serine residues. Silk of caddis flies has attracted the attention of biomedical engineers because its properties would make it well suited to serve as medical sutures.

Silk secretion in silkworms and spiders starts with production of silk proteins and builds up their concentration to levels of 25–30% protein when the solution has a viscosity about 3.5 million times greater than water. At this level, silkworms and spiders can produce fine streams that do not break into droplets. But it requires considerable pressure to pass this solution through the gland spigot (or spider spinneret). Shear forces align and order protein molecules to form a liquid crystal solution. This maintains the extensional viscosity of the solution while reducing its flow viscosity by an order of magnitude. The resulting solution can be discharged through the spigot as a droplet that is extensible into a thread by the movement of the insect or spider.

copepod crustaceans have escape responses coordinated by giant axons. Newer techniques, particularly patch clamping (Box 15.1), have now eliminated the need for giant axons in membrane research.

In any experiment with animals, the "control" group, which consists of individuals that were not treated or were treated with a placebo rather than an active ingredient, plays a pivotal role. Control groups give experimenters a picture of what is "normal," allowing them to assess deviations from the norm. Good experimental technique requires that research animals, both the experimental groups and the control groups, be humanely housed and treated. Whether the topic is related to human health or to identifying changes in physiological features (for example), if the animals used are not healthy, the experimental results will be compromised. An extension of this situation is the basis for arguments about the rights of animals and our right to exploit them. Ironically, perhaps, this leads us to the definition of "animal." The Canadian Council on Animal Care focuses mainly on vertebrates, thus on a very small proportion of the biodiversity of animals.

To get an impression of the importance of experimental animals to us, try an experiment. In Google® Scholar, type in the name of an animal and "medical research." When we tried this for beagles, we had 20 100 hits. The topics ranged from cartilage thickness in the knees of young beagles, to bone metabolism in ovariectomized aged beagles, to oral toxicity studies involving trinitrotoluene and Thallium-201.

Our dependence upon other species of animals and plants is part of our evolutionary heritage. The levels at which we consume other species will always depend upon the size of our population and our collective appetites.

STUDY BREAK

1. What causes overexploitation of animal resources?
2. Why are so many species of rats considered to be pests?
3. Would you be more accepting of using animals to advance the quest for a cure for cancer or for the development of a new cosmetic? Why?

18.4 Domestication

Domestication of animals refers to the process of selective breeding to enhance or conceal certain behavioural and/or physical traits. Domestication is more than just taming an animal, and the process has been key to the success of our species. Humans have domesticated many animals and plants, a process that started more than 10 000 years ago (Fig. 18.16).

Below we explore four examples of domestication in detail to emphasize the diversity of our activities in this area. Domestication was a key to avoiding some of the perils of overexploitation, and the process also helped to disconnect people from their dependence on local, naturally available foods that could not always be relied on. But domestication has brought other problems in its wake.

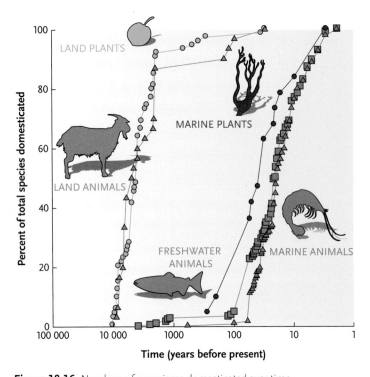

Figure 18.16 Numbers of organisms domesticated over time.

SOURCE: From Carlos M. Duarte, Nuria Marba, Marianne Holmer, "ECOLOGY: Rapid Domestication of Marine Species", *Science*, vol. 316, Apr 20, 2007, pp. 382–383. Reprinted with permission from AAAS.

18.4a Dogs

Genetic evidence indicates that about 40 000 years ago, somewhere in the Middle East, a group of our ancestors began to domesticate dogs. In 2007, the prevailing view was that dogs had been domesticated in eastern Asia, but new genetic data have changed this picture. The opportunity for domestication may have arisen because of an association between people and wolves (*Canis lupus* (1709)), whose descendants became dogs. Domesticated dogs could have been used to assist in the hunt and in protection. We can see how an original domestication event has spawned a huge variety of breeds of dogs, many of them specialized for different activities.

Domestic dogs show exceptional variation in both morphology and body size (Fig. 18.17). About 30% of the differences among the 85 breeds of dogs examined are genetic. Each of four genetic clusters contains mainly breeds with similar morphology, geographic origin, or role in human activities. Interestingly, small size in all dogs is controlled by a single insulin-like

Figure 18.17 Size extremes in dogs.

SOURCE: Eric Isselée / Shutterstock.com

growth factor 1 gene (*IFF1*), which is common in all small dogs and rare in giant breeds.

It is easy to be impressed by the morphological variety in domestic dogs and the range of tasks they perform for us. Behaviour may have been a more important determinant in domestication than any morphological feature. Dogs are better at reading human communication signals than other animals are, even great apes. Furthermore, wolves raised by humans do not show the same skill at reading human communication signals as do puppies of domestic dogs that are only a few weeks old. These observations suggest that selection for behavioural traits played a central role in the domestication of dogs.

Genetic evidence indicates that dogs arrived in the New World with the first people. By the time European explorers first visited, a genetically distinct lineage of dogs lived in the New World. Genetic data also reveal that European settlers appear to have minimized interbreeding between the dogs they brought with them and the dogs that were already here. The Salish people in northwestern North America bred the Salish wool dog, a distinct small dog, for its long fur that they used to make blankets. The Europeans brought blankets made from sheep's wool, which replaced the blankets made from dog's wool, and the last Salish wool dog is thought to have died early in the 20th century.

18.4b Chickens

Although our domestic chickens (*Gallus gallus domesticus* (1667)) appear to be descendants of red junglefowl (*Gallus gallus* (1666)), this is uncertain because several species of junglefowls are sympatric in South Asia. More than one domestication event may have involved more than one species of junglefowl (remember the turkeys). Charles Darwin was convinced that red junglefowl were the ancestral stock for chickens, but his view was not unanimously supported. Once again, genetic tools allow us to explore this history (see below).

Today we use chickens mainly as sources of meat and eggs. Birds that are good sources of meat ("broilers") are not necessarily as productive when it comes to eggs (Fig. 18.18). "Layers" are very prolific egg producers, but less so for meat. Early in the 20th century, specialized layer and broiler breeds were bred, avoiding the inherent problem of selecting for positive growth traits (meat production) and fertility (egg production) in the same individual. To date no domestic breeds of chicken are both good layers and good broilers.

All domestic chickens show selection at the locus for thyroid stimulating hormone receptor (TSHR) which, in vertebrates, plays a pivotal role in metabolic regulation and photoperiodic control of reproduction. Broiler strains share similar selection for genes associated with growth, appetite, and metabolic regulation (Fig. 18.18A).

Figure 18.20 A cave painting from the "Cave of the Spider" in Spain of a human figure climbing a vine to harvest honey. The picture is thought to be about 8000 years old.

SOURCE: Redrawn after Mesolithic rock painting of a honey hunter harvesting honey and wax from a bees nest in a tree. At Cuevas de la Araña en Bicorp. (Dating around 8000 to 6000 BC)

Figure 18.18 **A.** Broilers and **B.** layers are the two prevalent varieties of chickens today. They have been distinct since about 1900 and are raised in different settings.

Worldwide, most broilers and layers today are homozygous for the *yellow skin* allele—expressed as yellow legs in live birds. The general presence of yellow legs is genetically determined, but the amount of carotenoids, mainly xanthophylls, in the feed influences colour, with more intense yellows reflecting more carotenoids. Yellow skin reflects the activity of one or more cis-acting and tissue-specific regulatory mutation(s) inhibiting *BCDO2* (beta-carotene dioxygenase 2) in skin. But *yellow skin* does not originate in red junglefowl, instead in grey junglefowl (*Gallus sonneratii* (1668)). This evidence suggests that domestication of chickens involved more than one ancestor.

18.4c Bees

Honey bees (*Apis mellifera* (984), Fig. 18.19) have been associated with humans as long as recorded history. A painting from the Cueva de la Araña (spider's cave) in Spain shows a person harvesting honey from a bee colony (Fig. 18.20). At some point our ancestors began

Figure 18.19 Honey bee (*Apis mellifera* (984)) collecting nectar for honey.

beekeeping, long associated with churches and monasteries. Tenants on some church lands paid their rents in beeswax used to make the church candles. Fermented honey was converted to mead, the original liquor of the gods. Dionysus, the god of wine, was first worshipped with mead before viticulture developed in Greece. In northern Europe, the climate was too cool for grapes, and mead was the staple alcoholic drink.

Even simpler than mead is honey beer, which is common throughout Africa. Essentially this is honey and water, sometimes with added sugar or fruit, and some form of yeast typically obtained from germinating maize or millet. This mixture ferments rapidly in an open vessel in the sun but is short lived. Beekeeping and honey products are deeply embedded in the culture of many African tribes. Families owned sites where hollowed logs were hung from trees to attract bee swarms; these were included, rather like cattle, in bride prices.

African beekeeping practices have impinged on other areas of the world in an unexpected way. The African honeybee, *Apis adamsoni* (983), forms smaller colonies than its European counterpart *Apis mellifera* (984) and moves its colony location when pressured by a period of water shortage. African beekeepers take advantage of this mobility by using hollow logs, rubbed with beeswax, to attract swarms. When a colony is large enough, the beekeeper takes out honeycomb and larvae, destroying the colony to harvest the honey. Given that the beekeeper has little in the way of protective clothing, it is not surprising that docile colonies are completely destroyed while more aggressive colonies may drive off the beekeeper and survive. Thus, over time, African beekeepers inadvertently selected for aggressive bees. In 1956 a Brazilian apiarist imported a number of African queen bees for use in experimental crosses with *Apis mellifera*. An accident allowed swarms of the new crosses to escape from the experimental farm. They became established in the wild and since then, these "Africanized" bee colonies

BOX 18.2
Honeyguides

In many parts of Africa, birds known as Greater Honeyguides (*Indicator indicator* (**1629**)) and Lesser Honeyguides (*I. minor* (**1630**)) often guide people to bees' nests (Fig. 1). After the people have broken open the nest, the birds help themselves to the wax and larvae the human raiders leave behind. The experience of being "guided" by one of these birds is surely one of the more interesting events we can enjoy in nature. Honey gatherers among the Boran people in Kenya have developed a symbiotic relationship with Honeyguides. Boran honey gatherers that work with Greater Honeyguides are more efficient at finding bees' nests than those operating without the birds' help. The honey gatherers use a special whistle to call up their avian assistants (Fig. 2).

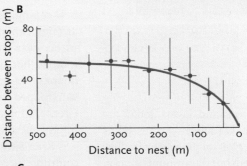

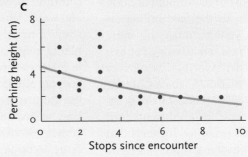

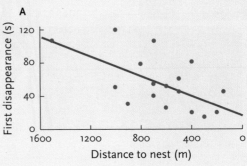

Figure 1
Lesser Honeyguide, Indicator minor *(1630).*

SOURCE: Courtesy of Naas Rautenbach

Figure 2
The behaviour of Honeyguides guiding Boran honey gatherers to a bee's nest. The birds keep the Boran honey gatherer in contact, stopping to wait for the human, and using perch height to indicate height of the nest above ground. **A.** *The time of contact (disappearance),* **B.** *stopping distance, and* **C.** *numbers of stops. Means are shown ± one standard deviation.*

SOURCE: From Isack, H.A. and H-U. Reyer, "Honeyguides and honey gatherers: interspecific communication in a symbiotic relationship," *Science* 10 March 1989: Vol. 243 no. 4896 pp. 1343–1346. Reprinted with permission from AAAS.

(the so-called killer bees) have steadily spread through Central America to California and Nevada.

The health benefits of honey are widely broadcast and sometimes exaggerated. Less well known is that honey is a very useful topical sterilizing agent for burns and ulcers because its osmotic concentration draws out water and dehydrates bacteria. But some honey can deadly. The so-called mad honey of Turkey and the Near East is produced by bees that have collected nectar from several of the native species of rhododendron (*Rhododendron* spp.). The nectar of these rhododendrons contains a potent neurotoxin, **grayanotoxin** (Fig. 18.21), also described from North American rhododendrons where it is known as andremodotoxin. Grayanotoxin is a diterpene hydrocarbon that binds to

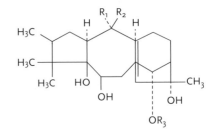

Figure 18.21
Grayanotoxin.

SOURCE: Based on http://www.answers.com/topic/grayanotoxin

GRAY	R_1	R_2	R_1R_1	R_3
GRAY1	OH	CH_3	–	H
GRAY2	–	–	$=CH_2$	H
GRAY3	OH	CH_3	–	Ac

open voltage-dependent sodium channels in cell membranes, blocking inactivation of the channel so that sodium ions readily enter and hyperpolarize the cell. At the whole body level, grayanotoxin causes hypotension (low blood pressure), brachycardia (slow heart rate), nausea, and vomiting—sometimes even death. Kateuas Mithridates IV, a king of Asia Minor fighting Pompey the Great in 67 BCE, left honeycombs containing toxic honey as he retreated. Advancing Roman troops ate the honey which effectively decimated their numbers.

The painful stings of bees are an adaptation for defending the colony and its honey store. A single bee sting, like a single wasp sting, is painful to a mammal and lethal to another insect. The most common victim of a bee sting is a robber bee from another hive. When a bee stings another insect, the barbs on the sting shaft do not catch in the broken exoskeleton, and the sting can be withdrawn, allowing the bee to sting again. But when a bee stings a mammal, the barbs catch in the skin and the sting apparatus is torn from the bee's body, rupturing a sac containing an alarm pheromone (iso-amyl acetate). The isolated sting continues to pump venom into the injection site, and the pheromone attracts other bees that also sting the threatening mammal. This ensures that the bear, honey badger (*Mellivora capensis* (1721)), or human will receive many stings and be driven off before the hive is severely damaged.

China is the world's largest honey producer with about 113 million kg of honey each year, compared to 68 million kg from the US. About 2.5 million colonies of bees live in the US, compared with 600 000 in Canada where annual production is about 29.4 million kg. In the US, the emphasis is not on honey production, but on pollination. An estimate of the value of crops pollinated by honey bees is around $15 billion, vastly exceeding the value of the honey crop (>US$200 million). Well over 100 North American crops are pollinated by bees, including all apples, pears, plums, peaches, and melons as well many species of vegetables. In California, pollination by about a million colonies of honey bees is essential for the almond (*Prunus dulcis*) crop and 810 000 acres of almond trees were pollinated between late-January and mid March of 2009. Beekeepers move colonies to California from as far away as Florida and the upper midwestern states to provide enough bees for pollination. California annually produces about 500 000 tonnes of almonds with a value of about US$900 million.

18.4d Aquaculture

If capture fisheries are equated to hunting, then aquaculture is the watery equivalent of farming. It can be defined as, "Man's attempt, through input of labour and energy, to improve the yield of useful aquatic organisms by deliberate manipulation of their rates of growth, mortality, and reproduction." An economist would add that the manipulation occurs "from a basis of site or stock ownership or leasehold." The economist's point is that the labour and investment in materials needed for the biological manipulations involved in aquaculture can rarely be justified in common property.

As we realize that most capture fisheries cannot be sustained at present levels (see "fishing down the food chain" on page 206), we pay more attention to aquaculture because some believe that it has the potential to supply protein to a hungry world. Projections from 2004 data suggest that the current annual worldwide production from aquaculture operations is about 40 million tonnes (Fig. 18.23), compared to capture fisheries producing about 100 million tonnes. Aquaculture production continues to grow rapidly. In North America, aquaculture is limited mainly to fresh salmon (*Salmo salar* (1316)) (Fig. 18.22), blue mussels (*Mytilus edulis* (536)), cultured oysters (*Ostrea edulis* (537), *Crassostrea virginica* (535), and *C. gigas* (534)), and bags of frozen penaeid shrimp. While catfish and tilapia (Fig. 18.24A) make occasional appearances at seafood counters, carp (Fig. 18.24B) and related cyprinids account for about half of global aquaculture productivity (Table 18.1). Over 90% of farmed freshwater fish come from Asia (Fig. 18.25).

About four times as much of the fish eaten in China comes from aquaculture (largely carp) than from the capture fishery, a ratio that is reversed in the rest of the world (Fig. 18.26). China produces about 70% of the world's aquaculture production (Fig. 18.25).

There are many concerns about the environmental impact and production efficiency of aquacultural operations. First is pollution, because a large salmon farm may produce as much nitrogenous waste as a small town. Second are the side effects of inorganic and organic compounds used by fish farmers to control pests, treat disease and inhibit net fouling. Third is the impact of animals that escape from the farm and enter the natural population. In an interesting modelling experiment, Japanese medaka (killifish, *Oryzias latipes* (1173)) were genetically modified with a growth hormone that enhanced their mating advantage but produced offspring with reduced viability. A modelled release of a small number of transgenic fish into a wild

Figure 18.22
Salmon cages in Haida Gwaii, British Columbia.

Figure 18.23 Trends in world aquaculture production: major species groups.

SOURCE: Food and Agriculture Organization of the United Nations, 2011, *World Aquaculture 2010. FAO Fisheries and Aquaculture Technical Paper 500/1*, http://www.fao.org/docrep/014/ba0132e/ba0132e.pdf

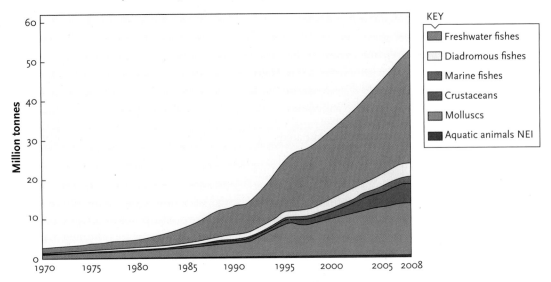

Trends in world aquaculture production: major species groups, 1970–2008

KEY
- Freshwater fishes
- Diadromous fishes
- Marine fishes
- Crustaceans
- Molluscs
- Aquatic animals NEI

population drove the population of wild fish to extinction within a few generations. At the same time the potential advantage of growth hormone treatment to the fish farmer is significant. Food production for cultured fish can have environmental consequences. Mussel and oyster farms depend on plankton in the water and involve no additional food supply, so these aquaculture operations have a low environmental impact. In contrast, farmed salmon are fed an artificial diet that includes fish from the capture fisheries. A wild salmon has a food conversion ratio (FCR) of about 10:1 (it takes 10 kg of food to produce 1 kg of salmon). This is roughly the same as the FCR of beef. Farmed pigs have an FCR of 5:1 and broiler chickens around 2:1. In aquaculture, grass carp (*Ctenopharyngodon*

idella (1196)) fed a soy-based diet grew from 775 g to 2.8 kg in 146 days and had an FCR of 1.7:1. Tilapia on the same diet achieved an FCR of 1.2:1 as they grew from 28 g to 525 g in 131 days. Some salmon farmers achieve an FCR of 1.5:1 by feeding dry soy-based food to their fish. But these figures ignore the reality that while salmon grow well on soy protein, they still require fish oil and fat in their diet. Three tonnes of wild-caught fish are needed to provide the oil for one tonne of dry feed for farmed salmon. This makes the effective FCR more than 4:1. While much of the capture fish used to make oil for salmon feed would not be used for human consumption, the removal of these fish from the food chain (see page 206) affects the population of wild fish used directly as food for people.

The benefits and costs of aquaculture must be considered on a case-by-case basis as aquaculture plays an increasing role in human nutritional economics.

STUDY BREAK

1. What do apiculture and aquaculture have in common? How do they differ?
2. What are the differences between domestication and taming? Give two examples of each.

A   discpicture / Shutterstock.com

B   Oleg_Z / Shutterstock.com

Figure 18.24 Freshwater aquaculture. **A.** Tilapia (*Tilapia natalensis* (1293)) and **B.** grass carp (*Ctenopharyngodon idella* (1196)).

18.5 Diseases

In this section we will consider two ways in which animals can affect human health:

1. as direct causative agents of human disease
2. as vectors that move disease-causing organisms between humans or between humans and other animals

Table 18.1 Top ten species groups in global aquaculture production.

	Quantity (millions tonnes)	Value (US$ billions)
Freshwater fishes	28.8 (55%)	40.5 (41%)
Molluscs	13.1 (25%)	13.1 (13%)
Crustaceans	5.0 (10%)	22.7 (23%)
Diadromous fishes	3.3 (6%)	13.1 (13%)
Marine fishes	1.8 (3%)	6.6 (7%)
Aquatic animals NEI	0.6 (1%)	2.4 (3%)

SOURCE: Food and Agriculture Organization of the United Nations, 2011, *World Aquaculture 2010. FAO Fisheries and Aquaculture Technical Paper 500/1*, http://www.fao.org/docrep/014/ba0132e/ba0132e.pdf

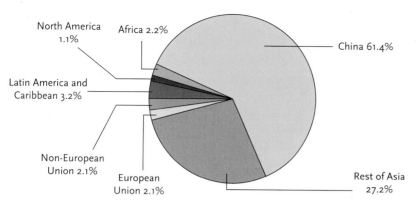

Figure 18.25 World aquaculture production by area.

SOURCE: Food and Agriculture Organization of the United Nations, 2012, *The State of World Fisheries and Aquaculture 2012*, http://www.fao.org/docrep/016/i2727e/i2727e.pdf

Not surprisingly, blood-feeding animals often serve as vectors. Mosquitoes are a prime example (Table 18.2), but many other animals that do not feed on blood also serve as vectors.

The list of human diseases that are caused or transmitted by animals is long and frightening. Diseases such as West Nile virus attract a lot of attention from the media at some times of the year in some parts of the world. Mosquitoes are the vectors and pick up the virus by eating the blood of infected birds. Although most people infected with West Nile virus experience mild flu-like symptoms or no symptoms at all, this virus can cause severe illness that requires hospitalization. West Nile virus has killed people, horses, and birds (mainly corvids, crows and ravens, and young geese).

Other mosquito-borne diseases are much more serious, for example, dengue and dengue hemorrhagic fever. These diseases are found in tropical and subtropical climates around the world. Dengue is caused by four distinct closely related viruses. People suffering from dengue exhibit flu-like symptoms that can be severe. In 2007 there were more than 890 000 cases of dengue in the Americas. Dengue hemorrhagic fever is a much more serious affliction causing illness and death, particularly among children. The statistics suggest that <10% of the cases of dengue turn into dengue hemorrhagic fever. Both diseases are now endemic in more than 100 countries in the tropics and subtropics.

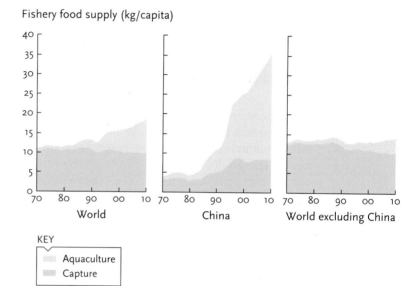

Figure 18.26 Relative contributions of aquaculture and capture fisheries to food fish consumption.

SOURCE: Food and Agriculture Organization of the United Nations, 2012, *The State of World Fisheries and Aquaculture 2012*, http://www.fao.org/docrep/016/i2727e/i2727e.pdf

Before 1970 they were reported from only nine countries, 36 countries by 1995.

Lymphatic filariasis is caused by *Wolbachia* bacteria that are endosymbiotic in the cells of thread-like nematode worms, *Wucheria bancrofti* and *Brugia malayi*. These worms lodge in the lymphatic system where they can live for 4–6 years while producing

Table 18.2	A sample of human diseases and their vectors.	
Disease	Causative agent	Vector
Plague (black death)	*Yersinia pestis* (Bacteria)	fleas
Malaria	*Plasmodium* spp. (Sporozoa)	*Anopheles* mosquitoes
Dengue haemorrhagic fever	viruses	mosquitoes
African trypanosomiasis (sleeping sickness)	*Trypanosoma* spp.	*Glossina* spp. (tsetse flies)
Trypanosomiasis (Chagas disease)	*Trypanosoma* spp.	*Rhodnius* (kissing bug, see page 355)
Lymphatic filariasis	*Wucheria bancrofti* and *Brugia malayi* (filarial worms)	mosquitoes
Onchocerciasis (river blindness)	*Onchocerca volvulus* (filarial worm)	black flies (Simuliidae)

millions of minute larvae (microfilariae) that, with their bacteria, circulate in the blood. The *Wolbachia* are released into the host's blood when the microfilariae die. The worms are associated with a disease known as **elephantiasis** (Fig. 18.27), which is spread by mosquitoes that pick up microfilariae when feeding on an infected person. In 7–21 days, microfilariae develop inside the mosquito and then migrate to the salivary glands. They are inoculated into a new human host when the mosquito takes her next blood meal. The most alarming symptoms of elephantiasis are enlarged legs, arms, genitals, and breasts due to accumulation of lymph that cannot circulate through blocked lymph ducts. In communities where lymphatic filariasis is endemic, up to 15% of men and 10% of women can be infected. While the symptoms are alarming and are socially stigmatizing, more dangerous is the damage done to the afflicted lymphatic system and kidneys.

Black flies (Simuliidae) can transmit filarial worms such as *Onchocerca volvulus* (589) where again *Wolbachia* endosymbiotic bacteria cause a disease, this time **onchocerciasis** (river blindness). In humans, the larvae form nodules in subcutaneous tissues where they mature into adults. Mated females can release up to 1000 microfilariae a day. Microfilariae move through the blood stream, and their endosymbionts cause conditions such as blindness, skin rashes, lesions, intense itching, and depigmentation. In addition to efforts to control black flies, available drugs (e.g., Ivermectin®) kill microfilariae, offering the possibility that this disease could be eradicated.

Wolbachia bacteria appear to infect over 60% of insect species. Perhaps ironically, infecting *Aedes aegypti* (953) with *Wolbachia pipientis* shortens the life span of this mosquito. The incubation period of viruses and parasites is relatively long, so a shorter lifespan renders the mosquito less likely to be a vector for human pathogens such as the one causing dengue fever.

African **trypanosomiasis**, also known as sleeping sickness (Fig. 18.28), is caused by the protist *Trypanosoma brucei*, which is spread by tsetse flies (*Glossina* spp. (963)). Sleeping sickness occurs only in sub-Saharan Africa and can appear in two forms. Infections by *Trypanosoma brucei gambiense* occur in west and central Africa and account for more than 90% of reported cases. Infections can be chronic and may take months or years before any symptoms develop. *Trypanosoma brucei rhodesiense*, in eastern and southern Africa, accounts for the remaining 10% of reported cases. In this case the infections are acute. The disease develops rapidly and affects the central nervous system. Other mammals, especially artiodactyls, can host *T. b. rhodesiense* so that both game and cattle can be reservoirs for the disease. The impact of sleeping sickness on cattle limits their use by people living in areas that harbour tsetse flies. To some, tsetse flies are the "saviours of Africa" because they have limited the spread of large ranching operations in tsetse fly territory.

Ticks—another blood-feeding arthropod—can also spread diseases. In the US and Canada, **Lyme**

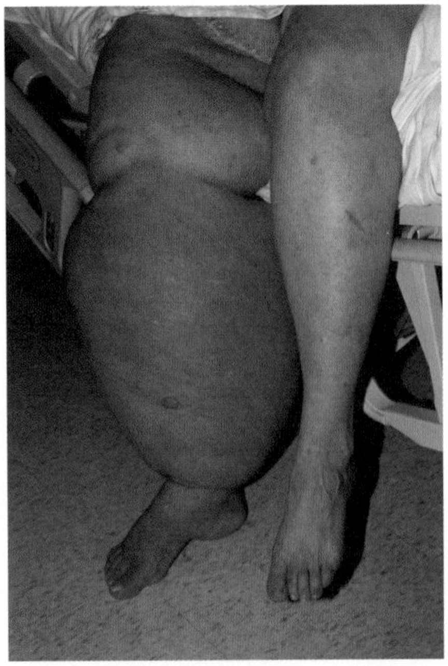

Figure 18.27 A victim of elephantiasis.

Figure 18.28 A sign warning of sleeping sickness and identifying traps designed to capture tsetse flies (*Glossina morsitans* (963)).

disease is caused by the bacterium *Borrelia burgdorferi*, while in Europe another species, *Borrelia afzelii*, causes a similar condition. Both bacteria are spread by tick bites. In eastern North America, *Ixodes scapularis* (651), the deer tick, is the common vector; *Ixodes pacificus* (650) is the carrier in western North America. In its early stages, Lyme disease is characterized by a rash, which, if untreated, is followed by more debilitating symptoms. Lyme disease was identified in the US in 1975 when mothers living near Lyme, Connecticut, contacted researchers because their children had all been diagnosed with rheumatoid arthritis. The researchers eventually identified the causative agent as the pathogenic bacterium transmitted by ticks, and the condition became known as Lyme disease.

People who walk in long grass should check themselves for (and remove) ticks, remembering that in their nymphal stages, ticks can be as small as the head of a pin.

We will now consider some diseases in more detail.

18.5a Malaria

Animals such as the mosquito *Anopheles gambiae* (954) are prime vectors for *Plasmodium* spp., a protist that causes malaria. The complex life cycle of the malarial parasite (Fig. 18.29) includes some stages that are completed in the mosquito and other stages completed in the human host.

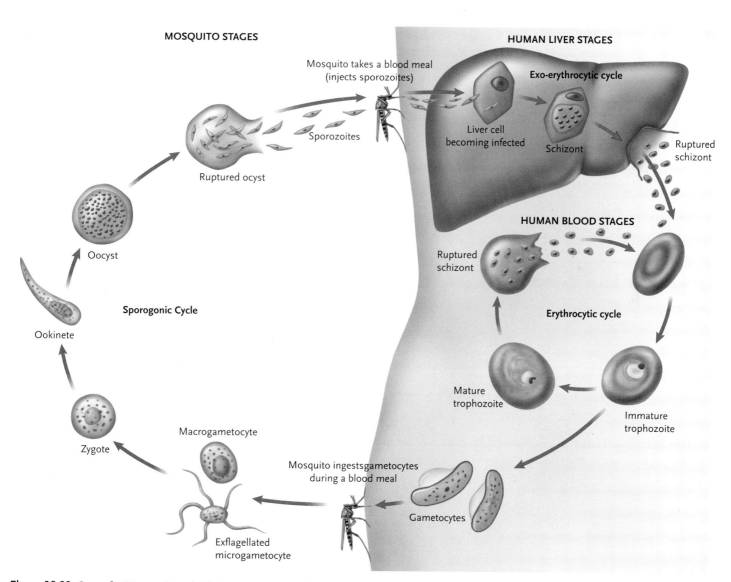

Figure 18.29 Center for Disease Control (CDC) representation of the malaria cycle. A biting mosquito (1) releases hundreds of sporozoites into the blood of a live host, where they infect hepatocytes (2) and form schizonts (3) that rupture (4), releasing merozoites that infect erythrocytes (5). There, some differentiate into gametocytes (7). Gametocytes (males = microgametocytes; female = macrogametocytes) are ingested by *Anopheles* mosquitoes (8). This begins the sporogenic cycle (C) where zygotes are formed when microgametocytes penetrate macrogametocytes (9). Zygotes become motile ookinetes (10) that invade the lining of the midgut wall of the mosquito and develop into oocysts (11). The oocysts grow. When they rupture, they release sporozoites (12) that move to the salivary glands of the mosquito. When the mosquito feeds on a host, it is infected by the sporozoites.

Like many other species of mosquito, *Anopheles gambiae* use olfactory cues to locate their hosts. The receptors on the olfactory neurons are controlled by a family of 79 *Or* genes. *A. gambiae* has a mixture of finely and broadly tuned olfactory receptor neurons that detect human-specific olfactory signatures.

Individual humans differ in their attractiveness to mosquitoes. For example, pregnant women are about twice as attractive to female *Anopheles* mosquitoes as women who are not pregnant. Research conducted in Burkina Faso (Africa) using human volunteers demonstrated that drinking beer consistently increased the attractiveness of people to mosquitoes (Fig. 18.30). Drinking water did not affect a person's attractiveness to mosquitoes.

These results provide evidence that people vary in their attractiveness to mosquitoes (and thus their exposure to malaria) and add another risk to the effects associated with drinking beer. Preference of blood-feeding insects for some individuals over others is a common experience for those working in areas where they are exposed to mosquitoes and other biting insects.

The most severe form of malaria is caused by *Plasmodium falciparum*, and other forms of the disease are caused by *P. vivax*, *P. ovale*, and *P. malariae*. Together these species account annually for about 515 million cases of malaria transmitted by bites of female

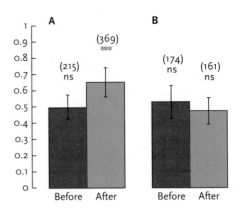

Figure 18.30 Beer consumption makes people more attractive to mosquitoes. Compared here are the numbers of mosquitoes caught in traps as they approached on of two groups: **A.** 25 volunteers who had consumed beer and **B.** 18 volunteers who had consumed water. Captures are expressed as percent of the total mosquitoes captured. *** = significant differences; ns = nonsignificant differences.

SOURCE: LeFevre et al. 2010. "Beer Consumption Increases Human Attractiveness to Malaria Mosquitoes." *PLoS ONE* 5:e9456. Doi: 10.1371/journalpone.0009546.

anopheline mosquitoes. Children under five are the most vulnerable to this disease, and many who survive it suffer impaired learning ability. Malaria caused by *P. falciparum* results from a recent bite by an infected mosquito. Malaria caused by *P. vivax* and *P. ovale* is

BOX 18.3
Mosquitoes and the Jenkins' Ear War

Diseases have played important roles in the outcome of many military operations. A good example is the Jenkins' Ear War which began in 1739 after a Spanish warship stopped an English merchant vessel, captained by Robert Jenkins, believed to be carrying goods looted from Spanish ships. The captain of the Spanish vessel cut off one of Jenkins' ears. The incident resulted in the assembly of an English armada with 18 000 troops that arrived on 15 March 1741 off Cartagena in Colombia. The English force blockaded Cartagena, but within two weeks of their arrival many troops were suffering from mosquito-borne diseases. By the time the English gave up their blockade, they had lost almost 9000 men to some combination of malaria and yellow fever.

Black vomit is one symptom of yellow fever, which begins to kill victims about five days after infection. Malaria takes longer to kill people, starting about 14 days after infection. Reports from the time suggest that both diseases could have been involved. Mosquitoes carrying malaria (*Anopheles* spp. (**956**)) and yellow fever (*Aedes aegypti* (**953**)) bred in cisterns in the areas around Cartagena. Earlier, in 1586, Sir Francis Drake captured Cartagena, but abandoned the prize after losing many of his men to disease.

After the 1741 debacle of Cartagena, the British navy took steps to protect their personnel from malaria by acquiring and using extracts of tree bark (*Cinchona* spp.), which was obtained from Bolivia and other areas of the Andes in South America. But even the bark extract was not always enough.

In 1809, King George III of England ordered a raid against Napoleon's forces in Europe and an English force landed at Walcheren, Belgium on 10 August 1809. Napoleon is said to have remarked that his forces need do nothing about the invasion because the mosquitoes would settle the matter. He ordered some of his men to break the dykes and flood the lowland where the English forces were camped. By the end of August, malaria had appeared in the English forces and by the end of September it had killed over 1000 troops. Several thousand others were evacuated to hospital ships. By the end of October 1809, the English forces were suffering 20–30 deaths a day. By the time the forces were withdrawn, malaria had killed over 4000 and another 11 000 were gravely ill. Yes, the English forces had extracts of *Cinchona* bark, but either there was not enough or it was the wrong bark.

also derived from the bite of an infected mosquito but can persist in a dormant hypnozoite form that may last for years after the victim has left the area of original exposure.

Traditionally, malaria was treated with alkaloids (quinine, quinidine, cinchonidine, and cinchonine) extracted from the bark of four species of trees in the genus *Cinchona*. Today malaria can be treated with drugs such as chloroquine and sulphadoxine-pyrimethamine, which are readily available and inexpensive, but increasing resistance of *Plasmodium* to these drugs raises concerns about the longer-term situation. As of 2010 there is no vaccine for malaria. A recombinant malaria protein fused to the surface of a hepatitis B protein and based on *P. falciparum* circumsporozoite protein could allow the development of a vaccine.

The genome sequence of *P. falciparum* was published in 2002, setting the stage for significant advances in the fight to eliminate malaria. This development, combined with advances in our knowledge about how *A. gambiae* choose their victims, offers opportunities for manipulating vectors to minimize their impact on humans. Knowledge of the behaviour and biology of vectors and disease-causing agents is necessary to advance our efforts to avoid malaria and other diseases.

18.5b Plague

Plague, also known as the Black Death, has killed many millions of people throughout history. The bacteria that causes plague, *Yersinia pestis*, infects small mammals, usually rodents, and their fleas. Fleas acquire the bacteria when they eat the blood of infected rodents, and an infected flea can pass the bacteria to humans (Fig. 18.31). Three to seven days after infection, the

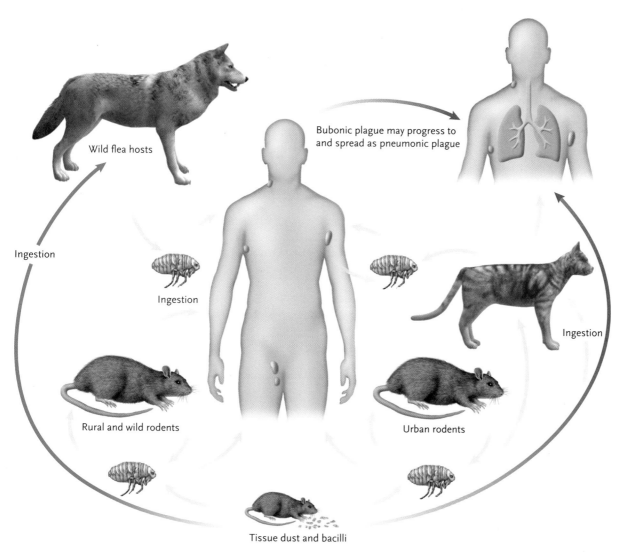

Bubonic plague may progress to and spread as pneumonic plague

Wild flea hosts

Ingestion

Ingestion

Ingestion

Rural and wild rodents

Urban rodents

Tissue dust and bacilli

Figure 18.31 The plague, caused by the bacteria *Yersina pestis*, is involved in several pathways shown here. Although we tend to associate the plague with rodents and their fleas, other animals can also be involved, depending upon the setting (rural or urban). Bubonic plague progresses to pneumonic plague when it is spread by droplets (sneeze or cough). Dark blue arrows show continuous pathways for bubonic plague, and red arrows show occasional pathways for pneumonic plague. Light blue arrows represent occasional pathways for bubonic plague.

person experiences flu-like symptoms such as a sudden fever, head and body aches, vomiting, and nausea.

There are three forms of clinical plague infections: bubonic, septicemic, and pneumonic. In the bubonic form, bacteria travel in the lymphatic system and accumulate in lymph node(s) close to the site of the bite. There they multiply and cause a "bubo" or swollen and painful lymph node. Bubos can suppurate as open sores, which are vulnerable to secondary bacterial infections. In the septicemic form, the bacterial infection spreads through the bloodstream and does not form bubos. This form of plague may result from a flea bite or through direct contact, perhaps through a cut or crack in the skin, with materials infected with *Y. pestis*. The pneumonic form of plague is the most virulent and least common, with a secondary spread of the bacterial infection to the lungs. The bacteria can move from human to human as an aerosol spray following a sneeze or cough, a mode of transmission that does not involve fleas or rodents.

Plague can be deadly and is mainly responsible for the drastic decline of about 20% of the human population between 1340 CE and 1400 CE. These declines were concentrated in Africa and Eurasia as the plague did not reach the New World until about 1600 CE. Today plague occurs in rodent populations in many countries in Africa, the former Soviet Union, the Americas, and Asia. In 2003 a total of 2118 cases of plague was reported worldwide (from nine countries), resulting in 182 deaths. More than 98% of these cases were in African countries.

Biologists working with live rodents in the field should be extremely careful to avoid bites from the fleas inhabiting their study animals.

18.5c Schistosomiasis

Schistosomiasis, also known as bilharziasis, is caused by infections by trematode flatworms. Larval cercaria stages emerge from freshwater snails (intermediate host) and penetrate the skin of people (definitive host), perhaps while bathing or swimming (Fig. 18.32). In the human hosts, the larvae become adult schistosomes that live in blood vessels. Female schistosomes produce and release eggs, some of which pass out of the host's body in urine or faeces. Eggs trapped in the host's body cause an immune reaction.

Adult schistosomes may live in blood vessels lining the bladder (urinary schistosomiasis) or the intestine. Symptoms of urinary schistosomiasis include damage to the bladder, ureter, and kidneys, while intestinal schistosomiasis causes enlargement of the spleen, damage to the intestines, and hypertension (high blood pressure) in the abdominal blood vessels (Table 18.3).

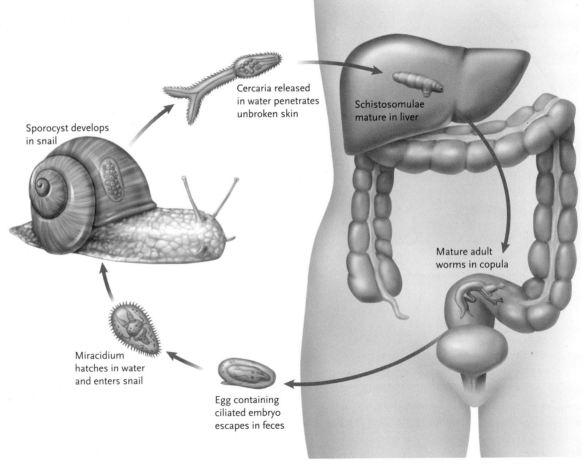

Sporocyst develops in snail

Cercaria released in water penetrates unbroken skin

Schistosomulae mature in liver

Mature adult worms in copula

Miracidium hatches in water and enters snail

Egg containing ciliated embryo escapes in feces

Figure 18.32 *Schistosoma* (265) life cycle.

Table 18.3	Schistosomiasis by species and geographic area.	
Disease	Species	Geographical distribution
Intestinal schistosomiasis	*Schistosoma mansoni*	Africa, Middle East, Caribbean, Brazil, Venezuela, Suriname
	S. japonicum	China, Indonesia, the Philippines
	S. mekongi	Areas of Cambodia, Laos
	S. intercalatum & S. guineansis	Rain forests of Central Africa
Urogenital schistosomiasis	*S. haematobium*	Africa, Middle East

Schistosomiasis is a chronic disease affecting >207 million people worldwide, with another 700 million at risk. The life cycle makes it clear that lack of hygiene and sanitation help spread this disease. The intermediate hosts are usually snails in the genus *Biomphalaria* (472), but other snails may also act as intermediate hosts (e.g., *Bulinus globosus* (474), *B. forskalii* (473), and *Australorbis (Biomphalaria) glabratus* (470)).

People are most apt to be exposed to *Schistosoma* cercaria if they enter stagnant or weakly flowing water that has been contaminated by feces from infected individuals. By slowing water flow, dams have increased the incidence of schistosomiasis—the Aswan dam in Egypt is a classic example. Among diseases associated with parasites, only malaria kills more people every year than schistosomiasis does.

18.5d Emerging Diseases

New infectious diseases appear at an astonishing rate of about one every eight months. Often the trade in wildlife is an important route for the appearance of new diseases in humans. Globally, the annual trade in wildlife involves ~40 000 live primates, 4 million live birds, 640 000 live reptiles, and 350 million tropical fish. In a single market in North Sulawesi, for example, up to 90 000 mammals are sold each year. In addition to trade in live wildlife, local caught wild animal meat ("bush meat") is consumed at astonishing rates, for example >1 billion kg per year in Central Africa, and 67–164 million kg per year in the Amazon Basin. Each animal that comes into contact with humans has the potential to pass to humans a disease usually only found in animals. These are called **zoonotic** diseases and include those caused by the viruses HIV, SARS-CoV, Nipah, and Hendra.

HIV may originally have been linked to humans that ate nonhuman primates and great apes as bush meat. African fruit bats (family Pteropodidae) appear to be the reservoir for the virus causing Ebola hemorrhagic fever. Severe acute respiratory syndrome (SARS) is caused by SARS-associated coronavirus (SARS-CoV) and had been associated with international trade in small carnivores, specifically palm civets (*Paguma larvata* (1725)). Further research indicated that palm civets were part of a market cycle, defined as wild-caught animals being held in a market before sale. These animals were kept in an animal market in Guangdon, China, where they were sold to locals. At the time, palm civets did not appear to be the natural reservoir for SARS-CoV. SARS severely affected the economy in Southeast Asia and had a large disruptive impact in Toronto, Canada, showing how such diseases can spread.

On the basis that bats are reservoir hosts for some zoonotic viruses, researchers sampled 408 bats (nine species, six genera, and three families) from four locations in China. Bats in the genus *Rhinolophus* (1813) tested positive for coronaviruses closely related to SARS-CoV. Bats also are known to be reservoirs for two other emerging viral diseases, namely Nipah and Hendra. People concerned about the conservation of bats did not welcome the news about bats and SARS and pointed out that the numbers of infected bats in the sample was small. Furthermore, lack of knowledge about the bats made it difficult to determine how bats contributed to the market cycle: what role could they have played in spreading SARS? The route of connection is clear in Nipah virus where chewed fruit fibres are spat out by the bats and then eaten by pigs. The fruit bats involved were species in the family Pteropodidae. Hendra virus is transferred in a similar manner from fruit bats to horses and thence to humans. Separating the fruit bats from the pigs or the horses minimized the chances of transfer of Nipah or Hendra from fruit bats to people. Fruit bats are also reservoirs for Ebola virus, and people who eat these bats may be exposed to this disease.

18.5e Other Diseases and Vectors

People who live in close contact with animals, and those that eat undercooked or raw animals, put themselves at risk of being exposed to disease-causing agents.

Some species of fish are intermediate hosts for parasitic worms (usually nematodes and cestodes) whose definitive hosts are fish-eaters. After eating raw fish, some people have experienced piercing stomach pains that are associated with a parasite that normally pierces the stomach of a seal to complete its life cycle. Knowing that this is a case of mistaken identity does not make the pain go away.

Figure 18.33
In an Asian rainforest, a tropical leech feeds on a human foot.

Other animals could be vectors for diseases that affect people. Leeches (class Hirudinidea) are often blood feeders and, as such, may be ideal vectors for spreading disease to people (and other animals). Leeches from Cameroon (Africa) were serologically positive for human immunodeficiency virus (HIV) and hepatitis B, while medicinal leeches (*Hirudo medicinalis* (367), Fig. 18.33) purchased from a pharmacy in Germany housed as many as 11 species of bacteria. Leeches are vectors for diseases of other animals, such as sea turtles. It is possible that land leeches are more often involved with human diseases than aquatic ones.

STUDY BREAK

1. What is an emerging disease? What factors are responsible for emerging diseases?
2. Compare malaria, schistosomiasis, and HIV. What do they have in common? How do they differ?

18.6 Recreation

It is difficult to overestimate the recreational value of animals and the associated benefits to individual humans and to society in general. Think of the spectrum, from the person whose house cat is their main day-to-day contact with another being, to the family whose dog gives them exercise and brings them into contact with other people and dogs. Consider the economic value of horse or dog races, rodeos, and contests between fighting fish, not to mention the social and economic impact of birdwatching (Fig. 18.34). The

economic impact of our dependence upon animals for recreation is astonishing, from pet products to veterinary services. In some ways this is the "other" side of the wildlife trade (see below).

Naturally, perhaps, there is tension around our recreational use of animals, with sport fishing, hunting, and dog racing providing familiar examples. To get an impression of the economic impact of sport fishing, visit a sporting goods store or the fishing tackle section of a department store during the "season." Look at the range of products, their costs, the numbers of customers, the numbers of people employed just in retail, and the places where the products are made. You can further your impression of the importance of sport fishing by visiting a town that has a fishing derby—and easily appreciate the social importance of the sport.

Hunting is a more difficult recreational activity for many people to accept, partly because of the use of firearms, but also because the quarry is usually mammals or birds. In countries such as Tanzania, the annual revenue from tourism is about US\$14 billion, reflecting the attraction of its wildlife to tourists who might be naturalists, photographers, or safari hunters. In 2010, Tanzania petitioned CITES (Convention for International Trade in Endangered Species) to make it legally possible to sell 90 tonnes of elephant ivory (Fig. 18.35). The CITES listing of African elephants forbids international trade in their ivory. Ironically, the revenue Tanzania could expect to receive from the sale of that ivory was <1% of their annual revenue from tourism, a reality used to emphasize the value of living versus dead wildlife. Another issue in this debate is the source of the ivory in question. The ivory could have been legally obtained as a result of management activities that included culling elephants to ensure that their populations did not exceed the carrying capacity of

Figure 18.35 Stored ivory in a National Parks and Wildlife vault in Zimbabwe. These tusks were obtained from control of problem elephants and confiscations from poachers.

Figure 18.34
Birdwatchers in Costa Rica.

their habitat. Elsewhere in Africa (e.g., Tsavo National Park), die-offs of elephants have resulted from over-population. In 2010 CITES did not approve the request from Tanzania to legalize the sale of ivory.

One important problem with legalizing the sale of the Tanzanian ivory is opening the market to other ivory that may have been poached and thus obtained illegally. Efforts to identify the geographic source of ivory by genetic or isotopic analysis have not been successful. Knowing the source of ivory (or any animal product) you want to buy is a widespread problem, and the labels in a market may not be accurate or truthful.

It is a mistake to think that so-called ecotourists have a less negative impact on habitat and animals than fishers or hunters. Bird watchers and photographers, as well as hikers, naturalists, and many other visitors, require infrastructure (roads, services, places to stay, toilet facilities, etc.) and may travel considerable distances to find animals. Their activities may interfere with the lives of the animals they are watching, particularly if the people are inconsiderate or unaware of the requirements of the animals.

STUDY BREAK

1. How does the sale of elephant ivory pose a threat to their conservation?
2. How do hunting, fishing, and birdwatching differ in their use of animals?

18.7 Humans and Environmental Change

18.7a Introduction

Exposure to emerging diseases (above) is one consequence of the trade in wildlife. Perhaps more important is the potential for introduction of alien species. The international Convention on Biological Diversity enjoins agencies to assess the risks associated with introducing alien species. Effective risk analysis requires input from all of those involved, from those who capture animals for trade, exporters, importers, people in the pet trade, and customers. In the US the Nonnative Wildlife Invasion Prevention Act (H.R. 669) is a recent effort to curb introductions and requires additional data about the animals to be imported. This includes information about the identity of the species, its biology, and its natural history. Some alien species may be more dangerous because of diseases they harbour. For example, the fungus *Batrachochytrium dendrobatidis* causes chytridiomycosis and is lethal to many amphibians. This fungus was unintentionally imported and has been responsible for drastic reductions in some amphibian populations.

Many introductions are accidental, as exemplified by the lengthy list of organisms that have invaded new regions by transport in ballast water of ships. Ocean-going ships transporting a light cargo pump seawater into a ballast hold to stabilize the ship during the ocean voyage. This ballast water is then released in a distant port when the light cargo is exchanged for a heavier load of goods. A survey of ballast water in 159 cargo ships at Coos Bay, Oregon revealed 367 taxa of alien species, including 16 animal and three protist phyla as well as three divisions of plants. Introduced species may out-compete native species. A classic example is the introduction of zebra mussels (*Dreissena polymorpha* (547)) into the Great Lakes, which coincided with reductions in populations of native species. Accumulations of zebra mussels also block water intake pipes vital to sewage treatment and hydro-electric facilities.

Other introductions have been intentional. Starlings (*Sturnus vulgaris* (1616)) and English sparrows (*Passer domesticus* (1608)) were introduced intentionally to the US and Canada, as were mongooses (*Herpestes auropunctatus* (1716)) to the West Indies. From an ecological standpoint, it is interesting to try to determine if the success of an immigrant species begins with a move into vacant niches or a displacement of a native species. The arrival of humans almost everywhere around the world heralded the release of domesticated animals (and plants) that wreaked havoc in many ecosystems. House cats (*Felis catus* (1713)) are particularly dangerous predators that have been as destructive as their counterpart accidental immigrants, rats (*Rattus norvegicus* (1903)); both have wiped out entire species in areas where they've been introduced (e.g., New Zealand, Australia). Pigs (*Sus scrofa* (1768)) and goats (*Capra aegagrus hircus* (1748)) are also notorious for the damage they do to local ecosystems.

Hybridization can be another impact of introduced species. Ruddy ducks (*Oxyura jamaicensis* (1660)) have been widely introduced into Europe, where they have interbred with the native white-headed ducks (*Oxyura leucocephala* (1661)). Hybrids between the two species of ducks are fertile, and the rapid expansion of European populations of ruddy ducks and hybrids threatens the survival of the native species.

The potential impact of political processes on the movement of wildlife is provided by data spanning the erection and dissolution of the Iron Curtain. Examining the records of bird introductions to countries on both sides of the Iron Curtain demonstrates that political reality can affect the movement of wildlife (Fig. 18.36).

In some cases the introduction of predators transforms habitats. Arctic foxes (*Alopex lagopus* (1706)) were introduced to islands in the Aleutian archipelago around 1900 to provide an additional source of furs for residents. About 100 years after the foxes had been introduced, the vegetation on islands with Arctic foxes had changed substantially: grassland communities had become dwarf shrub/forb ecosystems. Arctic foxes had reduced populations of seabirds and affected soil fertility by reducing deposits of bird dung (Fig. 18.37).

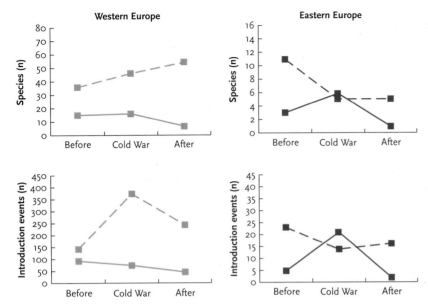

Figure 18.36 A comparison of the total numbers of European (solid lines) and non-European (dashed lines) exotic bird species introduced into Western and Eastern Europe (separated by the Iron Curtain) before, during, and after the Cold War. Note the difference in scale between Western and Eastern Europe.

SOURCE: Reprinted from *Biological Conservation*, 143, Chiron, F., S.M. Shirley, and S. Kark, "Behind the iron curtain: socio-economic and political factors shaped exotic bird introductions into Europe," Pages 352–356, Copyright 2010, with permission from Elsevier.

Once an alien species is established, it may be very difficult (or impossible) to eradicate. Steps to remove it may be counterproductive. Consider the case of Cook's Petrel (*Pterodroma cookii* (1635)), a seabird that nests on Little Barrier Island off the north coast of New Zealand's North Island. Both cats and Pacific rats (*Rattus exulans* (1902)) had been introduced to the islands some time before. Eradication of the cats in 2004 was expected to protect the petrels. However, the number of petrels declined, apparently because the populations of Pacific rats had been partly held in check by the cats (Fig. 18.38). The cats had been the top predator on the island, and their removal led to the expansion of the rat (a meso-predator) population. However, there was a clear effect of altitude (Fig. 18.38B). Worldwide, at least eight species of petrels are threatened or have been exterminated by introduced mammals (Table 18.4).

Alien species introductions can have dire consequences, and we will probably never succeed at extirpating them. Imagine the range of candidates—House Sparrows (*Passer domesticus* (1608)), Starlings (*Sturnus vulgaris* (1616)), Common Mynahs (*Acridotheres tristis* (1589)), Pigeons (*Columba livia* (1570))—just choosing birds gives us a seemingly endless list. And yet the international trade in wildlife continues both legally and illegally, often because of humanity's disregard for nature.

18.7b Land Use

The relentless increase in human population has enormous consequences for life on Earth. Increasing numbers of humans translate into growing demands for

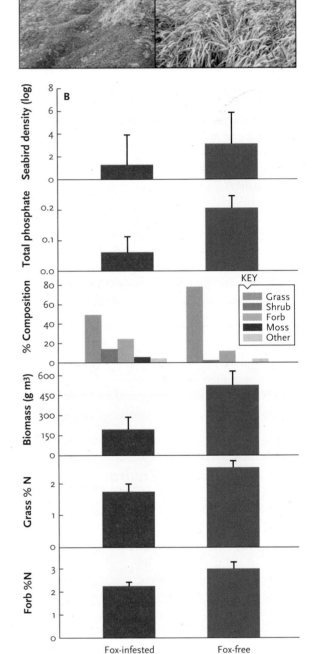

Figure 18.37 A. Photographs of the habitat in islands in the Aleutian archipelago with (left) and without (right) foxes. **B.** Some Aleutian islands have arctic foxes (red), others (blue) do not. The graphs show seabird density, total phosphate, composition, biomass, and grasses and forbs. In B, means are shown ± standard errors.

SOURCE: From Croll, D.A., J.L Maron, J.A. Estes, E.M. Danner and G.V. Byrd, "Introduced predators transform subarctic islands from grassland to tundra," *Science* 25 March 2005: Vol. 307 no. 5717 pp. 1959–1961. Reprinted with permission from AAAS.

food, space, and services, placing more and more pressure on the planet's ecosystems. Palm oil provides an interesting example. Global demand for oil from these palms (*Elaeis guineensis*) means that the land area used in its cultivation doubled in 15 years (Fig. 18.39),

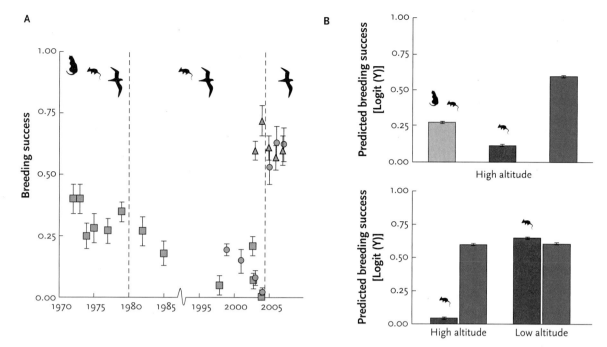

Figure 18.38 The situation involving cats, rats, and petrels on Little Barrier Island, New Zealand, between 1970 and 2006. **A.** The distribution of study burrows (of petrels) at high (squares) and low (circles or triangles) altitude sites (means ± standard error). **B.** The impact of two predators (cat and rat), one predator (rat), and no predators at high and low altitude sites. In the upper graph in B, the time period is 1972–2007; in the lower plot, 2003–2007.

SOURCE: Rayner, M.J., M.E. Hauber, M.J. Imber, R.K. Stamp and M.N. Clout. 2007. "Spatial heterogeneity of mesopredator release within an ocean island system," *PNAS*, 104:20862–20865. Copyright 2007 National Academy of Sciences, U.S.A.

particularly in southeast Asia which still houses about 11% of the world's remaining tropical forests.

Palm oil is used in cooking, as an additive to food, in cosmetics, as an industrial lubricant, and as a biofuel. Expansion of oil palm plantations threatens the habitat essential to many native animals, from orangutans (*Pongo pygmaeus* (1867)) to some 89 species of amphibians endemic to Malaysia. By 2007, three species of mammals, Verhoeven's giant tree rat (*Papagomys theodorverhoeveni* (1892)), Flores long-nosed rat (*Paulamys naso* (1893)), and Flores cave rat (*Spelaeomys florensis* (1905)) were listed as "extinct in the wild," partly due to habitat lost to the growth of oil palms.

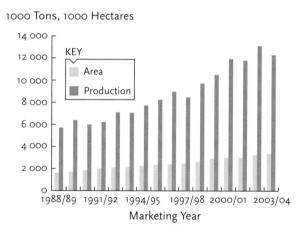

Figure 18.39 Expansion of palm oil production in Malaysia. The area under production is shown (green bar) along with the amount produced (blue bar).

SOURCE: USDA

Table 18.4	The world-wide distribution of petrels and the main threats to their survival.		
Common name	Latin name	Breeding location	Main causes of decline
Fiji Petrel	*Pseudobulweria macgillivrayi*	Fiji	Predation by cats and rats
Mascarene Black Petrel	*Pseudobulweria aterrima*	Reunion	Predation by cats and rats, urban light-induced mortality
Beck's Petrel	*Pseudobulweria becki*	Melanesia	Predation by cats and rats
Chatham Petrel	*Pterodroma axillaris*	Chatham Island	Predation by introduced predators; competition with another seabird for breeding burrows
Jamaica Petrel (possibly extinct)	*Pterodroma caribbaea*	Jamaica	Predation by rats and mongooses
Magenta Petrel	*Pterodroma magentae*	Chatham Island	Predation by cats and rats
Galapagos Petrel	*Pterodroma phaeopygia*	Galapagos Islands	Predation by introduced predators
Balearic Shearwater	*Puffinus mauretanicus*	Belearics	Predation by cats; bycatch in longline fisheries

One solution has been proposed, namely having nongovernmental organizations (NGOs) acquire small tracts of palm plantations and use the revenue to establish a system of privately owned protected areas. In 2005, palm oil plantations annually yielded about US$2000 per hectare, and the cost of acquiring plantations was about US$12 500 per hectare. To get into the business, NGOs would require funds up front to provide an opportunity for partnership ventures. The value of conservation is obvious and need not involve stopping palm oil operations.

Establishing protected areas and then enforcing their protection is an important element in conserving biodiversity. One problem with protected areas (some of them National Parks) is that they can appear to be recreational reserves for the well-to-do. Accelerated growth of human populations along the edges of protected areas (Fig. 18.40) could threaten the continuity of the protected areas and their effectiveness in conserving biodiversity. Ironically, the economic spin-offs for people living near protected areas may be largely associated with tourism. For developing countries this usually means visitors from wealthier countries, which translates into economic benefits and may explain the observed patterns of growth in human populations.

The encroachment of people into undeveloped areas is shown by the proliferation of the network of roads. No matter what the roads are built for—access for timber operations, mines, or communities—the net effect is the same. Roads provide convenient vehicular access to areas that had previously been relatively isolated. Figure 18.41 shows the situation in the contiguous US as represented by **roadless volume**, calculated based on the distance of any point from the nearest road. In 2006 there were 2.3 million km² of roadless volume in the United States, but its distribution was not uniform (Fig. 18.41).

To some species, road are physical barriers to dispersal. For small forest mammals in southern Ontario, a divided four-lane highway is a barrier because it is a wide stretch of inhospitable open habitat. Here the cleared right-of-way for the road is the barrier rather than the road surface or the traffic. Researchers came to

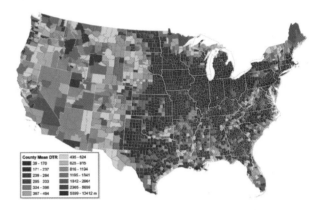

Figure 18.41 Variation of road volume in the US. Colours indicate the road density. DTR = distance to nearest road.

SOURCE: From Watts R. D., et al., "Roadless Space of the Conterminous United States," *Science* 4 May 2007: Vol. 316 no. 5825 pp. 736–738. Reprinted with permission from AAAS.

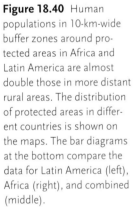

Figure 18.40 Human populations in 10-km-wide buffer zones around protected areas in Africa and Latin America are almost double those in more distant rural areas. The distribution of protected areas in different countries is shown on the maps. The bar diagrams at the bottom compare the data for Latin America (left), Africa (right), and combined (middle).

SOURCE: From Wittemyer, G., P. Elsen, W.T. Bean, A. Coleman, O. Burton and J.S. Brashares, "Accelerated Human Population Growth at Protected Area Edges," *Science* 4 July 2008: Vol. 321 no. 5885 pp. 123–126. Reprinted with permission from AAAS.

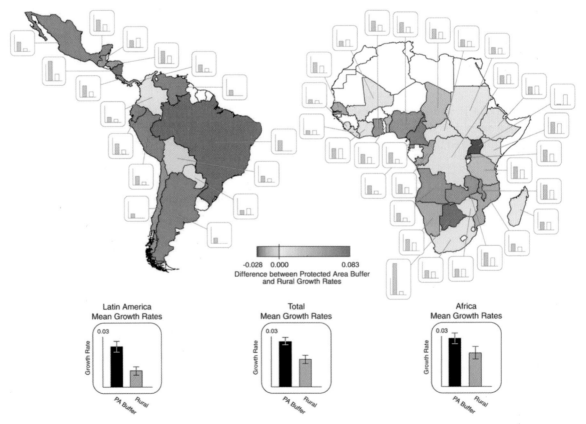

this conclusion when they noticed that white-footed mice (*Peromyscus leucopus* (1897)) did not cross a four-lane divided highway that was not yet open to traffic. The road itself, even without traffic, was the barrier to these mice. For larger mammals, increased chances of mortality through collisions with vehicles may be a more important factor than the roadway itself. Underpasses have been used to good effect to give mammals, salamanders, and turtles chances to cross roadways.

Woodland caribou (*Rangifer tarandus* (1767)) provide a picture of how roadways can severely disrupt the life cycle of a mammal, in this case one widely distributed in the Northern Hemisphere (from Europe, across Asia and North America, to Greenland) and of vital importance to communities of First Nations people. Caribou use traditional summer (calving) and wintering grounds connected by migratory routes. In this context, "traditional" means areas that have been repeatedly used by caribou for hundreds of years. In northwestern Ontario (Canada), woodland caribou were displaced 8–60 km from their traditional wintering areas in winters when a logging road was plowed and used to haul timber. Woodland caribou often abandon areas after road networks are established and in the wake of clearcuts.

Roadways and other linear features such as transmission lines, pipeline rights-of-way, survey lines, and snowmobile trails lead to increased mortality of caribou in two ways. First, these linear features are used as travel routes by wolves (*Canis lupus* (1709)) that are predators of caribou, and second, they are used as access routes by human hunters, whether legal or illegal. The obvious problem is that roads may be built for one purpose and then used for others. Woodland caribou provide a depressing glimpse of how human use of the landscape can have large negative impacts on animals.

18.7c Pollution

Pollution is a term that many people use and "everyone" understands is a direct or indirect consequence of human activity. Noise pollution is an example. In Hawaii, captive bottlenose dolphins (*Tursiops truncatus* (1772)) exposed to naval mid-frequency sonar showed temporary hearing loss and changes in behaviour. This information demonstrated a clear connection between background noise and cetacean behaviour. There is less clarity about the effects of the pervasive change in sound levels associated with the change from wind-powered to engine-powered vessels that occurred in the late 1800s (Fig. 18.42). This change turned the Earth's waterways and oceans into noisy places. Other anthropogenic noise sources include underwater explosions associated with exploration for oil and gas.

Noise pollution is not just a factor underwater. Male Nightingales (*Luscinia megarhynchos* (1607)) sing louder in areas with higher levels of background noise compared to quieter areas. This implies a higher cost

Figure 18.42 The noise associated with propeller-driven vessels changed the underwater soundscape.

M.B. Fenton

of territoriality for birds in noisy places such as those adjacent to heavily travelled roads. Common marmosets (*Callithrix jacchus* (1854)) also increased the strength and durations of their twitter calls as levels of background noise increased. Other birds, including Japanese Quail (*Coturnix coturnix japonica* (1665)) and King Penguins (*Aptenodytes patagonicus* (1640)) increase the number of syllables in their calls when confronted with higher levels of background noise. This may increase the chances of the transmitted signal being received.

Another source of pollution is the introduction of man-made chemicals into the environment. As of 2010, extracting petroleum products from the oil sands in Alberta, Canada, is a mainstay of Canada's economy and a central pillar in the energy policy of the US. Our society depends heavily on petroleum products, and many people are prepared to overlook the environmental costs of extracting oil from this source. One indication of pollution associated with the oil sands project is provided

by measuring the accumulation of polycyclic aromatic compounds deposited in and around the open pit mining and oil sands processing facilities (Fig. 18.43).

Polycyclic aromatic compounds at concentrations such those observed in Figure 18.43 are toxic to fish embryos, including species such as fathead minnows (*Pimephales promelas* (1199)) and white suckers (*Catostomus commersoni* (1195)) that are native to the watershed of the Athabasca River. The timing of fish spawning in the area means that their developing young will be exposed to toxic levels of contaminants. People and other animals that depend directly or indirectly on fish are also affected by the changes to fish in the Athabasca River.

18.7d Climate Change

There is ample evidence that climates change world-wide—they have changed in the past and continue to change today. Now, however, it is clear that the activities of our species are responsible for recent dramatic changes in concentrations of greenhouse gases (GHG). The resulting climate warming is reflected by the rapid retreat of sea ice, glaciers, and ice caps, demonstrating the global connectedness of systems.

The topic of climate change and global warming has been the focus of political and media attention that typically has more to do with points of view than about evidence or science. While this topic is controversial, the disagreement does not undermine the foundation of data supporting the reality of anthropogenic climate change. But messages about anthropogenic climate change, not to mention its impact on the planet and the life it supports (including our own species), often fall on deaf ears. One of the challenges of assessing climate change is to step

Figure 18.43 Snowpack accumulations of total polycyclic aromatic compounds (TPAC, particulate and dissolved) associated with the extraction of oil from the oil sands of northern Alberta (sites ARxx). This shows the spread of polycyclic aromatic compounds away from proposed and operating sites.

SOURCE: Kelly et al., "Oil sands development contributes polycyclic aromatic compounds to the Athabasca River and its tributaries," *PNAS* December 29, 2009 vol. 106 no. 52 22346–22351.

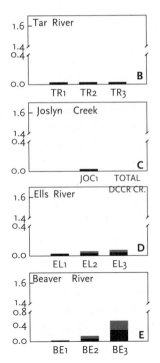

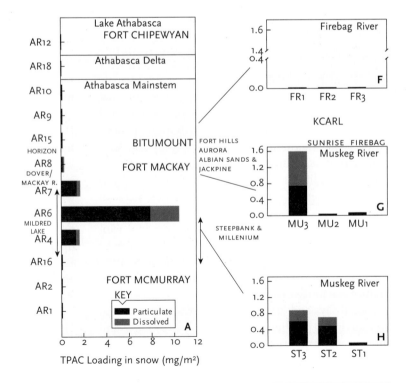

beyond the impression that it is a change in day-to-day weather. Another challenge is to make sense of the patterns of change that emerge when large and varied data sets are considered. Inevitably, researchers have turned to models to demonstrate what elements of climate are changing and to predict the eventual impacts.

Birds provide a good example. Daylength (photoperiod) has a strong influence on the timing of migration, but local environmental conditions may dictate the availability of food. In the Netherlands, populations of pied flycatchers (*Ficedula hypoleuca* (1601)) have declined by 90% because of reproductive failure. Normally, adult pied flycatchers return to the Netherlands in spring, establish territories, build nests, and lay eggs according to a time line that ensured that their eggs hatched at the same time as the caterpillars that adult pied flycatchers fed to their young. Caterpillars hatch when the temperature is warm enough, so now they hatch earlier. Because the flycatchers' hatching is now out of synch with that of the caterpillars, climate warming has resulted in the starvation of many young birds.

Humans are also at risk from climate change. By 2030, food crops will be at risk in 12 major regions of the world. The world's most malnourished individuals will live in groups of countries in which people have broadly similar diets. Projected numbers of malnourished people by 2030 include 158.5 million in China, 30.1% of the world's total population, compared to West Africa (27.5 million people, 3.2% of the world's total, or southern Africa, 33.3 million, 3.8% of the world's total. These human populations vary in their dependence on different crops, and not all crops are equally threatened by climate change (Fig. 18.43). Getting people to change from one food crop to another will involve changing social attitudes about foods as well as the use of different farming techniques.

Some animals change their patterns of distribution (their geographic ranges) in response to changes in climate, for example, the common opossum (*Didelphis marsupialis* (1684)). About the size of a house cat this marsupial occurs widely from southern Ontario in eastern North America and south through the eastern United States, Central America, and South America. Historical records indicate that opossums were found sporadically in southern Ontario between 1850 and 1860, from 1890 to 1910, and from 1930 to 1935, most often along the north shore of Lake Erie. By 2010, opossums were commonly seen, most often as road kills, over a wider area in southern Ontario. This extension of their range appears to reflect periods of warmer winters as the naked ears and tail of these animals are subject to frostbite.

During periods of rapid climate change, the normal dispersal of animals and plants may not keep pace with changing environments. Using spatial gradients (°C·km⁻¹) and models of temperature increase (°C·a⁻¹), one can calculate an index of the velocity of temperature change (km·a⁻¹) for different biomes. Topographic effects mean that the lowest velocities of climate change will occur in mountainous biomes, which include a range of habitats from tropical and subtropical forests to temperate coniferous forests and montane grasslands. The highest velocities of climate change will occur in deserts, flooded grasslands, and mangroves.

Modelling exercises, such as those involving birds or velocities of climate change, allow us to be proactive when trying to minimize negative effects of climate change and, most importantly, to consider and interpret the broader implications of the climate patterns for the places we live. It is important to distinguish science and data from value judgements. In the issue of climate change, scientists can provide, analyze, and interpret the data, but other people, often politicians, are usually the primary drivers of decisions about how to manage resources. Too often the value systems of the decision makers do not recognize or value the natural world and may even deny our connections to it.

For fiscal 2010, the US federal budget allocated US$40 million in support of landscape conservation cooperatives and climate science centres, allocations that had not been there before. Investments of this nature reflect the realities of planning for both plants and animals to live and even thrive in a warmer world.

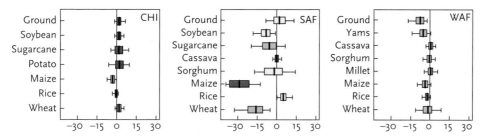

Figure 18.44 The impacts of climate change on major food crops projected for 2030, based on the effects of climate change on crop production between 1998 and 2002. CHI = China, SAF = southern Africa, and WAF = West Africa. Colours indicate degree of Hunger Importance Rating (HIR): red = 1–30, orange = 31–60 and yellow = 61–94, where lower scores indicate higher importance.

SOURCE: From Lobell, D.B., M.B. Burke, C. Tebaldi, M.D. Mastrandrea, W.P. Falcon and R.L. Naylor. 2008. Prioritizing climate change adaptation needs for food security in 2030. *Science* 319:607–610. Reprinted with permission from AAAS.

1. How are pollution and climate change interrelated?
2. Why are alien species so important? What can be done to reduce their impact?
3. How do roads affect habitats and animals? What can be done to ameliorate these effects?

18.8 Animals and People

We humans reveal our animal ancestry in almost every facet of our lives. As individuals and as a species we gain immeasurably from the other species with which we share the planet. We cannot afford to disregard our roots and how much other species mean to us.

Animals repeatedly provide us with astonishing examples of diversity and interactions with other species. A final example comes from the jungles of Borneo where tree shrews (*Tupaia montana* (1912)) visit aerial pitchers of a pitcher plant (*Nepenthes lowii*), crouching over the opening to the pitcher while drinking exudates from around the underside of the lid of the pitcher (Fig. 18.45). A close look revealed that the tree shrews often defecate into the pitcher and mark its lid with scent from its anal glands.

This pitcher plant produces two kinds of pitchers. One type, the typical pitchers of *Nepenthes* (Fig. 18.45A) grow on immature plants close to the ground where they catch insects, often ants. The pitchers visited by tree shrews are well above the ground, different in shape and structure (Fig. 18.45B), and do not contain the remains of insect prey. These pitchers often contain accumulations of tree shrew feces, which provides the pitcher plant with nitrogen to build proteins.

Tree shrews have a mutualistic relationship with the pitcher plants. The shrews are herbivores but

M.B. Fenton

© Ch'ien C. Lee

Figure 18.45 A. "Normal" insect-catching pitcher of a *Nepenthes* pitcher plant and **B.** a modified pitcher supporting a tree shrew (*Tupaia montana* (1912)). The tree shrew may defecate while drinking nectar produced around the lip of the pitcher. The feces provide nitrogen to the plant.

cannot digest cellulose, so the pitcher plants' nectar is an important source of energy. By providing the tree shrews a latrine and a sugar reward, the pitcher plants benefit by receiving tree shrew feces, which provide them with the nitrogen they require. The scent mark on the lid keeps other tree shrews away.

INTERACTIONS WITH HUMANS IN PERSPECTIVE

It should be completely apparent that humans are, by their ancestry, animals. Our lives are inextricably linked to other species, many of them animals. And yet, too often, too many of use choose (perhaps even refuse) to take actions to protect the very biological resources on which we depend. Use the examples and themes in this chapter to consider your place in nature and in the world.

Questions

Self-Test Questions

1. What interactions of humans with coral reefs represent sustainable use of these resources?
   a. increased ocean temperatures and acidification due to global human activities
   b. overfishing of reef environments and destructive fishing practices using cyanide and blast fishing
   c. establishment of marine protected areas (MPAs) for reef preservation
   d. nutrient enrichment through eutrophication in reef areas

2. How has domestication, the selective breeding of animal species to enhance or conceal certain traits, been reflected in human history?
   a. Domestication of dogs probably favoured behavioural traits rather than physical features.
   b. Domestication of chickens has favoured breeds that show both good growth characteristics (meat production) and fertility (egg production).
   c. Historically, beekeeping and the development of honey products were confined to northern Europe.
   d. Aquaculture provided an easy solution to the overfishing of natural habitats.

3. Which of the following illustrates an impact that animals have on human health?
   a. Blood-feeding insects such as mosquitoes transmit viruses only to human hosts.
   b. Animals can be the direct causative agents as well as the vectors of human disease.
   c. The effect of sleeping sickness in sub-Saharan Africa is limited to the physical effects on humans.
   d. Mosquitoes, which might transmit West Nile virus, are really the only major concern as vectors of human disease in North America.

4. How have human activities introduced animal species that impact other ecosystems?
   a. Trade in wild animal species has potentially exposed humans to emerging infectious diseases.
   b. Accidental introduction of zebra mussels into the Great Lakes has increased the diversity of species through competition in the local ecosystem.
   c. Human activities have accidentally and intentionally introduced alien species such as rats and Artic foxes into new habitats.
   d. Humans have introduced species to remove or eradicate other introduced species in biological control strategies that are usually successful.

5. Which of the following human activities has had a positive impact on other animal species?
   a. There are patterns of human land use in which roadways or other barriers or linear features subdivide large tracts of land.
   b. Noise pollution is present in areas where human activities increase, on land and in water.
   c. Anthropogenic climate change as a result of the proliferation of greenhouse gases has altered the distribution of many animal species.
   d. Humans have established protected areas and then enforced their protection.

Questions for Discussion

1. Why do humans domesticate some animals and not others?
2. Why do human activities sometimes aggravate their exposure to diseases for which animals are vectors?
3. How do the products or behaviours of animals influence the adoption of animals by people?
4. Using the electronic library, find a recent publication about interactions between animals and humans not covered in this book.

Appendix
Answers to Self-Test Questions

Chapter 1

1. b. Mandibles are not replaced, but rather the roles of the ants in the colony change, so d is incorrect. Choices a and c are not addressed in the story, so b is the correct answer.
2. c. Other animals obtain carotenoids from plants, fungi, and microorganisms, so a and b are incorrect. While the frequency of colours may change in a pea aphid population depending upon selection factors, a single individual cannot change colour, so d is incorrect.
3. a. The other answers illustrate structural or functional characteristics, but only a demonstrates the relationship between structure and function.
4. d. Plantarflexion and the reduced diaphysis angle between metatarsals and plantar surface are important in arboreal primates, rather than in bipedal organisms, so a and b are incorrect. The reduced angle in c would restrict dorsiflexion.
5. b. Structure is described in a and d, and function in c. Only b illustrates the relationship between structure and function.

Chapter 2

1. b. Transporter proteins on apical surfaces contribute to the regionalization of specialized functions within compartments. Option a combines structural features of muscle cells and physiology of neurons; c combines structural features of muscle and connective tissue, and d combines features of epithelial tissues and neural function.
2. c. The limits imposed on size are a direct consequence of the non-linear relationship between surface area and volume; larger animals solve the limitations of diffusion through multicellularity. While the other choices (a, b, and d) are related to increased size and complexity of animals, only c is directly a consequence of surface area:volume.
3. b. The question describes secondary body compartments, which can serve as hydrostatic skeletons, but not all hydrostatic skeletons are secondary body compartments. Options c and d are not secondary body compartments.
4. d. Development of the body axis is cued by different mechanisms in different animal groups.
5. b. While embryological development allowed for distinguishing some protostomes and deuterostomes (a and d), not all organisms fit into these groups. Similarly, molecular analysis of early clades (c) has yielded inconsistent results regarding phylogeny.

Chapter 3

1. d. The term *phylogeny* refers to the evolutionary history of an animal, rather than its placement in a taxonomic group, so a is incorrect because it doesn't answer the question. Many of the physical and molecular features of animals are present in organisms from widely diverse taxonomic groups, so b and c are incorrect, confounded by the mollusc DNA from gut contents!
2. b. Polyps, as well as some others, can be considered to have more than one plane of symmetry, so a is incorrect. Not all comb jellies have tentacles, and other features link them somewhat to the Cnidaria. Sea wasps, Scyphozoa, and Hydrozoa are currently considered separate classes in the Medusozoa.
3. a. See text for proposed phylogenies.
4. c. These features define all Ecdysozoa.
5. d. While these terms apply to the number of branches evident in the gut of Platyhelminthes, the specific anatomical feature is not evident. a. These terms describe the location of branchia = gills: proso, or in front; opistho, or posterior/dorsal. b. Gnathi = jaws. c. Echinoderm = spiny skin.

Chapter 4

1. b. All other answers contribute to the genetic variation within a population. Point mutations can introduce new variations. Genetic drift and gene flow can change the genetic variation within a population.
2. d. Geographic barriers may prevent gene flow (a), but so may differences in reproductive behaviours and other non-physical features (c). The dispersal of offspring distributes genes more uniformly across a wider geographic region than when offspring are more sedentary.
3. c. Genes that are selected for will become the most numerous, but this isn't the basis for selection, so option a is incorrect. Selection pressures determine which genes are favoured, so b describes the process by which a population can be divided by selection. Natural selection is not predictive, and so only those genetic traits in the current population that favour reproductive success will be selected for, so d is incorrect.
4. a. Phylogenies are constantly being revised based upon new information, including that provided by molecular clocks and the identification of organisms thought to be common ancestors.

5. c. Adaptive radiation describes the outcome when natural selection acts on the same pool of genetic variation in a variety of habitats, and rapid speciation occurs. Organisms that share a common ancestor will show different degrees of homology, independent of the forces of natural selection, so option a is incorrect. Pleiotropic genes influence many characteristics, and so selection is unlikely to result in any one specific adaptation. Species that coevolve must have some sort of mutualistic, predator–prey or host–parasite relationship, and so cannot be in unrelated ecosystems.

Chapter 5

1. b. Larval morphology differs significantly from that of adults, particularly through the absence of functional reproductive structures. Larvae typically have specific structures that disappear or are reduced in the adult form, but changes must occur prior to metamorphosis, regardless of the relationship to hatching, to be considered larval events.
2. d
3. d. Free-swimming nauplius larvae of crustaceans are well-designed for dispersal. The parasitic glochidium relies on nutrients from host tissues, whereas lecithotrophic larvae use egg resources to fuel their early development. The quest for nutrients places holometabolous insect larvae among some of the most destructive agricultural pests.
4. c. Metamorphosis in all organisms requires the elements listed in a and b.
5. c. Examples of viviparity occur in all but Aves and turtles, involving either parent, in a variety of body chambers, and while viviparity is associated with internal fertilization in many cases, it is not a requirement.

Chapter 6

1. c. Permineralization (a) and the replacement of structural materials by harder minerals (b) are mechanisms by which fossils are formed, but these don't give information about when they are formed. The variety of trace fossils recovered (d) similarly gives information about biological activity but not about the age of the fossils.
2. b. Check the text to determine the key distinguishing features of geological time periods.
3. c. Pachycormids (a) occupied the niche of the large planktivore from the middle Jurassic until the end of the Cretaceous. b. Multiplacophora of the Carboniferous (Devonian to Permian) had more valves than living and Cambrian Polyplacophora. d. *Eucritta* illustrates an early diversification of terrestrial vertebrates that combined characteristics usually found in lineages leading to living Amphibia and Amniota. Only horseshoe crabs (c)

retain the body form as seen in the fossil record of the Jurassic.
4. a. Extensive glaciation at high latitudes (b) is characteristic of Paleozoic–Carboniferous transitions in the tropics. c occurred about 2.5 billion years ago, prior to the appearance of the Ediacaran fauna (635–542 Mya). d is characteristic of the transition from Paleocene to Eocene.
5. d. The diversification of angiosperms is linked to the radiation of herbivores and nectarivores, whereas the radiation of terrestrial amphibians is linked to the diversification of insects. Radiation of modern sharks and rays (neoselachians) coincides with mass extinctions at the end of the Triassic period.

Chapter 7

1. a. Muscles can be paired with an elastic or other flexible element, so b is incorrect; c and d are simply not true.
2. c. Body undulations may also provide propulsion, without the action of appendages. Nematodes do not have circular muscle, and anguilliform locomotion is characteristic of fish with eel-like bodies only.
3. d. Surfactant is important for movement across the water's surface, not cursorial locomotion, which is terrestrial. Setae, microdroplets, and discs allow adhesion to surfaces, whereas changes in gait are important in the pace of cursorial locomotion.
4. b. While snail mucus serves as a glue, the bonds break under strain such that it is also a lubricant, so a is incorrect. Retrograde waves are typically generated for retrograde movement, and the morphology and mechanics associated with wave generation are different for movement in different directions, so c and d are incorrect.
5. d. Abductin is a hinge protein in scallops; b describes propulsion in salps, and c describes the mechanism of jet propulsion in hydrozoans.

Chapter 8

1. b. By eating the others, you are immediately at the next highest trophic level, but your trophic level will change depending upon the diet of your food organisms. Even as scavengers, the shrimp would be feeding on organisms that ate something else, so the shrimp are at least at trophic level 3.
2. a. Inertial forces are most important for larger organisms, but small animals likely to be captured by suspension feeding are greatly affected by the dominant influence of water viscosity. Small organisms increase their Reynolds number by increasing their speed of movement.
3. c. Ram feeders frequently show adaptations to reduce drag in water, like the drag of an open

mouth during ram feeding (b), because of the energetic cost of chasing prey. This is not true for suction feeders, which may remain more stationary in the water.

4. b. for Echinoidea. See text to identify other taxonomic groups and their characteristic jaws.

5. d. The fused upper jaw and mastication are mammalian features, but are not characteristic of other vertebrates. Cranial kinesis involves articulation between the upper jaw (not the mandible) and the neurocranium, and is not found in all vertebrates.

Chapter 9

1. a. Symbiotic bacteria are involved in digestion of cellulose in many animal groups, gastroliths may be found in digestive chambers of some tetrapods, and vacuoles are distributed throughout cells in which digestion occurs intracellularly. None of these reflect regional specialization.

2. b. Mechanical digestion is all about breaking apart food items so that more enzymes can work on more molecules to increase the rate of digestion and absorption.

3. b. Neural reflexes within the digestive tract coordinate motility. Chemical signals such as leptin provide information about energy stores, rather than gut contents, and sensory information about food in the environment can also stimulate digestive processes.

4. c. Glucose uptake into cells is by facilitated transport, with a transport protein coupled to sodium diffusion, taking advantage of the sodium gradient that is maintained across the cell membrane by active transport. Fatty acids can simply diffuse across the phospholipid bilayer. Amino acids, and other large polar molecules, are transported by the three-stage process of transcytosis. Animals with very high metabolic activity rapidly mobilize ingested sugars by paracellular absorption.

5. b. Antibiotic secretion by endosymbiont bacteria helps to limit the growth of other bacteria, as seen in the leech digestive tract. Endosymbiont bacteria in aphid guts aid in the synthesis of tryptophan, not its breakdown as stated in a. Similarly, microbial activity generates volatile fatty acids as well as vitamins and proteins that are essential to the host, and are essential for foregut fermentation of cellulose by herbivorous mammals.

Chapter 10

1. b. Parasitism is the most common form of life, and parasites range in size and association with their hosts, but these symbiotic associations involve very specific physical and biochemical interactions between parasite and host.

2. d. Bioluminescent reactions often depend upon a host's symbiotic relationship with a bioluminescent bacterium, common in marine organisms. Hydrothermal vent organisms are typically symbiotic with sulphide-oxidizing bacteria. Symbiotic algal cells in cnidarian polyps are known as zooxanthellae. Cytotoxins in free-swimming bryozoan larvae are produced by a symbiont bacterium, *Endobugula sertula,* and *T. welkeri* is characteristic of the symbiotic algae on sloths.

3. a. While b and c illustrate the symbiosis between ants and acacias, only answer a illustrates the interactive coevolution of this relationship.

4. c. While ectoparasites have an obvious free-living stage away from the host, the intermediate host in the life cycle of endoparasites represents the same stage of transmission. Both forms of parasites face the challenges of low probability of successful transmission through massive reproductive efforts, and both have enhanced features for attachment to the host. Only endoparasites are exposed to mechanical and immunological defences of the host.

5. d.

Chapter 11

1. c. Concentration gradients drives the rate of diffusion in all applications. While the relationship between surface area and volume is important in animal metabolism, Fick's Law describes factors that affect the rate of diffusion. Available diffusion area and the distance of diffusion are important but unrelated factors, and the coefficient of diffusion for any particular molecule remains unchanged.

2. d.

3. c. Respiratory pigments are widespread throughout aquatic and terrestrial animals, although they are not common in insects.

4. a. Parabronchi and air sacs are characteristics of avian lungs. Modifications of the water vascular system are found in echinoderms. Bipectinate gills are the generalized structures of molluscs, although some have monopectinate structures.

5. c. Bimodal respiration is characteristic of some fish and molluscs, to allow them to be successful in both water and on land.

Chapter 12

1. b. The surface area for exchange between circulating fluids and tissues is considerably greater in open systems, where the fluids bathe tissues directly. So, although fluids travel generally under lower pressure than in closed systems, the diffusion distance is negligible. Concentration gradients drive the rate of diffusion in all applications,

in both open and closed systems, and the coefficient of diffusion for any particular molecule remains unchanged.

2. c. Distribution vessels are larger vessels of longer length, which are typically more elastic, often with smooth muscle components, to help to pump blood and maintain pressure. At the tissues, lower pressure and flow facilitate exchange, and so b is also incorrect.

3. b. Not surprisingly, coelomic fluid circulates in the coelom, or body cavity. Tissue fluid is present in animals lacking a circulatory system, whereas the watery fluid between cells is interstitial fluid. More than half the volume of vertebrate blood is plasma, and all other body fluids are derived from plasma. Hemolymph of organisms with open circulatory systems is both blood and interstitial fluid.

4. a. Most hearts contain some sort of valves that prevent the backflow of blood that would otherwise occur during relaxation of the heart. Vertebrates that have dual circulation take advantage of the higher pressure generated by the heart for distribution through the systemic circuit, while still facilitating the drop of pressure in the pulmonary circuit essential for efficient gaseous exchange. The heart in a closed system will put the fluid under pressure, but this pressure is lost as blood flows through vessels better suited for diffusion (capillary beds).

5. c. Both groups of organisms have closed circulatory patterns, but these show relatively low blood volumes compared with other molluscs, and to organisms with open circulatory systems in general. Countercurrent exchange maximizes the concentration gradients for gas exchange in cephalopods, much like in aquatic vertebrates, and like most animals, cephalopods show greatly increased cardiac output during high levels of activity, driven by increased heart rate and stroke volume.

Chapter 13

1. c. The cells of other organisms (algae, bacteria, fungi, and plants) are reinforced by walls, but animal cells retain their ability to change shape by not having cell walls, despite the presence of aquaporins that allow passive diffusion of water across the cell membrane. The negative charge across the membrane maintained by Na^+/K^+-ATPase also maintains osmotic equilibrium.

2. a. The osmolarity of the body fluids of osmoconformers matches that of their environment so that there is no gradient. Mechanisms essential to organisms living in hyperosmotic and hypoosmotic environments are switched in answers c and d.

3. c. While excess amino acids can be put to other metabolic uses, the nitrogen-bearing amino group must first be removed, which generates nitrogenous waste. Urea is a large molecule unable to diffuse across membranes, and so must be removed by the excretory system (unless it is retained for osmotic functions or for a role in nitrogen cycling). Both uric acid and guanine crystallize under metabolic conditions that allow excretion of nitrogenous wastes as a solid, conserving water.

4. d. The pressure gradient generated in most nephridia involves muscular mechanisms to increase blood pressure, but in protonephridia this is accomplished by the beating of cilia. The terminal cells are important in filtration of protonephridia, a role performed by podocytes in annelids, molluscs, and vertebrates.

5. b. Increased filtration will increase water loss, which is a major challenge in terrestrial habitats, as is the buildup of nitrogenous wastes, which is reduced by down-regulating protein synthesis. While many marine crustaceans are isosmotic to seawater, this is not a possible osmoregulatory strategy in terrestrial habitats.

Chapter 14

1. b. The reproductive pattern described is hybridogenesis, an unusual form of unisexual reproduction that differs from asexual reproduction, where a new individual is produced from the preexisting cells of a diploid parent. Unisexual reproduction is derived from an ancestral, bisexual reproductive mode. Bisexual (or just "sexual") reproduction refers to the familiar combination of haploid gametes fusing during fertilization to form a diploid zygote.

2. d. ALL the terms apply to bisexual forms of reproduction. The differences are between dioecious (or gonochoristic) organisms, which possess only testes or ovaries, and monoecious (or hermaphroditic) animals, which produce both eggs and sperm within the ovotestis.

3. b. Entry of the sperm into the egg triggers blocks to polyspermy, which are important because of the role the sperm's proximal centriole plays in generating the mitotic spindle for the first embryonic division. c and d are important for species specificity of sperm binding, but do not prevent polyspermy.

4. c. While seasonal cues are important, the timing of reproductive events is linked typically to other, local environmental cues and social cues from conspecifics. The synchronization of reproductive events is important in all animals, so d is incorrect.

5. a. All these examples illustrate different aspects of sexual selection. Intersexual selection by females

favours the colourful plumage of males. Male red-back spiders "buy time" to increase the success of sperm transfer during female cannibalism. Mating activity can occur outside an establish pair bond by either sex. Only answer a is a direct result of male–male competition for females.

Chapter 15

1. d. All other answers are common features of all living cells. Only neurons have voltage-sensitive ion channels in the axonal membrane.
2. b. While giant axons have served as the basis for much research exploring nerve cell activity, this hardly provides an advantage for the animals in which they are found! Giant axons are also present in other invertebrates in which the speed of response is important for survival. Vertebrates make use of saltatory conduction because their energetic demands preclude the presence of giant axons.
3. c. Electrical synapses can conduct action potentials in either direction, which is extremely useful for the synchronization of rapid events. Both types of synapses are found in all animals, but neurotransmitters are employed only at chemical synapses.
4. d. Mauthner cells are sensory neurons important in the fish startle response "c-start." Mushroom bodies are prominent features of the brains of annelids and arthropods, and are involved in olfactory and visual processing and other cognitive functions. Nociceptors are sensory receptors that respond to noxious stimuli.
5. a. Vibratory cues arise external to the animal, whereas proprioception integrates external and internal cues. Vertebrate semicircular canals are continuous with the structures involved in hearing, but are sensitive to rotational movements of the animal's head. Tympanal and chordotonal organs are mechanoreceptors sensitive to sound but not to body position.

Chapter 16

1. c. All hormones are chemical messengers released into the circulatory system, and all act on the cells of target tissues, so a and b are not unique to peptide hormones. Only lipid-soluble hormones, such as steroids, are capable of crossing the cell membrane to bind to intracellular receptors.
2. d. The other answers outline synthesis, release and action of CHH. Negative feedback control is illustrated by answer d.
3. a. The integration of neural and chemical stimuli and responses is the key feature of neuroendocrine control. The other answers represent only chemical or only neural aspects of physiological processes.

4. b. Through the synthesis and release of gonadotropins, the HPG axis directs many different aspects of gamete production and release in animals. Physiological stress responses are mediated by the hypothalamus–pituitary–adrenal (HPA) axis, so answer a is incorrect. Answer c is true for vertebrate species only, whereas d is true primarily of invertebrate systems.
5. c. The main physiological role for vasopressins is an antidiuretic action, with a secondary role in elevation of blood pressure through effects on vascular smooth muscle and increase in the volume of body fluids.

Chapter 17

1. b. The behaviour of California mice increased the number of hormone receptors in brain cells and so is expected to alter future hormonal responses. The other examples show the impact of hormones on behaviour (a) or hormonal responses to sensory input (c and d).
2. d. The other answers do not describe defensive behaviours.
3. c. Migration has been redefined as directed movements of an animal away from its home range to exploit food or other resources not available in the home range. Long-distance movements (a) are not required. Not all bird species use the same routes southward and northward. Migratory birds typically have smaller brains than sedentary species.
4. a. Mixed-species social groups provide many defensive and foraging advantages to group members. The other responses (b, c, d) are true for single-species groups only.
5. d. Tools can be structures modified in many ways. Insects and other invertebrates demonstrate the use of tools. There are many examples of deferred benefit provided by tools, as with the empty coconut shells carried by the veined octopus that can be used as a shelter.

Chapter 18

1. c. Unfortunately, other human interactions with coral reef ecosystems are not usually beneficial.
2. a. Separate breeds of chickens have been selected for to separate egg-layers from broilers. Beekeeping is deeply embedded into the culture of African tribes. Aquaculture has its own challenges, including pollution produced by fish waste and the environmental consequences of food production for cultured fish.
3. b. Blood-feeding insects can also transmit bacteria, protozoans and other invertebrate parasites to human hosts. Because cattle serve as reservoirs for sleeping sickness, the use of cattle by humans is

limited in these regions. Ticks are also a vector of human disease, Lyme disease for example, and fleas can carry plague.

4. c. The exposure to emerging infectious diseases may affect humans, but has little impact on the environment. The introduction of species zebra mussels into the Great Lakes has eradicated most native species. Steps to remove introduced species are usually futile or even counterproductive, particularly when they involve the introduction of yet another new species into an ecosystem.

5. d. Protected areas, such as national parks, are important tools for preserving biodiversity, although increased human populations around the edges of these protected areas threaten their effectiveness.

Index

The Geological Time Scale and Major Evolutionary Events

Eons (Duration drawn to scale)	Eon	Era	Period	Epoch	Millions of Years Ago	Major Evolutionary Events
Phanerozoic: Cenozoic, Mesozoic, Paleozoic	Phanerozoic	Cenozoic	Quaternary	Holocene		
					0.01	
				Pleistocene		Origin of humans; major glaciations
					1.7	
				Pliocene		Origin of apelike human ancestors
					5.2	
			Tertiary	Miocene		Angiosperms and mammals further diversify and dominate terrestrial habitats
					23	
				Oligocene		Divergence of primates; origin of apes
					33.4	
				Eocene		Angiosperms and insects diversify; modern orders of mammals differentiate
					55	
Proterozoic				Paleocene		Grasslands and deciduous woodlands spread; modern birds and mammals diversify; continents approach current positions
					65	
		Mesozoic	Cretaceous			Many lineages diversify: angiosperms, insects, marine invertebrates, fishes, dinosaurs; asteroid impact causes mass extinction at end of period, eliminating dinosaurs and many other groups
					144	
			Jurassic			Gymnosperms abundant in terrestrial habitats; first angiosperms; modern fishes diversify; dinosaurs diversify and dominate terrestrial habitats; frogs, salamanders, lizards, and birds appear; continents continue to separate
					206	
			Triassic			Predatory fishes and reptiles dominate oceans; gymnosperms dominate terrestrial habitats; radiation of dinosaurs; origin of mammals; Pangaea starts to break up; mass extinction at end of period
					251	